Reindeer and Caribou

Health and Disease

Reindeer and Caribou

Health and Disease

Edited by
Morten Tryland
Susan J. Kutz

CRC Press
Taylor & Francis Group
Boca Raton London New York

CRC Press is an imprint of the
Taylor & Francis Group, an **informa** business

CRC Press
Taylor & Francis Group
6000 Broken Sound Parkway NW, Suite 300
Boca Raton, FL 33487-2742

First issued in paperback 2021

ISBN 13: 978-1-03-209433-5 (pbk)
ISBN 13: 978-1-4822-5068-8 (hbk)

Library of Congress Cataloging-in-Publication Data

Names: Tryland, Morten, editor. | Kutz, Susan J., editor.
Title: Reindeer and caribou : health and disease / editors, Morten Tryland and Susan J. Kutz.
Description: Boca Raton : Taylor & Francis, 2018. | Includes bibliographical references.
Identifiers: LCCN 2018005995 | ISBN 9781482250688 (hardback)
Subjects: LCSH: Caribou--Diseases. | Reindeer--Diseases. | MESH: Reindeer | Animal Diseases | Animal Husbandry
Classification: LCC SF997.5.C37 R45 2018 | NLM SF 997.5.C37 | DDC 636.2/948--dc23
LC record available at https://lccn.loc.gov/2018005995

Contents

Preface

Reindeer and caribou, distributed across arctic, subarctic and temperate latitudes of the northern hemisphere, are a keystone species and integral to the socio-ecological systems of these regions. Whether semi-domesticated or wild, *Rangifer* play critical roles in the ecosystems in which they live. Further, they are a focus for maintenance of cultural traditions and serve as a source of food and income for the northern Indigenous peoples of the Circumarctic.

Physiological and behavioural characteristics, together with the power of sheer numbers, have allowed *Rangifer* to adapt and persist across changing environments since the last Ice Age. Nevertheless, life in the North, with its short growing seasons and extremes in temperature, is often the story of life on the edge. Small perturbations to climate and habitat can tip animals over this edge, with consequences that manifest in many ways, including poor health and disease. Similarly, habitat degradation and increasingly intensive corralling and feeding of reindeer are leading to concentration of animals and emergence of new and re-emerging disease challenges.

Despite the importance of *Rangifer* around the Circumarctic, and despite a very broad literature on *Rangifer* ecology, conservation and husbandry, there exists no comprehensive book that covers health and diseases of *Rangifer*. The available information on *Rangifer* health is scarce, patchy, and often outdated, and has never before been collected and displayed as one story. With this book, we have engaged many experts in the field to provide updated knowledge on health and diseases of *Rangifer*. We hope this will serve as a resource for a broad audience of teachers, researchers, students, wildlife managers, herders and *Rangifer* users, including subsistence and recreational hunters, as well as others interested in *Rangifer* biology and ecology, and the role of health and diseases.

There is a vast literature on the biology and ecology of *Rangifer* that we simply could not capture with this book while still addressing health and disease. Rather, to orient the perhaps unfamiliar reader, we provide a general introduction to *Rangifer* as a species, with an overview of the ancient history of *Rangifer*, the genetics and diversity of the subspecies, and their physiology, behavior and management. We then move on to define what we mean by "health." We discuss the determinants for health in *Rangifer*, how stress can influence health and what health and disease mean for the individual and for the population. The subsequent chapters address a number of determinants of health in *Rangifer*, from nutrition and feeding to non-infectious and infectious diseases, and how these various determinants may act individually or in concert with others to influence *Rangifer* population dynamics. For a veterinarian, it is not enough to know the name of a disease; one must also know how to prevent and deal with it. Thus, hands-on knowledge of identification and treatment of many common health disorders, as well as capture and restraint, are covered. Important aspects of meat quality and hygiene are also addressed. Looking a little bit into the future, the last chapter deals with climate change scenarios and how they may affect the health and disease of *Rangifer*.

We hope that this book will generate increased interest in and knowledge of *Rangifer* health and diseases among new generations of scientists, biologists, veterinarians, managers and decision-makers, and in the longer term, a better understanding and improved management of the reindeer and caribou populations that inhabit our northern regions. *Rangifer* has, through history, played a decisive role for life in the Circumarctic, for Indigenous people herding or hunting them, for predators, for the plant communities and for northern ecosystems in general. At the same time, we know that reindeer herding and wild *Rangifer* populations are under pressure due to increasing habitat disturbance and fragmentation and that Arctic ecosystems are challenged by alarming climate change scenarios. These stressors will affect health and disease of *Rangifer* to a large extent, which will in turn affect the northern communities that depend on these animals for their social, physical, cultural and economic well-being. Hopefully this book will serve as a resource and provide some of the knowledge and tools to help understand and mitigate threats to this ancient species and see it thrive for the use and enjoyment of generations to come.

Morten Tryland
Susan J. Kutz

Acknowledgements

As with most books, this one is the result of many years of hard work by many dedicated people. It reflects the knowledge and experience of an international network of contributors that has been established over several decades through publications, meetings, workshops, and conferences. In fact, the work on the book itself has strengthened the circumpolar network among reindeer and caribou biologists, managers and veterinarians. A special thank you goes to Antti Oksanen, who contributed to the initial ideas and the drafting of the Table of Contents.

Each chapter has been peer reviewed by one or two experts on the topic, and we deeply appreciate their time and effort:

Lars Folkow, UiT – The Arctic University of Norway, Tromsø, Norway
Anne Gunn, Salt Spring Island, Canada
Rolf Egil Haugerud, Tromsø, Norway
Doug Heard, University of Northern British Columbia, Prince George, Canada
Kris Hundertmark, University of Alaska, Fairbanks, USA
Terje D. Josefsen, Nord University, Bodø, Norway
Rob McCorkell, University of Calgary, Calgary, Canada
Rob Mulley, Western Sydney University, Sydney, Australia
Antti Oksanen, Finnish Food Safety Authority, Oulu, Finland
Rolf-Arne Ølberg, Kristiansand Dyrepark, Kristiansand, Norway
Ulla Rikula, Finnish Food Safety Authority, Helsinki, Finland
Kathreen Ruckstuhl, University of Calgary, Calgary, Canada
Don Russell, Yukon College, Whitehorse, Canada
Satu Sankari, University of Helsinki, Helsinki, Finland
Anders Sirén, University of Helsinki, Helsinki, Finland
Timo Soveri, University of Helsinki, Helsinki, Finland
Michael A. Tranulis, Norwegian University of Life Sciences, Oslo, Norway

We are also deeply grateful for the many anonymous hands that contributed ideas, language corrections, translation, editing, information, experience, maps, figures and photos, and patience, and in other ways supported the project. Special recognition goes to Pat Curry, Michael Polterman, Sophie Scotter, Anne Cathrine Munthe, Lars Nesse, Sigurður Sigurðsson, Olav Strand, Sondre Myrvang and Kjell Bitustøyl.

Finally, we would like to thank the Taylor & Francis Group, who played the essential role of supporting and enabling the editors and authors to make this book a reality.

Editors

Dr. Morten Tryland is a professor in veterinary medicine – infection biology. He works in the Arctic Infection Biology research group at UiT – The Arctic University of Norway, and also holds a professor-II position at the Norwegian Polar Institute (NPI); both of these institutions are located in Tromsø, Norway. After a period of clinical veterinary practice, he transitioned to research on virology and wildlife diseases. He has been a member of the Norwegian Scientific Committee for Food Safety (VKM; Panel on Biological Hazards, 2007–2016) and the European Food Safety Authority (EFSA; Panel on Biological Hazards, 2016–2017), and he is currently the Norwegian editor of the scientific journal *Acta Veterinaria Scandinavica*. Dr. Tryland has spent most of his research career investigating infectious diseases and zoonoses in arctic wildlife and semi-domesticated reindeer, in close cooperation with Fennoscandian reindeer herders and research groups. He has led or participated in projects in Alaska, Canada, Iceland and Fennoscandia, including the Svalbard archipelago, and spent a sabbatical period at the University of Fairbanks, Alaska, USA. His research is focused on how pathogens impact humans and wildlife, both individuals and populations, of the vulnerable and changing northern ecosystems.

Dr. Susan J. Kutz is a professor in the Department of Ecosystem and Public Health at the University of Calgary Faculty of Veterinary Medicine, Alberta, Canada. She has devoted over two decades to wildlife health research in the Arctic and Subarctic. Her areas of expertise include wildlife parasitology, disease ecology, ecosystem health, arctic ecology, climate change and community-based wildlife health surveillance, with a focus on caribou and muskoxen. She is a member of the CircumArctic *Rangifer* Monitoring and Assessment Network and led the development and implementation of *Rangifer* health monitoring protocols during International Polar Year. She initiated and maintains the *Rangifer* Anatomy Website (2011: www.ucalgary.ca/caribou/index.html), an interactive website providing general and detailed information on caribou and reindeer anatomy, caribou hunting and caribou sampling, and co-produced the *Hunter Caribou Training Video* (2009), a 52-minute community-engaged training video containing four "chapters" on caribou health sampling, caribou disease and youth engagement. Working with local communities, Dr. Kutz has done extensive research on the impacts of a warming Arctic on the health of declining muskox and caribou populations and the consequent effects on food security in the Arctic.

Contributors

Jan Adamczewski, PhD
Ungulate Biologist
Wildlife Division
Government of Northwest Territories
Environment and Natural Resources
Yellowknife, Northwest Territories, Canada

Erik Ågren, DVM, Dipl. ECVP, Dipl. ECZM (WPH)
Head of Section of Wildlife Diseases
National Veterinary Institute
Department of Pathology and Wildlife Diseases
Uppsala, Sweden

Birgitta Åhman, PhD
Professor
Swedish University of Agricultural Sciences
Uppsala, Sweden

Jon M. Arnemo, DVM, PhD
Professor and Wildlife Veterinarian
Department of Forestry and Wildlife Management
Inland Norway University of Applied Sciences
Campus Evenstad
Koppang, Norway
and
Department of Wildlife, Fish and
    Environmental Studies
Swedish University of Agricultural Sciences
Umeå, Sweden

Kjetil Åsbakk, Cand. Scient., PhD
Professor
UiT – The Arctic University of Norway
Department of Arctic and Marine Biology
Tromsø, Norway

Kimberlee B. Beckmen, MS, DVM, PhD
Wildlife Veterinarian
Wildlife Health and Disease Surveillance Program
Division of Wildlife Conservation
Alaska Department of Fish and Game
and
Department of Veterinary Medicine
College of Natural Science and Mathematics
University of Alaska Fairbanks
Fairbanks, Alaska

John E. Blake, DVM, MVetSc
Director
Animal Resources Center
University of Alaska Fairbanks
Fairbanks, Alaska

Anja M. Carlsson, PhD
Senior Curator
Department of Environmental Research and
    Monitoring
Swedish Museum of Natural History
Stockholm, Sweden

Nigel A. Caulkett, DVM, MVetSci, DACVAA
Professor and Head
Department of Veterinary Clinical
    and Diagnostic Science
Faculty of Veterinary Medicine
University of Calgary
Calgary, Alberta, Canada

Steeve Côté, PhD
Professor
Caribou Ungava & Laval University
Québec City, Canada

Jim Dau, MS
Wildlife Biologist
Alaska Department of Fish and Game (Retired)
Kotzebue, Alaska

Carlos G. das Neves, DVM, PhD, Dipl. ECZM
Associate Professor
Norwegian Veterinary Institute
Oslo, Norway

Andy Dobson, DPhil
Professor
Ecology and Evolutionary Biology
Princeton University
Princeton, New Jersey

Alina L. Evans, DVM, MPH, PhD
Associate Professor and Wildlife Veterinarian
Department of Forestry and Wildlife
     Management
Inland Norway University of Applied Sciences
Campus Evenstad
Koppang, Norway

Gregory L. Finstad, PhD
Research Associate Professor
Reindeer Research Program
School of Natural Resources and Extension
University of Alaska
Fairbanks, Alaska

Bruce C. Forbes, PhD
Research Professor and Leader
Global Change Research Group
Arctic Centre
University of Lapland
Rovaniemi, Finland

Kerri Garner, MSc
Manager Lands Protection
Tłı̨chǫ Government
Behchokǫ̀, Northwest Territories, Canada

Maria Hautaniemi, PhD
Senior Researcher
Finnish Food Safety Authority Evira
Research and Laboratory Services Department
Virology Research Unit
Helsinki, Finland

Øystein Holand, PhD
Professor
Department of Animal and Aquacultural
     Sciences
Faculty of Biosciences
Norwegian University of Life Sciences
Ås, Akershus, Norway

Terje D. Josefsen, DVM, PhD
Associate Professor
Nord University
Bodø, Norway

Jörn Klein, PhD, MPhil
Graduated Engineer and Associate Professor
University College of Southeast Norway
Faculty of Health and Social Sciences
Kongsberg, Norway

Susan J. Kutz, DVM, PhD
Professor
Department of Ecosystem and Public Health
Faculty of Veterinary Medicine
University of Calgary
Calgary, Alberta, Canada

Sauli Laaksonen, DVM, PhD
Adjunct Professor
Department of Veterinary Biosciences
Faculty of Veterinary Medicine
University of Helsinki
Helsinki, Finland

Marianne Lian, DVM, MS
PhD Candidate and Wildlife Veterinarian
Departments of Veterinary Medicine and
     Chemistry and Biochemistry
College of Natural Science and Mathematics
University of Alaska Fairbanks
Fairbanks, Alaska

Bryan Macbeth, DVM, PhD
Postdoctoral Research Associate
Department of Ecosystem and Public Health
Faculty of Veterinary Medicine
University of Calgary
Calgary, Alberta, Canada

Gunnar Malmfors, PhD
Senior Consultant
Swedish University of Agricultural Sciences
     (Retired)
Uppsala, Sweden

Torill Mørk, DVM
Veterinary Researcher
Norwegian Veterinary Institute
Section of Pathology Oslo-Tromsø
Tromsø, Norway

Mauri K. U. Nieminen, PhD
Docent
University of Oulu and Kuopio
Finland

Arne C. Nilssen, Cand. Real.
Professor Emeritus
Zoology Department
Tromsø University Museum
UiT – Arctic University of Norway
Tromsø, Norway

Ingebjørg H. Nymo, DVM, PhD
Researcher
Norwegian Veterinary Institute
Section of Pathology Oslo-Tromsø
Tromsø, Norway

Åshild Ønvik Pedersen, PhD
Researcher
Norwegian Polar Institute
Tromsø, Norway

Virve Ravolainen, PhD
Researcher
Norwegian Polar Institute
Tromsø, Norway

Jan Åge Riseth, Cand. Agric. Dr. Scient.
Senior Research Scientist
Northern Research Institute (Norut)
Tromsø Science Park
Tromsø, Norway

Knut H. Røed, Cand. Real., Dr. Agric.
Professor
Department of Basic Sciences and Aquatic
    Medicine
Faculty of Veterinary Medicine
Norwegian University of Life Sciences
Oslo, Norway

Janice E. Rowell, PhD
Research Faculty
School of Natural Resources and Extension
and
Department of Veterinary Medicine
College of Natural Science and Mathematics
University of Alaska Fairbanks
Fairbanks, Alaska

Lavrans Skuterud, PhD
Biophysicist and Senior Research Scientist
Norwegian Radiation Protection Authority
Oslo, Norway

Owen M. Slater, BSc, DVM
Instructor
Department of Ecosystem and Public Health
    and Canadian Wildlife Health Cooperative
Faculty of Veterinary Medicine
University of Calgary
Calgary, Alberta, Canada

Timo Soveri, DVM, PhD
Professor
Department of Production Animal Medicine
Faculty of Veterinary Medicine
University of Helsinki
Saarentaus, Finland

Hans Tømmervik, Dr. Scient.
Senior Research Ecologist
Norwegian Institute for Nature Research
FRAM – High North Centre for Climate and
    the Environment
Tromsø, Norway

Morten Tryland, DVM, PhD
Professor
Arctic Infection Biology
Department of Arctic and Marine Biology
UiT – The Arctic University of Norway
Tromsø, Norway

Robert B. Weladji, PhD
Associate Professor
Department of Biology
Concordia University
Montréal, Québec, Canada

Jonas J. Wensman, DVM, PhD
Associate Professor
Swedish University of Agricultural Sciences
Department of Clinical Sciences
Uppsala, Sweden

Robert G. White, PhD
Professor Emeritus
Institute of Arctic Biology
University of Alaska Fairbanks
Fairbanks, Alaska

**Douglas P. Whiteside, DVM, DVSc,
    DACZM, DECZM (ZHM)**
Clinical Associate Professor
Faculty of Veterinary Medicine
University of Calgary
and
Senior Staff Veterinarian
Calgary Zoo
Calgary, Alberta, Canada

Eva Wiklund, PhD
Associate Professor
Swedish University of Agricultural Sciences
Uppsala, Sweden
and
University of Alaska Fairbanks
Fairbanks, Alaska

Glenn Yannic, PhD
Associate Professor
Université Savoie Mont Blanc
Le Bourget du Lac, France

1 INTRODUCTION

Knut H. Røed, Steeve Côté, Glenn Yannic,
Mauri K. U. Nieminen, Jan Åge Riseth, Hans Tømmervik,
Bruce C. Forbes, Jan Adamczewski, Jim Dau,
Åshild Ønvik Pedersen, Janice E. Rowell, John E. Blake,
Øystein Holand and Robert B. Weladji

CONTENTS

1.1   *RANGIFER TARANDUS*: CLASSIFICATION AND GENETIC VARIATION

Knut H. Røed, Steeve Côté and Glenn Yannic

Intraspecific genetic variation is the most fundamental level of biodiversity, provides the basis for evolutionary change and is crucial for maintaining the ability of species to adapt to new environmental conditions. Knowledge of the genetic variation and structure within reindeer and caribou thus provides essential insights to their future adaptability, including susceptibility to diseases and adaptation to climate change.

1.1.1   CLASSIFICATION AND DISTRIBUTION

Reindeer and caribou (*Rangifer tarandus*) are distributed throughout the northern Holarctic region (Figure 1.1.1); the species is a typical representative of the large mammalian fauna of this region. Reindeer have also been introduced to Iceland (Thorisson 1980) and to the southern hemisphere (e.g. the sub-Antarctic island of South Georgia [Leader-Williams 1988]), where an eradication has recently been conducted. The species belongs to the Cervidae family (the deer family) of ruminant mammals. Further classification and subdivision of *Rangifer* has been proposed and largely debated (Banfield 1961). Historically, there has been almost as much confusion in vernacular names for the species as in scientific names. The use of two vernacular names for this species – reindeer and caribou – has certainly contributed to the confusion. However, both refer to the same species, and caribou includes all wild specimens in North America, while reindeer refers to both wild and semi-domesticated animals in Eurasia. Domestic animals of Eurasian origin in North America are also called reindeer.

1.1.1.1   Subspecies Classification

The subspecies classification of *Rangifer* has been dominated by a high number of described subspecies without well-defined subspecific characteristics. Prior to the last formal taxonomic revision of *Rangifer* in 1961, 55 species and subspecies of caribou and reindeer had been described (Banfield 1961). Although outdated, Banfield's (1961) revision based mainly on craniometrical measurements is still widely used as a guideline for subspecies of reindeer and caribou.

1.1.1.1.1   Subspecies in Eurasia

The Eurasian tundra reindeer (*R. t. tarandus*) is distributed almost continuously across the tundra region and mountain areas of Eurasia (Figure 1.1.1). Most wild and domestic reindeer across Eurasia belong to this subspecies. Today, less than half of the approximately 3–4 000 000 reindeer of Eurasia are wild animals, and in many areas wild and domestic herds are managed in close coexistence (Syroechkovskii 1995; Baskin 2005). The largest population of wild reindeer in Eurasia inhabits the Taimyr Peninsula of northwest Siberia. During the last decades, this population has increased considerably and currently numbers approximately 750 000 animals (Kholodova et al. 2011). In the European part of Eurasia, wild *R. t. tarandus* are mainly distributed in the mountain regions of southern Norway (Reimers et al. 1980), but also in some scattered wild populations in the Kola Peninsula and in the Arkhangelsk and Komi regions of Russia (Baskin 2005).

It is difficult to draw a southern boundary for the distribution of the Eurasian tundra reindeer because of possible overlap with the Eurasian wild boreal forest reindeer (*R. t. fennicus*). The most typical form of *R. t. fennicus* is found in the taiga of the Karelia region near the border of Finland and Russia. The estimate for this subspecies ranges from 850 to 3 000 reindeer (Bisi and Härkönen 2007). However, similar boreal forest reindeer inhabit other areas in Russia and it is debatable whether these herds should be classified within the same subspecies or not (Banfield 1961). It has been argued that other populations in Russia may also be considered as subspecies (Banfield 1961). This applies in particular to reindeer of the Novaya Zemlya Island east of the Barents Sea (*R. t. pearsoni*), the reindeer living in southern taiga of Russian Asia (*R. t. valentinae*), and the population in Kamchatka in northeastern Siberia (*R. t. phylarchus*).

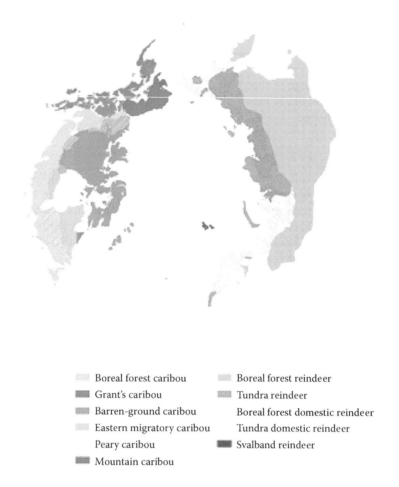

Boreal forest caribou   Boreal forest reindeer
Grant's caribou   Tundra reindeer
Barren-ground caribou   Boreal forest domestic reindeer
Eastern migratory caribou   Tundra domestic reindeer
Peary caribou   Svalband reindeer
Mountain caribou

FIGURE 1.1.1 Distribution of ecotypes of reindeer and caribou (Modified from CAFFs Arctic Flora & Fauna 2001).

The Svalbard reindeer (*R. t. platyrhynchus*) is isolated on the Spitsbergen Archipelago and represents a typical Arctic form of the species. Svalbard reindeer were heavily harvested about a century ago, but after banning of the harvest the population recovered, and currently there are around 10 000 reindeer (Tyler 1987; Hindrum et al. 1995).

1.1.1.1.2   Subspecies in North America

Contrary to Eurasia, the majority of *Rangifer* in North America are wild, and totaled approximately 2–3 000 000 caribou around the last millennium (updated from Miller 2003). The numbers of caribou in North America declined considerably in the early 2000s across most of the subspecies (Vors and Boyce 2009). The species is divided into four subspecies and, in Canada, the extant species are divided further into 11 Designatable Units by the Committee on the Status of Endangered Wildlife in Canada (COSEWIC). COSEWIC's Designatable Unit (DU) concept acknowledges that there are spatially, ecologically or genetically discrete and evolutionarily significant units that are irreplaceable components of biodiversity (www.cosewic.gc.ca/4E5136BF-F3EF-4B7A-9A79-6D70BA15440F /COSEWIC_Caribou_DU_Report_23Dec2011.pdf). In 2017, all the Canadian DUs of caribou were listed by COSEWIC with a conservation status of at least special concern or higher. The Alaska tundra or Grant's caribou, *R. t. granti*, is a migratory tundra subspecies that is distributed from western

Alaska through to the Yukon in Canada. The remaining migratory tundra herds, with the exception of those found in northern Québec, are classified as barren-ground caribou, *R. t. groenlandicus*, which are distributed east of the Alaska tundra caribou across northern Canada, including Baffin Island and Greenland (Miller 2003; Festa-Bianchet et al. 2011). The Canadian barren-ground caribou were listed as "Threatened" by COSEWIC in 2016 (www.cosewic.gc.ca/default.asp?lang=en&n=DD630173-1).

The woodland caribou, *R. t. caribou*, is distributed mainly in the taiga region south of the distribution of the Canadian barren-ground caribou from Newfoundland in the east through to British Columbia and southern Yukon in the west (Festa-Bianchet et al. 2011). The subspecies, as defined by Banfield (1961), contains both migratory barren-ground (Eastern Migratory DU), mountain (Northern, Central and Southern Mountain DUs) and boreal forest forms (Torngat, Boreal, Newfoundland, Atlantic-Gaspésie DUs) and consists of eight DUs. Within woodland caribou, herd size varies from tens of animals to hundreds of thousands. The migratory Rivière-George herd in northeastern Québec and Labrador numbered some 800 000 in the late 1980s, representing at the time the largest caribou herd in the world (Côté et al. 2012). In the early 2000s this herd declined by 98%, down to roughly 8 900 animals in 2016 (Ministère des forêts, de la faune et des parcs du Québec, mffp.gouv .qc.ca/the-wildlife/hunting-fishing-trapping/caribou-migration-monitoring-by-satellite-telemetry).

The Peary caribou (*R. t. pearyi*) is a typical Arctic type and is distributed on the high Arctic islands of northern Canada. Peary caribou have also declined from over 40 000 in the 1960s to about 4 000 in 2008 (Jenkins et al. 2011).

Two subspecies of caribou have gone extinct. The East-Greenland reindeer, *R. t. eogroenlandicus*, also belonged to an Arctic type (Gravlund et al. 1998), inhabited the north and east coast of Greenland and became extinct approximately 100 years ago (Banfield 1961). Dawson's caribou, *R. t. dawsoni*, inhabited Haida Gwaii, an isolated group of islands in the Pacific Northwest, and went extinct in the early 1920s (Byun et al. 2002).

Caribou have never been successfully domesticated and the reindeer husbandry in North America is based on reindeer introduced to Alaska from Eurasia in the 1890s/early 1900s to be managed as livestock (Jernsletten and Klokov 2002). Husbandry increased substantially and the population peaked at 640 000 reindeer in the 1930s, mainly on the Seward Peninsula (Stern et al. 1980). The population then declined sharply, probably due to combined effects of overstocked ranges, lack of care in herding, losses to migratory caribou and predation by wolves (*Canis lupus*). In the early 2000s the reindeer husbandry in Alaska numbered some 20–30 000 animals (Jernsletten and Klokov 2002).

1.1.2 ECOTYPES AND ADAPTATION

To cope with the lack of resolution of *Rangifer* taxonomy, ecotype designations have been increasingly applied (Figure 1.1.1). The ecotype (i.e. a population or group of populations adapted to a particular set of environmental conditions) is a convenient means of classifying caribou populations according to different life-history strategies and ecological conditions. Different terms have been used for *Rangifer* corresponding to ecological adaptations, i.e. the woodland or boreal forest or sedentary form, the barren-ground or tundra or migratory form, the mountain form and the Arctic form (Figure 1.1.2).

The woodland or boreal forest reindeer/caribou appears to be adapted to forest areas, characterized morphologically by large body size, relatively long legs and often with pelage generally dark chocolate brown (Banfield 1961). They are mostly solitary or form small isolated herds; they are rather sedentary and females do not aggregate during calving. The barren-ground or tundra form primarily inhabits open tundra habitat or mountain regions during the summer, with a medium body size, and is often distributed in large migratory herds where females aggregate in large concentrations for calving. The mountain form inhabits alpine areas, undertakes short to medium altitudinal migrations and does not aggregate for calving. Further distinctions, such as deep-snow mountain versus shallow mountain, and northern, central and southern mountain populations, have also been recognized in western Canada (COSEWIC 2014; see the discussion on p. 4 regarding DUs). The Arctic reindeer and caribou appear to be adapted to colder and harsher environments. They are

FIGURE 1.1.2 Ecotypes of reindeer/caribou: (a) American boreal forest caribou (*R. t. caribou*) represented by a female caribou from the Yukon, Canada (Photo: Martin-Hugues St-Laurent); (b) Eurasian tundra reindeer (*R. t. tarandus*) represented by a semi-domestic male from northern Finland (Photo: Steeve D. Côté); (c) Arctic ecotype represented by a group of female Peary caribou (*R. t. pearyi*) with calves from the Axel Heiberg Island, Canada (Photo: Debby Jenkins); (d) migratory tundra caribou (male) belongs to the Rivière-aux-Feuilles herd (*R. t. caribou*) in Canada (Photo: Steeve D. Côté); (e) mountain caribou: a female from the Gaspésie herd (*R. t. caribou*), Canada (Photo: Frédéric Lesmerises); (f) Arctic ecotype represented by a female Svalbard reindeer (*R. t. platyrhynchus*) with her calf from the Svalbard Archipelago (Photos: Nicolas Lecomte).

usually relatively small-bodied with short legs and rostrum and maintain thick insulating winter pelage. Small size reduces the surface:volume ratio of the animal, which suggests a selective pressure for small body size under the extreme environmental conditions of the high Arctic. The distinctiveness of the different ecotypes, however, is not absolute (Gravlund et al. 1998; Couturier et al. 2010), and a gradual transition between the different forms (or ecotypes) in a mainly south-north direction appears to exist in both Eurasia and North America.

1.1.3 COLONIZATION OF NEW PASTURES AFTER THE LAST GLACIAL PERIOD

As for most species of large mammals in northern Eurasia and America, the evolutionary history of *Rangifer* has been highly influenced by glacial and interglacial effects (Lorenzen et al. 2011; Yannic et al. 2014a). The last glacial period was particularly influential for the origin and colonization routes of present subspecies and populations. During a period of about 100 000 years, large parts of both northern Eurasia and America were covered by ice (Figure 1.1.3). During the last glacial maximum (25 000–19 000 years before present), the ice cover was extensive in North America, with most of Canada and the northern part of the United States covered by the massive Laurentide ice sheet (Dyke and Prest 1987). The ice extended south to about 40°N, while Alaska and part of the Yukon remained mostly ice free due to arid climatic conditions. In Eurasia complete ice cover was mostly limited to the northwest, including large parts of the Barents Sea, northwestern Russia, the Scandinavia Peninsula and passing through Germany and Poland, with most of the British Isles covered. Polar deserts and large continuous areas of tundra characterized large parts of Siberia and central Eurasia. At the time of the last glacial maximum, *Rangifer* was found south of the ice sheets in both Eurasia and North America, as well as in Beringia, an area encompassing most of Siberia as well as the Bering land bridge, including parts of Yukon and Alaska (Yannic et al. 2014a; Røed et al. 2014). In these areas reindeer/caribou lived side by side with several other large-bodied exotic mammals such as wooly mammoth (*Mammuthus primigenius*), wooly rhinoceros (*Coelodonta antiquitatis*), giant bison (*Bison latifrons*) and wild horse (*Equus ferus*), most of which are now extinct (Lorenzen et al. 2011). As the climate got warmer, the ice retreated and new *Rangifer* habitat became available (Bigelow et al. 2003; Yannic et al. 2014a). The herds isolated south of the large ice sheets in both Eurasia and North America then colonized areas towards the north, while those in Beringia could expand both west into northwest Europe and east to colonize most of northern Canada and west Greenland (Figure 1.1.3).

FIGURE 1.1.3 Possible routes of colonization of *Rangifer* from the three main glacier refugia at the end of the last glacial period approximately 12–14 000 years ago, the Beringia refugium covering Siberia and part of North America (in pink) and the refugia south of the glaciers in North America (in blue) and in Eurasia (in green). Light blue gives the approximate glacier distribution at the last glacial maximum.

1.1.4   Genetic Characterization and Origin of Subspecies

Genetic diversity has key ecological consequences for populations, communities and ecosystems, because it provides the basis for phenotypic plasticity, local adaptation or the ability to evolve in response to changing environmental conditions (Frankham et al. 2002). Consequently, knowledge of the genetic variation and structure within *Rangifer* is crucial for understanding the evolutionary and demographic history that has formed the contemporary genetic structure and for developing sustainable management strategies to ensure the long-term evolutionary potential of the species.

1.1.4.1   Genetic Markers Reveal Genetic Structure and Evolutionary History

Mitochondrial DNA (mtDNA) has been a popular marker because of characteristics such as its neutral and clock-like pattern of evolution, no recombination due to maternal inheritance, and an elevated mutation rate compared to nuclear DNA (Gissi et al. 2008). Comparing sequence variation in the control region of mtDNA in the different subspecies of reindeer and caribou revealed a set of different mtDNA haplotypes with a pattern indicating three haplotype groups (Figure 1.1.4), presumably representing three separate populations during the last glacial period (Flagstad and Røed 2003). The amount of genetic variability partitioned within and among populations, as well as groups of populations, revealed no relationship between the current subspecies designations and differentiation at the mtDNA level. The morphological differences among extant subspecies appear, therefore, not to have evolved in separate regions during glacial periods (Flagstad and Røed 2003; see also Weckworth et al. 2012; Yannic et al. 2014a). Similarly, weak relationships were found when grouping the populations according to the ecotypes of tundra-migratory, arctic and wood-land-sedentary types, or geographic distribution of Eurasia and North America (Flagstad and Røed 2003; Weckworth et al. 2012). However, the phylogenetic relationships among the different mtDNA haplotypes revealed three main haplotype groups corresponding respectively to *R. t. caribou*,

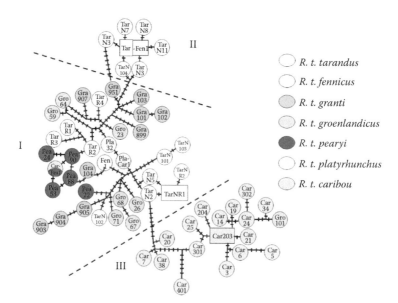

FIGURE 1.1.4   Network of the different mitochondrial DNA haplotypes obtained from subspecies of reindeer and caribou. Stippled lines separate haplotypes belonging to the three main groups. The network illustrates a pattern with two groups dominated by haplotypes found within III) American woodland caribou (*R. t. caribou*) and II) Eurasian tundra reindeer (*R. t. tarandus*), respectively, and I) a haplogroup characterized by a mix of haplotypes found in all the subspecies (Modified from Flagstad and Røed 2003).

R. t. tarandus, and a wide subnetwork with all subspecies represented (denoted III, II and I in Figure 1.1.4). Most recent phylogenetic reconstruction of mtDNA performed with the cytochrome *b* gene divided *Rangifer* into two well-supported and geographically structured lineages. The first lineage covers a vast portion of the species distribution range, from Eurasia to northwestern America, including high Arctic islands such as Greenland, Svalbard and the Canadian archipelagos. The second lineage has a more restricted distribution and includes herds distributed from the island of Newfoundland to the interior plains of Canada (Yannic et al. 2014a).

1.1.4.1.1 Caribou South of the Laurentide Ice Sheet

For both the control region and the cytochrome *b* gene, the most distinct mtDNA cluster comprised haplotypes distributed mainly among the different woodland caribou herds of North America (haplotype group III in Figure 1.1.4). This, and evidence based on genomic DNA (Røed 2005; Weckworth et al. 2012; Yannic et al. 2014a), provides support for an origin of this subspecies south of the Laurentide ice sheet (Figure 1.1.3). A thorough study of woodland caribou in North America revealed mtDNA lineages suggesting postglacial expansion south of the Laurentide ice sheet dating back 15–22 000 years (Klütsch et al. 2012; see also the climatic reconstruction in Yannic et al. 2014a). The putative center of these lineages indicates at least three disconnected regions located in the Rocky Mountains, east of the Mississippi, and the Appalachian Mountains in eastern North America.

1.1.4.1.2 Colonizing from the Beringia

The largest mtDNA group (I in Figure 1.1.4) constitutes a wide subnetwork of haplotypes with all subspecies of reindeer and caribou heavily represented except for those of the most easterly distributed woodland caribou in North America. This cluster appears to represent an ancestral population from the last glaciation period ranging across vast areas of tundra in Eurasia and extending into North America across the Beringian land bridge. As ice cover retreated by the end of the last glaciation period, representatives from this Euro-Beringia lineage appear to have re-colonized suitable habitat on the continental mainland in North America, Siberia and northern Europe, including Fennoscandia (Figure 1.1.3) (Flagstad and Røed 2003; Yannic et al. 2014a). The North American tundra forms (*R. t. granti* and *R. t. groenlandicus*) and the Arctic forms (*R. t. platyrhynchus* and *R. t. pearyi*) are almost exclusively composed by haplotypes that belong to this group.

1.1.4.1.3 Colonizing from a European Refugium

The mtDNA haplogroup II in Figure 1.1.3 exclusively consisted of haplotypes from herds in Fennoscandia. Extended analyses of Eurasian reindeer revealed that this cluster was particularly dominant among the domestic herds in Fennoscandia (cluster II in Figure 1.1.5). Haplotypes belonging to this group were absent from all the North American samples, suggesting a possible origin in southern Europe isolated from the large Euro-Beringian population. Recent modeling of combined mtDNA data from both extant and archaeological reindeer in Fennoscandia yielded an estimated mean separation time of this lineage more than 15 000 generations ago (Røed et al. 2014). This supports a scenario of an origin of the ancestors of the typical Fennoscandian domestic lineage in a glacial refugium different from the large Beringia refugium.

1.1.4.1.4 Increased Differentiation at the Margin of the Species Range

Over the last decades, DNA microsatellites have also been commonly used as genetic markers in studies of evolutionary history of populations and subspecies across the species range (Côté et al. 2002; 2005; Boulet et al. 2007; Røed et al. 2008; Weckworth et al. 2012; Mager et al. 2014; Yannic et al. 2014a). Most microsatellites are neutral loci and thus, their geographical distribution reflects the patterns of genetic drift and gene flow among populations (Slatkin 1995). As for the cytochrome *b*

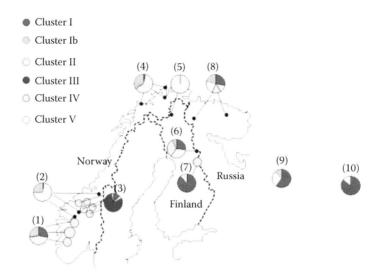

Distribution of mtDNA haplotype clusters in extant wild and domestic reindeer herds across Eurasia. The locations of the analyzed wild and domestic herds are given as green circles and black dots, respectively, and include: (1) the wild reindeer in Langfjella region in southern Norway (three herds); (2) three domestic herds in central Norway; (3) the wild herds of the Dovre/Rondane region in central Norway (five herds); (4) three domestic herds from northern Norway; (5 and 6) domestic herds from, respectively, northern and eastern Finland; (7) wild reindeer from eastern Finland; (8) two domestic herds from northwest Russia; (9) two domestic herds from north-central Russia and (10) two domestic herds from northeastern Russia (Modified from Røed et al. 2011).

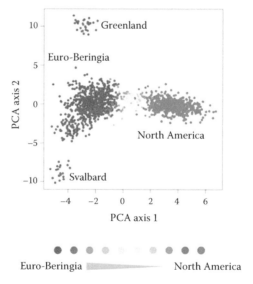

Plot of the first two coordinates from a principal component analysis on microsatellite variation of individual reindeer and caribou sampled across the species distribution. The blue and the red colour give their belonging to the North American or the Euro-Beringia mtDNA lineage (Modified from Yannic et al. 2014a).

mtDNA, the microsatellite data suggests a similar worldwide dichotomy of the genetic structure of *Rangifer*, with most woodland caribou separating from all the others (Figure 1.1.6) (Yannic et al. 2014a). There is also a higher genetic differentiation among herds both at the southern and northern margin of the species range (Yannic et al. 2014a). The isolation of such populations makes them more exposed to genetic drift, with subsequent increased differentiation and loss of genetic variation.

1.1.5 CONSERVATION AND FUTURE CHALLENGES

Marginally distributed populations can be of particular importance from a biodiversity conservation standpoint because they may have unique adaptive traits such as tolerance to extreme climates, or may be locally adapted to particular habitats such as boreal forest or high Arctic environments (Yannic et al. 2014b). However, in light of the ongoing habitat modification and climate change occurring in *Rangifer* habitat, such isolated populations may also be particularly vulnerable due to their reduced evolutionary potential to adapt to a changing environment. Combined climate change and modification/reduction of *Rangifer* habitat is an immense challenge for the maintenance of the different forms of the species. Warmer climate will increase human activities in the north, which will put extra pressure on the resources available for caribou and reindeer (Côté et al. 2012). During the last substantial climate change at the end of the last glacial period, *Rangifer* could respond to the warming climate by colonizing new habitats as the glaciers retreated. Little new *Rangifer* habitat, however, will be available during future climate warming (Yannic et al. 2014a). Consequently, the amount of genetic variation will be of great importance for the various populations of reindeer and caribou to be able to adapt to new environments, including increased pressure from various diseases, particularly with the potential increasing northern distribution of parasites and diseases (Altizer et al. 2013; Kutz et al. 2014).

REFERENCES

Altizer, S., R. S. Ostfeld, P. T. J. Johnson, S. Kutz, and C. D. Harwell. 2013. Climate change and infectious diseases: From evidence to a predictive framework. *Science* 341:514–19.

Banfield, A. W. F. 1961. A revision of the reindeer and caribou, genus *Rangifer*. National Museum of Canada. Bull. No. 177. *Biol. Ser.* 66.

Baskin, L. 2005. Number of wild and domestic reindeer in Russia in the late 20th century. *Rangifer* 25:51–7.

Bigelow, N. H., L. B. Brubaker, M. E. Edwards et al. 2003. Climate change and Arctic ecosystems: 1. Vegetation changes north of 55 degrees N between the last glacial maximum, mid-Holocene, and present. *J. Geophys. Res. Atmos.* 108, D19, 8170.

Bisi, J., and S. Härkönen. 2007. Status of the wild forest reindeer population. In Management plan for the wild forest reindeer population in Finland, 10–15. Ministry of Agriculture and Forestry, Finland.

Boulet, M., S. Couturier, S. D. Côté, R. Otto, and L. Bernatchez. 2007. Integrative use of spatial, genetic, and demographic analyses for investigating genetic connectivity between migratory, montane, and sedentary caribou herds. *Mol. Ecol.* 16:4223–40.

Byun, S. A., B. F. Koop, and T. E. Reimchen. 2002. Evolution of the Dawson caribou (*Rangifer tarandus dawsoni*). *Can. J. Zool.* 80:956–60.

COSEWIC 2014. COSEWIC assessment and status report on the Caribou *Rangifer tarandus*, Northern Mountain population, Central Mountain population and Southern Mountain population in Canada. Committee on the Status of Endangered Wildlife in Canada, Ottawa, Ontario.

Côté, S. D., J. F. Dallas, F. Marshall, R. J. Irvine, R. Langvatn, and S. D. Albon. 2002. Microsatellite DNA evidence for genetic drift and philopatry in Svalbard reindeer. *Mol. Ecol.* 11:1923–30.

Côté, S. D., A. Stien, R. J. Irvine et al. 2005. Resistance to abomasal nematodes and individual genetic variability in reindeer. *Mol. Ecol.* 14:4159–68.

Côté, S. D., M. Festa-Bianchet, C. Dussault et al. 2012. Caribou herd dynamics: impacts of climate change on traditional and sport harvesting. In *Nunavik and Nunatsiavut: From science to policy: An Integrated Regional Impact Study (IRIS) of climate change and modernization*, ed. M. Allard, and M. Lemay, 249–269. ArcticNet Inc., Québec City, Canada.

Couturier, S., R. D. Otto, S. D. Côté, G. Luther, and S. P. Mahoney. 2010. Body size variations in caribou ecotypes and relationships with demography. *J. Wildl. Manage.* 74:395–404.

Dyke, A. S., and V. K. Prest. 1987. Late Wisconsinan and Holocene history of the Laurentide Ice Sheet. *Géogr. Phys. Quatern.* 41:237–63.

Festa-Bianchet, M., J. C. Ray, S. Boutin, and S. D. Côté, A. Gunn. 2011. Conservation of caribou (*Rangifer tarandus*) in Canada: An uncertain future. *Can. J. Zool.* 89:419–34.

Flagstad, Ø., and K. H. Røed. 2003. Refugial origins of reindeer (*Rangifer tarandus* L.) inferred from mito-chondrial DNA sequences. *Evolution* 57:658–70.

Frankham, R., J. D. Ballou, and D. A. Briscoe. 2002. *An Introduction to Conservation Genetics*. Cambridge University Press, New York.

Gissi, C., F. Iannelli, and G. Pesole. 2008. Evolution of the mitochondrial genome of Metazoa as exemplified by comparison of congeneric species. *Heredity* 101:301–20.

Gravlund, P., M. Meldgaard, S. Pääbo, and P. Arctander. 1998. Polyphyletic origin of the small-bodied, high-arctic subspecies of tundra reindeer (*Rangifer tarandus*). *Mol. Phylogenet. Evol.* 10:151–9.

Hindrum, R., P. Jordhøy, O. Strand, and N. Tyler. 1995. Svalbardrein et nøysomt liv på tundraen. *Villreinen* 105–12.

Jenkins, D. A., M. G. Campbell, G. Hope, J. Goorts, and P. McLoughlin. 2011. Recent trends in abundance of Peary Caribou (*Rangifer tarandus pearyi*) and Muskoxen (*Ovibos moschatus*) in the Canadian Arctic Archipelago, Nunavut. Department of Environment, Government of Nunavut, *Wildl. Rep. No. 1*, Pond Inlet, Nunavut. 184 pp.

Jernsletten, J.-L. J., and K. Klokov. 2002. *Sustainable reindeer husbandry*. Centre for Saami Studies, University of Tromsø, Norway.

Kholodova, M. V., L. A. Kolpashchikov, M. V. Kuznetsova, and A. I. Baranova. 2011. Genetic diversity of wild reindeer (*Rangifer tarandus*) of Taimyr: Analysis of polymorphism of the control region of mitochon-drial DNA. *Biol. Bull. Russ. Acad. Sci.* 38:42–9.

Klütsch, C. F. C., M. Manseau, and P. J. Wilson. 2012. Phylogeographical analysis of mtDNA data indicates postglacial expansion from multiple glacial refugia in woodland caribou (*Rangifer tarandus caribou*). *PloS-One* 7, Issue 2:1–10.

Kutz, S. J., E. P. Hoberg, P. K. Molnár, A. Dobson, and G. G. Verocai. 2014. A walk on the tundra: Host-parasite interactions in an extreme environment. *Int. J. Parasitol. Parasites. Wildl.* 3:198–208.

Leader-Williams, N. 1988. Reindeer on South Georgia: *The Ecology of an Introduced Population*. Cambridge University Press, Cambridge.

Lorenzen, E. D., D. Nogués-Bravo, L. Orlando et al. 2011. Species-specific responses of Late Quaternary megafauna to climate and humans. *Nature* 479:359–65.

Mager, K. H., K. E. Colson, P. Groves, and K. J. Hundertmark. 2014. Population structure over a broad spatial scale driven by nonanthropogenic factors in a wide-ranging migratory mammal, Alaskan caribou. *Mol. Ecol.* 23:6045–57.

Miller, F. L. 2003. Caribou. In *Wild Mammals of North America: Biology, Management, and Conservation*, 2nd ed., ed. G. A. Feldhamer, B. C. Thompson, and J. A. Chapman, 965–97. Johns Hopkins University Press, Baltimore.

Reimers, E., L. Villmo, E. Gaare, V. Holthe, and T. Skogland. 1980. Status of *Rangifer* in Norway including Svalbard. In *Proc 2nd Int. Reindeer/Caribou Symp., Røros, Norway*, ed. E. Reimers, E. Gaare, and S. Skjenneberg, 774–785. Direktoratet for vilt og ferskvannsfisk, Trondheim, Norway.

Røed, K. H. 2005. Refugial origin and post-glacial re-colonization of Holarctic reindeer and caribou. *Rangifer* 25:19–30.

Røed, K. H. 2013. Genetisk variasjon hos rein som indikator for opprinnelse og innvandringshistorie. *Norsk veterinærtidsskrift* 2:72–8.

Røed, K. H., Ø. Flagstad, M. Nieminen et al. 2008. Genetic analyses reveal independent domestication origins of Eurasian reindeer. *Proc. R. Soc. B.* 275:1849–55.

Røed K. H., G. Bjørnstad, Ø. Flagstad et al. 2014. Ancient DNA reveals prehistoric habitat fragmentation and recent domestic introgression into native wild reindeer. *Conserv. Genet.* 14:1137–49.

Slatkin, M. 1995. A measure of population subdivision based on microsatellite allele frequencies. *Genetics* 139:457–62.

Stern, R. O., E. L. Arobio, L. L. Naylor, and W. C. Thomas. 1980. *Eskimos, Reindeer and Land*. Agricultural and Forestry Experimental Station, University of Alaska, Fairbanks. Bulletin 59, 205 pp.

Syroechkovskii, E. E. 1995. *Wild Reindeer*. Washington D.C. Smithsonian Institution Libraries.

Thorisson, S. 1980. Status of *Rangifer* in Iceland. In *Proc 2nd Int. Reindeer/Caribou Symp., Røros*, Norway, ed. E. Reimers, E. Gaare, and S. Skjenneberg, 766–770. Direktoratet for vilt og ferskvannsfisk, Trondheim, Norway.

Tyler, N. J. C. 1987. Natural limitation of the abundance of the High arctic Svalbard reindeer. Ph.D. thesis. University of Cambridge, Cambridge.

Vors, L. S., and M. S. Boyce. 2009. Global declines of caribou and reindeer. *Glob. Chang. Biol.* 15:2626–33.

Weckworth. B. V., M. Musiani, A. D. McDevitt, M. Hebblewhite, and S. Mariani. 2012. Reconstruction of caribou evolutionary history in Western North America and its implications for conservation. *Mol. Ecol.* 21:3610–24.

Yannic, G., L. Pellissier, J. Ortego et al. 2014a. Genetic diversity in caribou linked to past and future climate change. *Nat. Clim. Chang.* 4:132–7.

Yannic, G., L. Pellissier, M. Le Corre, C. Dussault, L. Bernatchez, and S. D. Côté. 2014b. Temporally dynamic habitat suitability predicts genetic relatedness among caribou. *Proc. R. Soc. B.* 281:20140502.

1.2 REINDEER AND MAN: FROM HUNTING TO DOMESTICATION

Mauri K. U. Nieminen

1.2.1 DESIGNATIONS

Human beings have had a rich and complex history of interaction with wild reindeer (*Rangifer tarandus*) in northern Eurasia, shifting from hunting to taming and domestication. People usually regard reindeer as a completely wild animal, so it is a surprise to many to know that many of them are domesticated. In the scientific literature, however, there is some confusion about how to classify herded reindeer. Often these reindeer are referred as "domesticated" or, especially during recent decades, as "semi-domesticated," and only sometimes as "tame" (Skarin and Åhman 2014).

Reindeer have been rather easy to domesticate, being docile, with a trusting nature, and allowing themselves to be groomed, milked, dehorned and even castrated, as well as being keen to work by pulling pulks or sleds and cooperating willingly with experienced drivers. Reindeer have been domesticated to a lesser extent than what we normally consider as "domestication" of an animal, in that they still freely roam on pastures year round. It is also one of the few species whose domestication occurred without taming and while the species was also hunted (Baskin 1974). The reindeer is, in fact, the only domesticated deer in the world, and it is also the last animal species to be domesticated.

1.2.2 EARLY HUNTING OF WILD REINDEER

Human-reindeer history goes back to the Paleolithic period and involves hunting of wild reindeer as well as later keeping tamed reindeer for transportation, clothing and food. There is archaeological evidence from the Combe Grenal and Vergisson caves in France that wild reindeer were hunted at least 45 000 years ago, but it is unlikely that reindeer were really domesticated until about 3 000 years ago. Many cave paintings and archaeological remains in Spain from the end of Pleistocene, 11 000–17 000 years ago, have led us to call that period the *Age of Wild Reindeer*. Wild reindeer constituted the main game of Magdalenian human groups of several geographical zones (Kuntz and Costamagno 2011), and today humans remain as the main "predator" of *Rangifer* (i.e. reindeer and caribou) in many areas. Norway and Greenland have unbroken traditions of reindeer hunting from the Ice Age until the present time, and hunting of wild forest reindeer remains common in Finland and Russia (Figure 1.2.1). In the non-forested mountains of central Norway it is still possible to find remains of stone trapping pits, guiding fences and rests, built especially for hunting wild mountain reindeer. Reindeer were one of the mainstays of the hunter-fisher-gatherer diet in many areas. Accounts describe the hunts during the spring and autumn before rutting season between the winter and summer pastures. The autumn reindeer were usually fat and strong, and their fur was also at its best for making warm clothes at this time of the year.

FIGURE 1.2.1   Reindeer hunting has been and still is important in many cultures. Hunting of wild reindeer at Nordmannslågen, Hardangervidda, Norway, about 1960 (Photo: Gunnhild Bitustøyl).

In addition to these accounts, there is plenty of material evidence of historical reindeer use, including the remains of bones and antlers, projective points, systems of hunting pits, corrals or enclosures and fences. The documented fences and systems of hunting pits are located along the migration routes of wild reindeer, especially at places where features in the terrain can lead reindeer to confined areas where they are easily trapped and killed. Good examples of historic hunting constructions are the large enclosures and systems of fences and pitfall traps in Varanger in northern Norway. Similar to the above, variation in hunting techniques was common in Eurasia, to the east from the Nansen hunters on the Taimyr Peninsula and to the Chukchi (Tegengren 1952, Popov 1966, Bogoras 1975). In northern Yukon, Canada, remains of old logs have been documented which formed "the caribou corrals" that were used to catch caribou thousands of years ago.

1.2.3   THEORIES OF DOMESTICATION

According to one much favored theory, domestication began at the end of the Ice Age with hunters following the wild reindeer herds. The hold of the hunters on the herds of wild reindeer gradually increased. Hunters consistently eliminated the "wildest" reindeer until the remaining animals gradually became tamed into a herd of semi-domesticated reindeer. This theory presupposed that reindeer husbandry began perhaps at the beginning of the Early Mesolithic Age about 11 500 years ago (9 500 BC). Another theory was based on the use of wild reindeer females as decoys in hunting during the rutting period. The aim was to get the male entangled in the antlers of the decoy reindeer while it was driving off competitors. Wild reindeer captured in this way would then be tamed by giving them human urine, and trained to be draft and pack reindeer. The use of wild reindeer as decoys and the employment of urine or salt to lure wild reindeer have been both historically and ethnographically documented among all the peoples of northern Eurasia who hunted wild reindeer. Today, Komi and Evenki people in Russia, Mongolia and China still continue to catch and herd reindeer with the use of urine and salt.

Reindeer have been domesticated by denizens of the Northern Hemisphere for some time, but exactly for how long and whether domestication occurred at different sites or only once have been the matter of some debate. Estimates indicate that some domestication took place earlier in Eurasia, ranging from 20 000 to 3 000 years ago. Finds of sledge runners may show that some reindeer might have been used as draught animals since the Mesolithic period, from 8 000 BC onwards

(Burov 1989). However, the semi-domestic transport reindeer were being adopted into the taiga hunting and fishing communities of northwest Siberia much later, during the 19th and early 20th centuries (Jordan 2012).

1.2.4 ANCIENT EVIDENCE OF DOMESTICATION

The evidence concerning the dating of domestication may be ascertained by anthropological studies and drawings in caves and upon rocks depicting domestication of reindeer. For example, caves near the river Lena in Russia contain paintings believed to be 3 000 years old depicting humans walking beside reindeer, without weapons. This suggests the early beginnings of reindeer domestication. In southern Siberia there are rock drawings dated about 2 000 years old that depict humans riding on reindeer. Reindeer have always been a part of the diet, and judging from parts of harnesses at the site Ust'Puloi in western Russia, using tamed reindeer as a means of transport existed in the early Iron Age in the 3rd century BC (Golovnev and Osherenko 1999). At this time hunters were building large reindeer sledges to travel long distances with their families in pursuit of the mobile wild reindeer herds (Jordan 2012). A number of sledge runners have also been recovered from bogs in northern Fennoscandia – some scholars have claimed these belong to sledges pulled by reindeer. The oldest dated remains of a sledge were found in Soukolojärvi, in Sweden and are dated back to the 13th century. Hunting pitfall systems in the interior of northern Scandinavia are suggested to have been used from the late Mesolithic, about 7 000 BC, while enclosures, such as corrals, have only been found depicted in rock art from about 4 300 to 3 700 BC (Helskog 2011). On the northern coast of Scandinavia near Alta fjord in Norway, there is a rock drawing dated to be 5 000–6 000 years old, depicting a circular corral like those still in use today for separating the semi-domesticated reindeer. Large groups of reindeer in enclosures have been found on these rock drawings and engravings. Probably, they are only drawings of hunting corrals. This is 6 000 years earlier than the 17th century AD, with the earliest direct historic evidence for reindeer corrals in northern Norway (Vorren 1958, Simonsen 1996). Considering the fact that humans have hunted reindeer for over 40 000 years, reindeer, compared with other domesticated animals, were tamed quite late. For example, sheep, the earliest domesticated animal, was tamed about 10 000 BC. The semi-domesticated reindeer only has a few thousand years behind it.

According to Adrianov (1888), one rock picture from the Sayan Mountains presented three animals with branching antlers, large ears and a short tail, which are typical of reindeer. In the *Kaia-Baji* picture, the reindeer was depicted as a domesticated animal used for riding; also, in another area, a rock picture was found which represented a reindeer hitched to a sledge. Reindeer saddles used by ethnographic Sayan appeared to be derived from horse saddles of the Mongolian steppes. Those were used by Evenkis, derived from Turkish cultures on the Altai steppe. Sledges drawn by draft reindeer also had attributes that appeared to be adapted from those used with cattle or horses. These contacts are estimated to have occurred no longer ago than about 1 000 BC. Mirov (1945) concluded from consideration of these rock pictures that reindeer were domesticated in the old Samoyed country of the upper Yenisei at the beginning of our era, a fact which also corroborated Laufer's conviction that the domestication of reindeer occurred about that time. It might then be possible that the Samoyed learned reindeer driving in the south, as an imitation of the horse and ox driving by the steppe nomads.

1.2.5 ARCHAEOLOGICAL AND ETHNOGRAPHICAL EVIDENCE

Both Aristotle and Theophrastus had short accounts of an ox-sized deer species, named *tarandos*, living in the land of the Bodines in Scythia. The descriptions have been interpreted as being of reindeer living in the southern Ural Mountains about 350 BC (Sarauw 1914). A deer-like animal described also by Julius Caesar in his *Commentarii de Bello Gallico* from the Hercynian Forest in 53 BC is most certainly to be interpreted as reindeer. In 98 AD, the Roman historian Tacitus wrote about a strange people in Thule, who used fur clothes, hunted reindeer and traveled on the snow with skis. Later the story of the Norwegian chieftain Ottar (Ohthere) appeared as an important information source, because he gave the first written description of domesticated reindeer herding in Fennoscandia. However, archaeological research has consistently pushed the date of domestication of reindeer and development of reindeer herding further back time. The northern peasant Ottar visited King Alfred the Great in England in 890, where he gave a description of his travels to the northern parts of Fennoscandia (see Bately and Englert 2007). He gave an account to the King about the Lapps and said that reindeer were domesticated and managed in herds.

Ottar had 600 unsold reindeer and six tame reindeer, and the latter were "*very valuable among the Finns* (Lapps)*, since they catch the wild reindeer with them*" (Simonsen 1996, Bately and Englert 2007). The interesting part here was not the 600 reindeer, but the six "valuable" ones. It was more than a qualified guess that Ottar might have tended 600 reindeer on an island for food supply and trade (Bratrein 1989), but the other six probably bear witness to the fact that draft reindeer have been a necessity for trade and communication for a very long time in northern Fennoscandia. Driving reindeer into corrals, as practiced among the present pastoral reindeer Sami, was similar to a hunting technique practiced by hunters in the Old as well as in the New World (Blehr 1990).

Information on the domestication of reindeer has existed for a long time. According to a Persian history book on Mongols from before the 14th century (Jami'al-Tarikh), Mongols raised reindeer instead of cattle and sheep. While migrating, they lived in the forests and stayed in birch bark tents. They believed that life in towns was unbearable and enjoyed hunting reindeer by skiing on the steppe or taiga (Shiwei 2011). Casual remarks about tamed reindeer can also be found in old Chinese annals, in Rashiduddin's *History of the Mongols*, in the Italian travels of Marco Polo and in later descriptions by some Scandinavians, such as Ottar in 898 AD (see Mirov 1945). Laufer (1917) was the first to investigate the information on domestic reindeer found in early Chinese documents. He told about a traveling Buddhist monk, Huei Shen, who in 499 AD visited a kingdom, Fu-sang, situated far off in the northeastern ocean. "*The people there have vehicles drawn by horses, oxen and stags. They raise deer ... and make cream from their milk.*" The geographic location of Fu-sang is ambiguous, but Laufer was thinking that the deer mentioned by Huei Shen were domesticated reindeer from the Baikal region. According to Maksimov (1928), "*they used wheeled carts, keep deer and from their milk they make kumis.*" However, vehicles drawn by the reindeer were, according to Laufer (1917), sledges.

The exact time of the domestication of reindeer is not known because domestication of reindeer may have begun at different times and in many places. The main problem with exact dating has also been the lack of firm archaeological evidence. The evidence relies mostly on some ethnographic observations, such as the development of reindeer saddles and sledges about 3 000 years ago, used by the nomadic pastoral people, such as Lapps, Sayans and Evenkis in Eurasian arctic or subarctic areas. Sledges have been used by archaeologists to identify the earliest evidence of reindeer domestication, and the first domesticators probably were the Samoyed (Donner 1933).

Concerning the people of the Sayan Mountains, a mountain range in southern Siberia in Russia, herding traditions are believed to be ancient and it is considered that the domestication of reindeer may have begun in this region. Originating from the Sayan Mountain region and Mongolia's northernmost province of Khovsgol, the Tsaatan, or ethnic Dukha peoples, are also credited as one of the world's earliest domesticators of reindeer (Keay 2006). This attribution rests on two types of evidence: (1) archaeological, largely consisting of 3 000-year-old rock carvings depicting reindeer and (2) ethnographic, comprising observations of the techniques of reindeer management employed by the Tuva. Evenki and Sayan reindeer herding is based on a closer relationship between the reindeer and the herder. As a result, their taiga reindeer are more tame than tundra reindeer.

Most reindeer in Evenki herds are used to being saddled and either ridden or burdened with a pack, and the female reindeer are also milked. This tamer character of Evenki taiga reindeer could be the result of a longer period of domestication. Evenki-type reindeer herding originated earlier than large-scale tundra herding. Probably, reindeer domestication took place in southern Siberia 2–3 000 years ago, either in the Sayan Mountains or around Lake Baikal, by the ancestors of today's Evenki and Eveny peoples. It seems likely that the species was first domesticated in small numbers to carry humans and their packs in order to hunt mainly wild reindeer. It was the ability to ride on reindeer that opened up much of north Asia, and it allowed the Evenki and Eveny to become among the most widely spread-out indigenous peoples anywhere in the world, transferring their discovery of reindeer domestication to other neighboring groups (Vitebsky and Alekseyev 2014).

Since the conquest of Siberia by the Russians, our knowledge of reindeer domestication has been greatly enriched. Early Siberian travelers contributed at least some information on either the wild or the tamed reindeer. But not until the beginning of the 19th century did the question of reindeer domestication become the subject of thorough investigation. Between 1916 and 1920, Sirelius published an account of the nature and time of reindeer domestication. Almost at the same time, Laufer's treatise (1917) on the subject appeared, followed by Wiklund's work (1918) and by Hatt's "*Notes on Reindeer Nomadism*" (1919). However, the subject was, in 1928, critically reviewed by Maksimov. Wiklund considered that there were four independent centers of reindeer domestication. This was later called the polycentric hypothesis (see Nelleman 1961, Gordon 2003). Hatt and Maksimov believed that taming of reindeer by the Lapps originated independently from a southern Siberian center of domestication, but Laufer supposed that reindeer husbandry started only in southern Siberia (in agreement with the monocentric hypothesis) and was transmitted to the Lapps by Samoyed, today called Nenets. The importance of southern Siberia as a center of domestication was, however, emphasized equally by all.

1.2.6 Development of Modern Reindeer Herding

The zones in which reindeer husbandry developed have been distinguished as follows: (1) the Sayan-Altai-western Siberia-northeastern Europe (the Nenets, the Khanty, the Mansi and the Komi), (2) central and eastern Siberia (the Evenki and the Chukchi) and (3) Fennoscandia (Sami and their neighbors). Our recent study on reindeer mitochondrial DNA (mtDNA) (Røed et al. 2008) identified at least two separate and apparently independent reindeer domestication events (polycentric hypothesis), one in eastern Russia and one in Fennoscandia. Reindeer could also have been domesticated and used as lures in hunting for wild reindeer in ancient times in Fennoscandia. They were tamed for milking and for pulling a skier or a sledge. The use of ear marks as an indication of ownership and the use of herding dogs indicate that reindeer husbandry in this region was originally a variant of Norwegian sheep and goat husbandry. Later, in the 16th and 17th centuries, reindeer husbandry started on a large scale in Norway. The semi-domesticated reindeer herds in Finnmark, which today play a key role in the culture of the indigenous Sami people, are still very similar to their wild relatives, but they have distinct genetic characteristics (see Chapter 1.1, *Rangifer tarandus* – Classification and Genetic Variation).

The Lapps adopted a nomadic lifestyle during the 16th century, and traditional reindeer herding today is practiced by approximately 10% of the Sami, with new technologies and for considerable profit. This has been a progressive process involving three stages of development.

The first of these was the hunter-gather phase, when they subsisted by living on berries, and by fishing and hunting, including of wild reindeer by setting traps. A small number of reindeer were tamed and sometimes used as draft animals and as decoys to trap others. With this system, reindeer and man lived in a state of equilibrium.

The second stage was pastoralism, mostly referred to as intensive reindeer herding, which started in central Norway during the 11th century, when the Lapps abandoned subsistence hunting of wild reindeer and began to gather reindeer in herds for domestication. Reindeer herding involved a nomadic life style – following the reindeer as they made their seasonal migrations. A symbiotic relationship was developed as the nomads protected the reindeer herds from predators and, in return, subsisted on reindeer for the provision of meat, milk, clothing and shelter. The reindeer herds were maximized by the control of reproductive increase, and by the 15th century, large-scale reindeer herding was in progress, and it also spread to Sweden and to northern part of Finland.

The third stage of reindeer domestication in this region was extensive reindeer herding. Extensive reindeer herding is characterized by the maximization of profits from selling reindeer for slaughter.

The reason for these presumed dramatic shifts is given in two ways. Vorren (1973, 1978) pointed to external explanations in terms of increased taxation, trade relations and the introduction of firearms, which led to the reduction and, in the end, depletion of the wild reindeer stock. As an alternative, people turned more to pastoralism. Hansen and Olsen (2004) referred to intrinsic reasons. Like Vorren (1978), their focus was the hunting pit and corral systems of the Varanger peninsula. The latest use of the corrals was dated to the early 17th century (Lillienskiold 1942). Such large-scale hunting must have depended upon sophisticated organizational forms and technology. To meet the external demands embedded in taxation and trade, an organizational elite developed to manage the seasonal hunting. Thus local hierarchies appeared, which established ownership over corrals and reindeer, and consequently pastoralism developed (Hansen and Olsen 2004). This transformation is dated to the 16th and 17th centuries (Vorren 1978, Hansen and Olsen 2004).

Referring to the so-called *Stallo* sites, some archaeologists have dated the emergence of pastoralism to the Viking Age, 800–1 000 AD (Aronsson 1991, Storli 1993, Hedman 2005, Bergman et al. 2008). According to this view, the *Stallo* sites proved that domestication had taken place and that pastoralism had become a reality. Others, however, defined these sites as seasonal camps used for wild reindeer hunting and suggested that a pastoral adaptation did not take place until the 16th to 17th centuries (Hansen and Olsen 2004, Sommerseth 2011). Common to both theories, whether reindeer husbandry was dated to 1 000 or 1 600 AD, is the idea that the management of domesticated reindeer represented a profound change in the Lapps culture and became a main strategy in their economic adaptation.

Writings after that time tell that the Lapps were using domesticated reindeer for transport and milking. In the 16th to 18th centuries, Sweden, which then included Finland, had imperial ambitions, which increased the tax burden on the Lapps' reindeer herding, which also may have stimulated a shift in reindeer herding practices. The Lapp reindeer herders were nomadic and moved with their reindeer herds between winter and summer pastures. In the mountain areas an intensive reindeer herding took shape, where the reindeer were monitored daily. The Lapps lived and worked in so-called *siiddat*, reindeer herding groups, and semi-domesticated reindeer were used for transport, milk and meat production. The *Siida* was an old community system, and could consist of several families and their reindeer herds. Reindeer were so important to nomadic people such as the Lapps that in their language there are many words describing reindeer, such as the variation of colours, size, texture of fur, shape of the antler, the ability to pull a sledge and the degree of tameness. Compared to Norway and Sweden, the reindeer husbandry developed in a different manner in Finland. Finnish settlers and peasants adopted reindeer herding as a livelihood from the Lapps, and Finnish reindeer herding became organized already in the 1700s. In 1898, state authorities obligated reindeer owners to establish geographically defined herding districts, *paliskunnat* (in Finnish) (Nieminen 2006). Reindeer are not currently watched year round, and reindeer move with relative freedom during certain periods, but because of poor lichen pastures, supplementary feeding of reindeer during winters has become common (see Figure 1.2.2; see also Chapter 4, Feeding and Associated Health Problems).

FIGURE 1.2.2   Semi-domesticated reindeer in a Finnish mountain area (Photo: Mauri Nieminen).

1.2.7   WILD AND DOMESTICATED HERDS TODAY

Wild and domesticated reindeer today show many of the same behavioural characteristics, including the need to migrate. Originally reindeer were herded and the annual migration was closely followed by nomadic herders. Today, semi-domesticated reindeer usually have an earlier onset of the breeding season, are smaller and have a less-strong urge to migrate than wild relatives. Nevertheless, semi-domesticated reindeer, unlike sheep and cattle, have retained much of their intrinsic similarity to their wild counterparts.

Because reindeer are well adapted to their environments, little selective breeding has taken place, and semi-domesticated reindeer can breed successfully with their wild counterparts and would soon return to their original natural state, which has been indicated through several translocations of semi-domesticated reindeer to other geographic regions. Semi-domesticated reindeer were introduced from Norway to East Iceland in the late 1700s (Skarphédinn 1984), and the population today consists of about 6 000 rather "wild" reindeer, which are not herded, but hunted, with an annual harvest of more than 1 000 animals. A few reindeer from Norway were also introduced to the South Atlantic island of South Georgia in the beginning of the 20th century. Because of ecological challenges with this invasive species, the whole population of about 5 000 "wild" semi-domesticated reindeer were eradicated during 2013–14. A total of 59 reindeer calves were translocated from South Georgia to the Falkland Islands in 2001 (Cameron and Dieterich 2010). Semi-domesticated reindeer have also been introduced to the French sub-Antarctic archipelago of Kerguelen Islands, today totalling a population of about 4 000 "wild" reindeer.

Rangifer originally had a circumpolar distribution in the mountain/tundra and forest/taiga zones of northern Europe, Siberia and North America (see Banfield 1961). No less than one-quarter of the land surface of the Earth is today used for reindeer herding, but most of this area is in the remote arctic and subarctic regions. Although it may be more accurate to consider them as semi-domesticated reindeer, many reindeer of this region remain entirely wild, such as the large herds of never-domesticated caribou which continue to freely roam in North America and Greenland and are mainly hunted by Inuit. There are as many as 3.5 million wild reindeer/caribou, and today there are about 60 major wild herds around the Northern Hemisphere. A total of 34 of these herds are declining, while no data exist for 16, and only eight herds are increasing in number. Many herds have been declining for a decade or more (see Solveig and Boyce 2009). About 1 million caribou live in

Alaska, a comparable number in northern Canada and about 100 000 in West Greenland. In total, more than 1 million wild reindeer live today in Russia, Finland (about 2 000), southern Norway (30 000) and Svalbard (10 000) (Nieminen 2014). Hunting of wild reindeer is strictly controlled in Norway and Russia, though poaching may still occur, and fencing has been erected in Finland to prevent the hybridization of wild forest reindeer with semi-domesticated reindeer.

Eurasia was the birthplace of reindeer herding, but it has spread around the world and is practiced today in ten countries and by 30 different ethnic and Arctic peoples. Approximately 100 000 people are engaged in reindeer husbandry, and wild and semi-domestic herds are managed in close coexistence in many areas. Alaskan semi-domesticated reindeer, originating from Russia and today counting about 10 000, have maintained a genetic variation comparable to that in Russia and differentiate from that of wild caribou, over 120 years after their introduction to Alaska (Cronin et al. 2006). About 3 000 semi-domesticated reindeer are in Canada today (Nieminen 2014). Further, about 3 000 semi-domesticated reindeer are herded in Greenland, 300 in Hokkaido, Japan and 150 in Scotland (Nieminen 2014).

Today, the semi-domesticated reindeer plays an important economic role for circumpolar peoples in Eurasia. In Russia there are about 1.6 million reindeer, which account for approximately 70% of the world's stock of semi-domesticated reindeer. Reindeer husbandry has great social significance in Russia, with 22 national minorities being involved in this branch of the economy (Baskin 2000) (Figure 1.2.3). The huge herds of modern times arose with Russian colonialism starting in the 17th century (see Krupnik 1993, Vitebsky and Alekseyev 2014).

The reindeer herding people in south Siberia and Mongolia include the Dukha of northwestern Mongolia, the Tozhu of the Republic of Tuva, the Tofa of Irkutsk Province, the Soyot of the Buryat Republic and the Evenki, who range throughout south Siberia and into the northern tip of China's Inner Mongolia Autonomous Region. The domestic reindeer in Mongolia, originally from Siberia's Tuva province, seem to be domesticated from Siberian forest reindeer (*R. t. valentinae*), and reindeer in Inner Mongolia, in China, are originally from Sakha, Russia, and are domesticated from Okhotsk forest reindeer (Baskin 1986; Nieminen 2011a,b). Inhabiting a fragile transition belt of taiga and alpine tundra between the Siberian boreal forest and the inner Asian steppes, these people represent the southernmost extreme of reindeer pastoralism in the world. These four groups, the, Dukha, Tozhu, Tofa and Soyot, all live in adjacent quadrants of the Eastern Sayan mountain range, but under different administrative regimens. Unlike large-scale reindeer "ranchers" of northern

FIGURE 1.2.3   Nenets men with reindeer in Jamalo-Nenets, Russia (Photo: Mauri Nieminen).

FIGURE 1.2.4   Riding reindeer in Mongolia (Photo: Mauri Nieminen).

Siberia, European Russia and Scandinavia, the south Siberian and northern Mongolian groups raise small herds of reindeer in the taiga. They use the reindeer predominantly as pack and riding animals to facilitate their hunting and as source of milk products, while fish and wild game are the principal sources of animal protein (Figure 1.2.4). Of these about 10 000 people, however, less than 1 000 are still actively involved in reindeer husbandry. About 3 500 semi-domesticated reindeer remain in the region, down from 15 000 just a decade ago.

There are about 5 million km^2 of pastures in the world, with a total semi-domesticated reindeer population of 2.5 million. Of this, forest-tundra is 3 million km^2, with a carrying capacity of 2.5 million reindeer, whereas taiga areas cover 1.9 million km^2, with a carrying capacity of 1 million reindeer. Mainly, meat production is practiced, by Sami (with about 500 000 reindeer), Finns (110 000), Nenets (800 000), Komi (115 000) Eveny (20 000), Chukchi (100 000) and Koryak (200 000) (Baskin 2005, Nieminen 2014) peoples. Reindeer supply high quality meat, skins and fur, and also young velvet antlers. In some parts of Siberia and Mongolia, reindeer are also milked (Fondahl 1989, Nieminen 2011a,b, 2014). Today, total reindeer meat production per year is about 50 million kg. Reindeer are very widely used as transport animals, but this becomes less important every year with the increasing use of mechanized vehicles.

REFERENCES

Adrianov, A. B. 1888. Journey to Altai and beyond Sayans in 1881. *Zapiski Imp. Russ. Geog. Ob.* 11:149–422.
Aronsson, K.-Å. 1991. *Forest reindeer herding A.D. 1–1800: An archaeological and palaeoecological study in northern Sweden*. Volume 10. University of Umeå, Department of Archaeology, Umeå, Sweden.
Banfield, A. W. F. 1961. A revision of the reindeer and caribou, genus *Rangifer*. Bulletin of the National Museum of Canada. Biological Services 177, 66 p.
Baskin, L. M. 1974. Management of ungulate herds in relation to domestication. In *The behavior of ungulates and its relation in management*, eds. V. Geist and F. Walther, 530–41, Volume 2. IUCN Publ., New Ser. 24. Morges, Switzerland. Int. Union for the Conservation of Nature.
Baskin, L. M. 1986. Differences in the ecology and behavior of reindeer populations in the USSR. *Rangifer*, Special Issue, No. 1:333–40.
Baskin, L. M. 2000. Reindeer husbandry/hunting in Russia in the past, present and future. *Polar Res.* 19(1):23–39.

Baskin, L. M. 2005. Number of wild and domestic reindeer in Russia in the late 20th century. *Rangifer* 25:51–8.

Bately, J. and A. Englert (eds.). 2007. Ohthere's voyages: A late 9th century account of voyage along the coast of Norway and Denmark and its cultural context. *Maritime cultures of the North, Roskilde.* Viking Ship Museum, 112–16.

Bergman, I., L. Liedgren, L. Östlund et al. 2008. Kinship and settlements: Sami residence patterns in the Fennoscandian alpine areas around A.D. 1000. *Arctic Anthropol.* 45(1):97–110.

Blehr, O. 1990. Communal hunting as a prerequisite for caribou (wild reindeer) as a human resource. In *Hunters of the Recent Past*, eds. L. B. Davies and B. O. K. Reeves, 304–326, Unwin Hyman, London.

Bratrein, H. D. 1989. Karlsøy og Helgøy bygdebok. Bd. 1. Karlsøy commune. (In Norwegian).

Bogoras, V. G. 1975. *The Chukchee.* AMS Press, New York.

Burov, C. M. 1989. Some Mesolithic wooden artifacts from the Mesolithic site visit in the European North East. In *The Mesolithic in Europe*, ed. C. D. Bonsall, 341–401, Edinburgh.

Cameron, M. B. and R. A. Dieterich. 2010. Translocation of reindeer from South Georgia to the Falkland Islands. *Rangifer* 30(1):1–9.

Cronin, M. A., M. D. MacNeil and J. C. Patton. 2006. Mitochondrial DNA and microsatellite DNA variation in domestic reindeer (*Rangifer tarandus tarandus*) and relationship with wild caribou (*Rangifer tarandus granti, Rangifer tarandus groenlandicus*, and *Rangifer tarandus caribou*). *J. Hered.* 97(5):525–30.

Donner, K. 1933. *Siperia, elämä ja entisyys.* Otava, Helsinki, 272 p. (In Finnish).

Fondahl, G. 1989. Reindeer dairying in the Soviet Union. *Polar Rec.* 25(155):285–94.

Golovnev, A. and G. Osherenko. 1999. *Siberian Survival: The Nenets and Their Story.* Cornell University Press, New York.

Gordon, B. 2003. *Rangifer* and man: An ancient relationship. *Rangifer,* Special Issue No. 14:15–27.

Hansen, L. I. and B. Olsen. 2004. *Samenes historie fram till 1750.* Cappelen Akademisk Forlag, Oslo. (In Norwegian).

Hatt, G. 1919. Rensdyrnomadismens elementer. *Geogr. Tidsskr.* 241–69. (In Norwegian).

Hedman, S. D. 2005. Renskötselns uppkomst in Övre Norrlands skogsområden. In *Fra villreinjakt til reindrift*, ed. O. Andersen. Skriftserie nr. 1. Drag, Arran Lulesamisk senter. (In Swedish).

Helskog, K. 2011. Reindeer corrals 4700–4200 BC: Myth or reality? *Quat Int.* 238:25–34.

Jordan, P. 2012. From hunter to herder? Investigating the spread of transport innovations in northwest Siberia. *Suomalais-Ugrilaisen Seuran toimituksia* 265. Helsinki 2012, 27–48.

Keay, M. G. 2006. The Tsaatan reindeer herders of Mongolia: Forgotten lessons of human-animal systems. *Encyclopedia of Animals and Humans*, p. 1–4

Krupnik, I. I. 1993. *Arctic Adaptations: Native Whalers and Reindeer Herders of Northern Eurasia.* University Press of New England, Hannover and London.

Kuntz, D. and S. Costamagno. 2011. Relationships between reindeer and man in southwestern France during the Magdalenian. *Quat. Int.* 238(1–2):12–24.

Laufer, B. 1917. The reindeer and its domestication. *Quarterly for the American Anthropological Association*, USA, 134 p.

Lillienskiold, H. 1942. Speculum boreale eller den Finnmarchiske beschriftwelsis. (In Norwegian). Bind 1. Nordnorske Samlinger utgitt av Etnografisk Museum IV. Oslo.

Maksimov, A. N. 1928. Origin of reindeer husbandry. *Inst. Istorii. Uchenile Zapiski* 6:3–37. (In Russian).

Mirov, N. T. 1945. Notes on the domestication of reindeer. *Am. Anthropol.* 47(3):393–408.

Nelleman, G. 1961. Theories on reindeer breeding. *Folk* 3:91–103.

Nieminen, M. 2006. History and development of Finnish reindeer husbandry. *Poromies* 5:26–9. (In Finnish).

Nieminen, M. 2011a. Reindeer husbandry, problems diagnosis: Visit and reindeer research in Mongolia. *Research Report 2011.* Finnish Game and Fisheries Research Institute, Reindeer Research Station, Kaamanen, Finland, 40 p.

Nieminen, M. 2011b. Reindeer herding and husbandry in Mongolia and Inner Mongolia. RKTL:n työraportteja 14/2011, 21 p. (In Finnish).

Nieminen, M. 2014. *Poro – Reindeer.* Books on Demand, Helsinki, Suomi, 244 pp.

Popov, A. A. 1966. *The Nganasan: The Material Culture of the Tavgi Samoyeds.* Indiana University Publications.

Røed, K. H., Ø. Flagstad, M. Nieminen et al. 2008. Genetic analyses reveal independent domestication origins of Eurasian reindeer. *Proc. R. Soc. Lond. B. Biol. Sci.* 275:1849–55.

Sarauw, G. 1914. Das Rentier in Europe zu den Zeiten Alexanders und Caesars. In *Mindeskrift in Anledning af Hundredeaaret for Japetus Steenstrups Födsel*, eds. H. F. E. Jungersen and E. Warming, 1–33, Copenhagen. (In German).

Skarphédinn, T. 1984. The history of reindeer in Iceland and reindeer study 1979–1981. *Rangifer* 4(2):22–38.

Shiwei, D. 2011. Looking for reindeer herders in Asia forest: Comparative studies in China, Mongolia and Russia. International Conference on "Cultural Diversity of Nomads," Ulan Baatar 2011, p. 87–91.

Simonsen, P. 1996. Ottar fra Hålogaland. In *Ultima Thule*, eds. S. Nesset and H. Salvesen, 27–42, Tromsø. (In Norwegian).

Skarin, A. and B. Åhman. 2014. Do human activity and infrastructure disturb domesticated reindeer? The need for the reindeer's perspective. *Polar Biol.* 37:1041–54.

Solveig, L. V. and M. S. Boyce. 2009. Global declines of caribou and reindeer. *Glob. Chang. Biol.* 15:2626–33.

Sommerseth, I. 2011. Archaeology and the debate on the transition from reindeer hunting to pastoralism. *Rangifer* 31:111–27.

Storli, I. 1993. Sami Viking Age pastoralism – or the "fur trade paradigm" reconsidered. *Norwegian Archaeol. Rev.* 26(1):1–48.

Tegengren, H. 1952. En utdöd lappkultur i Kemi lappmark: Studier i nordfinlands kolonisasjonshistoria. *Act. Acad. Ab., Humaniora* 19(4), Åbo. (In Swedish).

Vitebsky, P. and A. Alekseyev. 2014. What is a reindeer? Indigenous perspectives from northeast Siberia. *Polar Rec.* 51:413–21.

Vorren, Ø. 1958. Samisk villreinfangst i elder tid. *Ottar* 17(2). (In Norwegian).

Vorren, Ø. 1973. Some trends of the transition from hunting to nomadic economy in Finnmark. In *Circumpolar Problems: Habitat, Economy and Social Relations in the Arctic*, ed. G. Berg, 185–94.

Vorren, Ø. 1978. *Villreinfangst i Varanger fram til 1600–1700 årene*. Nordkalottforlaget, Stonglandseidet. (In Norwegian.)

Wiklund, K. B. 1918. Om renskötselns uppkomst. *Ymer* 249–73. (In Swedish).

1.3   SUSTAINABLE AND RESILIENT REINDEER HERDING

Jan Åge Riseth, Hans Tømmervik and Bruce C. Forbes

1.3.1   INTRODUCTION: SUSTAINABILITY AND RESILIENCE

The concept of *sustainability* has a long history in science. Its original meaning was to endure for a long time; cf. the German word *Nachhaltigkeit*. Though the concept, also due to its use in the political sphere (cf. WCED 1987), has different meanings, we will start out here with its original scientific meaning of *potential to endure in time* (cf. Salo et al. 2014). Basically, sustainability of natural production systems can be seen as dependent on two groups of factors: external and internal. Internal factors typically include harvesting and stocking rates, while external factors include encroachments on pastures, natural disasters and climate change effects. Well-known marine examples of how overharvesting can threaten a resource base to the verge of extinction are the North Sea herring fisheries and Antarctic whaling. For the discussion of sustainable harvesting and stocking rates in reindeer herding, a standard range management approach, originally developed in fisheries management (Gordon 1954), may be a useful starting point, not least due to its extensive use in public policy. Figure 1.3.1 has a revenue curve and a cost curve and a number of possible equilibria.

The revenue curve is based on pure logistic growth and has a maximum point at stock level MSY (maximum sustained yield), marking the maximum aggregate output per unit area. The revenue curve reaches zero at K, which represents what population biologists refer to as *ecological carrying capacity*. The cost curve is linear and crosses the revenue curve at OA, which resource economists refer to as *open access equilibrium*. This is the point of (average) zero profit, i.e. stocking rates beyond OA will provide negative profits. MP marks the level where ranchers will achieve *maximum profit*. The system depicted in Figure 1.3.1 is as simple as can be, but it illustrates a basic reasoning regarding intensity of resource use and output. These theoretical equilibria may seem objective;

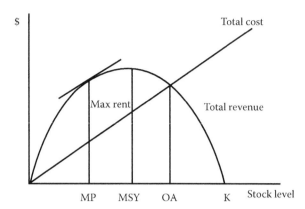

FIGURE 1.3.1 Economically and biologically optimal stocking rates (cf. Gordon 1954). MP = maximum profit, MSY = maximum sustained yield, OA = open access equilibrium, K = ecological carrying capacity.

however, and as discussed by Fox (1998), any specific definition of carrying capacity needs to be defined in relation to what are the goals of the manager(s). Critics consider the concept to be value-laden and elusive (Mysterud 2006; Keskitalo et al. 2016).

Indigenous peoples in the Northern Hemisphere traditionally use reindeer pastures in common, while the animals are private or family owned. This seeming contradiction has legitimized different political actions in the East as well as the West. With the rise of the Soviet Union, beginning in the late 1920s, much of the traditional herding was gradually collectivized into large state enterprises. However, one of the outcomes of the fall of the Soviet state was that many of the collective enter-prises were dissolved during the 1990s. In western countries, policies have been inspired by the allegory "Tragedy of the Commons" (Hardin 1968), suggesting that common resource use means adaptation at the stock level OA, i.e. depletion of the resource rent. The main political answer in Fennoscandia has been governmental top-down steering efforts. On the contrary, a large body of research substantiates that, when given the opportunity, users of common resource pools will tend to cooperate and agree upon less intensive resource use levels (Ostrom 2005), i.e. adaptations closer to the stock level MP. This also indicates that the standard model in Figure 1.3.1 can be too narrow to encompass a broad discussion of sustainability in reindeer herding. Further, the model is based on an assumption of stability. Unmanaged wild reindeer populations tend to follow cyclical behaviour (Syroechkovskii 1995), and domesticated reindeer populations often show cyclical behaviour as well (Moen and Danell 2003; Helle and Kojola 2006; Tømmervik and Riseth 2011). This requires that sources of instability need to be addressed (Riseth et al. 2016).

Though external impacts can make seemingly sustainable harvest levels unsustainable, an ecosys-tem's ability to maintain its integrity or to recover after a disturbance is important. This property is often referred to as *resilience* (Holling 1973; Ludwig et al. 1997). As sustainability is fundamentally a societal question, recent literature tries to build integrated theories to study *Social-Ecological Systems* (SESs) (Ostrom 2009). The historical and contemporary relationships between people and a changing Arctic can be seen as SESs. These systems for hunting/gathering, pastoralism, peasantry and local fisheries were, and still are, organized in local communities following natural divisions of landscape and its resources and nature's rhythm in their exploitation, organized through culturally based institutions, such as the *siida* among the Sámi (Solem 1970) and the *obschina* among the Nenets (Stammler 2005).

A crucial matter is what level of *robustness* SESs exhibit when encountering spatial and tempo-ral variability. One general finding is that even well-adapted SESs can become vulnerable to new types of disturbances (Janssen et al. 2007; Forbes 2013). Colonial regulations in Fennoscandia, national border establishments and closures as well as strict regulations of pasture use in time and space, are all typical examples of external impacts having become major long-term challenges for herding sustainability (Riseth et al. 2016; Brännlund 2015). More recently, the interface of several aspects of environmental change, including climate change, named as *global change* (Rasmussen et al. 2011), is discussed as a major challenge for herding sustainability (Rees et al. 2008) and resilience (Dong et al. 2011). The perspectives for reindeer herding can be very serious:

> The developments in the surrounding society are currently reducing the latitude for the reindeer industry at an accelerated rate and thereby also its capacity to handle new situations. In the complicated ecological, economic, social and institutional contexts, where reindeer husbandry is practiced, there [are] large risks for sudden and unpredicted disintegrations and collapses at different system levels. ... If it leads to a collapse of reindeer industry as mode of land-use, the risks of additional deterioration of the Sami indigenous rights is also apparent and thereby the scope for new solutions as well. ... The situation of reindeer husbandry has similarities with management crises in many other integrated socio-ecological systems, which have led to sustainability failures and unpredicted consequences. These insights seem to be deficient in the treatment of the problems, which [the] reindeer industry is facing.
>
> **Danell 2005:40**

Traditionally reindeer herding has been seen as the unification and a balance between the three factors *pasture, animals* and *humans* (Skum 1955; Ruong 1964; Paine 1972). Problems challenging sustainability tend to appear when imbalances between these factors become too large, as particularly stressed in Skum (1955), and may cause hunger and disease in reindeer herds. In Figure 1.3.2 we have suggested a model that may illustrate the total challenges of reindeer utilization practices in several layers of context.

On p. 23 we asserted that sustainability depends on both external and internal factors, i.e. sustainability problems can have external as well as internal origins. All four big arrows on the sides of the figure are examples of negative external influences, or what economists conceptualize as

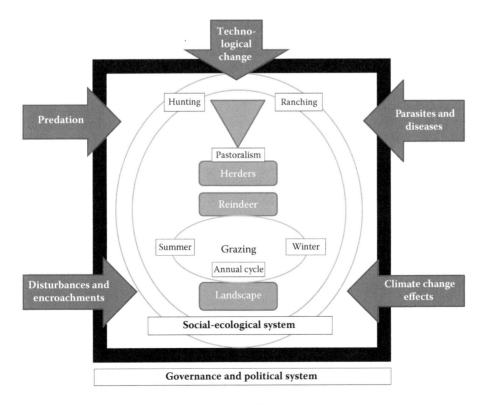

(1) *externalities*. However, we know that some of these hit animals directly, such as predation and disturbances, as well as most parasites and diseases; while others degrade or fragment pastures, such as encroachments; while even others, such as climate change effects, and some parasites and diseases, can harm both. The figure also has a vertical arrow displaying technological change; new technologies provide new control possibilities, but also incur extra costs.

The center of the figure, with an ellipse and an inverted triangle, represents the internal system. Starting with the ecological subsystem, which is the landscape: with seasonal grazing in an ellipse, reindeer's and herders' (2) *internal problems* can be the results of either animal-pasture-relations or can just be pasture problems. Typical (2a) *animal-pasture problems* are too-high utilization degrees, which are more serious when concerning winter (lichen) pastures than summer (green) pastures. (2b) *Pasture problems* can also be the outcome of inferior pasture balance and cause *out-of-season grazing*, i.e. grazing on lichen pastures in summer, thereby causing cause pasture destruction by trampling of dry lichens not protected by a snow cover (Riseth et al. 2004).

At the core of the model are the reindeer and the herders (above the ellipse). Two features of the herders are particularly important: their number and their abilities. To solve their (3) *control challenge*, the herders need to be numerous enough and have the necessary competence and tools to keep sufficient control over the herd and the individual animals, when required. The aim is to complete the annual cycle of grazing, life cycle events (such as mating and birth) and work operations (such as calf marking and slaughtering), with an optimal outcome. Practical forms of human utilization of reindeer have changed in space and time. Ingold (1980) has conceptualized this as the sequence of hunting, pastoralism and ranching, which we placed in an inverted triangle to mirror that pastoralism means close human-animal relations. These shifts in practice forms are clearly connected to technological change.

The circle around this core represents the social and institutional system around the ecological system and the practices – for the Sámi, with the *siida* as its core institution. It is included in the context of TEK, which Berkes (2012) describes as a *knowledge-practice-belief-complex* in four levels: (a) local empirical knowledge, (b) practice and management systems, (c) social institutions and (d) a worldview providing an ethical basis for the systems (cf. Riseth 2009, 2011; Sara 2013). Altogether, this is what has traditionally governed reindeer herding and also provided the necessary flexibility (Beach 1981; Brännlund and Axelsson 2011) for adaptation to the environment.

The outer rectangle in Figure 1.3.2, around the circle, represents the effects of state ambitions to govern herding societies on their territory (Scott 1998) as well as other effects of being embedded in a larger governance and political system. State policies may be supportive, but as both historical (Riseth et al. 2016; Brännlund 2015) and contemporary (Riseth 2009; Hausner et al. 2012) experiences show, they also can cause serious instability and even undermine resilience (O'Brien et al. 2009). As focused on by Janssen et al. (2007:308), there is abundant evidence that large-scale interventions in local SESs have frequently been counterproductive, while "*small-scale governance regimes that incorporate local knowledge, have clear rules that are enforced, and rely on high levels of trust frequently perform well.*" These authors also found that compared to slow, persistent change, top-down interventions affecting local coordination mechanisms have more severe effects since they may lead to maladaptations (op. cit.).

In an overall perspective there can be many sources of instability, and external and internal impacts may co-work, and reinforce or outbalance each other. As sustainability is basically about long-term SES survival, we need to find some criteria for sustainability in reindeer herding. Zijp et al. (2015) have explored selection of methods in sustainability assessment and found transparency between question and method, and consistency in methodological design, to be important. For our study, we think transparency and consistency is best achieved by turning to the basics and focus on sustenance of the three main factors: pasture, herd and humans, i.e.: *Are the pastures sufficient and in good condition? Are herd sizes adapted to the natural resource base? Is there a stable and competent work force? Do the herders have sufficient control over herding conditions to complete the annual cycle? Do the herders manage to make a good livelihood from reindeer herding? Do the governance and political system, including political authorities, bureaucracies and NGOs, contribute to support reindeer herding as a viable livelihood?* As for the impact of external effects, the main question will be: *How do different external forces, individually and together, influence the three main factors and the balance between them?*

On the basis of this introduction we will sketch an analysis of sustainability challenges in current reindeer herding. Figure 1.3.3 depicts the reindeer populations of Eurasia, both semi-domestic and wild.

We note that the northern parts of Russia encompass the majority of reindeer herding territories, while northern Fennoscandia constitutes a relatively minor proportion. Fennoscandia is Finland plus Scandinavia (Norway and Sweden). Outside the map there are populations of semi-domestic reindeer in Alaska and Greenland. We start with a general overview of Russia, and then focus the regions of Nenets and Yamal-Nenets, continue with an overview of Fennoscandia, and end with the three countries of Norway, Sweden and Finland.

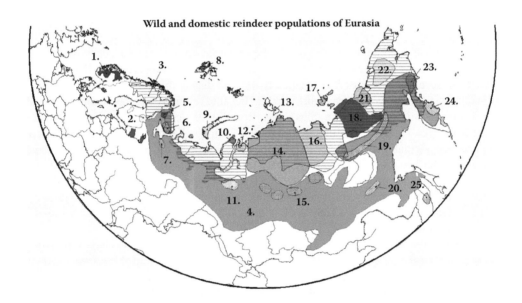

FIGURE 1.3.3   Geographical distribution of reindeer populations across Eurasia. Most of the numbers refer to the places (states, archipelagos, nature reserves, regions) where wild herds currently exist. Many of these are relatively well circumscribed spatially, based on discrete calving grounds, with the exception of the Taimyr herd that is extremely large and structurally complex (Kolpashchikov 2000). The ranges indicated for semi-domestic reindeer overlap in several places with those of so-called "'forest reindeer'" populations in Russia. The latter encompass both wild and semi-domestic herds, but these are fluid in space and time and not possible to differentiate at this scale. Semi-domestic populations are managed either extensively or intensively on subdivided territories that tend to be smaller and fenced in Fennoscandia versus larger and unfenced areas in Russia. (1) Wild herds in Norway; (2) Wild herds in Finland; (3) Range of semi-domestic Reindeer; (4) Range of forest Reindeer (mixed semi-domestic and wild); (5) Laplandskii Zapovednik (Kola); (6) Terskii Bereg (Kola); (7) Arkhangelsk Oblast; (8) Svalbard; (9) Novaya Zemlya; (10) Belyi; (11) Nadym-Pur (Yamal-Nenets Okrug); (12) Gydan; (13) Severnaya Zemlya; (14) Taimyr; (15) Evenkiya; (16) Lena-Olenek; (17) Novosibiriski Ostrova; (18) Yana-Indigirka; (19) Yakutsk; (20) Okhotsk; (21) Sudrunskaya; (22) Chukotka; (23) Parapolskii; (24) Kamchatka; (25) Amur (Source: Modified with permission from Circum-Arctic *Rangifer* Monitoring and Assessment Network (CARMA) (www.carmanetwork.com).

1.3.1.1   Russia

Russia has about two thirds of the world's semi-domestic reindeer kept by, altogether, 18 reindeer herding peoples (Jernsletten and Klokov 2002). In the North, the tundra type of herding is based on large herds and long migrations and is mainly meat-oriented. Further south, the taiga type is based on smaller herds and short migrations. Traditionally, taiga herds have been kept for transport and subsistence (Jernsletten and Klokov 2002:26). The semi-domestic reindeer stock in Russia today is about 1.5 million, but it was 1 million more as late as around 1970 (Klokov 2012). The 20th century was dramatic for reindeer herding societies in Russia, mainly due to Soviet and post-Soviet policies. Jernsletten and Klokov (2002:28) describe the collectivization as *"a very painful process for Northern indigenous peoples, especially at the initial stage. The size of private herds was strictly limited."* Further, the traditional way of life was partly disrupted and the aboriginal family life deteriorated. Total herd size was reduced from 2.2 million in 1926–27 to 1.4 million in 1934. In spite of passive opposition to the changes, the indigenous peoples of the North gradually got accustomed to the changes, and herd size slowly started to increase and reached 2.4 million around 1970 (op. cit.).

> It played an important role that the Soviet government paid great attention to the public reindeer husbandry....Various zootechnic and veterinary procedures were carried out...considerable work was done to overcome reindeer diseases, like anthrax, scabies, brucellosis, necrobacillosis.
>
> **(op. cit.:28)**

In addition, great attention also was paid to the scientific support of reindeer husbandry. While this developed large-scale reindeer herding of the tundra and some taiga regions, the taiga reindeer populations of several peoples were dramatically reduced during the 1970s and 1980s (op. cit.). An important reason for this is that transport reindeer became superfluous when helicopters and new means of transportation expanded in taiga areas. Recalling the close proximity of wild and semi-domestic herds (see Figure 1.3.3), in some cases (e.g. Taimyr, Yakutia and Chukotka) sharp rises in wild reindeer populations simultaneously depressed semi-domestic reindeer herding (Jernsletten and Klokov 2002; Klokov 2012). However, the wild reindeer population in Taimyr Peninsula has fallen by 40 percent since 2000 (Morelle 2016). This co-existence problem is in accordance with circumpolar-wide experience; semi-domestic reindeer cannot be sustained without natural borders to maintain functional separation from their wild counterparts. A contemporary example is the reindeer on Seward Peninsula in Alaska (see Jernsletten and Klokov 2002).

In the wake of the fall of the Soviet Union, the market reforms of the 1990s led to sharp declines in much of the large-scale semi-domestic tundra herds, particularly in the northeastern tundra zones. Reindeer population declines have ranged from marginal (Nenets Autonomous Okrug [NAO], Komi Republic) to steep (Chukotka, Yakutia) (Baskin 2000; Krupnik 2000; Jernsletten and Klokov 2002; Stammler 2005; Ulvevadet and Klokov 2004; Klokov 2012). In certain areas of Taimyr, Evenkia and Chukotka, reindeer herding has collapsed completely due to the termination of economic support and changes in institutional structures. Many enterprises lost most of their reindeer, and there are examples of serious social crises (Gray 2006), but reorganization and new economic support has been partly successful in some of the tundra regions (Jernsletten and Klokov 2002; Forbes and Kumpula 2009; Klokov 2012). In most Russian herding regions, general pasture resources are considered sufficient for further development of sustainable reindeer herding (Jernsletten and Klokov 2002).

1.3.1.2   Nenets and Yamal-Nenets

The most productive semi-domestic herds occur in the tundra Nenets regions of northwest Russia straddling the Ural Mountains (Forbes and Kumpula 2009). In contrast to Fennoscandia, even the largest semi-domestic herds in Russia are tended to by people and dogs around the clock (24/7), 365 days a year. This is what Ingold (1980) refers to as "close herding," and it is necessary to protect the animals from predators, prevent animals from mixing with neighboring herds, maintain them on more or less pre-determined migration routes, and manage pasture use and rotation to avoid excessive grazing and trampling. Environmental factors facilitating sustainability and resilience in Nenets regions include ample space and an abundance of resources, such as fish and game (e.g. geese), to augment the diet of not only the migratory herders, but also residents from coastal settlements (Forbes 2013). One of the key aspects of the Soviet-era administration of Yamal is that they did not restrict private ownership of animals. In NAO, privatization of *kolkhozy* and *sovkhozy* was mostly nominal, as they were renamed but their structure remained essentially intact (Tuisku 2002; Forbes 2013). Reindeer remain the most important means of transportation (Figure 1.3.4).

Although Nenets' rangelands are vast, mobility is not dependent on outside subsidies of energy, such as motorized transport and petroleum products, since herders continue to use their traditional reindeer-drawn sledges. There are probably more snowmobiles in the tundra in NAO than on Yamal, but reindeer remain the most important means of transportation during migrations, and hunting and fishing trips when away from settlements (Stammler 2002; Tuisku 2002; Forbes 2013). At the end of the 1950s a massive Soviet program to sedentarize the Nenets took place, albeit to a greater extent in NAO than in Yamal-Nenets Autonomous Okrug (YNAO) (Stammler 2005). On Yamal Peninsula, the tundra populations of people and herded reindeer have actually grown in recent years in spite of such efforts (Stammler 2008).

Herders in both NAO and YNAO have serious concerns about the progressive loss of pastures (Figure 1.3.5), campsites and sacred sites, poaching of reindeer and other wildlife, and wasteful fishing practices by gas and oil workers (Forbes 2013). In YNAO, a recent inventory around Yamal's Bovanenkovo Gas Field substantiates that the amount of territory visibly affected nearly doubled over a decade, from ca. 440 km^2 in 2001 to 836 km^2 in 2011 (Kumpula et al. 2012).

In spite of this loss of pastures, compared to some other regions in post-Soviet Russia, YNAO has experienced a steady increase in the number of semi-domestic reindeer and tundra nomads. Nevertheless, this implies impacts on the vegetation in YNAO. Persistent heavy use by large herds has frequently reduced or eliminated deciduous shrubs and fruticose or branched lichen species via summer trampling of dry lichen (Podkorytov 1995; Forbes and Kumpula 2009).

However, the lichen cover of the winter pastures in the forest tundra is still rich and productive.

FIGURE 1.3.4   Part of a caravan of draught animals from a large brigade crossing the Seyakha River in the midst of the Bovanenkovo Gas Field on Yamal Peninsula, West Siberia. Tundra Nenets migrate up to 1200 km each year using their reindeer as the main form of transportation (Photo: Bruce C. Forbes).

FIGURE 1.3.5   Nenets camp next to a new drill rig in Bovanenkovo Gas Field, Yamal Peninsula, West Siberia (Photo: Bruce C. Forbes).

Apart from management-related issues, there are potentially serious challenges presented by climate change. Firstly, warming summer air temperatures over the northwest Russian Arctic have been linked to increases in tundra productivity, longer growing seasons and accelerated growth of tall deciduous shrubs (Forbes et al. 2010; Macias-Fauria et al. 2012; Zeng et al. 2013). Secondly, sea ice loss is accelerating in the Barents and Kara Seas, probably due to warming summer air temperatures (Forbes et al. 2016) At the same time, autumn/winter rain-on-snow events (Figure 1.3.6) have become more frequent and intense, leading to record-breaking winter and spring mortality of reindeer (Forbes and Stammler 2009; Bartsch et al. 2010; Forbes et al. 2016).

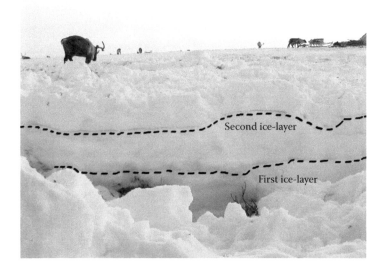

FIGURE 1.3.6   Ice layers formed during a major two-stage icing event on the southern Yamal Peninsula in November 2006. The affected area covered an estimated 60 * 60 km. The same area was affected by an even more extensive icing event a few months later in January 2007, covering approximately 60 * 100 km, and causing a large number of animals to perish. (Photo: Florian Stammler; Forbes and Stammler 2009.)

Two major rain-on-snow events during November 2006 and 2013 led to massive winter rein-deer mortality episodes on the Yamal Peninsula during the succeeding winters of 2006–07 and 2013–14. The latest led to the starvation of 61,000 reindeer out of a population of ca. 275,000 animals. Historically, this is the region's largest recorded mortality episode. Fieldwork with migratory herders has revealed that the ecological and socio-economic impacts from the catastrophic 2013–14 event will unfold for years to come. The suggested link between sea ice loss, more frequent and intense ROS events and high reindeer mortality has serious implications for the future of tundra Nenets nomadism (Forbes et al. 2016).

Thirdly, in July 2016 an outbreak of the *anthrax bacteria* killed a 12-year-old nomadic boy and sickened nearly 100 persons in YNAO, and also killed more than 2,300 reindeer. It is speculated that the outbreak was triggered by a record heat wave during the summer of 2016, which thawed the permafrost and brought a decades-old corpse of an infected reindeer to the surface, thus releasing the anthrax bacteria into the air. The real and potential effects of the combined sea ice retreat and permafrost thaw are cause for deep concern (Wang 2016). Initial governmental plans to cull up to 250,000 reindeer (Gertcyk 2016a), soon lowered to 100,000 (Gertcyk 2016b), were in part due to a somewhat unclear justification of "overgrazing." The recommendation to move towards a Finnish system of reindeer management (Gertcyk 2016a) is of further concern and would likely result in replacing one set of problems with another (Forbes 2016).

1.3.2 FENNOSCANDIA

Fennoscandia is relevant to consider as a joint region due to its political history and natural geography, which encompassed most of Sápmi. Fennoscandia has three major types of reindeer herding: (I) a mountain tundra type with intermediate to long seasonal migrations (Norway, Sweden, and historically also northernmost Finland), (II) a coastal-oriented type, with local seasonal migrations (Norway), and finally (III) a taiga type which is based on small-scale mosaic in forested landscapes (Sweden and Finland) (Figure 1.3.7).

Besides general challenges, each of these three adaptations faces its own distinct implications and challenges. The optimal land-use cycle for (I), the long-migrating tundra type, is to use the high mountains, with fresh leaves, herbs and grasses, along the mountain ridge and on the coast of northernmost Norway in summer, inland tundra with mushrooms and mires in fall, inland brushwood with lichens and mire plants in early winter, conifer forest with loose snow and accessible lichens in deep winter, and inland tundra again as spring transition. The short-migrating and stationary adaptations deviate mainly in that the coastal type (II) uses permanently or temporarily snow free coastal or sub-oceanic areas in winter, while the taiga type (III) uses mires and bogs in summer.

In total, today's populations of semi-domestic reindeer in Fennoscandia (ca. 664,000), Norway (212,000 in 2016), Sweden (261,000 in 2016) and Finland (191,000 in 2016) are nearly at the same level as the total of the YNAO (ca. 705,000 in January 2017).

Reindeer herding in Fennoscandia was exposed to major external political shocks during the late 19th and early 20th centuries (Riseth et al. 2016). As a long-term outcome of both global political changes and national politics, in particular the promotion of non-Sámi farming settlers in Sámi areas, the Nordic states during the 19th century dramatically changed their attitudes towards Sámi and reindeer herding (Pedersen 2007; Brännlund 2015). The shocks and their immediate effects include:

1. Border closures:
   a. Norway-Russia (in 1826),
   b. Norway-Finland (in 1852), and
   c. Sweden-Finland (in 1889),
      blocking cross-border herding and curtailing migrations, which forced much Sámi tundra type herding into a taiga pattern

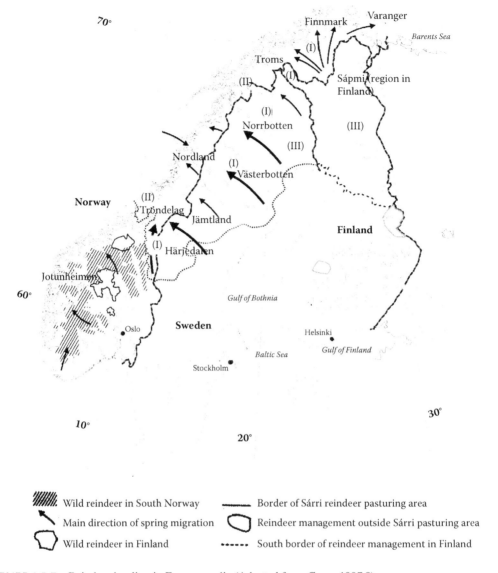

FIGURE 1.3.7   Reindeer herding in Fennoscandia (Adapted from Gaare 1997:8).

2. Joint Norwegian–Swedish legislation (1883), which
   a. limited the areas accessible for reindeer herding,
   b. led to much stricter control with use of areas in space and time, and
3. The first Norwegian–Swedish pasture convention (1922), which
   a   blocked access to much of Troms County in Norway for Swedish Sámi herders, and
   b. forced Sámi tundra types of herding into taiga patterns
   c. forced dislocating of herding Sámi from Norrbotten (Northernmost Sweden) and
       Troms (Northern Norway) southwards through Sweden and created new conflicts in
       southern regions

The border closures are considered to be the biggest upheaval reindeer herding in Sápmi has experienced and have had extensive and long-lasting repercussions all over northern Fennoscandia, i.e. nearly all Sámi herding was influenced (Sara 1997). A close study of herding adaptations in northern Scandinavia from the midst of the 19th century to now show that the political shocks aggravated the effects of severe climatic events and led to deteriorated pastures and major herd size fluctuations in both countries (Riseth et al. 2016). Helle and Kojola (2006) discuss synchrony in population trends in Fennoscandia for the last part of the 20th century and find that the three national curves seem to follow similar main trends, which were primarily an outcome of winter weather variability. We will comment on deviations from this under the paragraphs for the three countries.

The second part of the 20th century was also the period when anthropogenic encroachments started to significantly delimit reindeer herding land use by fragmenting it, typically by hydro power construction, spreading infrastructure and large-scale forestry (Beach 1981). As well, in recent decades, land use has been altered by urban peoples' recreation activities, both withdrawing land for recreation sites and cottages as well as creating disturbance (Vistnes 2008; Viken 2016). From the 1980s onward, new predator policies reinforced an old external effect in Fennoscandia:

> The increasing populations of large predators have developed to an extremely acute threat against the sustainability of reindeer husbandry. Via direct losses of animals, damage of herd production capacity, reduced space for herd improvement measures and obstructions in use of grazing lands as well as increase of management labor and costs, the consequences for the economy become generally very crushing.
>
> **(Danell 2010:79)**

The author states that herders frequently have traumatic experiences when they see injured and dead animals. Many herders have constant worries about whether they will be able to sustain their livelihood. The strong distrust and conflicts connected to predators cause immense social and mental discomfort. Sometimes this has tragic consequences. In many areas, the predator situation undoubtedly has caused population collapses.

1.3.2.1 Sweden

The map in Figure 1.3.7 shows that most of the reindeer herding in Sweden is based on adaptation type I, with long annual migrations between winters in continental woods and summers in the mountain ridge. This adaptation type is broadly intact, with two important exceptions. The first is that the border conventions with Norway dramatically reduced the access to summer pastures in Norway, mainly in Troms County, but also further south. The second is that land-use changes (which also apply to type II), the encroachments by industrial forestry in particular, have resulted in loss of important grazing resources, i.e. terrestrial and arboreal (old forest) lichens which are essential elements in the herding year. Further, as multilayered canopies offer more diverse patterns of snow hardness, climate change with winter warming spells increases the importance of old forest (Horstkotte 2013). Accordingly, climate change effects add to the severity of this challenge (Löf 2014) (Figure 1.3.8).

As for the predator situation, analyses of accessible data indicate that reindeer husbandry in the southern half of the Swedish reindeer herding area regionally appear to be in a phase where collapses can occur more or less at any time due to large populations of lynx, wolverine and bear, while the reindeer husbandry in Norrbotten is not yet in acute danger, although it is considerably affected (Danell 2010:78–79). The Arctic Council (2016) puts forward the case of Vilhelmina North Sameby in Västerbotten, which has been

> *forced to change its reindeer herding practices due to climate change, increased motorization, and reduced freedom of movement across the landscape. While technological innovations have enabled adaptation of reindeer herders' activities, they have also eroded Indigenous Knowledge.*
>
> **(Arctic Council 2016:105)**

Moen and Danell (2003) examined long-term data on population dynamics of reindeer in Sweden, and this analysis showed fluctuations, without any particular trends, around 225,000 animals for the

FIGURE 1.3.8 Traditional knowledge is included in climate science (cf. Riseth et al. 2011). Gustav Labba explains the importance of first snow conditions (Photo: Hans Tømmervik).

last century. They compared data on so-called "overgrazing" areas in parts of Sweden to the situation in parts of Finnmark (Norway), and in northern Finland where reindeer husbandry in recent decades seemed ecologically unsustainable. They concluded that large-scale overexploitation by reindeer in the Swedish mountains was not evident. However, strong grazing and trampling effects may be found around enclosures and fences.

Summing up history, Brännlund and Axelsson (2011) found that in spite of the clear differences between herding management a century ago and today, "*major adaptation strategies and constraining forces in the 19th century do not seem very different to those of today. The foremost adaptation strategy was, and remains, the flexible use of pasture areas*" (op. cit.:1103). In line with this, the current situation seems to be that land-use changes reduce herding flexibility year by year, while climate change effects create a need for increased flexibility (Löf 2014). An overall evaluation sums up:

> Institutional inconsistencies in Sweden have provided little protection for Sami reindeer herders, and often inhibit innovative responses. Technological, social, and governance change have decreased the diversity of herders' strategies and inhibited adaptive strategies, reducing the resilience of reindeer herding.

(Arctic Council 2016:105)

1.3.2.2 Finland

Reindeer herding in Finland has two different cultural roots; one is Finnish (in the southern and middle part of the reindeer herding area), while the other is Sámi (in the northernmost part of the reindeer herding area). The Finnish reindeer herding typically is developed as supplementary to farming, while the Sámi adaptation has been based on migration, like in Sweden and Norway. Both the border closure towards Norway, and Finnish legislation, have forced this adaptation into the Finnish frame despite the fact that this is contrary to the ecology, i.e. summer pasturing at the tundra removes the lichens due to trampling (Figure 1.3.9).

Kumpula et al. (2014) analyzed the changes in standing biomass of ground lichens from the period 1995–1996 to 2005–2008 in the 20 northernmost herding districts in Finland. The higher the long-term reindeer densities on the lichen pastures, the lower were the lichen biomass. On the basis of field site data, Colpaert and Kumpula (2012) found that the measured lichen biomass had declined significantly in 19 out of the 20 reindeer management districts and only one district showed slight improvement. Particularly, three districts exhibited notable reductions in lichen biomass, from over 1500 kg/ha to about 500 kg/ha. Based on satellite data they found that old growth forests with lush arboreal lichen cover had declined by 5 percent in the period 1995–2008 due to felling. The lichen biomass was also strongly affected by the grazing system; the lowest biomass values of

FIGURE 1.3.9   The image shows the contrast between intact lichen pastures in Norway compared with the areas in Finland. The pastures in Finland (back of the picture) are used during summer while the areas in Norway (in front of the picture) are only used in winter (Photo: Bernt Johansen).

lichens were measured in all grazing areas that were used in the snow-free seasons (Kumpula et al. 2014; Colpaert and Kumpula 2012). The lichen biomass also decreased as the proportion of arboreal lichen pastures within a district decreased. Further, it also decreased as the proportion of human infrastructure increased (Kumpula et al. 2014). The lichen biomass in old pine forests and in all mountain type lichen pastures was lower than that in mature and old pine forests (Colpaert and Kumpula 2012; Kumpula et al. 2014). Arboreal lichens also declined due to felling of old growth forests. Consequently, grass, shrub and sapling stands increased in felled areas (Colpaert and Kumpula 2012). Further, these authors assert:

> Our work therefore shows that it is very difficult to preserve lichen ranges in a moderate or good condition without implementing reasonable seasonal pasture rotation systems in semi-domesticated reindeer herding. This means that the most important lichen pasture areas should be protected from grazing and trampling by reindeer during the entire snow-free season...
>
> However, most of the reindeer herding districts in Finland still lack an appropriate seasonal pasture rotation system and therefore, it is very difficult to avoid the continuous, deterioration of lichen pastures. It is even more important to develop distinct seasonal pasture rotation systems in the districts located in mountainous areas where the present lichen pastures are heavily grazed and also more vulnerable to lichen reduction than in the districts located in coniferous forest areas. Although the grazing systems need considerable rearrangements in the reindeer management area, a recovery of lichen ranges is not probably possible without also making a marked reduction in the present number of reindeer on pastures. Especially in lichen pasture areas with low lichen biomass, a [considerable] reduction in grazing pressure is needed for starting the recovery process.

(Kumpula et al. 2014:550)

This evaluation confirms that the situation is more serious in tundra than taiga areas, but also that establishment of a reasonable pasture rotation system and a marked reduction in grazing pressure are necessary actions. Further, these studies also confirm that over time lichen pasture reductions made the herding dependent on supplementary feeding (Kumpula et al. 2002, 2014). Extensive supplementary feeding clearly impedes the regeneration of pastures but also means much higher herding costs and accordingly higher dependence on subsidies (Riseth et al. 2007). Supplementary feeding also increases animal density and animal-to-animal contact, which also increase the chance of contracting transmissible diseases (Tryland 2012). As for the predator situation, it seems still manageable in the North, although it is rapidly worsening due to expanding bear, wolf, lynx and wolverine populations. At the same time, in the southern parts of the Finnish reindeer herding area the expanding wolf population, in particular, has demonstrably resulted in reindeer population, as well as economic, collapses (Danell 2010:78–79). For Finland as a whole, the documented number of reindeer killed by large carnivores has been higher than the number slaughtered 3 years in a row since 2011–2012 (PY 2017).

1.3.2.3 Norway

The previously mentioned late 19th and early 20th century political shocks also influenced reindeer herding over large parts of Norway, both causing changing pasture balances and interregional relocations. For some regions World War II was another shock due to restrictions, German troops' forced slaughtering and practiced poaching (cf. Riseth et al. 2016). However, after a few decades of regeneration, pasture use was relatively balanced up to the 1980s, when the technological revolution (Pelto 1973), having started in the late 1960s, with the introduction and spread of snowmobiles, cars, ATVs and helicopters, changed both cost level and herd control possibilities. This resulted in different regional development patterns in Norway. The differences are clearest between, on the one hand, the tundra plateau Finnmarksvidda, winter pastures for about two-thirds of the semi-domestic reindeer stock in Norway, and on the other hand, the easternmost part of Finnmark Varanger/Polmak plus the South Sámi regions of Trøndelag and the concession herding in Jotunheimen in South Norway. While the former utilized the new opportunities to expand population size, the latter utilized husbandry knowledge about the animals' production potential to increase meat production and limit population size simultaneously, thereby improving their economy (Riseth 2009). The Norwegian government considered the Finnmarksvidda pasture situation to be unsustainable and tried, starting in the 1980s, to implement a command and control regime, with rather limited success. Figure 1.3.10 depicts population size at Finnmarksvidda, with the major regions, Kautokeino and Karasjok, contrasted against the region of Varanger/Polmak.

In the figure we observe that while the reindeer population in Varanger/Polmak show modest dynamics over seven decades, animal numbers beyond 1970 in Kautokeino and Karasjok vary with cycles with higher levels and amplitudes than ever before; post-1970, minima are higher than pre-1970 maxima (Tømmervik et al. 2009; Riseth and Lie 2016). The pasture situation at Finnmarksvidda has been closely monitored (Johansen et al. 2014; Tømmervik et al. 2009, 2012, 2014). Findings from the monitoring program in Figure 1.3.11 depict the dynamics of lichen biomass and reindeer population at Finnmarksvidda in the period 1957 to 2013.

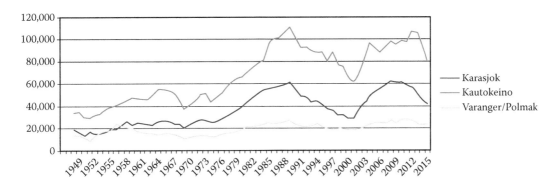

FIGURE 1.3.10   Reindeer herd size development in Finnmark subregions (Source: Reindriftsforvaltningen, Landbruksdirektoratet 1995–2015).

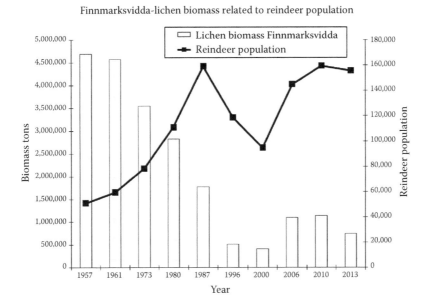

FIGURE 1.3.11   Finnmarksvidda – lichen biomass in tons related to reindeer population 1957–2013 (From Tømmervik et al. 2014).

We can observe that lichen biomass decreased considerably while the reindeer population increased in Finnmarksvidda during the period 1957–2000, hence with a subsequent increase of lichen biomass in the period 2000–2010 and a new decrease from 2010–2013 after a recent corresponding increase in the reindeer population. The post-millennium lichen recovery trajectory is particularly interesting, as the mean relative growth rate was at 8.3 percent, which is very rapid compared to previous studies (Tømmervik et al. 2012).

The explanation for the overall development involves several factors: a landscape without natural borders and the technological revolution providing new control possibilities, new governmental subsidies from the late 1970s, and favorable winter climate in the 1980s (Riseth 2009, 2013). The government has recently, by strict command and control methods, reduced the total reindeer population at Finnmarksvidda to a recommended level, but whether this will contribute to an increased long-term stability and which effects it will have in a broader sustainability perspective is an open question (Riseth and Lie, 2016). Over time, stricter top-down government has awakened everyday herder resistance (Johnsen and Benjaminsen 2017), and Ministry policy implementation is also challenged by a legal dispute raised by 25-year-old Jovsset Ante Sara, who refused to limit his herd to an imposed level of 75 reindeer, as he claimed that this would force him out of business. An appeal court verdict (LH-2016-92975) concludes that, according to international law (International Covenant on Civil and Political Rights (ICCPR), article 27), the imposed reduction is void because the governmental decision violates Sara's right to carry on his culture. The state has appealed the case to the Supreme Court. Further, the Ministry of Petroleum and Energy, in a complaint case, recently decided to deny concession to a wind farm complex on account of the same ICCPR article (OED 2016). These two cases are clear indications that international law has achieved a practical impact on the framing of Norwegian governmental policies towards reindeer herding. Nevertheless, a new Green Book (LMD 2017) pursues a unilateral focus on *ecological* sustainability, clearly contrasting Sámi policies pursuing social and cultural goals in concert with ecological ones (NTB 2017; Kveseth 2017).

In general, reindeer herding in Norway is exposed to much of the same challenges as in neighboring countries – in particular, external pressure created by infrastructure development, modernization and climate change. As for predation pressure, it seems fairly moderate in Finnmark County, but is considerably higher in Troms and Nordland Counties, in particular, and also in the Trøndelag Counties due to increasing predation by lynx, wolverine and golden eagle (Danell 2010). Norwegian governmental predator policy is based on a two-fold policy objective of preserving biodiversity and maintaining traditional local livelihoods. However, the current governance instruments in carnivore management do not address the spatial dynamics of carnivores (Risvoll et al. 2016). In practice, this increased the depredation pressure on reindeer herding and sheep farming. As in Sweden and Finland, both land-use change and climate change problems are probably the most serious, but are to a large extent overlooked by the government. In particular, inland winter pastures are threatened by expansion of woody vegetation, which tends to decrease their pasture potential considerably (Tømmervik et al. 2009; Karlsen et al. 2017; Käyhkö et al. 2014). Expansion of woody vegetation on the coast (Tombre et al. 2010) also risks facilitating the spread of diseases to reindeer, such as the tick-borne infection Lyme borreliosis (*Ixodes ricinus*), which has been detected in vegetation near the Arctic Circle (Hvidsten et al. 2015). Jore et al. (2011) and Nilssen (2010) concluded that *I. ricinus* reached coastal areas as far north as Harstad 69° N, and has been found sporadically in Finnmark County in northern Norway. It is not yet certain whether these ticks represent resident populations or introduced transient populations introduced by migratory birds or large mammals such as deer (Hvidsten et al. 2015). However, the increased populations of goose populations in spring staging sites along the coast of Norway (Tombre et al. 2010) may be a candidate for the future introduction of such infections on the coast.

1.3.3   Conclusion and Perspectives

We started out by defining sustainability as potential to endure in time, and put forward the standard range management approach and arguments for a wider scope and to see sustainability as connected to SESs and resilience. We introduced a model that places the factors – pasture, animals and humans – in a wider context, and used this as a basis for discussing challenges facing reindeer management in Russia and Fennoscandia. Seeing our findings in perspective, a case study from seven major pastoral regions across six continents found that vulnerability of pastoralism was very different for the cases across the globe, but that climate change and climate variability are driving fragile pastoral ecosystems into more vulnerable conditions. Further, socio-economic factors, such as changes in land tenure, agriculture, sedentarization and institutions, are fracturing large-scale pastoral ecosystems into spatially isolated systems (Dong et al. 2011).

The last century development of semi-domesticated reindeer herding in Russia is an outcome of policies by Soviet and post-Soviet authorities. Nenets herders are among those that have managed best through the large transitions, because their societal structures were kept intact. They also have managed to live with the impact of petroleum activities close by. Today they face major challenges caused by climate change, both sea ice withdrawal and an anthrax outbreak already having had dramatic effects. With increasing temperatures and other changing climatic conditions more major animal catastrophes are highly expected, and with melting permafrost the risk for other unexpected incidents also is increasing. An extra concern is that it is a fundamental issue that the authorities understand what the situation really is (Forbes 2016).

In Fennoscandia, Finland obviously has a challenge with sustaining high grazing pressures on limited lichen pastures as facilitated by the extensive use of supplementary feeding. This problem is most serious in tundra areas which have insufficient summer pastures, while in taiga areas it is more a question of establishing a reasonable system of pasture rotation. It is debatable how sustainable it is to go beyond the ecological niche of reindeer. The grazing pressure problem on Finnmarksvidda might seem to be solved, but the governmental policy has in practice challenged basic recommendations of co-management and building of trust (Riseth and Lie 2016). It follows that both mutual inter-siida agreements and compliance, as well as governmental respect for herders' own processes, are required to achieve a sustainable situation (cf. Ostrom 2005).

For Fennoscandia the predator situation in reindeer husbandry is generally a concern and in some regions is a serious threat to sustainability. A fundamental difficulty for the reindeer industries is also the low level of acknowledgement of the reality of the problems, both in society at large and within governmental authorities (Danell 2010).

An overall statement of concern is that: "*Potential threats facing reindeer populations of Eurasia, and reindeer herding as a livelihood, include rapid land use change, excessive predation, climate change, and ongoing institutional conflicts*" (Forbes 2010:86). Both land-use change and climate change are major global threats (UNEP 2001), while excessive predation and institutional conflicts are more typically regional challenges of a more variable significance. In some cases, human disturbances and fragmentation of the grazing grounds appear to be more important drivers of reindeer population dynamics than climate (Uboni et al. 2016). This is in agreement with Rees et al. (2008), who stated that the vulnerability to climate change for the Eurasian reindeer husbandry is comparatively small in relation to the influence of regional socio-economic dynamics.

In a long-term perspective, the combined effects of land-use change and climate change seem to be the more serious, and the recent experiences from Russia points in this direction. Climate challenges are furthermore the ones most difficult to foresee. Many effects are already well-known and recent extreme years probably will be characterized as more "normal" years within a few decades. However, we cannot yet project the various combinations of cumulative effects and the pace of change.

REFERENCES

Arctic Council. 2016. Arctic Council *Artic Resilience Report*. M. Carson and G. Peterson (eds.). Stockholm: Stockholm Environment Institute and Stockholm Resilience Centre.

Bartsch, A., T. Kumpula, B. C. Forbes and F. Stammler. 2010. Detection of snow surface thawing and refreezing using QuikSCAT: Implications for reindeer herding. *Ecol. Appl.* 20: 2346–58.

Baskin, L. M. 2000. Reindeer husbandry/hunting in Russia in the past, present and future. *Pol. Res.* 19 (1): 23–9.

Beach, H. 1981. Reindeer-herd management in transition: The case of Tuorpon Saameby in Northern Sweden. In *Uppsala Studies in Cultural Anthropology*. 3. Uppsala: Acta Universitas Uppsalensis.

Berkes, F. 2012. *Sacred Ecology*. 3rd Edition. New York: Routledge.

Brännlund, I. 2015. Histories of reindeer husbandry resilience: Land use and social networks of reindeer husbandry in Swedish Sápmi 1740–1920. PhD dissertation. Department of Historical, Philosophical and Religious Studies. Vaartoe. Centre for Sami research. Umeå: Umeå University.

Brännlund, I. and P. Axelsson. 2011. Reindeer management during the colonization of Sami lands: A long-term perspective of vulnerability adaptation strategies. *Glob. Environ. Chang.* 21: 1095–105.

Colpaert, A. and J. Kumpula. 2012. Detecting changes in the state of reindeer pastures in northernmost Finland, 1995–2005. *Polar Rec.* 48, Special Issue 01: 74–82.

Danell, Ö. 2005. The robustness of reindeer husbandry – Need for a new approach to elucidate opportunities and sustainability of the reindeer industry in its socio-ecological context. 13th Nordic Conference on Reindeer and Reindeer Husbandry Research, Røros, Norway, 23–25 Aug. 2005. *Rangifer Rep.* 10 (2005): 39–49.

Danell, Ö. 2010. Reindeer husbandry and the predators. 16th Nordic Conference on Reindeer and Reindeer Husbandry Research, Tromsø, Norway, 16–18 Nov 2010. *Rangifer Rep.* 14 (2010): 79–80.

Dong, S., L. Wen, S. Liu et al. 2011. Vulnerability of worldwide pastoralism to global changes and interdisciplinary strategies for sustainable pastoralism. *Ecol. Soc.* 16 (2): 10.

Forbes, B. C. 2010. Indicator # 18. Reindeer herding. In *Arctic Biodiversity Trends 2010 – Selected Indicators of Change*, 86–88. Akureyri, Iceland: CAFF International Secretariat.

Forbes, B. C. 2013. Cultural resilience of social-ecological systems in Nenets and Yamal-Nenets Autonomous Okrugs, Russia: A focus on reindeer nomads of the tundra. *Ecol. Soc.* 18 (4): 36.

Forbes, B. C. 2016. Beware of action that would put age old tundra nomadism at risk in Yamal, says expert. *Siberian Times*, 23 September 2016. siberiantimes.com/other/others/features/f0257-beware-of-action-that-would-put-age-old-tundra-nomadism-at-risk-in-yamal-says-expert/ (Accessed 9 Nov. 2017).

Forbes, B. C. and T. Kumpula. 2009. The ecological role and geography of reindeer (*Rangifer tarandus*) in Northern Eurasia. *Geography Compass* 3/4 (2009): 1356–80.

Forbes, B. C. and F. Stammler. 2009. Arctic climate change discourse: The contrasting politics of research agendas in the West and Russia. *Polar Res.* 28: 28–42.

Forbes, B. C., M. Macias-Fauria and P. Zetterberg. 2010. Russian Arctic warming and "greening" are closely tracked by tundra shrub willows. *Glob. Change Biol.* 16: 1542–54.

Forbes, B. C., T. Kumpula, N. Meschtyb et al. 2016. Sea ice, rain-on-snow and tundra reindeer nomadism in Arctic Russia. *Biol. Lett.* 12 : 20160466. dx.doi.org/10.1098/rsbl.2016.0466.

Fox, J. L. 1998. Finnmarksvidda: Reindeer carrying capacity and exploitation in a changing pastoral ecosystem – A range ecology perspective on the reindeer ecosystem in Finnmark. In *Commons in a Cold Climate: Coastal Fisheries and Reindeer Pastoralism in North Norway: The Co-Management Approach*, ed. S. Jentoft, 17–40. Paris and New York: UNESCO and Parthenon Publishers.

Gertcyk, O. 2016a. Huge cull of 250,000 reindeer by Christmas in Yamalo-Nenets after anthrax outbreak. *Siberian Times*, 19 September 2016, siberiantimes.com/other/others/news/n0738-huge-cull-of-250000-reindeer-by-christmas-in-yamelo-nenets-after-anthrax-outbreak/.

Gertcyk, O. 2016b. Yamal trims back reindeer cull to 100,000 as herders fear future of nomadic lifestyle. *Siberian Times*, 07 October 2016, siberiantimes.com/other/others/features/f0260-yamal-trims-back-reindeer-cull-to-100000-as-herders-fear-future-of-nomadic-lifestyle/.

Gordon, H. S. 1954. The economic theory of a common property resource. *J. Polit. Econ.* 62: 124–42.

Gray, P. A. 2006. Privatiseringen på Chukotka og reindriftens (foreløpige) skjebne. *Ottar* 259: 42–8.

Hardin, G. 1968. The tragedy of the commons. *Science* 162: 1243–7.

Hausner V. H., P. Fauchald and J.-L. Jernsletten. 2012. Community-based management: Under what conditions do Sámi pastoralists manage pastures sustainably? *PLoS ONE* 7(12): e51187. doi.org/10.1371/journal.pone.0051187.

Helle, T. and I. Kojola. 2006. Population trends of semi-domesticated reindeer in Fennoscandia – Evaluations of explanations. In *Reindeer Management in Northernmost Europe*, ed. B.C. Forbes et al. 319–39. Ecological Studies 184. Berlin: Springer.

Holling, C. S. 1973. Resilience and stability of ecological systems. *Annu. Rev. Ecol. Syst.* 4: 1–23.

Horstkotte, T. 2013. Contested landscapes – Social-ecological interactions between forestry and reindeer husbandry. PhD dissertation. Umeå: Umeå University.

Hvidsten, D., F. Stordahl, M. Lager et al. 2015. Borrelia burgdorferi sensu lato-infected *Ixodes ricinus* collected from vegetation near the Arctic Circle. *Ticks Tick Borne Dis.* 6: 768–73.

Ingold, T. 1980. *Hunters, Pastoralists and Ranchers: Reindeer Economies and Their Transformations.* Cambridge: Cambridge University Press.

Janssen, M. A., J. M. Anderies and E. Ostrom. 2007. Robustness of social-ecological systems to spatial and temporal variability. *Soc. Nat. Res.* 20: 307–22.

Jernsletten, J. L. and K. Klokov. 2002. *Sustainable Reindeer Husbandry.* Tromsø: Arctic Council.

Johansen, B., H. Tømmervik, J. W. Bjerke and S. R. Karlsen. 2014. *Vegetation and ecosystem transformation on Finnmarksvidda, Northern Norway, due to reindeer grazing pressure.* U.S. Geological Survey, Icel. Inst. Nat. Hist.

Johnsen, K. I. and T. A. Benjaminsen. 2017. The art of governing and everyday resistance: "Rationalization" of Sámi reindeer husbandry in Norway since the 1970s. *Acta Boreal.* doi:10.1080/08003831.2017.1317981.

Jore, S., H. Viljugrein, M. Hofshagen et al. 2011. Multi-source analysis reveals latitudinal and altitudinal shifts in range of *Ixodes ricinus* at its northern distribution limit. *Parasit. Vectors* 4: 84.

Karlsen S-R., H. Tømmervik, B. Johansen and J. Å. Riseth. 2017. Future forest distribution on Finnmarksvidda, North Norway. *Clim. Res.* Special Issue. Resilience in SENSitive mountain FORest ecosystems under environmental change. doi.org/10.3354/cr01459.

Käyhkö, J., L. K. Oksanen, T. Horstkotte et al. 2014. Can reindeer grazing stop expansion of woody vegetation in the warming tundra? *IGU Regional Conference*; 2014-08-18–2014-08-22 UiT.

Keskitalo, E. C. H., T. Horstkotte, S. Kivinen, B. Forbes and J. Käyhkö. 2016. "Generality of mis-fit"? The real-life difficulty of matching scales in an interconnected world. *Ambio*, 45 (6): 742–52.

Klokov, K. 2012. Changes in reindeer population numbers in Russia: An effect of the political context or of climate? *Rangifer* 32: 19–33.

Krupnik, I. 2000. Reindeer pastoralism in modern Siberia: Research and survival in the time of crash. *Polar Res.* 19 (1): 49–56.

Kumpula, J., A. Colpaert and M. Nieminen. 2002. Productivity factors of the Finnish semi-domesticated reindeer *(Rangifer t. tarandus)* stock during the 1990s. *Rangifer* 22: 3–12.

Kumpula, J., M. Kurkilahti, T. Helle and A. Colpaert. 2014. Both reindeer management and several other land use factors explain the reduction in ground lichens (*Cladonia* spp.) in pastures grazed by semi-domesticated reindeer in Finland. *Reg. Environ. Change* 14 (2): 541–59.

Kumpula, T., B. C. Forbes, F. Stammler and N. Meschtyb. 2012. Dynamics of a coupled system: Multi-resolution remote sensing in assessing social-ecological responses during 25 years of gas field development in Arctic Russia. *Remote Sens.* 4 (4): 1046–68. doi:10.3390/rs4041046.

Kveseth, M. 2017. Næringa kritisk til reindriftsmelding NRL-lederen ikke fornøyd med reindriftsmeldinga. *Altaposten*, 6 Apr. 2017. www.altaposten.no/nyheter/2017/04/06/N%C3%A6ringa-kritisk-til-reindrifts melding-14562813.ece (Accessed 9 Nov. 2017).

LH-2016-92975. *Hålogaland lagmannsrett* (Court of Appeal), Tromsø. lovdata.no/dokument/LHSIV/avg jorelse/lh-2016-92975?q=sp 27.

LMD. 2017. Meld. St. 32 (2016–2017) Reindrift—Lang tradisjon – unike muligheter. Oslo: Ministry of Agriculture and Food. www.regjeringen.no/contentassets/ffb8837d1f32425b962ceb23e5ccfc8e/no/pdfs /stm201620170032000dddpdfs.pdf (Accessed 9 Nov. 2017).

Löf, A. 2014. Challenging adaptability. Analysing the governance of reindeer husbandry in Sweden. PhD dissertation. Umeå University.

Ludwig, D., B. Walker and C. S. Holling. 1997. Sustainability, stability and resilience. *Conserv. Ecol.* 1 (1): 7.

Macias-Fauria, M., B. C. Forbes, P. Zetterberg and T. Kumpula. 2012. Eurasian Arctic greening reveals teleconnections and the potential for structurally novel ecosystems. *Nat. Clim. Change* 2: 613–8.

Moen, J. and Ö. Danell. 2003. Reindeer in the Swedish mountains: An assessment of grazing impacts. *Ambio* 32: 397–402.

Morelle, R. 2016. World's largest reindeer herd plummets. *BBC News.* Science & Environment. 13 December 2016. www.bbc.com/news/science-environment-38297464. (Accessed 9 Nov 2017).

Mysterud, A. 2006. The concept of overgrazing and its role in management of large herbivores. *Wildlife Biol.* 12 (2): 129–41.

Nilssen, A. C. 2010. Er skogflåtten i ferd med å innta Nord-Norge? *Ottar* 48–57.

NTB. 2017. ABC Nyheter. Reindriftsmeldingen – Sametinget er svært skuffet. www.abcnyheter.no/nyheter /norge/2017/04/05/195293321/reindriftsmeldingen-sametinget-er-svaert-skuffet (Accessed 9 Nov. 2017).

O'Brien, K., B. Hayward and F. Berkes. 2009. Rethinking social contracts: Building resilience in a changing climate *Ecol. Soc.* 14 (2): 12.

OED. 2016. Det Kongelige Olje- og Energidepartement. [Ministry of Petroleum and Energy] Fred. Olsen Renew ables AS-Kalvvatnan vindkraftverk i Bindal og Namsskogan kommuner-klagesak. 08/3602-11.11.2016.

Ostrom, E. 2005. *Understanding Institutional Diversity*. Oxford and Princeton: Princeton University Press.

Ostrom, E. 2009. A general framework for analyzing the sustainability of social-ecological systems. *Science* 325: 419–22.

Paine, R. 1972. The herd management of Lapp Reindeer Pastoralists. *J. Asian Afr. Stud.* 7 (1): 76–87.

Pedersen, S. 2007. Lappekodisillen av 1751 – "Samene – Det Grenseløse Folket." In E. G. Broderstad, E. Niemi and I. Sommerseth (eds.), *"Grenseoverskridende reindrift før og etter 1905"*. Skriftserie ,nr. 14. Senter for samiske studier, 9–20.Tromsø: Universitetet i Tromsø.

Pelto, P. J. 1973. *The Snowmobile Revolution: Technology and Social Change in the Arctic*. California: Cummings.

Podkorytov, F. M. 1995. *Reindeer Herding on Yamal* (in Russian). Sosnovyi Bor, Russia: Leningradskoi Atomoi Electrostantsii.

PY. 2017. Paliskuntain Yhdistys. paliskunnat.fi/reindeer-herders-association/reindeer-info/ (Accessed 25 May 2017).

Rasmussen, R. O. (ed.), A. Karlsdottir, C. Pellegatta et al. 2011. Megatrends. *Temanord* 2011: 527. Copenhagen: Nordic Council of Ministers.

Rees, W. G., F. M. Stammler, F. S. Danks and P. Vitebsky 2008. Vulnerability of European reindeer husbandry to global change. *Clim. Change* 87: 199–217.

Riseth, J. Å. 2009 [2000]. *Modernization and Pasture Degradation: A Comparative Study of Two Sámi Reindeer Pasture Regions in Norway 1960–1990*. Saarbrücken: VDM Verlag.

Riseth, J. Å. 2011. Can traditional knowledge play a significant role in nature management? Reflections on institutional challenges for the Sámi in Norway. In *Working With Traditional Knowledge: Communities, Institutions, Information Systems, Law and Ethics: Writings from the* Árbediehtu *Project on Sami Traditional Knowledge*, eds. J. Porsanger, Jelena and G. Guttorm. *Diedut* 1/2011: 127–61.

Riseth, J. Å. 2013. Reindrifta i Nord-Norge: Fra vikeplikt til bærekraft? In *Hvor går Nord-Norge? Politiske tidslinjer*, eds. S. Jentoft, J. I. Nergård and K. A. Røvik, 401–16. Orkana Akademisk.

Riseth, J. Å., B. Johansen and A. Vatn. 2004. Aspects of a two-pasture-herbivore model. *Rangifer*, Special Issue 1: 65–81.

Riseth, J. Å., L. Oksanen, N. Labba and B. Johansen. 2007. Macro policies of economy and border crossing: Opportunities and constraints for Sámi reindeer herd management in Northern Sapmi. *The workshop "Economic and ecological analysis of grazing systems" 5–6 June 2007*, Centre of Economic Research and Department of Economics at Norwegian University of Science and Technology, Trondheim, Norway.

Riseth, J. Å. and I. Lie. 2016. Reindrifta i Finnmarks betydning for næringsutvikling og samfunnsutvikling. In *Nordområdene i endring - Urfolkspolitikk og utvikling*, eds. E. Angell, S. Eikeland and P. Selle, 182–207. Oslo: Gyldendal Akademisk.

Riseth, J. Å., H. Tømmervik and J. W. Bjerke. 2016. 175 years of adaptation: North Scandinavian Sámi reindeer herding between government policies and winter climate variability (1835–2010). *J. For. Econ.* 24: 186–204.

Risvoll, C., G. E. Fedreheim and D. Galafassi. 2016. Trade-offs in pastoral governance in Norway: Challenges for biodiversity and adaptation. *Pastoralism* 6 (4): 1–15.

Ruong, I. 1964. Jåhkåkaska sameby. Särtrykk av: Svenska landsmål och svenskt folkliv.

Salo, M., A. Sirén and R. Kalliola. 2014. *Diagnosing Wild Species Harvest: Resource Use and Conservation*. Amsterdam: Academic Press.

Sara, M. N. 2013. Siida ja siidastallan. Being siida – On the relationship between siida tradition and continuation of the siida system. PhD. dissertation. UiT – The Arctic University of Norway.

Scott, J. C. 1998. *Seeing Like a State: How Certain Schemes to Improve the Human Condition Have Failed*. New Haven: Yale University Press.

Sara, O. K. 1997. Lovgivning og forvaltning av reindrift gjennom tidene. In *Reindrift før og nå*, ed. Sara, O. K. and I. Storli. Oslo: Landbruksforlaget.

Skum, N. N. 1955. Valla *R*enar. ed. E. Manker. Nordiska Museet: Acta Lapponica X. Stockholm: Almqvist and Wiksell/Gebers.

Solem, E., 1970 [1933]. *Lappiske Rettsstudier*, 2nd ed. Oslo: Universitetsforlaget.

Stammler, F. 2002. Success at the edge of the land: Present and past challenges for reindeer herders of the West-Siberian Yamal-Nenets Autonomous Okrug. *Nomadic Peoples* 6 (2):51–71.

Stammler, F. 2005. *Reindeer Nomads Meet the Market: Culture, Property and Globalization at the "End of the Land"*. Münster, Germany: Lit Verlag.

Stammler, F. 2008. Opportunities and threats for mobility: Reindeer nomads of the West Siberia coastal zone (Yamal) respond to changes (in Russian). *Environ. Plan. Man.* 3–4 (8–9): 78–91.

Syroechkovskii, E. E. 1995. *Wild Reindeer.* Washington: Smithsonian Institution Libraries.

Tombre, I. M., H. Tømmervik, N. Gullestad and J. Madsen. 2010. Spring staging in the Svalbard-breeding Pink-footed Goose (Anser brachyrhynchus) population: Site-use changes caused by declining agricultural management? *Wildfowl.* 60: 3–19.

Tømmervik, H., B. Johansen, J. Å., Riseth et al. 2009. Above ground biomass changes in the mountain birch forests and mountain heaths of Finnmarksvidda, Northern Norway, in the period 1957–2006. *For. Ecol. Man.* 257: 244–57.

Tømmervik, H. and J. Å. Riseth. 2011. Naturindeks. Historiske tamreintall i Norge fra 1800-tallet fram til i dag. *NINA Rapport* 672. Tromsø: NINA.

Tømmervik, H., J. W. Bjerke, E. Gaare, B. Johansen and D. Thannheiser. 2012. Rapid recovery of recently overexploited winter grazing pastures for reindeer in northern Norway. *Fungal Ecol.* 5: 3–15.

Tømmervik, H., J. W. Bjerke, K. Laustsen, B. E. Johansen and S. R. Karlsen. 2014. Monitoring of winter grazing areas for reindeer in inner parts of Finnmark 2013. Results of the field surveys – *NINA Rapport* 1066. Tromsø: NINA.

Tryland, M. 2012. Are we facing new health challenges and diseases in reindeer in Fennoscandia? *Rangifer* 32 (1): 35–47.

Tuisku, T. 2002. Transition period in the Nenets Autonomous Okrug: Changing and unchanging life of Nenets people. In *People and the Land: Pathways to Reform in Post-Soviet Siberia.* ed. E. Kasten, 189–206. Berlin, Germany: Dietrich Reimer Verlag.

Uboni, A., T. Horstkotte, E. Kaarlejärvi et al. 2016. Long-term trends and role of climate in the population dynamics of Eurasian reindeer. *PLoS ONE* 11(6): e0158359. doi:10.1371/journal. pone.0158359.

Ulvevadet, B. and K. Klokov (eds.). 2004. *Family-based reindeer herding and hunting economies, and the status and management of wild reindeer/caribou populations.* Tromsø, Norway: Centre for Saami Studies, UiT – Arctic University of Norway.

UNEP. 2001. C. Nellemann, L. Kullerud, I. Vistnes et al. GLOBIO Global methodology for mapping human impacts on the biosphere. UNEP/DEWA/TR.01-3.

Viken, A. 2016. Reiseliv i samiske områder-nyliberalisme og marginalisering? In *Nordområdene i endring - Urfolkspolitikk og utvikling,* eds. E. Angell, S. Eikeland and P. Selle, 235–57. Oslo: Gyldendal Akademisk.

Vistnes, I. I. 2008. Impacts of human development and activity on reindeer and caribou habitat use. Department of Ecology and Natural Resource Management, Norwegian University of Life Sciences, Ås, Norway. *Doctor philosophiae thesis* 2008:1.

Wang, A. B. 2016. Russia plans to kill a quarter-million Siberian reindeer amid anthrax fears. *The Washington Post,* 1 October 2016. www.washingtonpost.com/news/animalia/wp/2016/10/01/russia-plans-to-kill-a -quarter-million-siberian-reindeer-amid-anthrax-fears/ (Accessed 9 Nov 2017).

WCED. 1987. *Our Common Future.* Oxford: Oxford University Press.

Zeng, H., G. S. Jia and B. C. Forbes. 2013. Shifts in Arctic phenology in response to climate and anthropogenic factors as detected from multiple satellite time series. *Environ. Res. Lett.* 8 (2013) 035036.

Zijp, M. C., R. Heijungs, E. van der Voet, D. van de Meent et al. 2015. An identification key for selecting methods for sustainability assessments. *Sustainability – Basel* 7: 2490–512.

1.4 CARIBOU MANAGEMENT: AN OVERVIEW FROM NORTHERN CANADA AND ALASKA

Jan Adamczewski and Jim Dau

1.4.1 INTRODUCTION

Caribou (*Rangifer tarandus*) in North America range over most of Canada and Alaska. They occupy tundra, boreal and montane habitats and abundance ranges from remnant populations of southern mountain herds numbering fewer than 50 individuals to migratory tundra herds estimated at tens or hundreds of thousands of animals. Most of their northern ranges remain relatively intact, but the southern end of the geographic extent in Canada and the continental United States has contracted substantially. Management of caribou depends on the kind and abundance of caribou present in the region, goals of management agencies and the public and feasibility of monitoring and management as influenced by the remoteness of ranges and resource availability. Management is in many cases defined in plans or strategies for specific populations or kinds of caribou. This section provides a brief overview of current approaches to management of caribou in North America.

1.4.2 KINDS OF CARIBOU IN NORTH AMERICA

Caribou in North America have adapted to tundra, montane and boreal forest habitats in a variety of ways, with seasonal movements, group sizes and seasonal ecology shaped in various ways by their main predator, the wolf (Bergerud 1988, Seip 1991) and by weather and habitat characteristics (Skoog 1968). Four extant subspecies of caribou were identified by Banfield (1961) in North America: *R. t. granti* (Grant's caribou) in Alaska, *R. t. groenlandicus* (barren-ground caribou in Canada), *R. t. pearyi* (Peary caribou on the arctic islands) and *R. t. caribou* (woodland caribou). In addition, semi-domesticated reindeer (*R. t. tarandus*) were introduced to Alaska and a few locations in Canada. These subspecies designations pre-dated several decades of genetic research on caribou and reindeer, however (COSEWIC 2011), and do not recognize the full current range of caribou types and their genetics in North America. The diversity of caribou led the Committee on the Status of Endangered Wildlife in Canada (COSEWIC 2011) to identify 12 "Designatable Units" (DUs) of caribou in Canada (not including Alaska). The DUs identify ecological types by a combination of criteria that include genetics, behaviour, morphology, movements and distribution (Figure 1.4.1).

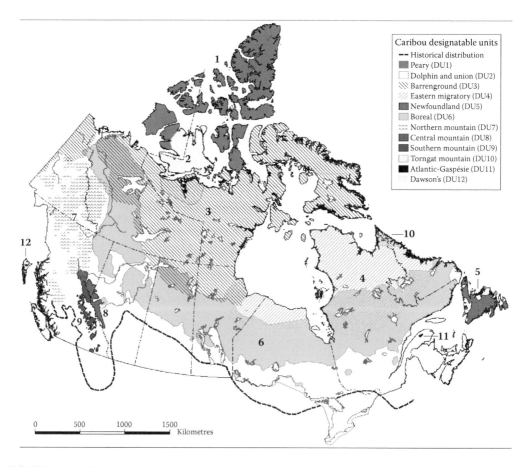

Caribou designatable units
- – – Historical distribution
- Peary (DU1)
- Dolphin and union (DU2)
- Barrenground (DU3)
- Eastern migratory (DU4)
- Newfoundland (DU5)
- Boreal (DU6)
- Northern mountain (DU7)
- Central mountain (DU8)
- Southern mountain (DU9)
- Torngat mountain (DU10)
- Atlantic-Gaspésie (DU11)
- Dawson's (DU12)

FIGURE 1.4.1 Designatable units of caribou in Canada (From COSEWIC 2011).

With the exception of one small mountain herd (Chisana) and some reindeer herds, Alaskan herds are considered to be *R. t. granti* (Figure 1.4.2). The larger northern migratory tundra herds in Alaska show many similarities to migratory herds in DU3 in Canada (barren-ground caribou).

The diversity of caribou has created challenges for caribou managers – among basic principles of wildlife management is the need to define populations and ranges, and to have some measure of relative abundance and trends. Where herds are identifiable, they are often named and are the basic unit of monitoring and management, e.g. migratory tundra caribou (Gunn and Miller 1986, Valkenburg 1998, Dau 2011). Boreal woodland caribou are sometimes named as herds, although studies in relatively large intact portions of their range, such as the Northwest Territories (NT), suggest that they form contiguous low-density distributions over large regions when not constrained by anthropogenic change. More than half the boreal caribou range in northern Ontario, Québec and Saskatchewan (SK), as well as all of the boreal caribou range in the NT are considered as large low-density contiguous populations (Environment Canada 2012a).

Broadly speaking, the range of caribou in North America is relatively intact in the north and the large migratory tundra herds still number thousands or, in a few cases, hundreds of thousands of individuals, although recent trends in Alaska, northern Canada and Québec/Labrador have been largely negative (Vors and Boyce 2009, Festa-Bianchet et al. 2011). Of the ten largest Alaskan caribou herds, only the Porcupine Herd was increasing as of 2015. The historic southern extent of caribou range extended across southern Canada and into several northern US states, but has retracted north over time as forests have been logged and turned into farmed and urban landscapes. Caribou are not currently found in New Brunswick, Nova Scotia or Prince Edward Island. A single small southern mountain herd has a portion of its range in northern Idaho and Washington (Environment Canada 2014a). Caribou occur throughout most of mainland Alaska. The relatively intact populations and habitat in the north reflect the lesser extent of human-caused landscape change at northern latitudes.

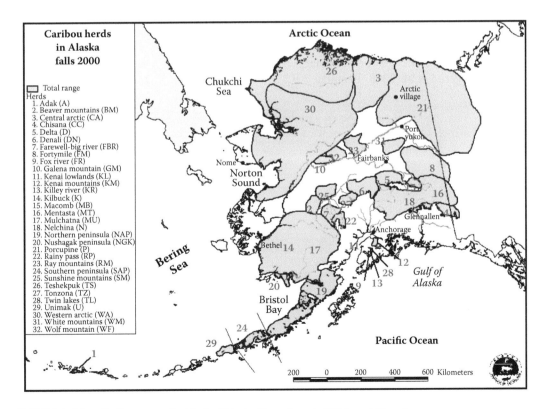

FIGURE 1.4.2   Caribou range in Alaska (Alaska Department of Fish and Game).

Each of the seven provinces and three territories and the state of Alaska have monitoring and management programs to ensure the conservation of this species and provide for the people who depend upon them. Where caribou are considered at risk of extirpation in Canada due to declining and/or low numbers, the federal government defines management at a national scale and the territories and provinces develop their individual recovery plans or strategies. In Alaska, management of caribou is shared between state (Alaska Department of Fish and Game) and federal (Bureau of Land Management, U.S. Fish and Wildlife Service, Bureau of Land Management) agencies. We will focus here mainly on caribou management for the following kinds of caribou as examples of management in Canada and Alaska: Peary, barren-ground, mountain and boreal. The largest section is devoted to barren-ground caribou, in part because they are the most abundant North American caribou and in part because they have had an extraordinary significance to Aboriginal cultures in northern Canada and Alaska for thousands of years (Giddings 1977, Gordon 2005, Burch 2006, Beaulieu 2012).

1.4.2.1 Barren-Ground Caribou

Barren-ground caribou are numerically the most successful of caribou types in Canada (DU3 in Figure 1.4.1), particularly if the George River and Leaf River herds in northern Québec and Labrador (Part of DU4, Eastern Migratory) are considered with DU3. These two herds have many attributes similar to those of migratory herds on the Canadian mainland included with DU3. Four migratory tundra herds in Alaska, including the Western Arctic and Porcupine herds, have a similar ecology and are the largest herds in the state (Alaska Department of Fish and Game 2011). Ranges occupied by the migratory herds are sometimes vast and may include multiple jurisdictions. For example, the overall range of the Bathurst herd in NT and Nunavut (NU) between 1996 and 2012 covered nearly 350,000 km^2 (Figure 1.4.3) and the range of the Western Arctic Herd between 1990 and 2014 covered about 370,000 km^2. Calving and summer ranges are generally on the tundra (Figures 1.4.4 and 1.4.5) and winter ranges south of the tree-line in the boreal forest. Barren-ground caribou also occur on several islands in Canada, including Baffin Island, and on the northeast mainland in NU, with their entire annual ranges on the tundra.

Wide variation in migratory caribou abundance on a time-scale of decades has been recognized from Traditional Knowledge of Aboriginal peoples across their North American range (Zalatan et al. 2006, Bergerud et al. 2008, Beaulieu 2012, Burch 2012). Large fluctuations in numbers have

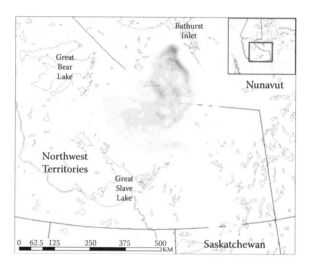

FIGURE 1.4.3 Annual range of the Bathurst caribou herd in northern Canada, based on locations of satellite radio-collared females, 1996–2012. Map A. D'Hont, GNWT. Darker (orange) areas were used more intensively than lighter (yellow) areas. The calving grounds at the north end of the range since 1996 have been southwest of Bathurst Inlet.

FIGURE 1.4.4   Barren-ground caribou on the Bathurst calving grounds in June. The sight of thousands of caribou cows with young calves is one of the most humbling sights in North America (Photo: Anne Gunn, GNWT).

FIGURE. 1.4.5   An immature male, mature male and mature female caribou (*R. t. granti*) of the barren-ground type from the Western Arctic Herd, northwest Alaska, October 2014 (Photo: Jim Dau).

continued since biologists first began surveying caribou in the 1950s and 1960s (Valkenburg 1998, Gunn 2003). Cycles or long-term fluctuations are complex but linked to large-scale weather trends such as the Arctic Oscillation and the Pacific Decadal Oscillation (Gunn 2003, Joly et al. 2011).

Migratory caribou herds have generally been defined and managed as discrete populations based on calving grounds used with high fidelity by female caribou (Skoog 1968, Gunn and Miller 1986, Valkenburg 1998). Status of the migratory herds of caribou and reindeer across the circumpolar north is monitored in part by the CircumArctic *Rangifer* Monitoring and Assessment (CARMA) network, and herd-specific ranges and data from the various agencies can be found on the CARMA website (carma.caff.is/index.php/herds).

Because of their high significance to northern Aboriginal cultures and communities, management boards have been set up to work toward collaborative management of a number of

North American herds. The boards are made up of government and community representatives to ensure a broad representation of interests. The Beverly and Qamanirjuaq Caribou Management Board (BQCMB) in NT, NU, Saskatchewan (SK) and Manitoba (MB) (www.arctic-caribou.com) was set up in 1982. The Porcupine Caribou Management Board (PCMB) in Yukon and NT (www .pcmb.ca) was set up in 1985 and the International Porcupine Caribou Board, which includes Alaska, was set up in 1987. In Alaska, the Western Arctic Caribou Herd Working Group was formed in 2000 (westernarcticcaribou.org). The mission statement of the WACH Working Group is *"To work together to ensure the long term conservation of the Western Arctic Caribou Herd and the ecosystem on which it depends, and to maintain traditional and other uses for the benefit of all people now and in the future."* Other boards have similar inclusive, long-term conservation goals.

Through the work of these boards and other co-management boards resulting from land claims, management plans have been developed for many herds. Recent examples are the Western Arctic Caribou Herd Cooperative Management Plan (WACHWG 2011) in Alaska, the BQCMB Management Plan for 2013–2022 (BQCMB 2014) for the Beverly and Qamanirjuaq herds in NT and NU, and the ACCWM management plan for the Cape Bathurst, Bluenose-West and Bluenose-East caribou herds in NT and NU (ACCWM 2014).

Although each herd has unique characteristics and plans vary, these management plans generally include the following elements, reflecting a desire from participants for a holistic, inclusive approach:

1. Goals and objectives
2. A summary of herd status and ecology
3. Monitoring, including demography and health
4. Harvest management
5. Predator management
6. Land use and habitat conservation
7. Education and awareness

Although some barren-ground caribou herds remain large, most have shown declining trends since the late 1990s and early 2000s (Alaska Department of Fish and Game 2011, Festa-Bianchet et al. 2011). In some cases (e.g. the Western Arctic Herd during the 1970s) severe declines have resulted in drastic short-term reductions in harvest, sometimes without a management plan to provide guidance. More recently, in December 2014, the Government of NU (GN) announced an interim moratorium on harvest of Baffin caribou when surveys showed about 5,000 caribou remaining after a 90–95% decline from population estimates in the 1990s (GN press release December 2014). In the Bathurst herd's range, an annual harvest estimated at 4,000–6,000 caribou was reduced in 2010 to 300 or less after this herd declined from an estimated 470,000 in 1986 to 32,000 in 2009 (Boulanger et al. 2011). These declines and the need for rapid action have created major challenges to collaborative caribou management in the Canadian North.

The threat of climate change looms ever larger on the horizon of caribou management in North America. Although caribou are an adaptable species and occupy a broad ecological niche, which has allowed them to survive the widespread extinctions that occurred at the end of the Pleistocene epoch, there are limits to which northern-adapted species can compete with herbivores, predators and parasites adapted for more temperate climes. A higher frequency of large-scale fire years driven by a warming climate may impact caribou negatively (Joly et al. 2012, Gustine et al. 2014). A positive aspect of this is that government agencies are increasingly funding environmental studies to investigate climate change, and these programs are substantially improving our understanding of weather, vegetation, hydrology and snow – all of which affect caribou.

Despite these challenges and the widespread declines in caribou numbers, the potential for conservation of migratory caribou in northern Canada and Alaska remains high for a number of reasons. First, the ancient relationship of caribou-hunting cultures and caribou means that the persistence of healthy caribou herds remains a high priority for northern people. Second, although there is increasing concern over the cumulative effects of mining, roads and other development on caribou, thus far the ranges of the great migratory herds are relatively intact (compared to southern Canada and the US states south of Alaska) and people can still make decisions to safeguard the integrity of caribou ranges. Third, methods used to monitor the size, demographics, movements and distribution of caribou herds have improved greatly over the past 40 years, and new technologies continue to develop. Finally, managers and the people who depend on caribou can be powerful advocates for the conservation of these herds and their habitat through the various collaborative management boards and comprehensive management plans that exist throughout Canada and Alaska.

1.4.2.2 Peary Caribou

Peary caribou are the smallest of North American caribou and are found on the northernmost islands in Canada (DU1 on Figure 1.4.1, Figure 1.4.6). These high arctic islands are generally sparsely veg-etated and some islands, particularly in the northeast (Ellesmere, Devon and Axel Heiberg Islands) have substantial montane glaciers. Much of the Peary caribou range is considered polar desert (SARA 2015). Peary caribou are known to move between neighboring islands, primarily in winter (Miller et al. 1977, Miller 1995, Environment Canada 2015). On the basis of caribou movements known from surveys and Traditional Knowledge, four groupings of islands and caribou have been proposed as "local populations": Banks Island & Northwest Victoria, Eastern Queen Elizabeth Islands, Western Queen Elizabeth Islands and Prince of Wales, Somerset, Boothia and Russell (Environment Canada 2015). Intervals between population surveys have varied, as have survey methods (Environment Canada 2015) and sometimes more than 10 years have elapsed between surveys; thus, population trends have been difficult to track. The total Peary caribou population was estimated at about 7,890 in 2015, a reduction of about 72% since 1980 (SARA 2015).

Large die-offs of Peary caribou have occurred repeatedly on the high arctic islands. Surveys of the Bathurst Island complex in the mid-1970s documented a catastrophic decline of over 90% from 1961, followed by recovery and then a 98% decline between 1993 and 1998 to less than 100 animals

FIGURE 1.4.6 Peary caribou (*R. t. pearyi*), the smallest and northernmost of North American caribou (Photo: John Nagy, GNWT).

(Miller and Gunn 2003). "Widespread, prolonged, exceptionally severe snow and ice conditions from 1994–95 to 1996–97 caused the die-off. Trends in snowfall are consistent with predictions for global warming in the western Canadian High Arctic" (Miller and Gunn 2003). Peary caribou and their cousins, the Dolphin-Union caribou that twice annually migrate across sea ice between Victoria Island and the Canadian mainland, are arguably the most sensitive of all North American caribou to climate change that is likely to include warmer winters, increased precipitation and shrinking sea ice (SARA 2015).

Peary caribou are considered endangered under the federal Canadian Species at Risk Act (SARA), and work led by Environment Canada was underway in 2015 to develop a national recovery strategy. Climate change, severe winter weather, icing events and loss of sea ice, reducing the ability of the caribou to move between islands, were collectively considered the greatest threat to Peary caribou (Environment Canada 2015). Changes to arctic weather are largely driven by global climate change and would need to be managed at a global scale. For some Peary caribou populations, hunter harvest may play a role if caribou are at very low numbers (Miller and Gunn 2003), but many Peary caribou ranges are far from human communities. Harvest of Peary caribou on Banks Island is limited under a quota system and harvest on northwest Victoria Island has been closed for a number of years; hunter and trapper organizations have established voluntary limits in a number of locations (Environment Canada 2015). Other possible threats to Peary caribou include industrial development and shipping between islands. Potential competition from muskoxen and predation by wolves have been proposed as threats to caribou (Environment Canada 2015). Overall, active management options for Peary caribou are limited, given the nature of the threats to the animals, especially the weather, as well as the remote nature of much of the range, and limited resources for monitoring.

1.4.2.3 Mountain Caribou

Mountain caribou occur in Yukon, NT, British Columbia (BC) and Alberta (AB). Identification of ecological types of mountain caribou has varied by jurisdiction and a consistent set of designations has not been universally adopted. COSEWIC (2011) identified three DUs: Northern Mountain (DU7), Central Mountain (DU8) and Southern Mountain (DU9).

Of these three groups, the northern mountain caribou found in northern BC and further north in Yukon and NT (Figure 1.4.7) have the most intact ranges, greatest abundance and generally stable

FIGURE 1.4.7   Female and male caribou of the northern mountain type (DU7) in northern British Columbia. Most of the populations of this type are healthy and their ranges relatively intact (Photo: Jan Adamczewski).

numbers. These populations are considered federally as Special Concern[1] and a management plan was completed in 2012 (Environment Canada 2012a). A total of 36 herds numbering about 45,000 caribou were estimated for 2008, although population status was unknown for 22 of these herds (Environment Canada 2012a; Figure 1.4.5). These caribou are generally found in alpine habitats through the snow-free season and either in lowland forested ranges with shallow snow or on windswept alpine plateaus during winter (Environment Canada 2012a). Several First Nations in the Yukon have a long history of depending on these caribou herds, and have contributed to successful recovery programs (Florkiewicz et al. 2007, Farnell 2009). Although most of the northern mountain herds have been stable, this most likely reflects the relatively remote nature of their habitat. For herds with accessible ranges, careful habitat management combined with harvest limitation will be needed for long-term persistence (Florkiewicz et al. 2007).

A recovery strategy was completed in 2014 for the Southern Mountain population of woodland caribou (Environment Canada 2014). This plan, somewhat confusingly, includes the central mountain (DU8) and southern mountain (DU9) groups identified by COSEWIC (2011), along with a few herds identified as northern mountain type (DU7). Regardless of the designations, the total estimated number of Southern Mountain caribou in 2014 was 5,800 distributed across 34 existing and four extirpated subpopulations; 18 of 34 subpopulations were estimated at fewer than 50 animals, and declining trends were near-universal (Environment Canada 2014). The main reason for the declines has been identified as habitat alteration from development leading to increased numbers of deer, elk and moose, increased wolf numbers supported by alternate prey and higher mortality rates of caribou (Wittmer et al. 2005a,b, 2007, Environment Canada 2014). The goals of the recovery strategy are to establish self-sustaining subpopulations in all cases and increased overall caribou numbers, and the strategy identifies objectives for adequate distribution of relatively undisturbed habitat (Environment Canada 2014). Intensive management has included "maternity penning," which involves capturing and confining caribou cows in temporary enclosures during late pregnancy and the first weeks of neonatal life to increase early calf survival (CBC News 2015). In February 2015, the BC government announced a 5-year program to cull wolves in the range of some of the most imperiled southern mountain herds, an approach that has often been highly controversial in Canada (CBC News 2015). Other measures to promote recovery of southern mountain caribou have included protection of key habitat and caribou trans-location to augment small populations (Kinley 2010). The long-term sustainability of these remnant herds remains questionable.

1.4.2.4 Boreal Woodland Caribou

Boreal woodland caribou have a wide geographic range across northern Canada in the boreal forest from Labrador to the northern NT, although their current southern limit is substantially further north than the historic limit (Figure 1.4.1). They are generally found at low densities in small groups in old forests where they are spatially separated from wolves and alternate prey for wolves (Bergerud 1988, Seip 1991, Environment Canada 2012b). Boreal caribou shift seasonally among habitats but do not undertake the long-range movements of migratory tundra caribou or the medium-range seasonal movements of some mountain herds. These caribou are difficult to count owing to their low densities and poor visibility in some forest habitats; there were an estimated 34,000 in Canada in 2012 (Environment Canada 2012b).

As with mountain caribou ranges, boreal caribou in the least disturbed tracts of forest are generally stable, but where oil and gas, logging and other development has substantially altered the range, population declines and range loss have been widespread (Environment Canada 2012b). The primary mechanism for boreal caribou declines is similar to the mechanism documented in southern mountain caribou.

Across most of the distribution of boreal caribou, human-induced habitat alterations have caused an imbalance in predator-prey relationships resulting in unnaturally high predation rates. This is the major factor affecting the viability of most boreal caribou local populations (Bergerud 1988, Stuart-Smith et al. 1997, Rettie and Messier 1998, Schaefer et al. 1999, James and Stuart-Smith, 2000, Wittmer et al. 2005, Chabot 2011) (Environment Canada 2012b).

The national recovery strategy set out objectives of maintaining the 14 local populations currently considered as self-sustaining in that state, and achieving self-sustaining status for the 37 local populations that are not currently considered self-sustaining (Environment Canada 2012b). A habitat objective of at least 65% undisturbed habitat for each local population was identified. Disturbed habitat is identified as habitat altered by human influences (primarily industrial development) and forest burned within the last 40 years. Canadian provinces and territories that have boreal caribou range are expected to develop range plans that will meet these objectives.

Because of the dire situation of these populations, management to preserve remnant boreal caribou populations has recently included intensive and controversial methods like shooting and poisoning wolves (Hervieux et al. 2014) and maternity penning. The ethics and rationale of some of these intensive methods have been questioned (e.g. Brook et al. 2015) and the long-term sustainability of these remnant herds remains in doubt, particularly in provinces like Alberta, where oil and gas development and functional loss of boreal caribou habitat has continued at a substantial pace.

REFERENCES

Advisory Committee for Cooperation on Wildlife Management (ACCWM). 2014. *Taking Care of Caribou: The Cape Bathurst, Bluenose-West, and Bluenose-East barren-ground caribou herds management plan.* C/O Wek'èezhìi Renewable Resources Board, 102A, 4504 – 49 Avenue, Yellowknife, Northwest Territories, Canada.

Alaska Department of Fish and Game. 2011. Caribou management report of survey-inventory activities 1 July 2008–30 June 2010. ed. P. Harper. Juneau, Alaska. 345 pp.

Banfield, A. W. F. 1961. A revision of the reindeer and caribou, genus *Rangifer.* National Museum of Canada, Bulletin No. 177. Queen's Printer, Ottawa, Ontario, Canada. 137 pp.

Beaulieu, D. 2012. Dene traditional knowledge about caribou cycles in the Northwest Territories. *Rangifer Special Issue* 20:59–67.

Bergerud, A. T. 1988. Caribou, wolves and man. *Trends in Ecology and Evolution* 3(3):68–72.

Bergerud, A. T., S. N. Luttich and L. Camps. 2008. *The Return of Caribou to Ungava.* McGill-Queen's University Press, Canada.

Beverly and Qamanirjuaq Caribou Management Board (BQCMB). 2014. Beverly and Qamanirjuaq Caribou Management Plan. www.arctic-caribou.com/PDF/bqcmb_managementplan_detailed2014.pdf.

Boulanger, J., A. Gunn, J. Adamczewski and B. Croft. 2011. A data-driven demographic model to explore the decline of the Bathurst caribou herd. *J. Wildl. Manag.* 75:883–96.

Brook, R. K., M. Cattet, C. T. Darimont, P. C. Paquet and G. Proulx. 2015. Maintaining ethical standards during conservation crises. *Can. Wildl. Biol. Manag.* 4:72–9.

Burch, E. S, Jr. 2006. *Social Life in Northwest Alaska: The Structure of Inupiaq Eskimo Nations.* University of Alaska Press, Fairbanks, Alaska. 478 pp.

Burch, E. S., Jr. 2012. *Caribou Herds of Northwest Alaska.* eds. I. Krupnick and J. Dau. University of Alaska Press, Fairbanks, Alaska. 203 pp.

CBC News. 2015. B.C. wolf cull will likely last 5 years, assistant deputy minister says. CBC News online, Feb. 11, 2015, British Columbia, Canada. www.cbc.ca/news/canada/british-columbia/b-c-wolf-cull -will-likely-last-5-years-assistant-deputy-minister-says-1.2952556.

Chabot, A. 2011. Suivi télémétrique et stratégie générale d'aménagement de l'habitat des caribous forestiers du Nitassinan de la Première Nation innue d'Essipit. Rapport du Groupe-Conseil AGIR inc., présenté au Conseil de la Première Nation innue d'Essipit. 43 p. et 1 annexe.

COSEWIC. 2011. *Designatable Units for Caribou* (Rangifer tarandus) *in Canada.* Committee on the Status of Endangered Wildlife in Canada. Ottawa, Ontario, Canada. 88 pp.

Dau, J. 2011. Western Arctic caribou herd. In *Caribou management report of survey-inventory activities 1 July 2008–30 June 2010,* ed. P. Harper, 187–251. Juneau, Alaska.

Environment Canada. 2012a. Management Plan for the Northern Mountain Population of Woodland Caribou (*Rangifer tarandus caribou*) in Canada. Species at Risk Act Management Plan Series. Environment Canada, Ottawa, Ontario, Canada. vii + 79 pp.

Environment Canada. 2012b. Recovery Strategy for the Woodland Caribou (*Rangifer tarandus caribou*), Boreal Population in Canada. Species at Risk Act Recovery Strategy Series. Environment Canada, Ottawa, Ontario, Canada. xi + 138 pp.

Environment Canada. 2014. Recovery Strategy for the Woodland Caribou, Southern Mountain Population (*Rangifer tarandus caribou*) in Canada. Species at Risk Act Recovery Strategy Series. Environment Canada, Ottawa, Ontario, Canada. viii + 103 pp.

Environment Canada. 2015. Recovery Strategy for Peary Caribou (*Rangifer tarandus pearyi*) in Canada. Species at Risk Act Recovery Strategy Series. Environment Canada, Ottawa Ontario, Canada (in prep., draft Feb. 2015).

Farnell, R. 2009. *Three decades of caribou recovery programs in Yukon: A paradigm shift in wildlife management.* Department of Environment, Government of Yukon, Whitehorse, Yukon, Canada. MRC-09-01.

Festa-Bianchet, M., J. C. Ray, S. Boutin, S. D. Côté and A. Gunn. 2011. Conservation of caribou (*Rangifer tarandus*) in Canada: An uncertain future. *Can. J. Zool.* 89:419–34.

Florkiewicz, R., R. Maraj, T. Hegel and M. Waterreus. 2007. The effects of human land use on the winter habitat of the recovering Carcross woodland caribou herd in suburban Yukon Territory, Canada. *Rangifer Special Issue* 17:181–97.

Giddings, J. L. 1977. *Ancient Men of the Arctic.* A. A. Knopf, New York, 391 pp.

Gordon, B. C. 2005. 8000 years of caribou and human seasonal migration in the Canadian Barrenlands. *Rangifer Special Issue* 16:155–62.

Gunn, A. 2003. Voles, lemmings and caribou – population cycles revisited? *Rangifer Special Issue* 14:105–12.

Gunn, A. and F. L. Miller. 1986. Traditional behavior and fidelity to caribou calving grounds by barrenground caribou. *Rangifer* 1:151–8.

Gustine, D. D., T. J. Brinkman, M. A. Lindgren et al. 2014. Climate-driven effects of fire on winter habitat for caribou in the Alaskan-Yukon Arctic. *PLoS ONE* 9(7): e100588. doi:10.1371/journal.pone.0100588.

Hervieux, D., M. Hebblewhite, D. Stepnisky, M. Bacon and S. Boutin. 2014. Managing wolves (*Canis lupus*) to recover threatened woodland caribou (*Rangifer tarandus caribou*) in Alberta. *Can. J. Zool.* 92:1029–37.

James, A. R. C., and A. K. Stuart-Smith. 2000. Distribution of caribou and wolves in relation to linear features. *J. Wildl. Manag.* 64:154–9.

Joly, K., D. R. Klein, D. L. Verbyla, T. S. Rupp and F. S. Chapin. 2011. Linkages between large-scale climate patterns and the dynamics of Arctic caribou populations. *Ecography* 34:345–52.

Joly, K., P. A. Duffy and T. S. Rupp. 2012. Simulating the effects of climate change on fire regimes in Arctic biomes: Implications for caribou and moose habitat. *Ecosphere* 3(5):36. dx.doi.org/10.1890/ES12-00012.1

Kinley, T. A. 2010. *Augmentation plan for the Purcells-South Mountain caribou population.* Prepared for British Columbia Ministry of Environment, Environmental Stewardship Division, Prince George, British Columbia, Canada.

Miller, F. L. 1995. Inter-island water crossings by Peary caribou, south-central Queen Elizabeth Islands. *Arctic* 48:8–12.

Miller, F. L., R. H. Russell and A. Gunn. 1977. Distributions, movements and numbers of Peary caribou and muskoxen on western Queen Elizabeth Islands Northwest Territories, 1972–1974. Canadian Wildlife Service Report Series No. 40. Edmonton, Alberta, Canada. 55 pp.

Miller, F. L. and A. Gunn. 2003. Catastrophic die-off of Peary caribou on the western Queen Elizabeth Islands, Canadian High Arctic. *Arctic* 56: 381–90.

Rettie, W. J., and F. Messier. 1998. Dynamics of woodland caribou populations at the southern limit of their range in Saskatchewan. *Can. J. Zool.* 76:257–9.

SARA (Species At Risk Act). 2015. Species at Risk Public Registry, Species Profile, Peary Caribou. Government of Canada, Ottawa, Ontario, Canada. www.sararegistry.gc.ca/species/speciesDetails_e.cfm?sid=823.

Schaefer, J. A., A. M. Veitch, F. H. Harrington et al. 1999. Demography of decline of the Red Wine Mountain caribou herd. *J. Wildl. Manag.* 63(2):580–7.

Seip, D. R. 1991. Predation and caribou populations. *Rangifer Special Issue* No. 11:46–52.

Skoog, R. O. 1968. Ecology of the caribou (*Rangifer tarandus granti*) in Alaska. PhD thesis, University of California, Berkeley, California.

Stuart-Smith, A. K., C. J. A. Bradshaw, S. Boutin, D. M. Hebert and A. B. Rippin. 1997. Woodland caribou relative to landscape pattern in northeastern Alberta. *J. Wildl. Manag.* 61:622–33.

Valkenburg, P. 1998. Herd size, distribution, harvest, management issues, and research priorities relevant to caribou herds in Alaska. *Rangifer Special Issue* 10:125–9.

Vors, L. S. and M. S. Boyce. 2009. Global declines of caribou and reindeer. *Global Change Biol.* 15:2626–33.

Western Arctic Caribou Herd Working Group (WACHWG). 2011. *Western Arctic Caribou Herd Cooperative Management Plan* – Revised December 2011. Nome, Alaska. 47 pp.

Wittmer, H. U., B. N. McLellan, D. R. Seip et al. 2005a. Population dynamics of the endangered mountain ecotype of Woodland Caribou (*Rangifer tarandus caribou*) in British Columbia, Canada. *Can. J. Zool.* 83:367–418.

Wittmer, H., A. R. E. Sinclair and B. McLellan. 2005b. The role of predation in the decline and extirpation of woodland caribou. *Oecologia* 144:257–67.

Wittmer, H., B. McLellan, R. Serrouya and C. Apps. 2007. Changes in landscape composition influence the decline of a threatened woodland caribou population. *J. Appl. Ecol.* 76:568–79.

Zalatan, R., A. Gunn and G. H. R. Hare. 2006. Long-term abundance patterns of barren-ground caribou using trampling scars on roots of *Picea mariana* in the Northwest Territories, Canada. *Arct. Antarct. Alp. Res.* 38:624–30.

ENDNOTE

1. Populations can be listed in Canada as Extirpated, Endangered, Threatened and Special Concern.

1.5 *RANGIFER* BIOLOGY AND ADAPTATIONS

Åshild Ønvik Pedersen

1.5.1 INTRODUCTION

The reindeer is a characteristic large key herbivore species of arctic and alpine regions and intensively herded by indigenous people. Reindeer have significant impacts on vegetation communities and nutrient cycling in these regions (Bernes et al. 2013). They also serve as important prey for large and medium sized predators (Tveraa et al. 2014) and provide carrion to the scavenger community (Killengreen et al. 2011) (Figure 1.5.1). Globally, many wild reindeer populations are declining and long-term monitoring and conservation efforts are intensified (Vors and Boyce 2009). This chapter will focus on reindeer and caribou biology and their adaptations to life in a highly seasonal environment. Several of the examples are given with

FIGURE 1.5.1   In northern ecosystems, wild and semi-domesticated reindeer provide prey and carrion to various predators and scavengers. The picture shows a dead Svalbard reindeer (*R. t. platyrhynchus*) (Photo: Eva Fuglei, Norwegian Polar Institute).

FIGURE 1.5.2   Winter co-feeding between Svalbard rock ptarmigan (*Lagopus muta hyperborea*) and Svalbard reindeer (*R. t. platyrhynchus*) (Photo: Nicolas Lecomte, Norwegian Polar Institute).

specific reference to the endemic wild Svalbard reindeer (*R. t. platyrhynchus*), which is one of the northernmost reindeer species, having developed many characteristic adaptations to arctic environments (Figure 1.5.2).

1.5.2   BODY CHARACTERISTICS

Reindeer and caribou are medium sized ruminants with a four-chambered stomach and a complex digestive system. The external features differ between the subspecies within their circumpolar geographic distribution ranges. Three distinct ecotypes are generally defined – woodland/forest, barren ground/tundra and Arctic reindeer/caribou – for which body characteristics and appearance are adapted to the environment in which they live. The trunk of a reindeer is generally elongated, the snout is long and the legs vary in size depending on the ecotype. For instance, the short-legged, bulky Svalbard reindeer, living in a virtually predator-free environment (but see Derocher et al. 2000), appears different from the long-legged, slender and elegant Eurasian forest reindeer (*R. t. fennicus*), which presumably need longer legs to escape predators and walk in deeper snow during winter (Syroechkovskii 1995, Bevanger and Jordhøy 2004). See also Chapter 1.1: *Rangifer tarandus* – Classification and Genetic Variation, and Figure 1.1.2.

The hair coat of the reindeer is thick and brownish during the summer and greyish during the winter months and appears almost white in late winter/spring. Colour variations are, however, common among herded reindeer. Their chest and underside are paler in colour and their rump and tail are white. Reindeer moult (lose fur) once a year, changing from winter to summer fur. The moult usually takes place in the spring, but may extend into the summer. Moulting starts at the head and extends to the neck, legs, dorsum and finally the flanks and abdomen. The summer fur undergoes a transition to winter fur in autumn. The coat of a newborn calf has various brownish shades and consists of soft elastic hairs that moult 2–3 weeks after birth and turn light greyish (Syroechkovskii 1995).

Reindeer are unique among cervids because both sexes have antlers. Those of males are larger and more complex than females, and males shed them after the rut in fall. In contrast, females generally shed their antlers in spring, shortly after the calves are born (Figure 1.5.3). The antlers give the females an advantage when competing for sparse winter and spring forage. The development of

FIGURE 1.5.3 Reindeer are unique among cervids because females also have antlers, although those of males are larger and more complex. The males shed the antlers after the rut in autumn, while the females shed them in spring before the calf is born. Picture shows a wild Svalbard reindeer (*Rangifer tarandus platyrhynchus*) (Photo: Bart Peeters, Norwegian University of Science and Technology).

antlers occurs in the first summer and well in advance of puberty (Holand et al. 2004). Antler size and status relate to both environmental conditions and individual characteristics (body size and physical condition) (e.g. Thomas and Barry 2005, Melnycky et al. 2013).

1.5.3 LIFE HISTORY

Reindeer breed at specific and highly synchronized times of the year in response to the interplay between endogenous clocks and day length (Figure 1.5.4; see also Chapter 1.6, *Rangifer* Reproduction Physiology). The calves are born at the end of spring or onset of summer (i.e. May–June).

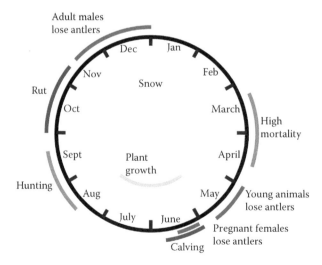

FIGURE 1.5.4 The annual life cycle of a reindeer, exemplified by the high-Arctic Svalbard reindeer (*R. t. platyrhynchus*), relates to the highly seasonal environment it inhabits. (Modified from Albon, S. D. et al. 2017, *Global Change Biology* 23:1374–1389; Graphic design by Jan Roald, Norwegian Polar Institute.)

FIGURE 1.5.5   Reindeer milk is rich in fat and protein and the lactation period can last up to 3–4 months, but varies depending on subspecies and geographic distribution. The calf suckles small quantities during each nursing, but frequently, which partly explains the rapid growth of a reindeer calf. Reindeer calves begin to consume plants a few days after birth. The picture shows a Svalbard reindeer (*Rangifer tarandus platyrhynchus*) (Photo: Bjørn Frantzen, Norwegian Polar Institute).

Females give birth, commonly in specific calving grounds, to a single calf after a gestation period ranging from 203 to 240 days (Ropstad et al. 2005, Rowell and Shipka 2009, Ropstad 2000 report common ranges from 225 to 235 in semi-domesticated reindeer). When the calf is only 1 day old, it is capable of reaching fast running speeds and can follow its mother and utilize newly emerging vegetation (Syroechkovskii 1995). A newborn reindeer calf weighs 5–6 kg, approximately 10% of the body mass of the mother, which is relatively larger than a moose calf that weighs only 3–5% of the mother's body mass. The lactation period completes around the rut, but varies among subspecies and populations. The fast growth and rapid maturation, much greater than for other ungulates, is essential for survival of the calves (Syroechkovskii 1995, Ropstad 2000) (Figure 1.5.5). Mortality is highest during the early life stages and peaks in mortality often occur during the autumn and winter. The life span of a reindeer may range up to 15–20 years, but large variations occur throughout the circumpolar ranges (Syroechkovskii 1995, Bevanger and Jordhøy 2004).

Reindeer puberty relates to body mass and they reach sexual maturity at age 1.5–3 years. Mature females normally conceive annually up to 12–14 years of age, but this can vary depending on geographic location and environmental conditions. As with other ungulates, there is a tight relationship between female body mass and reproductive output (Ropstad 2000). The variable climatic conditions in Svalbard results in high year-to-year variation in survival and reproductive rates, and the proportion of females that give birth to a calf in a given summer can vary from 10 to 90% (Øritsland 1986, Stien et al. 2012).

The reindeer is a polygynous species. Reindeer males improve their reproductive success by increasing the number of females to mate with, whereas females increase their reproductive success by mating with the male of best quality. The reindeer males become sexually mature at the same age as the female, but they are generally not competitive enough to mate until age 3–5 years (see also Chapter 1.7, *Rangifer* Mating Strategies).

1.5.4   HABITATS AND MIGRATION

Reindeer live in arctic and alpine open tundra landscapes and sub-arctic forests - spanning from the boreal taiga (Eurasian forest reindeer and North American woodland caribou *R. t. caribou*) to the high-Arctic sparsely vegetated and barren grounds (Svalbard reindeer, *R. t. platyrhynchus* and Peary caribou, *R. t. pearyi*). The spatial distribution of reindeer and their movements within and between seasons link to, among other things, snow and ground ice conditions, local changes in forage abundance and quality, weather variability, predators and insects. Generally, in early winter when the snow cover is limited, the reindeer graze in the lowlands (i.e. river valleys, near lakes). Later when the snow cover builds up they tend to move to the upland mountain plateaus. Even in summer, snow is important to reindeer because snow covered areas give protection from insects (mosquitos and parasitic flies) that influence movements and grazing time (Syroechkovskii 1995, Bevanger and Jordhøy 2004).

Reindeer and caribou are generally migratory animals that travel long distances between distinct summer and winter ranges (reviewed in Harris et al. 2009). Even when summer and winter ranges are contiguous, the reindeer may travel extensively, not only seasonally, but also within seasons in response to local environmental conditions. Generally, dietary needs (food availability and quality related to development of plants) have been the underlying cause for the migratory and seasonally nomadic behaviour in some populations, while in others such movements have been motivated by predator avoidance (e.g. wolves) or a combination of those factors and insect harassment (Syroechkovskii 1995, Bevanger and Jordhøy 2004) (Figures 1.5.6 and 1.5.7). For instance, the wild barren ground caribou in Alaska and northern Canada migrate northwards between late April and early June to find suitable calving areas on the arctic tundra that are less exposed to predators and insects. The Porcupine caribou herd migrate the longest distances – the annual distances travelled by caribou females between their winter range and the summer calving grounds may exceed 5000 km, which is the longest migration on land documented for any terrestrial mammal (Fancy et al. 1989). Human infrastructure is increasingly disturbing many migratory corridors and ranges, and having negative effects on reindeer populations (Vistnes and Nellemann 2008).

Among wild reindeer, the Svalbard reindeer is an outstanding example of a relatively stationary subspecies that does not undertake long seasonal migrations, nor does it appear nomadic within or between seasons. Females of this subspecies seem typically confined to small seasonal home ranges (Tyler and Øritsland 1989). However, unfavourable winter conditions, such as deep snowpack and ice that block access to food resources, cause range displacement and partial seasonal migrations to occur (Hansen et al. 2010b, Loe et al. 2016).

FIGURE 1.5.6   The high-Arctic Svalbard reindeer (*R. t. platyrhynchus*), along with the Peary caribou (*R. t. pearyi*), inhabit among the most hostile natural environments where ungulate species can live. The picture shows Svalbard reindeer in a typical wind-exposed winter pasture in the high-Arctic Svalbard (Photo: Ronny Aanes, Norwegian Polar Institute).

FIGURE 1.5.7 Many wild and semi-domesticated reindeer migrate over vast land areas between their seasonal distribution ranges. Eurasian wild reindeer (*R. t. tarandus*) at Hardangervidda, Norway (Photo: NINA Autocamera/Olav Strand, Norwegian Institute for Nature Research).

1.5.5 Diet

In arctic and alpine regions, plant species diversity and plant growth is limited due to the combination of low temperature, precipitation, soil-nutrient content and a short growing season. Limiting resources, particularly during the winter season (i.e. carrying capacity determined by the winter food availability), create strong pressure on the reindeer to use vegetation in an efficient manner to meet their energy requirements for survival, maintenance and reproduction (Moen et al. 2006). Forage availability is, in large parts of the year, determined by snow cover, which thus also impacts the spatial distribution of reindeer. Reindeer dig through the snow to access plants beneath and respond to difficult foraging conditions by switching diet or move to areas with less snow (e.g. Kumpula and Colpaert 2003, Tyler 2010).

The reindeer diet varies substantially between seasons and populations across their circumpolar range, and the seasonal grazing pattern commonly reflects the geographical distribution and abundances (biomass) of the most nutritional plants. The diet of wild reindeer consists mainly of those plants that are abundantly available in the pasture (Syroechkovskii 1995). The wide seasonal diet consists mainly of various grasses, herbs, sedges, mosses, shrubs (e.g. willows) and mushrooms and lichens (e.g. Bergerud 1972, Rominger and Oldemeyer 1990, Bjørkvoll et al. 2009) (see also Chapter 3, *Rangifer* Diet and Nutritional Needs).

Ruminants extract energy from the plant material by fermentation – through a symbiotic association with anaerobic microbes in their reticule-rumen. Reindeer depend on efficient physical degradation of forage through chewing to increase the surface area of the food particles presented to the microflora, and the main activity pattern of ruminants consists of a sequential series of foraging and rumination (Van Soest 1994). For instance, the high-Arctic Svalbard reindeer have season-dependent feeding-rumination intervals to optimize forage intake, closely linked to the changing light conditions in the Arctic (Loe et al. 2007) (Figure 1.5.8).

A characteristic feature of reindeer is the ability to feed on lichens – a vast resource with low nutritional value (i.e. low in nitrogen and vitamins) – that almost no other ungulate species feed on, except for the Siberian musk deer (*Moschus moschiferus*) (Nyambayar et al. 2015). Reindeer have special detoxifying microorganisms in their rumen that neutralize toxic compounds such as the usnic acid in lichens (Sundset et al. 2010). During winter, lichen may constitute up to 80% of the total intake by semi-domesticated reindeer, and reindeer can assimilate most of the carbohydrate in the lichens. Lichens mainly contain easily digestible carbohydrates, which can be stored as fat in the reindeer. However, lichens contain too little protein to allow growth and thus they contribute primarily to maintenance. In winter, reindeer generally prefer the lichen species *Cladonia stellaris* and *Cetraria nivalis*, but if these are not available, they feed on other types of lichen or plants (Syroechkovskii 1995, Bevanger and Jordhøy 2004).

In winter, the reindeer dig feeding craters in the snow, using their specialized hoofs as tools to access forage beneath the snowpack (Syroechkovskii 1995) (Figure 1.5.9). The feeding craters may cover large areas, and the reindeer use their olfactory senses to locate vegetation in ice-free micro-habitat beneath the snowpack (Hansen et al. 2010a). In the high-Arctic ranges, where reindeer feed in sparsely vegetated areas among rocks and gravel, increased tooth wear (i.e. reduced chewing efficiency) at earlier ages has been documented (Veiberg et al. 2007).

In summer, both forage biomass and quality are of importance to the reindeer, and the reindeer graze on tundra where grasses, sedges, herbs and shrubs are plentiful (Figure 1.5.10). Reindeer browsing has significant impacts on growth and expansion of the tall shrub communities and can thereby counteract the ongoing greening of arctic ecosystems (Bernes et al. 2013). Lichens are sensitive to grazing and trampling and at high reindeer population densities, preferred lichens quickly deplete. Recovery of lichen ranges after over-grazing is slow and almost non-existent in the high Arctic, but reduction of reindeer densities, as seen in northern Fennoscandia, can lead to faster recoveries (Hansen et al. 2007, Tømmervik et al. 2012).

FIGURE 1.5.8   The Svalbard reindeer (*R. t. platyrhynchus*) feed on low statured graminoids and herbs during the short high-Arctic summer (Photo: Tore Nordstad, Norwegian Polar Institute).

FIGURE 1.5.9   The reindeer, such as this Svalbard reindeer (*R. t. platyrhynchus*), has specialized hoofs for digging in snow to access food (Photo: Bjørn Frantzen, Norwegian Polar Institute).

FIGURE 1.5.10   In summer, the reindeer, such as this wild Eurasian tundra reindeer (*R. t. tarandus*) in Hardangervidda, Norway, graze on tundra where grasses, sedges, herbs and shrubs are plentiful (Photo: Anders Mossing, Norsk Villreinsenter Sør).

1.5.6   FACTORS AFFECTING POPULATION ECOLOGY

Caribou and reindeer populations fluctuate over time and space. In general, large herbivores living in seasonal environments are food-limited through density dependence and climatic variability (Klein 1991). Numerous studies from many of the circumpolar reindeer and caribou populations focus on both top-down (i.e. harvest and predation) and bottom-up (i.e. food limitation) factors and processes influencing population dynamics. Klein (1991) summarized the most important factors to be: climatic extremes and weather variability, food limitation, predation, hunting, insects, parasites,

FIGURE 1.5.11   Harvest is, in many reindeer populations, a primary factor controlling population growth rates and abundance. In many northern communities, semi-domesticated reindeer (*R. t. tarandus*) provide an important ecosystem service for indigenous people. The picture is from reindeer slaughtering in Finnmark, Norway (Photo: Manuel Ballesteros, Norwegian Institute of Nature Research).

disease, interspecific competition and human developments. In most populations of reindeer and caribou, several factors act together (e.g. Tveraa et al. 2014).

In Norway, large carnivores are effectively removed from extensive land areas to protect livestock, which leaves harvesting as the only significant top-down force on ungulate populations in the same regions to avoid food limitation. Intensive reindeer herding and restrictions on available pastures areas have resulted in a situation where some populations are unable to undertake their traditional seasonal migrations to escape unfavourable climatic winter conditions. The study by Tveraa et al. (2007), gathering data from 58 semi-domesticated reindeer populations in Norway, demonstrated the complexity of various factors and how they impact population dynamics differently across the herding districts. Their study revealed that top-down processes (harvest) appeared to be necessary to dampen the effect of harsh winters in populations with generally good winter conditions, and that bottom-up processes such as food limitation influenced populations subjected to poor winter conditions (Figure 1.5.11).

Another interesting example is the Svalbard reindeer that lives in an exceptionally simple high-Arctic food web characterized by absence of predators and insect harassment and limited hunting (Øritsland 1986). Here, the reindeer population dynamics are mainly shaped by limited food resources (i.e. direct density dependent growth), climatic variability (Aanes et al. 2003, Tyler et al. 2008, Stien et al. 2012, Hansen et al. 2013) and gastrointestinal nematodes (Albon et al. 2002). The Svalbard archipelago is currently experiencing a very fast rate of temperature change, resulting in mild winters with 'rain-on-snow' (ROS) events causing locked winter pastures (Putkonen and Roe 2003). Strong negative relations between ROS and reindeer population growth and reproduction are evident (Stien et al. 2012, Hansen et al. 2013).

1.5.7   ADAPTATIONS TO ARCTIC AND ALPINE ENVIRONMENTS

There are three basic types of adaptations to environmental conditions in birds and mammals - morphological, physiological and behavioural. The 'biological clock' of a reindeer is modified by environmental cues that govern temporal organization of physiological functions and behaviour, keeping the animal prepared for daily and seasonal biological events, such as moulting, antler growth, fat accumulation and rut.

The ability of reindeer and caribou to cope with Arctic winter environments and weather variability is truly outstanding. The reindeer is challenged by variable snow, ground ice, wind and temperature conditions during the winter season and by high temperatures, moisture and insects during the summer season. Seasonal variation in food resources and co-existence with predators are likely the most important ecological drivers of the adaptations of reindeer to life in arctic and alpine areas (see summaries in Blix 2005, Marchand 2013, Blix 2016).

1.5.8   MORPHOLOGICAL ADAPTATIONS

1.5.8.1   Body and Limbs

The body of the reindeer is adapted to life with snow. The large foot surface, relative to their size, gives reindeer superior mobility in snow - minimizing the extent to which they break through the crust and sink into the snow. Four deeply cleft hoofs, which act as 'snow shoes', enable reindeer to walk and dig through the snow. The medial claws, particularly on the foreleg, are strongly curved, broad and flat. Reindeer also have long and broad dewclaws that touch the ground even when the animal is standing. Dense, stiff and long hair around the hoof base and between the claws increase the contact surface of the feet with the ground. During the winter, the footpads shrink and tighten, which exposes the rim of the hoof, preventing them from slipping on ice and snow. During summer when the tundra is soft and wet, the footpads become spongy to provide extra traction when walking (Syroechkovskii 1995, Blix 2005).

1.5.8.2   Fur

Reindeer protect themselves from the cold by a well-developed fur, and unlike most other ungulates, the reindeer have no exposed skin. Even the ears and the nose are hair covered (Figure 1.5.12). The winter coat is extremely thick, dense and long, and has three times as many over-hairs, or guard hairs, as compared to other deer species. The thermal insulation of the fur depends on the

FIGURE 1.5.12   The thermal insulation of the reindeer fur is outstanding and, unlike other ungulates, the reindeer have no exposed skin. Picture shows a Svalbard reindeer (*R. t. platyrhynchus*) (Photo: Nicolas Lecomte, Norwegian Polar Institute).

inherent thermal conductivity of the individual hairs themselves and their collective ability to trap and maintain an air layer close to the skin. The high density of fibres on the skin and structural characteristics of the hair enhance the thermal insulation of the reindeer fur. The guard hairs on the flanks are approximately 50 mm long, on the back and sacrum up to 90–100 mm and on the neck up to 300 mm. They have a hollow core filled with air that occupies up to 90% of each hair's volume, which enhances insulation and makes reindeer excellent swimmers because the guard hairs serve as miniature floats, providing buoyancy.

Under the guard hair, the reindeer has a very dense, fine and crinkly fur that is not visible from the outside (Timisjarvi et al. 1984, Syroechkovskii 1995). Reindeer have a lower critical temperature in winter of about −50°C in Svalbard reindeer and −30°C in Eurasian tundra reindeer, which means that they can tolerate to rest at those low ambient temperatures without having to increase their metabolic rate (Nilssen et al. 1984).

1.5.9   PHYSIOLOGICAL ADAPTATIONS

1.5.9.1   Body Temperature and Thermoregulation

Besides having fur that offers excellent thermal insulation, physiological mechanisms reduce heat loss to the environment. Constriction of surface blood vessels in the extremities, which have a special vascular arrangement that facilitate the counter-current heat exchange, also increases the tissue insulation. In this way, peripheral tissues, such as extremities and the nose, cool under cold conditions while the core body temperature is kept constant, and the thermogradient between the core and the environment can be substantial. Such physiological adaptations increase the thickness of the insulating shell, while the core of the body is kept warm, and result in a lower foot temperature and thus reduced heat loss to the environment (Blix 2016).

The thick and insulating fur, which enables reindeer to survive under extremely cold conditions, potentially predisposes them to heat stress, e.g. in connection with physical exertion due to predators, insect harassment and human disturbances. Reindeer are, however, able to dissipate excess heat effectively through thermal panting, first, through the nose, then, when the heat load and the minute volume requirements increase due to exercise, primarily through the mouth. Eventually, they resort to selective brain cooling, a thermoregulatory effector mechanism that allows reindeer and other artiodactyls to selectively cool the brain tissue independently of the rest of the body (Folkow et al. 2002, Blix et al. 2011).

1.5.9.2   Evaporative Heat Loss

Reindeer also defend themselves against cold by restricting evaporative heat and water loss from the respiratory tract to the environment. An elaborate system of scrolled bone/cartilage structures (i.e. conchae) with a large surface, and coated with a vascularized mucosal membrane, fill the nasal cavities of the reindeer. The cold inhaled air passes over the warm mucosal membrane, is subsequently heated to body temperature and is saturated with water vapour on its way to the lungs. When the reindeer exhales, the warm, humid air passes the cold mucosal membrane and the water vapour condenses. The result of this process is that the reindeer expires cold, dry air while saving heat and water, in a process that can be regulated physiologically, depending on the animal's needs. Thus, under conditions of heat stress, less heat recovers in this system due to adjustments in the way the mucosal layer is perfused, and excess heat may be dissipated by means of panting (Blix and Johnsen 1983, Johnsen et al. 1985).

1.5.9.3   Body Mass and Composition

Caribou and reindeer undergo profound seasonal fluctuations in body mass and composition (i.e. amount of fat and protein) as an important adaptation to highly seasonal environments and food scarcity. During the short summer season when high-quality forage is abundant, the reindeer rapidly deposit fat, which compensates for losses during the prolonged winter season with limited food

FIGURE 1.5.13   The high-Arctic Svalbard reindeer (*R. t. platyrhynchus*) accumulate substantial amounts of subcutaneous fat in late summer and autumn. The black line represents approximately 5 cm. (Photo: Karstein Bye, in Blix, 2016).

resources (e.g. Reimers et al. 1982, Helle and Kojola 1994, Gerhart et al. 1996). Fat accumulation differs between the sexes due to differences in reproduction strategies. The females spend most of their fat reserves during birth and lactation, while males spend fat reserves during the rut. Thus, the ability to accumulate fat reserves is the ultimate key to reproductive success in reindeer. Particularly in female reindeer, the need for adaptive body composition strategies is magnified by high reproductive expenditure during food scarcity in winter (e.g. Bardsen et al. 2011).

For instance, at the end of the growing season in late fall, the fat content in the ingesta-free body of a Svalbard reindeer is as high as 27–40%. Most of the fat deposited is under the skin – over the back, flanks and rump and between the superficial skeletal muscles. At the end of winter, the decrease in body mass can be close to 50% of the autumn body mass (Reimers et al. 1982) (Figure 1.5.13). Accumulated fat reserves, together with muscle protein, contribute to about 25% of the energy requirements of the reindeer during winter, while the remaining energy requirements must be obtained through feeding (Tyler 1986). Absence of harassment from predators, insects and humans allows rapid restoration of body reserves in this subspecies during the summer.

1.5.9.4   Biological Rhythms

The daily and annual change in the photoperiod cycle synchronize the biological processes (i.e. termed circadian organization) of seasonally breeding animals. Their internal 'biological clock', situated in the brain of all vertebrates, is modified by environmental cues that govern temporal organization of physiological functions and behaviour. Reindeer adapt to the continuous Arctic summer light and winter darkness by having a seemingly free-running internal biological clock that allows them to be truly opportunistic during these seasons. Generally, this circadian organization and photoperiodic responsiveness in *Rangifer* decreases with latitude (van Oort et al. 2005). The temporal organization of seasonal and anticipatory adaptive changes in reindeer (e.g. winter pelage growth, appetite changes, timing of reproductive cycles, etc.) is of fundamental importance for the survival and successful existence of this species. Another interesting adaptation to Arctic light conditions is that the reindeer eyes are re-modelled between summer and winter so as to make optimal use of the prevailing light conditions to detect forage (Stokkan et al. 2013, Tyler et al. 2014).

1.5.10   Behavioural Adaptations

1.5.10.1   Herding

Reindeer are the most gregarious animal among the ungulates. They can appear in herds of up to several hundred thousand individuals, the largest herds being in North America and in Russia (Williams and Douglas 1986, Vors and Boyce 2009). Being in a herd gives the reindeer several advantages. Large herds cross snow far more easily due to the fact that large, strong animals lead the way, compacting the track. Further, feeding craters are numerous and higher animal densities reduce losses to predators. Generally, the belonging and co-existence in a herd relate to the contrasting needs between individuals' effective grazing search and forage availability and collective protection against predators in an environment where protection is difficult (Syroechkovskii 1995). The herd composition varies throughout the year and sexually segregated herds are common at certain times of the year. The herds generally become smaller throughout the winter due to depletion of grazing grounds, and in early summer the females and calves gather in smaller herds (Bevanger and Jordhøy 2004). However, isolated subspecies like the Svalbard reindeer operate in small groups or are even solitary, likely due to a general lack of predators.

1.5.10.2   Locomotor Activity

Reindeer can reduce their daily energy expenditure by temporarily reducing locomotor activity when exposed to low temperature and increased precipitation during winter. In general, the total distance the animal travels or climbs and the type of surface they move across (e.g. wet versus dry tundra, or hard snow-crust versus deep snow layers) determine the daily cost of locomotion. Reducing locomotor activity is one important strategy to save energy during harsh winter conditions (Blix 2005) (Figure 1.5.14). The Svalbard reindeer is an excellent example of how reduction in locomotor activity reduces energy expenditure. This subspecies appears to be almost tame, since it is sedentary and rarely runs unless provoked. The Svalbard reindeer have the ability to reduce their total daily energy cost of locomotion to become almost negligible (e.g. 2% of their daily winter energy expenditure). The resting metabolic rate (RMR) of a lying Svalbard reindeer is only 66–78% of the values for other reindeer/caribou, while the RMR for a standing animal is 44–88% of the values for other *Rangifer* subspecies. Also, by remaining lying more of the time, the Svalbard

FIGURE 1.5.14   Reindeer reduce their locomotor activity to save energy during the long Arctic winter. Eurasian wild reindeer (*R. t. tarandus*) at Hardangervidda, Norway (Photo: NINA Autocamera/Olav Strand, Norwegian Institute for Nature Research).

reindeer may conserve the equivalent of about 15 days' energy requirement over the winter. Their low energy expenditures for lying and standing and their sedentary activity budget illustrates the excellent energy saving strategies of reindeer (Cuyler and Øritsland 1993).

The truly outstanding adaptive capacity to their environment and their ecological and societal importance make reindeer and caribou a source of conservation concern in the highly changing environments of the far north.

REFERENCES

Aanes, R., B. E. Saether, E. J. Solberg et al. 2003. Synchrony in Svalbard reindeer population dynamics. *Canadian Journal of Zoology* 81:103–110.

Albon, S. D., A. Stien, R. J. Irvine et al. 2002. The role of parasites in the dynamics of a reindeer population. *Proceedings of the Royal Society of London Series B-Biological Sciences* 269:1625–1632.

Albon, S. D., R. J. Irvine, O. Halvorsen, R. Langvatn, L. E. Loe, E. Ropstad, V. Veiberg, R. Van Der Wal, E. M. Bjorkvoll, E. I. Duff, B. B. Hansen, A. M. Lee, T. Tveraa, and A. Stien. 2017. Contrasting effects of summer and winter warming on body mass explain population dynamics in a food-limited Arctic herbivore. *Global Change Biology* 23:1374–1389.

Bardsen, B. J., J. A. Henden, P. Fauchald, T. Tveraa, and A. Stien. 2011. Plastic reproductive allocation as a buffer against environmental stochasticity – Linking life history and population dynamics to climate. *Oikos* 120:245–257.

Bergerud, A. T. 1972. Food habits of Newfoundland caribou. *Journal of Wildlife Management* 36:913–923.

Bernes, C., K. A. Braathen, B. C. Forbes, A. Hofgaard, J. Moen, and J. D. M. Speed. 2013. What are the impacts of reindeer/caribou (*Rangifer tarandus* L.) on arctic and alpine vegetation? A systematic review protocol. *Environmental Evidence* 2:6.

Bevanger, K., and P. Jordhøy. 2004. *Reindeer: The Mountain Nomad.* Naturforlaget, Bokklubben Villmarksliv.

Bjørkvoll, E., B. Pedersen, H. Hytteborn, I. S. Jonsdottir, and R. Langvatn. 2009. Seasonal and interannual dietary variation during winter in female Svalbard reindeer (*Rangifer tarandus platyrhynchus*). *Arctic Antarctic and Alpine Research* 41:88–96.

Blix, A. S. 2005. *Arctic Animals and Their Adaptations to Life on the Edge.* Tapir Academic Press.

Blix, A. S. 2016. Adaptations to polar life in mammals and birds. *Journal of Experimental Biology* 219:1093–1105.

Blix, A. S., and H. K. Johnsen. 1983. Aspects of nasal heat-exchange in resting reindeer. *Journal of Physiology London* 340:445–454.

Blix, A. S., L. Walloe, and L. P. Folkow. 2011. Regulation of brain temperature in winter-acclimatized reindeer under heat stress. *Journal of Experimental Biology* 214:3850–3856.

Cuyler, L. C., and N. A. Øritsland. 1993. Metabolic strategies for winter survival by Svalbard reindeer. *Canadian Journal of Zoology* 71:1787–1792.

Derocher, A. E., O. Wiig, and G. Bangjord. 2000. Predation of Svalbard reindeer by polar bears. *Polar Biology* 23:675–678.

Fancy, S. G., L. F. Pank, K. R. Whitten, and W. L. Regelin. 1989. Seasonal movements of caribou in arctic Alaska as determined by satellite. *Canadian Journal of Zoology* 67:644–650.

Folkow, L. P., L. Walloe, and A. S. Blix. 2002. On how and when reindeer cool their brains. *Journal of Physiology–London* 543:7.

Gerhart, K. L., R. G. White, R. D. Cameron, and D. E. Russell. 1996. Body composition and nutrient reserves of arctic caribou. *Canadian Journal of Zoology* 74:136–146.

Hansen, B. B., R. Aanes, and B. E. Saether. 2010a. Feeding-crater selection by high-arctic reindeer facing ice-blocked pastures. *Canadian Journal of Zoology* 88:170–177.

Hansen, B. B., R. Aanes, and B. E. Saether. 2010b. Partial seasonal migration in high-arctic Svalbard reindeer (*Rangifer tarandus platyrhynchus*). *Canadian Journal of Zoology* 88:1202–1209.

Hansen, B. B., V. Grotan, R. Aanes, B. E. Saether, A. Stien, E. Fuglei, R. A. Ims, N. G. Yoccoz, and A. O. Pedersen. 2013. Climate events synchronize the dynamics of a resident vertebrate community in the High Arctic. *Science* 339:313–315.

Hansen, B. B., S. Henriksen, R. Aanes, and B. E. Sæther. 2007. Ungulate impact on vegetation in a two-level trophic system. *Polar Biology* 30:549–558.

Harris, G., S. Thirgood, J. G. C. Hopcraft, J. P. G. M. Cromsigt, and J. Berger. 2009. Global decline in aggregated migrations of large terrestrial mammals. *Endangered Species Research* 7:55–76.

Helle, T., and I. Kojola. 1994. Body-mass variation in semidomesticated reindeer. *Canadian Journal of Zoology* 72:681–688.

Holand, O., H. Gjostein, A. Losvar, J. Kumpula, M. E. Smith, K. H. Roed, M. Nieminen, and R. B. Weladji. 2004. Social rank in female reindeer (*Rangifer tarandus*): Effects of body mass, antler size and age. *Journal of Zoology* 263:365–372.

Johnsen, H. K., A. Rognmo, K. J. Nilssen, and A. S. Blix. 1985. Seasonal-changes in the relative importance of different avenues of heat-loss in resting and running reindeer. *Acta Physiologica Scandinavica* 123:73–79.

Killengreen, S. T., N. Lecomte, D. Ehrich, T. Schott, N. G. Yoccoz, and R. A. Ims. 2011. The importance of marine vs. human-induced subsidies in the maintenance of an expanding Mesocarnivore in the arctic tundra. *Journal of Animal Ecology* 80:1049–1060.

Klein, D. R. 1991. Limiting factors in caribou population ecology. *Rangifer* 7:30–35.

Kumpula, J., and A. Colpaert. 2003. Effects of weather and snow conditions on reproduction and survival of semi-domesticated reindeer (*R. t. tarandus*). *Polar Research* 22:225–233.

Loe, L. E., C. Bonenfant, A. Mysterud, T. Severinsen, N. A. Oritsland, R. Langvatn, A. Stien, R. J. Irvine, and N. C. Stenseth. 2007. Activity pattern of arctic reindeer in a predator-free environment: No need to keep a daily rhythm. *Oecologia* 152:617–624.

Loe, L. E., B. B. Hansen, A. Stien, S. D. Albon, R. Bischof, A. Carlsson, J. Irvine, M. Meland, I. M. Rivrud, E. Ropstad, V. Veiberg, and A. Mysterud. 2016. Behavioral buffering of extreme weather events in a high-Arctic herbivore. *Ecosphere* 7(6):1–11.

Marchand, P. J. 2013. *Life in the Cold: An Introduction to Winter Ecology*, fourth edition. University Press of New England.

Melnycky, N. A., R. B. Weladji, O. Holand, and M. Nieminen. 2013. Scaling of antler size in reindeer (*Rangifer tarandus*): Sexual dimorphism and variability in resource allocation. *Journal of Mammalogy* 94:1371–1379.

Moen, J., R. Andersen, and A. W. Illius. 2006. Living in a seasonal environment. In: Danell, K., Bergström, R., Duncan, P., Pastor, J., editors. *Large Herbivore Ecology, Ecosystem Dynamics and Conservation*. Cambridge University Press, pp. 50–70.

Nilssen, K. J., J. A. Sundsfjord, and A. S. Blix. 1984. Regulation of metabolic-rate in Svalbard and Norwegian reindeer. *American Journal of Physiology* 247:R837–R841.

Nyambayar, B., H. Mix, and K. Tsytsulina. 2015. Moschus moschiferus. The IUCN Red List of Threatened Species 2015. dx.doi.org/10.2305/IUCN.UK.2015-2.RLTS.T13897A61977573.en (accessed 18 May 2016).

Øritsland, N. A. (ed.). 1986. *Svalbardreinen og dens livsgrunnlag*. Universitetsforlaget AS.

Putkonen, J., and G. Roe. 2003. Rain-on-snow events impact soil temperatures and affect ungulate survival. *Geophysical Research Letters* 30:1188.

Reimers, E., T. Ringberg, and R. Sorumgard. 1982. Body-composition of Svalbard reindeer. *Canadian Journal of Zoology* 60:1812–1821.

Rominger, E. M., and J. L. Oldemeyer. 1990. Early-winter diet of woodland caribou in relation to snow accumulation, Selkirk mountains, British Columbia, Canada. *Canadian Journal of Zoology* 68:2691–2694.

Ropstad, E. 2000. Reproduction in female reindeer. *Animal Reproduction Science* 60:561–570.

Ropstad, E., V. Veiberg, H. Sakkinen, E. Dahl, H. Kindahl, O. Holand, J. F. Beckers, and E. Eloranta. 2005. Endocrinology of pregnancy and early pregnancy detection by reproductive hormones in reindeer (*Rangifer tarandus tarandus*). *Theriogenology* 63:1775–1788.

Rowell, J. E., and M. P. Shipka. 2009. Variation in gestation length among captive reindeer (*Rangifer tarandus tarandus*). *Theriogenology* 72:190–197.

Stien, A., R. A. Ims, S. D. Albon, E. Fuglei, R. J. Irvine, E. Ropstad, O. Halvorsen, R. Langvatn, L. E. Loe, V. Veiberg, and N. G. Yoccoz. 2012. Congruent responses to weather variability in high arctic herbivores. *Biology Letters* 8:1002–1005.

Stokkan, K. A., L. Folkow, J. Dukes, M. Neveu, C. Hogg, S. Siefken, S. C. Dakin, and G. Jeffery. 2013. Shifting mirrors: adaptive changes in retinal reflections to winter darkness in Arctic reindeer. *Proceedings of the Royal Society B-Biological Sciences* 280:9.

Sundset, M. A., P. S. Barboza, T. K. Green, L. P. Folkow, A. S. Blix, and S. D. Mathiesen. 2010. Microbial degradation of usnic acid in the reindeer rumen. *Naturwissenschaften* 97(3):273–278.

Syroechkovskii, E. E. 1995. *Wild Reindeer*. Smithsonian Institutions Libraries.

Thomas, D., and S. Barry. 2005. Antler mass of barren-ground caribou relative to body condition and pregnancy rate. *Arctic* 58:241–246.

Timisjarvi, J., M. Nieminen, and A. L. Sippola. 1984. The structure and insulation properties of the reindeer fur. *Comparative Biochemistry and Physiology* 79:601–609.

Tveraa, T., P. Fauchald, N. G. Yoccoz, R. A. Ims, R. Aanes, and K. A. Hogda. 2007. What regulate and limit reindeer populations in Norway? *Oikos* 116:706–715.

Tveraa, T., A. Stien, H. Broseth, and N. G. Yoccoz. 2014. The role of predation and food limitation on claims for compensation, reindeer demography and population dynamics. *Journal of Applied Ecology* 51:1264–1272.

Tyler, N. J. C. 1986. The relationship between the fat content of Svalbard reindeer in autumn and their death from starvation in winter. *Rangifer* 1:311–314.

Tyler, N. J. C. 2010. Climate, snow, ice, crashes, and declines in populations of reindeer and caribou (*Rangifer tarandus L.*). *Ecological Monographs* 80:197–219.

Tyler, N. J. C., M. C. Forchhammer, and N. A. Øritsland. 2008. Nonlinear effects of climate and density in the dynamics of a fluctuating population of reindeer. *Ecology* 89:1675–1686.

Tyler, N. J. C., G. Jeffery, C. R. Hogg, and K.-A. Stokkan. 2014. Ultraviolet vision may enhance the ability of reindeer to discriminate plants in snow. *Arctic* 67(2):159–166.

Tyler, N. J. C., and N. A. Øritsland. 1989. Why don't Svalbard reindeer migrate? *Holarctic Ecology* 12:369–376.

Tømmervik, H., J. W. Bjerke, E. Gaare, B. Johansen, and D. Thannheiser. 2012. Rapid recovery of recently overexploited winter grazing pastures for reindeer in northern Norway. *Fungal Ecology* 5:3–15.

Van Oort, B. E. H., N. J. C. Tyler, M. P. Gerkema, L. Folkow, A. S. Blix, and K. A. Stokkan. 2005. Circadian organization in reindeer. *Nature* 438:1095–1096.

Van Soest, P. J. 1994. *Nutritional Ecology of the Ruminant.* Cornell University Press.

Veiberg, V., A. Mysterud, E. Bjorkvoll, R. Langvatn, L. E. Loe, R. J. Irvine, C. Bonenfant, F. Couweleers, and N. C. Stenseth. 2007. Evidence for a trade-off between early growth and tooth wear in Svalbard reindeer. *Journal of Animal Ecology* 76:1139–1148.

Vistnes, I., and C. Nellemann. 2008. The matter of spatial and temporal scales: A review of reindeer and caribou response to human activity. *Polar Biology* 31:399–407.

Vors, L. S., and M. S. Boyce. 2009. Global declines of caribou and reindeer. *Global Change Biology* 15:2626–2633.

Williams, T. M., and C. H. Douglas. 1986. World status of wild *Rangifer tarandus* populations. *Rangifer* 1:19–28.

1.6   *RANGIFER* REPRODUCTIVE PHYSIOLOGY

Janice E. Rowell and John E. Blake

1.6.1   INTRODUCTION

Opportunities to gather detailed physiological information on reproduction in reindeer are limited and largely dependent on two primary sources: collection of tissues and blood from harvested free-ranging *Rangifer* and studies on captive or tame animals. The elements of reproductive physiology are highly conserved and similar among a diverse array of ruminants, providing a broader context for interpreting physiological features. While this allows us to extrapolate information from better studied species, it is still important to document species-specific anatomical and physiological details to enhance reproductive management strategies, to facilitate the development and application of reproductive technologies, and as a baseline for diagnosis and treatment of reproductive disorders and pathology. Presented here is a brief synopsis of our current knowledge on normal reproductive anatomy and physiology in *Rangifer*. The information stems primarily from harvested and captive reindeer and caribou studies, augmented with clinical observations from over 25 years of maintaining reindeer and caribou research herds at the University of Alaska Fairbanks (UAF).

1.6.2   ANATOMY

The anatomy of the reproductive tract of female *Rangifer* follows the typical ruminant pattern. Females have a bicornuate uterus characterized by short uterine body and coiled uterine horns. The ovaries, in the non-breeding season, are small, laterally flattened structures located close to the uterine body on the dorsal surface. Ovarian shape is distorted and ovarian weight increases with the seasonal development of Graafian follicles and corpora lutea (CL) (Leader-Williams and Rosser 1983). The endometrium (uterine lining) in non-pregnant females has approximately three to five caruncles per uterine horn. These are the specialized areas of the uterine lining that will form intimate maternal-foetal connections and become the sites of placental transfer between the growing foetus and the dam. In the non-pregnant state the caruncles form longitudinal ridge-like folds, blending with the mucous membrane at the margins, making identification difficult in prepubertal females. The cervix, also similar to small ruminants (sheep and goats in particular), consists of a variable number of circular folds projecting into the cervical canal and fitting closely together, making trans-cervical catheterization difficult, a consideration in artificial insemination protocols.

There is less published information on male reindeer anatomy. In general, reproductive anatomy of the male is very similar to descriptions in wapiti (Haigh 2007). The testes are held in a fur-covered scrotum relatively close to the body between the hind legs. They undergo a cyclical change in weight, increasing from a minimum in late winter to a maximum at the height of the breeding season. These cyclical changes become more dramatic with age and occur earlier in the season with older males (Leader-Williams 1988). There are paired ampulla and vesicular glands at the base of the bladder and a disseminate prostate gland covered by the urethral muscle. No mention is made of bulbourethral glands (Preobrazhenskii 1968). A dissection of the bladder and pelvic urethra can be found on the University of Calgary Caribou Anatomy website (www.ucalgary.ca/caribou/Bladder3.html). Similar to wapiti, there is no sigmoid flexure in the reindeer penis, making penile extrusion during semen collection under anesthesia more difficult.

1.6.3 PUBERTY

Like most ruminants, puberty in reindeer is related to body mass, with a body weight between 45 and 60 kg generally considered the threshold for sexual maturity among females (Ropstad 2000). Most females, especially in free-ranging situations, reach sexual maturity at 16 months of age. However, female calves can reach pubertal body weight at 6 months of age and mate successfully (Blake et al. 2007). A study in southern Norway reported that 35% of calves reached sexual maturity, with 20% of those becoming pregnant (Ropstad et al. 1991). This can be problematic, as yearling calf production can slow subsequent body weight gain for the dam and result in high newborn mortality (Ropstad et al. 1991). Alternatively, there were no apparent deleterious effects from yearling pregnancies among free-ranging reindeer on the Seward Peninsula, Alaska, a result attributed to their high plane of nutrition (Pritchard et al. 1999). The majority of female calves that become sexually mature do so much later in the season, as much as 1–3 months beyond peak rut, a factor contributing to lowered calving success.

Attempts to reduce calf pregnancies have produced variable results. Administration of prostaglandins to 7-month-old calves successfully terminated the corpus luteum (and presumably any early pregnancies) in only some of the treated calves. The percentage of non-responders was high (Ropstad and Lenvik 1991). Inducing abortion too early in the season could result in a return to oestrus and an even later pregnancy. Alternatively, administering prostaglandins later in the season risks complications associated with aborting a larger foetus. Puberty was delayed in only 31.6% of female calves treated with a synthetic progesterone (medroxyprogesterone acetate 75 mg i.m. administered 1 month prior to the breeding season), a result considered inadequate for general use (Ropstad et al. 1993). The use of long-acting hormones to delay puberty requires a lot more research into dose, time of administration, and clearance time in the animals, as well as acceptability in food animal operations. In captivity the general practice is to wean the calves at approximately 3 months of age and then separate male and female calves during the rut and keep them separated at least until the spring equinox.

Expression of puberty in male calves is harder to evaluate. Male calves do experience a rise in testosterone that roughly corresponds with the rut and can be influenced by nutrition and body weight. In one study, feed-restricted calves had lower amplitude testosterone concentrations that occurred later in the season compared to their *ad libitum* fed counterparts (Ryg 1984). There is also evidence of spermatogenic activity in the testes of calves, although among free-ranging reindeer calves, spermatozoa were not found until approximately 3–4 months after the period of rut

(Leader-Williams 1988). Leader-Williams (1988) suggests that the attainment of "physiological" puberty in male calves may be a mechanism that helps ensure complete spermatogenesis in yearlings, as opposed to an actual breeding strategy for calves. In large herds the social hierarchy normally inhibits male calves from gaining access to females and suppresses breeding tendencies. However, in farmed situations where high quality food stimulates rapid weight gain, male calves penned with only female reindeer are quite capable of breeding, usually later in the fall or early winter (November–January).

1.6.4 BREEDING SEASON AND MATING

Reindeer are seasonal breeders, breeding in response to shortening day length. As in most seasonal breeders, production of the hormone melatonin encodes the changing ratios of light and dark. Continuous light suppresses melatonin production (Eloranta et al. 1992), while continuous dark results in sustained, elevated melatonin with no evidence of circadian modulation (Stokkan et al. 2007). Melatonin secretion is highest in the fall and this fluctuating pattern of melatonin is associated with puberty in young animals, rutting in males and the initiation of oestrus among females (Eloranta et al. 1992). Changes in seasonal light:dark ratios are clearly a proximate cue for rut and oestrus, but genetic variables must also be playing a role in timing the breeding season. Initiation of the breeding season varies not only with latitude but also with different geographic and regional areas (Leader-Williams 1988). Whitehead and McEwan (1973) reported a 4–6 week difference in the onset of the breeding season between captive caribou and reindeer raised together in the same pen and on the same diet. This is similar to our observations in Fairbanks, where the reindeer breeding season is essentially over by the time the caribou show signs of rutting and breeding (Blake et al. 2007). Nevertheless, reindeer and caribou can and will successfully breed and produce fertile hybrids.

The first signs of impending rut are the hardening and cleaning of male antlers. The pre-rut endocrine events in males begin months earlier, in spring, with the pulsatile secretion of the pituitary hormones, follicle stimulating hormone (FSH) and luteinizing hormone (LH) (Bubenik et al. 1997). Systemic testosterone concentrations begin to rise and are responsible for rapid antler cleaning and development of associated secondary sexual characteristics. Thickening neck muscles, well developed neck mane, roaring and rutting odor are among the more prominent of these characteristics. The highest peaks of LH, FSH and, subsequently, testosterone pulses are correlated with dominance status and antler size in males (Stokkan et al. 1980, Bubenik et al. 1997). Physiological changes occurring in rutting males require special considerations in bull handling and husbandry during and immediately after rut (Blake et al. 2007). Voluntary feed intake decreases as rut progresses and ceases altogether among dominant males, representing a substantial energetic cost, with some bulls losing up to a third of their body mass (Barboza et al. 2004). Despite the rut-associated decrease in body condition these animals are not starving and most bulls transition reasonably well at the end of rut. However, after several weeks of animals not eating, we have encountered the whole spectrum – from individual reindeer who binge eat as rut wanes to others who refuse to start eating at all. During the breeding season, less dominant males and the females will continue to feed; however, in captive situations dominant males may guard feeding stations, preventing both sexes from gaining access to food.

Female reindeer are spontaneous ovulators and do not require bull stimulation to initiate seasonal ovulatory activity (Shipka et al. 2002, Shipka et al. 2007). However, the bull can have a pronounced stimulatory and synchronizing effect on ovulation. In farming situations this phenomenon (the male effect) can be used to synchronize and induce ovulation early in the breeding season by putting a rutting bull into a pen with female reindeer previously housed without males (Shipka et al. 2002).

Like many other ruminant species, small transient progesterone increases, lasting between 4 and 11 days, precede the first full-length estrous cycle (Eloranta et al. 1994, Ropstad 2000, Shipka et al. 2007). These short cycles may be accompanied by an LH peak and presumed ovulation

(Eloranta et al. 1994). Because increasing estradiol is responsible for female antler cleaning (Lincoln and Tyler 1994), cleaning and hardening of antlers can be a useful criterion for establishing the early onset of ovarian activity. Anecdotal information suggests that antlers are cleaned 2–3 weeks before the onset of oestrus, but there is no clear documentation on the interval between antler cleaning and first ovulation.

Females are polyoestrus when in season, with an estrous cycle length of approximately 24 days (range: 18–29 d: Alaska) (Shipka et al. 2007), although primiparous Norwegian reindeer exhibited greater estrous cycle variation (19 days; range: 13–33 d) (Ropstad 2000). Healthy, well fed females prevented from mating will continue to cycle, having between six and eight estrous cycles over the winter before entering anestrous. Among Alaskan reindeer, estrous cycle length did not differ between individuals or over the course of the winter (Shipka et al. 2007). The transition into anestrous occurred between late February and mid-April, presenting two different physiological pictures. In the first scenario, systemic progesterone dropped abruptly to baseline, where it remained, indicating a cessation of ovulatory activity, typical of most deer species. In the alternative mechanism, systemic progesterone remained elevated for a variable length of time (53–93 days), reminiscent of descriptions of persistent CL in red deer (Asher et al. 2000).

Aside from using ontlers as a broad indication of the impending breeding season, oestrus detection is difficult in reindeer. Behavioural oestrus is subtle (Ropstad 2000) and variable (Blake et al. 2007) and the use of oestrus detection aids has met with mixed results. Accuracy of radiotelemetric oestrus detection (a system developed for use in dairy cattle) was 70% and 42% effective in two separate studies, reflecting the difference in mounting behaviour between reindeer and domestic cattle. The limited data indicate that mounts are swift (3–9 secs) with females mounted only one to three times over the estrous period (Shipka et al. 2007).

1.6.5 PREGNANCY AND GESTATION LENGTH

Gestation length in reindeer ranges from 198 to 240 days, with the difference being almost twice the estrous cycle length (Rowell and Shipka 2009). A significant (P<0.001) negative correlation between conception date and gestation length has been described in Alaskan (Rowell and Shipka 2009) and Scandinavian reindeer (Holand et al. 2006). Essentially, female reindeer that breed early in the season have a longer gestation than those reindeer bred approximately 1 month later. The mean difference in gestation length between the early and late bred females is approximately 8–10 days. In Alaska, neither the sex of the calf or calf birth weight were associated with differences in gestation lengths (Rowell and Shipka 2009), although Holand et al. (2006) reported that reindeer bred early in the season produced more male calves and late conceiving females produced calves of lighter birth weight. While the physiological mechanism underlying gestation variability remains unclear, the end result is lack of precision in anticipating calving date or inferring breeding date from calving data. More information on the variability in gestation length and its potential impact on calf sex ratio, birth weight and calf long term viability is clearly warranted.

The endocrinology of pregnancy in reindeer is similar to ruminants in general and deer species more specifically. By 4 weeks post-conception, systemic progesterone levels rise from baseline to a mean of 5 ng/mL (range: 2.5–14 ng/mL, n=10) (Shipka et al. 2007). We have found progesterone concentrations remain elevated throughout pregnancy, although others report a dip or decline in levels during mid-pregnancy (Ropstad et al. 2005). While details may differ regionally, the progesterone profile is typical of many species that produce ovarian (luteal) progesterone throughout pregnancy (Bubenik et al. 1997, Shipka et al. 2007) and consistent with the fact that a functional CL is present in the ovaries for the duration of gestation (Leader-Williams and Rosser 1983). It is significant to note that progesterone levels from an estrous cycle and those of pregnancy overlap throughout the winter. For this reason, using elevated progesterone levels in a single blood sample for determining pregnancy is unreliable (Shipka et al. 2007). Estrogens (estradiol-17β, estrone and estrone sulphate) all followed similar patterns, remaining at baseline until the final 4–6 weeks of gestation (Ropstad et al. 2005, Shipka et al. 2007).

Pregnancy determination using pregnancy associated glycoproteins (PAGs) (Ropstad et al. 2005, Savela et al. 2009) and pregnancy specific protein B (PSPB) (Rowell et al. 2000) in blood samples collected at least 30 days post-conception has been successful. Once the proteins appear in maternal circulation they remain for the duration of pregnancy. In field trials, detectable levels of PAGs in non-pregnant females have been reported; however, it is unclear whether this phenomenon is related to foetal loss or assay error (Savela et al. 2009). Transrectal ultrasonography in reindeer has been done successfully between 3 and 6 weeks' gestation in both farmed and field situations. A field trial comparing pregnancy detection techniques (ultrasound, progesterone levels and PAG levels) found ultrasound to have the highest sensitivity and specificity. This, coupled with the fact that ultrasound is quick, provides an estimate of gestational age and gives results in real time, further enhances its utility (Savela et al. 2009).

Antler retention is not always an accurate indicator of pregnancy status. In our captive herds 89% of female reindeer retain their antlers until after calving, compared to only 32% of caribou, even though they are raised under identical conditions. If a female reindeer retains antlers into mid-April there is a very high probability that she is pregnant, although the converse is not true (Blake et al. 2007).

1.6.6   CALVING

Reindeer normally have a single offspring and, like most wild and semi-domestic species, calving difficulties are generally uncommon. Reindeer normally show few signs of impending parturition, although there are reports of cows separating themselves from the herd just prior to calving. In healthy females, calving occurs quickly. Skjenneberg and Slagsvold (1979) suggest calving usually occurs anywhere from a few minutes to half an hour after the first indications of labor have been observed, and this is similar to our experience with captive reindeer in Fairbanks. Dystocias (difficult births) do occur, most being associated with malpresentation. The normal presentation for reindeer calves, like most ruminants, is anterior with the front feet first, followed by the head (Figure 1.6.1a). At UAF, malpresentation (usually the head or a leg bent back towards the uterus) is most frequently associated with a dead calf. Posterior presentations (rear hooves presenting first in the birth canal) occur infrequently (Figure 1.6.1b) and intervention is a judgement call, based on prolonged calving or maternal distress. A true breech presentation (tail and hips presenting in the birth canal with the hind legs still deep in the uterus) is rare and does require assistance. There is only one documented case of a true breech at UAF and it was a result of twinning. The first calf was a breech presentation, with the twin in the normal anterior position. Knowing when to intervene is always difficult. Based on clinical experience at UAF, calving assistance is provided if a reindeer has been straining for more than an hour without delivering a calf or a portion of the calf, or membranes have been visible for over 30 minutes without evidence of delivery.

After a normal birth the cow immediately begins licking the calf and frequently eats the foetal membranes (Figure 1.6.2). The placenta is generally delivered within 1–2 hours. There have been no reported incidents of retained placentae in normal deliveries, although it is not an uncommon occurrence with dystocias. Healthy calves begin to stand within the first 10–15 minutes (Figure 1.6.3).

Twinning does occur (Figure 1.6.4), with a reported rate as high as 17.8% in a free-ranging population of Canadian reindeer (Godkin 1986). Two cases of pseudohermaphroditism have been documented in Fairbanks. This condition (free-martinism) results when a female foetus develops co-twin with a male. The female becomes masculinized *in-utero* and is infertile. Although we lack chromosome confirmation, a pseudohermaphrodite born in Fairbanks was co-twin with a normal male. The other case was identified in a reindeer calf captured from a free-ranging population in Western Alaska. Twinning clearly occurs both in captivity and in the wild, though given the propensity of this phenomenon to precipitate dystocia and produce sterile females, it is not a trait worth promoting in reindeer.

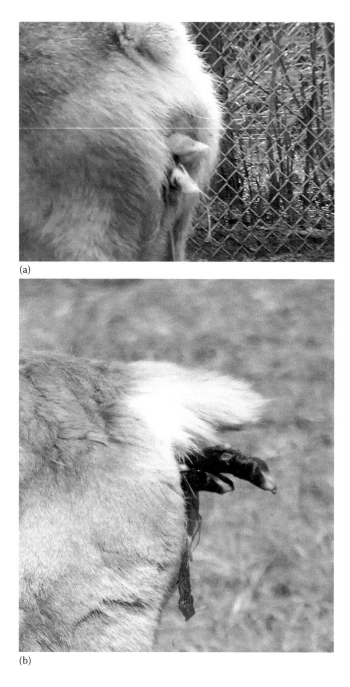

(a)

(b)

FIGURE 1.6.1   Normal foetal presentation in a reindeer. (a) In a proper anterior presentation the bottom of the hooves (i.e. front legs) should be facing the ground. (b) In a posterior presentation the calf hooves can be seen emerging with the bottom of the hoof facing up, i.e. hind leg presentation. This caribou female did require assistance in delivering her calf (Photo: John Blake).

FIGURE 1.6.2 Female reindeer eating the afterbirth. Her calf (foreground) has been licked clean, and has already suckled and is resting (Photo: John Blake).

1.6.7 REPRODUCTIVE TECHNOLOGIES

Many of the reproductive technologies available to the livestock industry can be applied to reindeer husbandry. A major constraint to their application is the ability to handle groups of reindeer repeatedly and with minimal stress. This can be both a logistic and economic barrier in extensive herding operations, reducing the advantages of using these technologies. However, in situations where reindeer are maintained on farms or held in temporary pens, techniques that synchronize estrous cycles among females and shorten the rut/breeding period, facilitate selective breeding programs, reduce rut related injuries among the animals and damage to facilities. At the other

FIGURE 1.6.3   In Fairbanks at the RG White Large Animal Research Station, UAF, calving normally occurs in early to mid April. With the warming of spring temperatures, calves are no longer born on snow pack but under conditions of melt and run-off (Photo: Christine Terzi).

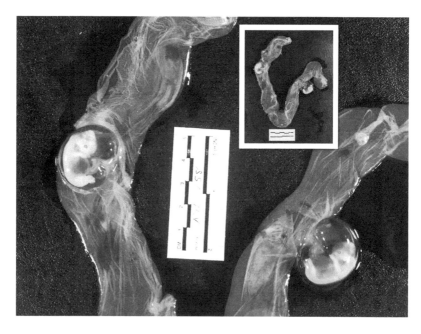

FIGURE 1.6.4   Twin reindeer embryos (6 weeks gestation) have been dissected from the uterus with the foetal membranes intact. The right side embryo and amniotic sac have been dissected out of the allantoic membrane. Inset A shows the connected allantoic membranes from both foetuses. DNA to determine gender was not available. A twinning rate of 0.4% has been reported in most reindeer populations (Godkin 1986) (Photo: Janice Rowell).

extreme, synchronizing oestrus with a reasonably precise knowledge of when ovulation occurs is necessary for artificial insemination (AI) techniques, especially using frozen/thawed semen. While AI appears more appropriate for large scale livestock farming, the technique is being used with increasing frequency as a means of transporting deer genetics across state and international boundaries (Shipka et al. 2010). In a place like Alaska, where distances between reindeer operations can be vast, AI technology offers the best opportunity for improving herd genetics through selective breeding.

Synchronizing estrous cycles among a group of females can be as simple as maintaining them in an all-female group away from rutting males for a period of weeks and then placing a mature rutting male in the pen with them. While the male effect can advance and synchronize oestrus (Shipka et al. 2002), it is not as precise as techniques that manipulate the endocrinology of the estrous cycle. Of the many techniques available, we have focused on evaluating simplified approaches that can be applied in more extensive husbandry situations. A two-injection schedule of prostaglandin $F_{2\alpha}$ (15 mg $PGF_{2\alpha}$ * i.m. administered 10 days apart) was 80% effective in synchronizing oestrus and was followed by an 88% conception rate. Prostaglandin is clearly luteolytic (Ropstad 2000) and a single injection at 6 weeks post-conception terminated pregnancies in both caribou and reindeer (Rowell et al. 2000). The fact that a functional CL must be present for $PGF_{2\alpha}$ to be effective limits its use, especially early in the breeding season (Rowell et al. 2000).

Progesterone has been used successfully in both Alaska and Scandinavia for oestrus synchronization. Sheep CIDRs (Controlled Internal Drug Releasing devices containing 0.3 grams progesterone) have successfully synchronized reindeer oestrus using a variety of different protocols. While some studies leave the vaginal inserts in place for 14 days (Lindeberg et al. 1999), we have had equal success with CIDRs inserted for 7 days (Rowell et al. 2004). Following CIDR removal the females are placed in harem with a single bull for 1 week. It should be stressed that this CIDR synchronization protocol has been used almost exclusively at the onset of the breeding season, usually at the first signs of antler cleaning among the females. Efficacy later in the breeding season, or outside of the normal breeding season, has not been evaluated. Using this synchronization protocol, mounting activity occurs approximately 48–60 hours later, with a resulting conception rate ranging between 90 and 100%. In Finland, eight of ten females conceived following a 14-day CIDR treatment with mating occurring 43 hours after CIDR removal. In this study, concurrent superovulation with FSH resulted in 20% embryo recovery (Lindeberg et al. 1999) and a live calf produced after embryo transfer to recipients (Lindeberg et al. 2003). Artificial insemination using frozen/thawed semen in nulliparous females (55 hours post-CIDR removal) resulted in a single live calf (Shipka et al. 2010). While these preliminary reports demonstrate the potential to use these technologies in reindeer, we need to develop reindeer specific protocols that are economically feasible and applicable to a wide range of husbandry practices.

The Reindeer Owners and Breeders Association (ROBA), a small reindeer-farming industry in the continental United States, has recently been using medroxyprogesterone acetate (Depo-Provera) to reduce rut related aggression in males, with variable success. Studies are currently underway at UAF to evaluate the effects of this drug on rutting male behaviour, semen quality and food intake. The ROBA members generally use their animals for display purposes and are not engaged in food production.

There is still much to learn about reproduction in healthy reindeer and caribou. Understanding normal reproductive patterns and how they fluctuate with season and nutrition provides an essential foundation for management of this species in farming and ranching situations. Captive and farmed *Rangifer* will play a key role in clarifying species-specific physiological details and in the development of reproductive technologies for genetic conservation and preservation of threatened populations.

REFERENCES

Asher, G. W., K. T. O'Neill, I. C. Scott, B. G. Mockett, and M. W. Fisher. 2000. Genetic influences on reproduction of female red deer (*Cervus elaphus*) (1) Seasonal luteal activity. *Anim Reprod Sci* 59:43–59.

Barboza, P. S., D. W. Hartbauer, W. E. Hauer, and J. E. Blake. 2004. Polygenous mating impairs body condition and homeostasis in male reindeer (*Rangifer tarandus tarandus*). *J Comp Physiol [B]* 174:309–17.

Blake, J. E., J. E. Rowell, and M. P. Shipka. 2007. Reindeer reproductive management. In *Large Animal Theriogenology*, ed. R. S. Youngquist and W. R. Threlfall, 970–4. Saunders Elsevier, St. Louis, Missouri.

Bubenik, G. A., D. Schams, R. G. White, J. E. Rowell, J. E. Blake, and L. Bartos. 1997. Seasonal levels of reproductive hormones and their relationship to the antler cycle of male and female reindeer. *Comp Biochem Physiol [B]* 116:269–77.

Eloranta, E., J. Timisjarvi, M. Nieminen, V. Ojutkangas, J. Leppaluoto, and O. Vakkuri. 1992. Seasonal and daily patterns in melatonin secretion in female reindeer and their calves. *Endocrinology* 130:1645–52.

Eloranta, E., J. Timisjarvi, M. Nieminen, J. Leppaluoto, and O. Vuolteenaho. 1994. Hormonal changes in reindeer (*Rangifer tarandus tarandus*) during silent heat. In *Recent Developments in Deer Biology*, ed. J. A. Milne, 151. Macaulay Land Use Research Institute.

Godkin, G. F. 1986. Fertility and twinning in Canadian reindeer. *Rangifer Special Issue* 1:145–50.

Haigh, J. C. 2007. Reproductive anatomy and physiology of male wapiti and red deer. In *Large Animal Theriogenology*, ed. R. S. Youngquist and W. Threlfall, 932–936. Saunders Elsevier, St. Louis, Missouri.

Holand, O., A. Mysterud, K. H. Roed et al. 2006. Adaptive adjustment of offspring sex ratio and maternal reproductive effort in an iteroparous mammal. *Proc R Soc B* 273:293–9.

Leader-Williams, N. 1988. *Reindeer On South Georgia: The Ecology of an Introduced Population*. Cambridge University Press, Great Britain.

Leader-Williams, N. and A. Rosser. 1983. Ovarian characteristics and reproductive performance of reindeer, *Rangifer tarandus*. *J Reprod Fertil* 67:247–56.

Lincoln, G. A. and N. J. C. Tyler. 1994. Role of gonadal hormones in the regulation of the seasonal antler cycle in female reindeer, *Rangifer tarandus*. *J Reprod Fertil* 101:129–38.

Lindeberg, H., J. Aalto, M. Oksman, M. Nieminen, and M. Valtonen. 2003. Embryo transfer in the semi-domesticated reindeer. *Theriogenology* 59:368.

Lindeberg, H., J. Aalto, S. Vahtiala et al. 1999. Recovery and cryopreservation of in vitro produced reindeer embryos after oestrous synchronization and superovulatory treatment. In *10th Arctic Ungulate Conference*, ed. R.E. Haugerud, Saskatoon, Canada. *Rangifer, Special Issue* 4:76.

Preobrazhenskii, B. V. 1968. Management and breeding of reindeer. In *Reindeer Husbandry*, ed. P. S. Zhigunov. Israel Program for Scientific Translation, Jerusalem.

Pritchard, A. K., G. L. Finstad, and D. H. Shain. 1999. Lactation in yearling Alaskan reindeer: Implications for growth, reproduction, and survival. *Rangifer* 19:77–84.

Ropstad, E. 2000. Reproduction in female reindeer. *Anim Reprod Sci* 60:561–70.

Ropstad, E. and D. Lenvik. 1991. The use of Clorprostenol and prostaglandin F2a to induce luteolysis in reindeer calves (*Rangifer tarandus*). *Rangifer* 11:13–16.

Ropstad, E., D. Lenvik, E. Bo, M. M. Fjellheim, and K. Romas. 1991. Ovarian function and pregnancy rates in reindeer calves (*Rangifer tarandus*) in southern Norway. *Theriogenology* 36:295–305.

Ropstad, E., G. Helland, H. Hansen, D. Lenvik, and E. T. Solberg. 1993. The effect of medroxy progesterone acetate on pregnancy rates in reindeer calves (*Rangifer tarandus*). *Rangifer* 13:163–7.

Ropstad, E., V. Veiberg, H. Sakkinen et al. 2005. Endocrinology of pregnancy and early pregnancy detection by reproductive hormones in reindeer (*Rangifer tarandus tarandus*). *Theriogenology* 63:1775–88.

Rowell, J. E., D. E. Russell, R. G. White, and R. G. Sasser. 2000. Estrous synchronization and early pregnancy. Proceedings of 8th North American Caribou Workshop: Whitehorse, Yukon, Canada, 1998. *Rangifer Special Issue* 12:66.

Rowell, J. E. and M. P. Shipka. 2009. Variation in gestation length among captive reindeer (*Rangifer tarandus tarandus*). *Theriogenology* 72:190–7.

Rowell, J. E., M. C. Sousa, A. M. Hirth, and M. P. Shipka. 2004. Reproductive management of reindeer in Alaska. In *Proceedings of the First World Deer Veterinary Congress*, ed. P. R. Wilson, 96–7. Deer Branch NZVA.

Ryg, M. 1984. Effects of nutrition on seasonal changes in testosterone levels in young male reindeer (*Rangifer tarandus tarandus*). *Comp Biochem Physiol [A]* 77:619–21.

Savela, H., S. Vahtiala, H. Lindeberg et al. 2009. Comparison of accuracy of ultrasonography, progesterone, pregnancy-associated glycoprotein tests for pregnancy diagnosis in semidomesticated reindeer. *Theriogenology* 72:1229–36.

Shipka, M. P., J. E. Rowell, and S. Bychawski. 2010. Artificial insemination in reindeer using frozen-thawed semen. *J Anim Sci* 88:124.

Shipka, M. P., J. E. Rowell, and S. P. Ford. 2002. Reindeer bull introduction affects the onset of the breeding season. *Anim Reprod Sci* 72:27–35.

Shipka, M. P., J. E. Rowell, and M. C. Sousa. 2007. Steroid hormone secretion during the ovulatory cycle and pregnancy in farmed Alaskan reindeer. *J Anim Sci* 85:944–51.

Skjenneberg, S. and L. Slagsvold. 1979. *Reindeer Husbandry and Its Ecological Principles*. U.S. Department of the Interior, Juneau, Alaska.

Stokkan, K.-A., K. Hove, and W. R. Carr. 1980. Plasma concentrations of testosterone and luteinizing hormone in rutting reindeer bulls (*Rangifer tarandus*). *Can J Zool* 58:2081–3.

Stokkan, K.-A., B. van Oort, N. J. C. Tyler, and A. S. I. Loudon. 2007. Adaptations for life in the Arctic: Evidence that melatonin rhythms in reindeer are not driven by a circadian oscillator but remain acutely sensitive to environmental photoperiod. *J Pineal Res* 43:289–93.

Whitehead, P. E. and E. H. McEwan. 1973. Seasonal variation in plasma testosterone concentration of reindeer and caribou. *Can J Zool* 51:651–8.

1.7 *RANGIFER* MATING STRATEGIES

Øystein Holand and Robert B. Weladji

1.7.1 INTRODUCTION

The genus *Rangifer* is a strongly polygynous species, with high sexual dimorphism. The *Rangifer* mating system is primarily driven by male ability to monopolize access to females in oestrus, but depends on the spatial and temporal distribution of receptive females, which is a function of their gregarious behaviour and synchronous ovulation. This can be expressed through three main strategies performed by dominant males: (1) harem defense, (2) tending and defending individual females and (3) a mixture between harem and territorial defense. The strategies adopted vary between *Rangifer*'s subspecies and this can be viewed as an adaptation to their variable environments. Females' major reproductive cost is related to gestation and lactation. Their strategy is therefore to maximize foraging time during the rut to maintain good body condition, while at the same time ensuring fertilization by a male during the first ovulation cycle. Here, we present current knowledge on *Rangifer* mating behaviour linked to intrasexual competition and intersexual mate choice, as well as intersexual conflicts. This overview is based on results from an ongoing long-term study of *Rangifer* reproduction in an experimental reindeer herd in northern Finland.

1.7.2 MATING SYSTEM THEORY

Sexual selection theory is fundamental for the development of mating system theory (Shuster and Wade 2003). Emlen and Oring (1977) described mating systems based on "the ecological and behavioural potential to monopolize mates". They suggested two measures for understanding mating system evolution and variation: the (1) spatial and (2) temporal distribution of estrous females, both of which influence males' mating strategy.

Sex-specific mating strategies are expressed in polygynous systems; males are risk prone and conspicuous, whereas females are risk aversive and cryptic. Variation in male mating strategy is particularly common in ungulates (Isvaran 2005). Dominant males may display a resource defense strategy when females aggregate at particular places, a harem defense strategy when groups of females are moving, or a tending strategy when females are spread out or when they form groups too large to defend. On the other hand, subordinate males may use a sneaking strategy instead. These male strategies would in turn influence female mating behaviours, underlining the dialectical intersexual relationship in mating strategies. A change in the operational sex ratio (OSR), defined as the ratio of fertile males to receptive females, will alter the strength of sexual selection because the greater the shortage of one sex, the more intense the sexual selection in the other (Emlen and Oring 1977).

Even if the ultimate process is the same, i.e. maximizing individuals' fitness, females' mating strategies cannot be classified in the same way as those of ungulate males. Males' reproductive investment is primarily related to rutting behaviour, whereas females have to partition their invest- ment into gestation, lactation and maternal care (Cézilly and Danchin 2008). In contrast to males, multiple mating opportunities do not necessarily increase females' mating success, and their repro- ductive skewness is less pronounced as compared to males in polygynous species (Andersson 1994). Indeed, the evolutionary forces that have shaped the breeding success of males are different for females. Firstly, female mate choice behaviour is related to the presence of males, and females can exert indirect mate choice by triggering male–male competition (Byers et al. 2010). This is the case, for example, when females attempt to move away, increasing male herding and/or tending activity (Byers et al. 1994). A dominant male will thus be more successful in his endeavour to con- trol females, thereby increasing the opportunity for sexual selection. Secondly, direct female mate choice requires assessment (i.e. sampling) of male quality followed by adjustment of their behaviour (Clutton-Brock and McAuliffe 2009). Quality validation is normally based on honest signals such as the males' physical appearance (e.g. body size, antler size), or behavioural and other pheno- typical characteristics (e.g. vocal performance, social rank and fighting ability) (Clutton-Brock and McAuliffe 2009), or their ability to provide protection against harassment (Bro-Jørgensen 2011).

1.7.3 *Rangifer* Mating Strategy

The genus *Rangifer* is an ice age relict (Geist 1999) which evolved in a harsh, stochastic and highly seasonal environment. This has molded *Rangifer* life history, including its reproductive strategy. The polygynous mating system in *Rangifer* is based on males' ability to control and mate with more than one female during the short and intense rut. However, the *Rangifer* mating strategy varies con- siderably between subspecies (i.e. across and within continents), demonstrating adaptations to the highly variable environments *Rangifer* inhabits, spanning from the high Arctic to the taiga zone. The forage resource distribution (both in space and time), the landscape and the risk of predation have resulted in variation in female group affinity and dynamic (Skogland 1989), with clear implica- tions for male mating strategies.

   In the non-migrating high arctic ecotype (e.g. Svalbard reindeer) where resources are clumped, female groups are spatially stable and dominant males defend harems at stable loca- tions (Heatta 2009). Migrating tundra/barren-ground *Rangifer* form large groups during fall migration, which coincide with the breeding season. Dominant males are therefore not able to control these mobile female bands and show a tending strategy, following one specific estrous female at a time (Lent 1965). When resources are more evenly distributed, the mating system is harem-based (Skogland 1989). These mobile harems are unstable due to female movement between groups, creating fission-fusion group dynamics (Body et al. 2015a). In addition, insta- bility of male hierarchies (Holand et al. 2012) may amplify the females' grouping dynamic. The forest ecotype of *Rangifer* forms small groups and dominant males show a combined strategy in defending a territory with a small harem (Kojola 1986); this is close to what is observed in the high arctic ecotype (Figure 1.7.1).

   Rangifer sexual dimorphism is one of the strongest observed among ungulates, with a male:female weight ratio among adults of different subspecies varying between 1 and 2.5 (Geist 1999), and male reproductive success being highly skewed towards high-ranked males (Røed et al. 2005). Supremacy, and thus priority of access to females, is manifested through dominance behaviour. Male body mass reflects physical maturity and their capacity to take an active part in the rut is correlated with their reproductive success (Røed et al. 2002). Fights are the ultimate expression of assessing competitive abilities among males. Although adult male combats contribute a minor part of their direct reproductive costs, fights may lead to severe injuries. The frequency and severity of fights will probably be higher in a harem setting, where dominant males are trying to monopolize groups of females, whereas in a tending situation, the stakes are not as high and the sexual selection

FIGURE 1.7.1   Svalbard reindeer males normally defend harems at stable location (Photo: Morten Tryland).

is expected to be less intense. There is, however, no indication that the sexual dimorphism is lower in populations where tending is the dominant male strategy as compared to a harem defense strategy (Geist 1999). Whether these different strategies result in differences in reproductive success and skewness, as well as reproductive costs, among adult males is not known. However, a certain level of plasticity in the strategy is expected, e.g. changing from tending to harem defense within a population and vice versa, depending on ecological and environmental conditions.

Males start rutting earlier and end it later than females. The adult males are eagerly fighting for dominance in the early phase of the rut. The fights, with clashing of antlers, probably trigger the females into entering mating mode. This leads to the next phase, in which dominant males start to control a female, or a group of females. Indeed, adult males are at the top of their mating activity during the highly synchronized peak rut (Holand et al. 2003), defined as the period during which 90% of the females are fertilized. This peak lasts about 10 days. As *Rangifer* females are receptive for only 1 to 2 days (Ropstad 2000), OSR during this peak period will average about five times the absolute adult sex ratio. Given a random distribution of females between mating groups, irrespectively of their reproductive phase, the local OSR will be equally skewed during the short peak rut, inducing intense local male–male competition. The female estrous synchrony and the length of the receptiveness "window" seem sufficient to maintain the environmental potential for polygyny in *Rangifer* Figure 1.7.2).

In contrast to male mating strategies, females' mate choice is less conspicuous and not well understood in *Rangifer*, as in most other polygynous ungulates. Indeed, it is difficult to demonstrate female choice, as male–male competition and male coercion and harassment may override female mate choice or limit its expression. Consequently, the mating strategy of females has been overlooked. Females' major reproductive investment is related to producing and raising their offspring. Accordingly, female reindeer/caribou will normally spend more time foraging during the rut to keep up their body condition necessary to overcome harsh winter conditions. Simultaneously, they have to avoid harassment by young males and capture the dominant males' attention to secure early fertilization by good quality males, suggesting there might be some female choice in the process.

FIGURE 1.7.2   Female reindeer give birth in early Spring before the vegetation growth has started and their major reproductive investment is related to producing and raising the calf (Photo: Olav Strand/Autocamera NINA, Norway).

1.7.4   THE KUTUHARJU REINDEER HERD

We have extensively studied alternative male mating tactics within the mobile harem defense strategy, in an experimental reindeer herd in Finland. The ultimate goal for adult males is to control and fertilize as many females as possible during peak rut. Obviously, not all males can be equally successful and several mating tactics have emerged. Three tactics have been observed: (1) *dominant*: adult males controlling as big a group as possible and tending individual females as they come into oestrus. These males are prepared for fights or any other agonistic interactions to maintain their supremacy. (2) *subdominant*: adult males willing to hang around waiting for the opportunity to take over mating groups by fighting, or by splitting up a group of females to establish their own harem. (3) *sneaker*: a tactic used by young males which were unable to keep their own groups. These males will either quickly enter into a group to attempt to mount females or sneak in and try to tend individual females when the dominant males are not paying attention to them (e.g. during fights, during prolonged courtships). The dominant tactic is obviously the most effective fitness-wise (Røed et al. 2002, Pintus et al. 2015). Indeed, the most successful dominant males are able to sire up to 30 calves in a single rutting season (Røed et al. unpublished data) given an extreme female-biased sex ratio. The two other tactics are probably employed by males to make the best out of a difficult situation, siring between zero and 20% of the females, depending on specific situations.

Males' mating tactic is plastic and conditional, as they may shift between a dominant and a subdominant tactic, or the other way around, within minutes, dependent on male-male competition levels (the local OSR) as well as their relative rank. This could happen when two dominant males (each having their own harem) meet and start fighting and the winner takes over the loser's harem. A subdominant male in a harem may also challenge and fight the dominant male (Holand et al. 2012), and if he wins, he will take over the group as the harem holder (Figure 1.7.3). The fact that males may reverse their tactic within a single rutting season makes it hard to relate reproductive success directly to mating tactic (Pintus et al. 2015). Nevertheless, the highest ranked males typically stick to the dominant tactic during peak rut, forming the biggest harems and fertilizing most of the females (Røed et al. 2002, Pintus et al. 2015). The sneaking tactic is more age-specific, and is not very successful (Pintus et al. 2015), siring less than 10% of the calves in any given year. However, sneaking males may express dominant behaviour if given the chance (e.g. when no adult males are around). Indeed, 1.5-year-old males, which often act as sneakers, are able to fertilize females. Their mating behaviour is less well developed, as their supremacy is not yet established. They have difficulties controlling a larger number of females. Accordingly, young males do not allocate much time and energy directly towards reproduction the first years of life. In contrast, adult males invest up to 25% of their pre-rut body mass into rutting activities (Mysterud et al. 2003). These efforts also reflect their age, independently of their mating tactic (dominant versus subordinate), and the overall intensity of male-male competition. Adult males seem pre-programmed to allocate all resources into this reproduction burst as they trade energy intake with staying alert and trying to get access to as many female as possible, in line with the foraging constraint hypothesis (i.e. bighorn sheep, *Ovis canadensis*, Pelletier et al. 2009). Indeed, the relative body mass loss of dominant males and agonistic behaviour towards other males increase with the number of females in the harem (Tennenhouse et al. 2011).

Adult males therefore become exhausted at the end of the rut. In comparison, young males spend most of their time foraging and are able to keep up their body mass during the rut (Tennenhouse et al. 2012). Their sneaking tactic allows them to obtain relatively more mating opportunities late in the rut (Røed et al. 2007), when less valuable females (i.e. young and light) enter oestrus. Also, in an experimental setting where we allowed females to re-ovulate (about 20 days after first ovulation), young males appeared to be relatively more successful as they sired up to 30% of calves from the second oestrus (Holand et al. 2006a). Re-ovulating is rare and late fertilization, which results in late parturition and decreased autumn calf weight (Holand et al. 2003), is highly unfavourable fitness-wise. Even though the immediate fitness gained by the sneaky males is low, being able to sire one offspring at the age of 1.5 years may have long lasting fitness benefits.

FIGURE 1.7.3   Adult males are fighting fiercely for dominance and hence priority of access to females (Photo: Ø. Holand).

During the breeding season, females preferentially aggregate. This behaviour may be a response to adult male herding or young male harassment, which is costly to females (Djaković 2012). Also, even without male presence, females do aggregate during the rut (Djaković et al. 2015). Daily group composition does vary and appears to follow a random pattern (Body et al. 2015a). However, females within mating groups seem more genetically related than expected by chance (Djaković et al. 2012). Within the 2 weeks preceding their oestrus, females visit several dominant males (L'Italien et al. 2012). Whether this is a sampling behaviour or is due to the fission-fusion group dynamics remains unknown (Body et al. 2015b). During heat, the females are rather stationary within one mating group (Body et al. 2015b), which secures fertilization during their first ovulation cycle and thereby an optimal time of parturition. The dominant male probably also restricts females' movements by way of herding or following (Body et al. 2014).

Females may choose their mates on a fine temporal scale. During their receptive phase, females court the dominant male to capture his attention (Djaković 2012). They also seem to seek shelter close to the dominant male to avoid harassment from young males. Indeed, the presence of younger males in the mating group increases body mass loss for both females (Holand et al. 2006b) and young males (Tennenhouse et al. 2011). If a harem gets too large, this will induce forage competition among females, suggesting a trade-off between being in a large group for access to good genes and protection, and the increased foraging disturbances in big groups (Uccheddu et al. 2015). Actually, by composing a female herd where half of the females were fertilized in the first oestrus and the other half not, we observed that the fertilized females stayed on the outskirts of the harem and partly formed smaller groups on their own, while the unfertilized females remained within the centre of the group (Holand unpublished data). This suggests that females fertilized in first oestrus find limited benefit in remaining within the harem, especially since they are less harassed by younger males so late in the rut.

Rangifer females show behaviour that could be coherent with direct and indirect mate choice, yielding both direct (e.g. protection from harassment) and indirect benefits (e.g. genetic). Their propensity to move between harems could be regarded as a form of indirect mate choice because females' departure attempts increase males' herding efforts. In addition, their foraging behaviour, as they are always on the move, enhances the encounter rate of other males and harem groups. In the absence of males during rut, females move even more, suggesting that males restrict their movement somewhat (Djaković et al. 2015) or that they are searching for males.

On a fine time scale, we have repetitively over the years observed females in oestrus guarding a male, often the dominant male, which may reflect courtship and preference for those dominant males, and may suggest a direct mate choice. However, females neither use relatedness (Holand et al. 2007) nor the major histocompatibility complex (MHC) (MHC heterogeneity plays an important role in the immune system to fight pathogens), to choose their mates (Djaković 2012). This suggests no inbreeding avoidance among females, probably due to high genetic variation, or that they are unable to express their choice.

Females spent as much as 80 to 90% of their active time feeding during the rut, even in the absence of males (Djaković et al. 2015). Indeed, foraging is their main activity and underlines the strong selection pressure in females for energy and nutrient acquisition. *Rangifer* females' body mass in autumn influences their pregnancy rate (Ropstad 2000) as well as the autumn weight of their calves (Holand et al. 2003), two important fitness traits. Indeed, there is a clear dominance hierarchy among females, and body mass as well as antler size and age influence their social rank (Holand et al. 2004a). Moreover, it has been shown that early pre-weaning growth rate and September body mass of calves increase with increasing females' social rank (Holand et al. 2004b). Keeping up their foraging time is therefore essential for their success. This is confirmed by their 2- to 3-hour foraging bouts intercepted by 1- to 2-hour resting and ruminating bouts throughout the day, independent of daylight (Body et al. 2012, Body unpublished data).

We have not been able to detect clear alternative female mating tactics, as we did for males, within the experimental herd. There are, however, some intimations towards alternative strategies,

FIGURE 1.7.4   Young dominant males do court females by flehmening and sniffing, but their courtship is less developed as compared to adults (Photo: Martin Dorber).

as we have observed females being copulated by multiple males, for reasons that are still to be investigated. Taborsky et al. (2008) argued that alternative tactics are expected to evolve more often in males, because of a generally higher reproductive investment outside the rut by females. Dominant males display an array of mating behaviours, the most conspicuous being grunting, bush trashing, urination while tripping their hind legs, herding females and chasing rivals. They stand watching their harem by being alert, always on the move looking for estrous females, testing females, flehmening while moving towards them, grunting with head low and sniffing their vaginal parts. When an estrous female is found, the dominant male approaches and tests her intensively; the frequency of grunts and flehmens increases while he steadily approaches her from the side and the back. This intensive courtship may go on for a whole day – depending on the female's state and willingness to accept the male. As she gets closer to ovulation, she starts following him closely. She gets alert towards sneakers and tries to keep them away using head treats, where her antlers play an active role. During the vorspiel, the male tests the female by lifting his front legs and jumping on her. Normally, there will be several attempts before she actually stands and accepts penetration by the male. The act is over within a few seconds. After successful copulation, the dominant male loses interest and the female is susceptible to chases, and copulation attempts from other (subdominant) males within the harem. She often tries to stay close to the dominant male, but he no longer seems interested in protecting her, suggesting that sperm competition might not be a driving force in evolution of the mating system in *Rangifer*. Surprisingly, rather many observed copulations do not fit with the fatherhood (Røed unpublished data). Therefore, multiple copulations must be common, although seldom observed (Figure 1.7.4).

1.7.5 IMPLICATIONS FOR REINDEER HUSBANDRY

Eurasian tundra *Rangifer* is the only cervid that has been domesticated (Røed et al. 2008). The reindeer herders have traditionally shaped their herd to meet multipurpose goals (i.e. meat, transportation, milk, fur, skin, etc.). Today, meat is the main product in a market-based economy. This modernisation has been supported by the introduction of new herd structures, culling practices and selection schemes to optimize production output (Lenvik 1988).

Assuming limited winter range resources, the reindeer herd should be composed of as high a proportion of reproductive females as possible, with a male segment big enough to serve the females successfully during the rut. Lenvik (1988) proposed to use only 1.5-year-old breeding males that can be slaughtered after the rut, as this will make room for the maximum number of prime females in the winter herd. As the growth is highest during the animals' first summer of life, the selection of animals for slaughter should primarily aim at calves, with removal of females reaching reproductive senescence. Selection of phenotypically superior female calves as herd recruits will normally secure high fecundity and low pre-weaning mortality and hence a high surplus of harvestable calves. Elucidating the maternal effect necessitates keeping track of the females' maternal pedigree, reproductive history, body condition and offspring performance. This strategy has been tested and successfully implemented in herds in all Nordic countries (Rönnegård 2003, Muuttoranta 2014). This strategy is also implemented in parts of Russia with a market oriented meat production. Actually, the application of the strategy had already started in the 1960s when Soviet Union scientists tried to realize the industry's meat production potential (Lenvik 1988).

Our findings suggest that a 1.5-year-old male proportion of ~10% will secure a high pregnancy rate, a synchronized rut and hence a concentrated parturition and acceptable autumn body mass of calves (Holand et al. 2003); this is in line with Lenvik's (1988) recommendations. Alternatively, a male proportion of only ~5% will normally be sufficient if using older males (Holand et al. 2003). Such a herd composition will reduce the male-male competition and hence the potential for injuries among males. The reduced intrasexual competition could partly be compensated for by selecting high quality breeding males. However, choosing breeding males based on calves' phenotype is not easy, as no study has confirmed that vigorous male calves will also be vigorous as adults. To keep a few adult males in the herd therefore seems beneficial. This may lead to monopolization of females by these few adult males, reducing the effective population size. Low male percentage in general may also contribute to loss of genetic heterogeneity since no inbreeding avoidance mechanism seems to be working (Holand et al. 2007) and females are partly organized in maternal lines (Djaković et al. 2012). This could be overcome by importing new genes from other herds (preferably as genetically different as possible, but from similar environmental conditions). Sampling of all calves and analyzing their genetic makeup by DNA-techniques would enable the herders to monitor the genetic structure within the herd and to take appropriate actions. DNA-techniques may be applied to establish a paternity pedigree. With such records, the selection of males for breeding could be substantially improved using estimated breeding values. Keeping up a robust and prime female segment, by a sustainable stocking rate and an appropriate selection scheme that is able to buffer environmental stochasticity, will still be a focus to secure the females' ability to express their maternal potential.

1.7.6 Implications for Wild *Rangifer* Management

In the 1970s and 1980s, the Fennoscandian wild ungulate management perspective was inspired by animal husbandry and range management to maximize meat production per unit area by composing a heavily female-biased winter population of "prime" age and condition (e.g. Haagenrud et al. 1987). Combined with hunting selection, i.e. offtake of the biggest and oldest animals within the licensed groups, even among the calf segment, concerns were raised about the ecological and evolutionary consequences of the practice (Skogland 1994), especially in herds where harvesting is the dominant mortality factor. The strongest effect was anticipated to be mediated through males, since the number of males was low and their reproductive success highly variable, and they were highly desired as trophies. This has lately been counteracted, by restricting harvesting of trophy males, through the number of licences issued. Today the Norwegian wild reindeer herds have a more balanced sex ratio, and the wildlife authority aims at 20% adults among the males in the winter herd (Andersen and Hustad 2004).

In the former Soviet Union, a commercial hunting regime was initiated in the 1970s, aiming at a high meat output by composing highly productive herds. This was especially well developed in the numerous migratory Taimyr herd(s). Since the fall of the Soviet Union, hunting has been deregulated and today, there is no clear management strategy (Kolpaschikov et al. 2015). Indeed, precise population estimates from Russia are lacking. However, there are no indices of highly skewed sex and age structures influencing the evolutionary processes beyond the direct effects of harvesting.

In North America, the big tundra migratory caribou herds typically experience substantial fluctuations in population size. However, the synchronous downward trend in many herds during the last decades (carma.caff.is/index.php/herds) has raised concerns about their resilience to climate change. The negative trend in many herds has resulted in restriction in harvesting, especially in outfitter-guided hunting, i.e. trophy hunting. The population size estimates are made uncertain by a lack of data reported by local hunters. However, it is hard to believe that the dynamics of these big herds could primarily be harvest-driven.

The situation for many of the small woodland caribou populations in Canada, and partly in the USA, is indeed different (Festa-Bianchet et al. 2011). Man-induced bottlenecks through harvesting, landscape exploitation and fragmentation have negatively affected their long-term viability (Festa-Bianchet et al. 2011). Strong management actions have therefore been taken to secure their survival, including a balanced harvesting regime. The current state of these populations resembles, in many ways, the situation of many of the small and fragmented populations of wild European tundra reindeer in Norway (Andersen and Hustad 2004), which need close monitoring and follow up.

REFERENCES

Albon, S. D., R. J. Irvine, O. Halvorsen et al. 2017. Contrasting effects of summer and winter warming on body mass explain population dynamics in a food-limited Arctic herbivore. *Global Change Biol.* 23:1374–89.

Andersen, R., and H. Hustad. 2004. Villrein & Samfunn. En veiledning til bevaring og bruk av Europas siste villreinfjell (in Norwegian). NINA Temahefte 27.

Andersson, M. 1994. *Sexual Selection.* Princeton, NJ: Princeton University Press.

Body, G., R. B. Weladji and Ø. Holand. 2012. The recursive model as a new approach to validate and monitor activity sensors. *Behav. Ecol. Sociobiol.* 66:1531–41.

Body, G., R. B. Weladji, Ø. Holand and M. Nieminen. 2014. Highly competitive reindeer males control female behavior during the rut. *PLoS ONE*, doi:10.1371/journal.pone.0095618.

Body, G., R. B. Weladji, Ø. Holand and M. Nieminen. 2015a. Measuring variation in the frequency of group fission and fusion from continuous monitoring of group sizes. *J. Mammal.* 96:791–9.

Body, G., R. B. Weladji, Ø. Holand and M. Nieminen. 2015b. Fission-fusion group dynamics in reindeer reveal an increase of cohesiveness at the beginning of the peak rut. *Acta Ethol.* 18:101–10.

Bro-Jørgensen, J. 2011. Intra- and intersexual conflicts and cooperation in the evolution of mating strategies: lessons learnt from ungulates. *J. Evol. Biol.* 38:28–41.

Byers, J. A., J. D. Moodie and N. Hall. 1994. Pronghorn females choose vigorous mates. *Anim. Behav.* 47:33–43.

Byers, J., E. Hebets and J. Podos. 2010. Female mate choice based upon male motor performance. *Anim. Behav.* 70:771–8.

Cézilly, F., and É. Danchin. 2008. Mating systems and parental care. In *Behavioural Ecology*, eds. É. Danchin, L. A. Giraldeau, and F. Cézilly, 429–465. Oxford: Oxford University Press.

Clutton-Brock, T., and K. McAuliffe. 2009. Female mate choice in mammals. *Q. Rev. Biol.* 84:3–27.

Festa-Bianchet, M., C. Ray, S. Boutin, S. D. Côté and A. Gunn. 2011. Conservation of caribou (*Rangifer tarandus*) in Canada: An uncertain future. *Can. J. Zool.* 89:419–34.

Djaković, N. 2012. Female mating strategy in reindeer (*Rangifer tarandus*): The importance of male age structure and availability in the social organisation, behaviour and mate choice. PhD diss., Norwegian University of Life Sciences.

Djaković, N., Ø. Holand, A. L. Hovland et al. 2012. Association patterns and kinship in female reindeer (*Rangifer tarandus*) during rut. *Acta Ethol.* 15:165–71.

Djaković, N., Ø. Holand, A. L. Hovland et al. 2015. Effects of males' presence on female behaviour during the rut. *Ethol. Ecol. Evol.* 27:148–60.

Emlen, S. T., and L. W. Oring. 1977. Ecology, sexual selection, and the evolution of mating systems. *Science* 197:215–23.

Geist, V. 1999. *Deer of the World: Their Evolution, Behaviour and Ecology.* Shrewsbury: Swan Hill Press.

Haagenrud, H., K. Morow, K. Nygren and F. Stålfelt. 1987. Management of moose in Nordic countries. *Swedish Wildl. Res. Suppl.*, Part 2: 635–42.

Heatta, M. J. 2009. The mating strategy of female Svalbard reindeer (*Rangifer tarandus platyrhynchus*). Master thesis, University of Tromsø.

Holand, Ø., K. H. Røed, A. Mysterud et al. 2003. The effect of sex ratio and male age structure on reindeer calving. *J. Wildl. Manage.* 67:25–33.

Holand, Ø., H. Gjøstein, A. Losvar et al. 2004a. Social rank in female reindeer (*Rangifer tarandus*) – effects of body mass, antler size and age. *J. Zool.* 263:365–72.

Holand, Ø., R. B. Weladji, H. Gjøstein et al. 2004b. Reproductive effort in relation to maternal social rank in reindeer (*Rangifer tarandus*). *Behav. Ecol. Sociobiol.* 57:69–76.

Holand, Ø., A. Mysterud, K. H. Røed et al. 2006a. Adaptive adjustment of offspring sex ratio and maternal reproductive effort in an iteroparous mammal. *Proc. Biol. Sci.* 273:293–9.

Holand, Ø., R. B. Weladji, K. H. Røed et al. 2006b. Male age structure influences females' mass change during rut in a polygynous ungulate: The reindeer (*Rangifer tarandus*). *Behav. Ecol. Sociobiol.* 59:682–8.

Holand, Ø., K. R. Askim, K. H. Røed et al. 2007. No evidence of inbreeding avoidance in a polygynous ungulate – The reindeer (*Rangifer tarandus*). *Biol. Lett.* 3:36–9.

Holand, Ø., L. L'Italien, R. B. Weladji et al. 2012. Shit happens – a glimpse into males' mating tactics in a polygynous ungulate – The reindeer. *Rangifer* 32:65–72.

Isvaran, K. 2005. Variation in male mating behaviour within ungulate populations: Patterns and processes. *Curr. Sci.* 89:1192–9.

Kojola, I. 1986. Rutting behaviour in an enclosured group of wild forest reindeer (*Rangifer tarandus fennicus* Lönnb.). *Rangifer* 6:173–9.

Kolpaschikov, L., V. Makhailov and D. E. Russell. 2015. The role of harvest, predators, and socio-political environment in the dynamics of the Taimyr wild reindeer herd with some lessons for North America. *Ecol. Soc.* 20(1):9. dx.doi.org/10.5751/ES-07129-200109.

Lent, P. 1965. Rutting behaviour in a barren-ground caribou population. *Anim. Beh.* 15:259–64.

Lenvik, D. 1988. Utvalgsstrategi i reinflokken (in Norwegian, with English summary). Dr. agric. diss., Norges landbrukshøgskole.

L'Italien, L., R. B. Weladji, Ø. Holand et al. 2012. Mating group size and stability in reindeer (*Rangifer tarandus*): The effects of male characteristics, sex ratio and male age-structure. *Ethol.* 118:783–92.

Muuttoranta, K. 2014. Current state of and prospects for selection in reindeer husbandry. PhD diss., University of Helsinki.

Mysterud, A., Ø. Holand, K. H. Røed et al. 2003. Effects of age, density and sex ratio on reproductive effort in male reindeer. *J. Zool.* 261:341–4.

Pelletier, F., J. Mainguy and S. Côté. 2009. Rut-induced hypophagia in male bighorn sheep and mountain goats: Foraging under time budget constraints. *Ethol.* 115:141–51.

Pintus, E., S. Uccheddu, K. H. Røed et al. 2015. Flexible mating tactics and associated reproductive effort during the rutting season in male reindeer (*Rangifer tarandus*, L. 1758). *Curr. Zool.* 61:802–10.

Røed, K. H., Ø. Holand, M. E. Smith et al. 2002. Reproductive success in reindeer males in a herd with varying sex ratio. *Mol. Ecol.* 11:1239–43.

Røed, K. H., Ø. Holand, H. Gjøstein and H. Hansen. 2005. Variation in male reproductive success in a wild population of wild reindeer. *J. Wildl. Manage.* 69:1163–70.

Røed, K. H., Ø. Holand, A. Mysterud et al. 2007. Male phenotypic quality influences offspring sex ratio in a polygynous ungulate. *Proc. Biol. Sci.* 274:727–33.

Røed, K. H., Flagstad Ø, Nieminen, M. et al. 2008. Genetic analyses reveal independent domestication origins of Eurasian reindeer. *Proc. Biol. Sci.* 275:1849–55.

Rönnegård, L. 2003. Selection, maternal effects and inbreeding in reindeer husbandry. PhD diss., Swedish University of Agricultural Sciences.

Ropstad, E. 2000. Reproduction in female reindeer. *Anim. Reprod. Sci.* 60–61:561–70.

Shuster, S. M., and M. J. Wade. 2003. *Mating Systems and Strategies.* Princeton, NJ: Princeton University Press.

Skogland, T. 1989. Comparative social organization of wild reindeer in relation to food, mates and predator avoidance. *Advances in Ethology* 29, Berlin: Paul Parey.

Skogland, T. 1994. *Villrein – fra urinnvåner til miljøbarometer* (in Norwegian). Oslo, Teknologisk forlag. 143 pp.

Taborsky M., R. F. Oliviera and H. J. Brockmann. 2008. The evolution of alternative reproductive tactics: Concepts and questions. In *Alternative Reproductive Tactics: An Integrative Approach*, eds. R.F. Oliveira, M. Taborsky, and H. J. Brockmann, 1–21. Cambridge: Cambridge University Press.

Tennenhouse, E. M., R. B. Weladji, Ø. Holand, K. H. Røed and M. Nieminen. 2011. Mating group composition influences somatic costs and activity in rutting dominant male reindeer (*Rangifer tarandus*). *Behav. Ecol. Sociobiol.* 65:287–95.

Tennenhouse, E. M., Weladji, R. B., Holand, Ø. and Nieminen, M. 2012. Timing of reproductive effort differs between young and old dominant male reindeer. *Ann. Zool. Fenn.* 49:152–60.

Uccheddu, S., G. Body, R. B. Weladji, Ø. Holand and M. Nieminen. 2015. Foraging competition in larger groups overrides harassment avoidance benefits in female reindeer (*Rangifer tarandus*). *Oecologia* 179:711–8.

2 *Rangifer* Health
A Holistic Perspective

Bryan Macbeth and Susan J. Kutz

CONTENTS

2.1 THE IMPORTANCE OF A HOLISTIC APPROACH TO REINDEER AND CARIBOU HEALTH

Wherever they occur, reindeer and caribou are ecologically, socially, and economically important. Despite this, wild *Rangifer* populations are declining globally (Gunn et al., 2016), and many semi-domesticated populations are now suffering from an array of health related issues (see Chapters 6–9). The number of programs seeking to evaluate, understand and safeguard *Rangifer* health is growing in response to these concerns; however, an operational definition of what precisely constitutes *Rangifer* "health" is lacking and different criteria are often employed to define or attempt to understand health based on the unique experiences, needs and priorities of different user groups. For example, biologists and ecologists may define *Rangifer* health in the context of endpoints such as reproductive or survival rates that can be quantified in the field and provide empirical evidence that may be used to inform wildlife management or conservation programs. In contrast, veterinarians may also consider the occurrence, prevalence, and effects of certain infectious diseases or body condition, or the influence of immunity and physiological stress (Stephen, 2014; Patyk et al., 2015). Semi-domesticated reindeer may be monitored and managed more closely than their wild counterparts and seasonal, annual or lifetime production targets such as weight gain, meat quality, calf production or even revenue may be used to define the health of reindeer kept as production animals. Similarly, traditional users of reindeer and caribou, such as indigenous peoples, may define *Rangifer* health using an array of behavioural, demographic and physical traits, or other factors related to the quality of meat or other products obtained from harvested animals. Although each definition is valuable in its own right, a more holistic and shared understanding of *Rangifer* health is urgently needed.

A cumulative effects model has recently been proposed as a useful strategy for defining and understanding the health of polar bears, another imperiled Arctic species (Patyk et al., 2015). A similar approach to *Rangifer* health would enhance communication among the many jurisdictions and diverse array of stakeholders involved in the management and conservation of reindeer and caribou across the species' distributional range. This would facilitate the development of clear priorities and quantifiable objectives for research that would increase our understanding of factors affecting reindeer and caribou health and of the precise role health plays in *Rangifer* population dynamics. In turn, a holistic understanding of *Rangifer* health would enable the development of management programs where the progress and efficacy of specific actions employed to protect or sustain the health of individual reindeer and caribou and *Rangifer* populations could be closely monitored and evaluated objectively. The relationship between animal health and welfare is well established (Nicks and Vandenheede, 2014) and a shared understanding of health would also benefit initiatives designed to preserve or improve the welfare of free-ranging *Rangifer* captured and handled for research, as well as reindeer and caribou kept in zoological collections or as production animals. Most importantly, a holistic approach to *Rangifer* health would make possible the development of more effective strategies to protect reindeer and caribou from those challenges that most threaten their long-term sustainability in a rapidly changing world.

2.2   DETERMINANTS OF HEALTH IN REINDEER AND CARIBOU: LOOKING BEYOND THE EFFECTS OF PATHOGENS AND DISEASE IN INDIVIDUAL ANIMALS

As in other wild and domestic species, "health" in reindeer and caribou has traditionally been defined as the absence of "disease." "Disease," in turn, has most often been described as a discrete physical or physiological state caused by infectious (e.g. bacteria, viruses, parasites) or non-infectious (e.g. toxins) agents and which directly cause significant and obvious dysfunction in, or the death of, affected animals (Wobeser, 2006; Wobeser, 2007). In the past, potentially important subclinical effects caused by these agents have not been recognized. The concept of *Rangifer* "health" has also been traditionally viewed primarily as a feature of individuals, although "herd" health is considered in semi-domesticated reindeer production.

In the field of human medicine, "health" has evolved past a concept focused only on the obvious effects of pathogens and disease in individuals. Instead, it is now recognized as a complex and dynamic condition that emerges from the cumulative effects of biological, environmental and socioeconomic pressures acting on and interacting with the intrinsic characteristics of both individuals and populations (Stephen, 2014). Wildlife health is increasingly recognized as a similar construct to be considered at the individual, population and ecosystem scales (Stephen, 2014; Patyk et al., 2015). An understanding of the importance of subclinical disease in reindeer and caribou is also growing (e.g. Chapter 10). A holistic definition of *Rangifer* health must reflect this contemporary understanding. Although pathogens and non-infectious agents are important, recognizing that many other factors may also affect the physiology, behaviour, growth, body condition, survival or reproduction of reindeer and caribou is a foundational step in developing a holistic understanding of *Rangifer* health.

Factors affecting *Rangifer* health may be either extrinsic or intrinsic. Extrinsic health determinants are those abiotic, biotic, or environmental conditions which annually influence reindeer and caribou. Extrinsic health determinants may include infectious and non-infectious agents, nutrition, habitat characteristics, climate and weather and sympatric species, as well as features such as population density and herd structure. Socioeconomic considerations related to tourism, subsistence hunting, outfitting, development, resource extraction and other human activities or beliefs that impact *Rangifer* or their habitat may also be considered as extrinsic determinants of reindeer and caribou health (Figure 2.1).

Intrinsic health determinants are those endogenous biological characteristics of *Rangifer* that may be influenced by, and/or affect the response to, extrinsic health challenges (Figure 2.1). Intrinsic health determinants include factors such as life history or production stage, genetic makeup,

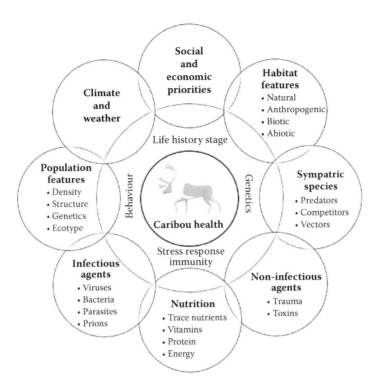

FIGURE 2.1   Determinants of health in *Rangifer*. Extrinsic determinants of health (outer circles) interact with each other and intrinsic biological characteristics (middle circle) to determine the health profile of reindeer and caribou (inner circle). This health profile is dynamic and may vary across time and space. It is directly linked to the fitness (reproduction and survival) of individual animals and the performance (reproductive and survival rates, juvenile recruitment) of *Rangifer* populations, as well as to the ability of each to cope with natural or anthropogenic challenges.

behaviour, the stress response (Hypothalamic-Pituitary-Adrenal [HPA] axis activity) and immunity (Figure 2.1).

The specific impacts of many extrinsic and intrinsic determinants of *Rangifer* health are considered in detail throughout this book. However, in the general context of understanding what influences and defines *Rangifer* health, it is most important to recognize that the cumulative effects of multiple health determinants act to shape the overall health profile of reindeer and caribou, interactions among and between extrinsic and intrinsic health determinants are common, and both types of determinants may act directly or indirectly and at the individual or population level. The occurrence and impact of both types of health determinants may also vary widely over time and space. As a result, their effects may be unequal between individuals or across different populations. *Rangifer* health can, therefore, be considered a dynamic or emergent state. Perhaps most importantly, the health profile of reindeer and caribou is directly linked to the fitness (i.e. survival and reproduction) of individual animals and the performance (i.e. survival and reproductive rates, juvenile recruitment) of *Rangifer* populations. These factors determine the long-term sustainability of *Rangifer* herds and, as such, the health status of *Rangifer* may also be considered as an index of resilience (or vulnerability) that reflects the ability of this species to cope with natural and anthropogenic challenges. In the face of increasing conservation threats and the recent widespread decline of *Rangifer* across their distributional range, the importance of considering health as an integral component of any reindeer and caribou management or conservation program is clear.

2.3   ALLOSTASIS AND PHYSIOLOGICAL STRESS

The concept of allostasis and its relationship to the physiological stress response provide a useful theoretical framework to demonstrate the value of considering a holistic approach to *Rangifer* health by highlighting potential pathways through which the effects of many health determinants, including natural or anthropogenic ecological change, may be linked with diminished health and population performance in this species (McEwen and Wingfield, 2003; Wingfield, 2005; McEwen and Wingfield, 2010). A more thorough understanding of these relationships will provide the enhanced knowledge necessary to develop more timely and effective management solutions to those threats that most threaten the long-term sustainability of reindeer and caribou populations (Landys et al., 2006).

2.3.1   ALLOSTASIS AND ALLOSTATIC LOAD

In any environment, reindeer and caribou are faced with a variety of challenges that may be related to relatively predictable life history processes (e.g. growth, migration, reproduction, feeding or predator avoidance behaviour, moult, seasonal environments) and unpredictable events (e.g. extreme weather, predator attacks, disease). The vital goal of any individual is not only to survive but also to reproduce successfully. In order to accomplish these goals, reindeer and caribou must maintain the stability of essential physiological systems (homeostasis) within a range optimal for their current life history stage. These goals are achieved by altering behaviour and physiology in order to acquire or re-allocate energy (Wingfield, 2005; Landys et al., 2006; Busch and Hayward, 2009; McEwen and Wingfield, 2010). Allostasis refers to this process of achieving stability in internal systems through physiological or behavioural change. Allostatic load may, therefore, be described as the cost to an individual of being forced to adapt to changing environments. It is the result of daily and seasonal routines to obtain food and survive plus additional energy needed for critical life cycle events or to deal with unexpected challenges (McEwen, 1998; Wingfield, 2005). More simply, allostatic load may be considered as the difference between an organism's available energy (from food or intrinsic energy stores) and that required to deal with energetic challenges related to normal life processes and unpredictable events (Figure 2.2). The physiological stress response and its effector hormones, including glucocorticoids (GCs) such as cortisol, play an important part in energy metabolism and an organism's response to changing energetic demands and are considered important mediators of allostasis (Landys et al., 2006; Busch and Hayward, 2009).

2.3.2   THE PHYSIOLOGICAL STRESS RESPONSE

The Hypothalamic-Pituitary-Adrenal (HPA) axis is the principal endocrine pathway of the mammalian stress response. Activation of the HPA axis occurs within seconds of an organism's identification of a stressful stimulus when signals from higher brain centres stimulate the synthesis and release of corticotropin releasing hormone (CRH) from the hypothalamus (e.g. Papadimitriou and Priftis, 2009). Upon release, CRH is transported to the anterior pituitary gland where it induces the production of adrenocorticotropic hormone (ACTH). ACTH is then released from the anterior pituitary into the blood stream and transported to the adrenal cortex where it stimulates adrenal steroidogenesis and the release of glucocorticoid (GC) effector hormones into the systemic circulation. In most mammals, cortisol is the primary GC secreted, and elevated levels of this hormone are a key component of the stress response. Within seconds to minutes following exposure to a stressful event, elevated cortisol levels act to mobilize and repartition energy, thus promoting changes in physiology or behaviour that facilitate immediate survival in the face of an unexpected challenge (Landys et al., 2006; Busch and Hayward, 2009; Papadimitriou and Priftis, 2009). These changes include transient permissive effects such as an increase in cardiac output, blood pressure, cerebral blood flow, and hepatic gluconeogenesis, along with enhanced mobilization of fat and protein stores, and

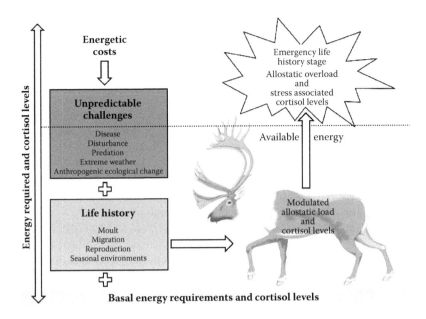

FIGURE 2.2   Allostasis and the relationship between allostatic load, allostatic overload, and cortisol levels. Allostasis refers to the process of achieving stability in internal systems through physiological or behavioural change. Allostatic load may be considered as the difference between an organism's available energy and the costs of dealing with challenges related to life history processes and unpredictable events. Glucocorticoids are primary mediators of allostasis, and as allostatic load increases cortisol levels rise within a modulated range. When energy required to meet demands exceeds that available, an organism may enter a state of allostatic overload. Allostatic overload is an emergency life history stage characterized by dysregulation of allostatic mediators and high (stress associated) levels of circulating cortisol. If allostatic load remains uncorrected, these changes may adversely impact the health and fitness of affected animals.

increased cognition (Sapolsky et al., 2000; Boonstra, 2005). Temporary suppressive effects such as the inhibition of energy storage, growth, immunity, reproductive behaviour, and physiology, along with decreased feeding and appetite, also occur (Charmandari et al., 2005). When noxious stimuli are removed, elevated levels of circulating cortisol normally act to terminate the stress response by means of negative feedback on the hypothalamus and pituitary and subsequent down regulation of CRH and ACTH secretion. As such, the effects of elevated cortisol levels may be modulated and the physiological response to short-term stress is generally considered to be adaptive (Sapolsky et al., 2000; Charmandari et al., 2005; McEwen and Wingfield, 2010).

Feedback inhibition of the stress response is not infallible, and when repeated or chronic activation of the HPA axis occurs over many weeks to months (chronic or long-term stress), the effects of persistently elevated cortisol may adversely impact many organ systems and physiological pathways (Sapolsky et al., 2000; Charmandari et al., 2005). For example, one of the effects of persistently elevated cortisol is enhanced protein breakdown and lipolysis to support glucose production. Elevated GCs also increase the amount of glucose available by inhibiting its uptake and use in peripheral tissues such as muscle, skin and bone. The production of bone is also directly suppressed by GCs. Ultimately, increased energy mobilization from protein and lipid stores in combination with a decrease in the maintenance of structural tissue (i.e. muscle, collagen and bone) may lead to diminished growth (body size, body mass or body condition) in chronically stressed animals.

Persistently elevated cortisol may also impact a variety of important endocrine pathways (Sapolsky et al., 2000; Tsigos and Chrousos, 2002; Choksi et al., 2003; Charmandari et al., 2005; Sheriff et al., 2009; Dunn et al., 2010; Sheriff et al., 2010). Through the direct actions of cortisol (and also indirectly via increased CRH), long-term stress may reduce secretion of growth hormone (GH) and other important growth factors (e.g. insulin-like growth factor, IGF-1) which may diminish long-term growth potential. Similarly, GCs may decrease production of thyroid stimulating hormone (TSH) along with the conversion of thyroxine (T4) to biologically active triiodothyronine (T3). T3 is an important regulator of carbohydrate and protein metabolism and diminished T3 levels may adversely impact growth, development, and reproduction in affected animals. Prolonged HPA axis activity may also inhibit the secretion or release of essential sex hormones such as gonadotropin releasing hormone (GnRH), luteinizing hormone (LH) and follicle stimulating hormone (FSH). Furthermore, long-term stress may render gonad tissue resistant to sex steroids, which may lead to gonadal atrophy. In aggregate, these changes may ultimately reduce reproductive behaviour, conception rates and pregnancy success in affected individuals. Chronic maternal stress during gestation has also been linked with lasting alterations in the development and function of the HPA axis in foetuses (HPA axis programming) that may negatively impact their growth and development along with their subsequent health, long-term viability, and lifetime fitness.

Virtually all components of the immune system are also inhibited by persistent HPA axis activity (McEwen et al., 1997; McEwen and Seeman, 1999; Sapolsky et al., 2000; Black, 2002; Charmandari et al., 2005). For example, GCs are lympholytic and promote T cell and B cell apoptosis (cell death) which, in turn, may be associated with atrophy of the thymus and other lymphoid tissue. Immunoglobulin production, along with cytokine synthesis, release, or efficacy, monocyte differentiation into macrophages and macrophage phagocytic activity, are also diminished. Likewise, elevated cortisol levels may reduce the inflammatory response, which is a vital component of efficient immune function. Through these mechanisms long-term stress may decrease an individual's resistance to infectious, autoimmune and allergic diseases, and cancer (McEwen et al., 1997; Black, 2002; Charmandari et al., 2005).

If uncorrected, the effects of chronic stress may eventually lead to the development of a pathological syndrome of distress characterized by poor health, reduced fitness (reproduction and survival), and diminished resilience in individual animals (Charmandari et al., 2005; Busch and Hayward, 2009; Linklater, 2010). In turn, distress occurring in many individuals may manifest as reduced rates of survival, reproduction, and recruitment at the population level and may ultimately culminate in the decline, extirpation or extinction of affected wildlife populations (Wikelski and Cooke, 2006; Ellis et al., 2012). Chronic physiological stress is increasingly recognized both as a consequence of natural and human-caused ecological change, and as a potential mechanism linking changing ecological conditions with impaired health and population performance in free-ranging wildlife, including reindeer and caribou (e.g. Freeman, 2008; Ashley et al., 2011; Wasser et al., 2011; Renaud, 2012; Joly et al., 2015; Ewacha, 2016; Hing et al., 2016).

Since cortisol and other glucocorticoids are important mediators of allostasis, the study of chronic stress in *Rangifer* may represent a particularly useful tool with which to increase our understanding of the importance of both natural and human-caused ecological conditions as determinants of health and population performance in this species. It may also have great potential to serve as an informative tool to monitor the response of *Rangifer* to management and conservation initiatives. Textbox 2.1 outlines methodologies for measurement of stress in *Rangifer*.

TEXTBOX 2.1   MEASURING STRESS

Quantitative measurements of physiological stress in reindeer and caribou have traditionally relied on the determination of cortisol concentrations in blood (Sheriff et al., 2011). Plasma or serum cortisol levels represent a point in time measure of short-term HPA axis activity and the utility of these media for the study of long-term stress may be limited by natural fluctuations in cortisol levels related to circadian and seasonal effects (Sheriff et al., 2011; Bubenik et al., 1998). Furthermore, blood collection itself may be stressful and may bring about an artificial increase in cortisol levels if samples are not obtained immediately after the initiation of handling (Sheriff et al., 2011; Romero and Reed, 2005). This may be especially important when considering large animals such as reindeer and caribou, where blood collection most often requires capture (usually involving pursuit and physical restraint) and/or chemical immobilization (Omsjoe et al., 2009; Säkkinen et al., 2004). Like handling stress, some sedative and immobilization drugs (e.g. alpha-2 adrenergic agonists, opioids) may directly increase blood cortisol levels in reindeer and caribou (Arnemo and Ranheim, 1999). Importantly, blood to be used in cortisol analyses must also be centrifuged, and derived plasma or serum kept cool (or immediately frozen) to minimize steroid metabolism and sample degradation (Sheriff et al., 2011). In many field-based studies of *Rangifer* these requirements may be difficult or impossible to achieve.

The measurement of cortisol in saliva has been explored as an alternate method with which to estimate HPA axis activity in captive cervids, including *Rangifer* (Rehbinder and Hau, 2005). Like blood, cortisol measured in saliva represents a point in time (short-term) measure of circulating free cortisol levels that may be influenced by circadian rhythms, season, and sample collection methods. Unlike blood, cortisol levels in saliva may be somewhat less affected by handling procedures and may remain relatively stable at room temperature for a few days to weeks so long as samples remain wet (Kobelt et al., 2003; Chen et al., 1992). Despite these potential advantages, the collection of saliva from free-ranging reindeer and caribou is usually impractical. It should also be recognized that GC levels can be evaluated in cerebrospinal fluid (CSF), milk, and adipose tissue (Sheriff et al., 2011). To date, the utility of these media have not been thoroughly evaluated in *Rangifer*.

In contrast, the measurement of GC metabolites in urine and faeces has emerged as a technique commonly employed to study HPA axis activity in both wild and domestic species (Sheriff et al., 2011). Glucocorticoids in blood are metabolized by the liver and excreted in the urine or faeces and both media may offer potential advantages over blood or saliva in that they may represent the accumulation of GC metabolites derived from biologically active free cortisol over several hours to several days (Wasser et al., 2000; Constable et al., 2006). Urine (from snow) and faeces can be collected opportunistically or remotely from captive or free-ranging *Rangifer*. The study of faecal GCs in particular has also demonstrated potential as a non-invasive technique with which to evaluate the effects of environmental stressors on *Rangifer* health and population performance (Heggberget et al., 2002; Ashley et al., 2011). Nonetheless, a number of confounding factors may diminish the utility of these techniques in field-based studies.

Under most field conditions, the collection of fresh, uncontaminated urine samples is difficult and may limit the practical use of this technique to captive *Rangifer* or reindeer and caribou that can be closely monitored in their natural habitat in winter. Although *Rangifer* faecal samples may be more easily collected [e.g. via winter-time aerial surveys to locate feeding (cratering) areas], levels of faecal GC metabolites may also be influenced by environmental exposure. Naturally occurring bacterial enzymes may begin to degrade steroid metabolites in as little as a few hours and, while the age of faecal samples is important, it is often unknown in field-based studies. As such, urine and faecal samples require immediate preservation once collected, yet GC levels in both media may also be influenced by preservation, storage and sample preparation protocols (Millspaugh et al., 2003; Sheriff et al., 2011).

The time delay in the excretion of cortisol metabolites corresponds to the time between urination events or gut transit time. Most species urinate several times per day, and circadian variations in levels of urinary GCs are commonly recorded; seasonal variations in urinary GC levels have also been observed in some species (e.g. Saltz and White, 1991; Bennett et al., 2008). While the detection of circadian changes in faecal GC levels is most often observed in small bodied animals with rapid gut transit, seasonal changes in faecal GC levels have been recorded in many mammals (e.g. Sheriff et al., 2011; Foley et al., 2001; Knojević et al., 2011). GC levels in both media may also vary with an animal's life history stage (e.g. Mooring et al., 2006; Owen et al., 2005). Where samples are collected remotely, and if additional analyses are not performed (e.g. DNA-based species and sex determination), these factors may introduce further bias into the interpretation of results generated from free-ranging reindeer and caribou. An animal's diet may also influence faecal GC levels. For example, fibre content may alter gut transit time and may make seasonal changes in the types or characteristics of forage plants consumed by reindeer and caribou particularly relevant (Keay et al., 2006; Drucker et al., 2010). The consumption of inorganic material during feeding (or mineral acquisition) may have similar effects. Dietary components or nutritional status may also influence GC metabolites measured in urine by altering urinary pH or other factors which may influence the GC stability. Most importantly, neither media can assess HPA axis activity occurring over weeks to months without the repeated sampling of individuals (Sheriff et al., 2011).

The evaluation of GC levels in hair has recently been proposed as alternate technique to evaluate chronic stress in both humans and animals (Mayers and Novak, 2012; Russell et al., 2012). The potential advantages of hair as a medium in which to assess chronic stress are closely tied to its structural and biological characteristics. Hair does not grow continually but instead follows a cyclical pattern of alternating periods of active growth and quiescence (Stenn and Paus, 2001). During the active growth phase the hair bulb is closely associated with the cutaneous blood supply. The opposite is true during the quiescent phase when the hair bulb regresses and this connection is terminated. The passive diffusion of circulating GCs (and other hormones and xenobiotics) from blood to cells in the zone of hair germination is considered to be the most important mechanism of incorporation into hair. This process is currently believed to be restricted to the period of active hair growth as the hair shaft is metabolically inactive tissue (Davenport et al., 2006). The duration of active hair growth is a species specific trait, and in *Rangifer* typically occurs from late spring through fall and lasts for many weeks. As such, cortisol levels measured in reindeer and caribou hair may represent an integrated measure of HPA axis activity occurring during this time period (however, see Di Francesco et al., 2017 for a discussion of effects of hair growth rate and possible integration of cortisol in hair post growth) and may, therefore, provide the longest record of cortisol levels (e.g. Macbeth, 2013).

In the last decade, the measurement of cortisol in hair has emerged as a promising indicator of long-term stress in humans as well as wild and domestic animals, including *Rangifer* (Ashley et al., 2011; Mayers and Novak, 2012; Renaud, 2012; Russell et al., 2012; Macbeth, 2013). Elevated hair cortisol levels have been associated with chronic exposure to a variety of stressful stimuli and adverse health effects in humans and animals (Mayers and Novak, 2012; Russell et al., 2012). Relationships between hair cortisol levels and large-scale environmental features (e.g. habitat disturbance) have also been recorded in a variety of wild species, including free-ranging wildlife including caribou (e.g. Renaud, 2012; Macbeth, 2013; Bourbonnais et al., 2013; Mastromonaco et al., 2014). Compared to blood, saliva, urine, or faeces, hair is also a relatively stable medium that can be transported and stored at room temperature and in which GCs may be measured many months to years after their deposition (Macbeth et al., 2010). In addition to potential for less invasive or opportunistic sampling (e.g. from harvested animals or mortality sites associated with predation), the collection and analysis of hair samples may also provide other relevant data for *Rangifer* health studies such as exposure to blood borne environmental contaminants, dietary analysis (via stable isotope analysis), and DNA-based identification and characterization of individual animals (e.g. Duffy et al., 2005). Other steroids hormones relevant to *Rangifer* biology, production and conservation (e.g. sex steroids) can also be measured in hair (e.g. Bryan et al., 2013; Terwissen et al., 2014).

Although promising, hair cortisol analysis should be considered an emerging technique. Recent research suggests that short-term stressors such as live-capture may also have the potential to influence hair cortisol levels in some species, and that GCs are likely to enter both the growing and quiescent hair shaft via diffusion from the external environment (contamination), tissues surrounding actively growing hair, or through glandular secretions in and around the follicle (Cattet et al., 2014; Di Francesco et al., 2017). The overall and relative importance of such factors in *Rangifer* and other species are currently unknown. Hair growth is also an energetically costly process and the incorporation of GCs may also be affected by an individual's nutritional or health status as well as its sex or life history stage (Bryan et al., 2013). In addition, trauma to the skin may stimulate a local stress response that may influence hair cortisol levels, while hair colour, along with the timing and duration of hair growth, may account for variation in hair cortisol levels recorded among body regions and hair types in *Rangifer* and other species (Sheriff et al., 2011; Bennett and Hayssen, 2010). As such, hair cortisol cannot be fully recommended as an explicit indicator of chronic stress until further validation studies are completed in *Rangifer*. Nonetheless, the potential utility of hair cortisol analysis in reindeer and caribou is great and continued research into this topic is encouraged.

2.3.3   Allostatic Overload Links the Physiological Stress Response and *Rangifer* Health in a Rapidly Changing World

Throughout their evolutionary history, *Rangifer* have developed a finely tuned set of physiological responses and behavioural strategies that have permitted the species to survive and thrive in challenging Arctic and Sub-Arctic ecosystems (see Chapter 1). A combination of human-caused changes in the global climate and the ecosystems in which reindeer and caribou live have recently disrupted this balance and now threaten the long-term sustainability of many populations (e.g. Racey, 2005; Vors and Boyce, 2009; Festa-Bianchet et al., 2011). The relationship between these factors, the physiological stress response, and allostasis highlights the multifactorial nature of pathways which converge to shape the health status of reindeer and caribou.

Allostasis and the physiological stress response are closely linked. As allostatic load increases, the stress response is activated and levels of circulating GCs begin to rise (Figure 2.2) (Wingfield, 2005; Landys et al., 2006; Busch and Hayward, 2009; McEwen and Wingfield, 2010). In an undisturbed animal at rest, relatively low (basal) levels of circulating GCs act to maintain glucose, salt and water balance within the minimum operating levels necessary to sustain life (Sapolsky et al., 2000; Landys et al., 2006; Busch and Hayward, 2009). Low to moderate increases in GC levels (compared to basal) are associated with an organism's response to increasing allostatic load associated with the added demands of daily (e.g. feeding, predator avoidance behaviour, general activity patterns) or seasonal (e.g. migration, reproduction, moult) life history events. At these levels, GCs enhance catabolic processes and promote behavioural changes (e.g. increased feeding or locomotion) that provide the extra energy necessary to counter the adverse effects of increasing allostatic load by maintaining physiological systems within an elevated (but modulated) operating range. These changes may be considered to be within the reaction norm of a particular species and may be observed as the circadian, seasonal, or life history stage associated alterations in levels of circulating GCs recorded in many species (Boonstra, 2005; Landys et al., 2006; Romero et al., 2008). As allostatic load continues to grow (and GC levels increase) GCs begin to bind low affinity cellular receptors associated with the suppressive effects that act to reduce allostatic load by mobilizing or repartitioning available energy through the temporary inhibition of feeding behaviour, energy storage, growth, immunity, and reproduction. If energy demands exceed energy available, an emergency physiological state of allostatic overload may occur which is characterized by dys-regulation of allostatic mediators, high levels of circulating GCs, and an associated increase in the potential for adverse health effects (McEwen, 1998; Wingfield, 2005; Charmandari et al., 2005; Landys et al., 2006; Busch and Hayward, 2009; McEwen and Wingfield, 2010) (Figure 2.2).

For *Rangifer*, unpredictable natural and anthropogenic challenges require extra energy compared to demands that fall within daily or seasonal norms typically encountered by this species. As such, exposure to unpredictable events superimposed on the energetic demands of regular life history processes may result in increased allostatic load in affected animals (McEwen, 1998; Wingfield, 2005; Landys et al., 2006; Busch and Hayward, 2009; McEwen and Wingfield, 2010; Wingfield et al., 2011) (Figure 2.2). In turn, *Rangifer* become less able to cope with additional energetic chal-lenges and more prone to entering a state of allostatic overload. In pristine ecosystems, reindeer and caribou may be able to escape from areas affected by unpredictable natural challenges such as forest fires or stochastic weather events (e.g. icing or deep snow) to unaffected areas, thus reducing energetic deficits and allostatic load back to levels within normal tolerances. In contrast, *Rangifer* unable to escape or inhabiting ecosystems significantly altered by human-caused climatic or eco-logical change may be forced to contend with persistently elevated allostatic loads. They may also be subject to repeated or prolonged bouts of allostatic overload in which persistently elevated GCs (chronic stress) directly threaten the health and fitness of affected animals. A wide range of extrinsic and intrinsic factors may contribute to allostatic load in reindeer and caribou (Figure 2.1) and dem-onstrate the importance of a holistic cumulative effects approach to understanding *Rangifer* health.

For example, current warming trends in *Rangifer* habitat may alter the quantity and type, nutri-tional quality, phenology or distribution of important forage species (Weladji and Holand, 2006; Cebrian et al., 2008; Post et al., 2008; Joly et al., 2009). As a result, energetic deficits associated with reduced foraging efficiency or long-distance travel to locate alternate food sources (McNeil et al., 2005; Sharma et al., 2009) may increase along with long-term stress levels in affected caribou populations. Warmer temperatures may also be associated with an increase in the intensity and duration of insect harassment, which may further reduce foraging efficiency and increase HPA axis activity during the summer period (Haegemon and Reimers, 2002; Weladji et al., 2003; Witter et al., 2011). These factors, along with direct energetic costs associated with heat stress (Soppela et al., 1986), may enhance allostatic load, making affected animals less able to deal with demands of daily life or additional anthropogenic challenges. In turn, the health status, productivity, and fitness of affected animals may decline.

Likewise, *Rangifer* habitat is threatened by increasing resource extraction (oil, gas, timber and mineral) activities along with escalating residential and recreational development in many regions (Festa-Bianchet et al., 2011). Reindeer and caribou are generally considered to be relatively intoler-ant of human disturbance (e.g. Vistnes and Nellemann, 2008) and their capacity to adapt to chang-ing conditions may be exceeded by the cumulative effects of natural and anthropogenic factors. Roads and trails through *Rangifer* habitat may increase the probability of contact with predators (Whittington et al., 2011; Boutin et al., 2012). These and other anthropogenic landscape features may also act as barriers to caribou movements and reduce habitat availability (Dyer et al., 2001; Gunn et al., 2009). For example, boreal and mountain caribou have been found to avoid habitat near roads, other anthropogenic linear landscape features, and petroleum extraction or forestry related disturbances (e.g. Dyer et al., 2002; Schaefer and Mahoney, 2007; Hins et al., 2009). Similarly, reindeer have also been found to avoid recreational development (e.g. resorts, cabins), even in the presence of high quality habitat in otherwise poor quality landscapes (Nellemann et al., 2000; Vistnes and Nellemann, 2008; Nellemann et al., 2010), while backcountry recreational activity (e.g. snowmobiling, skiing, road traffic) may displace *Rangifer* from preferred winter habitat, calving areas, or refugia from predators (Seip et al., 2007; Helle et al., 2012). Avoidance of human activity may have the potential to directly increase HPA axis activity in caribou (Cook, 1996; Bristow and Holmes, 2007; Busch and Hayward, 2009). This behaviour may also lead to the crowding of cari-bou in remaining areas of undisturbed habitat, which may indirectly increase levels of circulating cortisol in caribou through the density dependent effects of overgrazing (e.g. Mahoney et al., 2011) and social stress (Holand et al., 2004; Fauchald et al., 2007). Avoidance may also prevent caribou from migrating to circumvent the localized effects of stochastic weather events or may force caribou into sub-optimal habitat in which diminished foraging efficiency, intra and interspecific competition, or an increasing risk of predation or pathogen exposure, further enhance long-term stress levels and allostatic load.

2.4 A CONTEMPORARY DEFINITION OF REINDEER AND CARIBOU HEALTH

Given the wide range of health determinants that may impact reindeer and caribou, it is clear that *Rangifer* health, as with the health of any species (e.g. Stephen, 2014) cannot be defined in the context of any one feature. Likewise, the measurement of any one extrinsic or intrinsic parameter cannot fully describe the health status of an individual reindeer or caribou or a *Rangifer* population. A holistic definition of *Rangifer* health must, therefore, be multifaceted and multidisciplinary in nature. The criteria for defining healthy reindeer and caribou may also vary at the individual, population and ecosystem scales.

For example, at the individual level, a healthy reindeer or caribou may be defined as one which is relatively free of extrinsic and intrinsic health determinants that adversely impact physiology, behaviour, growth, body condition, survival or reproduction. Fitness is a key component of *Rangifer* health and to be considered healthy, a reindeer or caribou must also survive to reproduce successfully. Similar criteria may be used to describe a healthy *Rangifer* population. However, a holistic definition of *Rangifer* health must also recognize that, within populations, some individual reindeer and caribou may not meet this criterion. Accordingly, a free-ranging *Rangifer* population may be considered to be healthy despite the presence of unhealthy individuals within it. In free-ranging wildlife, the annual (finite) rate of population growth (λ) is often used to describe population dynamics (where $\lambda < 1$ population decline, $\lambda = 1$ population stable, $\lambda > 1$ population growing) and from a purely biological perspective, a healthy *Rangifer* population should be stable or growing.

A holistic definition must also consider *Rangifer* health at the ecosystem level, which includes the ecological, social and economic roles of reindeer and caribou (e.g. Stephen, 2014). Here, healthy reindeer and caribou individuals or *Rangifer* populations, must be able to meet the social, cultural, and economic needs of the multidisciplinary stakeholders in their consumptive and non-consumptive use, management or conservation. Similarly, a shared definition of *Rangifer* health must consider the relationship between *Rangifer* and their environment. A healthy reindeer, caribou, or *Rangifer* population must also have available an appropriate amount and type of habitat along with the food resources, water, and shelter necessary to meet the needs of the species' life history. In addition, reindeer and caribou must be able to fulfill their ecological role as a prey species, a keystone species, and a feature of biodiversity of ecological integrity in northern ecosystems. Most importantly, healthy *Rangifer* must be self-sustaining and resilient, persisting over time through their ability to cope with unpredictable natural and human-caused challenges. It is thus within this context that the incorporation of a broader concept of health as a desired outcome in cumulative effects models and *Rangifer* management is critical.

REFERENCES

Arnemo, J.M., and Ranheim, B. (1999). Effects of medetomidine and atipamezole on serum glucose and cortisol levels in captive reindeer (*Rangifer tarandus*). *Rangifer*, 19(2):85–89.

Ashley, N.T., Barboza, P.S., Macbeth, B.J. et al. (2011). Glucocorticosteroid concentrations in feces and hair of captive caribou and reindeer following adrenocorticotropic challenge. *General and Comparative Endocrinology*, 172(3):382–391.

Bennett, A., and Hayssen, V. (2010). Measuring cortisol in hair and saliva from dogs: Coat colour and pigment differences. *Domestic Animal Endocrinology*, 39(3):171–180.

Bennett, C., Fripp, D., Othen, L., Jarsky, T., French, J., and Loskutoff, N. (2008). Urinary corticosteroid excretion patterns in the Okapi (*Okapi johnstoni*). *Zoo Biology*, 10:1–13.

Black, P.H. (2002). Stress and the inflammatory response: A review of neurogenic inflammation. *Brain, Behavior and Immunity*, 16(6):622–653.

Boonstra, R. (2005). Equipped for life: The adaptive role of the stress axis in male mammals. *Journal of Mammalogy*, 86(2):236–247.

Bourbonnais, M.L., Nelson, T.A., Cattet, M.R.L., Darimont, C.T., and Stenhouse, G.B. (2013). Spatial analysis of factors influencing long-term stress in the grizzly bear (*Ursus arctos*) population of Alberta, Canada. *PLoS ONE*, 8(12): e83768. doi:10.1371/journal.pone.0083768.

Boutin, S., Boyce, M.S., Hebblewhite, M. et al. (2012). Why are caribou declining in the oil sands? *Frontiers in Ecology and the Environment*, 10(2):65–67.

Bristow, D.J., and Holmes, D.S. (2007) Cortisol levels and anxiety related behaviours in cattle. *Physiology and Behaviour*, 90(4):626–628.

Bryan, H.M., Darimont, C.T., Paquet, P.C., Wynne-Edwards, K.E., and Smits, J. (2013). Stress and reproductive hormones in grizzly bears reflect nutritional benefits and social consequences of a salmon foraging niche. *PLoS ONE*, 8(11): e80537. doi:10.1371/journal.pone.0080537.

Bubenik, G.A., Schams, D., White, R.G., Rowell, J., Blake, J., and Bartos, L. (1998). Seasonal levels of metabolic hormones and substrates in male and female reindeer (*Rangifer tarandus*). *Comparative Biochemistry and Physiology Part C*, 120:307–315.

Busch, D.S., and Hayward, L.S. (2009). Stress in a conservation context: A discussion of glucocorticoid actions and how levels change with conservation-relevant variables. *Biological Conservation*, 142:2844–2853.

Cattet, M., Macbeth, B.J., Janz, D.M. et al. (2014). Quantifying long-term stress in brown bears with the hair cortisol concentration: A biomarker that may be confounded by rapid changes in response to capture and handling. *Conservation Physiology*, 2:cou026. doi:10.1093/conphys/cou026.

Cebrian, M.R., Kielland, K., and Finstad, G. (2008). Forage quality and reindeer productivity: Multiplier effects amplified by climate change. *Arctic, Antarctic, and Alpine Research*, 40(1):48–54.

Charmandari, E., Tsigos, C., and Chrousos, G. (2005). Endocrinology of the stress response. *Annual Review of Physiology*, 67(1): 259–284.

Chen, Y.M., Cintron, N.M., and Whitson, P.A. (1992). Long-term storage of salivary cortisol samples at room temperature. *Clinical Chemistry*, 38:304.

Choksi, N.Y., Jahnke, G.D., St. Hilaire, C., and Shelby, M. (2003). Role of thyroid hormones in human and laboratory animal reproductive health. *Birth Defects Research Part B*, 68:479–491.

Constable, S., Parslow, A., Dutton, G., Rogers, T., and Hogg, C. (2006). Urinary cortisol sampling: A non-invasive technique for examining cortisol concentrations in the Weddell seal, *Leptonychotes weddellii*. *Zoo Biology*, 25:137–144.

Cook, C.J. (1996). Basal and stress response cortisol levels and stress avoidance learning in sheep (*Ovis ovis*). *New Zealand Veterinary Journal*, 44(4):162–163.

Davenport, M.D., Tiefenbacher, S., Lutz, C.K., Novak, M.A., and Meyer, J.S. (2006). Analysis of endogenous cortisol concentrations in the hair of rhesus macaques. *General and Comparative Endocrinology*, 147(3):255–261.

Di Francesco, J., Navarro-Gonzalez, N., Wynne-Edwards, K. et al. (2017) Qiviut cortisol in muskoxen as a potential tool for informing conservation strategies. *Conservation Physiology*, 5(1): cox052. doi:10.1093/conphys/cox052.

Drucker, D.G., Hobson, K.A., Oullet, J.P., and Courtois, R. (2010). Influence of forage preferences and habitat use on ^{13}C and ^{15}N abundance in wild caribou (*Rangifer tarandus caribou*) and moose (*Alces alces*) from Canada. *Isotopes in Environmental Health Studies*, 46(1):107–121.

Duffy, L.K., Hallock, R.J., Finstad, G., and Bowyer, R.T. (2005). Noninvasive environmental monitoring of mercury in Alaskan reindeer. *American Journal of Environmental Sciences*, 1(4):249–253.

Dunn, E., Kapoor, A., Leen, J., and Matthews, S.G. (2010). Prenatal synthetic glucocorticoid exposure alters hypothalamic-pituitary-adrenal regulation and pregnancy outcomes in mature female guinea pigs. *Journal of Physiology*, 58(5):887–899.

Dyer, S.J., O'Neil, J.P., Wasel, S.M., and S. Boutin, S. (2001). Avoidance of industrial development by woodland caribou. *Journal of Wildlife Management*, 65:531–542.

Dyer, S.J., O'Neil, J.P., Wasel, S.M., and S. Boutin, S. (2002). Quantifying barrier effects of roads and seismic lines on movements of female woodland caribou in northeastern Alberta. *Canadian Journal of Zoology*, 80:839–845.

Ellis, R.D., McWhorter, T.J., and M. Maron, M. (2012). Integrating landscape ecology and conservation physiology. *Landscape Ecology in Review*, 27:1–12.

Ewacha, M. (2016). Stress response of boreal woodland caribou, moose, and wolves to disturbance in eastern Manitoba. MSc. Thesis, University of Manitoba. 130 pp.

Fauchald, P., Rødven, R., Bårdsen, B.J. et al. (2007). Escaping parasitism in the selfish herd: Age, size, and density dependent warble fly infestations in reindeer. *Oikos*, 116:491–499.

Festa-Bianchet, M., Ray, J.C., Boutin, S., Côté, S.D., and Gunn, A. (2011). Conservation of caribou (*Rangifer tarandus*) in Canada: An uncertain future. *Canadian Journal of Zoology*, 89:419–434.

Foley, C.A.H., Papageorge, S., and Wasser, S.K. (2001). Noninvasive stress and reproductive measures of social and ecological pressures in free-ranging elephants. *Conservation Biology*, 15(4):1134–1142.

Freeman, N.L. (2008). Motorized backcountry recreation and stress response in mountain caribou (*Rangifer tarandus caribou*). MSc Thesis, University of British Columbia. 75 pp.

Gunn, A., Russell, D., White, R.G., and Kofinas, G. (2009). Facing a future of change: Wild migratory caribou and reindeer. *Arctic*, 62(3):iii–vi.

Gunn, A., Cuyler, C., Mizin, I. et al. 2016. *Rangifer tarandus*. The IUCN Red List of Threatened Species 2016: e.T29742A22167140. dx.doi.org/10.2305/IUCN.UK.2016-1.RLTS.T29742A22167140.en. Downloaded on 09 May 2017.

Haegemon, R.I., and Reimers, E. (2002). Reindeer summer activity pattern in relation to weather and insect harassment. *Journal of Animal Ecology*, 72:883–892.

Heggberget, T.M., Gaare, E. and Ball, J. P. (2002). Reindeer and climate change: Importance of winter forage. *Rangifer*, 22:13–31.

Helle, T., Hallikainen, V., Sarkela, M., Haapalehto, M., Niva, A., and Puiskari, J. (2012). Effects of a holiday resort on the distribution of semidomesticated reindeer. *Annales Zoologici Fennici*, 49(1–2):23–35.

Hing, S., Narayan, E.J., Thompson, R.C.A., and Godfrey, S.S. (2016). The relationship between physiological stress and wildlife disease: Consequences for health and conservation. *Wildlife Research*, 43:51–60.

Hins, C., Ouellet, J.P., Dussault, C., and St-Laurent, M.H. (2009). Habitat selection by forest dwelling caribou in managed boreal forest of eastern Canada: Evidence of a landscape configuration effect. *Forest Ecology and Management*, 257:636–643.

Holand, Ø., Gjostein, H., Losvar, A. et al. (2004). Social rank in female reindeer (*Rangifer tarandus*): Effects of body mass, antler size and age. *Journal of Zoology*, 263(4):365–372.

Joly, K., Jandt, R.R., and Klein, D.R. (2009). Decrease of lichens in Arctic ecosystems: The role of wildfire, caribou, reindeer, competition, and climate in north-western Alaska. *Polar Research*, 28:433–442.

Joly, K., Wasser, S.K., and Booth, R. (2015). Non-invasive assessment of the interrelationships of diet, pregnancy rate, group composition, and physiological and nutritional stress of barren-ground caribou in late winter. *PLoS One*, 10:e0127586.

Keay, J.M., Singh, J., Gaunt, M.C., and Kaur, T. (2006). Fecal glucocorticoids and their metabolites as indicators of stress in various mammalian species: A literature review. *Journal of Zoo and Wildlife Medicine*, 37(3):234–244.

Kobelt, A.J., Hemsworth, P.H., Barnett, J.L., and Butler, K.L. (2003). Sources of sampling variation in saliva cortisol in dogs. *Research in Veterinary Science*, 75:157–161.

Landys, M.M., Ramenofsky, M., and Wingfield, J.C. (2006). Actions of glucocorticoids at a seasonal baseline as compared to stress-related levels in the regulation of periodic life processes. *General and Comparative Endocrinology*, 148:132–149.

Linklater, W.L. (2010). Distress – An underutilised concept in conservation and missing from Busch and Hayward (2009). *Biological Conservation*, 143:1037–1038.

Macbeth, B.J. (2013). An evaluation of hair cortisol concentration as a potential biomarker of long-term stress in free-ranging grizzly bears (*Ursus arctos*), polar bears (*Ursus maritimus*), and caribou (*Rangifer tarandus* sp.). PhD Dissertation, Western College of Veterinary Medicine, University of Saskatchewan, Saskatoon. 298 pp.

Macbeth, B.J., Cattet, M.R.L., Stenhouse, G.B., Gibeau, M.L., and Janz, D.M. (2010). Hair cortisol concentration as a non-invasive measure of long-term stress in free-ranging grizzly bears (*Ursus arctos*): Considerations with implications for other wildlife. *Canadian Journal of Zoology*, 88:935–949.

Mahoney, S.P., Weir, J.N., Luther, J.G., Schafer, J.A., and Morrison, S.F. (2011). Morphological changes in Newfoundland caribou: Effects of abundance and climate. *Rangifer*, 31(1):21–34.

Mastromonaco, G.F., Gunn, K., McCurdy-Adams, H., Edwards, D.B., and Schulte-Hostedde, A.I. (2014). Validation and use of hair cortisol as a measure of chronic stress in eastern chipmunks (*Tamias striatus*). *Conservation Physiology*, 2: cou055. doi:10.1093/conphys/cou055.

Mayers, J.L., and Novak, M.A. (2012). Minireview: Hair cortisol: A novel biomarker of hypothalamic-pituitary-adrenocortical activity. *Endocrinology*, doi:10.1012/en.2012-1226.

McEwen, B.S. (1998). Stress, adaptation, and disease. Allostasis and allostatic load. *Annals of the New York Academy of Sciences*, 840:33–44.

McEwen, B.S., Biron, C.A., Brunson, K.W. et al. (1997). Neural–endocrine–immune interactions: The role of adrenocorticoids as modulators of immune function in health and disease. *Brain Research Reviews*, 23:79–133.

McEwen, B.S., and Seeman, T. (1999). Protective and damaging effects of mediators of stress: Elaborating and testing the concepts of allostasis and allostatic load. *Annals of the New York Academy of Sciences*, 896:30–47.

McEwen, B.S., and Wingfield, J.C. (2003). The concept of allostasis in biology and biomedicine. *Hormones and Behavior*, 43(1):2–15.

McEwen B.S., and Wingfield, J.C. (2010). What's in a name? Integrating homeostasis, allostasis and stress. *Hormones and Behavior*, 57(2):105–111.

McNeil, P., Russell, D.E., Griffith, B., and Kofinas, G.P. (2005). Where the wild things are: Seasonal variation in caribou distribution in relation to climate change. *Rangifer*, Special Issue 16:51–63.

Millspaugh, J.J., Washburn, B.E., Milanick, M.A., Slotow, R., and van Dyck, G. (2003). Effects of heat and chemical treatment on fecal glucocorticoid measurements: Implications for sample transport. *Wildlife Society Bulletin*, 31:399–406.

Mooring, M.S., Patton, M.L., Lance, V.A. et al. (2006). Glucocorticoids of bison bulls in relation to social status. *Hormones and Behavior*, 49(3):369–375.

Nellemann, C., Jordhoy, P., Stoen, O.G., and Strand, O. (2000). Cumulative impacts of tourist resorts on wild reindeer during winter. *Arctic*, 53:9–17.

Nellemann, C., Vistnes, I., Jordhoy, P. et al. (2010). Effects of recreational cabins, trails and their removal for restoration of reindeer winter range. *Restoration Ecology*, 18(6):873–881.

Nicks, B. and Vandenheede, M. (2014). Animal health and welfare: Equivalent or complementary? *Revue Scientifique et Technique Office International des Épizooties*, 3:97–101.

Omsjoe, E.H., Stien, A., Irvine, R.J. et al. (2009). Evaluating capture stress and its effects on reproductive success in Svalbard reindeer. *Canadian Journal of Zoology*, 87:73–85.

Owen, M.A., Czekala, N.M., Swaisgood, R.R., Steinman K., and Lindburg, D.G. (2005). Seasonal and diurnal dynamics of glucocorticoids and behaviour in giant pandas. *Ursus*, 16(2):208–211.

Papadimitriou, A., and Priftis, K.N. (2009). Regulation of the Hypothalamic-pituitary-adrenal axis. *Neuroimmunomodulation*, 16:265–271.

Patyk, K.A., Duncan, C., Nol, P. et al. (2015). Establishing a definition of polar bear (*Ursus maritimus*) health: A guide to research and management activities. *Science of the Total Environment*, 514:371–378.

Post, E., Pedersen, C., Wilmers, C.C., and Forchhammer, M.C. (2008). Warming, plant phenology and the spatial dimension of trophic mismatch for large herbivores. *Proceedings of the Royal Society B: Biological Sciences*, 275:2005–2013.

Racey, G.D. (2005). Climate change and woodland caribou in northwestern Ontario: A risk analysis. *Rangifer*, Special Issue 16:123–136.

Rehbinder, C., and Hau, J. (2005). Quantification of cortisol, cortisol immunoreactive metabolites and immunoglobulin A in serum, saliva, urine and feces for non-invasive assessment of stress in reindeer. *Canadian Journal of Veterinary Research*, 70:151–154.

Renaud, L.A. (2012). Stress-inducing landscape disturbances: Linking habitat selection and physiology in woodland caribou. *In*: Impacts de l'aménagement forestier et des infrastructures humaines sur les niveaux de stress du caribou forestier. MSc Thesis, Université du Québec à Rimouski. 97 pp.

Romero, L.M., and Reed, J.M., (2005). Collecting baseline corticosterone samples in the field: Is under three minutes good enough? *Comparative Biochemistry and Physiology Part A: Molecular and Integrative Physiology,* 140:73–79.

Romero, L.M., Meister, C.J., Cyr, N.E., Kenagy, G.J., and Wingfield, J. (2008). Seasonal glucocorticoid responses to capture in wild free-living mammals. *American Journal of Physiology: Regulatory, Integrative and Comparative Physiology*, 294:R614–R622.

Russell, E., Koren, G., Rieder, M., and Van Uum, S. (2012). Hair cortisol as a biological marker of chronic stress: Current status, future directions, and unanswered questions. *Psychoneuroendocrinology*, 37:589–601.

Säkkinen, H., Tornberg, J., Goddard, P.J. et al. (2004). The effect of blood sampling method on indicators of physiological stress in reindeer (*Rangifer tarandus*). *Domestic Animal Endocrinology*, 26:87–98.

Saltz, D., and White, G.C. (1991). Urinary cortisol and urea nitrogen responses to winter stress in mule deer. *Journal of Wildlife Management*, 55(1):1–16.

Sapolsky, R.M., Romero, L.M., and Munck, A.U. (2000). How do glucocorticoids influence stress responses? Integrating permissive, suppressive, stimulatory, and preparative actions. *Endocrine Reviews*, 21(1):55–89.

Schaefer, J.A., and Mahoney, S.P. (2007). Effects of progressive clear-cut logging on Newfoundland caribou. *Journal of Wildlife Management*, 71:1753–1757.

Seip, D.R, Johnson, C.J. and Watts, G.S. (2007). Displacement of mountain caribou from winter habitat by snowmobiles. *Journal of Wildlife Management*, 71(5):1539–1544.

Sharma, S., Couturier, S., and Côté, S.D. (2009). Impacts of climate change on the seasonal distribution of migratory caribou. *Global Change Biology*, 15:2549–2562.

Sheriff, M.J., C.J. Krebs, and Boonstra, R. (2009). The sensitive hare: Sublethal effects of predator stress on reproduction in snowshoe hares. *Journal of Animal Ecology*, 78:1249–1258.

Sheriff, M.J., Krebs, C.J., and Boonstra, R. (2010). The ghosts of predators past: Population cycles and the role of maternal programming under fluctuating predation risk. *Ecology*, 91(10):2983–2994.

Sheriff, M.J., Dantzer, B., Delehanty, B., Palme, R., and Boonstra, R. (2011). Measuring stress in wildlife: Techniques for quantifying glucocorticoids. *Oecologia*, 166:869–887.

Soppela, P., M. Nieminen, and Timisjàrvi, J. (1986). Thermoregulation in reindeer. *Rangifer*, Special Issue 1: 273–278.

Stenn, K.S., and Paus, R. (2001). Controls of hair follicle cycling. *Physiological Reviews*, 81(1):449–494.

Stephen, C. (2014). Toward a modernized definition of wildlife health. *Journal of Wildlife Diseases*, 50(3):427–430.

Terwissen, C.V., Mastromonaco, G.F., and Murray, D.L. (2014). Enzyme immunoassays as a method for quantifying hair reproductive hormones in two felid species. *Conservation Physiology*, 2: cou044. doi:10.1093/conphys/cou044.

Tsigos, C., and Chrousos, G.P. (2002). Hypothalamic-pituitary-adrenal axis, neuroendocrine factors and stress. *Journal of Psychosomatic Research*, 53:865–871.

Wingfield, J.C. (2005). The concept of allostasis: Coping with a capricious environment. *Journal of Mammalogy*, 86(2):248–254.

Vistnes, I. and Nellemann, C. (2008). The matter of spatial and temporal scales: A review of reindeer and caribou response to human activity. *Polar Biology*, 31:399–407.

Vors, L.S. and Boyce, M.S. (2009). Global declines of caribou and reindeer. *Global Change Biology*, 15:2626–2633.

Wasser, S.K., Hunt, K.E., Brown, J.L. et al. (2000). A generalized fecal glucocorticoid assay for use in a diverse array of nondomestic mammalian and avian species. *General and Comparative Endocrinology*, 120(3):260–275.

Wasser, S.K., Keim, J.L., Taper, M.L., and Lele, S.R. (2011). The influences of wolf predation, habitat loss, and human activity on caribou and moose in the Alberta oil sands. *Frontiers in Ecology and the Environment*, 9(10):546–551.

Weladji, R.B., Holland, Ø., and Almøy, T. (2003). Use of climatic data to assess the effect of insect harassment on the autumn weight of reindeer calves. *Journal of the Zoological Society of London*, 260:79–85.

Weladji, R.B., and Holand, Ø. (2006). Influences of large scale climatic variability on reindeer population dynamics: implications for reindeer husbandry in Norway. *Climate Research*, 32:119–127.

Whittington, J., Hebblewhite, M., DeCesare, N.J. et al. (2011). Caribou encounters with wolves increase near roads and trails: A time to event approach. *Journal of Applied Ecology*, 48(6):1535–1542.

Wikelski, M., and Cooke, S.J. (2006). Conservation physiology. *Trends in Ecology and Evolution*, 21(2):38–46.

Wingfield, J.C. (2005). The concept of allostasis: Coping with a capricious environment. *Journal of Mammalogy*, 86(2):248–254.

Wingfield, J.C., Kelley, J.P., and Angelier, F. (2011). What are extreme environmental conditions and how do organisms cope with them? *Current Zoology*, 57(3):363–374.

Witter, L.A., Johnson, C.J., Croft, B., Gunn, A., and Gillingham, M.P. (2011). Behavioural trade-offs in response to external stimuli: Time allocation of an Arctic ungulate during varying intensities of harassment by parasitic flies. *Journal of Animal Ecology*, doi: 10.1111/j.1365-2656.2011.01905.x.

Wobeser, G.A. (2006). *Essentials of Disease in Wild Animals*. Blackwell Publishing, Ames, Iowa, USA. 256 pp.

Wobeser, G.A. (2007). *Disease in Wild Animals: Investigation and Management*. Springer-Verlag Berlin Heidelberg, Germany. 393 pp.

3 *Rangifer* Diet and Nutritional Needs

Birgitta Åhman and Robert G. White

CONTENTS

This chapter discusses the importance of nutrition and adaptation to the seasonally changing availability and quality of reindeer and caribou diets, and the subsequent importance this may have for *Rangifer* health and performance. The topic includes the seasonal diet and nutrient intake, the function of the gastrointestinal tract, nutritional needs, deficiencies and starvation and the effect of nutrition on growth, reproduction and survival, with implications for population dynamics and management.

3.1 SEASONAL HABITAT CHOICE, DIET AND NUTRIENT INTAKE

In order to support nutritional needs for survival and reproduction, reindeer and caribou use a set of foraging strategies at different scales. Migrations between seasonal ranges are typical for most *Rangifer* herds, wild as well as semi-domesticated. These migrations are several hundred km for many populations of tundra and barren-ground *Rangifer* subspecies. Forest/woodland and high arctic *Rangifer* types, as well as some of the tundra-type herds, are more sedentary, although they also perform relocations and select habitat in relation to season.

Migrations from winter to summer ranges can be made along a south–north gradient, from lower to higher altitudes or from inland to coastal areas. This makes it possible for reindeer and caribou to follow snow-melt and the succession of spring green-up and to find areas in winter where the snow is not too deep or dense. Winter ranges are often situated either on open lichen heaths or in forests with a high abundance of terricoulous lichens (Bergerud et al. 2008). Habitats with access to winter-green graminoids and forbs may offer additional forage of high nutritive value (Warenberg 1982).

Semi-domesticated reindeer grazing on summer ranges in Sweden (July) (Photo: Birgitta Åhman).

Summer ranges on open tundra (Figure 3.1) generally consist of a mosaic of vegetation types with large variation in plant composition (Gaare and Skogland 1975). Short-term movements between vegetation types and foraging sites make it possible for the animals to select for high available biomass and high nutritional quality of forage plants.

Rangifer ranges are similar over the northern hemisphere with respect to plant composition. Even if species are not altogether the same, genera are usually similar within vegetation types. A large variety of plant species have been identified as being eaten by *Rangifer* through analyses of rumen content, faeces, samples collected from esophageal fistulas or direct observations of grazing animals (Bergerud 1972, Bjune 2000, Russell et al. 1993, Scotter 1967). Research by Skjenneberg et al. (1975) revealed that semi-domesticated reindeer in Norway consumed most plant species that were available (in this case close to 100 species), although lichens, at least when scarce, were clearly selected and mosses were avoided.

Lichens typically dominate the *Rangifer* diet in winter, but there are large variations between different *Rangifer* populations, and lichens are also eaten during the snow-free period if available (Bergerud 1972, Finstad and Kielland 2011, Gaare and Skogland 1975) (Figure 3.2, Table 3.1). In contrast to mainland *Rangifer* subspecies, the high arctic types (Peary caribou and Svalbard reindeer) often have little access to lichens in their habitats and thus few or no lichens in their diets (Bjune 2000, Larter and Nagy 2004). Instead forbs, deciduous shrubs and graminoids typical for these habitats are grazed in varying proportions throughout the year.

Lichens contain carbohydrates with high digestibility for reindeer. In vitro dry matter (DM) digestibility for different *Cladina* species, using inoculum from reindeer adapted to a lichen diet, have been reported to range from about 40% up to 70% (higher for *C. rangiferina* and *arbuscula* than for *C. stellaris*) (Danell et al. 1994, Storeheier et al. 2002, White et al. 1975). The digestibility of *Cetraria islandica* and *nivalis* seems to be similar to that of the most digestible *Cladina* species, while the less favored *Stereocaulon pascale* was shown to be less digestible (Storeheier et al. 2002). The digestibility of arboreal *Bryoria* spp. has been found to be as high as 88% (Danell et al. 1994). Although a good source of energy, most lichens grazed by *Rangifer* are low in protein and macro minerals (calcium, phosphorous, potassium and magnesium) (Ophof et al. 2013). The *Cladonia/ Cladina* species contain only around 2% crude protein (on a DM basis), and some arboreal lichens even more (*Bryoria*, 5%). The chemical composition of lichens does not change with season, but

FIGURE 3.2   Reindeer winter forage – lichens, *Cladina rangiferina* (grey), *Cladina stellaris* (round/tubby and yellowish white) and some small fractions of *Cladina mitis* (yellowish white and similar shape as *C. rangiferina*) (Photo: Birgitta Åhman).

mat-forming lichens (e.g. reindeer lichens) are typically more nutritious in the upper growing part than in the lower part of the lichen thalli (Ophof et al. 2013, Storeheier et al. 2002).

Reindeer and caribou cannot survive the winter on lichens alone, due to the low content of nitrogen and macro minerals. Graminoids and other vascular plants are thus part of the winter forage of most *Rangifer* herds in winter, adding important nutrients to the diet. In line with this, Helle (1984) observed that, throughout winter, reindeer in northern Finland used not only lichen heaths, but also adjacent feeding sites where they had access to grasses and sedges. For some herds (Russell et al. 1993, Scotter 1967), shrubs dominate the non-lichen part of the diet; however, for many *Rangifer* populations, graminoids seem to constitute the major fraction of the non-lichen part of the winter diet (Finstad and Kielland 2011, Gaare and Skogland 1975, Thomas et al. 1996).

The content of crude protein in most shrubs, graminoids and forbs varies from 4% to about 10% in winter (Heggberget et al. 2002), which is at least two to three times that of most lichens, and the mineral content may be several times higher than in lichens (Ophof et al. 2013). Green shoots of vascular plants and buds of deciduous shrubs are highly digestible and may contain over 20% crude protein (Ophof et al. 2013). Even in small amounts, they may provide a substantial part of necessary protein and minerals, especially in late winter and early spring before snow-melt.

Bryophytes are consumed by *Rangifer* and in some cases constitute 10–20% of the diet (Boertje 1984), even though they seem to be less palatable and of low nutritional value (Danell et al. 1994). Moss forms a large part of the diet (35–49%) of the Porcupine caribou herd during late spring, prior to green-up and along migration routes that contained few lichens (Russell et al. 1993).

Limitation not only in quality of forage, but also quantity, is a challenge to reindeer and caribou in winter. Snow is a key factor in the winter-feeding patterns of *Rangifer*. In winter, forage intake is limited by the time it takes to search and dig for food, and there is a trade-off between forage intake and the energy costs associated with searching and cratering. It is crucial for the animal to be able to distinguish sites with enough forage from those with little or no forage. In fact, reindeer can use smell to distinguish good lichen patches from poor patches through soft snow over 90 cm thick (Helle 1984). Accordingly, feeding craters are often made at patches with the highest lichen biomass (Helle 1984, Johnson et al. 2001). Hard snow increases the time needed for cratering, and reindeer and caribou typically choose to crater at locations with the lowest depth and hardness of snow (Johnson et al. 2001). Arboreal lichens are chosen when deep and hard snow prevents the reindeer from feeding on terrestrial lichens, and are therefore mainly grazed in late winter.

There is a positive relationship between the availability and the intake of lichens (Trudell and White 1981), and snow is thus a key factor for total biomass intake. The availability of forage seems to overrule the nutritional quality of forage in the choice of habitat or feeding sites in winter (Johnson et al. 2001).

The importance of snow for reindeer foraging is demonstrated by traditional classification of winter pastures within reindeer herding, where factors affecting snow properties may be as important as lichen abundance in categorizing a certain land area as good or poor winter pasture (Roturier and Roué 2009).

As snow melts and new green vegetation emerges, vascular plants become increasingly important. Graminoids, deciduous shrubs and forbs will usually dominate the summer diet (Boertje 1984, Finstad and Kielland 2011, Gaare and Skogland 1975), substantially enhancing the nutritional quality of the diet. The early summer diet is characterized by high digestibility and high protein and mineral content. However, the nutritional value of forage varies between plant species, plant parts, climatic conditions and phenological development (Johnstone et al. 2002, Mårell et al. 2006). Peak nitrogen content of between 3 and 4% (corresponding to 19–25% crude protein) has been recorded for several alpine plant species in early growth stages (Mårell et al. 2006). Several of the macro minerals (phosphorus, magnesium and potassium) follow a similar pattern (Mårell et al. 2006, Staaland and Sæbø 1993).

Access to habitats with high plant biomass production and plants of high digestibility makes it possible for reindeer and caribou to increase forage intake considerably in summer compared to winter. For *Rangifer*, being a ruminant, high digestibility of the diet is crucial to enable the consumption of sufficient biomass of forage. Being able to select the most digestible and protein rich plants and plant parts is therefore crucial if the animals are to obtain enough nutrients for growth (young animals) and for regaining body reserves during the relatively short summer.

In late summer and autumn, reindeer and caribou eagerly eat mushrooms (fruiting bodies of fungi) and often search for habitats where they are available (Boertje 1984, Helle 1981). Mushrooms provide a late summer supply of highly digestible, high protein forage. Caribou and reindeer continue to eat graminoids as well as some woody plants and herbs, but gradually increase the lichen component typical of a winter diet (Boertje 1984, Gaare and Skogland 1975).

TABLE 3.1

Seasonal Diet of Reindeer and Caribou, Average and Range (Minimum-Maximum) Values Reported from Different *Rangifer* Herds

	Lichen	Shrub[a]	Graminoid	Forb	Fungi[b]	Moss	Other
Spring (Apr-May)	62%	8%	12%	1.6%	0%	9%	8%
	(56–69%)	(5–14%)	(10–14%)	(1–2%)	–	(7–10%)	(4–17%)
Summer (June-Aug)	32%	29%	23%	8%	5%	3%	4%
	(10–62%)	(13–49%)	(10–51%)	(0–41%)	(0–25%)	(0–6%)	(0–12%)
Autumn (Sep-Oct)	58%	9%	19%	4%	5%	3%	2%
	(34–87%)	(2–21%)	(5–47%)	(0–12%)	(0–12%)	(1–5%)	(0–7%)
Winter (Dec-Mar)	65%	11%	12%	1.0%	0.3%	6%	4%
	(34–88%)	(5–26%)	(0–28%)	(0–7%)	(0–2%)	(1–13%)	(0–11%)

Source: Extracted from Bergerud, A.T., *J. Wildl. Manage.*, 36:913–23, 1972; Boertje, R.D., *Arctic*, 37:161–5, 1984; Finstad, G.L., and K. Kielland, *Arct. Antarct. Alp. Res.*, 43:543–54; 2011; Gaare, E., and T. Skogland, Wild reindeer food habits and range use at Hardangervidda, in *Fennoscandian Tundra Ecosystems, Part 2. Animals and System Analysis*, ed. F.E. Wielgolaski, 195–205, Berlin, Heidelberg, New York, Springer Verlag, 1975.

[a] Woody plants, from which mainly leaves and shoots are eaten.

[b] Fungi content is strongly influenced by annual and seasonal availability.

3.2 THE GASTROINTESTINAL TRACT

Both the anatomy and function of the gastrointestinal tract in reindeer and caribou are adapted to a diet with large variations in structure and nutrient content over the year. The species *Rangifer tarandus* is generally regarded as an intermediate mixed feeder on a scale from grazers to browsers or concentrate selectors (Hofmann 1989). Grazers are represented by, for example, cattle and sheep, which are able to digest high amounts of fibrous grass, while browsers (e.g. moose and roe-deer) are adapted to a diet with high digestibility. As opposed to grazers, concentrate selectors are characterized by a narrow mouth, relatively small rumen and reticulum, small omasum, shorter intestine and a large distal fermentation chamber (caecum). The retention time for digesta is shorter than for grazers, and the microorganisms in the rumen are less capable of digesting cellulose (Mathiesen 1999).

Most studies on the gastrointestinal function on *Rangifer* are made on semi-domesticated Eurasian tundra reindeer (*R. t. tarandus*) or on the high arctic Svalbard reindeer (*R. t. plathyrynchus*). The sub species are generally alike with regard to the gross anatomy of the digestive tract (Mathiesen et al. 2005), although Staaland et al. (1979) point out the shorter intestines and larger caecum – important for mineral absorption – as characteristics of *R. t. plathyrynchus* as compared to *R. t. tarandus*.

Investigations by Mathiesen et al. (2000) show that reindeer have the narrow and pointed muzzle and rather low crowned molariform teeth considered typical for selective feeders. They also have large lips, and prominent first incisor teeth, which help to select for the most digestible and nutritious parts of the vegetation.

A major part of the processing of food in *Rangifer*, as in other ruminants, occurs in the rumen-reticulum (hereafter referred to as rumen) where less digestible carbohydrates, like cellulose and hemicellulose, are degraded by various microorganisms. Rumination, the chewing of regurgitated forage, will break the food into smaller particles and facilitate further processing by microorganisms.

Animals ruminate in bouts following feeding. Daily feeding and ruminating patterns of captive reindeer and caribou differ depending on time of year. Eriksson et al. (1981) and Maier and White (1998) recorded daily feeding and ruminating patterns of captive reindeer during different seasons. Both studies found four to six activity peaks per 24 hours in winter (6 hours daylight) and six to nine peaks in summer (with continuous or near-continuous daylight), representing three more events than in free-ranging reindeer (van Oort et al. 2007) and caribou (Maier and White 1998). Reindeer were fed grain-based pellets and lichens that were freely available, while caribou had free-grazing and a grain-based diet. Much of their active time was spent eating, and rumination took place when the animals were resting between feeding sessions. For reindeer, time spent ruminating between meals was shorter in summer (28 minutes per ruminating session) than in winter (47 and 54 minutes in December and March, respectively). Since the frequency of activity and rest was higher in summer, the resulting total time for ruminating (not reported) probably did not differ much between seasons. Captive reindeer offered different diets ad libitum (Nilsson et al. 2006b) ruminated 16–26% of their time, and this did not differ between diets (lichens, silage and pellets in different combinations) once the animals had been adapted.

Experiments with reindeer equipped with esophageal fistulas (Trudell and White 1981) and foraging on different natural vegetation showed that about 10 g dry matter is usually regurgitated each time and chewed for about a minute before being swallowed again.

The utilization of forage in ruminants depends on a range of anaerobic bacteria, ciliate protozoa, fungi and methanogenic archaea (previously classified as bacteria) that are present in the rumen and hindgut. These microorganisms have the capacity to digest the different carbohydrate fractions in the forage and make them available to the host. The microbial flora differs between environments in both quantity and composition, and changes with season and diet (Mathiesen et al. 2005). Bacteria that generally dominate in summer are more adapted to using soluble carbohydrates, while cellulolytic bacteria dominate more in winter.

It is well documented that the digestive system of *Rangifer* is poor at handling large quantities of fibrous forage (Aagnes et al. 1996). Rapidly increasing live body mass in animals fed hay or silage may actually be a sign of failure of cellulolysis and subsequent accumulation of forage in the rumen (Olsen et al. 1995). Svalbard reindeer, which need to survive on a very poor-quality diet in winter, are known to host relatively more cellulolytic bacteria compared to *Rangifer* in other environments (Mathiesen et al. 2005, Orpin and Mathiesen 1990).

Animals of the *Rangifer* species are unique in their ability to utilize lichens as their main forage at times when the availability and quality of vascular plants is poor. The polysaccharides of lichens contain glucose linkages that differ from those in starch and cellulose (Culberson 1969), and specific bacterial enzymes are required for breaking these. In addition, secondary phenolic compounds, including usnic acid, which is toxic to fungi and many Gram-positive bacteria, are found in several of the lichen species that are eaten by reindeer (Ingolfsdottir 2002). However, rumen microbes in reindeer seem to tolerate usnic acid (Palo 1993), and novel species of usnic acid resistant bacteria (*Eubacterium rangiferina*) have been identified by Sundset et al. (2008). Sundset et al. (2010) showed that even when reindeer were provided with relatively large amounts of usnic acid, no traces were found in rumen fluid, urine or faeces. The authors conclude that usnic acid is degraded by rumen microbes, explaining the ability of reindeer and caribou to fully utilize lichens. Detoxification of usnic acid, alternatively microbial resistance, was further confirmed by research showing that rumen bacteria and methanogenic archaea populations did not change significantly when usnic acid was added to the rumen content (Glad et al. 2014). Rumen bacteria from reindeer have also been shown to be resistant to other phenolic secondary compounds that are present in lichens (Glad et al. 2009).

The adaptation to lichens has been further demonstrated by measurements of in vitro digestibility of lichens, showing that digestibility (of *Cladina* spp.) was higher (70%, compared to 63%) in rumen fluid from lichen-fed reindeer than from reindeer on other diets (Wallsten 2003). It was also evident that lichens were poorly digested in rumen fluid from cattle (11% digestibility), while the digestibility of vascular plants did not differ depending on rumen fluid donor (cattle or reindeer). The importance of adaptation to lichens was confirmed by Nilsson et al. (2006a), who found that rumen content from lichen-fed reindeer contained more bacteria that could be cultured on media with lichens as the only energy source, compared to rumen content from reindeer fed diets without lichens (pellets and grass silage).

Bacterial species that are attached to the rumen wall differ from those in the rumen fluid (Cheng et al. 1979), and include, for example, ureolytic bacteria, which may be important for recycling nitrogen and producing ammonia that can be further used for protein synthesis by other rumen bacteria.

The concentrations of total protozoa in the rumen of reindeer has been reported to range from 130 to 250×10^4, which is higher than in other domestic livestock investigated ($5.2 - 44.5 \times 10^4$) (De la Fuente et al. 2006). Westerling (1970) investigated the ciliate fauna of semi-domesticated reindeer in Finland and found that it differed substantially from that found in domestic livestock like cattle and sheep. The composition of ciliate species was also shown to vary with season (Orpin and Mathiesen 1990, Westerling 1970). Dehority (1975) reported that domesticated reindeer in Alaska had similar ciliate species to domestic livestock, while the ciliate flora of wild caribou was more similar to that found in wild and semi-domesticated Eurasian reindeer. Based on several studies, Mathiesen et al. (2005) conclude that ciliate fauna is probably not primarily linked to host species, but rather to diet and possible isolation of host populations.

Methanogenic archaea are a group of microorganisms that produce methane and other by-products, and have received growing attention during recent years. Archaea were earlier defined as bacteria but have their own evolutionary history and are now regarded as a separate domain. Methanogens have not been investigated much in reindeer. Thus far, studies on Svalbard and Norwegian reindeer (Sundset et al. 2009a, Sundset et al. 2009b) suggest that their numbers are relatively small (10^7–10^8 cells/g rumen content) compared to that reported for cattle (up to $1.34 \ 10^9$ cells/g) (Denman et al. 2007). This might indicate lower methane emission from reindeer compared to cattle, at least with regard to reindeer free-grazing on natural ranges.

In addition to bacteria, methanogens and protozoa, the rumen also hosts anaerobic fungi with the capacity to utilize, for example, cellulose (Mathiesen et al. 2005). It seems that these fungi have a higher capacity to invade plant tissue than many bacteria, and thus that they are important for the breakdown of fiber in the rumen.

Microorganisms utilize carbohydrates in the food eaten by the animal to produce carbon dioxide, methane and energy-rich volatile fatty acids (VFA): mainly acetic, butyric and propionic acid. All are highly absorbed across the rumen wall (72–82%) (White and Gau 1975) and their net energies may constitute up to 60–80% of metabolizable energy intake (Annison and Armstrong 1970). The production and concentration of VFA in the rumen can be used as an indicator of rumen activity and forage quality, and they are considerably reduced in poor diets or during starvation. Nilsson et al. (2006a) found a 56% reduction in the VFA concentration in the rumen (from 110 to 48 mmol/l) of reindeer calves after a period of 9 days with restricted feed intake. Due to reduced volume of rumen content, the total VFA decreased to 34% of the original level. Reindeer calves that had been starved for four days showed even lower VFA concentrations of 14–37 mmol/l (Aagnes et al. 1995).

The absorption of VFA depends on the total rumen surface. Westerling (1975) found that the reindeer rumen mucosa changed according to diet. This was later confirmed by studies on reindeer calves by Josefsen et al. (1997), who found that the size and density of rumen papilla changed with season, so that the total rumen surface (or surface enlargement factor, SEF) in late summer (September) was almost double that in late winter (April). Most of the observed decline occurred by November. The effect of diet quality (and not quantity) was demonstrated in a feeding experiment, using silage with different digestibility (content of soluble carbohydrates versus cellulose), where high digestibility was shown to be positively correlated with high SEF. The mechanism behind this difference is the concentration of VFA in the rumen content that stimulates the growth of rumen papilla (Sakata and Tamate 1979).

Food needs to be degraded into small enough particles before it can proceed further through the digestive system. The threshold size seems to be general for most ruminants and ranges from 1 to 4 mm (Lechner et al. 2010). The rumen retention time of particles and rumen liquor declines with increasing intake of digestible dry matter (Lechner et al. 2010, White and Trudell 1980). Particle retention time was reported to be 23–69 hours in lichen-fed reindeer (Aagnes and Mathiesen 1994), while it was only 10.8 hours in Alaskan reindeer consuming early summer pasture (White and Trudell 1980). Turnover time of the liquid component of rumen contents is less than that for particles and is inversely related to dry matter intake (Lechner et al. 2010, White and Trudell 1980). Lechner et al. (2010) found that the rumen retention time in reindeer was shorter than in muskoxen (a grazer), but longer than in moose (a typical browser), though all species showed a similar relative pattern with regard to retention of different sized particles.

The caecum, together with the proximal part of the colon, represent the distal fermentation chamber (DFC) where cellulolytic bacteria take care of undigested cellulose that escaped the rumen, thereby further improving the digestibility of the consumed forage. As in the rumen, VFAs are absorbed into the blood to be used as an energy source by the animal. It has been observed that as much as 17% of the total VFA is produced in the DFC of Svalbard reindeer in winter (Sørmo et al. 1997), and the corresponding production rate of reindeer fed commercial pellets or lichen-based diets was 20% of the rumen production rate (White and Gau 1975).

In the distal colon, water is absorbed and faecal pellets are formed. *Rangifer* produce small faecal pellets that are usually relatively dry in winter when food is limited, but are more soft and "clotted" on lush summer pastures, or when the reindeer are fed grain-based diets (pellets).

Reindeer and caribou can obviously be sustained on poor ranges during winter, where their ability to digest lichens and tolerate anti-digestive substances like usnic acid give them an advantage compared to other herbivores in the same environment. However, in summer they need to compensate for the restricted nutrient intake with rapid growth and replenishment of body reserves. As their ability to digest fiber is limited, they require a highly digestible diet that enables them to process a large enough volume of forage with the necessary nutrients for synthesis of fat and protein. The possibility for selective grazing is then a key issue, as demonstrated by White (1983) and stressed later in this chapter. The limited capability of *Rangifer* to digest fiber is an important consideration when feeding reindeer, as accumulation of undigested forage in the rumen may subsequently lead to starvation.

Although reindeer and caribou are clearly very well adapted to utilizing lichens, there seems to be a limitation in their ability to use lichens as the only, or major, dietary component. Several feeding experiments show that reindeer have a lower total DM intake when fed only lichens or a major share of lichens, compared to when fed other diets (Jacobsen et al. 1981, Nilsson et al. 2000). These experiments also revealed that reindeer fed lichen-diets did not eat enough to maintain their body mass, even though the diets were offered ad libitum, while reindeer fed other diets increased in body mass. The reason for this is unclear, though it might be linked to a lack of necessary nutrients (nitrogen and minerals) for the gastrointestinal microbes, or may indicate that some substances in lichens (e.g. the secondary compounds discussed on p. 112) reduce the animals' appetite.

3.3   NUTRITIONAL PHYSIOLOGY AND NEEDS

Reindeer and caribou are highly adapted to the seasonal variations in availability and nutrient content of forage typical of their Arctic and Sub-Arctic ranges. Long winters with limited daylight, constant snow cover and limited access to forage through occasional icing events, is a challenge that the species has evolved to cope with through evolution, and which has shaped life history timing and anatomical, physiological and metabolic adaptations. *Rangifer* have, for example, several adaptations for saving energy during winter (Tyler and Blix 1990). Heat loss is minimized by thick fur and heat exchange mechanisms in the legs and nose (Folkow and Mercer 1986). The energy necessary for locomotion is reduced by anatomical attributes like large hooves, long legs and cost effective gaits (Fancy and White 1987). The sharp edges and size of the hooves makes digging in snow effective (Fancy and White 1985a). Antlers on females is a unique feature among *Cervidae*, enhancing the social rank of *Rangifer* females in late winter, when the males have already lost their antlers, making the females superior in the competition for forage (Barrette and Vandal 1986, Espmark 1964) and allowing their offspring to share feeding sites.

Seasonality of food availability and quality has selected for a high capacity to store energy and protein as fat and muscle tissue. Fat and protein are regained during periods with higher forage abundance (summer and autumn) in response to body reserves lost during periods of low nutrient intake (winter). Negative energy balance and consequent weight loss in winter is often considered a regulated process for *Rangifer* (Tyler et al. 1999, White et al. 2014) and should be regarded as dependent on nutrition but not necessarily be described as malnutrition. In most environments with low or moderate *Rangifer* densities, available forage in winter is sufficient to maintain the reproductive capacity and survival of populations, often through millennia. Reproductive output is, however, seasonally variable via adjustment of conception rate (Cameron and Ver Hoef 1994, Reimers 1983), rate of foetal growth in the final trimester (Bergerud et al. 2008, Reimers 2002) and variability in weaning date (Russell et al. 2005). Body fat and muscle tissue serve as supplements to available forage and in females provide a regulated reserve for terminal gestation and the initiation of lactation. In addition, fat and muscle reserves provide insurance against mortality during potential periods of acute starvation. Adaptations that minimize N loss in winter assist in this process.

Timing of the reproductive cycle of *Rangifer* follows annual patterns of snow-melt and forage availability, demonstrating plasticity with respect to seasonal variation in forage availability and quality (Cameron et al. 1993, Flydal and Reimers 2002).

3.3.1   ENERGY AND PROTEIN FOR MAINTENANCE, THERMOREGULATION AND ACTIVITY

The allocation of energy and protein (nitrogen) for maintenance, activity, reproduction, growth and body reserves in *Rangifer* has been described in detail by White, Russell and co-workers

(Russell 2011, White et al. 2014), with most based on the NRC *Nutrient Requirements of Small Ruminants* (NRC 2007).

Irrespective of season, a certain minimum of energy and nutrients is needed for base requirements to keep the animal alive (maintenance). This includes the basal metabolism, plus added cost for maintenance activity (e.g. foraging) and sustaining body temperature. It has been speculated whether reindeer and caribou lower their basal metabolic rate (BMR) in winter as an adaptation to save energy. However, observed seasonal differences in resting metabolic rate are related to food intake before measurement of energy expenditure, confusing estimation of seasonal effect (Nilssen et al. 1984b, Tyler and Blix 1990). White et al. (2014) assumed a constant daily BMR of 293 kJ/kg$^{0.75}$ for *Rangifer* in winter when modeling energy expenditure. Taking the efficiency of use of metabolizable energy in forage into consideration, daily maintenance requirements for sedentary reindeer and caribou were calculated to be 493 kJkg$^{0.75}$ and 560 kJ/kg$^{0.75}$, respectively.

Nitrogen requirements for maintenance are associated with urinary losses and faecal output of metabolic N, and increase with increasing energy intake. The nitrogen requirements of reindeer have been little investigated, but for growing lambs N requirements have been estimated to be about 1 g N ($\approx$6 g protein) per MJ energy intake (Black and Griffiths 1975), while larger and older animals need relatively less.

Due to effective mechanisms for reducing heat loss, the excess heat produced by digestive processes and metabolism is usually sufficient to maintain body temperature even at low ambient temperatures. Apparent lower critical temperatures have been estimated to be as low as −50°C for Svalbard reindeer and −30°C for Norwegian reindeer in winter (Cuyler and Øritsland 1993, Nilssen et al. 1984b). However, these critical temperatures require a certain level of heat-producing basic metabolism and activity (Parker and Robbins 1985) and may be different in situations of starvation, for example. Extra energy will be required at ambient temperatures below the critical temperature, and Dryden (2011) suggests a net energy cost of 14.6 kJ/kg$^{0.75}$ BW for each degree (°C) below the critical temperature, based on research on temperate deer species (red and white-tailed deer). The thermal costs for *Rangifer* are expected to be lower because of their thick fur and other effective heat conserving mechanisms. Accordingly, costs of conductance of 5.9 and 7.7 kJ/kg$^{0.75}$ BW and °C for reindeer and caribou at low temperatures were reported by Parker and Robbins (1985).

The energetic costs of standing, foraging or walking on plane ground are just slightly higher than the costs associated with resting, while activities like running, walking uphill, and walking or cratering in snow may increase energy requirements substantially (Fancy 1986). White et al. (2014) used estimates of energy requirements for activity from different sources for modeling *Rangifer* energetics. Lying or standing were assumed to require less than 0.5 kJ/kg body mass (BM) per hour, while the energetic costs for eating, pawing and walking were assumed to be 2–4 kJ/kg BM per hour. The total daily energy requirement for maintenance in summer was estimated at 660 kJ/kg$^{0.75}$, or about 17 MJ for a 75 kg reindeer, and was the sum of energy required for basal metabolism, minimum activity for foraging, for growth of hair and antlers and for hosting parasites. The requirement of N for maintenance was assumed to be proportional to the energy requirements and was estimated to be about 0.73 g N/MJ, which equals 12 g N (or 77 g protein) for a 75 kg reindeer. Daily N requirements for maintenance determined by McEwan and Whitehead (1970), 0.82 g/kg$^{0.75}$, is theoretically achievable using a crude protein concentration in the diet of about 8.3% and a daily dry matter intake of 50 g/kg$^{0.75}$ (NRC 2007, p. 98), though these levels are not always achieved in the wild.

The cost of walking, additional to that of standing, is 1.6 kJ/kg body BM per km (Fancy and White 1987), corresponding to 120 kJ/km for a 75 kg reindeer. Similarly, Nilssen et al. (1984a) showed that the energetic cost for locomotion increases linearly with speed, irrespective of season and ambient temperatures. Only when temperatures were below the lower critical temperature did energy expenditure not increase with movement rate, indicating that heat from exercise substituted for the extra energy otherwise needed for the animal to keep warm. Walking uphill increases energy expenditure. As an example, lifting one kg one meter was shown to cost 23 kJ (Fancy and White 1987). Part of this cost could, however, be recovered when the animal walks downhill.

Walking and cratering in snow result in increased energy requirements for activity in winter, compared to summer. The cost of walking in snow increases exponentially with sinking depth (Fancy and White 1985a), with a 100% increase (doubling) in the energy needed for locomotion at 30–40 cm snow depth (depending on crust or not). At 50 cm, the energetic cost may be five times as high as that of walking on bare ground.

The cost of cratering in snow increases with snow density and hardness. According to Fancy and White (1985b), the mean cost for cratering was 118 J per stroke in light snow without crust and 219 J per stroke in denser snow with a thin, hard crust. The resulting daily cost for cratering was estimated to be 1.2–1.5 kJ/kg and 2.3–2.9 kJ/kg BM for a caribou cratering in soft and dense snow, respectively.

Gotaas et al. (2000) summarized total daily energy expenditures in winter, which ranged from 617 to 830 kJ/kg$^{0.75}$ BM for free-living reindeer, barren-ground caribou and woodland caribou (recalculated from W/kg). A similar value of 837 kJ/kg$^{0.75}$ BM was reported by McEwan and Whitehead (1970) for reindeer in winter. Considerably lower daily energy expenditure for maintenance in summer (232 kJ/kg$^{0.75}$) was reported for penned non-lactating females (Chan-McLeod et al. 1994). The low energy requirements in the latter case may, however, be partly explained by the fact that the reindeer were fenced and fed, thus not having to expend much energy on foraging.

3.3.2 Costs for Reproduction and Growth

Similar to many other large herbivores, *Rangifer* females depend on body stores of fat and protein for satisfying the extra energy and nitrogen needed for foetal growth and milk production (Barboza and Parker 2008, Taillon et al. 2013). This makes their reproductive success less sensitive to the irregular, and sometimes dramatic, variations in forage availability that are typical of their environment.

The energy costs for pregnancy are low during the first months. Boertje (1985) calculated (from published data) that the daily costs were 0.15 and 0.38 MJ in September–October and November–December, respectively, for a 110 kg caribou. This corresponded to 1% of the estimated total energy expenditure for a pregnant female. In January–February the daily cost increased to 0.97 MJ and in March–April to 2.41 MJ. This is in line with the rate of foetal growth reported by Roine et al. (1982), at only 1 g/d in mid-December, around 20 g/d in January–February and 100 g/d a month before parturition. Although foetal growth is remarkably linear at approximately 49 g/d between day 120 and 190 of gestation, both Reimers (2002) (reindeer) and Bergerud et al. (2008) (caribou) report large differences in terminal foetal growth (40–113 g/d) attributable to year and population effects, that likely act through the mother's protein reserves. In line with this, White at al. (2013) estimated the tissue deposition for foetal growth at term gestation to be 14 g/d for protein and 2.1 g/d for fat. The gross energy content of this amount of fat and protein is about 300 kJ. Most of the extra energy required for gestation is lost as heat, however, and only a minor part (10–20%) is retained as new tissue (Robbins 1993, p. 200). Likewise, most maternal protein used over winter is not captured by the concept and White et al. (2013) speculated that extra protein must be mobilized to meet maternal and foetal glucose requirements through gluconeogenesis.

The daily energy needs for lactation are substantially higher than those for gestation. *Rangifer* calves in the Arctic are born before the start of green-up and females have very limited access to nutritious forage. Consequently, much of the energy and protein required for early milk production has to be mobilized from the female's own body reserves. Experiments with reindeer and caribou fed a low nitrogen diet confirmed that lactating females continued to rely on body fat and protein during the first weeks of lactation (Barboza and Parker 2008). Similarly, Taillon et al. (2013) found that 88–91% of the nitrogen in milk protein was derived from female body reserves (Figure 3.3).

Similar to many other cervids (Robbins 1993, pp. 205–6), reindeer and caribou produce concentrated milk with a high protein and fat content (around 8–12% protein and 10–20% fat) that changes

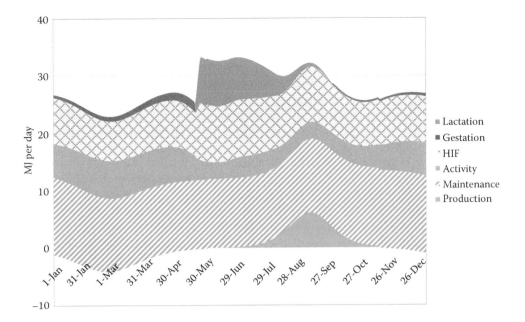

FIGURE 3.3 Simplified example of energy allocation of a reproducing *Rangifer* female under satisfactory grazing conditions using figures derived from articles referred to in the text, mainly White et al. 2014, Russel et al. 1993, and Fancy 1986. Maximum female BM is set to 75 kg, and it is assumed that the female loses 15% of her BM from 15 November until 1 June. Birth date of the calf is set to 20 May, and birth weight to 5.5 kg. The maximum daily milk production is assumed to be 1.2 liters and occurs 1 week after calving. Weaning is set to 1 October. "Production" includes regaining of body reserves (fat and protein) as well as coat and antler production. Increased cost for activity during winter is mainly attributed to extra costs for walking and cratering in snow, while slightly increased costs from late June to early August are caused by higher activity because of insect harassment. Bottom line below zero implies that the animal uses body reserves and loses body mass. HIF is the heat increment of feeding (heat produced from digestion and metabolic processes), which may vary depending on diet and how the energy is used. Here a default value (30% of total metabolizable energy intake) has been used to estimate HIF.

during the lactation cycle (Gjøstein et al. 2004, Luick et al. 1974). Milk yield – determined using machine milking – was shown to vary substantially between individual female semi-domesticated reindeer, ranging between 595 and 1239 g/d (corresponding to an average of 8 MJ/d) at peak lactation around 3 weeks postpartum (Gjøstein et al. 2004). Milk intake in bottle-fed caribou calves was about 100 g/kg BM per day during the first month after birth, declining until weaning at the age of around 4 months (Parker and Barboza 2013). It should be noted that milk production is highly dependent on female condition and nutrient intake, and may be substantially reduced in years with poor forging conditions (White and Luick 1984). Likewise, in free-grazing reindeer milk production varies between individuals and between years, which is largely attributable to the female's body reserves (White and Luick 1984). As the lactation period progresses, milk production declines, fat and protein content increases, and there is a decline in the concentration of lactose (Chan-McLeod et al. 1994, Gjøstein et al. 2004). However, the calf continues to suckle even when milk production is low, and complete weaning does not usually occur until the rut in autumn, 22–26 weeks postpartum (Gjøstein et al. 2004, Lavigueur and Barrette 1992). There seems to be rather large variations, however, and lactation can sometimes be terminated early, or be extended through winter (Skjenneberg and Slagsvold 1968). Five weaning strategies have been documented for caribou (Russell et al. 2005, White et al. 2014).

Lactation imposes high energetic costs on a female with a calf. The daily energy requirements for maintenance were reported to be twice as high (457 kJ/kg$^{0.75}$ BM) for a lactating reindeer or caribou, compared to a non-lactating (232 kJ/kg$^{0.75}$ BM) (Chan-McLeod et al. 1994), resulting in a daily lactation cost of 7.6 MJ for a female weighing 110 kg (about three times the cost for late pregnancy). The amount of energy and protein allocated for lactation is, however, limited by the female's own needs (White 1983, White and Luick 1984). In situations with low access to forage (e.g. at high population densities) *Rangifer* females are expected to adopt a "selfish cow strategy" (Russell et al. 1996) and use relatively more resources for regaining their own body reserves and for future reproduction, than for current milk production (Chan-McLeod et al. 1999). Since *Rangifer* calves are normally born before, or at the start of, vegetation growth, more body resources are required for gestation and early lactation in early breeders compared to late breeders. On the other hand, the calf will have more time to grow before autumn if born early.

Calf growth rates of around 300 g/d from birth to weaning have been reported for captive semi-domesticated reindeer (Gjøstein et al. 2004, Rognmo et al. 1983). This is in accordance with results from semi-domesticated reindeer on natural mountain ranges in Sweden (Åhman, unpublished results). White et al. (1981) report growth rates of between 295 and 350 g/d for caribou and reindeer in captivity (compiling results from several investigations); however, these values can be higher and more variable in caribou, e.g. between 380 and 440 g/d during the first 3 weeks and 193–344 g/d at 3 to 6 weeks for the Porcupine caribou herd (Griffith et al. 2002). Bottle-fed caribou gained 355 g/d on average until peak milk intake at age 30–40 days. This did not differ from the subsequent weight gain (351 g/d) that proceeded until weaning (Parker and Barboza 2013).

Higher weight gains (on average 428 g/d during the first month, and 454 g/d during the following months) were recorded for calves of caribou females kept on a high plane of nutrition (Parker et al. 1990). Daily energy costs were reported to vary substantially, but the energy for daily existence was reported to be 439 kJ/kg BM on average, corresponding to 923 kJ/kg$^{0.75}$ between 20 and 40 days of age. This was 1.7 times the estimated fasting metabolism for caribou calves at 3 months of age (536 kJ/kg$^{0.75}$ BM per day) (Luick and White 1983). Daily milk intake during the same period corresponded to about 10 MJ and 140 g protein per day, on average.

The costs of growth in *Rangifer* calves have been summarized by White et al. (1981), with an overall value reported to be 30 MJ per kg weight gain. As the calf grows, gradually more energy will be needed for maintenance, while at the same time milk production and intake decline. More of the nutrient intake will therefore have to be supported by foraging. *Rangifer* calves have been observed foraging from the available vegetation within a few days of birth (Espmark 1971, Skjenneberg and Slagsvold 1968). Even if milk is the main source of energy and nutrients during the calf's first month of life, the size of the rumen gradually increases and the calf adapts to the utilization of forage. Knott et al. (2005) show that the relative size of the rumen of a 60-day-old *Rangifer* calf was almost as large as in the adult animal, and this coincided with increasing content of the gastrointestinal tract, similar to the content of the adult animal.

3.3.3 MINERAL REQUIREMENTS

Rangifer's requirements have not been established for all essential minerals, but are generally believed to be similar to those of other deer species and ruminants in general. Skeletal growth, antler formation, hair growth and milk production may require specific nutrients, and dietary requirements thus vary with season, age and reproductive stage.

Calcium (Ca) is the main component of bone and teeth and is commonly discussed together with phosphorus (P), because of their ratio in the skeleton (Ca:P = 2:1). According to NRC (2007, p. 115) Ca balance in cervids is achieved at a daily intake of 73 mg/kg BM. The daily requirement of P is reported to be 20 mg/kg BW. Growing animals need substantially more, the recommended daily intake for a calf weighing 20 kg and growing 300 g per day being 5.6 g Ca and 4.0 g P. Antler growth increases the need for both Ca and P, and the requirements are related to antler size and growth rate. Moen and Pastor (1998) estimated the daily requirement for male reindeer at peak antler growth to be at least 25 g for Ca and 12 g for P.

Potassium (K) and magnesium (Mg) requirements are not available for cervids, but values for goats are recommended (NRC 2007, pp. 123–24). Daily maintenance requirements for these species can be calculated to be around 0.15 g K/kg BM and 0.02 g Mg/kg BM, respectively. For growing, pregnant and lactating animals, the requirements are higher. For animals on natural pastures the supply of K is generally sufficient. The natural winter forage for *Rangifer* is generally low in Mg, however, and there are several documented cases of deficiency (see section 3.4 of this chapter). Most vegetation is also low in sodium (Na) and salt hunger is consequently common among ungulates. Robbins (1993, p. 47) recommended 9 mg/kg BM of sodium, but *Rangifer* research indicates that a daily requirement at 3.2 mg/kg is probably more relevant for reindeer and caribou (NRC 2007, p. 121).

Trace elements are generally not limiting for free-grazing reindeer, and deficiencies have only been reported for copper. However, trace element requirements are relevant with regard to supplementary feeding and the composition of feed rations for *Rangifer*, and are therefore discussed in relation to feeding in Chapter 4, Feeding and Associated Health Problems.

3.3.4 Retention and Use of Body Reserves

Reindeer and caribou often lose around 10–20% of their body mass during winter, with some island or high arctic types losing even more (Couturier et al. 2009, Tyler and Blix 1990). The corresponding loss of protein may be as high as 29% (Gerhart et al. 1996a). However, *Rangifer* females seem to allocate as much of their body reserves as possible for foetal growth and instead use dietary energy and nitrogen for maintenance (Parker et al. 2005).

Protein and fat reserves have to be regained during the relatively short snow-free period, which requires sufficient quantity and quality of summer forage. The chemical energy in dry body protein is 23.6 kJ/g on average, while the energy in fat is about 39.5 kJ/g (Robbins 1993, p. 10). Protein generally contains about 16% nitrogen, yet muscle and other soft tissue contain a large proportion of water (normally at least 70%), which means that the nitrogen content will be about 50 mg and the energy content will be 7 kJ per gram of fat free muscle tissue.

The allocation of energy, protein and other nutrients from forage or body reserves at a given time depends not only on the supply of nutrients but also on the animal's reproductive state and the time of year. Energy for maintenance, including, for example, the necessary activity involved in foraging, as discussed on pp. 116–117, evidently must be given a high priority in order to keep the animal alive, while there is a "choice" in the degree of retention or mobilization of energy and nitrogen for activity, body tissue or reproduction.

As earlier discussed, there is a trade-off between using resources for lactation and using them for direct growth and regaining body reserves. Although lactating females gain body mass considerably slower in summer than do non-lactating females (Chan-McLeod et al. 1994), there is evidence that milk production is limited when food resources are insufficient (White and Luick 1984).

Adult males lose a significant part of both their fat and protein (muscle) reserves in autumn, due to increased activity and little time spent eating during the rut. Some of this may be regained after the rut if foraging conditions are favorable, but there may be a risk that the body reserves will be insufficient for the animal to survive periods of poor foraging conditions during the following winter. Apparently, this is still an advantageous strategy for lifetime reproductive success (Barboza et al. 2004), and similar strategies are common for males of other polygamous species (Pelletier et al. 2009).

3.3.5 Association between Nutrition and Parasites

Reindeer and caribou can be plagued by a range of external and internal parasites (discussed in detail in Chapter 6, Parasitic Infections and Diseases). It is well known from research on domestic ruminants that, even in cases where they do not induce clinical symptoms, gastrointestinal parasites may depress appetite, impair gastrointestinal function and increase endogenous protein loss, thereby negatively affecting the animal's body condition (van Houtert and Sykes 1996). At the same time, poor nutrition may make the animal particularly susceptible to parasitic infections.

As pointed out in a review by Gunn and Irvine (2003), the nutritional implications of parasitism for wild and semi-domesticated ungulates have rarely been acknowledged in research. Their review focuses on gastrointestinal nematodes and confirms that findings regarding the importance of parasites for domestic ruminants also have relevance for wild ungulates, and that parasites should be considered further in relation to ungulate nutrition. Impaired appetite, and therefore poor weight gain, has been demonstrated as an effect of abomasal nematode infection in semi-domesticated reindeer in Norway (Arneberg et al. 1996), and experiments on Svalbard reindeer have shown that body condition was negatively affected by gastrointestinal parasites (Stien et al. 2002). Furthermore, anthelmintic treatment was shown to have positive effects on fecundity, and the effects were large enough to influence reindeer population density (Albon et al. 2002).

Hughes et al. (2009) found a negative effect of abomasal nematodes on body mass of barren-ground caribou, even at substantially lower parasite burdens than those observed in Svalbard reindeer (Stien et al. 2002). Caribou in Greenland are typified by high burdens of abomasal parasites as well as both warble and nasal bot fly larvae (Oestridae) in late winter/spring (Cuyler et al. 2012). However, pregnancy rates were high and body fat reserves were not fully depleted despite the high estimates of energy and protein costs imposed by these parasites. In addition, Hughes et al. (2009) found that oestrid larvae infections (measured in April) did not affect body mass significantly, but had a negative effect on back fat and were associated with a reduced probability of pregnancy. This might be a result of direct action of the parasite larvae in the body of the host, but may also be a delayed effect of heavy harassment by oestrid flies during the preceding summer, reducing the time that the animals can spend foraging. Ballesteros et al. (2012) found a significant effect of anthelmintic treatment of female reindeer in autumn on their body mass the following summer. Reproductive success of treated females did not, however, differ from that of untreated controls. It was assumed that the treated reindeer would be re-infected by intestinal parasites after the treatment and that the experiment would capture mainly the effect of oestrid larvae. Whether this was the case, however, is unclear.

3.4 NUTRITIONAL DEFICIENCIES AND STARVATION

3.4.1 DEFICIENCIES

As already mentioned, a negative energy and nitrogen balance is common for reindeer in winter. This is, however, something to which reindeer and caribou are adapted to via their capacity to accumulate body reserves in summer for later use, combined with their ability to effectively recycle endogenous nitrogen (Barboza and Parker 2006, Valtonen 1979). Moderate under-nutrition is thus not expected to cause significant loss of muscle tissue, although Soppela et al. (2000) found that a lichen-dominated winter diet may lead to a reduction in important serum lipids.

More specific deficiencies, in terms of lack of certain vitamins or minerals, are not very well documented among reindeer and caribou, magnesium (Mg) being an exception. Lichens contain only minor amounts of Mg compared to most vascular plants, and clinical illness due to Mg deficiency has been observed in reindeer grazing on lichen ranges in inner Finnmark, Norway (Hoff et al. 1993). Reindeer calves with clinical signs (they were paretic and ataxic) had extremely low levels of Mg in their blood (0.19 ± 0.18 mmol/l in blood serum, compared to an average of 0.82 in a healthy reference group), accompanied by a low calcium/phosphorus ratio (1.21, compared to 1.85 in the reference group) and greatly elevated levels of aspartate aminotransferase, creatine kinase and lactate dehydrogenase. Exceptionally low Mg levels (below 0.3 mmol/l) in blood serum have also been reported for Swedish reindeer in poor body condition in late winter (Åhman et al. 1986). In line with this, Ropstad et al. (1997) observed large variations both between herds and between years in the concentrations of Mg (0.16–1.39 mmol/l), together with a large variation in calcium (Ca, 0.9–3.6 mmol/l) and phosphorous (P, 0.2–3.4 mmol/l), in blood plasma collected from reindeer in Northern Norway in winter. They also observed that the probability of a female being pregnant was positively correlated with the concentrations of Mg and Ca in plasma. Moderately low serum-Mg levels (0.67 mmol/l), alongside low serum levels of Ca and copper (Cu, 0.42 μmol/l), were observed in a reindeer herd with a subsequent high mortality rate (40%) in late spring, soon after sampling (Hyvärinen et al. 1977). The average Cu level in this herd was less than half of that observed in well-nourished herds, and Cu deficiency was believed to have contributed to the high mortality rate.

Judging from concentrations determined in the liver, trace elements may vary substantially between *Rangifer* herds. Vikøren et al. (2011) found dramatic variations in liver concentrations of some trace elements (cobalt, copper, manganese, molybdenum, selenium and zinc) in wild reindeer from Norway, in which Cu varied more than 100 times (min 2.3 and max 289 μg/g dry weight). However, neither low nor high levels could be linked to any obvious clinical signs of deficiency or toxicity. Unlike moose and roe-deer from the same investigation, reindeer had sufficient concentrations of selenium in the liver.

Staaland et al. (1980) observed low levels of sodium (Na) compared to potassium (K) in reindeer saliva on winter ranges, suggesting insufficiency in Na that is a reason for the well-known salt hunger among *Rangifer* and many other herbivore species. Animals are clearly attracted to natural salt licks as well as, for example, roads that are salted for deicing, and Staaland and Hove (2000) point out evidence for a reproductive response to Na supplementation in an interior reindeer herd in Norway that exhibited Na deficiency.

Other than for magnesium and the above examples for copper and sodium, there are few documented cases of clinically significant mineral deficiencies in *Rangifer*. In many cases, even animals that have starved to death have not been reported as showing signs of specific vitamin or mineral deficiency, and, for example, the content of vitamin E and selenium in the liver of reindeer that died of starvation showed satisfactory levels (Josefsen et al. 2007). However, it is uncertain if lack of evidence for deficiencies is because they are generally uncommon, because the symptoms are difficult to detect or because deficiencies are simply not commonly investigated.

3.4.2 STARVATION

Starvation is a relatively customary cause of death in *Rangifer* during severe winters, despite the fact that the species is adapted to large seasonal variation in nutrient intake. Lack of food will initially result in emptying of the gastrointestinal tract, depletion of glycogen reserves in the liver and use of easily accessible protein sources (Robbins 1993, p. 234). Body mass will drop as an effect of decreased gastrointestinal content and loss of water. In a second stage the animal will start to utilize body fat, at the same time reducing basic metabolism to save energy. This stage may continue for quite some time in an animal with plenty of fat reserves, and the decline in body mass will be rather slow. If the animal has no intake of nitrogen it will also have to utilize some body protein for maintenance and glucose synthesis. When most fat reserves have been used, protein will start to be utilized as an energy source. As lean muscle tissue contains only about 25% dry matter and protein contains about half as much energy as fat, the weight loss will be dramatic compared to when fat is used as the energy source. If depletion of protein reserves continues the animal will soon die from emaciation.

Results regarding the depletion of body resources in Svalbard reindeer demonstrated that both reindeer that survived and those that died had lost a substantial part of their total and lean body mass, and almost all their fat reserves during winter (Reimers 1984). However, although surviving reindeer had lost up to 97% of their body fat during winter, they still had a small amount remaining, while those that died had lost everything, including all bone marrow fat – the last fraction to be utilized.

Roffe (1993) investigated the cause of perinatal mortality in caribou calves, basing the diagnosis of emaciation on the appearance of fat tissue (reduced, firm and dark red, with histology indicating lack of lipids in adipose tissue) and on lack of full functional development indicated by a missing, or poorly formed, urethra tube.

Josefsen et al. (2007) investigated 23 completely emaciated reindeer from different free-ranging herds in Northern Norway. The reindeer were found dead in late winter (March to May). All animals had very little lichen or leafy parts of grass in their rumen. Most of the rumen content was either stem parts of grasses, woody plants or mosses and litter. The animals had depleted fat reserves (shown by lack of visible fat in the abdomen), extremely low fat content in bone marrow (0.2–1.0%), and muscle mass also seemed considerably reduced. Eyes were sunken, a sign of dehydration. Common pathological findings were, for example, abomasal lesions (in 19 animals, 68%), a dark liver, excessive hemosiderosis in the liver and spleen and liver lipofuscinosis. The authors concluded that emaciation had arisen because of inadequate quantity or quality of food, although some of the animals had actually been offered supplementary feeds prior to death. In contrast to results by other authors (Westerling 1972, Ågren and Rehbinder 2000), Josefsen et al. (2007) did not observe that diarrhoea was indicative of starvation.

3.5 BODY CONDITION INDICATORS

There is often an interest in assessing the body condition of individual animals or whole populations in *Rangifer* research and management. A range of body condition indicators have been used or suggested for this purpose. Practical and logistical considerations often limit the methods that can be used.

Body mass is a straightforward measure of body condition, but must be related to age, sex and reproductive stage. Body mass is evidently affected by size, which complicates its use as a general indicator of body condition. Nonetheless, body mass may often be the most suitable indicator for measuring change in body condition in individual animals or within a population. For comparing individuals or *Rangifer* populations, a body size variable (e.g. back length or length of a specific bone) is generally needed as covariate in statistical analyses. Alternately, or in addition, a body condition score (1–5) based on that used in animal husbandry has been used to assess fat level (Gerhart et al. 1996b). When using body mass, it is also important to consider the amount of digesta. The rumen content represents a large part of the total body mass (typically 10–20%) and may vary considerably depending on feed intake. Ingesta-free body mass is consequently a better indicator of body reserves than body mass as such.

In Norway, Sweden and Finland, carcass weights of slaughtered animals are recorded in national databases for reindeer husbandry. This data has been used for describing variations and changes in general body condition of reindeer in relation to, for example, management, lichen resources and climatic factors, yet suffers from the same limitations related to animal size as does whole body mass or ingesta-free body mass. Fat and conformation classifications are standardized measures at reindeer slaughter in Sweden and are independent of size, therefore may be used in addition to carcass weight for accessing body condition (Olofsson et al. 2011).

Bone marrow fat, kidney fat and back fat are commonly used indicators of body fat reserves. Fat around the heart could be used in addition (Figure 3.4). Kidney fat, or kidney fat index, is probably the most well-used indicator of body reserves in ungulates. Kidney fat index is calculated by relating the amount of fat to the weight of the kidney. However, Serrano et al. (2008a) point out several weaknesses with using these kinds of indexes or quotas for statistical analyses, and recommend using the measure as such, while including the size measurement (in this case kidney weight) as a covariate in the statistical model. A muscle index – using the weight of a certain muscle in relation to bone length – was suggested by Tyler (1987b), and bone length was used in a multiple regression model to predict pregnancy rate in caribou (Gerhart et al. 1997). Such an index could have statistical limitations as shown for the kidney fat index.

As discussed in relation to starvation, bone marrow fat is a good indicator for assessing degree of emaciation. However, since bone marrow fat is one of the last fat reserves to be mobilized,

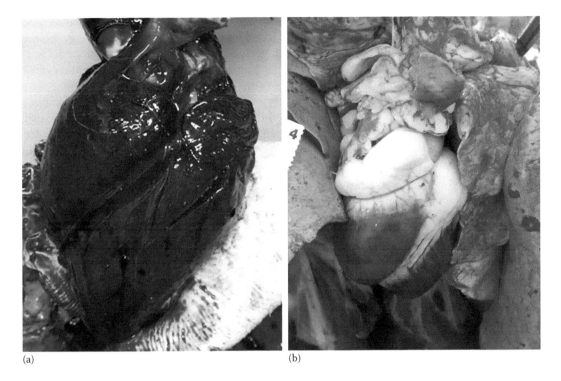

(a)   (b)

FIGURE 3.4   Fat around the heart is a useful indicator of *Rangifer* body condition: (a) Heart of an emaciated reindeer. (b) Heart of a reindeer in good body condition (Photo: Ulrika Rockström).

its concentration is not a very suitable measure of total fat reserves in animals of adequate body condition (Chan-McLeod et al. 1995). Nonetheless, Rehbinder and Nikander (1999, p. 66) suggested use of the following bone marrow fat contents for estimating the nutritional state of animals: >70% "good," 30–70% "less good," 5–30% "poor," 2–5% "emaciated but possible to save," <2% "inanition and the animal has no possibility of surviving." Chan-McLeod et al. (1995) compared different variables for body mass and body composition that could be used for harvested caribou without access to specialized laboratories. They concluded that water content in indicator muscles, together with kidney fat index (left kidney), were the indicators of fat reserves in the caribou body that worked best across all seasons. Total marrow fat was suggested as a preferable alternative to percentage of fat in bone marrow, and back fat proved to be a good index in autumn only, when the animals were relatively fat.

Taillon et al. (2011) used a range of body condition indices (including body mass, a number of length measurements, chest girth, peroneus muscle mass, rump fat, kidney fat and bone marrow fat) for multivariate analyses in order to find a combined measure of body condition. For females, they identified a "body bulkiness index" that separated the heavy, round-bodied females from the light and lean ones. Whereas body condition of newborn calves was best described by body mass, a combination of mass and size measurements worked best for calves at weaning.

It is not only total fat that may change as result of poor nutrition, but also the proportions of individual fatty acids. Soppela and Nieminen (2001) observed a decline in oleic and linoleic acid in bone marrow fat as an effect of under-nutrition in adult female reindeer, while there was a decline only in linoleic acid in reindeer calves.

Thomas and Barry (2005) tested whether variation in antler mass correlated with changes in body mass, body condition and pregnancy rate, but concluded that this measure varied too much and was inferior compared to other body condition indicators.

Blood analyses can be used for measuring the nutritional state of live animals, and Soveri et al. (1992) found that blood serum parameters varied with diet and season in reindeer calves in winter.

Säkkinen et al. (2005) measured blood constituents from free-ranging semi-domesticated reindeer and concluded that plasma total protein and albumin, and to some extent also globulin, may serve as nutritional biomarkers for reindeer, and that they reflect the nutritional composition of the animal's diet. Total serum protein and serum triglycerides have also been found to reflect physical condition of Iberian wild goats (Serrano et al. 2008b).

Sample size often inhibits the assessment of temporal and spatial differences in body condition. In North America, initiatives have been conducted that involve aboriginal hunters providing their own assessment of body condition based on a number of quantifiable and qualitative indicators (Kofinas et al. 2004, Lyver and Gunn 2004). Protocols have been incorporated into monitoring activities in a number of jurisdictions.

3.6 EFFECTS OF NUTRITION ON POPULATION DYNAMICS AND MANAGEMENT

3.6.1 Body Condition, Survival and Reproductive Success

It has been suggested that foraging conditions during summer are the main factor in determining growth rate and body size in *Rangifer*, and that conditions on winter ranges are important for survival, and thus for population density (Klein 1968, Reimers 1997). However, as discussed previously, enough summer forage is needed to accumulate sufficient body reserves to ensure survival during adverse winter periods with, for example, severe snow conditions.

It is frequently assumed that forage shortage, as such, is rarely problematic in summer. High animal density may, however, reduce the possibility for an individual animal to select plants with the highest nutritional quality and may thereby limit growth and the regaining of body reserves. For an individual animal, plant biomass can limit food intake on an instantaneous and daily basis, which is frequently the situation during spring green-up on Arctic ranges. Many plant species are of low palatability, preference or nutrient quality and thus selective feeding among the forage sward typifies the *Rangifer* feeding strategy. Selective grazing not only optimizes digestibility, but also boosts nutrient intake and retention to produce a multiplier effect. Consequently, White (1983) estimated a doubling in the rate of weight gain as result of a 14% increase in forage digestibility. At low or moderate animal densities, the limitations in forage intake and the possibility for selective grazing in summer are usually an effect of disturbances, the most important in many habitats being insect harassment. Detrimental effects of insect harassment have been described in several papers (Fancy 1986, Reimers 1980, Witter et al. 2012). Biting, stinging and parasitic insects are abundant in *Rangifer* habitats, where the most significant are the oestrid (warble and nasal bot) flies that force reindeer and caribou to move and seek windy and elevated areas (e.g. snow patches, if available) with little or no forage.

The rate of insect activity depends on the effective temperature, a combination of temperature and wind (Mörschel 1999), with oestrids requiring higher temperatures for activity than, for example, mosquitos and black flies. Fancy (1986) estimated the energetic loss and subsequent negative effects on weight gain in *Rangifer* caused by warm summer days and the resulting insect harassment. Using data on observed temperature and wind during a single month (July 1981), he calculated that a lactating female was in a negative energy balance during most days within the observed period. Similarly, Reimers (1980) estimated negative effects of increased activity – because of insects – on reindeer productivity. This confirms that insect harassment is probably a key factor explaining temporal and geographical variations in *Rangifer* autumn body condition. It also implies that one of the more severe effects of global warming on *Rangifer* populations might be that of insect harassment.

As already mentioned, body condition at the onset of winter is expected to affect winter survival and be important for the animal's ability to cope with periods of adverse snow conditions. Body reserves are especially crucial for calves during their first winter, and autumn body condition is also crucial for reproduction. Research on wild as well as domesticated *Rangifer* shows significant effects of female autumn body mass and fat reserves on conception rate and reproductive success the following year (Cameron and Ver Hoef 1994, Gerhart et al. 1997, Reimers 1983, Rönnegård et al. 2002). Eloranta and Nieminen (1986) found large effects of female autumn body mass on calving rate and calf survival in semi-domesticated reindeer, especially for young females. Differences in autumn body condition may thus explain a considerable part of the variation in reproductive rate that can be seen among and within reindeer and caribou populations.

Most *Rangifer* females over 1 year of age are fertile and ovulate at rut, and in sufficient nutritional conditions even juveniles may get pregnant (Tyler 1987a). However, raising a calf in the first year of life is expected to negatively affect a female's future reproductive success, since she will probably not be able to invest enough in her own growth. Data from Norwegian wild reindeer showed that over a certain body weight threshold, most females were likely to get pregnant (Reimers 2002). Based on results from the Porcupine caribou herd, all females with a fat reserve of 12 kg or more would be expected to be pregnant. For non-lactating females, the corresponding fat content for a 100% pregnancy rate was around 7 kg (Gerhart et al. 1997). This is also in accordance with findings for captive caribou (Crête et al. 1993).

Rangifer females typically produce one offspring every year. Nevertheless, there are regions where female lifetime reproduction is very high and twins seem to be more common (Cuyler and Ostergaard 2005), at least in some years (Godkin 1986). Poor environments, on the other hand, where lactating females do not always succeed in regaining enough body mass during summer, may be associated with frequent reproductive pauses and alternate year breeding (Cameron 1994). It has also been shown that both the youngest and the oldest females are at higher risk of failing to produce a calf during a certain year compared to those at prime age (5–8 years old) (Rönnegård et al. 2002).

Some females exhibit a prolonged lactation (over the breeding season), which has been shown to be a factor that may reduce the probability of pregnancy at a given body mass (Gerhart et al. 1997). This might be a mechanism to support the survival of a calf when food resources are restricted, at the expense of not calving the following summer.

The timing of calving varies among *Rangifer* population and between years, and could be affected by nutrient supply and female body condition via effects on either conception date or gestation length. Cameron et al. (1993) observed that females that calved before a certain date (June 7) were significantly heavier both in the previous autumn and in the summer after calving. It has been observed that older, heavier females mate earlier than young and light females (Mysterud et al. 2009). There seems, however, not to be a direct effect of female autumn body condition on gestation length, and late breeders are shown to have generally shorter gestations than early breeders (Rowell and Shipka 2009). This actually counteracts some of the effects of female autumn body mass and timing of mating on calving date. On the other hand, poor nutrition during the latter part of pregnancy can extend the length of gestation and thus delay calving date (Cameron et al. 1993, Skogland 1983).

Female nutrition during pregnancy affects foetal growth and early calf survival in *Rangifer*. Loison and Strand (2005) confirm that foetal growth rate is positively correlated with female body mass in wild reindeer. Skogland (1984) observed large differences in calf birth weight, calving date and calf survival depending on animal density and female body condition. Enhanced feed intake and an increased content of protein in the diet (13.7% crude protein compared to a lichen diet with 3.1% crude protein) were observed to have positive effects on foetal growth rate in domesticated reindeer (Rognmo et al. 1983). The same experiment also revealed a strong negative effect of a low protein diet on early calf survival, as five out of 17 calves on this diet died within 2 days of birth. This agrees with results from an experimental herd of semi-domesticated reindeer in Finland, where calves that were stillborn or died during the first day of life weighed 2 kg less at birth, than those that survived until autumn (Eloranta and Nieminen 1986). A similar effect has been deduced for caribou (Bergerud et al. 2008, Roffe 1993).

Energy and protein intake during pregnancy seem not to significantly affect the composition or energy content of the milk (Jacobsen et al. 1981, Luick et al. 1974, Rognmo et al. 1983), while milk volume may be substantially affected (Jacobsen et al. 1981). Calf growth rate during the first weeks of life may vary between 150 and 400 g/d depending on the female's milk production (White and Luick 1984). Delayed effects of female winter nutrition on early calf growth was also observed by Rognmo et al. (1983), who recorded an initial (first 3 weeks) weight gain of 290 g/d for calves of females fed only lichens prior to parturition, compared to 350 g/d for calves of females on an improved diet. However, the effect of winter nutrition disappeared by the time the calves were about 2 months old.

3.6.2   NUTRITION AND POPULATION DYNAMICS

Fluctuations in population size regularly occur in wild as well as in domesticated *Rangifer* populations, and changes in forage availability are often seen as a main driver (Post and Klein 1999, Skogland 1983). High winter mortality, especially in young animals, sometimes occur because of icing and locked pastures (Chan et al. 2005). Even if reindeer and caribou have been shown to survive winter in habitats without lichens, persistent or long-term fluctuations in *Rangifer* population size are often linked to a gradual decline in lichen biomass due to high animal density on the ranges. Lichens grow slowly and overgrazed lichen ranges need to be protected from grazing over several years (maybe decades) in order to recover to optimal productivity. Lichens grow from the top of their branches and also need enough light and humidity to be able to grow. A heavily grazed and trampled lichen mat will have few branches to grow from, therefore having low density and height, and drying out quickly. Reindeer lichens (*Cladina* spp.) in northern Finland were shown to have a maximum annual production at 175 kg dm per ha (10 000 m^2) on stands with about 2600–2800 kg dry matter of living lichen per ha (Kumpula et al. 2000). Substantially higher maximum growth rates, around 35%, were recorded at optimal conditions by Cabrajic et al. (2010).

Dramatic population crashes have been observed in *Rangifer* populations the world over (Tyler 2010), one of the more spectacular being that of the introduced caribou population on St. Matthew Island, Alaska, happening in the winter of 1963–1964, when almost the entire population of 6 000 caribou died, and only about 40 animals survived (Klein 1968). During a 20-year period, after the introduction of 28 caribou on the island in 1944, the population had grown rapidly. Lichens, that had been abundant, were by that time almost eliminated, and during the years prior to the crash body mass and reproduction rate declined substantially.

Winter foraging conditions are important and complex with respect to forage abundance, snow characteristics and the variability of rain-on-snow and icing events. Nevertheless, summer forage and autumn body condition are crucial for determining the capacity of reindeer and caribou to cope with periods of food deficiency in winter. Regain of body reserves in summer decreases with age and is linked to tooth wear (Kojola et al. 1998), making older animals more sensitive to inadequate forage abundance and quality in summer. In winter, tooth wear is enhanced if the lichen availability is low and the animals have to eat alternative forage that is coarser than lichens. Worn teeth will reduce the animal's ability to process and utilize both summer and winter forage, thereby limiting the potential for fat and protein accumulation.

It has been discussed to what extent body condition affects the risk that a reindeer or caribou will be killed by a predator. Top-down (predation) and bottom-up (food resources) mechanisms may, however act, and interact, in diverse ways in different environments and over time (Bastille-Rousseau et al. 2016, Mahoney et al. 2016). Fat reserves (marrow fat) did not affect the risk of being killed by predators among animals older than 1 year in a declining population of mountain caribou (McLellan et al. 2012), while there seems to be a negative correlation between calf body mass and risk of predation (Jenkins and Barten 2005) – although this relationship may differ depending on predator species (Nieminen et al. 2013).

In general, the most important influence of foraging conditions on population demographics is via the effect on recruitment rate. As previously stated, female body condition in autumn and over winter has significant effects on conception rate, foetal survival, timing of parturition, calf birth weight and neonatal survival, while adult survival in herbivores generally shows little variation or density dependence (Gaillard et al. 2000). This is confirmed by findings on wild reindeer in Norway (Skogland 1990), where density-dependent limitation of winter forage did not influence adult survival, although calf survival the following summer was significantly affected. An outstanding difference for arctic barren-ground caribou is the finding that for some herds – after effects of recruitment by younger cohorts are accounted for – population declines can been linked to adult survival, especially as affected by icing events (Griffith et al. 2002).

3.6.3 NUTRITION AND MANAGEMENT OF *RANGIFER*

Controversy exists over whether factors limiting caribou and reindeer populations are related to predation and harvest, or to winter and summer habitat, and therefore if the population size of *Rangifer* should be regulated, rather than the number of predators. These contrasting approaches tend to reflect objectives and management efforts in North America, and particularly Alaska, where predator control represents a major tool used by management (Bergerud et al. 2008, Boertje et al. 1996, Mosnier et al. 2008), despite habitat fragmentation and industrial development having been shown to threaten survival of woodland caribou in Canada (Wittmer et al. 2005). In contrast, in Fennoscandia, long-term concerns over winter habitat loss and habitat fragmentation for both herded and wild reindeer have emphasized bottom-up approaches to management; much of it emphasizing the role of nutritional ecology (Colpaert et al. 1995, Olofsson et al. 2011). Predation is generally regarded to affect harvest rate (Hobbs et al. 2012) rather than population size, except in some exceptional cases (Åhman et al. 2014). Evolving approaches to management use resource selection analysis to assess habitat use (Johnson et al. 2004), and when combined with determination of cumulative effects, this can integrate top-down and bottom-up, as well as behavioural, factors. These may, either separately or in combination, affect recruitment and can be the basis for determining, for example, the influence of climate change (Bastille-Rousseau et al. 2016, Tyler et al. 2008). The future of reindeer and caribou management will require a thorough understanding of the interactions between environment and population dynamics through nutrition, as well as a wide-scale collation of data sets. "A glimpse of the future needs (to include) a different approach to sharing data and information to avoid fragmentation and to take advantage of electronic data such as networking the MERRA climate data sets" (Anne Gunn, 16th North American Caribou Workshop, Thunder Bay, Ontario, Canada, May 2016).

REFERENCES

Aagnes, T.H., A.S. Blix, and S.D. Mathiesen. 1996. Food intake, digestibility and rumen fermentation in reindeer fed baled timothy silage in summer and winter. *J. Agric. Sci.* 127:517–23.

Aagnes, T.H., and S.D. Mathiesen. 1994. Food and snow intake, body mass and rumen function in reindeer fed lichen and subsequently starved for 4 days. *Rangifer* 14:33–8.

Aagnes, T.H., W. Sørmo, and S.D. Mathiesen. 1995. Ruminal microbial digestion in free-living, in captive lichen-fed, and in starved reindeer (*Rangifer tarandus tarandus*) in winter. *Appl. Environ. Microbiol.* 61:583–91.

Ågren, E.O., and C. Rehbinder. 2000. Case report: Malnutrition and undernutrition as cause of mortality in farmed reindeer (*Rangifer tarandus tarandus* L.). *Rangifer* 20:25–30.

Åhman, B., A. Rydberg, and G. Åhman. 1986. Macrominerals in free-ranging Swedish reindeer during winter. *Rangifer* Special Issue No. 1:31–8.

Åhman, B., K. Svensson, and L. Rönnegård. 2014. High female mortality resulting in herd collapse in free-ranging domesticated reindeer (*Rangifer tarandus tarandus*) in Sweden. *PLoS One* 9:e111509.

Albon, S.D., A. Stien, R.J. Irvine et al. 2002. The role of parasites in the dynamics of a reindeer population. *Proc. R. Soc. Lond. B* 269:1625–32.

Annison, E.F., and D.G. Armstrong. 1970. Volatile fatty acid metabolism and energy supply. In *Physiology of Digestion and Metabolism in the Ruminant*, ed. A.T. Phillipson, 422–37. Newcastle upon Tyne: Oriel Press.

Arneberg, P., I. Folstad, and A.J. Karter. 1996. Gastrointestinal nematodes depress food intake in naturally infected reindeer. *Parasitology* 112:213–9.

Ballesteros, M., B.J. Bårdsen, K. Langeland et al. 2012. The effect of warble flies on reindeer fitness: A parasite removal experiment. *J. Zool.* 287:34–40.

Barboza, P.S., D.W. Hartbauer, W.E. Hauer, and J.E. Blake. 2004. Polygynous mating impairs body condition and homeostasis in male reindeer (*Rangifer tarandus tarandus*). *J. Comp. Physiol. B* 174:309–17.

Barboza, P.S., and K.L. Parker. 2006. Body protein stores and isotopic indicators of N balance in female reindeer (*Rangifer tarandus*) during winter. *Physiol. Biochem. Zool.* 79:628–44.

Barboza, P.S., and K.L. Parker. 2008. Allocating protein to reproduction in arctic reindeer and caribou. *Physiol. Biochem. Zool.* 81:835–55.

Barrette, C., and D. Vandal. 1986. Social rank, dominance, antler size and access to food in snow-bound wild woodland caribou. *Behaviour* 97:118–46.

Bastille-Rousseau, G., J.A. Schaefer, K.P. Lewis et al. 2016. Phase-dependent climate-predator interactions explain three decades of variation in neonatal caribou survival. *J. Anim. Ecology* 85:445–56.

Bergerud, A.T. 1972. Food habits of the Newfoundland caribou. *J. Wildl. Manage.* 36:913–23.

Bergerud, A.T., S.N. Luttich, and L. Camps. 2008. *The Return of Caribou to Ungava.* Montreal, Quebec: McGill-Queen's University Press.

Bjune, A.E. 2000. Pollen analysis of faeces as a method of demonstrating seasonal variations in the diet of Svalbard reindeer (*Rangifer tarandus platyrhynchus*). *Polar Res.* 19:183–92.

Black, J.L., and D.A. Griffiths. 1975. Effects of live weight and energy-intake on nitrogen-balance and total N requirement of lambs. *Br. J. Nutr.* 33:399–413.

Boertje, R.D. 1984. Seasonal diets of the Denali Caribou Herd, Alaska. *Arctic* 37:161–5.

Boertje, R.D. 1985. An energy model for adult female caribou of the Denali herd, Alaska. *J. Wildl. Manage.* 38:468–73.

Boertje, R.D., P. Valkenburg, and M.E. McNay. 1996. Increases in moose, caribou, and wolves following wolf control in Alaska. *J. Wildl. Manage.* 60:474–89.

Cabrajic, A.V.J., J. Moen, and K. Palmqvist. 2010. Predicting growth of mat-forming lichens on a landscape scale – comparing models with different complexities. *Ecography* 33:949–60.

Cameron, R.D. 1994. Reproductive pauses by female caribou. *J. Mammal.* 75:10–13.

Cameron, R.D., W.T. Smith, S.G. Fancy, K.L. Gerhart, and R.G. White. 1993. Calving success of female caribou in relation to body weight. *Can. J. Zool.* 71:480–6.

Cameron, R.D., and J.M. Ver Hoef. 1994. Predicting parturition rate of caribou from autumn body-mass. *J. Wildl. Manage.* 58:674–9.

Chan, K.S., A. Mysterud, N.A. Oritsland, T. Severinsen, and N.C. Stenseth. 2005. Continuous and discrete extreme climatic events affecting the dynamics of a high-arctic reindeer population. *Oecologia* 145:556–63.

Chan-McLeod, A.C.A., R.G. White, and D.F. Holleman. 1994. Effects of protein and energy intake, body condition, and season on nutrient partition and milk production in caribou and reindeer. *Can. J. Zool.* 72:938–47.

Chan-McLeod, A.C.A., R.G. White, and D.E. Russell. 1995. Body mass and composition indices for female barren-ground caribou. *J. Wildl. Manage.* 59:278–91.

Chan-McLeod, A.C.A., R.G. White, and D.E. Russell. 1999. Comparative body composition strategies of breeding and nonbreeding female caribou. *Can. J. Zool.* 77:1901–7.

Cheng, K.J., R.P. McCowan, and J.W. Costerton. 1979. Adherent epithelial bacteria in ruminants and their roles in digestive-tract function *Am. J. Clin. Nutr.* 32:139–48.

Colpaert, A., J. Kumpula, and M. Nieminen. 1995. Remote sensing, a tool for reindeer range land management. *Polar Rec.* 31:235–44.

Couturier, S., S.D. Cote, J. Huot, and R.D. Otto. 2009. Body-condition dynamics in a northern ungulate gaining fat in winter. *Can. J. Zool.* 87:367–78.

Crête, M., J. Huot, R. Nault, and R. Patenaude. 1993. Reproduction, growth and body composition of Riviere George caribou in captivity. *Arctic* 46:189–96.

Culberson, C.F. 1969. *Chemical and botanical guide to lichen products.* Chapel Hill: University of North Carolina Press.

Cuyler, L.C., and N.A. Øritsland. 1993. Metabolic strategies for winter survival by Svalbard reindeer. *Can. J. Zool.* 71:1787–92.

Cuyler, C., and J.B. Ostergaard. 2005. Fertility in two West Greenland caribou *Rangifer tarandus* groenlandicus populations during 1996/97: Potential for rapid growth. *Wildl. Biol.* 11:221–7.

Cuyler, C., R.R. White, K. Lewis et al. 2012. Are warbles and bots related to reproductive status in West Greenland caribou? *Rangifer* 32:243–57.

Danell, K., P.M. Utsi, T. Palo, and O. Eriksson. 1994. Food plant selection by reindeer during winter in relation to plant quality. *Ecography* 17:153–8.

De la Fuente, G., K. Skirnisson, and B.A. Dehority. 2006. Rumen ciliate fauna of Icelandic cattle, sheep, goats and reindeer. *Zootaxa* 1377:47–60.

Dehority, B.A. 1975. Rumen ciliate protozoa in Alaskan reindeer and caribou (*Rangifer tarandus* L.). In *Proc 1st Int Reindeer/Caribou Symposium, Fairbanks (1972). Biological Papers of the University of Alaska*, eds. J.R. Luick, P.C. Lent, D.R. Klein, and R.G. White, 241–50. Fairbanks: University of Alaska.

Denman, S.E., Tomkins, N.W., and C.S. McSweeney. 2007. Quantitation and diversity analysis of ruminal methanogenic populations in response to the antimethanogenic compound bromochloromethane. *FEMS Microbiol. Ecol.* 62:313–22.

Dryden, G.M. 2011. Quantitative nutrition of deer: Energy, protein and water. *Anim. Prod. Sci.* 51:292–302.

Eloranta, E., and M. Nieminen. 1986. Calving of the experimental reindeer herd in Kaamanen during 1970–85. *Rangifer* Special Issue No. 1:115–21.

Eriksson, L.-O., M.-L. Källkvist, and T. Mossing. 1981. Seasonal development of circadian and short-term activity in captive reindeer, *Rangifer tarandus* L. *Oecologia* 48:64–70.

Espmark, Y. 1964. Studies in dominance-subordination relationship in a group af semi-domestic reindeer (*Rangifer tarandus* L.). *Anim. Behav.* 12:420–6.

Espmark, Y. 1971. Mother-young relationship and ontogeny of behaviour in reindeer (*Rangifer tarandus* L.). *Z. f. Tierpsycol.* 29:42–81.

Fancy, S.G., 1986. *Daily Energy Budgets of Caribou: A Simulation Approach*. University of Alaska, Fairbanks.

Fancy, S.G., and R.G. White. 1985a. Energy expenditures by caribou while cratering in snow. *J. Wildl. Manage.* 49:987–93.

Fancy, S.G., and R.G. White. 1985b. The incremental cost of activity. In *Bioenergetics of Wild Herbivores*, eds. R.J. Hudson, and R.G. White, 143–159. Boca Raton, Florida: CRC Press, Inc.

Fancy, S.G., and R.G. White. 1987. Energy expenditures for locomotion by barren-ground caribou. *Can. J. Zool.* 65:122–8.

Finstad, G.L., and K. Kielland. 2011. Landscape variation in the diet and productivity of reindeer in Alaska based on stable isotope Analyses. *Arct. Antarct. Alp. Res.* 43:543–54.

Flydal, K., and E. Reimers. 2002. Relationship between calving time and physical condition in three wild reindeer *Rangifer tarandus* populations in southern Norway. *Wildl. Biol.* 8:145–51.

Folkow, L.P., and J.B. Mercer. 1986. Partition of heat loss in resting and exercising winter-and summer-insulated reindeer. *Am. J. Physiol.* 251:R32–R40.

Gaare, E., and T. Skogland. 1975. Wild reindeer food habits and range use at Hardangervidda. In *Fennoscandian Tundra Ecosystems, Part 2. Animals and System Analysis*, ed. F.E. Wielgolaski, 195–205. Berlin, Heidelberg, New York: Springer Verlag.

Gaillard, J.-M., M. Festa-Bianchet, N.G. Yoccoz, A. Loison, and C. Toïgo. 2000. Temporal variation in fitness components and population dynamics of large herbivores. *Annu. Rev. Ecol. Syst.* 31:367–93.

Gerhart, K.L., D.E. Russell, D. Van DeWetering, R.G. White, and R.D. Cameron. 1997. Pregnancy of adult caribou (*Rangifer tarandus*): Evidence for lactational infertility. *J. Zool.* 242:17–30.

Gerhart, K.L., R.G. White, R.D. Cameron, and D.E. Russell. 1996a. Body composition and nutrient reserves of arctic caribou. *Can. J. Zool.* 74:136–46.

Gerhart, K.L., R.G. White, R.D. Cameron, and D.E. Russell. 1996b. Estimating fat content of caribou from body condition scores. *J. Wildl. Manage.* 60:713–8.

Gjøstein, H., Ø. Holand, and R.B. Weladji. 2004. Milk production and composition in reindeer (*Rangifer tarandus*): Effect of lactational stage. *Comp. Biochem. Physiol. A* 137A:649–56.

Glad, T., P. Barboza, R.I. Mackie et al. 2014. Dietary supplementation of unic acid, an antimicrobial compound in lichens, does not affect rumen bacterial diversity or density in reindeer. *Curr. Microbiol.* 68:724–8.

Glad, T., A. Falk, P. Barboza et al. 2009. Fate and effect of usnic acid in lichen on the bacterial population in the reindeer rumen. *Microb. Ecol.* 57:570–1.

Godkin, G.F. 1986. Fertility and twinning in Canadian reindeer. *Rangifer* Special Issue No. 1:145–50.

Gotaas, G., E. Milne, P. Haggarty, and N.J.C. Tyler. 2000. Energy expenditure of free-living reindeer estimated by the doubly labelled water method. *Rangifer* 20:211–9.

Griffith, B., D.C. Douglas, N.E. Walsh et al. 2002. Section 3: The Porcupine Caribou Herd. In *Arctic Refuge Coastal Plain Terrestrial Wildlife Research Summaries*, eds. D.C. Douglas, P.E. Reynolds, and E.B. Rhode, 8–37. Anchorage, Alaska: U.S. Geological Survey, Biological Resources Division.

Gunn, A., and R.J. Irvine. 2003. Subclinical parasitism and ruminant foraging strategies – A review. *Wildl. Soc. Bull.* 31:117–26.

Heggberget, T.M., E. Gaare, and J.P. Ball. 2002. Reindeer (*Rangifer tarandus*) and climate change: Importance of winter forage. *Rangifer* 22:13–32.

Helle, T. 1981. Studies on wild forest reindeer (*Rangifer tarandus fennicus* Lönn.) and semi-domestic reindeer (*Rangifer tatandus tarandus* L.) in Finland. *Acta Universitatis Ouluensis. Series A 107 Biol.* 12:1–283.

Helle, T. 1984. Foraging behaviour of the semi-domestic reindeer (*Rangifer tarandus* L.) in relation to snow in Finnish Lapland. *Rep. Kevo Subarctic Res. Stat.* 19:35–47.

Hobbs, N.T., H. Andrén, J. Persson, M. Aronsson, and G. Chapron. 2012. Native predators reduce harvest of reindeer by Sámi pastoralists. *Ecol. Appl.* 22:1640–54.

Hoff, B., A. Rognmo, G. Havre, and H. Morberg. 1993. Seasonal hypomagnesemia in reindeer on Kautokeino winter pasture in Finnmark county, Norway. *Rangifer* 13:133–6.

Hofmann, R.R. 1989. Evolutionary steps of ecophysiological adaptation and diversification of ruminants: A comparative view of their digestive system. *Oecologia* 78:443–57.

Hughes, J., S.D. Albon, R.J. Irvine, and S. Woodin. 2009. Is there a cost of parasites to caribou? *Parasitology* 136:253–65.

Hyvärinen, H., T. Helle, M. Nieminen, P. Väyrinen, and R. Väyrinen. 1977. The influence of nutrition and seasonal conditions on mineral status in the reindeer. *Can. J. Zool.* 55:648–55.

Ingolfsdottir, K. 2002. Molecules of interest: Usnic acid. *Phytochemistry* 61:729–36.

Jacobsen, E., K. Hove, R.S. Bjarghov, and S. Skjenneberg. 1981. Supplementary feeding of female reindeer on a lichen diet during the last part of pregnancy. Effects on plasma composition, milk production and calf growth. *Acta Agr. Scand.* 31:81–6.

Jenkins, K.J., and N.L. Barten. 2005. Demography and decline of the Mentasta caribou herd in Alaska. *Can. J. Zool.* 83:1174–88.

Johnson, C.J., K.L. Parker, and D.C. Heard. 2001. Foraging across a variable landscape: Behavioral decisions made by woodland caribou at multiple spatial scales. *Oecologia* 127:590–602.

Johnson, C.J., D.R. Seip, and M.S. Boyce. 2004. A quantitative approach to conservation planning: Using resource selection functions to map the distribution of mountain caribou at multiple spatial scales. *J. Appl. Ecol.* 41:238–51.

Johnstone, J., D.E. Russell, and B. Griffith. 2002. Variations in plant forage quality in the range of the Porcupine caribou herd. *Rangifer* 22:83–91.

Josefsen, T.D., T.H. Aagnes, and S.D. Mathiesen. 1997. Influence of diet on the occurrence of intraepithelial microabscesses and foreign bodies in the ruminal mucosa of reindeer calves (*Rangifer tarandus tarandus*). *J. Vet. Med. A* 44:249–57.

Josefsen, T.D., K.K. Sørensen, T. Mørk, S.D. Mathiesen, and K.A. Ryeng. 2007. Fatal inanition in reindeer (*Rangifer tarandus tarandus*): Pathological findings in completely emaciated carcasses. *Acta Vet. Scand.* 49:27.

Klein, D.R. 1968. The introduction, increase, and crash of reindeer on St. Matthew Island. *J. Wildl. Manage.* 32:350–67.

Knott, K.K., P.S. Barboza, and R.T. Bowyer. 2005. Growth in Arctic ungulates: Postnatal development and organ maturation in *Rangifer tarandus* and *Ovibos moschatus*. *J. Mammal.* 86:121–30.

Kofinas, G., P. Lyver, D. Russell et al. 2004. Towards a protocol for community monitoring of caribou body condition. *Rangifer* Special Issue No. 14:43–52.

Kojola, I., T. Helle, E. Hauta, and A. Niva. 1998. Foraging conditions, tooth wear and herbivore body reserves: A study of female reindeer. *Oecologia* 117:26–30.

Kumpula, J., A. Colpaert, and M. Nieminen. 2000. Condition, potential recovery rate, and produictivity of lichen (*Cladonia* spp.) ranges in the Finnish reindeer management area. *Arctic* 53:152–60.

Larter, N.C., and J.A. Nagy. 2004. Seasonal changes in the composition of the diets of Peary caribou and muskoxen on Banks Island. *Polar Res.* 23:131–40.

Lavigueur, L., and C. Barrette. 1992. Suckling, weaning, and growth in captive woodland caribou. *Can. J. Zool.* 70:1753–66.

Lechner, I., P. Barboza, W. Collins et al. 2010. Differential passage of fluids and different-sized particles in fistulated oxen (*Bos primigenius f. taurus*), muskoxen (*Ovibos moschatus*), reindeer (*Rangifer tarandus*) and moose (*Alces alces*): Rumen particle size discrimination is independent from contents stratification. *Comp. Biochem. Physiol. A* 155:211–22.

Loison, A., and O. Strand. 2005. Allometry and variability of resource allocation to reproduction in a wild reindeer population. *Behav. Ecol.* 16:624–33.

Luick, B.R., and R.G. White. 1983. Indirect calorimetric measurements of the caribou calf. *Acta Zool. Fennica* 175:89–90.

Luick, J.R., R.G. White, A.M. Gau, and R. Jenness. 1974. Compositional changes in the milk secreted by grazing reindeer. 1. Gross composition and ash. *J. Dairy Sci.* 57:1325–33.

Lyver, P.O.B., and A. Gunn. 2004. Calibration of hunters' impressions with female caribou body condition indices to predict probability of pregnancy. *Arctic* 57:233–41.

Mahoney, S.P., K.P. Lewis, J.N. Weir et al. 2016. Woodland caribou calf mortality in Newfoundland: Insights into the role of climate, predation and population density over three decades of study. *Popul. Ecol.* 58:91–103.

Maier, J.A.K., and R.G. White. 1998. Timing and synchrony of activity in caribou. *Can. J. Zool.* 76:1999–2009.

Mårell, A., A. Hofgaard, and K. Danell. 2006. Nutrient dynamics of reindeer forage species along snowmelt gradients at different ecological scales. *Basic Appl. Ecol.* 7:13–30.

Mathiesen, S.D. 1999. *Comparative Aspects of Digestion in Reindeer*, Department of Arctic Biology and Institute of Medical Biology, University of Tromsø, Tromsø.

Mathiesen, S.D., R.I. Mackie, A. Aschfalk, E. Ringø, and M.A. Sundset. 2005. Microbial ecology of the digestive tract in reindeer: Seasonal changes. In *Microbial Ecology in Growing Animals*, eds. W.H. Holzapfel, and P.J. Naughton, 75–102. Edingburgh: Elsevier Limited.

Mathiesen, S.D., W. Sørmo, Ø.E. Haga et al. 2000. The oral anatomy of Arctic ruminats: Coping with seasonal changes. *J. Zool.* 251:119–28.

McEwan, E.H., and P.E. Whitehead. 1970. Seasonal changes in the energy and nitrogen intake in reindeer and caribou. *Can. J. Zool.* 48:905–13.

McLellan, M.L., R. Serrouya, B.N. McLellan et al. 2012. Implications of body condition on the unsustainable predation rates of endangered mountain caribou. *Oecologia* 169:853–60.

Moen, R., and J. Pastor. 1998. Simulating antler growth and energy, nitrogen, calcium and phosphorus metabolism in caribou. *Rangifer* Special Issue No. 10:85–97.

Mörschel, F.M. 1999. Use of climatic data to model the presence of oestrid flies in caribou herds. *J. Wildl. Manage.* 63:588–93.

Mosnier, A., D. Boisjoly, R. Courtois, and J.P. Ouellet. 2008. Extensive predator space use can limit the efficacy of a control program. *J. Wildl. Manage.* 72:483–91.

Mysterud, A., K.H. Roed, O. Holand, N.G. Yoccoz, and M. Nieminen. 2009. Age-related gestation length adjustment in a large iteroparous mammal at northern latitude. *J. Anim. Ecol.* 78:1002–6.

Nieminen, M., H. Norberg, and V. Maijala. 2013. Calf mortality of semi-domesticated reindeer (*Rangifer tarandus tarandus*) in the Finnish reindeer-herding area. *Rangifer* Special Issue No. 21:79–90.

Nilssen, K.J., H.K. Johnsen, A. Rognmo, and A.S. Blix. 1984a. Heart rate and energy expenditure in resting and running Svalbard and Norwegian reindeer. *Am. J. Physiol.* 246:R963–R967.

Nilssen, K.J., J.A. Sundsfjord, and A.S. Blix. 1984b. Regulation of metabolic rate in Svalbard and Norwegian reindeer. *Am. J. Physiol.* 247:R837–R841.

Nilsson, A., B. Åhman, M. Murphy, and T. Soveri. 2006a. Rumen function in reindeer (*Rangifer tarandus tarandus*) after sub-maintenance feed intake and subsequent feeding. *Rangifer* 26:73–83.

Nilsson, A., B. Åhman, H. Norberg et al. 2006b. Activity and heart rate in semi-domesticated reindeer during adaptation to emergency feeding. *Physiol. Behav.* 88:116–23.

Nilsson, A., Ö. Danell, M. Murphy, K. Olsson, and B. Åhman. 2000. Health, body condition and blood metabolites in reindeer after sub-maintenance feed intake and subsequent feeding. *Rangifer* 20:187–200.

NRC. 2007. *Nutrient Requirements of Small Ruminants: Sheep, Goats, Cervids, and New World Camelids*. Washington, D.C.: National Academies Press.

Olofsson, A., Ö. Danell, B. Åhman, and P. Forslund. 2011. Carcass records of autumn-slaughtered reindeer as indicator of long-term changes in animal condition. *Rangifer* 31:7–20.

Olsen, M.A., T.H. Aagnes, and S.D. Mathiesen. 1995. Failure of cellulolysis in the rumen of reindeer fed timothy silage. *Rangifer* 15:79–86.

Ophof, A.A., K.W. Oldeboer, and J. Kumpula. 2013. Intake and chemical composition of winter and spring forage plants consumed by semi-domesticated reindeer (*Rangifer tarandus tarandus*) in Northern Finland. *Anim. Feed Sci. Technol.* 185:190–5.

Orpin, C.G., and S.D. Mathiesen. 1990. Microbiology of digestion in Svalbard reindeer (*Rangifer tarandus platyrhyncus*). *Rangifer* Special Issue No. 3:187–99.

Palo, R.T. 1993. Usnic acid, a secondary metabolite of lichens and its effect on *in vitro* digestibility in reindeer. *Rangifer* 13:39–43.

Parker, K.L., and P.S. Barboza. 2013. Hand-rearing wild caribou calves for studies of nutritional ecology. *Zoo Biol.* 32:163–71.

Parker, K.L., P.S. Barboza, and T.R. Stephenson. 2005. Protein conservation in female caribou (*Rangifer tarandus*): Effects of decreasing diet quality during winter. *J. Mammal.* 86:610–22.

Parker, K.L., and C.T. Robbins. 1985. Thermoregulation in ungulates. In *Bioenergetics of Wild Herbivores*, eds. R.J. Hudson, and R.G. White, 161–213. Boca Raton, Florida: CRC Press, Inc.

Parker, K.L., R.G. White, M.P. Gillingham, and D.F. Holleman. 1990. Comparison of energy metabolism in relation to daily activity and milk consumption by caribou and muskox neonatales. *Can. J. Zool.* 68:106–14.

Pelletier, F., J. Mainguy, and S.D. Cote. 2009. Rut-induced hypophagia in male bighorn sheep and mountain goats: Foraging under time budget constraints. *Ethology* 115:141–51.

Post, E., and D.R. Klein. 1999. Caribou calf production and seasonal range quality during a population decline. *J. Wildl. Manage.* 63:335–45.

Rehbinder, C., and S. Nikander. 1999. *Ren och rensjukdomar.* Lund: Studentlitteratur.

Reimers, E. 1980. Activity pattern: The major determinant for growth and fattening in *Rangifer*? In *Proceedings of the 2nd International Reindeer/Caribou Symposium, Røros, Norway, 1979*, eds. E. Reimers, E. Gaare, and S. Skjenneberg, 466–74. Trondheim: Direktoratet for vilt og ferskvannsfisk.

Reimers, E. 1983. Reproduction in wild reindeer in Norway. *Can. J. Zool.* 61:211–17.

Reimers, E. 1984. Body composition and population regulation of Svalbard reindeer. *Rangifer* 4(2):16–21.

Reimers, E. 1997. *Rangifer* population ecology: A Scandinavian perspective. *Rangifer* 17:105–18.

Reimers, E. 2002. Calving time and foetus growth among wild reindeer in Norway. *Rangifer* 22:61–6.

Robbins, C.T. 1993. *Wildlife Feeding and Nutrition.* 2nd ed. San Diego, London: Academic Press.

Roffe, T.J. 1993. Perinatal-nortality in caribou from the Porcupine Herd, Alaska. *J. Wildl. Dis.* 29:295–303.

Rognmo, A., K.A. Markussen, E. Jacobsen, H.J. Grav, and A. Skytte Blix. 1983. Effects of improved nutrition in pregnant reindeer on milk qualiy, calf birth weight, growth, and mortality. *Rangifer* 3(2):10–18.

Roine, K., M. Nieminen, and J. Timisjärvi. 1982. Foetal growth in the reindeer. *Acta Vet. Scand.* 23:107–17.

Rönnegård, L., P. Forslund, and Ö. Danell. 2002. Lifetime patterns in adult female mass, reproduction and offspring mass in semidomestic reindeer (*Rangifer tarandus tarandus*). *Can. J. Zool.* 80:2047–55.

Ropstad, E., O. Johansen, K. Halse, H. Morberg, and E. Dahl. 1997. Plasma magnesium, calcium and inorganic phosphorus in Norwegian semi-domestic female reindeer (*Rangifer tarandus tarandus*) on winter pastures. *Acta Vet. Scand.* 38:299–313.

Roturier, S., and M. Roué. 2009. Of forest, snow and lichen: Sámi reindeer herders' knowledge of winter pastures in northern Sweden. *For. Ecol. Manage.* 258:1960–7.

Rowell, J.E., and M.P. Shipka. 2009. Variation in gestation length among captive reindeer (*Rangifer tarandus tarandus*). *Theriogenology* 72:190–7.

Russell, D.E., 2011. *Energy-Protein Modeling of North Baffin Caribou in Relation to the Mary River Mine Project*, Shadow Lake Environmental Consultants, Whitehorse, Yukon, pp. 1–48.

Russell, D.E., A.M. Martell, and W.A.C. Nixon. 1993. Range ecology of the Porcupine caribou herd in Canada. *Rangifer* Special Issue No 8:168 pp.

Russell, D.E., D. Van de Wetering, R.G. White, and K.L. Gerhart. 1996. Oil and the Porcupine Caribou Herd – Can we quantify the impacts? *Rangifer* Special Issue No. 9:255–7.

Russell, D.E., R.G. White, and C.J. Daniel, 2005. *Energetics of the Porcupine Caribou Herd: A Computer Simulation Model*, Technical Report Series No. 431, Ottawa, Ontario, p. 64.

Sakata, T., and H. Tamate. 1979. Rumen epithelium cell proliferation accelerated by propionate and acetate. *J. Dairy Sci.* 62:49–52.

Säkkinen, H., A. Tverdal, E. Eloranta et al. 2005. Variation of plasma protein parameters in four free-ranging reindeer herds and in captive reindeer under defined feeding conditions. *Comp. Biochem. Physiol. A* 142:503–11.

Scotter, G.W. 1967. The winter diet of barren-ground caribou in northern Canada. *Can. Field.-Nat.* 81:33–9.

Serrano, E., R. Alpizar-jara, N. Morellet, and A.J.M. Hewison. 2008a. A half a century of measuring ungulate body condition using indices: Is it time for a change? *Eur. J. Wildl. Res.* 54:675–80.

Serrano, E., F.J. Gonzalez, J.E. Granados et al. 2008b. The use of total serum proteins and triglyserides for monitoring body condition in the Iberian wild goat (*Capra pyrenaica*). *J. Zoo Wildl. Med.* 39:646–9.

Skjenneberg, S., P. Fjellheim, E. Gaare, and D. Lenvik. 1975. Reindeer with esophageal fistula in range studies: A study of methods. In *Proc 1st Int Reindeer/Caribou Symposium, Fairbanks (1972). Biological Papers of the University of Alaska*, eds. J.R. Luick, P.C. Lent, D.R. Klein, and R.G. White, 528–45. Fairbanks: University of Alaska.

Skjenneberg, S., and L. Slagsvold. 1968. *Reindriften og dens naturgrunnlag.* Oslo/Bergen/Tromsø: Scandinavian University Books - Universitetsforlaget.

Skogland, T. 1983. The effects of density dependent resource limitation on size of wild reindeer. *Oecologia* 60:156–68.

Skogland, T. 1984. The effects of food and maternal conditions on fetal growth and size in wild reindeer. *Rangifer* 4(2):39–46.

Skogland, T. 1990. Density dependence in a fluctuation wild reindeer herd; maternal vs. offspring effects. *Oecologia* 84:442–50.

Soppela, P., U. Heiskari, M. Nieminen et al. 2000. The effects of a prolonged undernutrition on serum lipids and fatty acid composition of reindeer calves during winter and spring. *Acta Physiol. Scand.* 168:337–50.

Soppela, P., and M. Nieminen. 2001. The effect of wintertime undernutrition on the fatty acid composition of leg bone marrow fats in reindeer (*Rangifer tarandus tarandus* L.). *Comp. Biochem. Physiol. B* 128:63–72.

Sørmo, W., Ø.E. Haga, R.G. White, and S.D. Mathiesen. 1997. Comparative aspects of volatile fatty acid production in the rumen and distal fermentation chamber in Svalbard reindeer. *Rangifer* 17:81–95.

Soveri, T., S. Sankari, and M. Nieminen. 1992. Blood chemistry of reindeer calves (*Rangifer tarandus*) during the winter season. *Comp. Biochem. Physiol. A* 102:191–6.

Staaland, H., and K. Hove. 2000. Seasonal changes in sodium metabolism in reindeer (*Rangifer tarandus tarandus*) in an inland area of Norway. *Arct. Antarct. Alp. Res.* 32:286–94.

Staaland, H., E. Jacobsen, and R.G. White. 1979. Comparison of the digestive tract in Svalbard and Norwegian reindeer. *Arct. Alp. Res.* 11:457–66.

Staaland, H., and S. Sæbø. 1993. Forage diversity and nutrient supply of reindeer. *Rangifer* 13:169–77.

Staaland, H., R.G. White, J.R. Luick, and D.F. Holleman. 1980. Dietary influences on sodium and potassium metabolism of reindeer. *Can. J. Zool.* 58:1728–34.

Stien, A., R.J. Irvine, E. Ropstad et al. 2002. The impact of gastrointestinal nematodes on wild reindeer: Experimental and cross-sectional studies. *J. Anim. Ecology* 71:937–45.

Storeheier, P.V., S.D. Mathiesen, N.J.C. Tyler, and M.A. Olsen. 2002. Nutritive value of terricolous lichens for reindeer in winter. *Lichenologist* 34:247–57.

Sundset, M.A., P.S. Barboza, T.K. Green et al. 2010. Microbial degradation of usnic acid in the reindeer rumen. *Naturwissenschaften* 97:273–8.

Sundset, M.A., J.E. Edwards, Y. Cheng et al. 2009a. Rumen microbial diversity in Svalbard reindeer, with particular emphasis on methanogenic archaea. *FEMS Microbiol. Ecol.* 70:553–62.

Sundset, M.A., J.E. Edwards, Y.F. Cheng et al. 2009b. Molecular diversity of the rumen microbiome of Norwegian reindeer on natural summer pasture. *Microb. Ecol.* 57:335–48.

Sundset, M.A., A. Kohn, S.D. Mathiesen, and K.E. Præsteng. 2008. *Eubacterium rangiferina*, a novel usnic acid-resistant bacterium from the reindeer rumen. *Naturwissenschaften* 95:741–9.

Taillon, J., P.S. Barboza, and S.D. Cote. 2013. Nitrogen allocation to offspring and milk production in a capital breeder. *Ecology* 94:1815–27.

Taillon, J., V. Brodeur, M. Festa-Bianchet, and S.D. Cote. 2011. Variation in body condition of migratory caribou at calving and weaning: Which measures should we use? *Ecoscience* 18:295–303.

Thomas, D., and S. Barry. 2005. Antler mass of barren-ground caribou relative to body condition and pregnancy rate. *Arctic* 58:241–6.

Thomas, D.C., E.J. Edmonds, and W.K. Brown. 1996. The diet of woodland caribou populations in west-central Alberta. *Rangifer* Special Issue No. 9:337–42.

Trudell, J., and R.G. White. 1981. The effect of forage structure and availability on food intake, biting rate, bite size and daily eating time of reindeer. *J. Appl. Ecol.* 18:63–81.

Tyler, N.J.C. 1987a. Fertility in female reindeer: The effects of nutrition and growth. *Rangifer* 7(2):37–41.

Tyler, N.J.C., 1987b. *Natural Limitations of the Abundance of the High Arctic Svalbard Reindeer*, University of Cambridge.

Tyler, N.J.C. 2010. Climate, snow, ice, crashes, and declines in populations of reindeer and caribou (*Rangifer tarandus* L.). *Ecol. Monogr.* 80:197–219.

Tyler, N.J.C., and A.S. Blix. 1990. Survival strategies in arctic ungulates. *Rangifer* Special Issue No. 3:211–30.

Tyler, N.J.C., P. Fauchald, O. Johansen, and H.R. Christiansen. 1999. Seasonal inappetence and weight loss in female reindeer in winter. *Ecol. Bull.* 47:105–16.

Tyler, N.J.C., M.C. Forchhammer, and N.A. Oritsland. 2008. Nonlinear effects of climate and density in the dynamics of a fluctuating population of reindeer. *Ecology* 89:1675–86.

Valtonen, M. 1979. Renal responses of reindeer to high and low protein diet and sodium supplement. *J. Sci. Agric. Soc. Finl.* 51:381–419.

van Houtert, M.F.J., and A.R. Sykes. 1996. Implications of nutrition for the ability of ruminants to withstand gastrointestinal nematode infections. *Int. J. Parasit.* 26:1151–67.

van Oort, B.E.H., N.J.C. Tyler, M.P. Gerkema, L. Folkow, and K.-A. Stokkan. 2007. Where clocks are redundant: Weak circadian mechanisms in reindeer living under polar photic conditions. *Naturwissenschaften* 94:183–94.

Vikøren, T., A.B. Kristoffersen, S. Lierhagen, and K. Handeland. 2011. A comparative study of hepatic trace element levels in wild moose, roe deer, and reindeer from Norway. *J. Wildl. Dis.* 47:661–72.

Wallsten, J., 2003. *In Vivo and In Vitro Digestibility of Lichens and Silage for Reindeer*, MSc Thesis, Swedish University of Agricultural Sciences, Uppsala.

Warenberg, K. 1982. *Reindeer Forage Plants in the Early Grazing Season. Growth and Nutritional Content in Relation to Climatic Conditions*. Uppsala: Svenska växtgeografiska sällskapet.

Westerling, B. 1970. Rumen ciliate fauna of semi-domestic reindeer (*Rangifer tarandus* L.) in Finland: Composition, volume and some seasonal variations. *Acta Zool. Fennica* 127:1–76.

Westerling, B. 1972. Överutfodring som orsak till svältdöd bland renkalvar [Overfeeding as a starvation cause amongst reindeer calves]. *Finsk Veterinärtidskrift.* 78:131–8.

Westerling, B. 1975. Effects of changes in diet on the reindeer mucosa. In *Proc 1st Int Reindeer/Caribou Symposium, Fairbanks (1972). Biological Papers of the University of Alaska*, eds. J.R. Luick, P.C. Lent, D.R. Klein, and R.G. White, 278–83. Fairbanks: University of Alaska.

White, R.G. 1983. Foraging patterns and their multiplier effects on productivity of northern ungulates. *Oikos* 40:377–84.

White, R.G., F.L. Bunnell, E. Gaare, T. Skogland, and B. Hubert. 1981. Ungulates on arctic ranges. In *Tundra Ecosystems: A Comparative Analysis. Int. Biol. Progr.*, eds. L.C. Bliss, O.W. Heal, and J.J. Moore. 397–483. Cambridge, New York, Melbourne: Cambridge University Press.

White, R.G., and A.M. Gau. 1975. Volatile fatty acid (VFA) production in the rumen and cecum of reindeer. In *Proc 1st Int Reindeer/Caribou Symposium, Fairbanks (1972). Biological Papers of the University of Alaska*, eds. J.R. Luick, P.C. Lent, D.R. Klein, and R.G. White, 284–9. Fairbanks: University of Alaska.

White, R.G., and J.R. Luick. 1984. Plasticity and constraints in the lactational strategy of reindeer and caribou. *Symp. Zool. Soc. Lond.* 51:216–32.

White, R.G., D.E. Russell, and C.J. Daniel. 2013. Modeling energy and protein reserves in support of gestation and lactation: Glucose as a limiting metabolite in caribou and reindeer. *Rangifer* Special Issue No. 21:167–72.

White, R.G., D.E. Russell, and C.J. Daniel. 2014. Simulation of maintenance, growth and reproduction of caribou and reindeer as influenced by ecological aspects of nutrition, climate change and industrial development using an energy-protein model. *Rangifer* Special Issue No. 22:1–126.

White, R.G., B.R. Thomson, T. Skogland et al. 1975. Ecology of caribou at Prudhoe Bay, Alaska. In *Ecological Investigations of the Tundra Biome in the Prudhoe Bay Region, Alaska*, ed. J. Brown, 151–87. Fairbanks: University of Alaska.

White, R.G., and J. Trudell. 1980. Habitat preference and forage consumption by reindeer and caribou near Atkasook, Alaska. *Arct. Alp. Res.* 12:511–29.

Witter, L.A., C.J. Johnson, B. Croft, A. Gunn, and M.P. Gillingham. 2012. Behavioural trade-offs in response to external stimuli: Time allocation of an Arctic ungulate during varying intensities of harassment by parasitic flies. *J. Anim. Ecology* 81:284–95.

Wittmer, H.U., A.R.E. Sinclair, and B.N. McLellan. 2005. The role of predation in the decline and extirpation of woodland caribou. *Oecologia* 144:257–67.

4 Feeding and Associated Health Problems

Birgitta Åhman, Gregory L. Finstad and Terje D. Josefsen

CONTENTS

The extent to which reindeer are fed varies across different management systems in the world. This chapter focuses on common systems of supplementary feeding of semi-domesticated reindeer in traditional (mainly Sami) reindeer husbandry in Fennoscandia, and feeding of semi-domesticated reindeer introduced to North America. These systems include supplementary feeding of free-ranging reindeer as well as temporary feeding of fenced reindeer within free-range management systems. Feeding in situations when reindeer or caribou are temporarily or permanently kept in captivity for research or other purposes are also discussed. Various commonly observed health problems and diseases related to feeding are described in the second part of the chapter.

4.1   FEEDING STRATEGIES

Birgitta Åhman and Greg Finstad

4.1.1   FEEDING AS PART OF REINDEER HUSBANDRY IN FENNOSCANDIA

Within the reindeer herding industry in Norway, Sweden and Finland, semi-domesticated reindeer generally range freely, feed on natural forage and migrate between summer and winter ranges in basically the same way as wild herds. The use of supplementary feeding was previously (50–100 years ago) restricted mainly to reindeer used for transportation. Winter-feeding has, however, become increasingly common during the last decades (Helle and Jaakkola 2008, Staaland and Sletten 1991, Turunen and Vuojala-Magga 2014), though artificial feeds are still only a minor part of the nutrient intake for most reindeer in this part of the world.

The use of supplementary feeding in reindeer herding has increased for several reasons. A major reason is the loss of land to other human activities. In Sweden and Finland, there has been a great loss of winter pasture due to modern forestry, with the loss of lichen-rich forests in Sweden during the last 60 years estimated to be over 70% (Sandström 2015). Hydropower development has destroyed important migration routes, and roads and infrastructure have created barriers and fragmentation of the landscape (Kivinen et al. 2012). National borders and state regulations have also limited the available land. All this has made it more difficult for the reindeer (and herders) to find access to suitable land in different seasons and in various weather conditions. It has also created a situation where some areas are overused or used at non-optimal times of the year. This is especially serious for lichen pastures, where the biomass production dramatically decreases as lichen height declines (Cabrajic et al. 2010). Moreover, lichens are sensitive to trampling, and therefore lichen pastures may easily deteriorate if they are used during the snow-free season.

Severe winters with icing over large areas have always occurred intermittently, irrespective of access to land, and lead to starvation and high reindeer mortality. There has thus been the motivation, yet not always the possibility, to support reindeer nutrition. However, the general interest in supplementary feeding arose as this need became more apparent. Development of new feedstuffs suitable for reindeer started during the 1960s, and knowledge about how to feed reindeer grew. Availability of better feedstuffs and the gradual increase of competence among herders has contributed to the intensification of feeding.

Feeding during autumn and spring migration, and during reindeer gathering events, is now a commonly used practice in many parts of Fennoscandia. Reindeer may be fed at stopovers during the night in order to keep the herd together, but are also fed if they have to be fenced for days in connection with, for example, separation into sub-herds or selection for slaughter.

Feeding is a widespread management action when natural forage is unavailable due to hard snow and ice and there is therefore a risk for acute starvation (this is commonly referred to as "emergency feeding"). In some regions however, winter-feeding has become a regular routine. This is the case in the southern part of the Finnish reindeer herding area, where many herders keep pregnant females fenced in and fed in late winter and spring until the calves are born (Turunen and Vuojala-Magga 2014). Feeding of reindeer prior to slaughter is used in areas of Sweden and Norway for ranges contaminated by radioactive fallout after the nuclear power plant accident in Chernobyl in 1986 (Åhman 1998). Radioactive cesium (the significant part of the fallout with respect to contamination of food) is excreted from the body relatively fast if the animals are provided non-contaminated feed. One to 2 months of "clean feeding" is usually sufficient to remove the major part of the radioactive cesium from the reindeer body. A few herders also feed reindeer prior to slaughter in order to be able to deliver high-quality reindeer meat to the slaughter companies at times when free-ranging reindeer are normally in reduced body condition and therefore not suitable for slaughtering (e.g. in late winter). However, reindeer meat is generally marketed as a "product of nature," and feeding the animals compromises this concept. Furthermore, feeding grain-based feeds changes the fatty

FIGURE 4.1   Reindeer winter-feeding in Sweden (Photo: Birgitta Åhman).

acid composition of the meat, affecting its taste and making it less healthy compared to meat from reindeer foraging on natural pastures (Wiklund et al. 2001, Wiklund et al. 2003).

In Fennoscandia, reindeer are either fed on the natural ranges ("field feeding") or taken into enclosures for feeding ("pen feeding," Figure 4.1). In field feeding, reindeer mostly have access to some natural forage, while in pen feeding most, or all, of their feed is provided by the herders. Either system can be used for winter-feeding, and practical considerations determine which system is preferable. Reindeer that are fed prior to slaughter to improve their body condition or get rid of radioactive contamination, are usually fed in pens. Finnish publications reviewed by Turunen and Vuojala-Magga (2014) show that both field feeding and pen feeding have been practiced in several reindeer herding districts in Finland since the 1970s.

As mentioned, pen feeding of females prior to calving is a relatively common practice in parts of Finland. The pregnant female is usually kept fenced in only until the calf is born. The calf is marked directly after birth (while it can still be captured by hand), after which the female is let out for free grazing on natural pastures along with her calf. In Sweden and Norway, pen feeding of females during calving has only occasionally been practiced; the reason is usually to protect the calves from predation. In this situation, the females and their newborn calves need to be kept within fences until calving is completed (beginning of June) and the calves have become mobile enough that they are no longer as vulnerable to predation. Although this practice is uncommon in Sweden, there are documented cases when it has led to increased calf mortality due to the spread of infectious diseases (discussed later in this chapter).

4.1.2   Feeding Reindeer in North America

Eurasian reindeer were introduced to Alaska in 1892, and subsequently to Canada in 1929, as a foundation for social management of Alaska's Native populations and red meat production. Reindeer were originally raised using a free-range system where there was little or no use of supplements or milled rations.

During the 1980s, reindeer were being moved off the tundra into intensively managed operations across the continental U.S. states in order to support animal leasing, commercial viewing and live sale operations. Also during the 1980s, Alaska began clearing and developing large tracts of agricultural land for the production of forage and cereal grain products, primarily barley. Reindeer producers on the Seward Peninsula were having difficulties delivering a government inspected and approved meat product to markets because of limited accessibility to slaughter and processing facilities, and lack of well-developed transportation and shipping systems. Increasing losses of reindeer to the Western Arctic Caribou Herd, combined with a high demand for reindeer meat, added to the problem. This led some reindeer producers with large herds of reindeer on the Seward Peninsula to consider translocating and raising reindeer along the road system, behind fences, using various combinations of locally produced cereal grains and forage. Because Alaska imports most of its food from the continental U.S., there continues to be a strong interest and push for locally produced reindeer meat, regardless of management or feeding regime.

The feeding of reindeer can occur under a number of different scenarios. Supplementary feeding aims to make better use of available feed by supplying those nutrients in which the pasture is deficient. Free-ranging reindeer can be given a supplementary feed, concentrate or forage to provide a temporary boost in nutrition that offsets inadequate winter forage (lichen) or severe weather events. As stated previously, this strategy is becoming more common in Europe, and has just recently been tested in North America, as most reindeer ranges still provide enough year-round nutrition.

Many Seward Peninsula reindeer herders were losing their herds due to changing migratory patterns of the Western Arctic Caribou Herd during the late 1990s and early 2000s. Some reindeer herders wanted to re-establish their herds; however, the possibility of caribou returning to their rangelands called for herders to assume increased control of their animals. Use of enclosures to temporarily hold reindeer captive was suggested as a way to restrict comingling and loss to the caribou herds. This was also tested with good results by a herder from Koyuk in 2006 (Finstad 2007). The reindeer quickly socialized to humans and habituated to intensive feeding and husbandry practices, while the enclosure acted as a refuge for the reindeer. When threatened by predators or disturbed, the reindeer attempted to return and re-enter the enclosure. Placing free-ranging animals under a temporary captive feeding regime changed the human/animal relationship and allowed herders the opportunity to adapt their reindeer to environmental disturbances like migrating caribou or icing events.

There are hundreds of small reindeer operations scattered across the United States, holding groups of three to four to 100 animals. Fenced reindeer can also be found in other U.S. states, as far south as Arizona and Tennessee. There are currently (2016) ten to 12 fenced populations in Alaska, including a research herd at the University of Alaska Fairbanks. The reindeer raised in these operations are completely dependent upon a milled ration, although some operations may have small supplemental pastures. Reindeer rations are produced using locally available feed components. A wide variety of ration configurations is therefore being used across the U.S.

The temporary capture and holding of wild caribou has also been employed as a management strategy to recover small, nearly extirpated herds. The first experiment was conducted on the Chisana herd in the Yukon, a herd that was in danger of disappearing due to low calf survival as a result of predation. Cows were captured in spring, a couple of months prior to calving, and kept behind large enclosures to eliminate predation of their newborn calves. Cow-calf pairs were held for nearly a month after parturition before being released. Supplemental feed, comprising a pelleted ration and lichens, was provided throughout the captive period (Chisana Caribou Recovery Team 2010).

4.1.3   COMMONLY USED FEEDSTUFFS FOR REINDEER

Before commercial feeds for reindeer were developed in Fennoscandia, the most common way to support semi-domesticated reindeer in adverse winter situations was to cut down branches or whole trees, providing arboreal lichens that the reindeer could feed on. This practice is still used in areas with old forest, where arboreal lichens are readily available. Valuable reindeer (e.g. those used for transport) were offered fodder that had been gathered and stored for the winter by the herders. This could be reindeer lichens (*Cladonia* spp.), arboreal lichens, leaves (mainly *Salix*), hay made from naturally growing grasses, sedges or horsetail (*Equisetum* spp.).

Even if pellets and grass silage are currently the main feedstuffs offered to reindeer, it is generally considered desirable to provide some lichens (*Cladonia* spp.) when reindeer shift from natural forage to a cereal-based feed. Lichens can be mixed with other feedstuffs to make them more palatable. Also, if animals show symptoms of digestive disturbances, the best cure is usually to feed them lichens for some time.

4.1.3.1 Hay and Grass Silage

In Finland, supplementary feeding with hay started in the late 1960s. It soon developed into a normal practice in many regions, especially in the south (Helle and Jaakkola 2008, Turunen and Vuojala-Magga 2014). Common (dry) hay is still used to feed reindeer but is not recommended in large quantities since it generally contains more fiber than the digestive system can handle. Therefore, hay has gradually been replaced by grass silage in combination with grain-based rations.

Grass for making silage is cut and dried, usually to contain up to 40–50% dry matter (DM), then wrapped, airtight, in plastic. Initially, the pH of forage is approximately 6.0. The grass will undergo anaerobic fermentation, producing lactic acid, which will drive pH as low as 4.2 (higher in dry silages). When the forage drops to this pH, the acidic environment will halt the fermentation process. This low pH also prevents the growth of harmful bacteria and fungi (mold) and thereby preserves the quality of the forage.

Grass silage was introduced as winter forage for reindeer in Fennoscandia during the 1990s. Silage is also referred to as haylage (either generally, when made into round bales, or, more specifically, for relatively dry types of silage of >50% DM). This conservation technique allows for the harvesting and preserving of forages when the quality is at its peak, thereby maintaining a high percent of total digestible nutrients that will more closely meet the unique dietary requirements of *Rangifer*. The plants are harvested at a relatively early vegetation stage, when they contain more protein and less fiber, and a smaller fraction of the nutrients are lost in the process compared to when it is dried to make hay. Drawbacks compared to hay are that the commonly produced bales (Figure 4.2) are heavy (about 200 kg DM, plus, usually, at least as much water), and if too wet, the bales will freeze in cold weather.

Timothy (*Phleum pratense*) is commonly used for making hay and silage for reindeer in both Fennoscandia and the U.S., with good results (Moen et al. 1998). In North America, slender wheatgrass (*Elymus trachycaulus*) has also been used and was found to be very palatable for reindeer. Grass harvested for feeding reindeer should be leafy, and the forage could also contain some herbs (e.g. *Trifolium*, *Taraxacum* and *Alchemilla* spp.). The silage should preferably contain at least 45%

FIGURE 4.2   Semi-domesticated reindeer bull feeding on grass silage (Photo: Birgitta Åhman).

DM, both because a small water content reduces the risk of growth of unwanted microorganisms (bacteria and mold), and because it makes the bales lighter and less difficult to handle.

4.1.3.2 Reindeer Pellets Used in Fennoscandia

Feeds for reindeer resembling those used for cattle and sheep began to be developed in Fennoscandia during the 1960s. The feeds mainly contained different cereal grains like barley, wheat and oats, as well as their bran products. Sugar-beet pulp and molasses were usually added. Present-day commercial reindeer feeds contain the same main ingredients, although feed composition and structure have been further developed. They are mainly intended for winter-feeding, and are composed to provide the nutrients reindeer need during this time of the year (Table 4.1). All commercial feeds are thus supplemented with vitamins and minerals (the same premix rations that are used for other ruminant livestock). Some feeds also contain additives, with the purpose of alleviating digestive problems. The technique used to make pellets from milled rations (Figure 4.3) resolves a previous problem with dusty feeds and feed particles that sometimes cause irritation or small wounds in, for example, the mouth or eyes, which create a gateway for bacterial infections (Aschfalk et al. 2003a, Rehbinder and Nilsson 1995).

There are several companies in Fennoscandia that produce reindeer pellets. The production is highest in Finland, where around 18 million kg are produced annually (Nieminen 2010). This corresponds to about 100 kg per reindeer per year in Finland. In Sweden, the production is lower (corresponding to 20–40 kg per reindeer per year), and even less is produced in Norway.

Data from two Swedish feed manufacturers shows that, in Sweden, feeding of reindeer was not very widespread in the beginning of the 1980s. In 1986, the annual production of reindeer pellets corresponded to around 1.5 kg per reindeer in the winter herd. Due to the need for so-called "clean feeding" after radioactive fallout from the Chernobyl accident of 1986, the commercial production of reindeer feed in Sweden rapidly grew. The need for clean feeding has now ceased; however, the production of reindeer pellets in Sweden has continued to grow. One probable reason is that by being forced to feed, herders became practiced in feeding and consequently continued to use supplementary feeding in regular reindeer management. Other reasons may include the increasing loss of land to competing human activities (mentioned on p. 136), as well as milder winters with more icing events. Nevertheless, as the market for reindeer pellets increased, producers also became more experienced and interested in improving feeds. Thus, present-day reindeer pellets are certainly better than those available in the 1980s.

4.1.3.3 Reindeer Feeds in North America

Similar to other feeds used for domestic animals, the main energy components of reindeer pellets used in North America are grains, which are fed as part of a total mixed ration. Cereal grains, primarily barley and oats, are used most commonly, although corn has been incorporated in some reindeer rations. The nutritional composition of different locally produced feed rations for reindeer is highly variable with regard to crude protein and carbohydrate fractions (Table 4.2), as well as mineral content (Table 4.3). This may be due to the choice of ingredients, as further described below (information for the individual rations is, however, not available). Some feeds may also be designed for special purposes and adapted to nutritional needs, depending on season or category of reindeer (e.g. lactating females or growing calves in summer).

Barley is the most used grain in U.S. reindeer diets because it can be grown in cooler climates, is relatively inexpensive and has a higher crude protein content than corn and wheat. The energy content of barley is slightly lower than the energy content of corn and wheat, partially due to its higher fiber content (neutral detergent fiber, NDF, and acid detergent fiber, ADF). Barley has a relatively high starch content, which may lead to ruminal acidosis and bloat if fed too abruptly or if proportions in the ration are too high. Oats are generally lower in energy and have more fiber content than other grains, and therefore are often used to balance energy and fiber levels, in conjunction with barley. Cereal grains are also high in phosphorous, so calcium must be added to the ration to balance

TABLE 4.1

The Nutritional Content per kg DM in Some Commercially Available Swedish and Finnish Reindeer Pellets According to Manufacturers' Declaration (Obtained in May 2016)

	CP	Dig CP	Crude Fat	Crude Fiber	NDF	Starch	Energy	Ca	P	K	Mg	Na	Cu	Se	Vit A	Vit D_3	Vit E
	%	%	%	%	%	%	MJ/kg	%	%	%	%	%	mg/kg	mg/kg	IU/kg	IU/kg	mg/kg
Renfoder standard	12.6	9.2	4.0	16.1	–	–	11.7	0.8	0.57	–	0.57	–	13.8	0.8	12069	4023	80
Renfoder start	10.9	8.0	4.0	16.7	–	–	10.3	0.8	0.57	–	0.57	–	13.8	0.8	12069	4023	80
Renfoder stand. P	12.1	–	5.7	16.0	–	–	–	–	–	–	0.46	0.23	8.0	0.5	7586	2529	51
Renfor bas	11.3	8.0	3.5	–	44	17	11.4	1.0	0.40	0.80	0.25	–	15.0	0.6	6000	–	60
Renfor nära	11.3	8.0	3.5	–	44	15	11.2	1.0	0.40	0.80	0.25	–	15.0	0.6	6000	–	60
Tähtiporo-Basic	12.4	–	4.3	13.0	–	–	12.2	1.2	0.47	–	0.35	0.35	15.1	0.4	7442	2326	44
Tähtiporo-Arctic	13.1	–	4.5	11.9	–	–	13.1	1.3	0.47	–	0.35	0.35	15.1	0.5	7442	2326	44
Tähtiporo-Balans	13.7	–	3.4	10.1	–	–	13.6	1.2	0.58	–	0.35	0.35	26.7	0.4	6977	2326	47

FIGURE 4.3   Local production of reindeer pellets in Finland. The pellets contain grains, grass meal and beet pulp (Photo: Birgitta Åhman).

TABLE 4.2

The Content (% in DM) of Crude Protein (CP) and Carbohydrate Fractions in Reindeer Rations Produced in Alaska and Other Parts of North America and Analyzed at the University of Alaska Fairbanks (Codes Are Those Used by UAF to Keep the Producer Confidential)

	CP	NDF	ADF	Hemicellulouse	Cellulouse	Lignin
	%	%	%	%	%	%
RRPS05	16.1	25.1	10.0	15.1	–	–
RRPW12	13.6	40.0	12.3	27.8	9.0	3.3
RRPS12	15.2	38.2	12.3	26.0	9.1	3.2
RRPS16	18.0	23.6	10.4	13.2	–	–
ROBA-1	16.3	45.1	25.9	19.2	21.7	4.1
ROBA-2	17.1	35.2	14.8	20.3	12.6	2.2
ROBA-3	13.1	23.7	9.1	14.7	7.5	1.6
ROBA4W	14.8	30.8	15.1	15.7	11.7	3.4
ROBA4S	17.1	29.8	14.0	15.8	9.2	4.8
ROBA-5	15.3	46.2	28.4	17.8	19.3	7.0
ROBA-6	9.7	27.0	10.8	16.2	8.6	2.1
ROBA-7	17.6	34.0	14.4	19.6	12.0	2.4
ROBA-8	15.2	33.3	13.1	20.3	9.2	3.9
ROBA-9	9.4	51.2	26.4	24.8	19.7	4.5
ROBA-11	15.1	30.4	12.1	18.4	9.3	2.8
ROBA-12	11.8	35.2	13.0	22.2	9.6	3.3
ROBA-13	20.4	23.0	6.3	16.7	4.4	2.0
ROBA-14	9.8	42.5	26.3	16.2	21.1	5.1
ROBA15	19.4	29.2	16.2	13.0	–	–
MIX-BHR	18.5	23.4	12.8	10.6	–	1.7
MIX-JBS	17.6	21.8	10.5	11.3	–	1.8
MIX-KNP	16.6	25.6	10.4	15.2	–	1.6
Median	15.7	30.6	12.9	16.45	9.45	3.2
Min	9.4	21.8	6.3	10.6	4.4	1.6
Max	20.4	51.2	28.4	27.8	21.7	7.0

TABLE 4.3

Mineral Content (g/kg or mg/kg DM) in Reindeer Rations Produced in Alaska and Other Parts of North America and Analyzed at the University of Alaska Fairbanks (Codes Are Those Used by UAF to Keep the Producer Confidential)

	Ca	P	K	Mg	S	Na	Cu	Zn	Mn	Fe	Co	Mo
	%	%	%	%	%	%	mg/kg	mg/kg	mg/kg	mg/kg	mg/kg	mg/kg
RRPS05	0.6	0.4	0.8	0.2	0.2	0.10	13	129	115	184	0.7	0.9
RRPW12	0.8	0.5	0.6	0.2	0.2	0.25	17	139	103	329	0.3	1.1
RRPS12	0.8	0.4	0.8	0.2	0.2	0.24	15	122	119	144	0.7	1.3
RRPS16	1.1	0.6	1.0	0.2	–	0.40	22	188	115	228	–	1.6
ROBA-1	0.9	0.3	1.1	0.3	0.5	0.42	65	200	167	466	0.9	0.6
ROBA-2	1.9	0.7	1.2	0.3	0.5	0.22	60	329	211	589	5.5	1.7
ROBA-3	0.4	0.3	0.8	0.2	0.4	0.12	27	122	108	154	0.7	0.3
ROBA4W	0.8	0.5	1.0	0.2	0.2	0.25	21	176	109	531	0.8	1.2
ROBA4S	1.0	0.6	1.1	0.2	0.3	0.19	22	165	111	441	0.5	1.5
ROBA-5	1.4	0.5	0.8	0.4	0.2	0.14	22	110	141	725	1.1	1.1
ROBA-6	0.7	0.4	0.4	0.2	0.1	0.06	20	124	84	477	1.0	1.0
ROBA-7	1.1	0.6	0.9	0.3	0.3	0.16	26	162	125	257	3.2	2.9
ROBA-8	1.4	0.6	1.0	0.4	0.3	0.30	26	144	107	131	1.6	1.2
ROBA-9	0.9	0.1	0.7	0.2	0.3	0.13	6	23	58	1012	0.4	0.0
ROBA-11	0.8	0.5	0.8	0.2	0.2	0.17	19	128	100	485	0.8	0.7
ROBA-12	0.3	0.3	0.7	0.2	0.2	0.04	9	52	40	119	0.2	0.6
ROBA-13	1.1	0.8	0.8	0.3	0.2	0.004	164	176	208	296	0.4	1.4
ROBA-14	1.4	0.3	0.5	0.3	0.3	0.10	43	112	113	573	4.2	0.3
ROBA15	1.4	0.7	1.3	0.3	–	0.40	59	225	316	209	3.7	1.5
MIX-BHR	1.2	0.6	1.4	0.3	0.3	0.49	53	163	124	293	1.0	1.3
MIX-JBS	1.0	0.4	1.2	0.2	0.2	0.37	9	41	48	328	0.2	1.8
MIX-KNP	1.4	0.9	1.2	0.4	0.2	0.43	53	191	159	245	1.6	1.2
Median	1.0	0.5	0.9	0.2	0.2	0.20	22	141	114	312	0.8	1.2
Min	0.3	0.1	0.4	0.2	0.1	0.004	6	23	40	119	0.2	0.0
Max	1.9	0.9	1.4	0.4	0.5	0.49	164	329	316	1012	5.5	2.9

calcium-phosphorus (Ca:P) ratios. Molasses is a common ingredient added to promote appetite and is a readily available energy supply for rumen microbes. It is often added as a binder as well as a dust and particle suppressor.

Canola, or corn oil, is used as an energy source as well as to promote absorption of fat-soluble vitamins, and to support lactation. Care must be taken in developing a reindeer ration, since the concentration of fat in a reindeer ration should not exceed 4%. This is because unsaturated fatty acids, such as those found in canola oil, are relatively toxic to rumen microbes, particularly forage-fiber-digesting species. This doesn't mean that canola oil and other unsaturated vegetable oils should not be used in a reindeer ration, as most rumen microbes have the ability to detoxify and reduce the toxic effects of unsaturated fats through "bio-hydrogenation" (hydrogen is added to the unsaturated fats, which turns them into rumen-protected saturated fats). However, concentrations of unsaturated fats and oil exceeding 4% can overwhelm this process and, as a result, can interfere with rumen fermentation.

Chopped hay (fiber length ~2 cm) and hulls from oats, rice, cottonseed and soybeans are often used to add fiber to the ration and therefore slow down the rate of fermentation. Reindeer readily consume beet pulp and this is often used as a ration component, as a top dressing or given *ad libitum* on the side. Typically, soybean meal is used as a ration component to adjust crude protein concentrations, although it has been discovered that whitefish meal used as a protein source in a reindeer ration significantly increases growth efficiency over that of soybean meal (Finstad et al. 2007). Similar to the Fennoscandian feeds, vitamin and mineral premixes are often added to a ration to ensure that the nutritional needs of reindeer are met.

4.1.4 MINERAL SUPPLEMENTATION

As mentioned, commercial feeds for *Rangifer* are supplemented with minerals and vitamins, and premixes designed for other ruminants are also often used in these feeds. Minerals can also be provided in the form of mineral stones (Figure 4.4). The following guidelines are generally based on NRC (2007) recommendations for *Rangifer* or cervids in general, though recommendations for sheep or goats are used when recommendations for cervids are not available. Mineral requirements are also discussed in Chapter 3, *Rangifer* Diet and Nutritional Needs.

The requirements for calcium (Ca) and phosphorus (P) are usually expressed together, and proportions of 0.8–1.2% calcium and 0.6–0.9% potassium, on a dry matter (DM) basis, are used in Alaskan rations and support good skeletal and antler growth. Contents of 0.8–1.3% Ca and 0.4–0.6% P in feed DM are found in Fennoscandian reindeer feeds (Table 4.1). Di-calcium phosphate can be added to a ration to adjust concentrations of Ca and P. Growth, lactation and antler formation increase the requirements substantially in summer; however, the above concentrations of Ca and P in reindeer feeds are still expected to be sufficient for the summer needs.

A daily sodium (Na) intake of 9.0 mg/kg BW is generally recommended for cervids, but is probably too high for *Rangifer* (NRC 2007, p. 121). The content of Na in most *Rangifer* feeds is 0.2–0.4% of DM (0.5–1.0% NaCl). Salt (NaCl) can also be used as an attractant and as an enticement (treat).

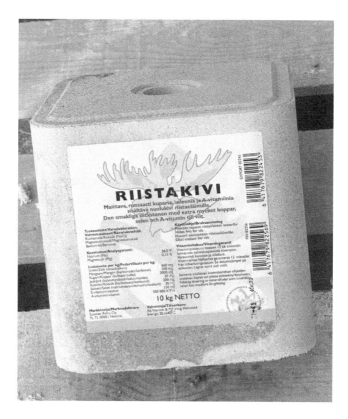

FIGURE 4.4   Mineral stone made for game and reindeer (produced in Finland). The stone contains salt (NaCl), magnesium, trace elements and vitamins E and A (Photo: Birgitta Åhman).

Dietary recommendations for potassium (K) and magnesium (Mg) are not available for *Rangifer* or other cervids, and therefore recommendations for sheep and goats are used, but can vary with age, growth or lactation (NRC 2007, pp. 122–4). Contents of potassium in feed DM of 0.5 and 0.8% are valid for growing and lactating animals, respectively. Magnesium concentrations for maintenance should be 0.15% of the feed DM, and 0.20% is recommended to support fast growth or lactation. Additional potassium or components containing potassium may have to be added to grain-based diets, as grains are typically low in potassium. Many *Rangifer* feeds in Alaska and Fennoscandia contain around 0.8–0.9% K and 0.2–0.5% Mg, which should be sufficient in most situations. Potassium content of cultured pastures that are grazed by reindeer should be evaluated, as some grasses may have concentrations exceeding 2.5%, which may lead to grass tetany if heavily grazed.

Sulfur (S) is a component of protein and many enzymes. The sulfur proportion of protein averages around 1.0% and approximates a nitrogen:sulfur ratio of 16:1, which can be used as a basis for estimating sulfur requirements (NRC 2007, p. 125). Dietary needs for cervids are not available, though based on recommendations for sheep and goats, a dietary content of 0.20% of feed DM should be adequate.

Dietary recommendations regarding trace elements are not available for cervids and the following is based on NRC (2007) recommendations for sheep and goats. Copper (Cu) is a very important mineral since it is essential for many enzyme, nervous and immune systems. Deficiency in Cu has been implicated in poor reindeer calf survival in some U.S. states. Copper absorption seems to be more important than the actual dietary content, and secondary copper deficiency can be induced by unbalanced rations or incorrect feeding methods. Dietary excess of molybdenum, sulfur or iron has been clearly identified as a suppressant of copper uptake in ruminants (Davis and Mertz 1987). Dietary Cu requirements for *Rangifer* have not been established, but many reindeer producers in the U.S. use around 20 mg/kg DM of the diet in their ration configuration (Table 4.3), while contents in most Fennoscandian reindeer feeds are lower (8–15 mg/kg, Table 4.1).

Iron (Fe) is the most abundant trace mineral in the body since it is needed for the formation of hemo- and myoglobin. Iron deficiencies in reindeer with access to pasture or hay are unlikely, as grasses and forage have high iron contents. However, reindeer fed only concentrate (grain) diets may be susceptible to iron deficiencies. Fe concentrations should be 150–200 mg/kg DM of the diet, but may have to be higher if reindeer are being affected by parasites and suffering significant blood loss.

Manganese (Mn) is involved in the formation of cartilage, antler growth and blood clotting. Manganese requirements have not been established for cervids, but a general guideline using sheep and goat recommendations is 60–80 mg/kg DM of the diet. However, reindeer fed barley or corn may need a diet higher in Mn, since high levels of Ca, Fe, P, or K may induce Mn deficiency (NRC 2007, p. 133).

Molybdenum (Mo) is an essential trace mineral in the diet because of its role in enzymatic reactions. Again, there are no established dietary Mo requirements for reindeer, but 0.1 to 1.0 mg/kg DM of the diet is recommended based on sheep and goat standards (NRC 2007, p. 134).

Sodium bicarbonate (NaHCO) has been used in ruminant diets containing cereal grains or during transition from a high fiber to highly digestible grain in order to buffer rumen pH and thereby avoid rumen acidosis. NaHCO was used in the Agricultural and Forestry Experiment Station Reindeer Research Program (AFES-RRP) reindeer herd at the University of Alaska Fairbanks when a number of animals were exhibiting signs of peritonitis and laminitis, suggesting low-grade rumen acidosis. Results showed that a barley-based diet containing 1.5% NaHCO resulted in a decrease in DMI, along with a decrease in weight gain. There was no decrease in DMI or weight gain when the percentage was dropped to 1.0% and laminitis in the herd disappeared. NaHCO in the diet can also be effective in maintaining normal milk-fat percentages when high grain, milk-fat-depressing diets are fed.

With regard to vitamins, A and E are those that need attention. According to NRC (2007, p. 152) the daily intake of vitamin A for maintenance should be 31 RE/kg body mass (BM) (where 1 RE corresponds to 3 IU). Growing animals need 100 RE per kg BM, while lactating females require 54 RE/kg BM. Vitamin E requirement is difficult to establish in isolation since it is linked to other compounds in the diet. More vitamin E is required if selenium is deficient. A recommended daily intake of vitamin E for maintenance in sheep is 5.3 IU/kg BM (1 IU = 0.025 mg) (NRC 2007, p. 156), while growing and lactating animals need more. Fennoscandian reindeer feeds typically contain 2000–4000 RE (6000–12000 IU) vitamin A and 1600–3200 IU (40–80 mg) vitamin E per kg DM (Table 4.1), which should be sufficient, as these feeds are intended for winter-feeding. A higher content of vitamin A is, however, needed for growing reindeer during summer.

4.1.5 Feed Intake

A daily log of the amount of feed consumed by the reindeer in the AFES-RRP herd at the University of Alaska Fairbanks facility indicated that a 100 kg adult reindeer will consume 1.6 kg DM per day in winter and 3.1 kg during summer. This corresponds to 51 and 98 g DM/kg$^{0.75}$ BM (that is, per kg metabolic weight) per day in winter and summer, respectively. This seems low compared to results from Sweden, where reindeer calves (BM 30–60 kg) fed pellets during two months in winter consumed an average of 1.64 kg DM (88 g DM/kg$^{0.75}$ BM) per day, while adult reindeer (average BM 70 kg) in the same experiment consumed 4.4 kg per day (Åhman, unpublished results). In another experiment, reindeer calves (40–45 kg BM) fed pellets-based diets (80% pellets combined with either lichens or silage) ad libitum during winter consumed 1.8 kg DM (about 100 g DM per kg$^{0.75}$ BM) per day, while those fed a diet with 80% lichens consumed less, 1.3 kg DM per day (Nilsson et al. 2000).

The consumption of hay or silage is usually limited by the fiber content and the subsequent rate of passage through the digestive system. For example, reindeer calves fed first-cut silage containing 30.4% cellulose and 6.2% water soluble carbohydrates (WSC) had a daily consumption of 0.6–0.7 kg DM (34 g per kg$^{0.75}$ BM), while those fed a leafy regrowth silage containing 18.7% cellulose and 30.0% WSC consumed significantly more, around 1 kg (53 g per kg$^{0.75}$) DM per day (Aagnes et al. 1996). Similarly, Nilsson et al. (1996) found that reindeer calves offered a mixed barley-silage diet had a considerably lower total DM intake (0.49 kg per day, or 65 g/kg$^{0.75}$ BM per day) when they received 40% silage in their diet, compared to when they received only 20% (0.82 kg per day, or 78 g/kg$^{0.75}$ BM per day).

4.1.6 Rules of Thumb

Based on experiences feeding reindeer within AFES-RRP at the University of Alaska Fairbanks, the following "rules of thumb" for *Rangifer* feeding have been established:

1. Grain dry matter (DM) should be 65% or less of total DM intake, or not more than 2% of BM.
2. Minimum recommended ADF in ration DM is 16–17%, or higher if fat is higher than 4% in the diet.
3. NDF should be at least 26% of the ration DM, but not more than 33%.
4. CP requirements change significantly across cohorts and season: 8–11% CP in winter and 15–16% in summer, though lactating females, yearlings and calves require higher concentrations (16–20%).
5. The total fat in rations should be a maximum of 4% of the DM. A guideline is no more than 2% added fat from any one of three sources: animal, vegetable or rumen inert fats. When feeding a diet with greater than 4% fat, calcium should be increased to 0.9–1%, magnesium to 0.3% and ADF to 17% or more in the ration DM.
6. Salt should be included in the ration at 1%.

7. A calcium-phosphorus mineral source should be included in the grain mix at 1–2%.
8. Supplemental vitamins (A, D and E) and trace minerals should be included in the ration as per requirements.
9. NDF of forage (hay or silage/haylage) should be 55–60% DM, though it can be higher if the percentage of hemicellulose is greater than that of cellulose.
10. ADF of forage (hay or silage/haylage) should be 30% or less of DM.

Pellet-fed reindeer need access to water. Clean fresh snow may be sufficient, or the feeding site could be a place close to an open stream with clean running water or water offered in a tank (Figure 4.5).

4.1.7   CULTURED PASTURES FOR REINDEER

Reindeer will utilize a wide variety of native graminoids when grazing in a free-range setting; however, most of these forages cannot be grown in a farm environment. In Fennoscandia, sufficient natural pastures are generally available for summer grazing and therefore there has been little interest in providing cultured pastures for reindeer in this part of the world. In Norway, however, old abandoned meadows were reported to be grazed by reindeer in late spring (Eilertsen et al. 1999), where dominating grasses were *Festuca rubra*, *Poa alpigenea*, *Agrostis tenuis*, *Descampsia caespitosa* and *Phleum commutatum*. Herbs such as *Achillea millefolium*, *Rumex acetosa*, *Ranuculus repens* and *Alchemilla subcrenata* were also available. The crude protein content and digestibility were high, and the forage seemed to be well utilized by the reindeer. Reindeer have occasionally been found grazing on cultured land intended for harvest of forage in Sweden. Old lays comprising mixed species of grass and "weeds" were preferentially consumed over first-year growth of timothy and clover. In Alaska, smooth bromegrass (*Bromus inermis*) and Kentucky Bluegrass (*Poa pratensis*) are grass species grown for pasture and hay production, as well as for private and public lawns. These species have been shown to be grazed by reindeer during summer and could be used as pasture in a reindeer production system (Luick and Blanchard 1980).

4.1.8   BOTTLE-MILK FEEDING OF *RANGIFER* CALVES

Bottle-feeding of reindeer calves has been studied at the University of Alaska Fairbanks, where Parker (1989) used a mixture of homogenized cow's milk and lamb's milk replacer for feeding caribou neonates. The calves were initially fed seven times per day and milk was the major food source during their first month of life. After peak intake of around 1200 g/d, the milk volume and feeding frequency was gradually reduced until weaning at 100 days of age. Pellets then replaced milk feeding and the calves also had access to pasture. The calves developed well and their growth rate was similar to that of calves in a free-ranging herd.

A commercial milk replacer (Zoologic® Milk Matrix) designed specifically for *Rangifer* calves was used in a more recent experiment (Parker and Barboza 2013). The nutrient composition closely resembled the milk composition observed in different *Rangifer* populations over the world, with 8.9% protein, 10.8% fat, 3.4% lactose and 26.3% DM. The calves were introduced to the diet slowly by being given only 75 g per feeding event, after which the daily ration was gradually increased to 10% of the calves' body mass. As in the previous experiment, the calves were initially fed seven times per day. Milk was replaced with a buffer electrolyte (Re-Sorb; Pfizer, Animal Health, New York, NY) on a few occasions when calves got diarrhoea. If the diarrhoea did not stop after 1 day, rice cereal was used in addition to slow down the passage through the digestive system. This proved to be a good strategy, since the growth rate of the bottle-fed calves did not differ from that of maternally raised calves at the experimental station, and by autumn the calves were as big as free-ranging caribou calves.

(a)

(b)

FIGURE 4.5   Pellet-fed reindeer need access to water. Feeding site could be a place close to an open stream with clean running water (a) or offered water in a tank with electric heating (b) (Photo: Birgitta Åhman).

4.2   DISEASES RELATED TO FEEDING

Terje D. Josefsen and Birgitta Åhman

Reindeer, whether they are wild or domesticated, are basically adapted to finding their own food in summer as well as in winter. Most feedstuffs that are used for supplementary feeding differ substantially from natural forage. The feeding situation, with a surplus of feed served once or twice per day, is also very different from natural foraging. Adaptation to a new diet and a new situation may pose several challenges to reindeer health. Most problems are related to dysfunction of the digestive system, though high animal density, stress and increased concentration of pathogens represent additional risk factors with respect to animal health.

4.2.1   ACCUMULATION OF GRASS IN THE RUMEN

The term "gressbuk" (Norwegian) or "gräsbuk" (Swedish), can be translated as "grass belly" and is used by reindeer herders to describe grass accumulation in the reindeer rumen. The condition is associated with feeding on hay or silage, and occurs when rumen digestion fails and undigested grass accumulates in the rumen. The reindeer retain their appetite and continue to eat, gradually increasing the rumen volume. At the same time, the animal does not get enough energy from the diet and therefore has to use its body reserves to meet energy demands. If the diet is not changed, the animal will eventually die from emaciation, with its rumen full of undigested feed.

Grass accumulation is an example of what can happen when the reindeer, as a selective mixed feeder, is forced to eat only roughage with high contents of cellulose and lignin. "Grass belly," as a disease diagnosis, describes the advanced state where the rumen has become noticeably large and bulky. However, the underlying physiology, i.e. the reindeer's limited ability to utilize high-fiber grasses, is not an either/or condition, with rumen volume increasing to varying degrees when reindeer are fed high-fiber diets.

Many reports of emaciation and starvation in reindeer, in spite of free access to hay or silage, can probably be attributed to the consumption of high-fiber roughage (Josefsen et al. 2007, Syrjälä-Qvist 1985). In a description of 12 calves that all died of emaciation during a period of feeding with hay (Westerling 1972), the most plausible explanation seemed to be an accumulation of undigested grass in the rumen. However, in most situations the condition is reversible and can be reversed by supplying a more easily digestible diet.

Feeding experiments carried out in Norway (Aagnes et al. 1996, Aagnes and Mathiesen 1996) provide examples of grass accumulation in reindeer. Three reindeer calves in each experiment were fed silage from a late first cut (FC) of timothy (*Phleum pratense*), and three were fed silage comprising regrowth (RG) of the same timothy. The silages differed substantially in their stem to leaf content, where FC had a substantially higher fiber content (57.8%) than RG (38.7% fiber). The more fiber-rich diet resulted in a lower feed intake (0.6–0.7 kg/d, on a dry matter basis, compared to 0.9–1.1 kg/d for RG), larger rumen and poorer nutritional status than the more easily digestible feed, and this confirms the limited ability of reindeer to utilize high-fiber roughages.

4.2.2   RUMINAL ACIDOSIS

Ruminal acidosis is a disease in which rumen digestion stops because the content has become too acidic. Ruminal acidosis occurs due to high intake of easily digestible carbohydrates. The high intake may be due to a sudden event of overeating, or an intentional but overly rapid increase in carbohydrate consumption, exceeding the adaptive capacity of the rumen and its microbial flora.

In an adequately functioning reindeer rumen the pH varies between 6 and 7 (Nilsson et al. 2006). A large intake of easily digestible carbohydrates (e.g. starch, which is the main component in grains and thus in most commercial reindeer feeds) changes the rumen microbial flora. In the worst case, this may result in a flora dominated by lactobacilli (*Lactobacillus* spp.), which produce lactic acid and consequently cause rumen pH to drop to between 4 and 5. This is a life-threatening condition since the absorption of acid from the rumen may exceed the buffering capacity of the blood, leading to general acidosis and possibly resulting in death. In addition, the high concentration of lactic acid in the rumen content causes water from the general circulation to enter rumen by osmosis, resulting in severe dehydration.

Ruminal acidosis is a well-known condition in ruminant livestock (cattle, sheep and goats) and often occurs when animals gain unintentional access to feed stores, hence the term "grain overload." In reindeer, this disease is always associated with feeding, be it the use of pelleted feed mixtures or other feedstuffs with high carbohydrate content, such as bread or cereal grains.

A severe case of ruminal acidosis was observed in a group of female reindeer that had been fed reindeer pellets from February until the first week of April (unpublished results from the Norwegian Veterinary Institute). The owner had then lost six females and two calves without them showing any clear symptoms of illness beforehand. A pregnant female in seemingly good body condition (87 kg BM) was found in a critical state, and the rumen seemed inflated. She was then euthanized and submitted for necropsy. The main finding was a large, dilated rumen with a bright green watery content (Figure 4.6). Rumen pH was 4.5–5.0. The rumen mucosa was firmly attached to the rumen wall. The content of the colon was bright green and watery, while the content of the distal rectum

FIGURE 4.6   Rumen from an adult female reindeer diagnosed with ruminal acidosis. The rumen was filled with highly fluid content. In such cases the sloshing sound of moving fluid may be heard when the reindeer walks, giving raise to the Swedish nickname "skvalpmage" ("rippling belly"). The acidity (pH) of the rumen content was between 4.5 and 5.0 (Photo: Terje D. Josefsen).

was more viscid and dark green. Histological examination of the mucosa showed vacuolation of the rumen epithelium and infiltration of inflammatory cells (neutrophilic granulocytes). In this case, ruminal acidosis was diagnosed with certainty. It was, however, not obvious what had caused the condition, since the feeding had already proceeded for 6 weeks without problems. Most cases of ruminal acidosis occur within 2–3 weeks of the introduction of grain-based feeds. The best suggestion was that the amounts of pellets provided over time had been a bit too large, and therefore perhaps only minor changes were needed to shift the rumen flora towards lactobacilli and, consequently, lactic acid production, causing the disease.

Common symptoms of ruminal acidosis in reindeer include lack of appetite, increased thirst, lethargy, dehydration, decreased or ceased rumen contractions and eventually diarrhoea (Rehbinder and Nikander 1999, p. 55). The rumen content becomes fluid and, in many cases, one can hear the sloshing sound of moving fluid from the stomach of the reindeer, giving the disease its Swedish nickname "skvalpmage" ("rippling belly"). However, this term has also been used for other conditions with watery rumen content (Rehbinder and Nikander 1999, p. 56).

A diagnosis can be made in live animals from their dietary history (large intake of pellets, bread or cereal) and clinical signs. Necropsy findings include a liquid rumen content with pH of between 4 and 5. If the fodder that caused the disease contained whole grains, these may be found in the rumen and abomasum. The contents of the colon will be softer than normal and sometimes diarrhoea is evident. However, if only a short time has passed from the start of the disease until the reindeer dies, diarrhoea may not have developed and faecal pellets in the rectum may still be normal.

A way to diagnose ruminal acidosis upon field necropsy may be to investigate whether or not the rumen mucosa detaches from the rumen wall. In a rumen with normal pH, the mucosa starts loosening only a few hours after death. In reindeer with ruminal acidosis, the rumen mucosa will not peel off (Norwegian Veterinary Institute, Terje D. Josefsen, own experience). This phenomenon is also known in other ruminants (Brown et al. 2007). The diagnosis may be verified by histological examination of the rumen mucosa. The acidic rumen content causes chemical rumenitis, which is not visible macroscopically but evident histologically as a vacuolated rumen epithelium with numerous small aggregates of neutrophilic lymphocytes (Brown et al. 2007).

The treatment of ruminal acidosis in reindeer should follow the same principles as for other ruminants. The outcome is, however, uncertain. Severe cases often end fatally (Rehbinder and Nikander 1999, p. 55), while mild cases may recover after a few days. However, there is a risk of complications: lesions in the rumen mucosa may provide entry points for bacterial or fungal infections. Bloat (ruminal tympany – see p. 152) can occur as a secondary condition. If the animal is already very lean, there is a risk that failing ruminal digestion and loss of appetite results in death by emaciation before the rumen fermentation recovers.

Different strategies have been tested to avoid pH drop and subsequent rumen acidosis when introducing grain-based diets to reindeer. By replacing part of the barley and wheat bran in a standard reindeer diet with seaweed (15%), a pH drop could be avoided (Sletten and Hove 1990). The addition of seaweed proved to be more effective than the inclusion of 4% sodium bicarbonate in the original diet.

4.2.3 Diarrhoea

Diarrhoea in reindeer occurs almost exclusively in connection with feeding. It is mostly associated with the use of pellets (Johansson 2006) and therefore may be a symptom of ruminal acidosis (see above). However, diarrhoea has also been observed in association with consumption of high-fiber feed mixtures (Jacobsen and Skjenneberg 1979) or grass silage (Josefsen et al. 2007).

Occurrence of diarrhoea in reindeer that eventually die of emaciation has been interpreted as hunger by some, and that emaciation in itself can cause diarrhoea (Westerling 1971). However, results by Josefsen et al. (2007) indicate that diarrhoea in such cases was caused by the feed or other conditions, and that emaciation due to food shortage does not trigger diarrhoea. Diarrhoea may also be caused by bacterial infections in the digestive tract (e.g. by *Salmonella* spp. or *Clostridium perfringens*).

4.2.4 BLOAT

Bloat (ruminal tympany) denotes a state in ruminants where the rumen is filled with gas or foam, without the animal being able to eructate (burp up) the gas. Bloat may have a range of different causes, though irrespective of this, the pressure in the abdomen may become so large that it obstructs breathing and blood circulation, and the animal may die of suffocation. In cattle, bloat is generally associated with grazing on pastures with very lush grass or large amounts of leguminous plants (e.g. alfalfa or clover), or with feeding on high amounts of concentrates (Radostits et al. 2000). Bloat may occasionally affect reindeer in connection with feeding (Ippen and Henne 1985, Luick 1977) but is far less common among reindeer than in domestic livestock in more highly productive systems.

4.2.5 WET BELLY

So called "wet belly" is a disease that manifests itself in the reindeer's fur becoming wet, primarily at the axilla, then down the legs and along the lower parts of the thorax and abdomen, and sometimes also on the lower part of the neck (Åhman et al. 2002) (Figure 4.7). Hair loss may occur in the affected areas (Rydberg 1969), possibly because the animals lick themselves. The disease is exclusively observed in relation to feeding and has been repeatedly reported since feeding experiments with reindeer started during the 1960s (Jacobsen and Skjenneberg 1979, Persson 1967, Westerling 1971, Åhman et al. 2002).

The disease has, to our knowledge, not been reported in any other herbivores. Wet belly is considered rare in Norway, which may be related to winter-feeding not being so widespread. In interviews and questionnaire surveys in Sweden and Finland (Johansson 2006, Sainmaa 1998), about half of the participating reindeer herders had observed the condition among their reindeer, and several stated that wet belly was common (that it had occurred several times). The herders all agreed that the condition only occurred during feeding. If they had a cure, it was to change the feed (irrespective of what feed they used when the condition became apparent).

Other symptoms are linked to this condition besides the typical wetting of the fur. Reindeer with wet belly are reported to have high feed intake and to appear hungrier than healthy reindeer in the same situation (Jacobsen and Skjenneberg 1979, Westerling 1971, Åhman et al. 2002). In addition, reindeer with wet belly have been reported to curl up when they lie down, which has been interpreted as a sign of freezing, while Åhman et al. (2002) also found that the rectal temperature was slightly lower in reindeer with wet belly than in healthy reindeer from the same group.

The reason for wet belly is unknown. It was previously unclear whether the moisture came from within the animal – as sweat – or whether it emerged because the skin was warm and melted the snow where the reindeer was lying, thus making the coat wet (as proposed by Rehbinder and Nikander 1999, p. 55). Åhman et al. (2002) ended this debate by placing a reindeer with wet belly indoors in a dry environment and drying the coat with a hair dryer, after which it was observed that the reindeer became wet again within a few hours.

It has been theorized that wet belly appears when the reindeer are fed hay, especially timothy. Jacobsen and Skjenneberg (1977, 1979) experienced the disease when feeding reindeer a high-fiber diet containing straw. However, Åhman et al. (2002), showed that the disease also occurred when reindeer were fed restricted amounts of a "natural" diet containing reindeer lichens (80%), bilberry shrubs and willow leaves, without access to grass forage. It was established that wet belly, although obviously arising during feeding, is not tied to a specific feedstuff or feeding regime. Despite this, the common recommendation is still to change the feed.

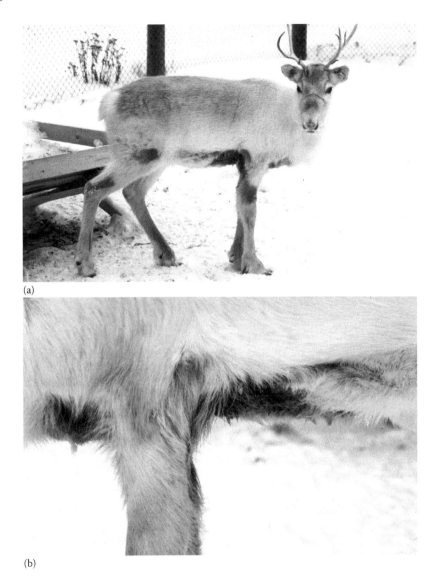

FIGURE 4.7   Reindeer calf affected by "wet belly" (a), with the typical wetness at the axilla region, between the forelegs (b) (Photo: Birgitta Åhman).

Reindeer may die from wet belly, but the prognosis is uncertain. In the experiment by Åhman et al. (2002) a total of 11 reindeer got sick, of which four died and one was euthanized because of the disease. Four reindeer recovered after 10 to 28 days, one was excluded from the experiment (because of an injury) and one was slaughtered according to the study design. All reindeer that recovered had changed to pellet-based diets, while those that died and the one that had to be euthanized were first fed mainly silage before gradually changing to a pellet-based diet. Necropsies of the four that died and the one that was euthanized showed emaciation as well as combinations of several other pathological findings, e.g. mycotic pneumonia in two reindeer, abomasal ulcers in two, rumen ulcers with abscess formation and local peritonitis in one, enteritis and hepatitis in one. Ulceration in the abomasum has also been described in other cases of wet belly (Jacobsen and Skjenneberg 1979). The findings give reason to speculate whether wet belly is associated with fungal contamination and/or fungal infection of the feed, as mycotic pneumonia was detected in two of the animals, and ruminal and abomasal ulcers – though not diagnosed in the above experiment – are common signs of fungal infection.

4.2.6   OTHER FEEDING-RELATED DISORDERS

Constipation may sometimes occur during feeding. Five out of 40 Swedish reindeer herders interviewed (Johansson 2006) reported that they occasionally observed reindeer suffering from constipation during feeding. Observations by one of the authors of this chapter (Josefson) indicate that constipation (or obstruction) in the omasum may occur as a result of feeding on high-fiber fodder. However, constipation in reindeer is inadequately documented both clinically and pathologically. A physical blockage of the passage between the stomach compartments or in the intestine may occur due to foreign bodies ingested with the feed – for example, pieces of string, plastic or other foreign objects (Rehbinder and Nikander 1999, p. 57).

Accumulation of grass in the rumen (described on p. 149) may be seen as functional constipation. There is no physical obstruction preventing the passage, yet malfunction in rumen digestion and contraction results in rumen content not passing at a normal rate out of the rumen.

4.2.7   INFECTIOUS DISEASES

Feeding generally results in aggregation of animals, increasing the possibility of transmission of bacteria, viruses and parasites. Contagious eye inflammation and ecthyma have been reported in association with feeding (Aschfalk et al. 2003a, Tryland et al. 2001, Tryland et al. 2009). *Clostridium perfringens* and *Salmonella* spp. have been reported to occur more commonly among reindeer that have been subject to corralled feeding than among free-ranging reindeer (Aschfalk et al. 2003b, Aschfalk and Thórisson 2004). Infection by *Clostridium perfringens* was identified in several reindeer neonates that had died during fenced-in calving in Sweden (Åhman and Ågren, unpublished results). In connection with another case of corralled calving in Sweden, neonate reindeer were infected with *Fusobacterium necrophorum*, causing a severe outbreak of oral necrobacillosis where many calves died shortly after the herd had been let out of the corral (Ågren, Åhman et al. unpublished results). (See also Chapter 16, Climate Change – Potential Impacts on Pasture Resources, and Health and Diseases of Reindeer and Caribou.)

Sufficient space for the animals, good hygiene around feeding sites and water sources, as well as frequent changes of feeding site, are needed in order to reduce the risk of outbreak of contagious disease. Supplemental feeding of free-ranging reindeer usually means lower animal density and considerably less risk of infection and disease.

4.2.8   CONCLUDING REMARKS ON HEALTH RISKS AT FEEDING

The feeding of reindeer has always been considered problematic, especially feeding in emergency situations when the pasture is locked and the reindeer are already nutritionally stressed. However, the situation today is better than before. Well-tested pelleted reindeer feeds are commercially available, and knowledge of the use of different feedstuffs has increased significantly. If herders choose to feed pellets, they can be confident that the reindeer are receiving a feed that covers all nutritional needs. The challenge is to regulate the quantity, so that every animal gets to eat but no single animal eats too much. If the herder chooses to feed roughage, there is little risk of overeating. It is, however, important that the nutritional quality of roughage is sufficient (high enough digestibility) in order to avoid accumulation of grass in the rumen, and it is generally not recommended to feed only roughage over longer time periods (several weeks).

Although feeding may cause disease among reindeer, practical experience has shown that far more reindeer die from lack of feed than from feeding-related diseases. This is also why winter-feeding has become increasingly common in Fennoscandia over the past 20 years. If feeding starts in time, before the reindeer are emaciated, and suitable feedstuffs are used, there is a high probability of a successful outcome.

Reindeer seem to be capable of adapting well to long-term feeding, provided they are given a suitable diet. At the University of Tromsø (Norway), reindeer have been fed for years with commercial reindeer feed (formerly R-80, now FK Reinfôr) without showing signs of health problems or disease caused by the feeding. Also, reindeer that are kept permanently in zoos or research institutions usually seem to do well on the feeds that are available.

REFERENCES

Aagnes, T.H., A.S. Blix and S.D. Mathiesen. 1996. Food intake, digestibility and rumen fermentation in reindeer fed baled timothy silage in summer and winter. *J. Agric. Sci.* 127: 517–23.

Aagnes, T.H. and S.D. Mathiesen. 1996. Gross anatomy of the gastrointestinal tract in reindeer, free-living and fed baled timothy silage in summer and winter. *Rangifer* 16: 31–9.

Åhman, B. 1998. Contaminants in food chains of arctic ungulates: What have we learned from the Chernobyl accident? *Rangifer* 18: 119–26.

Åhman, B., A. Nilsson, E. Eloranta and K. Olsson. 2002. Wet-belly in reindeer (*Rangifer tarandus tarandus*) in relation to body condition, body temperature and blood constituents. *Acta Vet. Scand.* 43: 85–97.

Aschfalk, A., T.D. Josefsen, H. Steingass, W. Muller and R. Goethe. 2003a. Crowding and winter emergency feeding as predisposing factors for kerato-conjunctivitis in semi-domesticated reindeer in Norway. *Dtsch. Tierarztl. Wochenschr.* 110: 295–8.

Aschfalk, A., N. Kemper and C. Holler. 2003b. Bacteria of pathogenic importance in faeces from cadavers of free-ranging or corralled semi-domesticated reindeer in northern Norway. *Vet. Res. Commun.* 27: 93–100.

Aschfalk, A. and S.G. Thórisson. 2004. Seroprevalence of Salmonella spp. in wild reindeer (*Rangifer tarandus tarandus*) in Iceland. *Vet. Res. Commun.* 28: 191–5.

Brown, C.C., D.C. Baker and I.K. Barker. 2007. Alimentary system, in *Jubb, Kennedy & Palmer's Pathology of Domestic Animals*, ed. Maxie, M.G. 1–298. London: Saunders Elsevier.

Cabrajic, A.V.J., J. Moen and K. Palmqvist. 2010. Predicting growth of mat-forming lichens on a landscape scale – comparing models with different complexities. *Ecography* 33: 949–60.

Chisana Caribou Recovery Team, 2010. *Recovery of the Chisana Caribou Herd in the Alaska/Yukon Borderlands: Captive-Rearing Trials. Yukon Fish and Wildlife Branch Report TR-10-02*, Yukon Department of Environment, Whitehorse.

Davis, G.K. and W. Mertz. 1987. Copper, in *Trace Elements in Human and Animal Nutrition*, ed. Mertz, W. 301–64. San Diego: Academic Press.

Eilertsen, S.M., I. Schjelderup and S.D. Mathiesen. 1999. Utilization of old meadow by reindeer in spring in Northern Norway. *Rangifer* 19: 3–11.

Finstad, G. 2007. Reindeer in Alaska: Under new management. *Agroborealis* 38: 22–8.

Finstad, G., E. Wiklund, K. Long, P.J. Rincker, A.C.M. Oliveira and P.J. Bechtel. 2007. Feeding soy or fish meal to Alaskan reindeer (*Rangifer tarandus tarandus*) – effects on animal performance and meat quality. *Rangifer* 27: 59–75.

Helle, T.P. and L.M. Jaakkola. 2008. Transitions in herd management of semi-domesticated reindeer in northern Finland. *Ann. Zool. Fenn.* 45: 81–101.

Ippen, R. and D. Henne. 1985. A contribution on the diseases of Cervidae, in *Erkrankungen der Zootiere. Verhandlungsbericht des 27. Internationalen Symposiums uber die Erkrankungen der Zootiere, 9–13 Juni, 1985, St. Vincent/Torino.* 7–16. Berlin: Akademie-Verlag Berlin.

Jacobsen, E. and S. Skjenneberg. 1977. Resultater og erfaringer fa foring av rein med høy, surfor, "sauenøtt" og "komplettfor m/halm" [Results and experiences from feeding reindeer with hay, silage, sheep pellets and "completed straw feed"]. *Forskning og forsøk i landbruket* 28: 651–60.

Jacobsen, E. and S. Skjenneberg. 1979. Forsøk med ulike forblandinger till rein, Forverdi av reinfor (RF-71) [Experiment with different diets for reindeer, Feeding value of reindeer feed (RF-71)]. *Meldinger fra Norges Landbrukshøgskole* 58: 1–11.

Johansson, A., 2006. *Djurägarbehandling inom renskötseln*, MSc thesis, Swedish University of Agricultural Sciences, Faculty of Veterinary Medicine and Animal Science, Uppsala.

Josefsen, T.D., K.K. Sørensen, T. Mørk, S.D. Mathiesen and K.A. Ryeng. 2007. Fatal inanition in reindeer (*Rangifer tarandus tarandus*): Pathological findings in completely emaciated carcasses. *Acta Vet. Scand.* 49:27.

Kivinen, S., A. Berg, J. Moen, L. Ostlund and J. Olofsson. 2012. Forest fragmentation and landscape transformation in a reindeer husbandry area in Sweden. *Environ. Manage.* 49: 295–304.

Luick, J. 1977. Diets for captive reindeer, in *CRC Handbook Series in Nutrition and Food. Section G. Diets, Culture, Media, and Food Supplements.*, ed. Rechcigl, M. 279–94. Cleveland, Ohio: CRC Press.

Luick, J.R. and J.M. Blanchard. 1980. Seasonal changes in grazing preferences of reindeer for six grasses, in *Proc. 2nd Int. Reindeer/Caribou Symp., Røros, Norway, 1979*, eds. Reimers, E., E. Gaare and S. Skjenneberg. 78–83. Trondheim: Direktoratet for vilt och ferskvannsfisk.

Moen, R., M.A. Olsen, Ø.E. Haga, W. Sørmo, T.H. Aagnes-Utsi and S.D. Mathiesen. 1998. Digestion of timothy silage and hay in reindeer. *Rangifer* 18: 35–45.

Nieminen, M. 2010. Why supplementary feeding of reindeer in Finland? *Rangifer Report* No. 14: 41.

Nilsson, A., B. Åhman, M. Murphy and T. Soveri. 2006. Rumen function in reindeer (*Rangifer tarandus tarandus*) after sub-maintenance feed intake and subsequent feeding. *Rangifer* 26: 73–83.

Nilsson, A., Ö. Danell, M. Murphy, K. Olsson and B. Åhman. 2000. Health, body condition and blood metabolites in reindeer after sub-maintenance feed intake and subsequent feeding. *Rangifer* 20: 187–200.

Nilsson, A., I. Olsson and P. Lingvall. 1996. Evaluation of silage diets offered to reindeer calves intended for slaughter. I. Feeding of silage and barley from September to March. *Rangifer* 16: 129–38.

NRC. 2007. *Nutrient Requirements of Small Ruminants: Sheep, Goats, Cervids, and New World Camelids.* Washington, D.C.: National Academies Press.

Parker, K.L. 1989. Growth rates and morphlogical measurements of Porcupine caribou calves. *Rangifer* 9: 9–13.

Parker, K.L. and P.S. Barboza. 2013. Hand-rearing wild caribou calves for studies of nutritional ecology. *Zoo Biol.* 32: 163–71.

Persson, S. 1967. Studier av renarnas energibehov [Energy requirements of reindeer]. *Kungliga Lantbruksstyrelsen, Meddelanden – Serie B. Lantbruksavdelningen. Rennäringen* Nr 75: 23 pp.

Radostits, O.M., C.C. Gay, D.C. Blood and K.W. Hinchcliff eds., 2000. *Veterinary Medicine: A Textbook of the Diseases of Cattle, Sheep and Horses.* London: Saunders.

Rehbinder, C. and S. Nikander. 1999. *Ren och rensjukdomar.* Lund: Studentlitteratur.

Rehbinder, C. and A. Nilsson. 1995. An outbreak of kerato-conjunctivitis among corralled, supplementary fed, semidomestic reindeer calves. *Rangifer* 15: 9–14.

Rydberg, A. 1969. Utfodringsförsök med renar. *Röbäcksdalen meddelar* 1996: 1–6.

Sainmaa, S. 1998. *Blötbuk hos renar (*Rangifer tarandus tarandus *L.) i Finland och Sverige [Wet belly in reindeer in Finland and Sweden]*, Master thesis, Swedish University of Agricultural Sciences, Faculty of Veterinary Medicine, Uppsala.

Sandström, P. 2015. *A toolbox for co-production of knowledge and improved land use dialogues*, PhD thesis, Acta Universitatis Agriculturae Sueciae 2015: 20, 80 pp.

Sletten, H. and K. Hove. 1990. Digestive studies with a feed developed for realimentation of starving reindeer. *Rangifer* 10: 31–7.

Staaland, H. and H. Sletten. 1991. Feeding reindeer in Fennoscandia: The use of artificial food. In *Wildlife Production: Conservation and Sustainable Development*, eds. Renecker, L.A. and R.J. Hudson. 227–42. Fairbanks: Agricultural and Forestry Experiment Station, University of Alaska Fairbanks.

Syrjälä-Qvist, L. 1985. Hö som foder till renar [Hay as the feed to reindeer]. *Rangifer* 5: 2–5.

Tryland, M., C.G. das Neves, M. Sunde and T. Mork. 2009. Cervid herpesvirus 2, the primary agent in an outbreak of infectious keratoconjunctivitis in semidomesticated reindeer. *J. Clin. Microbiol.* 47: 3707–13.

Tryland, M., A. Oksanen, A. Aschfalk, T.D. Josefsen and K.A. Waaler. 2001. Smittsom munnskurv påvist hos tamrein i Norge [Contagious ecthyma in semidomesticated reindeer]. *Norsk Veterinærtidskrift* 113: 9–13.

Turunen, M. and T. Vuojala-Magga. 2014. Past and present winter feeding of reindeer in Finland: Herders' adaptive learning of feeding practices. *Arctic* 67: 173–88.

Westerling, B. 1971. Den svältande renen. *Kungl. lantbruksstyrelsen. Meddelanden – Serie B* Nr 88: 75–82.

Westerling, B. 1972. Överutfodring som orsak till svältdöd bland renkalvar [Overfeeding as a starvation cause amongst reindeer calves]. *Finsk Veterinärtidskrift*: 131–8.

Wiklund, E., L. Johansson and G. Malmfors. 2003. Sensory meat quality, ultimate pH values, blood parameters and carcass characteristics in reindeer (*Rangifer tarandus tarandus* L.) grazed on natural pastures or fed a commercial feed mixture. *Food Qual. Prefer.* 14: 573–81.

Wiklund, E., J. Pickova, S. Sampels and K. Lundstrom. 2001. Fatty acid composition of M-longissimus lumborum, ultimate muscle pH values and carcass parameters in reindeer (*Rangifer tarandus tarandus* L.) grazed on natural pasture or fed a commercial feed mixture. *Meat Sci.* 58: 293–8.

5 Non-Infectious Diseases and Trauma

Erik Ågren and Torill Mørk

CONTENTS

Non-infectious diseases in *Rangifer* sp. are, in general, rarely described in scientific literature, as they mainly affect individuals (e.g. malformations) and do not affect or threaten the overall populations. Being free-ranging animals (mainly, or most of the time), caribou and wild or semi-domesticated reindeer are well adapted to their natural habitat, and cases of disease may only be noted during hunter harvest or meat inspection at slaughter. When herds are managed, other risks and diseases appear when herds are in corrals, fed or transported. In this chapter, there is an emphasis on findings in Fennoscandian semi-domesticated reindeer, with an overview of the major causes of death, such as predation and trauma, as well as descriptions of non-infectious diseases affecting *Rangifer* species, many of which are not commonly documented in the scientific literature.

5.1 MALFORMATIONS

Malformations in reindeer and caribou are only infrequently noted or reported. Hereditary malformations due to inbreeding should not be expected to be a problem in large populations of free-ranging animals. Selection for breeding is limited in semi-domesticated reindeer herding, as usually the females are evaluated by the individual owners, who often cull older, barren females. Selection for certain traits such as hair colour, antler shapes or large body size is sometimes done in traditional reindeer herding (Kuhmunen 2000). Life-threatening malformations occurring in populations are likely under-reported due to scavenging. In contrast, malformations in reindeer calves have been noted during herding operations where calving takes place in corrals or fenced areas associated with research stations or management efforts to avoid predation on the calving grounds. Malformations in adult reindeer are typically reported as opportunistic observations during scientific studies for various other reasons. Table 5.1 lists the types of malformations reported in reindeer.

In juvenile calves (up to 1 month of age), facial and oral defects seem to be the most commonly reported malformations. Other defects, mainly reported as single or few cases, involve the spine, joints, heart, skin, limbs, or brain. Rehbinder and Nikander (1999) listed some sporadic cases of malformations in semi-domesticated reindeer calves diagnosed at the Swedish National Veterinary Institute. Brachygnathia inferior (overbite or undershot jaw), ichthyosis (fish scale disease), umbilical hernia, amelia (congenital lack of limb), and acquired or congenital hydrocephalus ("water on the brain") are listed. One case of brachygnathia inferior and palatoschisis (cleft palate) from Sweden was published by Elvestad and Nordkvist (1985), with references to Schmit (1937), who reported a prevalence of 4.6% of "short jaws" in calves in certain reindeer herds at Chutkotka peninsula, USSR. The affected calves were all entirely white, lacking all pigmentation, and the changes were considered hereditary in these small herds. Also, Skjenneberg and Slagsvold (1968) mention that calves with undershot jaws are called "hare-mouth" in Scandinavia (Norwegian: haremunn, Sámi: njoammelnjalbme), and that deformities of the head are more common in white calves. Renecker and Blake (1992) described two newborn albino caribou calves from Alaska with 2 cm overshot jaws, pale blue eyes with abnormal pupils, domed skulls, malformed nasal bones and trachea, and ears placed lower than normal. Traditional Nordic reindeer herders do select to keep completely white reindeer (Sámi: gabba) for esthetic reasons, although they are considered to be weaker and to have a shorter life span than normal coloured reindeer (Kuhmunen 2000).

Studying perinatal mortality (usually defined as deaths during delivery or within the first month of life) in caribou in Alaska, Roffe (1993) found severe congenital defects in two of the 13 stillborn calves, in a total of 61 dead calves examined. One calf had a heart malformation with a high ventricular septal defect, and an overriding, dextraposed aorta, as well as renal dysplasia. The other calf had hypoplasia of the intestine, with a markedly shortened intestine.

Ågren (2001) reported six calves with malformations based on necropsy of 158 perinatal calf carcasses collected in fenced calving grounds in a semi-domesticated herd in Sweden. Two of the

---

TABLE 5.1

Malformations Reported in Reindeer and Caribou, with Reference Example

Amelia (lack of limb)	Rehbinder and Nikander (1999)
Aplasia (non-development) of kidney	Ågren (2001)
Aplasia (non-development) of rectum and anus	Ågren (2001)
Arthrogryposis (joint contractions)	Ågren (2001)
Brachygnathia inferior (overshot jaw)	Ågren (2001)
Brachygnathia superior (undershot jaw)	Ågren (2001)
Cheilo-palato-gnathoschisis (cleft palate, lip and upper jaw)	Ågren (2001)
Cryptorchid (retained testicle)	Leader-Williams (1979)
Dental anomalies: supernumerary teeth, missing teeth	Miller and Tessier (1971)
Dermoid cyst	Wobeser et al. (2009)
Heart malformations	Roffe (1993)
Hydrocephalus (water on the brain)	Rehbinder and Nikander (1999)
Palatoschisis (cleft palate)	Elvestad and Nordkvist (1985)
Polydactylism (supernumerary toes)	Miller and Broughton (1971)
Scoliosis (lateral deviation of spine)	Josefsen et al. (2014)
Tail vertebra malformation (kinked tail)	Ågren (2001)
Umbilical hernia	Rehbinder and Nikander (1999)

affected calves had multiple malformations. Cleft palate (palatoschisis) was the most common malformation, seen in four calves (Figure 5.1). A cleft palate invariably leads to a very short life span for the calf, as it cannot create normal suction when suckling from the dam's teat, with an abnormal passage between the mouth and the nasal cavity. If the calf does receive some milk, there is a high risk of aspiration of milk into the lungs, causing secondary gangrenous pneumonia. Two of these four calves had multiple defects, one with cheilo-palato-gnathoschisis (cleft lip, as well as cleft palate and upper jaw), a defect in a tail vertebra (kinked tail), a ventricular septal defect (heart malformation with an incomplete wall between the right and left ventricle), as well as aplasia (non-development) of the left kidney, rectum, and anus. The second calf had both cleft palate and brachygnatia maxilla (underbite or undershot jaw) (Figure 5.2). Another calf was diagnosed with arthrogryposis (congenital joint contractions), featuring torticollis, scoliosis, and ankyloses in both hind limbs. Arthrogryposis can be hereditary in livestock, but is also caused by viral infections (e.g. Schmallenberg virus or bluetongue virus) or intake of toxic and teratogenic plants. Josefsen and coworkers (2014) reported three malformations among 30 examined Norwegian semi-domesticated juvenile reindeer calves submitted for necropsy over a period of 13 years: two brachygnathia inferior and one case of scoliosis (abnormal, sideways curvature of the spine). One additional case of brachygnathia inferior was diagnosed at the Norwegian Veterinary Institute in an 8-month-old reindeer calf which died of emaciation, in 2017. The calf did not have a white colour and had obviously managed to feed and suck milk from its mother despite the malformation.

Polydactylism (supernumerary toes) has occasionally been reported in cervid species. Only one case in caribou has been published, describing a complete extra digit in the place of the normally absent first digit, on the right front foot of a 5-year-old caribou female in Canada (Miller and Broughton 1971). In a study of 1 226 caribou skulls from Canada, Miller and Tessier (1971)

FIGURE 5.1 Oral malformation (cheilo-palato-gnathoschisis) in a reindeer calf. The calf cannot suckle milk, and will starve, or contract aspiration pneumonia (Photo: Erik Ågren).

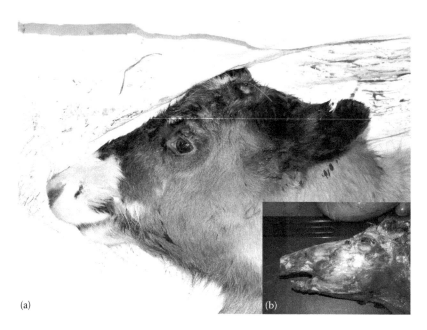

(a)   (b)

FIGURE 5.2   (a) Reindeer calf with shortened lower jaw, also called overbite or undershot jaw (*brachygnathia inferior*). (b) Skinned head of the calf (Photos: Torill Mørk).

noted 43 dental anomalies. Supernumerary teeth were noted in 13 skulls, with both extra incisors and premolars, especially peg-shaped first premolars, PM1. Twenty-five skulls were determined to have missing teeth, based on lack of indications of previous trauma or previous loss of a tooth. Five skulls had at least one tooth with extreme morphologic variation in root development or abnormal dental pattern. These dental changes in this latter study were considered to have a genetic background.

Dermoid cysts are congenital hair-filled cysts found in or just under the skin, often in the neck or throat area. The origin is thought to be an abnormal entrapment of a part of the epidermis during embryonic development. Wobeser and coworkers (2009) found dermoid cysts in eight of 557 caribou specimens sent to diagnostic labs from various geographic regions of Canada. The samples were collected as abnormal findings from hunter harvested caribou.

Studying reproductive organs, Leader-Williams (1979) found four males with abnormal testes out of 111 harvested male reindeer which were the descendants of reindeer introduced to the island of South Georgia from Norway in 1911 and 1925. Three males had cryptorchid testes, two of which were unilateral cryptorchid testes, while the third was bilateral cryptorchid. The fourth male had one vestigial testis (not further examined) of two scrotal testes. All four males had hard polished antlers, but the antlers were smaller in size in all the cryptorchid males compared to males with normal testes. Testicular abnormalities or lesions have been associated with antler abnormalities in cervids, which is also reflected among castrated male semi-domesticated reindeer. Castrated males are traditionally used as draught reindeer to pull sleds (Sámi: ackja) for the nomadic reindeer herding people. Among these animals, antlers in permanent velvet or abnormal antler shedding patterns are commonly observed.

In a study of 315 female reproductive organs collected from semi-domesticated Swedish reindeer slaughtered during the autumn and winter, within the expected pregnancy period, only ten adult females were not pregnant (Mossing and Rydberg 1982). Two of these ten barren females were noted to have morphologic malformations of the reproductive organs, but no details on the nature of the malformations were given.

5.2 TRAUMA

Trauma was found to be the second most common category of diagnosis in semi-domesticated reindeer in Norway when causes of death were documented by necropsy of fallen stock, second only to emaciation (Josefsen et al. 2014). However, only a few cases of predation were included in that study, since animals suspected to have been killed by predators in Norway usually are only investigated in the field. In another study using radio-collared animals, accidental trauma was the second most common cause of death after predation (Nybakk et al. 2002).

The most common type of trauma causing mortality can vary, depending on geographical region, landscape, presence of infrastructure such as road and railways, and lastly, but of major importance if present, large predators within the reindeer herding area or grazing lands. Josefsen and coworkers (2014) listed, in descending order, the causes of 96 trauma diagnoses (not including predation) in 300 necropsies of reindeer from northern Norway: traffic (69 cases), trauma occurring during moving or transport of reindeer herds (12), falling off steep cliffs (3), rifle or shotgun wounds (3), hit by snowmobiles (2), and unknown cause (probably traffic) (10). In the same study, 35 of the 300 animals (12%) were killed by predators. In the annual account for the reindeer herding industry in Norway, predation is by far the most commonly reported cause of losses (Ressursgrunnlaget for reindriftsnæringen 2016).

5.2.1 PREDATORS

Caribou and wild or free-ranging reindeer are natural prey for large predators present in their habitat. Losses to predation can be especially high in regions where there are multiple predatory species established; however, the effect of the presence of different combinations of predator species is not well documented. Winter condition is an important risk factor concerning predation, since some predators will take advantage of weather and snow conditions. In spring and early summer, calf losses due to predation can be severe. Several studies with radio-collared calves have been conducted. The loss of caribou calves to predation in Newfoundland was between 65 and 90% (Trindale et al. 2011, Lewis and Mahoney 2014). In studies of semi-domesticated reindeer in Finland, up to 54% of the calf mortality was caused by predation, with golden eagles (*Aquila chrysaetos*) being the main predator (Nieminen et al. 2011, Norberg et al. 2006). In a Norwegian study, it was estimated that 75.5% of the calf mortality was due to predation, with lynx (*Lynx lynx*) as the main predator (Nybakk et al. 2002). The total number of reindeer killed by predators is difficult to verify, but estimates can be as high as 70% to 83% of the total mortality (Nieminen et al. 2011, Nieminen et al. 2013). In semi-domesticated reindeer herds, fewer calves are available for predators after slaughtering in the autumn, and the risk of predation of females will therefore increase during winter. If this predation is high, the reproduction in the herd decreases and, in extreme cases, the herd can collapse (Åhman 2013).

In Fennoscandia, protection of predators and restrictive rules for hunting predator species has generally contributed to an increase of these populations, which has had a negative effect on reindeer herding production (Hobbs et al. 2012). Seip (1991) reports that wolves (*Canis lupus*) are the main predators that limit or decrease some populations of non-migratory caribou herds, especially if there is alternate prey present in the area, as wolf numbers then can be maintained even when caribou are scarce. Migratory caribou herds can sustain higher densities as they spatially separate from predators by annual migration and are to a larger extent limited by factors such as feed limitation or human harvest.

5.2.1.1 Evaluation of Predatory Trauma

It is necessary to do a partial or full necropsy to determine if a predator, including what species of predator, has killed a reindeer or caribou, and to differentiate predation from death from other causes followed by scavenging. Experience is important in evaluating the lesions found on a carcass, but there are field guides and wildlife forensic investigation textbooks available that can aid the examiner (e.g. Acorn and Dorrance 1990, Levin et al. 2008, Cooper and Cooper 2013). A kill site examination is an important part of the investigation. Predators can leave tracks and clues around the kill, as there are species specific patterns in killing, eating, and handling or hiding the remains of a carcass.

5.2.1.1.1   Wolf

Wolves have a powerful bite and their long jaws can cause large bite wounds as they often tear and pull when biting, crushing soft tissues and bone. The typical lesion pattern of a wolf kill differs, depending on the size of the prey. Large animals such as moose and adult reindeer are often killed by forceful bites in the throat or neck region, while small animals such as young reindeer calves are bitten over the back and neck, as well as on the head and nasal bridge. Bite wounds can, during the initial parts of an attack, occur high up on the side of the body, the chest area and high up on the hind limbs or forelimbs, and occasionally in the udder or tail region in adult reindeer (Skåtan and Lorentzen 2011, Rehbinder and Nikander 1999). Wolf teeth do not always penetrate the skin, but the clamping force of the bites cause severe subcutaneous damage and hemorrhage, which is notable when skinning a wolf-killed reindeer. When eating, wolves can crush all larger bones, and consume meat, bone and inner organs (Bjärvall et al. 1990). Only the forestomachs, jaws and claws may be left of the carcass, as the wolf may repeatedly return to a kill site, if undisturbed. Wolves do not usually remove parts of the carcass but may cover the body to hide it.

Wolves tend to make multiple kills in herds or corralled animals. As wolves also scatter reindeer herds by their attacks, the Sami reindeer herders traditionally consider the wolf to be the worst of all predators, and the presence of wolves is not considered compatible with reindeer husbandry (Kuhmunen 2000). Based on what reindeer husbandry sees as acceptable levels of predation of reindeer, active political decisions have been made in both Sweden and Norway to disallow establishment of reproducing wolf pairs within the semi-domesticated reindeer herding areas, which is roughly the northern half of both countries.

5.2.1.1.2   Domestic Dog

Unattended domestic dogs can kill reindeer, and the lesions may be similar to those made by wolves, making it difficult to determine the species of the attacking predator. Swabbing skin bite wounds on the prey for genetic analysis of the saliva left by the predator can aid in resolving these cases. Most dogs have less powerful bites compared to a wolf. When dogs attack reindeer there are more often multiple bite wounds that vary in size depending on the size of the dog. Dogs tend to bite and get hold of a skin fold, which creates wounds with torn skin, but less damage compared to wolf bites. Lesions can be found all over the body, but some larger dogs may bite mainly in the neck and throat area, as wolves do. In general, domestic dogs often bite low on the body or on the legs when chasing prey, and injuries may be found mainly on the hind legs, with multiple bites and subcutaneous hemorrhage (Skåtan and Lorentzen 2011) (Figure 5.3).

5.2.1.1.3   Brown Bear

Brown bears (*Ursus arctos*) predate on reindeer when they are available, especially on newborn calves in the calving grounds during early spring, when the bears wake up hungry after the winter denning period (Karlsson et al. 2012). A single bear can cause high losses of newborn calves within a herd. The calves are swiftly killed by a bite over the head or back, or just crushed to the ground by a paw. Bears use their paws to catch and keep their prey pinned down, using bites to make their kill, but can also start to feed on a live animal. Larger prey is killed with multiple bites, usually over the back, neck, or the head, and sometimes over the nose, but more over the skull compared to the wolf. Bears typically cause deep and severe bite wounds, but this varies depending on the size and age of the bear (Skåtan and Lorentzen 2011). Bear claw injuries can sometimes be noted in the skin and fur of larger prey that survive attacks. After a kill, bears may drag the

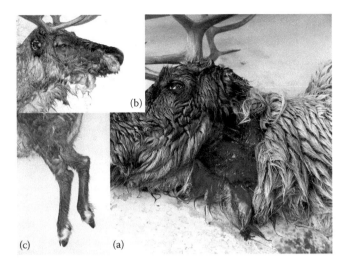

FIGURE 5.3 (a) Reindeer with bite wound in the throat from of a domestic dog. (b) Bite wound with skin torn off the nose. (c) Bite wound on the hind leg (Photos: Terje D. Josefsen).

FIGURE 5.4 Nine reindeer calves killed by one brown bear during one night in a calving area (Photo: Erik Ågren, SVA).

carcass away from the killing site, and cover the remainder with vegetation or snow, or place it in water. Bears hibernate during winter, and they also use a variety of other food sources, so they cause less overall damage to the reindeer population than the other large predators (Sikku and Torp 2008) (Figure 5.4).

5.2.1.1.4 Lynx

Lynx (*Lynx lynx* and *Lynx canadensis*) hunt by stalking and attacking prey from cover, leaping onto the back and biting at the throat or neck, killing efficiently with one or very few bites (Krofel et al. 2009). The bite wounds over the throat are typical, and can be found after skinning the carcass. The teeth are sharp, and distinct penetrating wounds are normally found in the skin. However, there are some cases when the teeth do not penetrate the skin, maybe due to the

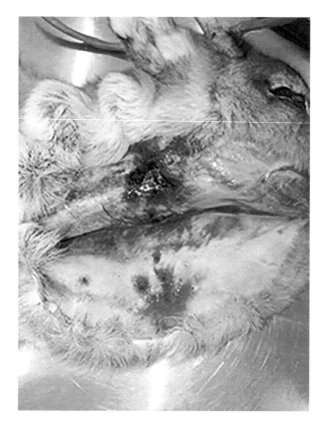

FIGURE 5.5   Bite wound from a lynx in the throat of female reindeer. Minimal local tissue damage, with hemorrhage from the sharp canine teeth (Photo: Terje D. Josefsen).

thick layer of winter fur, leaving only less distinct bleeding in the subcutis and throat (Skåtan and Lorentzen 2011). Measuring the distance between obvious canine teeth bite marks can aid in identifying the lynx as a predator, but can be difficult if there are several bites and tearing of the skin. The sharp claws can cause parallel lacerations in the fur and skin, or discrete puncture wounds on the sides and back of the body. Lynx prefer to prey on calves but also kill adult reindeer. After pulling away wads of the thick fur, they mainly eat the muscle tissue, usually first from a hind leg or a shoulder. The abdomen of a fresh kill is rarely opened. Large bones are rarely broken, and lynx do not dismember or bite off the head of carcasses. A lynx can stay and eat from a carcass for several days if undisturbed. The remains of the carcass may be removed into a nearby thicket, or covered with leaves, moss, or snow, but not as carefully as a bear does. Lynx do not scavenge on other predator-killed prey and may have trouble eating from a frozen carcass (Bjärvall et al. 1990). A female lynx with cubs living in areas with reindeer herds is estimated to kill 30 reindeer in one winter season (Pedersen et al. 1999) (Figure 5.5).

5.2.1.1.5   Wolverine

Wolverines (*Gulo gulo*) can be very efficient in hunting reindeer when there is deep snow with an icy crust that carries the weight of the wolverine but not a reindeer. The wolverine can then easily pursue a reindeer until it is exhausted from running in the deep snow. In such snow conditions, multiple reindeer in a herd may be killed (Bjärvall et al. 1990). The wolverine's most important prey is reindeer where the two species co-exist (Haglund 1966). In autumn, a wolverine may kill several animals and store the carcasses for feeding on during the winter. Wolverines tend to prey on calves but can also kill adult reindeer. Despite its short jaws, the bite of a wolverine is very powerful and can crush thick

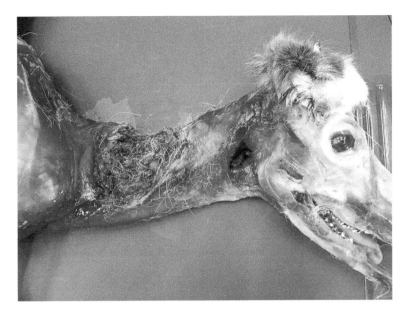

FIGURE 5.6   Wolverine bite wound in the neck of a reindeer. The wolverine has a powerful bite that can crush bone tissue, but in this case the shown lesion is not a fresh wound, and the reindeer survived the attack but was put down for animal welfare reasons (Photo: Terje D. Josefsen).

bones such as neck and thoracic vertebrae (Skåtan and Lorentzen 2011). Wolverines typically attack by jumping up on the back of the reindeer. Powerful bites damage the neck and dorsal back region, until the reindeer is immobilized and recumbent. The wounds are seldom instantly fatal, and the wolverine may feed on the muscle of the still live prey (Bjärvall et al. 1990). Larger carcasses are often dismembered and a carcass, or parts of it, is often removed and hidden away, buried, or placed in water sources. When other predators are present, the wolverine will scavenge the remains of kills of other predators in the same area, being a typical scavenger (Skåtan and Lorentzen 2011) (Figure 5.6).

5.2.1.1.6   Golden Eagle

Golden eagles (*Aquila chrysaetos*) typically kill young calves in spring and early summer, but yearlings or adults are also occasionally killed (Nybakk et al. 1999). Reindeer herds seem more at risk in open landscape and highlands where the eagles can kill a large number of calves during summer (Nieminen et al. 2011). Eagles only make stab wounds in the tissues (Skåtan and Lorentzen 2011). The long talons cause deep stab wounds in the skull and brain of young calves, or in the throat when the eagle strikes at the head, or when striking from above at the chest, making perforations into the lungs, leading to collapse of the lungs (pneumothorax) and hemorrhage (Bjärvall et al. 1990). They typically start feeding by making openings between the ribs, and do not crush bones when eating, although ribs may break, especially in calves (Bergo 1990). Golden eagles are also scavengers, and are often seen on carcasses. White-tailed eagles (*Haliaeetus albicilla*) may also be observed as scavengers on carcasses, but they have not been documented to actively hunt larger prey such as reindeer (Figure 5.7).

5.2.1.1.7   Red Fox

Red foxes (*Vulpes vulpes*) normally kill young calves with multiple bite wounds in the throat, but bite wounds can also be found on the neck or skull, and on the side or hindquarters of larger calves. The small, sharp canine teeth penetrate the skin and multiple perforating bite wounds are often found at necropsy. Larger bones are not crushed, but the head or a complete leg can be dismembered and dragged away (Bjärvall et al. 1990). Coyotes (*Canis latrans*) in North America, similar to the red fox, may predate on young calves and scavenge on kills from other predators (Lewis and Mahoney 2014).

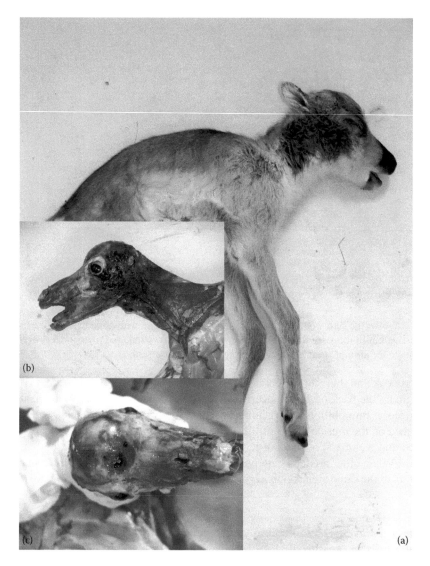

FIGURE 5.7   (a) Reindeer calf killed by golden eagle. (b) Bleeding in the neck and throat. (c) Puncture wounds in the skull from the sharp talons (Photos: Siri Knudsen).

5.2.1.1.8   Raven

Ravens (*Corvus corax*) are mainly scavengers but may attack and kill newly born calves. The eyes are typically missing, as well as the tongue and anal region of the prey. Several ravens can collaborate in attacking and killing weak adult reindeer by puncturing the thorax and abdomen of recumbent animals (Rehbinder and Nikander 1999).

5.2.2   TRAFFIC

Although northern Norway is sparsely populated, 66 of 96 trauma cases reported by Josefsen and coworkers (2014) were due to road traffic, with slightly more killed (65% of the road kills) during the winter period. In the yearly account for reindeer herding in Norway, 1,038 reindeer were reported killed by traffic in 2015 (Ressursregnskapet for reindriften 2016). In Sweden, which has a similar number of semi-domesticated reindeer as Norway (about 250,000), there were, on average, about 1,500 road-killed

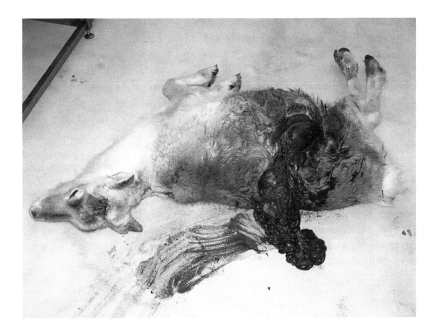

FIGURE 5.8   Road-killed reindeer calf with massive blunt trauma (Photo: Terje D. Josefsen).

and 1,100 train-killed reindeer reported every year between 2012 and 2016 (statistics retrieved in December 2017 from www.viltolycka.se, The National Wildlife Traffic Accident Council [Nationella Viltolycksrådet], and www.sametinget.se) (Figure 5.8). Reindeer sometimes prefer to move along plowed roads when there is deep snow cover elsewhere, increasing the risk of being hit by vehicles. Loss of animals can be especially high if a herd is moving on a railroad track when a train comes at full speed. Secondary train-strike fatalities of scavengers such as golden eagles, wolves, or brown bears occur regularly when they are feeding on train-struck cervid carcasses, as it often takes some time before staff can clear the tracks of animal remains (unpublished: wildlife case diagnosis database, SVA, Sweden).

5.2.3   ENVIRONMENTAL TRAUMA

The hairs of the reindeer's winter coat contain insulating air, which keeps their bodies very buoyant when they are swimming. In calm waters, drowning is unlikely. In Norway, reindeer swim out to islands to graze in early spring, while still in their winter coat. There is greater risk of drowning during spring migration on lakes with rotten ice, or on frozen regulated rivers where the water level has been lowered during winter, creating a space between the ice and the underlying water. Other types of "drowning" occur when neonatal calves find themselves in deep, soft snow. When it sinks into the snow, the calf can suffocate, or if it is unable to move to its mother, the calf quickly dies of hypoglycemia/starvation (Rehbinder and Nikander 1999). Females usually choose to calve on dry and snow-free areas of hills or slopes, but there is usually still snow cover present at the time of calving. Young calves can also get stuck in bogs or deep mud, fall into crevices, and die from exposure when the weather is both cold and very wet during the calving period (unpublished, Erik Ågren).

Chasing reindeer and other herd animals over the edge of a cliff is a known ancient hunting method. Even nowadays, herds of reindeer can be accidentally frightened or chased by predators into areas of steep slopes or cliffs, resulting in falls and mass mortality. Reindeer are also at risk of falling down and impacting on boulders when grazing on steep rocky mountain slopes with snow-free patches, especially in winters with heavy snow (Skjenneberg and Slagsvold 1968). Both semi-domesticated and wild reindeer herds have been accidentally caught in snow avalanches, with several examples of 100 to 200 animals found dead in the snow masses, as noted by Sami reindeer herders and official rangers in Norway (unpublished). Occasional incidents have been reported, one from 1977 in the Bunnerfjällen, a mountain area in the county of Jämtland, in Sweden. Fifteen reindeer carcasses were found spread out on a steep and very icy slope with an incline of 20–30°. The carcasses were up to 200 meters apart and were scavenged, but it was concluded that the reindeer herd had been frightened onto the slope and succumbed by tumbling down the mountain side (Bjärvall et al. 1990). In another report from April 2017, almost 100 reindeer died in northern Norway after falling down steep mountain cliffs when being herded to summer pasture. The animals likely slipped on icy snow, and one herder almost slid down as well, but managed to cling onto the hillside (Rensberg 2017).

Lightning strikes can kill animals – with four contact points with the ground, there are multiple pathways for the electrical current to pass through the body. Lightning strike has not been recorded as a cause of death in semi-domesticated reindeer, but it probably occurs occasionally. An unusual mass mortality due to lightning strike was recently documented (2016) when a flock of 323 Norwegian wild Eurasian tundra reindeer were found dead on a mountain top after a summer thunder storm in the area. The carcasses were found grouped close together, without any external signs of trauma (Miljødirektoratet 2016).

5.2.4 MANAGEMENT TRAUMA

If semi-domesticated reindeer are crowded or stressed in handling pens or during loading into transports, it can result in trauma from sharp antler tines. Lesions that arise during loading can often be found in the rump or genital areas, as one reindeer butts on the rump of a hesitating animal in front, in the narrow driveways or loading ramps. In the corrals, a reindeer herd always rotates in the same direction (often a counter-clockwise direction), and the rotating herd necessitates using rounded corrals, without angled corners. When driving a reindeer herd, calves can get trampled if the herd panics for any reason. Calves trying to escape through fencing without a tarp cover may push their head or legs through the fence mesh and contract serious skin, leg or antler (if in velvet) lesions when thrashing to get freed from the fence. Driving reindeer in forested areas can occasionally result in impalement on sharp dry branches (Rehbinder and Nikander 1999).

Mechanical trauma to eyes by antler tines can be the result of dominance conflicts at crowded feeding troughs if there are not enough troughs or if the feed is not spread out well enough to allow simultaneous feeding of the herd at the site (Figure 5.9). Also, feed particles from dusty and broken pelleted reindeer fodder can cause eye irritation and traumatize the cornea and conjunctiva, creating lesions that can subsequently be infected by bacteria. Similarly, dust from dry bare ground in handling areas can be a mechanical contributor, in addition to other factors, resulting in outbreaks of eye infections, when the reindeer are milling around during marking or sorting in the corrals (Rehbinder and Nikander 1999). Other contributing factors during such outbreaks are stress, Cervid herpesvirus 2, and various bacteria (see also Chapter 7, Bacterial Infections and Diseases and Chapter 8, Viral Infections and Diseases).

Antler fractures can easily occur if a lasso is caught on an antler in velvet (Skjenneberg and Slagsvold 1968), which is to be avoided. Calves that are caught with a lasso over both antlers may contract a skull fracture in the antler base area. As the lasso tightens, the antlers are pressed together. Even with apparently minor skull fractures, the following subdural hemorrhage causes pressure on the brain and may lead to death shortly afterwards (Rehbinder and Nikander 1999). Pelvic lesions from a splay-leg injury can occur if the lasso is caught in a hind leg and there is too much struggling. Using a lasso causes the whole corralled herd to run, milling around. An alternative to using a lasso, avoiding the stress of continuous milling around in a corral to catch calves during calf marking in the summer, is to use a small plastic snare loop at the end of a long pole to carefully snare a hind foot, a method used by many reindeer herders nowadays.

Antlers may be cause of traumatic death. However, the thick winter fur of reindeer acts as a very good padding and protection against a lot of trauma, and serious wounds do not seem very common,

FIGURE 5.9 Reindeer crowding around trough to feed, with increased risk for antler trauma to eyes or face (Photo: Erik Ågren).

despite sometimes quite rough handling (Skjenneberg and Slagsvold 1968). Sharp antler tines may cause penetrating wounds into the thorax or abdomen during handling of reindeer herds in corrals or for truck transports, with limited space for the reindeer to retreat in a conflict situation. These accidents happen when it gets too crowded and an aggressive dominant reindeer uses its antlers to move away other reindeer. A penetrating wound may lead to a generalized infection, with peritonitis, pleuritis or pneumothorax. Occasionally, when males fight during the rut, the pair may entangle the tines of their antlers so that they are solidly locked to each other (Figure 5.10). Both antagonists may die of exhaustion, stress and dehydration, and even only skeletal remains provide obvious evidence of what has happened (Rehbinder and Nikander 1999).

FIGURE 5.10 Reindeer bulls with interlocked antlers. This occasionally happens when bulls fight during the rut, leading to the death of both animals (Photo: Erik Ågren).

5.3 POISONING

There are few reports of poisoning in reindeer. This may be explained by the animal's being well adapted to its natural pasture and being a selective grazer. However, there are most likely undiagnosed and unreported cases of poisoning, since reindeer mostly roam freely and also close to human settlements, where they could encounter contaminants or imported poisonous garden plants.

5.3.1 Plant Poisoning

Reindeer grazing close to human settlements can be at risk of getting poisoned by eating toxic garden plants. Yew (Swedish: idegran, Norwegian: barlind) (*Taxus* sp.) are evergreen trees and shrubs, several species of which are used as ornamental shrubs. The taxine alkaloids are cardiotoxic and can cause acute heart failure and cardiac arrest; acute intoxication results in death within hours after ingestion. Most poisonings are reported in horses and cattle; however, there are reports of cases in sheep, deer, poultry, and game birds (Burrows and Tyrl 2004). Rhododendron (*Rhododendron*), another evergreen plant commonly found in gardens, along with other species in the Ericaceae family, contains gryanotoxins. This toxin binds to sodium channels in cell membranes of nerves, and heart and skeletal muscles. Rhododendron poisoning has been reported in goats, sheep and cattle, as well as in some exotic animal species (Burrows and Tyrl 2004). A possible case of poisoning from English Yew (*Taxus baccata*) and a case of rhododendron poisoning in reindeer were reported in northern Norway. The animals were found dead in residential areas. The autopsy findings were unspecific, but leaves and plant remains of the respective plants were found in the ruminal contents (Josefsen et al. 2014) (Figure 5.11).

Water hemlock (*Cicuta virosa*) is well known for causing deadly poisoning in several animals as well as in humans. The root and stem base contain cicutoxin, which affects the CNS and causes seizures. The toxic dose is very low, and animals usually die within a few hours after consuming a lethal amount of the plant (Burrows and Tyrl 2004). The plant grows mainly in wetlands and is widespread in North America and in Europe. There are single reports of poisoning of reindeer in

FIGURE 5.11 Poisonous plants: Yew (*Taxus* sp.) found in rumen content of a reindeer (Photo: Terje D. Josefsen).

northern Norway and Sweden in such landscapes (Skjenneberg and Slagsvold 1968, Rehbinder and Nikander 1999).

Witch's hair (*Alectoria sarmentosa*) is a lichen found on tree branches. It is traditionally used as forage for semi-domesticated reindeer, and herders cut down branches that are out of reach for the reindeer when other feed is scarce. Reindeer herders have reported toxic effects in reindeer when large amounts had been given, causing a condition where the head is bent back towards the withers (opistotonus) (Skjenneberg and Slagsvold 1968). The description resembles clinical signs of central nervous system lesions, such as cerebrocortical necrosis, which can be caused by thiamine deficiency in ruminants. However, this condition has never been confirmed with diagnostic methods.

Reindeer do not naturally feed on branches and bark that contain tannins. They lack the salivary tannin-binding protein, found in red deer (*Cervus elaphus*) and moose (*Alces alces*), that binds toxic plant tannins. Tannin poisoning of reindeer is reported by zoos that have fed reindeer with pelleted fodder produced for farmed deer (Rehbinder and Nikander 1999).

Sodium nitrite poisoning, known to cause methemoglobinemia, has been documented experimentally in reindeer. The purpose of the experiment was to evaluate the risk of poisoning in reindeer that grazed in fertilized forests. However, the reindeer seem to avoid drinking water contaminated with fertilizers (Nordkvist et al. 1984).

5.3.2 Radioactivity

In 1986, the nuclear plant accident in Chernobyl, Russia, caused the spread of radioactive pollution throughout Europe, and the Nordic countries in particular. This caused pollution of reindeer pastures in central Norway and Sweden, and grazing animals were contaminated with radioactive particles. Reindeer were especially exposed by grazing lichens, which do not have roots, as they receive water and nutrition, as well as radioactivity, from the air. The contamination of lichens has gradually been reduced as the plants have slowly grown in size. Presently, radioactive cesium is still present in the soil and contaminates mushrooms, which are browsed by reindeer in the fall season. After 30 years, control of radioactivity in live animals before slaughter is still done in herds within the known contaminated areas. Reindeer selected for slaughter in these areas are corralled and fed with cesium-free pelleted fodder or a cesium-binding substance for a period to eliminate all cesium in the muscle tissue before slaughter (Statens Strålevern 2016). There are no reports of clinical effects of elevated cesium in reindeer from these areas.

5.4 MISCELLANEOUS NON-INFECTIOUS DISEASES

5.4.1 Tying-Up

Lameness of the hind quarters has been described in draught reindeer, used for pulling sleds or doing hard work, especially after a period of rest. The background and symptoms are similar to "tying-up" in horses, known as equine exertional rhabdomyolysis, azoturia or Monday morning disease. A breakdown of muscle tissue results in pain and dark red urine, stained by myoglobin (Skjenneberg and Slagsvold 1968, Rehbinder and Nikander 1990).

Historically, arthritic joints causing symptoms of stiff joints (Sámi: Roavve-vikke) that were swollen and fluid-filled were seen in hard-working draught reindeer (Skjenneberg and Slagsvold 1968).

5.4.2 Deficiencies

Specific deficiencies such as lack of vitamins or minerals are, in general, not well documented among reindeer and caribou, except for magnesium deficiency (grass tetany) and possible copper deficiency. Nutritional needs and deficiencies are dealt with in Chapter 3, *Rangifer* Diet and Nutritional Needs.

5.4.3 Hypomagnesemia

Seasonal hypomagnesemia has been reported in reindeer in Norway and Sweden (Åhman et al. 1986, Hoff et al. 1993). Acute hypomagnesemia in ruminants is usually caused by feeding on plants with low levels of Mg in combination with high nitrogen contents, as in rapidly growing grass in spring (Peek and Divers 2008). A more chronic, slowly developing deficiency can occur during winter when feeding on pastures low in magnesium, such as lichen (Hoff et al. 1993).

Clinical symptoms are correlated to the degree of deficiency and vary from mild symptoms of ataxia and weakness to light nervousness and excitability and, in severe cases, seizures and death if untreated (Zelal 2017). Subclinical hypomagnesemia may occur without symptoms, other than unthriftiness and weight loss, but may change suddenly to acute disease when the animal is exposed to stress (Hoff et al. 1993, Zelal 2017). Hoff and coworkers (1993) reported hypomagnesemia in a reindeer herd connected to a special winter pasture area in Norway. Reindeer herders had reported unexpected deaths of reindeer grazing in this area during winter. The animals showed no clinical symptoms but died when exposed to physical activity. Blood analyses showed low levels of magnesium compared to other herds grazing in nearby regions and with similar winter conditions.

5.4.4 Hormonal Disturbances

5.4.4.1 Perruque Antlers

Tumour-like growths on deer antlers are variously called perruque antlers, antleromas or cactus antlers, and are likely due to hormonal disturbances. They are seen in cervid males, most commonly in roe deer (*Capreolus capreolus*), and even roe deer females are infrequently noted to have perruque antler formation (Ågren et al. 2016). Occasionally, perruque antlers are seen in other deer species such as red deer (*Cervus elaphus*), white tailed deer (*Odocoileus virginianus*), and moose (*Alces alces*), and more seldom in reindeer. In males, the condition is seen when there are lesions or disease damaging the testicles, resulting in disturbed hormone production. The hormonal cycling of cervid antler shedding is shut down after castration and the velvet continues to grow when sex hormones are missing (Bubenik et al. 1975). In reindeer this is apparent if males are castrated when the antlers are in velvet, as nodular fibromatous masses of velvet may form over time (Foster et al. 2013). When extremely overgrown, the velvet masses can cover the eyes, there may be infections after trauma to the velvet, or fly strikes can reduce fitness and even lead to death (Figure 5.12).

5.5 NEOPLASIA

Most cases of neoplasia are reported through meat control registries and the incidence seems low (Rehbinder and Nikander 1999). Most cases are reported in older females and there is a wide variety of tumour types found. There are a few reports of lymphomas (Kummeneje and Poppe 1978, Järplid and Rehbinder 1995, Josefsen et al. 2014) (Figure 5.13). Other types of tumours are mainly reported as single cases, such as squamous cell carcinoma in the eye ("cancer eye"), malignant melanoma, and different types of liver tumours: hepatic carcinoma, hemangiocarcinoma, hepatocellular adenoma and bile duct adenoma (Rehbinder and Nikander 1999) (Figure 5.14). Dermoid cysts in reindeer have been described by Rehbinder and Nikander (1999) and Josefsen et al. (2014). Wobeser and coworkers (2009) reported eight cases of dermoid cysts in caribou, all found in the cervical region (Figure 5.15). Fibropapillomas caused by papilloma virus are occasionally reported in reindeer and include some severe cases of generalized papillomatosis (Josefsen et al. 2014). See also Chapter 8, Viral Infections and Diseases.

FIGURE 5.12 Perruque antlers of reindeer, due to hormonal disturbance, resulting in loss of normal shedding of antler velvet. This may be a result of testicular trauma or following castration of males while antlers are in velvet (Photo: SVA, Sweden).

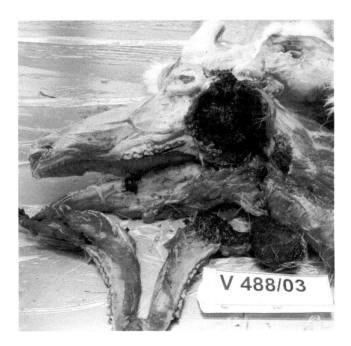

FIGURE 5.13 Neoplasia: Malignant melanoma in an aged white-coated female reindeer. The black pigmented tumour had eroded soft tissues in the left jaw area, and metastastic tumour tissue was found in the neighboring lymph node (Photo: Erik Ågren).

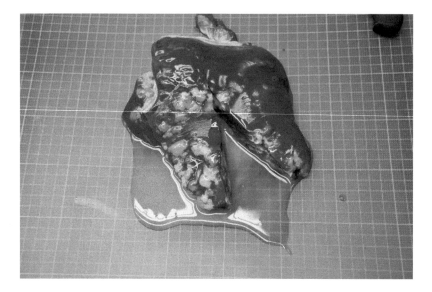

FIGURE 5.14   Reindeer liver, multiple bile duct cysts. A benign lesion of the bile ducts, regularly found as single or sometimes multiple cysts with clear liquid contents. Larger monocystic cysts need to be examined to be differentiated from possible *Echinococcus* parasite cysts (Photo: SVA).

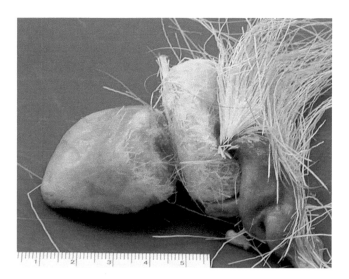

FIGURE 5.15   Reindeer haired skin from the cervical region, with a 4 cm large dermoid cyst packed with white hairs, growing down into the subcutaneous tissue. This is a benign malformation occasionally found in reindeer and caribou (Photo: Erik Ågren).

REFERENCES

Acorn, R., and M. Dorrance. 1990. Methods of investigating predation on livestock. Alberta Agriculture and Rural Development, Alberta, Canada. www1.agric.gov.ab.ca/$Department/deptdocs.nsf/all/agdex44/$FILE/684-14 .pdf (Accessed 15 July 2016).

Ågren, E. 2001. Malformations in reindeer calves. Paper presented at the NOR 11th Nordic Research Conference on Reindeer and Reindeer Husbandry, 18–20 June, in Kaamanen, Finland. p. 101. septentrio.uit.no/index .php/rangifer/article/viewFile/1540/1446 (Accessed 13 July 2016).

Ågren, E.O., H. Uhlhorn, C. Bröjer, and D. Gavier-Wideén. 2016. Studies on peruke antlers in roe deer (*Capreolus capreolus*). Conference abstract. *J Comp Path* 154:98.

Åhman, B. 2013. Breakdown/collapse of a reindeer herd. [Renhjord i kollaps.] *Report 285*. Department of Animal Nutrition and Management, Swedish University of Agricultural Science, Uppsala [In Swedish].

Åhman, B., A. Rydberg, and G. Åhman. 1986. Macrominerals in free-ranging Swedish reindeer during winter. *Rangifer* 1:31–8.

Anonymous. 2016. [30 år siden Tsjernobyl: Kor mange fleire år med tiltak i reindrifta?] Norwegian Radiation Protection Authority. Stråleverninfo 4, 16:1–2.

Anonymous. 2016. [Ressursregnskapet for reindriftsnæringen]. *Report 14/2016*. Norwegian Agricultural Authority.

Anonymous. The National Wildlife Traffic Accident Council [Nationella Viltolycksrådet]. Retrieved on Dec 15, 2017. www.viltolycka.se.

Anonymous. The Sami parliament [Sametinget]. Retrieved on Dec 15, 2017. www.sametinget.se.

Bergo, G. 1990. Eagle damage on livestock and deer. [Ørneskader på småfeog hjortedyr.] *NINA Scientific Report No. 09:1–37*. Trondheim, Norway. ISBN 82-426-0089-9. [In Norwegian, abstract in English].

Bjärvall, A., R. Franzén, M. Nordkvist, and G. Åhman. 1990. Renar och rovdjur. Naturvårdverket förlag, Solna, Sweden, 296 pp. [In Swedish].

Bubenik, G.A., A.B. Bubenik, G.M. Brown, and D.A. Wilson. 1975. The role of sex hormones in the growth of antler bone tissue: I: Endocrine and metabolic effects of antiandrogen therapy. *J Exp Zool* 194:349–58.

Burrows, G., and R.J. Tyrl. 2004. Plants. In: *Veterinary Toxicology*, Ed: Plumlee, K.H. Mosby Inc.

Cooper, J.E., and M.E. Cooper. 2013. *Wildlife Forensic Investigation: Principles and Practice*. CRC Press, Boca Raton, Florida, USA.

Elvestad, K., and M. Nordkvist. 1985. A case of brachygnathia inferior and palatoschisis in a reindeer calf. *Rangifer* 5:22–7.

Foster, A.P., A.M. Barlow, L. Nasir et al. 2013. Fibromatous lesions of antler velvet and haired skin in reindeer (*Rangifer tarandus*). *Vet Rec* 172:452.

Haglund, B. 1966. De stora rovdjurens vintervanor I. In: *Viltrevy 4*, Almquist & Wiksells Boktryckeri AB, Uppsala, pp. 81–299.

Hobbs, N.T., H. Andrén, J. Persson, M. Aronsson, and G. Chapron. 2012. Native predators reduce harvest of reindeer by Sámi pastoralists. *Ecol Appl* 22:1640–54.

Hoff, B., A. Rognmo, G. Havre, and H. Morberg. 1993. Seasonal hypomagnesemia in reindeer on Kautokeino winter pasture in Finnmark county, Norway. *Rangifer* 13(3):133–6.

Järplid, B., and C. Rehbinder. 1995. Lymphoma in reindeer (*Rangifer tarandus tarandus* L.). *Rangifer* 15:37–8.

Josefsen, T., T. Mørk, K.K. Sørensen, S.K. Knudsen, H.J. Hasvold, and L. Olsen. 2014. Findings at necropsy or by examination of organs from reindeer (*Rangifer tarandus tarandus*) in Norway 1998–2011. *Norsk veterinærtidsskrift*, 126:174–84. [In Norwegian, abstract in English].

Karlsson, J., O.-G. Støen, P. Segerström et al. 2012. *Bear predation on reindeer and potential effects of three preventive measures*. [Björnpredation på ren och potentiella effekter av tre förebyggande åtgärder]. Report from Viltskadecenter 2012:6, ISBN: 978-91-86331-50-4.

Krofel, M., T. Skrbinsek, F. Kljun, H. Potocnik, and I. Kos. 2009. The killing technique of Eurasian lynx. *Belg J Zool* 139:79–80.

Kuhmunen, N. 2000. *Reindeer herding in Sweden, then and now* [Renskötseln i Sverige förr och nu]. Sápmi 11/2000. Sámiid Riikasearvi/SSR. ISBN: 91-8820-406-5. [In Swedish or Sámi].

Kummeneje, K., and T.T. Poppe. 1978. A case of multiple cutaneous malignant lymphoma in reindeer, probably of immunologic-parasitic origin. *Acta Vet Scand* 19:129–32.

Leader-Williams, N. 1979. Abnormal testes in reindeer, *Rangifer tarandus*. *J Reprod Fertil* 57:127–30.

Levin, M., J. Karlsson, L. Svensson, M. HansErs, and I. Ängstig. 2008. *Inspection of predator-killed livestock*. Edita Sverige AB, Västerås, Sweden. ISBN: 978-91-977318-0-5. [In Swedish].

Lewis, K.P., and S.P. Mahoney. 2014. Caribou survival, fate, and cause of mortality in Newfoundland: A summary and analysis of the patterns and causes of caribou survival and mortality in Newfoundland during a period of rapid population decline (2003–2012). *Tech Bull No. 009*. Sustainable Development and Strategic Science, Department of Environment and Conservation, Government of Newfoundland and Labrador, St. John's.

Miller, F.L., and E. Broughton. 1971. Polydactylism in a barren-ground caribou from northwestern Manitoba. *J Wildl Dis* 7(4):307–9.

Miller, F.L., and G.D. Tessier. 1971. Dental anomalies in caribou, *Rangifer tarandus*. *J Mammal* 52:164–74.

Miljødirektoratet. 2016. 323 villrein drept av lynet. http://www.miljodirektoratet.no/no/Nyheter/Nyheter/2016/August-2016/322-villrein-drept-av-lynet/.

Mossing, T., and A. Rydberg. 1982. Reproduction data in Swedish domestic forest reindeer (*Rangifer tarandus* L.). *Rangifer* 2:22–7.

Nieminen, M., H. Norberg, and V. Maijala. 2011. Mortality and survival of semi-domesticated reindeer (*Rangifer tarandus tarandus* L.) calves in northern Finland. *Rangifer* 31:71–82.

Nieminen, M., H. Norberg, and V. Maijala. 2013. Calf mortality of semi-domesticated reindeer (*Rangifer tarandus tarandus*) in the Finnish reindeer-herding area. *Rangifer* 33:79–90.

Nordkvist M., C. Rehbinder, S.C. Mukherjee, and K. Erne. 1984. Pathology of acute and subchronic nitrate poisoning in reindeer (*Rangifer tarandus* L). *Rangifer* 4:9–15.

Norberg, H., I. Kojola, P. Aikio, and M. Nylund. 2006. Predation by golden eagle *Aquila chrysaetos* on semi-domesticated reindeer *Rangifer tarandus* calves in Northeastern Finnish Lapland. *Wildl Biol* 12: 393–402.

Nybakk, K., O. Kjelvik, and T. Kvam. 1999. Golden eagle predation on semidomestic reindeer. *Wildlife Soc B* 27(4):1038–42.

Nybakk, K., O. Kjelvik, T. Kvam, K. Overskaug, and P. Sunde. 2002. Mortality of semi-domestic reindeer *Rangifer tarandus* in central Norway. *Wildl Biol* 8(1):63–8.

Pedersen, V.A., J.D.C. Linnell, R. Andersen, H. Andrén, M. Lindén, and P. Segerström. 1999. Winter lynx *Lynx lynx* predation on semi-domestic reindeer *Rangifer tarandus* in northern Sweden. *Wildl Biol* 5:203–11.

Peek, S.F., and T.J. Divers. 2008. Metabolic diseases. In: *Rebhun's Diseases of Dairy Cattle*. Ed: Divers, T.J. and Peek, S.F., Elsevier Inc.

Rehbinder, C., and S. Nikander. 1999. Reindeer and reindeer diseases. [Ren och rensjukdomar.] Studentlitteratur Lund. [In Swedish].

Renecker, L., and J.E. Blake. 1992. *Congenital Defects in Reindeer: A Production Issue*. University of Alaska Fairbanks, Circular 87.

Rensberg V. 2017. [Nærmere 100 reinsdyr stupte i døden]. NRK Sápmi, 19.04.2017. www.nrk.no/sapmi /naermere-100-reinsdyr-stupte-i-doden-1.13479229.

Roffe, T.J. 1993. Perinatal mortality in caribou from the Porcupine herd, Alaska. *J Wildl Dis* 29:295–303.

Schmit, E.V. 1937. The case of an inherited deformity in reindeer. *Arctic Inst USSR, The Soviet Reindeer Industry*. 9:103–6. Quoted in Elvestad and Nordkvist, 1985.

Seip, D.R. 1991. Predation and caribou populations. *Rangifer* Special Issue No 7:46–52.

Sikku, O.J., and E. Torp. 2008. *The wolf is worst: Traditional Saami knowledge on predators*. [Vargen är värst: Traditionell samisk kunskap om rovdjur] in Swedish. Jamtli förlag/Jämtlands läns museum. 152 pp. ISBN: 978-91-7948-216-9

Skåtan, J.E., and M. Lorentzen. 2011. *Drept av rovvilt? Håndbok for dokumentasjon av rovviltskade på husdyr og tamrein*. Statens naturoppsyn, ISBN: 978-82-7072-887-9.

Skjenneberg, S., and L. Slagsvold. 1968. *Reindeer husbandry and its ecological principles*. Translated from Norwegian [Reindriften og dens naturgrunnlag. Universitetsforlaget, Oslo, Norway]. Pub. by U.S. Dept. Interior, Bureau of Indian Affairs, Juneau, Alaska.

Trindale, M., F. Norman, K.P. Lewis et al. 2011. Caribou calf mortality study. A summary and analysis of the patterns and causes of caribou calf mortality in Newfoundland during a period of rapid population decline: 2003, 2007. *Sustainable Development and Strategic Science*, Government of Newfoundland and Labrador, St. John's, NL.

Wobeser, G., T. Bollinger, A. Neimanis, K.B. Beckmen. 2009. Dermoid cysts in caribou. *J Wildl Dis* 45:505–7.

Zelal, A. 2017. Hypomagnesemia tetany in cattle. *J Adv Dairy Res* 5:178.

6 Parasitic Infections and Diseases

Susan J. Kutz, Sauli Laaksonen,
Kjetil Åsbakk and Arne C. Nilssen

CONTENTS

6.1 WHAT IS A PARASITE?

The classic definition of a parasite, established by Crofton in 1971 (Crofton, 1971), consists of the following:

1. A parasitic relationship is an ecological relationship between **two different organisms/ species**, one designated the parasite, the other the host.
2. The parasite is physiologically or metabolically dependent upon its host.
3. Heavily infected hosts may be killed by their parasites or harm will be done to the host.
4. The **reproductive potential** of the parasites exceeds that of their hosts.
5. There is an **overdispersed frequency distribution** of parasites within the host population. In other words, the parasite population is not evenly or randomly distributed among the host population; it is clumped, such that some hosts have a lot of parasites and most have very few.

By definition, parasites depend on other organisms to complete their lifecycles, and they can harm their hosts, particularly if present in high numbers. Parasites tend not to be evenly distributed among a population of hosts; rather, a few animals usually harbor the majority of the organisms (clumped distribution). This means that under "normal" conditions most animals in a host population will be unaffected or marginally affected, while a few will be more severely affected, resulting in reduced fitness of these individuals. In a wild population exposed to predation, a range of habitat quality and severe weather events, these heavily infected animals will be eliminated or populations will be regulated by parasites (see Chapter 10); thus, there is natural control of parasite abundance. However, under semi-domestic conditions with shelter, protection from predators, supplementary feeding, and frequent overcrowding, the overall abundance of parasites in a population may rise such that the shape of their distribution more closely resembles a "normal curve," with a greater proportion of the population harboring moderate to high parasite intensities. In the absence of anti-parasitic treatment, this may mean that a larger segment of the population will suffer ill health due to parasites (Figure 6.1).

Parasites are important components of the community of organisms that live in or on *Rangifer*. They can influence survival, reproduction, behaviour, interspecific interactions, and overall population dynamics (see Chapter 10). Globally, more than 45 species of helminth, protozoal and arthropod parasites have been reported in wild and semi-domesticated *Rangifer* (Table 6.1).

FIGURE 6.1   Under semi-domestic conditions where reindeer are held at high densities, the overall abundance of parasites may increase (Photo: Sauli Laaksonen).

TABLE 6.1

Common Parasites of *Rangifer* by Organ System and Their Impacts on the Host
P = Protozoa, A = Arthropod, N = Nematode, C = Cestode, T = Trematode

Organ	Parasite	Type	Impact
Skin	*Besnoitia tarandi*	P	Skin thickening, hair loss
	Hypoderma tarandi	A	Damage to hide, energetic cost
	Dermacentor albipictus	A	Damage to hide, hair loss, anaemia, energetic cost
	Solenopotes tarandi	A	Hair loss, anaemia, energetic cost
	Bovicola tarandi	A	Hair loss, energetic cost
	Chorioptes	A	Hair loss
	Lipoptena cervi	A	Hair loss, heat loss, energetic cost
Respiratory system			
Nose/throat	*Cephenemyia trompe*	A	Irritation, disturbance
Sinuses	*Linguatula arctica*	A	Sinusitis, increased mucous secretion, and sneezing
Lungs	*Dictyocaulus eckerti*	N	Respiratory distress, young animals
	Varestrongylus eleguneniensis	N	Unknown
	Echinococcus canadensis	C	Possible increased susceptibility to predation/accident
Gastrointestinal tract			
Rumen	*Paramphistomum*	T	Unknown
Abomasum	*Ostertagia gruehneri*	N	Weight loss, inflammation, decreased pregnancy
	Marshallagia marshalli	N	Weight loss, decreased pregnancy
	Haemonchus contortus[a]	N	Blood loss, anaemia
	Teladorsagia boreoarcticus	N	Weight loss, inflammation, decreased pregnancy?
	Teladorsagia circumcincta[a]	N	Not described for *Rangifer*
	Trichostrongylus axei	N	Not described for *Rangifer*
Small Intestine	*Nematodirus* spp.	N	Not described for *Rangifer*
	Nematodirella spp.	N	Not described for *Rangifer*
	Cooperia oncophora[a]	N	Not described for *Rangifer*
	Eimeria spp.	P	Diarrhoea, young animals
	Giardia	P	None reported
	Cryptosporidium	P	Diarrhoea, young animals
Large Intestine/Caecum	*Skrjabinema tarandi*	N	Not described for *Rangifer*
	Trichuris	N	Hemorrhagic diarrhoea, young animals
Abdomen			
Peritoneal cavity	*Setaria* spp.	N	Peritonitis
	Taenia hydatigena	C	None reported
Liver	Fascioloides magna	T	Hepatic damage
	Taenia hydatigena	C	None reported
	Paramphistomum	T	Unknown
Neuromuscular, skeletal, lymphatic and connective tissues system			
Heart and skeletal muscle	*Sarcocystis* spp.	P	Lethargy, pneumonia, reduced growth rates or weight loss

(Continued)

TABLE 6.1 (CONTINUED)

Common Parasites of *Rangifer* by Organ System and Their Impacts on the Host

P = Protozoa, A = Arthropod, N = Nematode, C = Cestode, T = Trematode

Organ	Parasite	Type	Impact
Musculature	*Taenia krabbei*	C	Myositis in severe cases
Central nervous system	*Parelaphostrongylus tenuis*	N	Fatal neurologic disease
	Elaphostrongylus rangiferi	N	Neurologic disease
	Toxoplasma gondii	P	Foetal loss, enteritis
	Neospora caninum	P	Foetal loss
Bone/periosteum	*Besnoitia tarandi*	P	Reduced mobility
Connective tissue	*Onchocerca* spp.	N	Poor condition
	Lappnema auris	N	Ear nodules and "hot ear"
Haematopoietic	*Babesia* spp.	P	Anaemia, mortality
	Trypanosoma	P	None reported
Lymphatics	*Rumenfilaria andersoni*	N	Possible skin inflammation

[a] Indicates parasites that have spilled over from domestic species.

Given this great diversity, it is not possible to cover all of these species in detail; rather, in this chapter we provide a general overview of the most common parasites of *Rangifer*. More in-depth discussions of parasites of *Rangifer* are available elsewhere (Kutz et al., 2012; Josefsen et al., 2014; Laaksonen et al., 2017).

6.2 HELMINTHS

Rangifer are infected by a variety of helminth parasites. These organisms encompass the "worms" and include roundworms (nematodes), flatworms/flukes (trematodes), and tapeworms (cestodes). All occur as various life stages in *Rangifer*, parasitizing virtually every organ system, and their impacts can vary from undetectable to severe disease and death. Helminths have diverse lifecycle strategies that include direct transmission, indirect transmission with vectors or intermediate hosts and transmission through predator-prey interactions.

6.2.1 Nematodes

6.2.1.1 Nematodes of the Gastrointestinal Tract

Gastrointestinal nematodes are well-known causes of production loss in domestic livestock and there is growing evidence that they have impacts on condition and reproduction in *Rangifer* (Albon et al., 2002). The literature on *Rangifer* is limited but several experimental and long-term studies on gastrointestinal nematodes of Soay sheep (*Ovis aries*), red grouse (*Lagopus lagopus*) and snowshoe hare (*Lepus americanus*) have shown important population level impacts of parasites, as well as interactions with climate and predators (see Chapter 10). It is reasonable to assume similar ecological patterns for host-parasite interactions in *Rangifer*.

Three orders of nematodes are found in the gastrointestinal tract of *Rangifer*: the Strongylida, Oxyurida, and Trichocephalida. Of these, the most important and best studied are those of the Strongylida, which include members of the sub-families Ostertagiinae and Nematodirinae, and these are typically found in mixed infections in *Rangifer* (Kutz et al., 2012) (Figure 6.2). *Ostertagia gruehneri* is by far the most abundant species, and occurs in the abomasum. Other species found in the abomasum of *Rangifer* throughout the Circumarctic include *Teladorsagia boreoarcticus*, *Marshallagia marshalli* and *Trichostrongylus* spp.. *Nematodirus* and *Nematodirella* species are

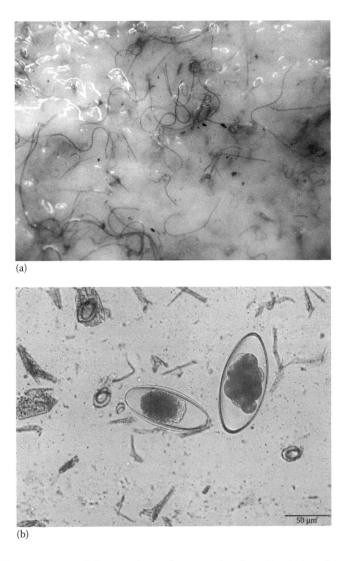

(a)

(b)

FIGURE 6.2   (a) Numerous strongylid nematodes on the mucosal surface of a reindeer abomasum (Photo: Sauli Laaksonen). (b) Eggs of *Nematodirus* sp. (right) and *Marshallagia marshalli* (left) (Photo: Alejandro Aleuy).

the most common nematodes of the small intestines. Experimental infections in semi-domesticated reindeer have demonstrated susceptibility to parasite species of domestic livestock, including *Teladorsagia circumcincta* and *Haemonchus contortus* from sheep and *Ostertagia ostertagi* from cattle. To a lesser extent, semi-domesticated reindeer may also be susceptible to *Cooperia oncophora* (Hrabok et al., 2006). The Oxyurida in *Rangifer* consist of the pinworm *Skrjabinema* sp., found in the large intestine, and the Trichocephalida consist of *Capillaria* sp. in the small intestine and the whipworm *Trichuris* sp. in the cecum and large intestine. Pinworms do not appear to cause significant production loss in domestic livestock and have not been reported as a significant pathogen in *Rangifer. Capillaria* sp. were found in 60% of studied semi-domesticated reindeer calves in their first winter, with much lower abundance in adult animals (Hrabok et al., 2006). The impact of this parasite is not known and its distribution in North America appears limited to eastern Canada. *Trichuris* spp. are also found more frequently in young animals. They can cause bloody diarrhoea in cervids (Hoeve et al., 1988) and likely increase in intensity in captive or farming situations (Lankester and Samuel, 2007).

All of these nematodes have direct lifecycles. Eggs are shed in the faeces. They then either hatch as first-stage (L1) larvae (*O. gruehneri, T. boreoarcticus, Trichostrongylus* spp., *Skrjabinema, M. marshalli*) or third-stage (L3) larvae (*Nematodirus/Nematodirella*), or they develop to infective larvae but remain in the egg (*Trichuris*). In all cases, the infective stage is ingested by the host and subsequently develops to an adult parasite in the gastrointestinal tract. Climate, and more specifically temperature and humidity, influence the development and survival rates of these parasites in the environment. Eggs of some of these parasites, such as *M. marshalli* and *Nematodirus*, are resistant to freezing, whereas eggs of others, such as *O. gruehneri*, cannot survive freezing, but their larval stages can (Hoar et al., 2012b; Kutz et al., 2012). Pathology caused by this group of parasites is commonly associated with the migration of larvae through the mucosa as they transition to the adult stage.

Below are some details about the main nematode parasites known to affect *Rangifer*.

6.2.1.1.1   Ostertagia gruehneri

Ostertagia gruehneri is the best studied of the gastrointestinal nematodes of *Rangifer*. It is the most common gastrointestinal parasite of *Rangifer* globally and is found in virtually all *Rangifer* populations that have been investigated, with the exception of those in Iceland and the Kangerlussuaq-Sisimiut caribou herd in Greenland (Kutz et al., 2012; Josefsen et al., 2014). This parasite can also infect domestic sheep that share pasture with infected *Rangifer* (Manninen et al., 2014).

Ostertagia gruehneri has two strategies for development in the host; it can develop immediately to an adult nematode after ingestion, or else remain in an arrested larval stage in the abomasal mucosa until the following spring when environmental conditions are better for egg survival. The strategy taken varies depending on the geographic location. In more temperate environments, L3 will typically complete development to adults in the same season as they are ingested, in which case the next generation of eggs is produced that same year. In contrast, at higher latitudes, ingested L3 are more likely to enter arrested development, delaying maturity and egg production to the following year when conditions are more likely to support egg survival and subsequent larval development (Hoar et al., 2012a; Kutz et al., 2014). These different strategies may be reflected in two different disease syndromes. In the case where parasites continually mature (i.e. no arrested development), impacts may be seen throughout the summer season and manifest as poor-doing animals and weight loss. The equivalent syndrome in domestic animals infected with related parasites is termed "Type I Ostertagiasis." In the case of arrested development, the larvae tend to emerge from the abomasal mucosa and mature to adult worms synchronously in the spring. In domestic livestock, this is termed "Type II Ostertagiasis" and results in substantial inflammation, hemorrhage, edema and protein and blood loss depending on the numbers of larvae. This latter syndrome is also suspected to occur in *Rangifer*. The lifecycle strategy and generation time will vary according to geography, climate, behaviour and management system. For example, in reindeer in Fennoscandia and on South Georgia Island in the sub-Antarctic, *O. gruehneri* appears to maintain an annual cycle, with calves infected and shedding parasite eggs in the first summer (Hrabok et al., 2006; Morgan and Kutz, unpubl. data). In contrast, caribou of the Bathurst caribou herd on the arctic tundra of Canada, and likely any herds ranging north of this, maintain a cycle where calves that are infected the first year do not shed eggs until the subsequent year (Hoar et al., 2012a,b). In this case, the majority of larvae will enter arrested development in the abomasal mucosa and remain there over winter. Larvae mature to adult worms the following spring and produce eggs. These eggs may or may not develop to infective larvae that summer and infect another caribou; alternatively, they may overwinter in larval stages. Owing to this arrested development and time delays in subsequent development of free-living larvae, and to exposure dependent on caribou behaviour (i.e. migration and spatial distribution on the landscape), the lifecycle takes at least two summers, and perhaps three, depending on climate and exposures (Hoar et al., 2012a).

Ostertagia gruehneri can have subclinical to severe effects depending on the context, disease syndrome, and infection intensity. In wild *Rangifer* from Svalbard (a Norwegian archipelago in the Arctic Ocean) and Greenland, *O. gruehneri* has been associated with poorer body condition and reproduction, and may regulate populations, with a two-year time delay (Albon et al., 2002; Steele et al., 2013). Reduced food intake and pregnancy rates, as well as weight loss, have been reported in reindeer with infection intensities exceeding 5,000 adult nematodes (Arneberg et al., 1996; Arneberg and Folstad, 1999; Albon et al., 2002). Semi-domesticated reindeer generally have high prevalence and low intensity of gastrointestinal nematodes, and it is assumed that these infections are subclinical and do not contribute to significant disease or productivity losses (Hrabok et al., 2006; Josefsen et al., 2014).

6.2.1.1.2   Marshallagia marshalli

Marshallagia marshalli is a generalist parasite that infects caribou, muskoxen (*Ovibos moschatus*), Dall's and Stone's sheep (*Ovis dalli*), bighorn sheep (*Ovis canadensis*), and Saiga antelope (*Saiga tatarica*). Despite its host generalism, this nematode appears to have specific habitat requirements, as it is found in drier regions and is more common in the alpine and on the high arctic islands than on low-lying tundra (Kutz et al., 2012). The eggs of *M. marshalli* can develop and hatch after prolonged freezing, and egg production is seasonal: high in the winter/spring and declining over summer (i.e. essentially the reverse of the pattern for *O. gruehneri*) (Samuel and Gray, 1974; Irvine et al., 2000; Kutz et al., 2012). Infective larvae are also resistant to freezing, and winter transmission has been reported for reindeer (Irvine et al., 2000) and Saiga antelope (Morgan et al., 2006).

The impact of *M. marshalli* in *Rangifer* is not well understood; however, in the Kangerlussuaq-Sisimiut caribou herd in West Greenland, the intensity of infection with *M. marshalli* was negatively associated with protein mass index, carcass weight index, and kidney fat index (Steele, 2012). Negative impacts of this parasite on pregnancy and condition in Dall's sheep have also been reported (Kutz et al., 2012; Aleuy et al., 2018).

6.2.1.1.3   Nematodirus/Nematodirella

Nematodirella longissemispiculata, *N. longispiculata* and *Nematodirus tarandi* are small intestine parasites in caribou and reindeer (Kutz et al., 2012). *Nematodirus skrjabini* has also been reported but may be a synonym for *N. tarandi* (Hoberg et al., 2001). Nematodirines tend to primarily parasitize young animals; however, they have been seen in adult males during rut (Fruetel, 1987), and unusually high prevalence of Nematodirine eggs (approximately 50%) was reported for adult female caribou in the Kangerlussuaq-Sisimiut herd of West Greenland (Steele et al., 2013). In caribou calves, the freeze and desiccation tolerant eggs of *Nematodirella longissemispiculata* are produced throughout winter, but by spring, calves tend to clear the infection (Fruetel, 1987). In Finland, *N. tarandi* and *N. longispiculata* were detected only in reindeer calves, with 65% of calves slaughtered in October through January infected (Hrabok et al., 2007).

6.2.1.2   Nematodes of the Lungs, Muscles, Tissues, and Body Cavities

Rangifer are host to a variety of nematodes that parasitize the lungs and tissues. These can be divided into three major groups: (1) a group that includes the lungworm *Dictyocaulus* of the Dictyocaulidae family, (2) the lung, muscle, and nervous system nematodes of the Protostrongylidae family, and (3) organisms of the family Filaroididae that parasitize fascia, the lymphatics, body cavities and other tissues.

6.2.1.2.1   Dictyocaulidae: Dictyocaulus eckerti

Dictyocaulus spp. are large white nematodes commonly found with a dorsal caudal distribution in the bronchi and bronchioles of a variety of ungulate species (Figure 6.3). A number of different species of *Dictyocaulus* have been reported in wild and domestic ungulates across the Holarctic; *D. eckerti* is considered to be the species in *Rangifer* (Höglund et al., 2003) and has also been reported in moose and muskoxen (Divina et al., 2002; Höglund et al., 2003).

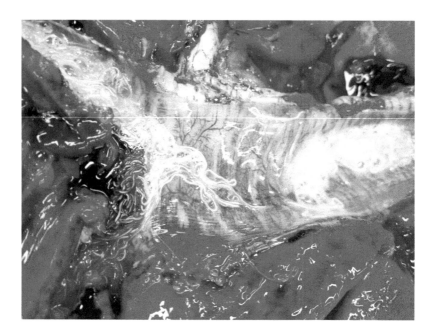

FIGURE 6.3 Numerous adults of *Dictyocaulus eckerti* in the trachea and bronchi of a reindeer (Photo: Antti Oksanen).

Dictyocaulus spp. have a direct lifecycle. Eggs are shed by adult worms in the lungs and are moved up the airways to the mouth, swallowed, and then larvated eggs or hatched L1 are passed in the faeces. Larvae undergo temperature-dependent development in the environment to infective L3. There are no data for *D. eckerti* with respect to this development rate; however, development of *D. filaria* from domestic sheep in northeast England took 4–9 days in spring and summer, but 5.5–7 weeks in winter (Gallie and Nunns, 1976). Eggs and L1 of *D. eckerti* do not seem to survive freezing (Kutz, unpubl. obs. based on examination of frozen faeces). *Dictyocaulus* spp. have been reported in muskoxen as far as 75°N on Bathurst Island, Canada, which demonstrates that they are able to complete their lifecycle in a high arctic environment. A related species from cattle, *D. viviparous*, has a fascinating mechanism of dispersal facilitated by the faecal fungus *Pilobolus*. Infective *D. viviparous* larvae will crawl up on the sporangium of the fungus and be dispersed up to meters from the faecal pat when the spores eject from the stalk (Eysker, 1991). It is not known whether this fungus is present and the phenomenon occurs for *D. eckerti* in the Arctic. Some *Dictyocaulus* spp. can undergo arrested development, where ingested L3 will enter an arrested stage in the lungs over winter and not mature to adults until the following spring. This phenomenon appears to also occur for *D. eckerti* in muskoxen (Kutz et al., 2012) and in reindeer (Josefsen et al., 2014).

In domestic livestock, *Dictyocaulus* spp. tends to cause disease in calves and yearlings, with some degree of age-related immunity observed. The infective larvae, adults, eggs and hatched larvae in the lungs can all cause significant lung pathology that results in respiratory disease (Panuska, 2006). Clinical disease tends to be more common during warm, wet years. In reindeer, lung pathology associated with *Dictyocaulus* includes acute inflammation and chronic changes with fibrotic thickening, exudate, and eosinophilic infiltration, and may predispose animals to *Pasteurella* infections (Rahko et al., 1992).

6.2.1.2.2 Protostrongylidae

The protostrongyid family of nematodes are found as adults in the lungs, musculature, or neurological tissue of ungulates and lagomorphs. Members of this group require gastropods (slugs or snails) as intermediate hosts to complete their lifecycles. Depending on where the adult nematodes are found, clinical symptoms can range from pulmonary disease only through to neurological signs, muscular wasting, and weight loss, with particularly severe disease occurring in aberrant hosts (e.g., *Parelaphostrongylus tenuis* in *Rangifer*) (Anderson, 2000; Lankester, 2001). The life-cycle starts with adult nematodes releasing eggs (directly in the lungs if a true lungworm [e.g. *Varestrongylus*] or in the lymphatics or vasculature if adults in the musculature or nervous system [e.g. *Parelaphostrongylus* and *Elaphostrongylus*]). Eggs travel to the lungs via the lymphatics and vasculature, hatch to L1 in the parenchyma or airways, travel up the respiratory tree, and are then swallowed and passed in the faeces. L1 in faeces will then penetrate the foot tissue of slugs or snails and undergo temperature-dependent development to the infective L3 while in the gastropod (Anderson, 2000). Development success and rates in the gastropod depends on species of parasite, species of gastropod, and ambient temperatures (i.e. not all gastropod species are good hosts for all protostrongylid species). The final host (caribou) is infected by ingesting a gastropod containing infective L3. For some protostrongylid species (e.g. *Varestrongylus eleguneniensis*), infective L3 may emerge from the gastropod and remain free in the environment, and this may be another source of infection for the final host. In the gastrointestinal tract of the final host, the L3 will be digested from an infested gastropod the animal has eaten, and will migrate to its predilection site (e.g. lung, muscle or nervous system). At least three genera and five protostrongylid species have been identified in *Rangifer*. All five species (*Parelaphostrongylus tenuis*, *P. odocoilei*, *P. andersoni*, *V. eleguneniensis*, *Elaphostrongylus rangiferi*) are present in North America, and only one (*E. rangiferi*) is commonly reported in Eurasia. Protostrongylids have not been reported in caribou in Greenland (Steele et al., 2013).

First-stage larvae of species of *Parelaphostrongylus*, *Elaphostrongylus* and *Varestrongylus* are morphologically similar, with kinked tails with small dorsal spines (and are therefore referred to as "dorsal spined larvae," or DSL); however, subtle features may allow morphological differentiation between *P. andersoni* and *V. eleguneniensis* (Kafle et al., 2017) and possibly others. The L1 of these genera are environmentally resistant, and are able to survive desiccation and to overwinter at arctic latitudes.

6.2.1.2.3 Varestrongylus eleguneniensis

This parasite was first found in caribou and muskoxen in North America in 2002 (Kutz et al., 2007) and was subsequently described as a new species (Verocai et al., 2014). It is a small lung nematode (approximately 1–2 cm) found in the terminal bronchioles of caribou, muskoxen, and rarely moose in Canada and Alaska, but has not been reported in Eurasia. This parasite generally appears to cause minimal pathology in the lungs of caribou; however, unusual mortality rates of caribou in the northern Alaskan peninsula in the early 2000s were attributed to heavy pulmonary infections with an unidentified protostrongylid. *Varestrongylus eleguneniensis* is now known to occur in this area but was undescribed at the time. It has a typical protostrongylid lifecycle, and the widespread meadow slug, *Deroceras laeve*, is one natural intermediate host. This parasite expanded its range to the Arctic Archipelago of Canada in or prior to 2010 (Kutz et al., 2013) and continues to expand its range further north.

Pulmonary lesions consistent with infection by *Varestrongylus alces* (from moose) and *V. capreoli* (from roe deer) (see Verocai et al., 2014b) are frequently seen in slaughtered Finnish reindeer; however, the identity of the parasite remains unknown (Laaksonen, unpublished; Figure 6.4).

6.2.1.2.4 Parelaphostrongylus *spp.*

Parelaphostrongylus spp. are native to North America and are not found in Eurasia. Three species infect *Rangifer* in North America: *P. andersoni*, *P. tenuis*, and *P. odocoilei*.

Parelaphostrongylus andersoni, a small 1–3.5 cm long muscle-dwelling nematode, is the most common species of this genus and is found in caribou across their mainland range in Canada and Alaska (Kutz et al., 2007; Verocai, 2015). In the Arctic Archipelago, Canada, *P. andersoni* has been found only in caribou of the Dolphin and Union herd (Kafle et al., 2017). Reindeer are susceptible to

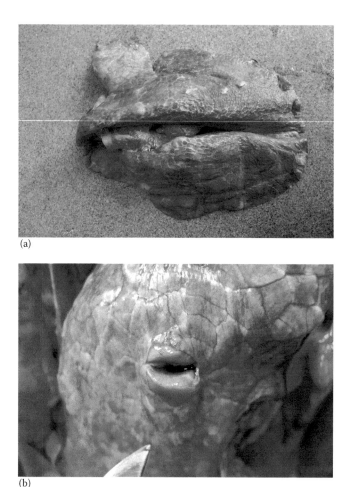

(a)

(b)

FIGURE 6.4   Lungs from a reindeer showing (a) the surface and (b) a cut section of the multifocal lesions in the lung parenchyma that are consistent with *Varestrongylus* spp. infection (Photos: Sauli Laaksonen).

this parasite under natural conditions, as demonstrated by its occurrence in an introduced population of semi-domesticated reindeer in Alaska (Verocai et al., 2013).

Most of what is known about *P. andersoni* comes from natural and experimental infections in white-tailed deer, another common host in North America. However, in one experiment on a caribou calf, the prepatent period (i.e. the interval from infection to shedding larvae) was 66 days (Lankester and Hauta, 1989). The adult parasites have a predilection for skeletal muscles and are often found in the longissimus dorsus and psoas, but they may be found elsewhere and in fat; post mortem findings generally include diffuse hemorrhage and fasciitis (Nettles and Prestwood, 1976; Lankester, 2001). Clinical signs described in deer include weakness, reluctance to stand, and falling to the ground when light pressure is put on the loin muscles (Nettles and Prestwood, 1976). Pulmonary disease can also occur. Eggs released from adult female worms are carried to the lungs through the lymphatics or blood; the migratory path has not yet been described. There, the eggs and hatched L1 can cause significant parasitic granulomatous pneumonia (Lankester, 2001). *Parelaphostrongylus andersoni* is widespread and abundant in caribou, and causes muscular and pulmonary pathology that likely translates to poor mobility and exercise tolerance;

however, the impacts on the ecology and survival of *Rangifer* at the individual and population level are unknown.

In contrast, *Parelaphostrongylus tenuis*, also known as meningeal worm, causes debilitating neurologic disease in caribou, moose, mule deer, and other domestic and wild ungulate species (Lankester, 2001). This parasite's primary host is white-tailed deer, and it is widely distributed in this host throughout eastern North America. In caribou, infection is almost invariably fatal. Failure of caribou re-introduction events in eastern North America is in part attributed to spillover of *P. tenuis* from white-tailed deer to the introduced caribou, and the parasite is increasingly considered to be a significant threat for woodland caribou and moose populations across much of the Canadian boreal forest (Kutz et al., 2012). Disease modeling suggests that current and future climatic conditions will be suitable for its range expansion west and north (Pickles et al., 2013) with expanding white-tailed deer populations.

Parelaphostrongylus odocoilei is a 2–5.6 cm long muscle-dwelling nematode found primarily in mule deer (*Odocoileus hemionus*) and also in Dall's, Stone's and bighorn sheep (Kutz et al., 2012). Despite an extensive geographic survey of caribou across North America (Verocai, 2015), there remain only two confirmed reports in *Rangifer*, one of which documents a single L1 from a woodland caribou near Hay River, Northwest Territories, Canada (Kutz et al., 2007), and the other from an experimental infection (Gray and Samuel, 1986). In experimentally infected Stone's-Dall's sheep cross, *P. odocoilei* caused weight loss, muscle atrophy and pulmonary disease (Kutz et al., 2012). It has also been responsible for sporadic mortality events in wild Dall's sheep (Jenkins et al., 2007). This parasite has not been found in barrenground caribou and its significance in woodland caribou remains unknown.

In Eurasia, *Elaphostrongylus rangiferi*, also referred to as meningeal worm, is common and widespread in reindeer and is the only protostrongylid described in reindeer in Fennoscandia (Josefsen et al., 2014; Laaksonen, 2016). This parasite can infect a variety of ruminant hosts, such as reindeer and moose (*Alces alces*) (Steen et al., 1997), and goats and sheep, and its lifecycle in these hosts has been extensively studied.

The infective larva penetrates the abomasal wall, moves through the bloodstream to the liver and lungs and then spreads to all tissues via the general circulation. Nematodes mature in the central nervous system (CNS), and then migrate to the skeletal muscles (SM) via the spinal nerve roots (Figure 6.5a, b). Eggs are deposited into veins and carried by the venous blood to the lungs, where the L1 hatch and enter the airways (Handeland, 1994) (Figure 6.5c, d). Although the typical infection is symptomless, in heavily infected animals the development of larvae to adults in the central nervous system (CNS) may cause disease (Josefsen et al., 2014). The typical CNS symptoms, known as cerebrospinal nematodiasis, appear 4–8 weeks after infection in late autumn or early winter. Symptoms manifest as ataxia and posterior paresis (Handeland, 1994). Owing to temperature-dependent development in gastropods, clinical disease occurs at higher frequencies after warm summers (Figure 6.5e) (Handeland and Slettbakk, 1995). *Elaphostrongylus rangiferi* can sporadically infect and cause disease in other animal species, most commonly in goats, sporadically in sheep (Handeland et al., 1993), and rarely in moose (Steen et al., 1997) and muskoxen (Holt et al., 1990).

Elaphostrongylus rangiferi was introduced to Newfoundland, Canada, in 1908 with infected reindeer from Norway, and has since expanded its range across the island of Newfoundland (Lankester, 2001). Reindeer from Newfoundland have been moved across Canada, and several other translocations of Eurasian reindeer into Canada and Alaska have occurred (Lankester and Fong, 1998); however, widespread geographic surveys for Protostrongylidae in caribou and reindeer have not detected *E. rangiferi* outside of the island of Newfoundland (Kutz et al., 2007; Verocai, 2015). Considering that the appropriate hosts (caribou and several terrestrial gastropod species) and climate exist to maintain *E. rangiferi* on mainland Canada and Alaska, appropriate biosecurity measures, including ongoing surveillance and thorough health assessments prior to any translocations, should be done to prevent invasion of this parasite to mainland North America.

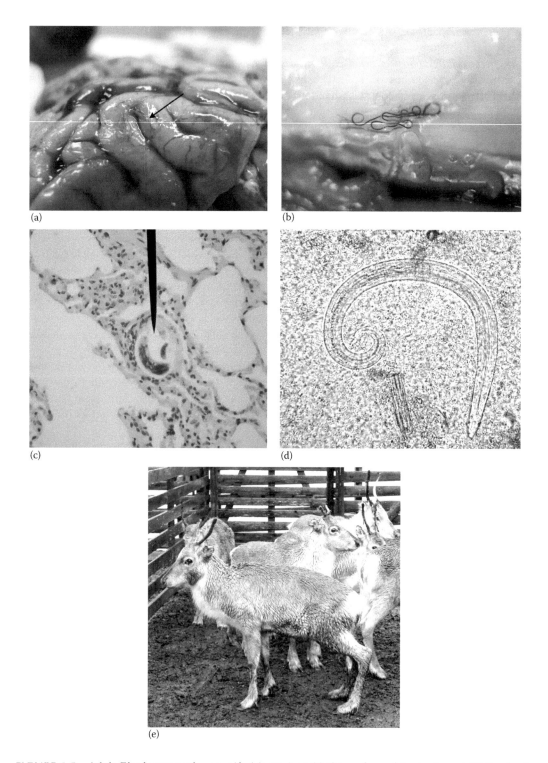

FIGURE 6.5   Adult *Elaphostrongylus rangiferi* (arrow) on (a) the surface of the brain and (b) the fascia of skeletal muscle. (c) Histological section of first-stage larva (L1) of *E. rangiferi* in the lungs. (d) L1 isolated from the faeces. (e) A reindeer calf showing signs of clinical disease caused by *E. rangiferi* (Photos: Sauli Laaksonen).

Several species of filarioids, the tissue and body cavity nematodes, have been found in *Rangifer* in North America and Eurasia, but none have been reported in Greenland. This group of nematodes is of increasing interest as their emergence in Fennoscandia, with major mortality events and meat condemnation, together with links to unusually warm weather conditions and climate change, presents a significant concern for animal health and food security (Figure 6.6).

Worldwide, filarioid nematodes parasitize a variety of species and have important medical, veterinary and economic implications. Until recently, these parasites were a neglected component of the helminth fauna of arctic ungulates. At northern latitudes, species of several filarioid genera are now known to circulate among ungulate definitive hosts, with various haematophagous insects as vectors. In *Rangifer*, these belong to the family Onchocercidae and include species in the genera *Setaria*, *Rumenfilaria*, and *Onchocerca* (reviewed by Laaksonen et al., 2017). The adult parasites are found in the abdominal cavity (*Setaria*), subcutaneous and facial tissue (*Onchocerca*) (Anderson, 2000), or lymphatics (*Rumenfilaria*) (Laaksonen et al., 2010).

Each adult female filarioid worm can produce thousands of microfilariae (mf) daily; for example, more than 200,000 *Setaria tundra* mf were found in the uterus of a single female worm (Nikander et al., 2007). Microfilariae occur in the circulatory system or in the skin of their ungulate host, where they are available for uptake by arthropod intermediate hosts (vectors) during blood meals (Figure 6.7a). In the arthropod host, the microfilaria develops to an infective stage (Figure 6.7b–d). Vectors of different filarioid nematodes include most of the major arthropod groups known to feed on the blood of higher vertebrates (e.g. biting midges, blackflies, horse and deer flies, mosquitoes, lice, fleas, mites and ticks) (Anderson, 2000).

In their normal definitive hosts, most species of filarioid nematodes are often well-tolerated. The nematodes *Setaria tundra*, *Onchocerca* spp. and *Rumenfilaria andersoni* appear to have emerged in Fennoscandian reindeer during the latter half of the 20th century, most likely as a result of climate warming and colonization of reindeer from white-tailed deer, roe deer, and red deer (*Cervus elaphus*).

FIGURE 6.6   A reindeer calf with severe *Setaria tundra* infection. Infection rates increase during warm and wet summers (Photo: Sauli Laaksonen).

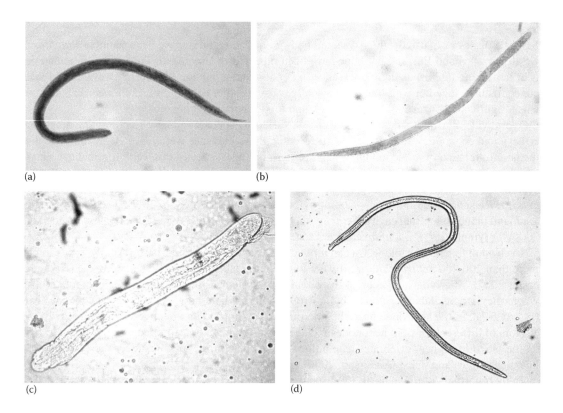

FIGURE 6.7   (a) An *Onchocerca* spp. mf from the skin of a reindeer. The stages of *S. tundra* in *Aedes* sp. mosquito: (b) L1; (c) L2 (sausage state); (d) L3 (long and slender) (Photos: Sauli Laaksonen).

Clinical signs have included peritonitis, necrotic granulomas, and tarsitis. Patterns of emergence in Fennoscandia suggest a direct relationship between temperature (and probably humidity) and incremental warming over the past 40 to 50 years (Laaksonen et al., 2017). It is likely that climate change will continue to favor the northward expansion of filarioid nematodes, and this could make them an even greater threat to the health of arctic cervids as well as food safety and security.

6.2.1.2.6   Setaria spp.

Setaria tundra was first described in semi-domesticated reindeer in the Arkhangelsk region of Russia in 1928, and has since been reported in reindeer in the Baikal region (1980), in roe deer across much of Europe, and in moose and wild forest reindeer (reviewed by Laaksonen et al., 2007; 2009a; 2017). Outside of northern Eurasia, other species of *Setaria* have been reported in reindeer, including *S. yehi* from Alaska (Dieterich and Luick, 1971) and Canada (Fruetel and Lankester, 1989) and *S. labiatopapillosa* from China (Wang et al., 1989). White-tailed deer, mule deer, moose, caribou, fallow deer (*Cervus dama*) and bison (*Bison bison*) have been described as hosts for *S. yehi* by Sonin (1977), and reindeer, elk and roe deer have been reported hosts for *S. tundra* in Europe. The distinction between *S. tundra* and *S. yehi* is based on minor morphological differences and Nikander et al. (2007) have questioned the validity of *S. yehi* as a species.

6.2.1.2.7   Setaria tundra

Adult, white, slender *S. tundra* worms (females 67 mm and males 35 mm long) inhabit the host's abdominal cavity (Figure 6.8). Female *S. tundra* produce up to 200,000 mf into the host's blood circulation (Nikander et al., 2007), where they are available to arthropod vectors. Microfilaremia peaks from the beginning of June to mid-September, with a mean of 950 smf/mL blood

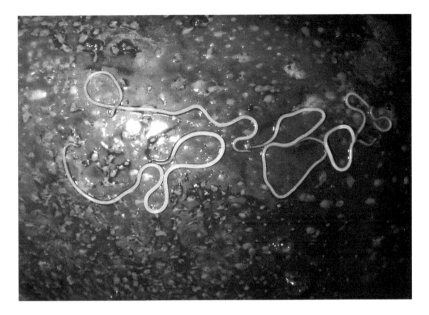

FIGURE 6.8   Adult *Setaria tundra* worms on the surface of the liver. Notice the multiple white spots on the liver associated with irritation and inflammation caused by the nematodes (Photo: Sauli Laaksonen).

(range, 62–4,000 smf/mL blood). The prepatent period of *S. tundra* is approximately 4 months and the life span of female *S. tundra* in the definitive host is at least 14 months, and probably much longer. Low prevalence and density of *S. tundra* in definitive hosts can maintain the infection in reindeer populations (Laaksonen et al., 2009a).

Mosquitoes, particularly *Aedes* spp. and to a lesser extent *Anopheles* spp., play an important role in the transmission of *S. tundra* in Finland (Laaksonen et al., 2009b). It is suggested that *S. tundra* is not a highly vector-specific parasite, and that this may enhance its ability to expand its geographic range (Laaksonen et al., 2017). Development of *S. tundra* larvae in mosquitoes is temperature-dependent; at mean temperature 21°C, larvae develop in the mosquito to the infective L3 in approximately 2 weeks, whereas at mean temperature 14.1°C, development is not completed. Warm summers apparently promote transmission and genesis of disease outbreaks by favoring the development of *S. tundra* in its mosquito vectors, by improving the development and longevity of mosquitoes, and by forcing the reindeer to flock on mosquito-rich wetlands (Laaksonen et al., 2009b).

Initial recognition of *S. tundra* in Scandinavia appears to coincide with geographic expansion of roe deer from more southern latitudes of western Europe (Laaksonen et al., 2007). Both wild forest reindeer and semi-domestic reindeer serve as competent hosts for *S. tundra*, and the parasite likely circulates between these cervids (Laaksonen et al., 2009a).

Clinical signs and pathology include peritonitis with ascites, green fibrin deposits, adhesions and living and dead *S. tundra* worms in the peritoneal cavity, as well as decreased body condition (Figure 6.9). Clinical signs are strongly correlated with intensity of infection (Laaksonen et al., 2007). Histopathological changes indicate granulomatous peritonitis with lymphoplasmacytic and eosinophilic infiltration. Infections can cause significant meat condemnation and production loss.

Setaria tundra has caused severe disease outbreaks in reindeer in Fennoscandia and Finland. Outbreaks are associated with mean summer temperatures exceeding 14°C for 2 consecutive years. If these conditions occur in two consecutive summers, disease emergence occurs in the second summer (Laaksonen et al., 2010). Four major outbreaks of parasitic peritonitis caused by *Setaria* have been reported in Finland: one in reindeer in 1973 (also observed in Sweden and Norway), one in

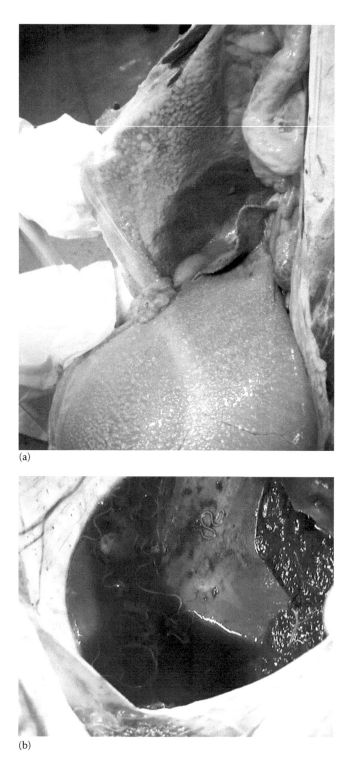

(a)

(b)

FIGURE 6.9 Abdominal cavity of a reindeer with a severe infection of *Setaria tundra*. Note (a) the roughened surfaces of the organs and body walls (granulomatous peritonitis) and (b) the abundant fluid (ascites), green fibrin deposits, adhesions and living and dead *S. tundra* worms in peritoneal cavity (Photos: Sauli Laaksonen).

moose in 1989, another in reindeer in 2003 and another in reindeer in 2014 (Laaksonen et al., 2007; 2009a; 2017). In the 2003 outbreak, the prevalence and intensity of infection were very high in calves and caused substantial economic losses for reindeer herders in the region (Laaksonen et al., 2007; 2009a). The most recent outbreak occurred during 2014 and, as with the previous events, followed two consecutive summers of elevated temperatures (Laaksonen et al., 2017).

6.2.1.2.8   Onchocerca spp.

In North America, two species of *Onchocerca* are reported to infect cervids: *Onchocerca cervipedis* and a recently identified, but not yet named, species in white-tailed deer; additional diversity of this genus in North America is suspected (Verocai et al., 2012; McFrederick et al., 2013). The geographic range of *O. cervipedis* extends into subarctic regions of western North America, where hosts include caribou (*Rangifer tarandus granti*) and Yukon-Alaska moose (*A. a. gigas*) (Verocai et al., 2012). Authors have suggested a continuous distribution of *O. cervipedis* in the northern boreal forest regions of western North America.

Onchocerca sp. have also been observed in reindeer across most of Eurasia, including Russia, Sweden, and Finland (reviewed by Laaksonen et al., 2017). The species was identified as *Onchocerca tarsicola*, which was later synonymized with *O. skrjabini* by Yagi et al. (1994). Bylund et al. (1981) reported 35% prevalence of these worms among reindeer in Finnish Lapland. The parasite has never been reported in Norway (Josefsen et al., 2014), but this likely reflects lack of search effort.

Species of *Onchocerca* have generally been considered to have low pathogenicity, and consequently there has been little veterinary interest. These parasites are primarily observed in nodules in subcutaneous tissues. Dead worms in the subcutaneous tissues usually become calcified and surrounded by dense fibrous tissue, causing little damage, but they may also act as a focus for bacteria, and abscesses may develop (reviewed by Laaksonen et al., 2017). Different species localize and develop in specific anatomical sites of predilection in the host (reviewed by Laaksonen et al., 2017).

Onchocerca cervipedis, commonly referred to as "legworm," is transmitted by blackflies and generally affects subcutaneous tissues of the hindquarters from the tibio-tarsal joint to the hoof. While this species is found in caribou in North America, it appears to be more common in moose, where it rarely causes clinical signs (Verocai et al., 2012; McFrederick et al., 2013). In caribou, however, substantial pathology was observed, including periosteitis, and cellulitis and granulation associated with larvae and adult worms. Infections can also cause swelling and hoof damage in species of *Odocoileus*, which may increase susceptibility to predation (Verocai et al., 2012).

Onchocerca tarsicola is the main species described in reindeer in Eurasia. The adult nematodes (females 20–25 cm long, in contrast to males only 2.5 cm, and 0.2–0.3 mm thick) are typically found in or associated with nodules in the subcutaneous tissues of the muzzle, the metacarpal and carpal bones, and occasionally elsewhere (e.g. shoulder, brisket). The microfilariae are present in the skin, concentrated at a considerable distance away from the adults. For example, they are common in the skin of the outer parts of the ears and the nose (Schulz-Key, 1975). The microfilariae are transmitted by blackflies, in which they develop to infective L3 larvae in 23–25 days at 17–18°C (Schulz-Key and Wenk, 1981) (Figure 6.10). The prepatent period in the host is 6 months (Schulz-Key, 1975).

In reindeer, the worms are most often found in flat swellings or nodules of connective tissue in membranes surrounding the tendons of the tibio-tarsal and radiocarpal joints (Bylund et al., 1981). In heavy infections, *O. tarsicola* can manifest as granulomatous nodules in most organs, especially the liver. In Finland, heavy infections, presumed to be *O. tarsicola*, have been associated with severe hemorrhagic tarsitis, which likely causes pain and lameness for reindeer and also for moose (Figure 6.11) (Laaksonen et al., 2017).

Greenish granulomatous nodules have also often been found between muscle fasciae and surrounding the hip joint, and have been observed in other organs (e.g. the abdominal wall, diaphragm

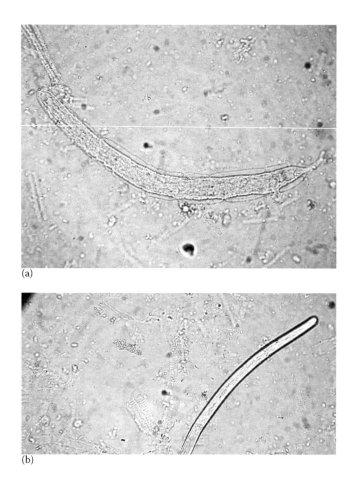

(a)

(b)

FIGURE 6.10   (a) Developing *Onchocerca* sp. larvae (sausage stage) and (b) infective larvae from the black fly (Photo: Sauli Laaksonen).

and rumen) (Laaksonen, 2016; Laaksonen et al., 2017). It is postulated that reindeer are not the "normal" host for *O. tarsicola*; rather, it is suspected to have colonized from red deer as a result of range expansion of this putative primary host in Sweden during the late 1960s (Laaksonen et al., 2017). Recent genetic evidence suggests that there is even greater species diversity of *Onchocerca* circulating in Finnish reindeer (Laaksonen et al., unpublished).

6.2.1.2.9 Rumenfilaria andersoni

Rumenfilaria andersoni is a lymphatic-dwelling filarioid nematode found in moose, reindeer, caribou, and other deer species. It was originally described in a single moose in Ontario (Lankester and Snider, 1982) and more recently has been identified in all cervid species in Finland: reindeer (0% to 90%) and wild forest reindeer (41% to 100%), moose (0% to 12%), white-tailed deer (15% to 22%) and roe deer (3%) (Laaksonen et al., 2010; 2015). It is also present in moose in Alaska and appears to be widespread in moose and white-tailed deer in the United States (Grunenwald et al., 2016). Its distribution in Canada is not well documented, but it is likely prevalent given the distribution in Alaska and the United States. Adult *R. andersoni* (length: males 33–35 mm, females 40–100 mm) inhabit the lymphatic vessels of the subserosal rumen and mesenteries, and are associated with a high abundance of microfilariae (rmf) (mean, 11,089 rmf/mL blood during summer months) (Figure 6.12).

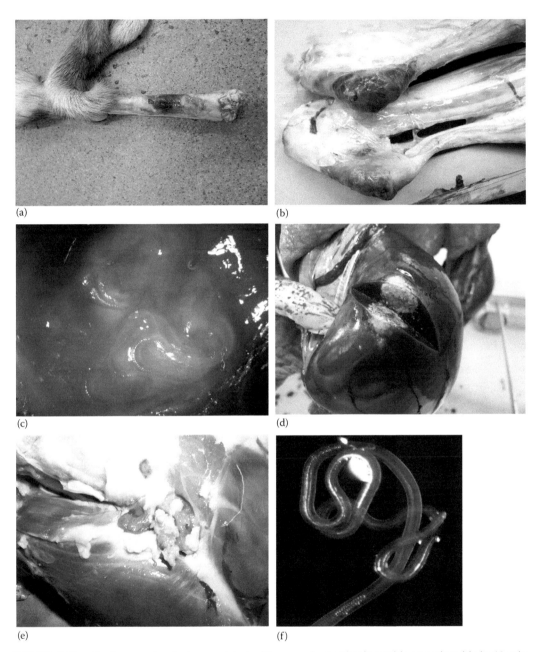

FIGURE 6.11 *Onchocerca tarsicola* associated with severe hemorrhagic tarsitis or periosteitis in (a) reindeer and (b) moose. (c) Live *O. tarsicola* embedded in fibrous tissue; These nematodes can cause granulomatous nodules in (d) the liver and (e) between muscle fasciae; (f) Adult *O. tarsicola*.

In Finland in 2004–2006, overall, rmf were detected in 64% of reindeer blood samples, with a mean density of 452 rmf/mL blood (range, 1–19,400 rmf/mL blood). The prevalence and density of rmf was higher in adults than in calves (Laaksonen et al., 2015). The parasite has expanded its geographic range northward in Finland, coinciding with the outbreak of *S. tundra* (Laaksonen et al., 2007). During reindeer health monitoring in 2015, 71% of adult Finnish reindeer were found to be infected with *R. andersoni* (Laaksonen et al., 2017). The parasite has not been reported in Norwegian cervids, but one of 15 Swedish moose (sampled near the Finnish border) was positive in 2010 (Laaksonen et al., 2015). The insect vector of *R. andersoni* is not known, but it is likely not a species of mosquito (see Laaksonen et al., 2009a).

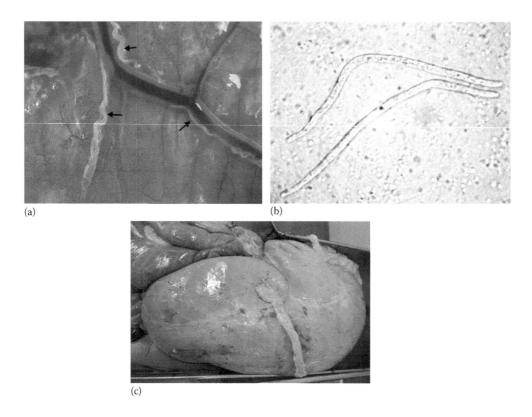

(a)   (b)

(c)

FIGURE 6.12   (a) Adult *Rumenfilaria andersoni* in the lymphatic vessels of the subserosal rumen (arrows). (b) Stained microfilariae in the blood. (c) Typical presentation of a rumen with a severe infection with *Rumenfilaria*. Note the granular surface of the rumen, the dilatation and swelling of the vessel walls around the living worms, and greenish or grayish granulomatous reaction in the foci containing dead worms (Photos: Sauli Laaksonen).

The typical gross pathological changes associated with the infection are dilatation of the vessels, lympho-edematous swelling of the vessel walls around the living worm, and greenish or grayish granulomatous or fibrotic reaction in the foci containing dead worms (Figure 6.12).

Nematodes may be seen through the wall of the dilated vessels as thin, white, winding threads occluding the vessel lumen. They are readily visible in emaciated reindeer calves with scant or no fat around the lymphatic vessels on the serosa of the rumen (Laaksonen et al., 2010). The impact of *R. andersoni* on cervid health and well-being remains unclear, however, the eosinophilic reaction in skin and lymph nodes suggests that high counts of rmf in blood would have a negative impact on overall cervid health (Laaksonen et al., 2015).

In contrast to the emergence of *Setaria* and *Onchocerca*, which appear to be "natural invasion," it is likely that *Rumenfilaria* established in Finland as a result of the introduction of infected white-tailed deer from North America in 1935 (Laaksonen et al., 2015; Grunenwald et al., 2016). Observations of *R. andersoni* in other Scandinavian countries or across Eurasia are lacking, and the parasite is considered to be absent from these regions (Laaksonen et al., 2015). However, considering that *R. andersoni* is found in all four cervid species from Finland, it can be anticipated to have considerable potential for geographic expansion and host colonization in Eurasia.

6.2.1.2.10 Lappnema auris

Lappnema auris is an unusual nematode of the family Robertdollfusidae. It occurs in subcutaneous capillaries of the ears or eyelids of reindeer in Fennoscandia. It induces the formation of large nodules where it occurs and is called "hot ear" by reindeer herders (Figure 6.13). Females are small

FIGURE 6.13   Reindeer infected with *Lappnema auris*. Note the large swelling at the base of the ear. This is called "hot ear" by reindeer herders (Photo: Antti Oksanen).

(5–6 mm long and 20–25 µm wide) and males have not been observed; it is postulated that these parasites may reproduce by parthenogenesis. The females give birth to living infective larvae, with development from L1 to L3 taking place in the worm's uterus. The vector is unknown but tabanids have been suggested (Bain and Nikander, 1982). It is also thought that the parasite could be spread via uncleaned knives during ear marking (Lisitzin, 1976).

The parasite has been reported only in Finland (Rehbinder, 1990), where historically it was a common finding during round-ups (Lisitzin, 1976); however, the species has become rare, perhaps because of regular endectocide treatment.

6.2.2   TREMATODES

Trematodes or "flukes" are hermaphroditic flatworms that require at least one intermediate host, which is often a snail. Three species have been reported in *Rangifer*. In North America, these include *Fascioloides magna*, the giant liver fluke (Figure 6.14), and *Paramphistomum cervi*, the rumen fluke, and in Norway, *Dicrocoelium dendriticum*, the small liver fluke, or lancet fluke, has been reported in reindeer.

6.2.2.1   *Fascioloides magna*

Fascioloides magna has only been reported from the Rivière-George and Rivière-aux-Feuilles (George River and Leaf River) herds in Quebec and Labrador in Canada (Kutz et al., 2012), where the prevalence and intensity in adult caribou approach 83% and 12.6 flukes/animal (Simard et al., 2016), and in the now extirpated mountain woodland caribou in Banff National Park, Alberta, Canada (N. DeBruyn, unpubl. data). However, the distribution may be more widespread. The fluke is transmitted through an aquatic snail intermediate host in the family Lymnaeidae, and can overwinter in these snails (Pybus, 2001; Králová-Hromadová et al., 2011). Eggs hatch and miracidia are released and invade the aquatic snail, where they develop to numerous cercariae that are eventually released as metacercariae. These remain in the water or encyst on the vegetation and continue their lifecycle once ingested by a suitable definitive host. Development in the snail is temperature-dependent; however, the distribution of *F. magna* well into Arctic Quebec suggests that temperature may not be limiting its northern distribution.

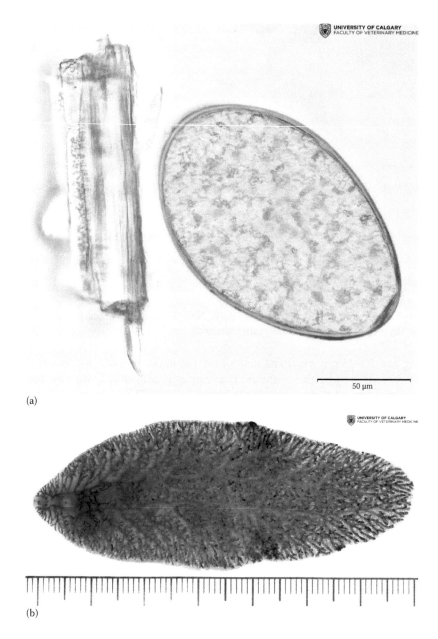

FIGURE 6.14   (a) Egg and (b) adult of *Fascioloides magna* (Photos: Paul Gajda).

Not all ungulates are competent hosts (e.g. moose are dead-end hosts), but caribou are suitable hosts in which *F. magna* can complete its lifecycle. Migrating juvenile flukes cause damage in the liver, leaving blood-filled tunnels. Adult flukes are found in capsules in the liver, each of which contains fluid and usually two flukes. In caribou, prevalence of *F. magna* tends to increase with age; for example, in the Rivière-George and Rivière-aux-Feuilles herds, calves had a lower prevalence (19%) than yearlings (70%) and adults (83%) (Simard et al., 2016). The liver fluke also has a seasonal pattern of abundance, with intensity peaking in spring. At a herd level, in one study, abundance of liver fluke was highest when the herds were at their peak size, and it remained high as the herds declined, perhaps reflecting a combination of parasite longevity in the definitive and intermediate hosts as well as the environment (Simard et al., 2016). While there is no evidence that *F. magna* has negative effects on body condition of caribou (Lankester and Luttich, 1988; Pollock et al., 2009), carefully designed experimental and/or longitudinal studies (see Chapter 10) are required to truly evaluate their impact.

Fascioloides magna was introduced to Europe from North America, with the earliest reports from Italy in 1875 (Pybus, 2001). While it is maintained in red deer and fallow deer, there are no reports in the literature of *F. magna* in reindeer. However, range expansion of the fluke historically has occurred through translocations or natural migration (Pybus, 2001; Malcicka, 2015), and that, together with warming climatic conditions, may enable this parasite to expand its range into reindeer habitat (e.g. Pickles et al., 2013).

6.2.2.2   *Dicrocoelium dendriticum*

Dicrocoelium dendriticum, referred to as the lancet liver fluke or small liver fluke, occurs in more than 30 countries, including some in Europe, Asia, Africa, South and North America and Australia. Its main reservoirs are grazing ruminants, mainly sheep, goat and cows. In reindeer, this fluke has been reported only in Norway (reviewed by Josefsen et al., 2014).

Adult flukes, 6–10 mm long and 1.5–2.5 mm wide, live in the bile ducts of ruminants and produce eggs, which are passed through the bile ducts and ultimately end up in the faeces. Eggs are ingested by a land snail, where they hatch to miracidia that undergo multiple generations of asexual reproduction (sporocysts) in the snail, finally producing cercariae. These cercariae are released from the snail in slime balls that are excreted through the respiratory pore of the snail. The second intermediate host is an ant. The ants, attracted by the moisture left behind in the snail trail, will feed on the slime ball, which contains hundreds of cercariae. Metacercariae then develop in the ants, with one locating in the brain of the ant and forcing it to remain clamped on the tops of vegetation where it will be more likely to be consumed by the next herbivore host. Once a grazing animal consumes the ant in grass, the larvae migrate to the biliary tree, where they mature into adults and produce eggs, continuing the cycle of the species.

The symptoms of *D. dentricum* infection appear to be mild, causing symptomless fibrotic thickening of the bile ducts. However, in animals with heavy infections, it can be a serious problem, as they can accumulate large numbers in bile ducts, eventually leading to cirrhosis (Figure 6.15). Symptoms include cachexia, weakening, anaemia, icterus, edema in the head, abortions and death (Kimberling, 1988).

Dicrocoelium dentricum is present in white-tailed deer in Finland, and in high density deer populations, clinical disease, including liver cirrhosis and death, is reported (Laaksonen and Paulsen, 2015). In Norway, the parasite was reported in reindeer during meat inspection during the 1970s and 1980s, when 1.2–2.8% of livers (in some areas over 50%) were condemned because of changes caused by the fluke (reviewed by Josefsen et al., 2014). The prevalence in Norway is likely connected to the intensive sheep farming and reindeer sharing the same pastures with sheep.

6.2.2.3   *Paramphistomum* spp.

Paramphistomum cervi is a rumen fluke reported in Alaskan caribou and moose (Dieterich, 1981; USNPC, 2011). Surveys for *Paramphistomum* spp. in *Rangifer* elsewhere in North America and in Greenland are limited, although a *Paramphistomum* sp. has been reported in moose in Ontario, Canada (Lankester et al., 1979). *Paramphistomum leydeni* is documented in reindeer in Eurasia, and in one study in Fennoscandia this parasite was found in 4.8% of 731 reindeer (Laaksonen, 2016). Similar to *F. magna*, these parasites require aquatic snails (Lymnaeidae and Planorbidae) as intermediate hosts, and the entire portion of the lifecycle outside the *Rangifer* host is temperature-dependent (De Waal, 2010). Lankester et al. (1979) suggested that the parasite has a one-year life-cycle in moose in Ontario, Canada, with infection occurring in the fall, adult flukes present by the following spring, and fluke death in autumn. There is limited information on the effects of rumen flukes in *Rangifer*, but heavy infections have been documented in severely debilitated adult moose

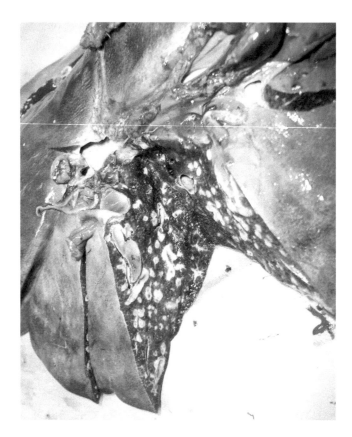

FIGURE 6.15   Liver heavily infected with *Dicrocoelium dendriticum* (Photo: Sauli Laaksonen).

and caribou in Alaska (K. Beckmen, unpubl. obs.). Pathology, including denudation of rumen villae, was observed in heavily infected moose calves (Lankester et al., 1979). In domestic livestock, juvenile parasites in the small intestine have been associated with diarrhoea and enteritis (De Waal, 2010), but while some pathological changes have been observed in reindeer, hemorrhagic duodenitis caused by the migrating flukes has not been diagnosed in reindeer calves in Finland (Figure 6.16).

6.2.3   CESTODES

Cestodes, or tapeworms, are hermaphroditic flatworms with complex lifecycles that require two hosts, often in a predator-prey relationship. Depending on the cestode family, *Rangifer* are either definitive hosts (family Anoplocephalidae) or intermediate hosts (family Taeniidae).

6.2.3.1   Anoplocephalidae

Rangifer are host to at least two or three genera of tapeworms in the family Anoplocephalidae, and these are found across most of the *Rangifer* range. These include *Avitellina arctica* from caribou in the Northwest Territories and Quebec, Canada (Gibbs, 1960; USNPC, 2011), as well as *Moniezia benedeni*, *M. taimyrica* and *M. rangiferina* in reindeer from Russia, Norway and South Georgia Island, and possibly *Thysanosoma* (Semenova, 1967; Zelinskii, 1973; Leader-Williams, 1980; Bye, 1985). These identifications remain suspect, however, as detailed morphological and molecular analyses have not been done, and hidden diversity and cryptic species are suspected.

Anoplocephalid tapeworms live as adults, ranging up to several meters in length, in the intestines of *Rangifer* (Figure 6.17). Tapeworm segments (proglottids) are passed in the faeces and eggs are then ingested by arthropod intermediate hosts (oribatid mites or members of collembola), in which

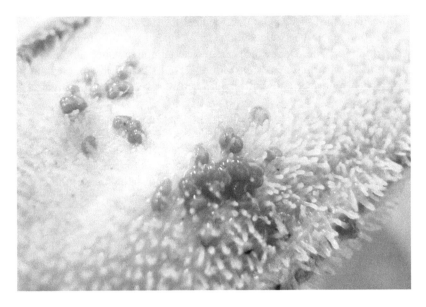

FIGURE 6.16   Multiple *Paramphistomum* flukes among villi of the rumen. (Photo: Sauli Laaksonen).

(a)

(b)

(c)

FIGURE 6.17   (a) A severe infection with tapeworms in a reindeer. (b) A single tapeworm pulled out of the small intestine on post mortem. (c) Tapeworm segments (proglottids), which contain hundreds of eggs, drop off the main tapeworm regularly and are passed in the faeces (Photo: Sauli Laaksonen).

the eggs undergo temperature-dependent development (e.g. 27–97 days for *M. expansa* [Narsapur and Prokopic, 1979]) to reach the cysticercoid stage. *Rangifer* are infected when they accidentally ingest infected arthropods. These tapeworms are most common in young animals (Kirilenko, 1975; Bye, 1985), and this is where clinical signs are observed. Weight loss/poor gain, diarrhoea and unthriftiness has been reported, and can be sufficiently severe to require treatment (Kirilenko, 1975) (Figure 6.17).

6.2.3.2 Taeniidae

The second family of tapeworms that infect *Rangifer* are the Taeniidae, which include the genera *Taenia* and *Echinococcus*. *Rangifer* are intermediate hosts for these parasites. Both genera are transmitted through a predator-prey lifecycle, with wild canids being the definitive hosts (Jones and Pybus, 2001).

6.2.3.2.1 Taenia *spp.*

Taenia hydatigena and *T. krabbei* have both been reported in *Rangifer* across their holarctic range (Loos-Frank, 2000; Haukisalmi et al., 2011; Kutz et al., 2012); however, ongoing molecular and phylogenetic assessment of the genus *Taenia* in high-latitude ungulates has revealed hidden diversity (e.g. *T. krabbei* in moose and brown bears was recently recognized as a new species, *T. arctos*; see Haukisalmi et al., 2011); thus, until they are definitely identified, we will refer to these species in *Rangifer* as *T.* cf. *hydatigena* and *T.* cf. *krabbei*.

The larval forms of both species are found as cysticerci in the liver, omentum or elsewhere in the peritoneal cavity (*T.* cf. *hydatigena*), or in the skeletal, esophageal, or cardiac musculature, and rarely the brain (*T.* cf. *krabbei*) (see Kutz et al., 2012; Lavikainen et al., 2011) (Figure 6.18 and Figure 6.19). These are transmitted to the canid definitive hosts through predation or access to slaughter offal. In the intestines of the definitive host, they develop to adult tapeworms that can be meters long, and the host sheds proglottids containing thousands of eggs in its faeces. The eggs are tough and can survive in the environment for extended periods of time. *Rangifer* are re-infected by ingesting eggs from the environment. The prepatent period ranges from 34–37 days (*T.* cf. *krabbei*) to 51–76 days (*T.* cf. *hydatigena*), and eggs may be shed for several months. Proglottids of *T.* cf. *hydatigena* are highly mobile and can move up to 3 feet away from the faeces of the definitive host, effectively separating them from the faeces and potentially increasing the likelihood that they will be ingested by the next caribou or reindeer (reviewed in Kutz et al., 2012).

Taenia cf. *hydatigena* prevalence and intensity increases with age (Pollock et al., 2009; Simard et al., 2016). There may also be a seasonal pattern of occurrence; in one study, prevalence decreased from December to March (Thomas, 1996); however, in another study, this was not observed (Simard et al., 2016). As well, no association has been found between number of cysts in the liver and kidney fat index (Pollock et al., 2009). In northern Norway, the prevalence of *T.* cf. *hydatigena* recorded in meat inspections from 1976–1980 was 7.4%, but the current prevalence is likely lower due to anti-parasitic treatment of dogs (reviewed by Josefsen et al., 2014). In Finland, during official meat inspections conducted from 2004 to 2015, the parasite was found in 0.01% of 700,000 reindeer (Laaksonen, unpublished).

For *T.* cf. *krabbei*, no age or seasonal patterns of infection have been described. While these parasites are very common in *Rangifer*, infection intensities tend to be low and clinical impacts are rare. However, hunters have reported high-intensity infections with *T.* cf. *krabbei* associated with poor body condition (Kutz, 2007). In Norway, the parasite was previously detected during meat inspection, with muscle cysticerci found in 1% of reindeer inspected from 1976 to 1980 (reviewed by Josefsen et al., 2014). In contrast, Bye (1985) found cysticerci in 38% of wild Svalbard reindeer. In official Finnish meat inspection data, cysticerci were reported in only four of 700,000 inspected reindeer (Laaksonen, unpublished).

Rangifer are also intermediate hosts for *Echinococcus canadensis*. These parasites live as larvae in hydatid cysts of up to several centimeters' diameter in the lungs, and less commonly in

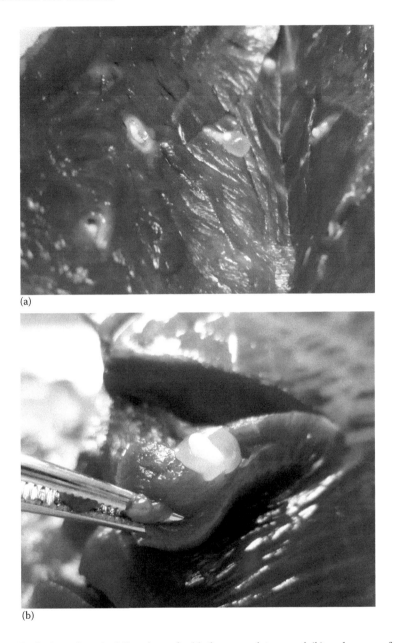

(a)

(b)

FIGURE 6.18   Typical cysticerci of *Taenia* sp. in (a) the musculature and (b) a close-up of a cysticercus dissected from the muscle, showing the single larva inside (Photo: Sauli Laaksonen).

sterile cysts in the liver, and they are widespread across the range of *Rangifer*, with the exception of Greenland (OIE, 1998; Jenkins et al., 2013; Kutz et al., 2012) (Figure 6.20).

The lifecycle is predator-prey and at high latitudes the sylvatic cycle includes wolves as the definitive host; however, dogs are also suitable hosts and are an important source of human infection (reviewed by Oksanen and Lavikainen, 2015) (Figure 6.20c). Adult tapeworms are very small (4–10 mm long) and the definitive host may have infection intensities in the tens of thousands (Rausch, 1993). In experimentally infected dogs, the time from ingestion to eggs in feces is 56–65 days and the adult worms may live as long as 8–12 months. The eggs are very environmentally resistant and may persist for many months (Gemmel, 1977). *Rangifer* are infected by eating

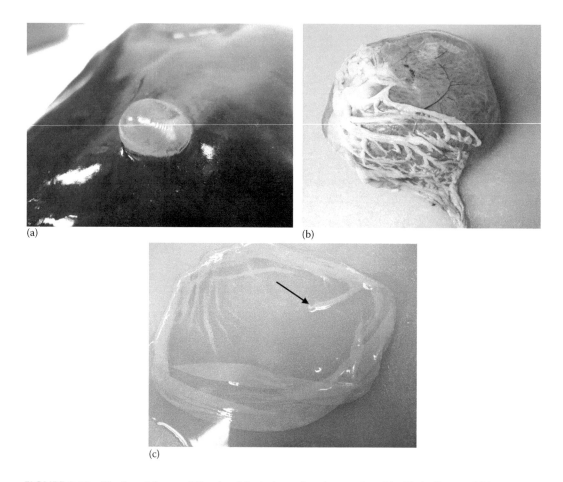

(a)

(b)

(c)

FIGURE 6.19 The larval forms of *Taenia* cf. *hydatigena* found as cysticerci in (a) the liver and (b) omentum. (c) Cysticercus from the omentum opened up showing the one infective larvae (arrow) (Photo: Sauli Laaksonen).

the eggs. Infection intensity in this intermediate host increases with age and may be associated with poorer body condition and lower pregnancy rates (Thomas, 1996). In moose, it increases susceptibility to predation and hunting (Rau and Caron, 1979; Joly and Messier, 2004) and, although not proven, similar impacts may be suspected for *Rangifer*.

6.2.3.2.2 Echinococcus *spp.*

Echinococcus canadensis is zoonotic, and causes cystic echinococcosis (CE) in people. The cysts primarily develop in the lungs (though they may be found less frequently in the liver and brain), are less pathogenic than those of *E. granulosus*, and are often asymptomatic. In fact, historically, *E. canadensis* infection was diagnosed radiographically during screening of aboriginal peoples for tuberculosis (Rausch, 1993). Transmission to people typically occurs through a semi-synanthropic cycle (i.e. via sled dogs and wild cervids) or a synanthropic cycle (i.e. via herding dogs and semi-domesticated reindeer), where people are exposed through contact with eggs from the faeces of dogs (Figure 6.20c).

Aggressive public health campaigns, burial of slaughter waste, anthelmintic treatment of dogs and replacement of herding and sled dogs with snowmobiles has resulted in far fewer human cases in both North America and Fennoscandia, as well as a substantial decline in occurrence in semi-domesticated reindeer (Oksanen and Lavikainen, 2015; Jenkins et al., 2013). In Norway, this parasite has not been found since 1990, and in Sweden the last finding was in the slaughter season of 1996–97

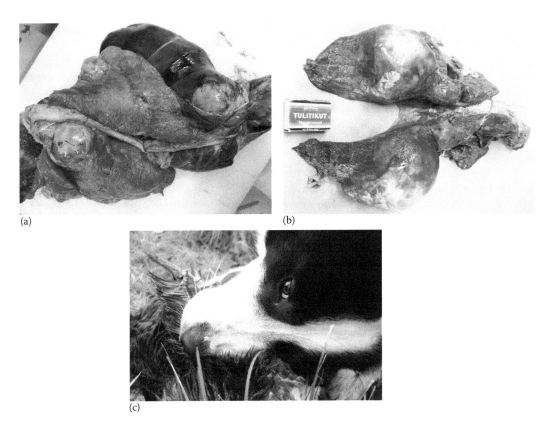

(a)

(b)

(c)

FIGURE 6.20 (a, b) Hydatid cysts of *Echinococcus canadensis* in the lungs and liver of *Rangifer*. (c) Transmission of *E. canadensis* to people typically occurs through contact with eggs from the faeces of dogs which have fed on slaughter waste of infected cervids (Photos: Sauli Laaksonen).

(reviewed by Josefsen et al., 2014). Currently, Fennoscandia appears to be free of *E. canadensis*, excluding the eastern Finnish reindeer herding area, where 62 findings have been registered during reindeer meat inspection over the last 10 years (Laaksonen, unpublished). The parasite remains widespread in wild caribou throughout its range in North America.

6.3 PROTOZOA

Protozoa are single-celled organisms and include numerous species that can parasitize the gastrointestinal tract, blood, and tissues in *Rangifer*. Depending on the species, protozoans have a diversity of life-history strategies.

6.3.1 GASTROINTESTINAL PROTOZOA

Gastrointestinal protozoa of importance in *Rangifer* include *Eimeria* spp., *Giardia* and *Cryptosporidium*. *Eimeria* spp. are by far the most common and include at least eight species across their holarctic distribution of *Rangifer* (reviewed by Kutz et al., 2012 and Josefsen et al., 2014; Skirnisson and Cuyler, 2016). These parasites have a direct lifecycle. Relatively tough oocysts that are shed in *Rangifer* faeces must first sporulate (a temperature-dependent process) in the environment before they are infective to the next *Rangifer* host. Once ingested, the parasites undergo several asexual replication events in cells of the mucosa or lamina propria of the small intestine, and then a final sexual cycle in which oocysts are produced and then passed in the faeces. Each asexual and sexual cycle results in extensive destruction of intestinal cells, and it is this that may cause clinical disease. *Eimeria* spp. are most common in calves, where disease results in diarrhoea (Figure 6.21).

Prevalence of *Emeria* spp. in wild adult *Rangifer* is low; however, this is a common parasite in semi-domesticated reindeer calves. In Finland, a high prevalence of infection (35–50%) has been observed in reindeer calves (Oksanen et al., 1990). For semi-domesticated reindeer, infection with these direct lifecycle parasites is favored by crowding in enclosures (Laaksonen, 2016). Because each coccidian oocyst destroys at least one epithelial cell when it emerges into the intestinal lumen, the peak faecal oocyst counts, which can be up to 800,000 oocysts per gram (opg), cannot be considered apathogenic (Oksanen et al., 1990).

Giardia duodenalis and *Cryptosporidium* spp. are directly transmitted protozoa that are often waterborne, and these are relatively rare parasites in *Rangifer* (Kutz et al., 2009b; 2012). Both genera have zoonotic species and strains that can cause diarrhoea in susceptible hosts, usually the young or immunocompromized. *Giardia* has been reported in reindeer in Norway (reviewed by Josefsen et al., 2014), and Oksanen et al. (unpublished) found *G. duodenalis* Assemblage E (livestock assemblage) in 22% of 54 reindeer calves younger than 3 weeks old. *Giardia duodenalis* Assemblage A is also common in at least one arctic muskox population that is sympatric with caribou, but may not be widespread in muskoxen (Kutz et al., 2008; 2012). *Giardia* is an important cause of diarrhoea in a number of species, including domestic livestock (mainly young) and humans; however, it has not been documented as an important cause of disease in *Rangifer*. *Cryptosporidium* spp. are sporadically detected in *Rangifer* faeces, but this parasite is uncommon and has not been associated with clinical disease in free-ranging *Rangifer* (reviewed by Kutz et al., 2012). It has been observed in semi-domesticated reindeer calves at 1 month old and can be associated with diarrhoea (Josefsen et al., 2014). A new genotype discovered in caribou in Alaska is most closely related to *C. andersoni*, a cattle species that parasitizes the abomasum and may cause decreased weight gains and milk production (reviewed by Kutz et al., 2012). This suggests some potential as-yet-unknown clinical significance in *Rangifer*.

6.3.2 PROTOZOA OF BLOOD AND TISSUE

Several blood and tissue protozoa are of importance at the individual and population level for *Rangifer*. These are members of three main families: the Sarcocystidae (*Besnoitia tarandi*, *Neospora caninum*, *Sarcocystis* spp. and *Toxoplasma gondii*), the Babesiidae (*Babesia odocoilei* and *B. divergens*, *B. venatorum*, *B. capreoli*, *Theileria* sp.) and the Trypanosomatidae. Species of the Sarcocystidae are typically transmitted through a predator-prey lifecycle with *Rangifer* as the intermediate host, whereas the Babesiidae and Trypanosomatidae are transmitted via arthropod vectors.

6.3.2.1 *Besnoitia tarandi*

Besnoitia tarandi is common and widespread in *Rangifer* in North America, but does not occur in Greenland (Kutz et al., 2012; Ducrocq et al., 2013). *Besnoitia* occurs in Fennoscandia but appears to be less common than in North America; during reindeer meat inspection in Finland over the past 10 years, *Besnoitia* was detected in 31 of approximately 700,000 animals (Laaksonen, unpublished). The true prevalence is likely higher, though, because mild or recent infections are difficult to detect during routine meat inspection (Laaksonen, 2016).

Besnoitia is most commonly seen in the chronic state, as tissue cysts in the skin and subcutaneous tissues of the lower legs and in the ocular conjunctiva; however, it can be found in a variety of body tissues (Ducrocq et al., 2012) (Figure 6.22). The acute course of infection is not well described but may be similar to that reported for *B. besnoiti*. While *B. tarandi* is common across its range and there is focal pathology associated with the cysts, this parasite has rarely been implicated as a cause of significant disease.

Exceptions to this include findings in the Rivière-George and Rivière-aux-Feuilles caribou herds of Quebec, Canada, where *B. tarandi* emerged as a disease-causing agent in 2007 (Kutz et al., 2009a). As well, an outbreak at the Winnipeg Zoo in Manitoba, Canada, led to extensive morbidity and

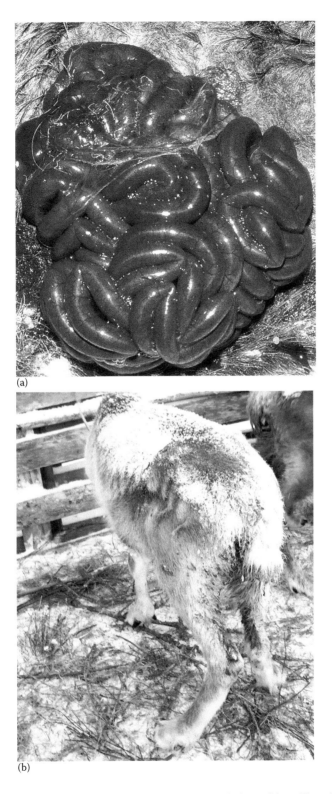

FIGURE 6.21 (a) Intestines with severe *Eimeria* infection. (b) Reindeer with an *Eimeria* induced diarrhoea (Photos: Sauli Laaksonen).

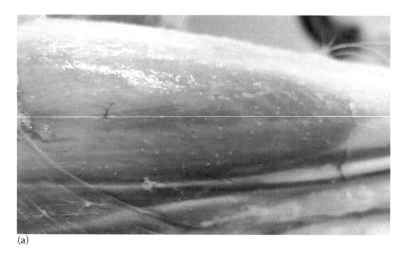

(a)

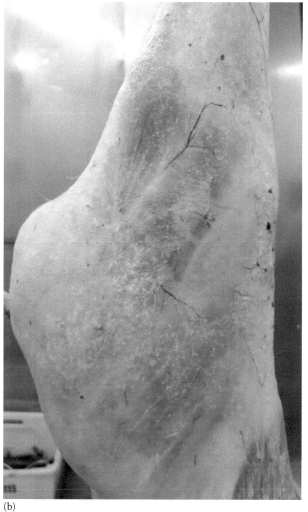

(b)

FIGURE 6.22   (a, b) *Besnoitia tarandi* is most commonly seen as tissue cysts (the size of salt grains) in the skin and subcutaneous tissues of the legs of *Rangifer*.   *(Continued)*

FIGURE 6.22 (CONTINUED)   (c) In the acute state, *Besnoitia* causes a state preferred as "hot foot," with itching and biting leading to local hair loss. In chronic conditions, tissue cysts are seen in (d) the conjunctiva of the eye. (e) on the mucous membranes of nasal turbinate, and (f) trachea and in a variety of tissues throughout the body. (g) Reindeer showing clinical signs of chronice besnoitiosis poor condition, alopecia, skin thickening, decreased mobility, resistance to movement, and mortality (Photos: Sauli Laaksonen).

mortality (Glover et al., 1990). Clinical signs observed in both these cases included poor condition, alopecia, skin thickening, decreased mobility and resistance to movement (Ducrocq, 2011). In addition, *Besnoitia* spp. have been implicated as a cause of infertility, and *B. tarandi* may have similar impacts. In caribou, it can cause substantial pathology in the pampiniform plexus of the testicles, which could also lead to reduced testosterone levels and abnormal growth of antlers (Ducrocq, 2011).

While infection with *B. tarandi* has been associated with broken antlers and velvet retention in *Rangifer* (Rehbinder et al., 1981; Ducrocq, 2011), it is not yet known whether this is cause and effect.

The *B. tarandi* transmission cycle remains unknown but is thought to be similar to other species of *Besnoitia* in ungulates, with transmission through carnivore definitive hosts as well as arthropods. Experimental studies exposing dogs, domestic cats, raccoons and an arctic fox to *B. tarandi* have not proven these species to be definitive hosts (reviewed by Kutz et al., 2012); however, transmission by arthropods is supported in part by epidemiological investigations in zoo and wild caribou (Glover et al., 1990; Ducrocq et al., 2013). Prevalence of infection with *B. tarandi* is higher in male caribou than in females, and increases over the first year of life and then declines. There is also a seasonal pattern, with infection intensity in the spring lower than that in the previous fall, which may suggest overwinter mortality associated with higher infection levels (Ducrocq et al., 2013).

Besnoitia tarandi can be diagnosed by direct observation of the conjunctiva and observation and palpation of the skin of the metatarsus; however, more sensitive methods include histopathology of skin from the mid-cranial metatarsus (Ducrocq et al., 2012), and a multi-species ELISA test for *Besnoitia* spp. has also been developed (Gutiérrez-Expósito et al., 2012).

6.3.2.2 *Toxoplasma gondii* and *Neospora caninum*

Toxoplasma gondii and *Neospora caninum* are two important tissue protozoa of *Rangifer* that may cause reproductive loss (Figure 6.23). Serological evidence of both these parasites is common across most herds in North America (Kutz et al., 2012). In Fennoscandian reindeer, the overall prevalence of antibodies against *T. gondii* has generally been low, documented at 0.9–1.0% (Oksanen et al., 1997; Vikøren et al., 2004). However, there is a clear association between seroprevalence and degree of domestication, and in some areas seroprevalence exceeds 20% (Oksanen et al., 1997). Svalbard reindeer and sibling voles on Svalbard are free from *T. gondii* infection despite the fact that this parasite is common in sympatric geese and foxes (Prestrud et al., 2007).

Toxoplasma and *Neospora* are transmitted through carnivore definitive hosts, felids and canids, respectively. They undergo asexual and sexual cycles in the intestines of these hosts, and then shed oocysts in their faeces. *Rangifer* primarily become infected by ingesting sporulated oocysts from the

FIGURE 6.23 Aborting reindeer. *Neospora caninum* and *T. gondii* are transmitted across the placenta to the foetus and can result in abortion, mummified foetuses, or the birth of stillborn or weak calves (Photo: Sauli Laaksonen).

environment; however, transplacental transmission, as occurs with domestic livestock, is suspected (Stieve et al., 2010). Both *T. gondii* and *N. caninum* are well-known causes of abortion, foetal abnormalities and neonatal mortality in sheep and cattle, respectively (Figure 6.23) (Dubey et al., 2007).

For *N. caninum*, dogs, wolves and coyotes are all suitable definitive hosts, and for *T. gondii*, lynx (seroprevalence 15% to 44% in northern Canada and Alaska) and cougars are considered likely definitive hosts in the wild. While neither parasite has been explicitly linked to reproductive loss in the wild, they are both known pathogens in captive *Rangifer* (reviewed in Kutz et al., 2012).

An abortion storm in a captive reindeer herd was associated with an incursion of coyotes onto the pasture the previous summer/fall and exceptionally high *N. caninum* seroprevalence in adult females the following spring (Kutz et al., 2012). In wild herds, *N. caninum* seroprevalence can be high, and for many of these herds, poor productivity/early calf survival is often implicated as the cause for declines. In a study of reindeer experimentally infected with *T. gondii*, one animal developed clinical signs of anorexia, depression and fatal enteritis (Oksanen et al., 1996). As well, transplacental transmission and birth of a stillborn calf has been documented in a captive reindeer (reviewed in Kutz et al., 2012). For both *T. gondii* and *N. caninum*, the impacts at a population level in free-ranging *Rangifer* remain unknown; however, the effects of an acute infection, as documented by Oksanen et al., (1996), together with abortions/stillbirths/neonatal mortalities, would have a significant impact on lifetime reproductive success and, thus, important consequences for population dynamics.

Infection of people with *Toxoplasma* can cause abortion or severe foetal abnormalities if the mother is exposed for the first time during pregnancy (Jenkins et al., 2013). Human exposure with respect to *Rangifer* can occur from handling or consuming undercooked meat. *Neospora* is not zoonotic.

6.3.2.3 Sarcocystis

Sarcocystis spp. are common protozoa of *Rangifer* throughout their range. They are found as small rice-grain-like cysts or microscopic lesions in muscle (Figure 6.24). Several species have been described in reindeer in Norway and Iceland, including *S. gruehneri*, *S. rangi*, *S. tarandivulpes*, *S. hardangeri*, *S. rangiferi* and *S. tarandi* (reviewed by Dahlgren et al., 2008), but the species identities in North American *Rangifer* are unknown (Kutz et al., 2012). As with *Toxoplasma* and *Neospora*, *Rangifer* are intermediate hosts and the parasite is typically transmitted through a predator-prey cycle, although other definitive hosts may include avian scavengers, such as ravens and crows. Canids are definitive hosts for at least three species in *Rangifer*, but bears, wolverine, cougar and

FIGURE 6.24 Two species of *Sarcocystis*, *S. rangiferi*, and *S. hardangeri* can be seen with the naked eye, as shown here. The small rice-grain-like white stripes are the sarcocysts (Photo: Peter Paulsen).

lynx are also potential hosts (reviewed in Kutz et al., 2012). "Sarcocysts," the form of the parasite in *Rangifer*, can be found in the skeletal and cardiac musculature and nervous system.

Sarcocysts are transmitted to carnivore definitive hosts through ingestion of infected tissue, and then they undergo a sexual phase in the gastrointestinal tract of these hosts. The effects of *Sarcocystis* spp. on *Rangifer* are not well understood. In other cervid hosts, natural and experimental infections with *Sarcocystis* spp. have been associated with lethargy, pneumonia, reduced growth rates/weight loss, weakness and even death. Sarcocysts in the muscle is a cause of meat condemnation in Fennoscandia (Dahlgren and Gjerde, 2007), and may result in altered meat quality and increased bacterial contamination (Daugschies et al., 2000).

6.3.2.4 *Babesia*

Babesia spp. are blood protozoa transmitted by Ixodid ticks. These parasites occur naturally and cause disease in *Rangifer* in eastern Russia (Rehbinder, 1990). In Eurasia, infections caused by several *Babesia* spp. have been reported in captive or corralled reindeer (Nilsson et al., 1965; Langton et al., 2003; Wiegmann et al., 2015), but not in free-ranging reindeer in Fennoscandia.

In North America, *Babesia* is not found naturally cycling in wild *Rangifer*. However, *Babesia odocoilei* has caused severe disease and mortality in captive caribou and reindeer (reviewed in Kutz et al., 2012), but clinically normal carriers of *Babesia* species have also been identified in zoos in Germany (Wiegmann et al., 2015). *Babesia* causes disease by invading and destroying red blood cells, resulting in hemolysis and anaemia, and renal disease (Petrini et al., 1995). Recent outbreaks of *Babesia* in zoos in eastern Canada suggest that that *B. odocoilei* may be expanding its range northwards and may become a pathogen of increasing concern under current climate change scenarios.

6.3.2.5 *Trypanosoma*

Trypanosoma spp. are also arthropod-borne blood protozoa with high prevalence (72% to 84%) in *Rangifer* across North America and Greenland (summarized in Kutz et al., 2012), and are reported in reindeer in Fennoscandia as well (Dirie et al., 1990; Kingston and Nikander, 1985).

Trypanosoma cervi has been described morphologically in a number of North American ungulates; however, differences in biology and infectivity across these hosts (e.g. *T.* cf. *cervi* from Alaskan reindeer did not establish when administered experimentally to two elk) suggests that further characterization is required to establish species identity. Transmission of *Trypanosoma* sp. in *Rangifer* is considered to occur through bites from tabanid flies (e.g. horse flies, deer flies) and ticks, and peak prevalence in summer for *Rangifer* may support this mode of transmission. Transplacental transmission has been reported in deer but the significance of this is unknown, and it is also not known whether this applies to *Rangifer*. Infection with *Trypanosoma* in North American or Fennoscandian wild cervids, including *Rangifer*, has not been associated with disease (reviewed in Kutz et al., 2012). Theileriasis, anaplasmosis, babesiosis and mixed infections have caused high mortality among reindeer of the Amur region of Russia (Tashkinov, 1976).

6.4 ARTHROPODS

Rangifer are subject to harassment by a number of ectoparasites, some of which can significantly affect the animals' welfare and energetic balance. Here we discuss three main groups: lice, arachnids (mites, ticks), keds and oestrids (bot flies), and one unusual taxa, a pentastomid (sinus worm). Flies, including horse flies, black flies and mosquitoes, are also important ectoparasites of *Rangifer* but, with the exception of the moose ked, these will not be discussed here.

6.4.1 LICE

Chewing lice (*Bovicola tarandi*) and blood-feeding lice (*Solenopotes tarandi*) have been reported in *Rangifer* across much of their holarctic range, including in Sweden (Mjöberg, 1915),

Alaska (Weisser and Kim, 1973), Finland (Laaksonen, 2016) and Canada (Kashivakura, 2013). Lice have highly specialized legs with tarsi that are adapted to attach to the hairs of their specific host species; thus, they are highly host-specific. These parasites spend their entire lifecycle on the *Rangifer* host and transmission among hosts is by direct contact; young animals become infected from their dams shortly after birth. Chewing lice, and hair loss and itch caused by them, are not uncommon findings in corralled reindeer during late winter in Finland (Laaksonen, 2016) (Figure 6.25). Sucking lice have been described in barrenground caribou in Alaska and in Northwest Territories, Canada, with an apparent predilection for the head and anterior neck (Weisser and Kim, 1973; Kashivakura, 2013).

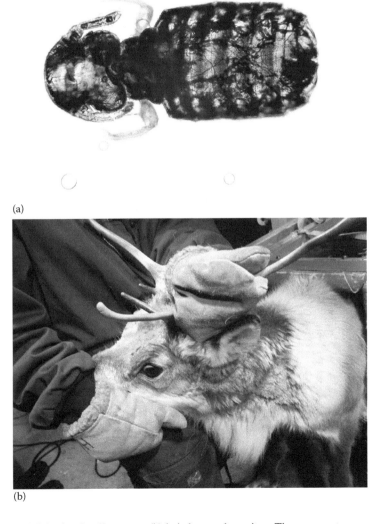

(a)

(b)

FIGURE 6.25   (a) Adult chewing lice cause (b) hair loss and pruritus. These are not uncommon findings in corralled reindeer during late winter in Finland (Photos: Sauli Laaksonen).

6.4.2   ACARIDAE

6.4.2.1   Mites

Mites are very small ectoparasites, approximately 0.5 mm in length, and rarely visible to the naked eye (Figure 6.26). Similar to lice, they complete their lifecycle on the host and are transmitted through direct contact with an infected host. In Russia, reindeer are known to be parasitized by both sarcoptic and chorioptic mange mites (Saval'ev, 1968). *Sarcoptes scabiei* initially causes small vesicles to develop on the skin, but later the affected area may become covered by thick scabs. This is considered an important reindeer health problem in Russia. Reindeer suffering from sarcoptic mange gradually lose body condition. During winter, the weight loss is rapid and the affected reindeer may perish (Saval'ev, 1968). The disease can be transmitted to humans as well.

Chorioptic mange mites cause local bald spots with no thick scabs and are not considered to be as serious of pathogens as sarcoptic mange mites (Saval'ev, 1968). *Chorioptes texanus* has been detected on ears of reindeer in Canada (Kutz et al., 2012), in Norway (Josefsen et al., 2014 referring to unpublished data), and in Finland (Nikander, unpublished) with no significant clinical disease reported. *Chorioptes bovis* infestation was detected on the body and legs of reindeer with hair loss in Finland (Laaksonen, unpublished).

6.4.2.2   Ticks

6.4.2.2.1   Dermacentor albipictus

Dermacentor albipictus, also known as the winter tick or moose tick, is an emerging tick of concern in *Rangifer* in North America. This is a medium-sized, one-host tick that can infect a wide range of ungulates and other mammals. It attaches to the host as a larva in the autumn and remains on that host, blood feeding and undergoing two molts to reach the adult stage. Reproduction occurs on the host in late winter and engorged adult females drop onto the ground the following spring and lay hundreds to thousands of eggs (6.27). Eggs develop and hatch to larvae, which then quest for the next host the following autumn (Drew and Samuel, 1985). Some animals show no

FIGURE 6.26   *Chorioptes bovis* mites found on the body and legs of reindeer with hair loss in Finland (Photo: Sauli Laaksonen).

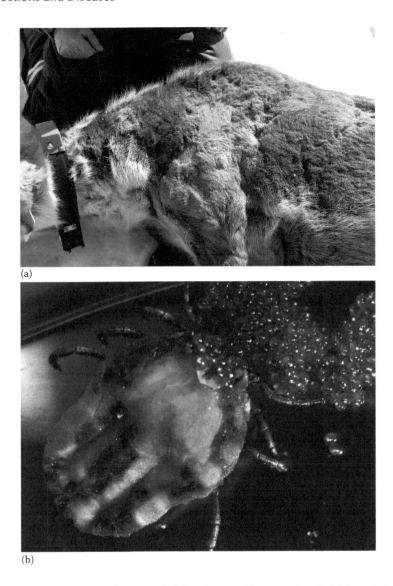

(a)

(b)

FIGURE 6.27   (a) Boreal caribou with severe hair breakage and loss associated with the winter tick, *Dermacentor albipictus* (Photo: Diane and Brad Culling). (b) Female winter tick laying eggs (Photo: Cyntia Kashivakura).

clinical signs; in others, the signs range from mild to severe hair loss to anaemia. This parasite is expanding its range north in Canada and is increasingly reported as a cause of morbidity in woodland caribou (Kutz et al., 2009a; Kashivakura, 2013) (Figure 6.27). To date, there is no evidence that it has invaded barrenground caribou herds; however, this is considered a potential risk under current climate warming scenarios (Kutz et al., 2009a).

6.4.3 KEDS

The deer ked, *Lipoptena cervi* (L. 1758) (Diptera, Hippoboscidae), is a blood-sucking ectoparasitic fly that parasitizes several deer species, including *Rangifer* (Figure 6.28) (Haarløv, 1964). All species of deer keds, including *L. cervi*, are viviparous and produce one larva at a time. Winged young adults of *L. cervi* emerge from pupae on the ground from late summer through autumn, at which time they find a host and then lose their wings. Both sexes remain on the host over winter and feed

FIGURE 6.28 *Lipoptena cervi*. After finding a host, the *L. cervi* fly (a) drops its wings and, after a blood meal, (b) its abdomen expands. (c) Females extrude white mature larvae, the prepupa, one at a time. The skin of prepupa hardens, darkens and drops to the ground where it remains until it hatches to an adult fly in late summer/autumn. Deer ked cause notable harm and acute behavioural disturbances leading to (d) hair breakage. (e) heat loss (Photos: Sauli Laaksonen).

on blood and interstitial fluid. Females extrude white mature larva, the prepupa, one at a time, from autumn until the following summer (Härkönen et al., 2010). The skin of prepupa fattens and hardens to the pupa and drops to the ground, where it remains until it hatches to an adult fly in late summer/ autumn (Haarløv, 1964).

The distribution of *L. cervi* is bimodal in Fennoscandia. A western population is found in Norway and Sweden, with the respective northern limits of the geographical ranges being ca. 61°N and 62°N in those countries. An eastern population in Finland occupies more northern latitudes, with the northern range reaching as far as 65°N. The parasite has expanded its geographical range throughout Fennoscandia in recent decades (Välimäki et al., 2010), and is considered an invasive species in the southern reindeer herding area of Fennoscandinavia.

In Finland, the main host of deer ked is moose (*Alces alces*), with abundances up to 17,000 parasites per moose (Paakkonen, 2010). The ked also uses wild forest reindeer (*Rangifer tarandus fennicus*) and semi-domestic reindeer (*Rangifer tarandus tarandus*) as hosts (Kaunisto et al., 2009; Kynkäänniemi et al., 2010). In the southern reindeer herding area of Finland, deer ked infestations are commonly recognized in reindeer (Laaksonen, 2016). Kynkäänniemi et al. (2014) demonstrated that deer ked cause notable harm and acute behavioural disturbances in reindeer, even when the infection level is relatively low (300 deer keds) (Figure 6.28). Heat loss during winter and weakening of moose through loss of winter hair has been reported in Norway and Sweden by Madslien et al. (2011). Openings and chafes in the fur were also observed in reindeer by Kynkäänniemi et al. (2014), and this would, without doubt, lead to detrimental effects, especially during hard winter conditions. Deer keds are sensitive to ivermectin (Kynkäänniemi et al., 2010) and the presence of this fly in the southern reindeer herding area of Finland is one indication for early ivermectin treatment in autumn (Laaksonen, 2016).

Bites by wingless parasitic adults have been reported on various other mammals, including humans (Rantanen et al., 1982), dogs (Hermosilla et al., 2006), horses and even birds (Johnsen, 1946). The known Central European breeding hosts that support reproduction of *L. cervi* are the red deer (*Cervus elaphus*), the roe deer (*Capreolus capreolus*) and, to a lesser extent, the fallow deer (*Dama dama*) (Haarløv, 1964). In Fennoscandia, the main host is the moose (*Alces alces*). In addition to the red deer and the roe deer, a few records on successful reproduction have been reported for reindeer (*Rangifer tarandus*) as well (Kaunisto et al., 2009; Kynkäänniemi et al., 2010; Kynkäänniemi et al., unpublished). The white-tailed deer (*Odocoileus virginianus*) is also among the breeding hosts of *L. cervi* in North America, but the prevalence of infection seems relatively low (Matsumoto et al., 2008; Samuel et al., 2012).

In addition to being a nuisance and an obstacle to traditional human outdoor activity, *L. cervi* may cause serious health problems and symptoms, including chronic deer ked dermatitis (Rantanen et al., 1982) and occupational allergic rhinoconjunctivitis in people (Laukkanen et al., 2005). The species is also a potential vector of various diseases (Rantanen et al., 1982; Duodu et al., 2013).

6.4.4 OESTRIDS: BOTFLIES

Two fly species of the family Oestridae (botflies) parasitize reindeer/caribou (*Rangifer tarandus* and subspecies): the reindeer warble fly (*Hypoderma* [= *Oedemagena*] *tarandi*) and the reindeer throat botfly (*Cephenemyia trompe*) (Zumpt, 1965). Both species belong to the order Diptera (two-winged). With most of the body covered by golden and black hairs, and a body length 15–18 mm for *H. tarandi* and 14–16 mm for *C. trompe*, the flies resemble bumblebees (Figure 6.29a, b). They also resemble the moose throat botfly, *Cephenemyia ulrichii* (Figure 6.29c). Both reindeer botflies have a similar 1-year lifecycle, with the larvae living as obligate parasites in reindeer/caribou. There is only a short time in June–July (Figure 6.30) when the animals are free of larvae. The term myiasis (Greek myia = fly) denotes the state of being infested by fly larvae. The female *H. tarandi* can oviposit on mammalian species unsuitable for complete larval development (accidental hosts), and reported cases include red deer (Nilssen and Gjershaug, 1988), moose (Ågren and Chirico, 2005), muskoxen (Samuelsson et al., 2013), and humans (e.g. Lagacé-Wiens et al., 2008; Landehag et al., 2017).

The reindeer botflies have a northern-hemisphere circumpolar distribution. They are found where reindeer/caribou live in Alaska, Canada, Greenland, Fennoscandia and Russia (Zumpt, 1965), but not in the northernmost high arctic regions of Canada and the high Arctic Archipelago of Svalbard. The flies are also absent from Iceland, although the reindeer present there are

FIGURE 6.29   (a) *Hypoderma tarandi* – the reindeer warble fly, male (Photo: Arne C. Nilssen). (b) *Cephenemyia trompe* – the reindeer throat botfly, male (Photo: Arne C. Nilssen). (c) *Cephenemyia ulrichii* – moose throat botfly (Photo: Hanna Halonen).

descendants of reindeer that were imported from Finnmark (northernmost county of Norway) four times between 1771 and 1787 (Sigurdarson and Haugerud, 2004). Visual inspection of hides from slaughtered reindeer in Finnmark in 1984–1985 revealed *H. tarandi* larvae in 99.9% (Folstad et al., 1989), and recent reports (Fauchald et al., 2007; Rødven et al., 2009) still showed high *H. tarandi* prevalence (counts of larvae) in reindeer in northern Norway. For reindeer calves in Finland in 2005–2007, there was a significant increase in *H. tarandi* prevalence and intensity from south (11%) to north (100%) in the reindeer management area (Åsbakk et al., 2014). Nilssen and Haugerud (1995) demonstrated an overall prevalence of 65% of throat botfly larvae in reindeer in northern Norway in 1983–1985/1987–1988, with densities of up to 200 larvae per animal.

6.4.4.1   *Hypoderma tarandi* – Warble Fly
The reindeer warble fly is closely related to *Hypoderma lineatum* and *Hypoderma bovis*, both warble fly species in cattle that are considered practically extinct in Fennoscandia and most other countries in Europe and America after campaigns using effective anti-parasitic agents (Colwell et al., 2004). The mature *H. tarandi* larvae leave the host mainly in May–June, and they burrow superficially into the ground and pupate. Adult flies emerge after a few weeks, usually in early July (Nilssen and Haugerud, 1994; studies in northern Norway), but the first ones may appear in June if the early summer is warm (Nilssen, 1997a). The flies mate shortly after they emerge.

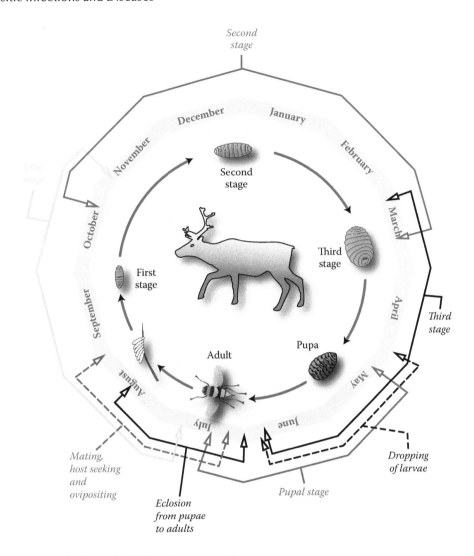

FIGURE 6.30   The life cycle of *Hypoderma tarandi*.

Mating and oviposition activities depend on good weather conditions for flying, and are greatest in warm weather (>13–15°C) and when there is little wind (Anderson et al., 1994). On the vidda in northern Norway (i.e. the biome above the tree line), mating activities typically occur in rocky areas along rivers and streams and arid riverbeds, whereas in forest areas mating typically occurs at certain topographical sites along gravel roads and trails (Anderson et al., 1994). The males die after mating, and only females seek reindeer. In experiments with female flies on a flightmill, some flew continuously for 15 hours, with the maximum distance flown through life estimated at 600–900 km (Nilssen and Anderson, 1995). With assumed normal airspeed of 20–30 km/h, females thus have great ability to seek reindeer over long distances (Nilssen, 1997b). The period for oviposition is mainly July and August (Anderson and Nilssen, 1996). The female attaches the eggs to thin summer hairs, close to the skin, using an extendable ovipositor organ (Figure 6.31a). The eggs attach firmly to the hair (Figure 6.31b), typically in rows of four to six, and hatch after 4–7 days, depending on the temperature in the haircoat (Karter et al., 1992). The larva must migrate quickly to the skin surface and penetrate it. At this stage, L1 (the larva has three developmental stages separated by molting, L1-L3), the larva is 0.6 mm long (Zumpt, 1965) (Figure 6.31c), sticky and dries out easily. Larvae of

FIGURE 6.31   (a) Female *Hypoderma tarandi* during oviposition (Photo: Arne C. Nilssen). (b) Eggs of *Hypoderma tarandi* on a reindeer summer hair. Most are open (slit) in the distal end after the larva has left (Photo: Kjetil Åsbakk). (c) Scanning electron microscopy picture of the anterior end of a newly hatched larva of *Hypoderma tarandi* (Photo: Kjetil Åsbakk). (d) A reindeer with many *Hypoderma tarandi* larvae under the skin in May (Photo: Arne C. Nilssen).

H. lineatum that were unable to find a suitable penetration site died from desiccation on the bovine skin surface after a few minutes (Nelson and Weintraub, 1972). It is thus reasonable to believe that if eggs are laid on animals with a haircoat thicker than reindeer, the animal may remain uninfested because the distance between hatching site and skin surface is too great. Also, L1 larvae have a high mortality rate (>70%) in reindeer even after they penetrate the skin successfully (Breyev, 1961). The larvae migrate within the reindeer (Tashkinov, 1976), and by October–November the majority end up under the skin and are especially abundant on the rear portion of the host's back, now as L2 (Figure 6.31d). The larvae develop during the winter in capsules of connective tissue with a breathing hole that opens to the outside air (Figure 6.32). In heavily infested reindeer, larvae can be seen also on the neck.

The larvae leave through the breathing hole as L3, 2.5–3 cm long and weighing approximately 1.5 grams. Mouthparts are vestigial in adult flies, so they do not take nourishment (Zumpt, 1965). Domesticated reindeer that normally can be touched by hand avoid such contact when the larvae are large under the skin, apparently because it is painful. This may protect the larvae from mechanical damage since the reindeer probably will avoid bumping into tree trunks and rocky outcrops.

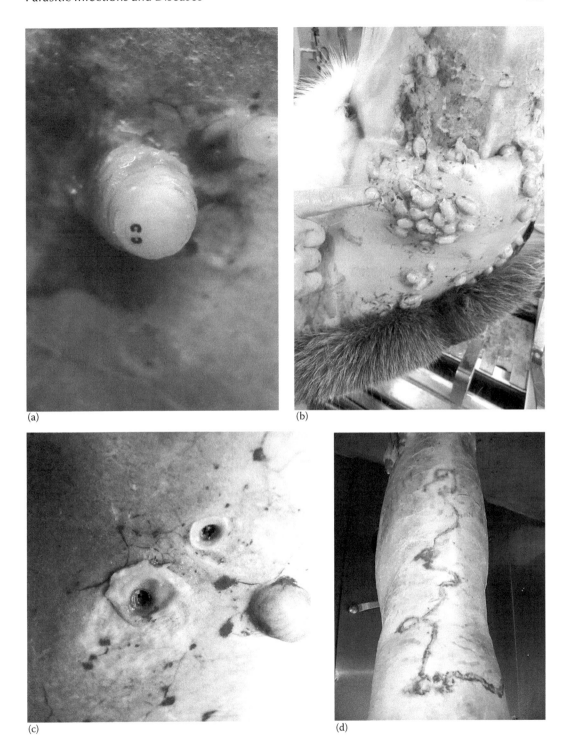

(a)

(b)

(c)

(d)

FIGURE 6.32 (a, b) The larvae of *H. tarandi* develop during the winter in capsules of connective tissue under the skin with (c) a breathing hole to the outside. (d) Tracts of a migrating larva on the back of a reindeer (Photos: Sauli Laaksonen).

Accordingly, larvae are absent from regions of the body where there are bone structures just beneath the skin (Skjenneberg and Slagsvold, 1968).

During migration in the host body, the larvae excrete enzymes (serine proteases; Moiré et al., 1994) digesting host tissues so that they become available as nourishment for the larva (Figure 6.32d). Some of the enzymes also have an inhibitory effect on the host's immune system (Otranto, 2001), which is probably evident in the host as a general lack of, or low grade of, inflammation around intact larvae under the skin.

Serological assays detecting antibodies against one of the enzymes secreted by larvae, hypodermin C (HyC), have proved useful for serodiagnosis of *H. tarandi* infestation (hypodermosis). HyC is common to several *Hypoderma* species, and the enzyme from one species (e.g., *H. lineatum*) is applicable for serodiagnosis of hypodermosis caused by other species (Boulard et al., 1996). Most reindeer calves are born in May (Fennoscandia), and if the mother is seropositive for antibodies against HyC, the calf receives such antibodies with the colostrum (the first mother's milk) (Åsbakk et al., 2005). By the onset of the major *H. tarandi* oviposition period in mid-July, however, these antibodies have declined to almost undetectable levels. Thus, elevated anti-HyC antibody levels in calves-of-the-year in fall/winter signals acquired infestation during the preceding summer. Following the first year of infestation, anti-HyC antibodies again decline to very low levels by the subsequent summer after the larvae have left the 1-year-old calf (yearling). Antibodies against HyC in adult reindeer give no information as to whether the animal acquired an *H. tarandi* infestation during the previous summer, since elevated levels seem to persist after repeated infestations (Åsbakk et al., 2005). The number of larvae per animal is usually higher in calves than in adult reindeer (Helle, 1980; Folstad et al., 1989), and the lack of antibodies in calves and yearlings during the summer may be a contributing factor. The difference may also be due to behavioural differences between juveniles and adults (hierarchy and dominance conditions), and whether the animals occur in herds or as single individuals during the infestation period (Folstad et al., 1989; Fauchald et al., 2007). In addition to flying and landing on reindeer for oviposition, the females may approach reindeer resting on the ground by walking or jumping up from the ground and, as such, be able to lay eggs unnoticed (Skjenneberg and Slagsvold, 1968). Strategies not involving flying may be especially important on days of low temperature, wind, and rain, and body height of calves may make them easy targets. If reindeer move a long distance after the larvae have dropped to the ground, the flies emerging from the pupae may have a challenge finding reindeer hosts, especially if conditions are suboptimal for flying (Nilssen and Haugerud, 1994; Landehag et al., 2017).

6.4.4.2   *Cephenemyia trompe* – Throat Botfly

The time at which the larvae drop from the host and the period at which the adult females attack reindeer are approximately the same for the reindeer throat botfly and the reindeer warble fly (Nilssen and Haugerud, 1994). Like the warble fly adults, the throat botfly adults take no nourishment and instead live only on fat reserves acquired from the host during the larval stage. The throat botfly female does not oviposit; instead, the eggs hatch inside the female body, whereupon the fly sprays the newly hatched larvae onto the nose and mouth of the reindeer. From there they end up in the nasal sinuses, where they live during the first few months; they grow very little during this time. Later in the winter, they begin to grow more rapidly while moving further back in the pharynx, where they attach themselves with two hooks (Figure 6.33). From late April to June/July, the larvae are ejected through the mouth or nostrils by the coughing of the host (Nilssen and Haugerud, 1994).

The mature throat botfly larvae are slimmer than the warble fly larvae and up to 4 cm long. The larvae rapidly pupate and eclose as adult flies after 4–5 weeks or more, depending on temperature. Mating takes place at distinct peaks in the terrain; males congregate at these locations and approaching females are caught in the air and pulled to the ground, where mating occurs. After a few days, females start to seek reindeer, guided in part by carbon dioxide exhaled from the reindeer and possibly by other chemical substances, such as octenol. Sight also plays a role. If the female manages to make a direct hit with larvae on the reindeer nose, the animal reacts strongly by sneezing and hiding

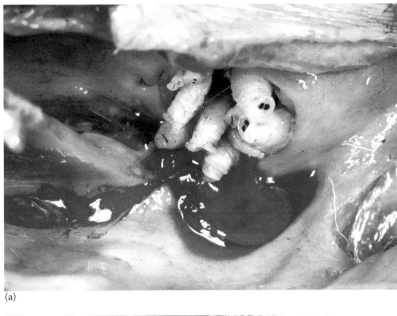

(a)

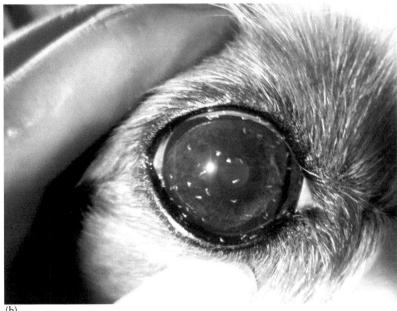

(b)

FIGURE 6.33   (a) *Cephenemyia trompe* larvae in the pharynx of a reindeer (Photo: Arne C. Nilssen). (b) Throat botfly larvae (*Cephenemyia ulrichii* or *C. trompe*) on the surface of the eye of a dog in Finland (Photo: Anette Brockmann).

its muzzle close to the ground. In an attempt to reduce the insect harassment, the reindeer seek mountaintops and patches of snow, if present. Like the warble fly, the throat botfly can fly long distances.

6.4.4.3   Impact of Botflies on Reindeer and Reindeer Management

Reindeer react with panic when attacked by botflies, and much more strongly than when harassed by bloodsuckers like mosquitoes and horse flies (Nilssen, 1997a). Combined with the activity of other insects, botfly activity reduces grazing and resting time (Skjenneberg and Slagsvold, 1968; Hagemoen and Reimers, 2002), and this, in turn, affects the animals' ability to build body reserves for winter. Heavy infestation, particularly in late winter when the larvae are large, imposes a strong physiological load on the animal, which can reduce body weight of individuals and, in turn, affect reindeer at the population level (Ballesteros et al., 2012). The breathing holes of the warble fly larvae also reduce leather quality (Nieminen, 1992). Consequently, botflies create economical and animal-welfare problems and are a production-limiting factor in reindeer husbandry (Weladji et al., 2003).

6.4.4.4   Treatment of Reindeer against Botflies

In the former Soviet Union, attempts were made to reduce reindeer botfly populations by spraying areas with insecticides (briefly summarized in Åsbakk and Oksanen, 2001). In the 1980s, ivermectin fully arrived on the commercial market and became the main drug used to combat reindeer botflies. Ivermectin is a broad-spectrum anti-parasitic drug and is considered very efficient against the larval stages of reindeer botflies and various nematodes in reindeer (Oksanen, 1999; Laaksonen et al., 2008). The most effective recommended treatment with ivermectin is once a year, in early winter (October–February, during the reindeer winter round-up), with a standard dose of 200 μg/kg body weight injected subcutaneously (Oksanen et al., 1992; 2014). Even when large larvae die under the skin after such treatment, the body of the host resorbs the larvae without apparent adverse effects to the host, as shown in cattle. A herd-by-herd and district-by-district (reindeer herding cooperatives) treatment is practiced. Among the Nordic countries, Finland has particularly extensive (>20 years) experience in treating reindeer with ivermectin, with a large proportion of the population treated annually. For example, in 2002–2004, the proportion treated per herding district was 76–80% (Laaksonen et al., 2008), with the aim of helping the individual animals through the winter (Oksanen et al., 1998; Laaksonen et al., 2008). There is, however, a desire within the reindeer husbandry industry (Fennoscandia) to keep reindeer as natural a product as possible, and many reindeer owners are reluctant to treat with anti-parasitic drugs. In addition, little is known about the importance of botflies in the ecosystem (Colwell et al., 2009).

Experiments with reindeer calves showed that treatment with ivermectin during the first autumn of life resulted in substantial growth gains a year later compared to untreated calves (Heggstad et al., 1986). Treatment of reindeer calves in early July, however, resulted in no difference in autumn slaughter weight between groups of treated and untreated animals (Oksanen et al., 1998, Laaksonen et al., 2008). In pregnant females, ivermectin treatment had a positive effect on weight gain (Oksanen et al., 1992). For adult females treated with ivermectin in autumn/winter, there was a positive effect on body weight until the next summer (Ballesteros et al., 2012). Attempts to eradicate *H. tarandi* on isolated islands using anti-parasitic drugs failed (Kummeneje, 1980; Nilssen et al., 2002). After reduction of the warble fly population by 94% to 99% (Kummeneje, 1980), the population increased again to pre-treatment levels after just 2 years. With 667 as the average number of eggs per *H. tarandi* female (Anderson and Nilssen, 1996), only a few untreated reindeer are required to enable a rapid increase of the warble fly population, especially after favorable summers (Nilssen, 1997b) and with high reindeer density (Fauchald et al., 2007). Although Finland has many years of experience with ivermectin treatment that removes the larvae from the individual reindeer, this treatment does not appear to have reduced the warble fly problem (Åsbakk et al., 2014).

Possible environmental impacts of the use of ivermectin are a cause of concern (Herd, 1995). Most of the treatment dose given to an animal is excreted with the faeces in an unchanged and still-active form, regardless of the route of administration (Sommer et al., 1992). Ivermectin is fat-soluble and, when present as thin films on water surfaces, it is rapidly broken down by sunlight to less bioactive compounds (Halley et al., 1993). After treatment of reindeer by subcutaneous injection, the ivermectin concentration in faeces increases to a maximum on around day four after treatment, and residues are detectable in the stool for more than 30 days (Nilssen et al., 1999). Thus, faeces excreted immediately after treatment are a source of high local concentrations of ivermectin on reindeer pastures, and the drug can be detected in faeces on pastures for more than 2 years (Åsbakk et al., 2006).

Because the time for treating reindeer is usually in winter and the dry faecal pellets from reindeer excreted at this time of the year are of little interest to dung-degrading insects (e.g. flies, beetles) the following summer, it is likely that the treatment has minimal impact on this segment of the decomposer-fauna (Nilssen et al., 1999). However, the main dung-degrading fauna and the fauna with the greatest biomass on reindeer pastures consists of springtails (Collembola), mites (Acari), and small earthworm-like oligochaetes (Enchytraeidae). Whether ivermectin residues on pastures have any impact on these fauna is unknown. Ivermectin residues in reindeer dung on pasture had no apparent adverse effect on nematode communities under the dung (Yeates et al., 2007).

6.4.4.5   Reindeer Botfly Larvae in Humans

Reports on myiasis caused by *H. tarandi* larvae in humans include cases in Norway (e.g. Syrdalen et al., 1982), Sweden (Chirico et al., 1987), Canada (Lagacé-Wiens et al., 2008), and Greenland (Bangsø et al., 2016). Most of the cases reported before 2012 involved ophthalmomyiasis interna, a condition in which the larva invades the eyeball and, in the worst case, can damage the eye or even cause loss of vision. A case of myiasis caused by *H. tarandi* in a boy in Sweden in 2009 was diagnosed by a serological test for antibodies against HyC, and the boy was successfully treated with ivermectin (Kan et al., 2010). Eggs found in the boy's hair were first mistaken as head lice eggs, and the case demonstrated a lack of knowledge and awareness of *H. tarandi* myiasis as a potential medical problem. The reporting resulted in diagnosis of additional cases in Sweden (Kan et al., 2013). These recent reports provide information on signs and symptoms of human *H. tarandi* myiasis and ophthalmomyiasis. The report in 2010, together with media coverage, led to the diagnosis of 39 new cases in Norway in 2011–2016 (Landehag et al., 2017). These patients were diagnosed based on typical clinical signs (i.e. migratory dermal swellings of the head region, enlarged lymph nodes, periorbital edema) combined with the serological test for antibodies against HyC, and in some cases also by isolation and identification of larva. Most patients were youth (3–13 years old) living in Finnmark, the northernmost county of Norway, where there is extensive reindeer husbandry practice. The patients were treated with ivermectin, and no ophthalmic complications occurred. Since most of the cases reported before 2012 involved ophthalmomyiasis, it is reasonable to believe that the ophthalmic complication was the primary reason for the diagnoses. The many new cases identified after 2012 suggests that *H. tarandi* myiasis is likely under-diagnosed in humans. Previous cases may have remained unidentified; for example, the larvae could have died without causing significant health problems (humans are unsuitable hosts for complete larval development), or there may have been ophthalmic complications of unknown origin. Personal experiences of the authors and others show that the warble fly is highly capable of entering the hair of humans for oviposition without making the prospective host aware of its presence. This is easily prevented by covering the hair (with a hat or cap) when in areas where there are reindeer, not only on hot summer days (>13–15°C) when the botflies can fly, but also on days that are suboptimal for flying, when the flies may be able to walk or jump onto potential hosts. There are also reported cases of throat botfly larvae in the eyes of humans (Jaenson, 2011) and dogs (Figure 6.33b).

6.4.5   Pentastomid

6.4.5.1   *Linguatula arctica*

Linguatula arctica is an unusual parasite of *Rangifer* found in North America and Eurasia. It belongs to the pentastomids. Of the roughly 130 species of pentastomids globally, only one is known to infect *Rangifer*. The majority are obligatory parasites of the respiratory passages of reptiles in tropical regions, and they are often referred to as tongue worms due to the resemblance of some species to a tongue (Nikander and Saari, 2006).

Early reports of the sinus worm (*L. arctica*) in reindeer were erroneously recorded as *L. serrata*, a sinus worm in canids. The parasite was reported in introduced reindeer on Unimak Island,

Alaska as early as 1926, suggesting that it may have been introduced to North America. It has also been reported in caribou on Baffin Island, Canada, however, where traditional knowledge suggests that it may have been present prior to reindeer introductions (Ferguson, 2003). High prevalence of *L. arctica* (31%, *n* = 130) has been reported in reindeer in Taimyr, Russia, and, at least historically, was also common in Norwegian, Swedish and Finnish wild and semi-domesticated reindeer (reviewed by Nikander and Saari, 2006).

The morphology and lifecycle of this parasite was described by Riley et al. (1987). This organism is called the "sinus worm" because it lives in the paranasal sinuses of reindeer, especially calves. Worms are paddle-shaped, transparent, pale yellow, dorso-ventrally flattened, and pseudo-segmented with a long, tapering end (Figure 6.34). Adult female parasites are 8–14 cm long, with a flat, broad (1.3–2 cm) oval front end. The male is smaller, at 3.2–4 cm. The lifecycle of *L. arctica* differs from all other known pentastomids in that it is a direct lifecycle with *Rangifer* as the definitive host (Haugerud, 1988), although the possibility of an invertebrate intermediate host cannot be excluded (Riley et al., 1987). Fertilized females in the sinuses produce thousands of eggs – up to 10,000/day – from April to October, after which the females die (Haugerud and Nilssen, 1985; Haugerud, 1988). The eggs are released to the environment by sneezing, by nasal secretions, or sometimes via faeces, and are immediately infective (Riley, 1986). The lifecycle and development in the host is not known, but it is suggested that the larvae migrate from the lungs to the paranasal sinuses (Haugerud, 1988).

Linguatula arctica is predominantly parasitic in first- and second-year animals, with most hosts becoming infected in their first year on the summer coastal grazing areas (Riley, 1986). With the help of an oral papilla, the sinus worm produces an ulcer on the epithelium and feeds on the blood that pools on the ulcer. Usually the infestation is symptomless. Sometimes this parasite causes mild inflammation on the mucous membranes of sinuses, with eosinophilic and neutrophilic infiltration and increased mucous secretion and sneezing (Nikander and Rahko, 1989). A conspicuous ulcerative lesion was a consistent finding in the studies conducted by Nikander and Saari (2006), and this was located at the attachment site of the parasite in the sinusoid mucosa of infected reindeer. *Linguatula arctica* is very rare in older animals, and it is suggested that reindeer can develop immunity to it (Haugerud, 1988).

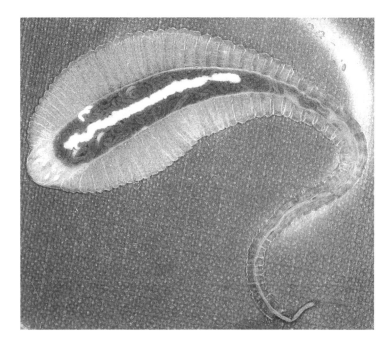

FIGURE 6.34   Adult female *Linguatula arctica*, or "sinus worm" (Photo: Arne C. Nilssen).

REFERENCES

Ågren E., and J. Chirico. 2005. Reindeer warble fly larvae (*Hypoderma tarandi*) in a moose (*Alces alces*) in Sweden. *Acta Vet Scand* 46(1–2):101–3.

Albon, S.D., A. Stien, R.J. Irvine, R. Langvatn, E. Ropstad, and O. Halvorsen. 2002. The role of parasites in the dynamics of a reindeer population. *Proc Biol Sci* 269(1500):1625–32.

Aleuy, O.A., K. Ruckstuhl, E.P. Hoberg, A. Veitch, and S.J. Kutz. 2018. The abomasal nematode *Marshallagia marshalli* is negatively associated with fitness indicators in Dall's sheep. In press, *PLoS* 1.

Anderson, J.R., and A.C. Nilssen. 1996. Trapping oestrid parasites of reindeer: The response of *Cephenemyia trompe* and *Hypoderma tarandi* to baited traps. *Med Vet Entomol* 10:337–46.

Anderson, J.R., A.C. Nilssen, and I. Folstad. 1994. Mating behavior and thermoregulation of the reindeer warble fly, *Hypoderma tarandi* L. (Diptera: Oestridae). *J Insect Behav* 7:679–705.

Anderson, R.C. 2000. *Nematode Parasites of Vertebrates: Their Development and Transmission*. Wallingford Oxon, UK: CAB International.

Arneberg, P., and I. Folstad. 1999. Predicting effects of naturally acquired abomasal nematode infections on growth rate and food intake in reindeer using serum pepsinogen levels. *J Parasitol* 85(2):367–9.

Arneberg, P., I. Folstad, and A.J. Karter. 1996. Gastrointestinal nematodes depress food intake in naturally infected reindeer. *Parasitology* 112(2):213–19.

Åsbakk, K., J.T. Hrabok, A. Oksanen, M. Nieminen, and J.P. Waller. 2006. Prolonged persistence of fecally excreted ivermectin from reindeer in a sub-arctic environment. *J Agric Food Chem* 54:9112–8.

Åsbakk, K., J. Kumpula, A. Oksanen, and S. Laaksonen. 2014. Infestation by *Hypoderma tarandi* in reindeer calves from Northern Finland – Prevalence and risk factors. *Vet Parasitol* 200:172–8.

Åsbakk, K., and A. Oksanen. 2001. Oestrid fly parasites of reindeer and control efforts – A review. In: *Mange and Myiasis of Livestock*. Luxembourg: Office for Official Publications of the European Commission.

Åsbakk, K., A. Oksanen, M. Nieminen, R.E. Haugerud, and A.C. Nilssen. 2005. Dynamics of antibodies against hypodermin C in reindeer infested with the reindeer warble fly, *Hypoderma tarandi*. *Vet Parasitol* 129:323–32.

Bain, O., and S. Nikander. 1982. [An aphasmid nematode in the ear capillaries of the reindeer, *Lappnema auria* n. gen., n. sp.(Robertdollfusidae)]. *Annales de parasitologie humaine et comparee* 58(4):383–90.

Ballesteros, M., B.J. Bårdsen, K. Langeland, P. Fauchald, A. Stien, and T. Tveraa. 2012. The effect of warble flies on reindeer fitness: A parasite removal experiment. *J Zool* 287:34–40.

Bangsø, J., K.F. Thøgersen, P. Nejsum, and C.R. Stensvold. 2016. The first case of *Hypoderma tarandi*-associated human myiasis in Greenland. *Ugeskr Laeger* 178(1):V10150796. (In Danish).

Boulard, C., C. Villejoubert, and N. Moiré. 1996. Cross-reactive, stage-specific antigens in the Oestridae family. *Vet Res* 27:535–44.

Breyev, K. 1961. Biological principles of the control of warble-flies. *Entomol Rev* 40:36–45.

Bye, K. 1985. Cestodes of reindeer (*Rangifer tarandus platyrhynchus* Vrolik) on the Arctic islands of Svalbard. *Can J Zool* 63(12):2885–7.

Bylund, G., H.P. Fagerholm, C. Krogell, and S. Nikander. 1981. Studies on *Onchocerca tarsicola*. Bain and Schulz-Key, 1974 in reindeer (*Rangifer tarandus*) in northern Finland. *J Helminthol* 55(1):13–20.

Chirico, J., S. Stenkula, B. Eriksson et al. 1987. Renkorm, en styngflugelarv, orsak till tre fall av human myiasis [The reindeer warble fly larva, a botfly larva, the cause of three cases of human myiasis]. *Läkartidningen* 84:2207–8. (In Swedish).

Colwell, D.D., D. Otranto, and J.R. Stevens. 2009. Oestrid flies: Eradication and extinction versus biodiversity. *Trends Parasitol* 25:500–4.

Colwell, D.D., P.J. Scholl, B. Losson et al. 2004. Management of myiasis: Current status and future prospects. *Vet Parasitol* 125:93–104.

Crofton, H.D. 1971. A model of host-parasite relationships. *Parasitology* 63(3):343–64.

Dahlgren, S., and B. Gjerde. 2007. Genetic characterisation of six *Sarcocystis* species from reindeer (*Rangifer tarandus tarandus*) in Norway based on the small subunit rRNA gene. *Vet Parasitol* 146(3–4):204–13.

Dahlgren, S., B. Gjerde, K. Skirnisson, and B. Gudmundsdottir. 2008. Morphological and molecular identification of three species of *Sarcocystis* in reindeer (*Rangifer tarandus tarandus*) in Iceland. *Vet Parasitol* 149:191–8.

Daugschies, A., J. Hintz, M. Henning, and M. Rommel. 2000. Growth performance, meat quality and activities of glycolytic enzymes in the blood and muscle tissue of calves infected with *Sarcocystis cruzi*. *Vet Parasitol* 88(1–2):7–16.

De Waal, T. 2010. *Paramphistomum* – A brief review. *Irish Vet J* 63(5):313–15.

Dieterich, R.A. 1981. *Alaskan Wildlife Disease*. Fairbanks, Alaska: University of Alaska.

Dieterich, R.A., and J.R. Luick. 1971. The occurrence of *Setaria* in reindeer. *J Wildl Dis* 7(4):242–5.

Dirie, M.F., S. Bornstein, K.R. Wallbanks, D.H. Molyneux, and M. Steen. 1990. Comparative studies on Megatrypanum trypanosomes from cervids. *Trop Med Parasitol* 41(2):198–202.

Divina, B.P., E. Wilhelmsson, T. Morner, J.G. Mattsson, and J. Hoglund. 2002. Molecular identification and prevalence of *Dictyocaulus* spp. (Trichostrongyloidea: Dictyocaulidae) in Swedish semi-domestic and free-living cervids. *J Wildl Dis* 38(4):769–75.

Drew, M.L., and W.M. Samuel. 1985. Factors affecting transmission of larval winter ticks, *Dermacentor albipictus* (Packard), in moose *Alces alces* L. in Alberta, Canada. *J Wildl Dis* 21:274–82.

Dubey, J.P., G. Schares, and L.M. Ortega-Mora. 2007. Epidemiology and control of neosporosis and *Neospora caninum. Clin Microbiol Rev* 20(2):323–CP3.

Ducrocq, J. 2011. Écologie de la besnoitiose chez les populations de caribous (*Rangifer tarandus*) des régions subarctiques. Sciences Cliniques, Université de Montréal, St-Hyacinthe, Québec, Canada. MSc. dissertation. (In French).

Ducrocq, J., G. Beauchamp, S. Kutz et al. 2012. Comparison of gross visual and microscopic assessment of four anatomic sites to monitor Besnoitia tarandi in barren-ground caribou (*Rangifer tarandus*). *J Wildl Dis* 48(3):732–8.

Ducrocq, J., G. Beauchamp, S. Kutz et al. 2013. Variables associated with *Besnoitia tarandi* prevalence and cyst density in barren-ground caribou (*Rangifer tarandus*) populations. *J Wildl Dis* 49(1):29–38.

Duodu, S., K. Madslien, E. Hjelm et al. 2013. *Bartonella* infections in deer keds (*Lipoptena cervi*) and moose (*Alces alces*) in Norway. *Appl Environ Microbiol* 79(1):322–7.

Eysker, M. 1991. Direct measurement of dispersal of *Dictyocaulus viviparus* in sporangia of *Pilobolus* species. *Res Vet Sci* 50:29–32.

Fauchald, P., R. Rødven, B.-J. Bårdsen et al. 2007. Escaping parasitism in the selfish herd: Age, size and density-dependent warble fly infestation in reindeer. *Oikos* 116:491–9.

Ferguson, M.A. 2003. Evolutionary and global change implications of the occurrence of two nasal parasites in caribou on Baffin Island? *Rangifer* Report No. 7, 61.

Folstad, I., A.C. Nilssen, O. Halvorsen, and J. Andersen. 1989. Why do male reindeer (*Rangifer t. tarandus*) have higher abundance of second and third instar larvae of *Hypoderma tarandi* than females? *Oikos* 55:87–92.

Fruetel, M. 1987. The biology of the gastro-intestinal helminths of woodland and barren-ground caribou (*Rangifer tarandus*). Lakehead University, Thunder Bay, Ontario, Canada. Dissertation.

Fruetel, M., and M.W. Lankester. 1989. Gastrointestinal helminths of woodland and barren ground caribou (*Rangifer tarandus*) in Canada, with keys to species. *Can J Zool* 67:2253–69.

Gallie, G.J., and V.J. Nunns. 1976. The bionomics of the free-living larvae and the transmission of *Dictyocaulus filaria* between lambs in North-East England. *J Helminthol* 50(2):79–89.

Gemmel, M.A. 1977. Taeniidae: Modification to the life span of the egg and the regulation of tapeworm populations. *Exp Parasitol* 4(2):314–18.

Gibbs, H.C. 1960. A redescription of *Avitellina arctica* Kolmakov, 1938 (Anoplocephalidae: Thysanosominae), from *Rangifer arcticus arcticus* in northern Canada. *J Parasitol* 46:624–8.

Glover, G.J., M. Swendrowski, and R.J. Cawthorn. 1990. An epizootic of besnoitiosis in captive caribou (*Rangifer tarandus caribou*), reindeer (*Rangifer tarandus tarandus*) and mule deer (*Odocoileus hemionus hemionus*). *J Wildl Dis* 26(2):186–95.

Gray, J.B., and W.M. Samuel. 1986. *Parelaphostrongylus odocoilei* (Nematoda, Protostrongylidae) and a protostrongylid nematode in woodland caribou (*Rangifer tarandus caribou*) of Alberta, Canada. *J Wildl Dis* 22(1):48–50.

Grunenwald, C.M., M. Carstensen, E. Hildebrand et al. 2016. Epidemiology of the lymphatic-dwelling filarioid nematode *Rumenfilaria andersoni* in free-ranging moose (*Alces alces*) and other cervids of North America. *Parasit Vectors* 9(1):450.

Gutiérrez-Expósito, D., L.M. Ortega-Mora, A.A. Gajadhar, P. García-Lunar, J.P. Dubey, and G. Alvarez-García. 2012. Serological evidence of *Besnoitia* spp. infection in Canadian wild ruminants and strong cross-reaction between *Besnoitia besnoiti* and *Besnoitia tarandi*. *Vet Parasitol* 190(1–2):19–28.

Haarløv, N. 1964. Life cycle and distribution pattern of *Lipoptena cervi* (L.) (Dipt., Hippobosc.) on Danish deer. *Oikos* 15(1):93–129.

Hagemoen, R.I.M., and E. Reimers. 2002. Reindeer summer activity pattern in relation to weather and insect harassment. *J Anim Ecol* 71:883–92.

Halley, B.A., W.J. Vanden Heuvel, and P.G. Wislocki. 1993. Environmental effects of the usage of avermectins in livestock. *Vet Parasitol* 48:109–25.

Handeland, K. 1994. Experimental studies of *Elaphostrongylus rangiferi* in reindeer (*Rangifer tarandus tarandus*): Life cycle, pathogenesis, and pathology. *Zentralbl Veterinarmed B* 41(5):351–65.

Handeland, K., A. Skorping, and T. Slettbakk. 1993. Experimental cerebrospinal elaphostrongylosis (*Elaphostrongylus rangiferi*) in sheep. *Zentralbl Veterinarmed B* 40(3):181–9.

Handeland, K., and T. Slettbakk. 1995. Epidemiological aspects of cerebrospinal elaphostrongylosis in small ruminants in northern Norway. *Zentralbl Veterinarmed B* 42(2):110–7.

Härkönen, L., S. Härkönen, A. Kaitala et al. 2010. Predicting range expansion of an ectoparasite – The effect of spring and summer temperatures on deer ked *Lipoptena cervi* (Diptera: Hippoboscidae) performance along a latitudinal gradient. *Ecography* 33(5):906–12.

Haugerud R.E. 1988. A life history appoach to the parasite-host interaction *Linguatula arctica* (Riley, Haugerud and Nilssen, 1987) – *Rangifer tarandus* (Linnaeus, 1758). Ecology, Zoology, University of Tromsø, Norway. Thesis.

Haugerud, R.E., and A.C. Nilssen, 1985. *Linguatula* sp. (Pentastomida) in reindeer. A new species with a direct life-cycle? In *Proceedings of the 12th Scandinavian Symposium of Parasitology*, Tromsø, Norway, June, 1985. – Information 18:51. Institute of Parasitology, Åbo Akademi, Åbo, Finland.

Haukisalmi, V., A. Lavikainen, S. Laaksonen, and S. Meri. 2011. *Taenia arctos* n. sp. (Cestoda: Cyclophyllidea: Taeniidae) from its definitive (brown bear *Ursus arctos* Linnaeus) and intermediate (moose/elk *Alces* spp.) hosts. *Syst Parasitol* 80:217–30.

Heggstad, E., E. Bø, and D. Lenvik. 1986. Behandling av reinkalver med ivermectin første levehøst. Effekt på levendevekter andre levehøst [Treatment of reindeer calves the first autumn of living. Effect of live-weights the second autumn of living]. *Rangifer* 1:77–9. (In Norwegian).

Helle, T. 1980. Abundance of warble fly (*Oedemagena tarandi*) larvae in semi-domestic reindeer (*Rangifer tarandus*) in Finland. *Rep Kevo Subarctic Res Stat* 16:1–6.

Herd, R. 1995. Endectocidal drugs: Ecological risks and counter-measures. *Int J Parasitol* 25:875–85.

Hermosilla, C., N. Pantchev, R. Bachmann, and C. Bauer. 2006. *Lipoptena cervi* (deer ked) in two naturally infested dogs. *Vet Rec* 159(9):286–7.

Hoar, B.M., A.G. Eberhardt, and S.J. Kutz. 2012a. Obligate larval inhibition of *Ostertagia gruehneri* in *Rangifer tarandus*? Causes and consequences in an arctic system. *Parasitology* 139(10):1339–45.

Hoar, B.M., K. Ruckstuhl, and S. Kutz. 2012b. Development and availability of the free-living stages of *Ostertagia gruehneri*, an abomasal parasite of barrenground caribou (*Rangifer tarandus groenlandicus*), on the Canadian tundra. *Parasitology* 139(8):1093–100.

Hoberg, E.P., A.A. Kocan, and L.G. Rickard. 2001. Gastrointestinal strongyles in wild ruminants. In *Parasitic Diseases of Wild Mammals*. 193–220. W.M. Samuel, M.J. Pybus, and A.A. Kocan (eds). Ames, Iowa: Iowa State University Press.

Hoeve, J., D.G. Joachim, and E.M. Addison. 1988. Parasites of moose (*Alces alces*) from an agricultural area of Eastern Ontario. *J Wildl Dis* 24(2):371–4.

Höglund, J., D.A. Morrison, B.P. Divina, E. Wilhelmsson, and J.G. Mattsson. 2003. Phylogeny of *Dictyocaulus* (lungworms) from eight species of ruminants based on analyses of ribosomal RNA data. *Parasitology* 127(Pt 2):179–87.

Holt, G., C. Berg, and A. Haugen. 1990. Nematode related spinal myelomeningitis and posterior ataxia in muskoxen (*Ovibos moschatus*). *J Wildl Dis* 26(4):528–31.

Hrabok, J.T., A. Oksanen, M. Nieminen, A. Rydzik, A. Uggla, and P.J. Waller. 2006. Reindeer as hosts for nematode parasites of sheep and cattle. *Vet Parasitol* 136(3–4):297–306.

Hrabok, J.T., A. Oksanen, M. Nieminen, and P.J. Waller. 2007. Prevalence of gastrointestinal nematodes in winter-slaughtered reindeer of northern Finland. *Rangifer* 27(2):133–9.

Irvine, R.J., A. Stien, O. Halvorsen, R. Langvatn, and S.D. Albon. 2000. Life-history strategies and population dynamics of abomasal nematodes in Svalbard reindeer (*Rangifer tarandus platyrhynchus*). *Parasitology* 120(Pt 3):297–311.

Jaenson, T.G.T. 2011. Larver av nässtuyngfluga i ögat – Ovanligt men allvarligt problem. Fall av human oftalmomyiasis från Dalarna och Sydöstra Finland redovisas [A case of throat botfly larvae in the eye – an uncommon but serious problem. Cases of human ophthalmomyiasis from Dalarne and Southeastern Finland recognized]. *Läkartidningen* 108:928–30. (In Swedish).

Jenkins, E.J., L.J. Castrodale, S.J. de Rosemond et al. 2013. Tradition and transition: Parasitic zoonoses of people and animals in Alaska, northern Canada, and Greenland. *Adv Parasitol* 82:33–204.

Jenkins, E.J., A.M. Veitch, S.J. Kutz et al. 2007. Protostrongylid parasites and pneumonia in captive and wild thinhorn sheep (*Ovis dalli*). *J Wildl Dis* 43(2):189–205.

Johnsen, P. 1946. Bidrag til kundskaben on den danske Ixodidaefauna. *Entomologiske Meddelelser* 24:397–401.

Joly, D.O., and F. Messier. 2004. The distribution of *Echinococcus granulosus* in moose: evidence for parasite-induced vulnerability to predation by wolves? *Oecologia* 140(4):586–90.

Jones, A., and M.J. Pybus. 2001. Taeniasis and Echinococcosis. In *Parasitic Diseases of Wild Mammals*. W.M. Samuel, M.J. Pybus, and A.A. Kocan (eds). Ames: Iowa State University Press.

Josefsen, T.D., A. Oksanen, and B. Gjerde. 2014. Parasiter hos rein I Fennoskandia – en oversikt. *Norsk Veterinærtidsskrift* 2:185–201.

Kafle, P., L.M. Leclerc, M. Anderson, T. Davison, M. Lejeune, and S. Kutz. 2017. Morphological keys to advance the understanding of protostrongylid biodiversity in caribou (*Rangifer* spp.) at high latitudes. *Int J Parasitol Parasites Wildl* 6(3):331–9. doi:10.1016/j.ijppaw.2017.08.009.

Kan, B., K. Åsbakk, K. Fossen, A. Nilssen, R. Panadero, and D. Otranto. 2013. Reindeer warble fly-associated human myiasis, Scandinavia. *Emerg Infect Dis* 19(5):830–2.

Kan, B., C. Åsen, K. Åsbakk, and T.G.T. Jaenson. 2010. Misstänkte lusägg i pojkes hår avslöjade farlig parasit [Suspected head-lice eggs in the hair of a boy revealed dangerous parasite]. *Läkartidningen* 107:1694–7. (In Swedish).

Karter, A.J., I. Folstad, and J.R. Anderson. 1992. Abiotic factors influencing embryonic development, egg hatching, and larval orientation in the reindeer warble fly, *Hypoderma tarandi*. *Med Vet Entomol* 6:355–62.

Kashivakura, C.K. 2013. Detecting *Dermacentor albipictus*, the winter tick, at the northern extent of its range: Hunter-based monitoring and serological assay development. Graduate Studies, University of Calgary, Calgary, Alberta, Canada. M.Sc. Dissertation.

Kaunisto, S., R. Kortet, L. Harkonen, S. Harkonen, H. Ylonen, and S. Laaksonen. 2009. New bedding site examination-based method to analyse deer ked (*Lipoptena cervi*) infection in cervids. *Parasitol Res* 104 (4):919–25.

Kimberling, C.V. 1988. *Jensen and Swift's Diseases of Sheep*. Philadelphia: Lea & Febirger. Infectious Diseases: Dicrocoeliasis: Worldwide distribution. 2009. GIDEON. 20 Feb. 2009.

Kingston, N., and S. Nikander. 1985. Poron, *Rangifer tarandus*, *Trypanosoma* sp. Suomessa. (*Trypanosoma* sp in reindeer, *Rangifer tarandus*, in Finland.) *Suomen Eläinlääkärilehti* 91:1–3.

Kirilenko, A.V. 1975. *Avitellina arctica* in *Rangifer tarandus*. *Veterinariya* 7:58–9.

Králová-Hromadová, I., E. Bazsalovicsová, J. Stefka, M. Spakulová, S. Vávrová, T. Szemes, V. Tkach, A. Trudgett, and M. Pybus. 2011. Multiple origins of European populations of the giant liver fluke Fascioloides magna (Trematoda: Fasciolidae), a liver parasite of ruminants. *Int J Parasitol* 41(3–4):373–83. doi:10.1016/j.ijpara.2010.10.010.

Kummeneje, K. 1980. Some treatment trials for eradication of the reindeer grub fly (*Oedemagena tarandi*). In: *Proceedings of the 2nd International Reindeer/Caribou Symposium*. E. Reimers, E. Gaare, and S. Skjenneberg (eds). 459–61. Røros 1979. Trondheim: Direktoratet for vilt og ferskvannsfisk.

Kutz, S.J. 2007. *An Evaluation of the Role of Climate Change in the Emergence of Pathogens and Diseases in Arctic and Subarctic Caribou Populations*. Prepared for the Climate Change Action Fund, Government of Canada.

Kutz, S.J., I. Asmundsson, E.P. Hoberg et al. 2007. Serendipitous discovery of a novel protostrongylid (Nematoda: Metastrongyloidea) in caribou, muskoxen, and moose from high latitudes of North America based on DNA sequence comparisons. *Can J Zool* 85(11):1143–56.

Kutz, S.J., S. Checkley, G.G. Verocai et al. 2013. Invasion, establishment, and range expansion of two parasitic nematodes in the Canadian Arctic. *Glob Chang Biol* 19(11):3254–62.

Kutz, S.J., J. Ducrocq, G.G. Verocai et al. 2012. Parasites in ungulates of Arctic North America and Greenland: A view of contemporary diversity, ecology, and impact in a world under change. *Adv Parasitol* 79:99–252.

Kutz, S.J., E.P. Hoberg, P.K. Molnar, A. Dobson, and G.G. Verocai. 2014. A walk on the tundra: Host-parasite interactions in an extreme environment. *Int J Parasitol Parasites Wildl* 3(2):198–208.

Kutz, S.J., E.J. Jenkins, A.M. Veitch et al. 2009a. The Arctic as a model for anticipating, preventing, and mitigating climate change impacts on host-parasite interactions. *Vet Parasitol* 163(3):217–28.

Kutz, S.J., R.C.A. Thompson, and L. Polley. 2009b. Wildlife with *Giardia*: Villain, or Victim and Vector? In *Giardia* and *Cryptosporidium*: from molecules to disease. Ortega-Pierres, G., Cacciò, S., Fayer, R., Mank, T.G., Smith, H.V., and Thompson, R.C.A. (eds). CAB International

Kutz, S.J., R.C.A. Thompson, L. Polley et al. 2008. *Giardia* assemblage A: human genotype in muskoxen in the Canadian Arctic. *Parasit Vectors* 1(32):1–4.

Kynkäänniemi, S.-M., M. Kettu, R. Kortet et al. 2014. Acute impacts of the deer ked (*Lipoptena cervi*) infestation on reindeer (*Rangifer tarandus tarandus*) behaviour. *Parasitol Res* 113(4):1489–97.

Kynkäänniemi S.-M., R. Kortet, L. Härkönen, A. et al. 2010. Threat of an invasive parasitic fly, the deer ked (*Lipoptena cervi*), to the reindeer (*Rangifer tarandus tarandus*): Experimental infection and treatment. *Ann Zool Fenn* 2010, 47:28–36.

Laaksonen, S. 2016. *Tunne poro – poron sairaudet ja terveydenhuolto*. Riga, Latvia: Livonia print.

Laaksonen, S., J. Kuusela, S. Nikander, M. Nylund, and A. Oksanen. 2007. Outbreak of parasitic peritonitis in reindeer in Finland. *Vet Rec* 160(24):835–41.

Laaksonen, S., A. Oksanen, and E. Hoberg. 2015. A lymphatic dwelling filarioid nematode, *Rumenfilaria andersoni* (Filarioidea; Splendidofilariinae), is an emerging parasite in Finnish cervids. *Parasit Vectors* 8:228.

Laaksonen, S., A. Oksanen, S. Kutz, P. Jokelainen, A. Holma-Suutari, and E. Hoberg. 2017. Filarioid nematodes, threat to arctic food safety and security. In *Game Meat Hygiene: Food Safety and Security*. Paulsen, P., Bauer, A., and Smulders, F.J.M. (eds). Netherlands: Wageningen Academic Publishers.

Laaksonen, S., A. Oksanen, T. Orro, H. Norberg, M. Nieminen, and A. Sukura. 2008. Efficacy of different treatment regimes against setariosis (*Setaria tundra*, Nematoda: Filarioidea) and associated peritonitis in reindeer. *Acta Vet Scand* 50:49.

Laaksonen S., and P. Paulsen. 2015. *Hunting Hygiene*. Netherlands: Wageningen Academic Publishers.

Laaksonen, S., J. Pusenius, J. Kumpula et al. 2010. Climate change promotes the emergence of serious disease outbreaks of filarioid nematodes. *Ecohealth* 7(1):7–13.

Laaksonen, S., M. Solismaa, R. Kortet, J. Kuusela, and A. Oksanen. 2009b. Vectors and transmission dynamics for *Setaria tundra* (Filarioidea; Onchocercidae), a parasite of reindeer in Finland. *Parasit Vectors* 2:3.

Laaksonen, S., M. Solismaa, T. Orro et al. 2009a. *Setaria tundra* microfilariae in reindeer and other cervids in Finland. *Parasitol Res* 104(2):257–65.

Lagacé-Wiens P.R.S., R. Dookeran, S. Skinner, R. Leicht, D.D. Colwell, and T.D. Galloway. 2008. Human ophthalmomyiasis interna caused by *Hypoderma tarandi*, Northern Canada. *Emerg Infect Dis* 14:64–6.

Landehag J., A. Skogen, K. Åsbakk, and B. Kan. 2017. Human myiasis caused by the reindeer warble fly, *Hypoderma tarandi* – Case series from Norway, 2011-2016. *Euro Surveill* 22(29):pii=30576. DOI: http:// dx.doi.org/10.2807/1560-7917. ES.2017.22.29.30576.

Langton, C., J.S. Gray, P.F. Waters, and P.J. Holman. 2003. Naturally acquired babesiosis in a reindeer (*Rangifer tarandus tarandus*) herd in Great Britain. *Parasitol Res* 89(3):194–8.

Lankester, M.W. 2001. Extrapulmonary lungworms of cervids. In *Parasitic Diseases of Wildlife Mammals*. W.M. Samuel, M.J. Pybus, and A.A. Kocan (eds). Ames, Iowa: Iowa State University Press.

Lankester, M.W. and D. Fong. 1998. Protostrongylid nematodes in caribou (*Rangifer tarandus caribou*) and moose (*Alces alces*) of Newfoundland. *Rangifer* Special Issue No. 10, 73–83

Lankester, M.W., and P.L. Hauta. 1989. *Parelaphostrongylus andersoni* (Nematoda: Protostrongylidae) in caribou (*Rangifer tarandus*) of northern and central Canada. *Can J Zool* 67(8):1966–75.

Lankester, M.W., and S. Luttich. 1988. *Fascioloides magna* (Trematoda) in woodland caribou (*Rangifer tarandus caribou*) of the George River herd, Labrador. *Can J Zool* 66:475–9.

Lankester, M.W., and W.M. Samuel. 2007. Pests, parasites and diseases. In *Ecology and Management of the North American Moose*. 479–518. A.W. Franzmann and C.C. Schwartz (eds). Boulder, Colorado: Colorado University Press.

Lankester M., and J.B. Snider. 1982. *Rumenfilaria andersoni* n. gen., n. sp. (Nematoda: Filarioidea) in moose from northwestern Ontario, Canada. *Can J Zool* 60:2455–8.

Lankester, M.W., J.B. Snider, and R.E. Jerrard. 1979. Annual maturation of *Paramphistomum cervi* (Trematoda: Paramphistomatidae) in moose, *Alces alces* L. *Can J Zool* 57(12):2355–7.

Laukkanen, A., Ruoppi, P., and S. Mäkinenkiljunen. 2005. Deer ked-induced occupational allergic rhinoconjunctivitis. *Ann Allerg Asthma Im* 94:604–8.

Lavikainen, A., S. Laaksonen, K. Beckmen, A. Oksanen, M. Isomusrsu, and S. Meri. 2011. Molecular identification of *Taenia* spp. in wolves (*Canis lupus*), brown bears (*Ursus arctos*) and cervids from Europe and Alaska. *Parasitol Int* 60:289–95.

Leader-Williams, N. 1980. Observations on the internal parasites of reindeer introduced into South Georgia. *Vet Rec* 107(17):393–5.

Lisitzin, P. 1976. Loisetko poron korvanseudun tulehduksen aiheuttajia? Suomen eläinlääkärileht 82:198–9.

Loos-Frank, B. 2000. An up-date of Verster's (1969) "Taxonomic revision of the genus Taenia Linnaeus" (Cestoda) in table format. *Syst Parasitol* 45:155–83.

Madslien, K., B. Ytrehus, T. Vikøren et al. 2011. Hair-loss epizootic in moose (*Alces alces*) associated with massive deer ked (*Lipoptena cervi*) infestation. *J Wildl Dis* 47(4):893–906.

Malcicka, M. 2015. Life history and biology of *Fascioloides magna* (Trematoda) and its native and exotic hosts. *Ecol Evol* 5(7):1381–97.

Manninen, S.M., S.M. Thamsborg, S. Laaksonen, and A. Oksanen. 2014. The reindeer abomasal nematode (*Ostertagia gruehneri*) is naturally transmitted to sheep when sharing pastures. *Parasitol Res* 113(11):4033–8.

Matsumoto, K., Z.L. Berrada, E. Klinger, H.K. Goethert, and S.R. Telford III. 2008. Molecular detection of *Bartonella schoenbuchensis* from ectoparasites of deer in Massachusetts. *Vector Borne Zoonotic Dis* 8(4):549–54.

McFrederick, Q.S., T.S. Haselkorn, G.G. Verocai, and J. Jaenike. 2013. Cryptic *Onchocerca* species infecting North American cervids, with implications for the evolutionary history of host associations in *Onchocerca*. *Parasitology* 140(10):1201–10.

Mjöberg, E. 1915. Über eine neue Gattung und Art von Anopluren. [On a new sucking louse genus and species]. *Entomol Tidjskr* 36:282–5. (In German).

Moiré, N., Y. Bigot, G. Periquet, and C. Boulard. 1994. Sequencing and gene expression of hypodermins A, B, C in larval stages of *Hypoderma lineatum*. *Mol Biochem Parasitol* 66:233–40.

Morgan, E.R., M. Lundervold, G.F. Medley, B.S. Shaikenov, P.R. Torgerson, and E.J. Milner-Gulland. 2006. Assessing risks of disease transmission between wildlife and livestock: The Saiga antelope as a case study. *Biol Cons* 131(2):244–54.

Narsapur, V.S., and J. Prokopic. 1979. The influence of temperature on the development of *Moniezia expansa* (Rudolphi 1810) in oribatid mites. *Folia Parasitologica* 26:239–43.

Nelson, W.A., and J. Weintraub. 1972. *Hypoderma lineatum* (De Vill.) (Diptera: Oestridae): Invasion of the bovine skin by newly hatched larvae. *J Parasitol* 58:614–24.

Nettles, V.F., and A.K. Prestwood. 1976. Experimental *Parelaphostrongylus andersoni* infections in white-tailed deer. *Vet Path* 13(5):381–93.

Nieminen, M. 1992. Increasing the quality and value of reindeer hides. II. Export of warble-free leather and suede. *Poromies* 59(6):26–7. (In Finnish).

Nikander, S., S. Laaksonen, S. Saari, and A. Oksanen. 2007. The morphology of the filaroid nematode *Setaria tundra*, the cause of peritonitis in reindeer *Rangifer tarandus*. *J Helminthol* 81(1):49–55.

Nikander, S., and Rahko, T. 1989. Sinusitis in reindeer caused by *Linguatula arctica*. Poster presentation at the 13th WAAVP conference (P 3–15), Berlin 1989.

Nikander, S., and S. Saari. 2006. A SEM study of the reindeer sinus worm (*Linguatula arctica*). *Rangifer* 26(1):15–24.

Nilssen, A.C. 1997a. Reinbremsene. En oversikt over nye forskningsresultater [The reindeer botflies. An overview of recent research results]. *Fauna* 50:122–37. (In Norwegian).

Nilssen, A.C. 1997b. Effect of temperature on pupal development and eclosion dates in the reindeer oestrids *Hypoderma tarandi* and *Cephenemyia trompe* (Diptera: Oestridae). *Environ Entomol* 26:296–306.

Nilssen, A.C., and J.R. Anderson. 1995. Flight capacity of the reindeer warble fly, *Hypoderma tarandi* (L), and the reindeer nose bot fly, *Cephenemyia trompe* (Modeer) (Diptera, Oestridae). *Can J Zool* 73:1228–38.

Nilssen, A.C., K. Åsbakk, R.E. Haugerud, W. Hemmingsen, and A. Oksanen. 1999. Treatment of reindeer with ivermectin – Effect on dung insect fauna. *Rangifer* 19:61–70.

Nilssen, A.C., and J.O. Gjershaug. 1988. Reindeer warble fly larvae found in red deer. *Rangifer* 8(1):35–6.

Nilssen, A.C., and R.E. Haugerud. 1994. The timing and departure rate of larvae of the warble fly *Hypoderma* (=*Oedemagena*) *tarandi* (L.) and the nose bot fly *Cephenemyia trompe* (Modeer) (Diptera: Oestridae) from reindeer. *Rangifer* 14:113–22.

Nilssen, A.C., and R.E. Haugerud. 1995. Epizootiology of the reindeer nose bot fly, *Cephenemyia trompe* (Modeer) (Diptera, Oestridae) in reindeer, *Rangifer tarandus* (L), in Norway. *Can J Zool* 73:1024–36.

Nilssen, A.C., W. Hemmingsen, and R.E. Haugerud. 2002. Failure of two consecutive treatments with ivermectin to eradicate the reindeer parasites (*Hypoderma tarandi*, *Cephenemyia trompe* and *Linguatula arctica*) from an island in northern Norway. *Rangifer* 22:115–22.

Nilsson, O., M. Nordkvist, and L. Ryden. 1965. Experimental *Babesia divergens* infection in reindeer (*Rangifer tarandus*). *Acta Vet Scand* 6(4):353–9.

OIE. 1998. *World Animal Health in 1997–Part 2*. Paris, France: Office International des Épizooties.

Oksanen, A. 1999. Endectocide treatment of the reindeer. Tromsø. *Rangifer* Special Issue 11. University of Helsinki, Finland. Doctoral Dissertation.

Oksanen, A., K. Åsbakk, M. Nieminen, H. Norberg, and A. Nareaho. 1997. Antibodies against *Toxoplasma gondii* in Fennoscandian reindeer – Association with the degree of domestication. *Parasitol Int* 46(4):255–61. doi.org/10.1016/S1383-5769(97)00033-0

Oksanen, A., K. Åsbakk, M. Raekallio, and M. Nieminen. 2014. The relative plasma availabilities of ivermectin in reindeer (*Rangifer tarandus tarandus*) following subcutaneous and two different oral formulation applications. *Acta Vet Scand* 56:76–82.

Oksanen, A., K. Gustafsson, A. Lund, J.P. Dubey, P. Thulliez, and A. Uggla. 1996. Experimental *Toxoplasma gondii* infection leading to fatal enteritis in reindeer (*Rangifer tarandus*). *J Parasitol* 82(5):843–5.

Oksanen, A., and A. Lavikainen. 2015. *Echinococcus canadensis* transmission in the North. *Vet Parasitol* 213(3–4):182–6.

Oksanen, A., M. Nieminen, T. Soveri, and K. Kumpula. 1992. Oral and parenteral administration of ivermectin to reindeer. *Vet Parasitol* 41:241–7.

Oksanen A., M. Nieminen, T. Soveri, K. Kumpula, U. Heiskari, and V. Kuloharju. 1990. The establishment of parasites in reindeer calves. *Rangifer*, Special Issue 5:20–1.

Oksanen, A., H. Norberg, and M. Nieminen. 1998. Ivermectin treatment did not increase slaughter weight of first-year reindeer calves. *Prev Vet Med* 35:209–17.

Otranto, D. 2001. The immunology of myiasis: Parasite survival and host defense strategies. *Trends Parasitol* 17:176–82.

Paakkonen, T., A.M. Mustonen, H. Roininen, P. Niemelä, V. Ruusila, P. Nieminen. 2010. Parasitism of the deer ked, *Lipoptena cervi*, on the moose, *Alces alces*, in eastern Finland. *Med Vet Entomol* 24(4):411–7. doi:10.1111/j.1365-2915.2010.00910.x.

Panuska, C. 2006. Lungworms of ruminants. *Vet Clin North Am Food Anim Pract* 22(3):583–93.

Petrini, K., P.J. Holman, J.C. Rhyan, J.J. Sharon, and G.G. Wagner. 1995. Fatal babesiosis in an american woodland caribou (*Rangifer tarandus caribou*). *J Zoo Wildl Med* 26(2):298–305.

Pickles, R.S., D. Thornton, R. Feldman, A. Marques, and D.L. Murray. 2013. Predicting shifts in parasite distribution with climate change: A multitrophic level approach. *Glob Chang Biol* 19(9):2645–54.

Pollock, B., B. Penashue, S. McBruney et al. 2009. Liver parasites and body condition in relation to environmental contaminants in caribou (*Rangifer tarandus*) from Labrador, Canada. *Arctic* 62(1):1–12.

Prestrud, K.W., K. Åsbakk, E. Fuglei et al. 2007. Serosurvey for *Toxoplasma gondii* in arctic foxes and possible sources of infection in the high Arctic of Svalbard. *Vet Parasitol* 150(1-2):6–12.

Pybus, M.J. 2001. Liver flukes. In *Parasitic Diseases of Wild Mammals*. W.M. Samuel, M.J. Pybus, and A.A. Kocan (eds). Ames: Iowa State University Press.

Rahko, T., S. Saari, and S. Nikander. 1992. Histopathological lesions in spontaneous dictyocaulotic pneumonia of the reindeer (*Rangifer tarandus tarandus* L.) *Rangifer* 12(2):115–22.

Rantanen, T., T. Reunala, P. Vuojolahti, and W. Hackman. 1982. Persistent pruritic papules from deer ked bites. *Acta Derm Venereol* 62(4):307–11.

Rau, M.E., and F.R. Caron. 1979. Parasite-induced susceptibility of moose to hunting. *Can J Zool* 57:2466–8.

Rausch, R. 1993. The biology of *Echinococcus granulosus*. In *Compendium on Cystic echinococcosis with Special Reference to the Xinjiang Uygur Autonomous Region, the People's Republic of China*. F.L. Anderson, J.-J. Chai, and F.-J. Liu (eds). Provo, Utah: Brigham Young University.

Rehbinder, C. 1990. Some vector borne parasites in Swedish reindeer (*Rangifer tarandus tarandus* L). *Rangifer* 10(2):67–73.

Rehbinder, C., M. Elvander, and M. Nordkvist. 1981. Cutaneous besnoitiosis in a Swedish reindeer (*Rangifer tarandus* L.). *Nordisk Veterinaer* 33:270–2.

Riley, J. 1986. The biology of pentastomids. *Adv Parasitol* 25:45–128.

Riley, J., R.E. Haugerud, and A.C. Nilssen. 1987. A new species of pentastomid from the nasal passage of the reindeer (*Rangifer tarandus*) in northern Norway, with speculation about its life-cycle. *J Nat Hist* 21:707–16.

Rødven, R., I. Männikkö, R.A. Ims, N.G. Yoccoz, and I. Folstad. 2009. Parasite intensity and fur coloration in reindeer calves – Contrasting artificial and natural selection. *J Anim Ecol* 78:600–7.

Samuel, B., K. Madslien, and J. Gonynor-McGuire. 2012. Review of deer ked (*Lipoptena cervi*) on moose in Scandinavia with implications for North America. *Alces* 48:27–33.

Samuel, W.M., and D.R. Gray. 1974. Parasitic infection in muskoxen. *J Wildl Manage* 38(4):775–82.

Samuelsson F., P. Nejsum, K. Raundrup, T.V.A. Hansen, and C.M.O. Kapel. 2013. Warble infestations by *Hypoderma tarandi* (Diptera; Oestridae) recorded for the first time in West Greenland muskoxen. *Int J Parasitol: Parasites Wildl* 2:214–16.

Saval'ev, D.V. 1968. Control of warble flies and bloodsucking Diptera. In *Reindeer Husbandry*. 294–311. P. S. Zhigunov (ed). (Translated from Russian). Israel: Israel Program for Scientific Translations.

Schulz-Key, H. 1975. [Studies on the Filariidae of Cervidae in southern Germany. 2. Filariidae of the red deer (*Cervus elaphus*)]. (In German). *Tropenmed Parasitol* 26(3):348–58.

Schulz-Key, H., and P. Wenk. 1981. The transmission of *Onchocerca tarsicola* (Filarioidea: Onchocercidae) by *Odagmia ornata* and *Prosimulium nigripes* (Diptera: Simuliidae). *J Helminthol* 55(3):161–6.

Semenova, N.S. 1967. Study of *Moniezia* (Moniezia) *taimyrica* Semenova, 1967 in reindeer. (In Russian). *Materialy Nauchnykh Konferentsii Vsesoyuznogo Obshchestva Gel'-mintologov 1971–1972* 25:199–202.

Sigurdarson, S., and R.E. Haugerud. 2004. "Wild reindeer" in Iceland. In *Family-based Reindeer Herding and Hunting Economies, and the Status of Management of Wild Reindeer/Caribou Populations*. 159–62. B. Ulvevadet and K. Klokov (eds). Tromsø: Centre for Sami Studies, University of Tromsø.

Simard, A.A., S. Kutz, J. Ducrocq, K. Beckmen, V. Brodeur, M. Campbell, B. Croft, C. Cuyler, T. Davison, B. Elkin, T. Giroux, A. Kelly, D. Russell, J. Taillon, A. Veitch, and S.D. Côté. 2016. Variation in the intensity and prevalence of macroparasites in migratory caribou: A quasi-circumpolar study. *Can J Zool* 94(9):607–17.

Skirnisson, K., and C. Cuyler. 2016. A new *Eimeria* species (Protozoa: Eimeriidae) from caribou in Ameralik, West Greenland. *Parasitol Res* 115(4):1611–15.

Skjenneberg, S., and L. Slagsvold. 1968. *Reindriften og dens naturgrunnlag* [The reindeer husbandry and its foundation]. Oslo: Universitetsforlaget. (In Norwegian).

Sommer, C., B. Steffansen, N.B. Overgaard et al. 1992. Ivermectin excreted in cattle dung after subcutaneous injection or pour-on treatment: Concentrations and impact on dung fauna. *Bull Entomol Res* 82:257–64.

Sonin, M.D. 1977. *Setaria tundra*, *Setaria yehi* In *Filariata of Animals and Man and the Disease Caused by Them*. Vol. 27. 120–24. Sonin, M.D. (ed). Moscow: Izdatelstvo Nauka. (In Russian).

Steele, J., 2012. The Devil's in the Diversity: Divergent parasite faunas and their impacts on body condition in two Greenland caribou populations. Veterinary Medical Sciences. MSc thesis, University of Calgary, Calgary, p. 139.

Steele, J., K. Orsel, C. Cuyler, E.P. Hoberg, N.M. Schmidt, and S.J. Kutz. 2013. Divergent parasite faunas in adjacent populations of west Greenland caribou: Natural and anthropogenic influences on diversity. *Int J Parasitol: Parasites Wildl* 2:197–202.

Steen, M., C.G. Blackmore, and A. Skorping. 1997. Cross-infection of moose (*Alces alces*) and reindeer (*Rangifer tarandus*) with *Elaphostrongylus alces* and *Elaphostrongylus rangiferi* (Nematoda, Protostrongylidae): Effects on parasite morphology and prepatent period. *Vet Parasitol* 71(1):27–38.

Stieve, E., K. Beckmen, S.A. Kania, A. Widner, and S. Patton. 2010. *Neospora caninum* and *Toxoplasma gondii* antibody prevalence in Alaska wildlife. *J Wildl Dis* 46(2):348–55.

Syrdalen, P., T. Nitter, and R. Mehl. 1982. Ophthalmomyiasis interna posterior: Report of case caused by the reindeer warble fly larva and review of previous reported cases. *Br J Ophthalmol* 66:589–93.

Tashkinov, N.I. 1976. Migration of *Oedemagena tarandi* L. at the artificial infection in reindeer. *Parazitologiia* 10(1):56–60. (In Russian).

Thomas, D.C. 1996. Prevalence of *Echinococcus granulosus* and *Taenia hydatigena* in caribou in north-central Canada. *Rangifer* 16(4):331–5.

USNPC. 2011. United States National Parasite Collection.

Välimäki, P., K. Madslien, H. Malmsten, L. Härkönen, S. Härkönen, A. Kaitala, R. Kortet, S. Laaksonen, R. Mehl, L. Redford, H. Ylönen, and B. Ytrehus. 2010. Fennoscandian distribution of an important parasite of cervids, the deer ked (*Lipoptena cervi*), revisited. *Parasitol Res* 107(1):117–25. doi:10.1007/s00436-010-1845-7.

Verocai, G.G. 2015. Contributions to the biodiversity and biogeography of the genus *Varestrongylus* Bhalerao, 1932 (Nematoda: Protostrongylidae), lungworms of ungulates, with emphasis on a new nearctic species. University of Calgary, Calgary, Alberta, Canada. Ph.D. Dissertation.

Verocai, G.G., E.P. Hoberg, T. Vikøren et al. 2014b. Resurrection and redescription of *Varestrongylus alces* (Nematoda: Protostrongylidae), a lungworm of the Eurasian moose (*Alces alces*), with report on associated pathology. *Parasit Vectors* 7(1). doi:10.1186/s13071-014-0557-8.

Verocai, G.G., S.J. Kutz, M. Simard, and E.P. Hoberg. 2014. *Varestrongylus eleguneniensis* sp. n. (Nematoda: Protostrongylidae): A widespread, multi-host lungworm of wild North American ungulates, with an emended diagnosis for the genus and explorations of biogeography. *Parasit Vectors* 7:556.

Verocai, G.G., M. Lejeune, K.B. Beckmen et al. 2012. Defining parasite biodiversity at high latitudes of North America: New host and geographic records for *Onchocerca cervipedis* (Nematoda: Onchocercidae) in moose and caribou. *Parasit Vectors* 5:242.

Verocai, G.G., M. Lejeune, G.L. Finstad, and S.J. Kutz. 2013. A Nearctic parasite in a Palearctic host: *Parelaphostrongylus andersoni* (Nematoda; Protostrongylidae) infecting semi-domesticated reindeer in Alaska. *Int J Parasitol: Parasites Wildl* 2:119–23.

Vikøren, T., J. Tharaldsen, B. Fredriksen, and K. Handeland. 2004. Prevalence of *Toxoplasma gondii* antibodies in wild red deer, roe deer, moose, and reindeer from Norway. *Vet Parasitol* 120(3):159–69.

Wang, A., Y.M. Kang, W.C. Peng, Z.Y. Cui, S.C. Li, Z.Y. Li. 1989. Investigation of parasites in reindeer. *Chinese Journal of Veterinary Science and Technology* 3:14–15. (In Chinese with English abstract).

Weisser, C.F., and K.C. Kim. 1973. Rediscovery of *Solenopotes tarandi* (Mjöberg, 1915) (Linognathidae: Anoplura), with ectoparasites of the barren ground caribou. *Parasitology* 66(1):123–32.

Weladji, R.B., Ø. Holand, and T. Almøy. 2003. Use of climatic data to assess the effect of insect harassment on the autumn weight of reindeer (*Rangifer tarandus*) calves. *J Zool* 260:79–85.

Wiegmann, L., C. Silaghi, A. Obiegala et al. 2015. Occurrence of *Babesia* species in captive reindeer (*Rangifer tarandus*) in Germany. *Vet Parasitol* 211(1–2):16–22.

Yagi, K., O. Bain, and C. Shoho. 1994. *Onchocerca suzukii* n. sp. and *O. skrjabini* (= O. tarsicola) from a relict bovid, *Capricornis crispus*, in Japan. *Parasite* 1(4):349–56.

Yeates, G.W., J.T. Hrabok, A. Oksanen, M. Nieminen, and P.J. Waller. 2007. Soil nematode populations beneath faeces from reindeer treated with ivermectin. *Acta Agric Scand Sect B* 57:126–33.

Zelinskii, L.M. 1973. [Identification of *Moniezia rangiferina* from reindeer on the Kamchatka Peninsula, Siberia, Russia]. *Sbornik Rabot. Leningradskii Veterinarnyi Institut* 32:43–4. (In Russian).

Zumpt, F. 1965. *Myiasis in Man and Animals in the Old World: A Textbook for Physicians, Veterinarians and Zoologists.* London: Butterworth.

7 Bacterial Infections and Diseases

Terje D. Josefsen, Torill Mørk and Ingebjørg H. Nymo

CONTENTS

7.1   INTRODUCTION

Bacterial diseases occur in reindeer, just as they occur in all animals. However, bacterial diseases in wild reindeer are not always easily detected. Reindeer roam in remote areas, where few people are present to observe disease. Diseased animals are often quickly taken care of by predators; however, if this is not the case, carcasses of deceased animals will readily be consumed by scavengers. Thus, many cases of bacterial disease may go unnoticed in wild reindeer populations.

It is easier to detect diseases in semi-domestic reindeer, and much knowledge of reindeer disease comes from reindeer husbandry. Semi-domestic reindeer are called "semi-domestic" because they lie somewhere between wild and domestic animals. They are wild in the sense that they are allowed to roam more or less freely in a natural environment all through the year; nevertheless, they are domestic in the sense that they are owned by people. Semi-domestic reindeer are regularly gathered for different purposes (ear marking, slaughtering, dividing mixed herds, supplemental feeding), and therefore the chance of detecting diseased or dead animals is increased relative to that in wild reindeer.

The statement that "much knowledge of reindeer disease comes from reindeer husbandry" is in some ways true; however, it could also be changed to "many diseases in reindeer are due to reindeer husbandry." It is a recurring theme in this chapter that conditions created by humans to exploit reindeer also create stress, crowding and unsanitary conditions that promote bacterial disease. This relationship between reindeer, bacteria and the environment is well illustrated by several bacterial diseases.

The most exotic and strange diseases in reindeer often occur in zoos and parks where reindeer are kept in a climate that they are not adapted to, and where they come into contact with infectious agents that they would not normally be exposed to. There is a huge difference between the infectious diseases reindeer have the possibility of contracting if exposed to the infectious agent under unfavorable conditions, and the diseases reindeer normally contract in their natural environment.

In this chapter, we introduce the main bacterial diseases recorded in reindeer and also mention some of those less common.

7.2 PASTEURELLOSIS

Torill Mørk and Terje D. Josefsen

7.2.1 AETIOLOGY

The term "pasteurellosis" is used for any disease caused by infection with bacteria within the genera *Pasteurella, Mannheimia and Bibersteinia*, all previously included in genus *Pasteurella*. In veterinary medicine, this concerns three relevant species: *Pasteurella multocida, Mannheimia hemolytica and Bibersteinia trehalosi*. These bacteria belong to the family Pasteurellaceae, which is a large group of Gram-negative Gammaproteobacteria prevalent as commensals, opportunistic pathogens, and primary pathogens among both humans and animals worldwide (Wilson and Ho 2013). In reindeer, only *P. multocida* is known to cause disease, and hence "pasteurellosis" in reindeer refers to disease caused by this species.

Pasteurella multocida causes diseases like hemorrhagic septicemia in bovids and cervids, respiratory disease or septicemia in rabbits and hares, avian cholera in birds and atrophic rhinitis in swine. In reindeer, two forms of pasteurellosis are seen: hemorrhagic septicemia and bronchopneumonia.

Pasteurella multocida is a small Gram-negative, non-motile, facultative anaerobic rod. Within *P. multocida*, three subspecies are differentiated based on their ability to ferment the sugars D-sorbitol and trehalose: *P. m. multocida, P. m. septica* and *P. m. gallicida*, Furthermore, five serogroups, A, B, D, E and F, have been identified based on differences in capsular polysaccharides. Those within serogroups B and E are involved in hemorrhagic septicemia in cervids and bovids, while those in serogroup A are involved in avian cholera and pasteurellosis in rabbits and hares (Quinn et al. 2011, Ferroglio 2012). Hemorrhagic septicemia caused by *P. multocida* serogroup B or E is a notifiable disease in some countries. Pasteurellosis in reindeer is caused by *P. m. multocida*, but further typing and phylogenetic comparison of isolates obtained from reindeer has not been reported.

7.2.2 PATHOGENESIS

Pasteurella multocida is found as a commensal on mucous membranes in the upper respiratory tract in many animal species. Transmission among animals is mainly through direct contact with nasal secretions. The bacterium acts as an opportunistic pathogen that may cause disease when various types of stress, including other infections, reduce the immune status of the host. A disease outbreak may also occur due to the introduction of a new *Pasteurella* strain to the population, or a change in the virulence of an existing strain. Despite extensive research, very little is known about the mechanisms involved in the transition from an upper airway commensal to a severe pathogen. (Wilson and Ho 2013).

In reindeer, outbreaks of hemorrhagic septicemia are often reported during summer, especially during periods of dry and warm weather. Heat stress, insect harassment and drought are stress factors that may contribute to disease outbreaks (Nordkvist and Karlsson 1962).

However, Skjenneberg (1957) and Kummeneje (1976) reported pasteurellosis in 1-year-old calves in late winter/early spring. These disease outbreaks were characterized by respiratory disease, and were associated with lungworm (*Dictyocaulus* sp.) and throat bot fly (*Cephenemyia trompe*), affecting mostly animals in poor condition after a harsh winter.

In buffalo and fallow deer, it has been shown that the bacterium may lodge in the tonsils after primary infection (De Alwis et al. 1990, Aalbæk et al. 1999), and in cattle the bacteria have been isolated from the nasopharynx of healthy animals (Hotchkiss et al. 2010). *Pasteurella multocida* is most likely a commensal of the nasopharynx/upper respiratory tract of adult reindeer (Kummeneje 1976), but this has not been confirmed.

The predominance of young reindeer developing pasteurellosis indicates some kind of immunity or resistance to the infection in older animals. The factor(s) responsible for activating the latent *P. multocida* in nasal secretions are still unknown (Shivachandra et al. 2011).

7.2.3 Clinical Signs and Pathology

Pasteurella multocida infection in reindeer is associated with both peracute hemorrhagic septicemia and acute or subacute lung disease. In most large outbreaks, the peracute form has been the dominating clinical presentation (Magnusson 1913, Brandt 1914, Horne 1915, Nordkvist and Karlsson 1962).

Clinical symptoms of hemorrhagic septicemia in ruminants typically include fever, subcutaneous edema of the head and neck, salivation, lacrimation and nasal discharge. Respiratory distress, septic shock and death follow (Ferroglio 2012, Wilson and Ho 2013). In reindeer, most reports describe pasteurellosis as a peracute disease with no signs of disease before the animal is found dead. During the epizootic in Sweden in 1913 (Brandt 1914), a herder saw that a flock of animals, apparently healthy, came to rest quietly on a snow drive in the middle of the day. When the animals stood up and started moving again, one or more dead or dying reindeer were left behind.

In cases with a longer disease course there are symptoms in the upper respiratory tract, such as nasal discharge and coughing. Calves with clinical symptoms often show weakness and apathy, and die (Horne 1915, Kummeneje 1976).

The pathological findings are usually general signs of septicemia, such as acute congestion in the liver and spleen, acute lung edema and petheciae in serosal surfaces. Inflammatory changes in the lungs are rare in such cases, but do occur (Figure 7.1). In cases of lung disease, the most common finding is fibrinous pneumonia or bronchopneumonia, and occasionally fibrinous pleuritis and pericarditis (Figure 7.2). *Pasteurella multocida* has also been isolated from apparently healthy slaughtered animals with small focal areas of bronchopneumonia, indicating a chronic and subclinical course (Kummeneje 1976).

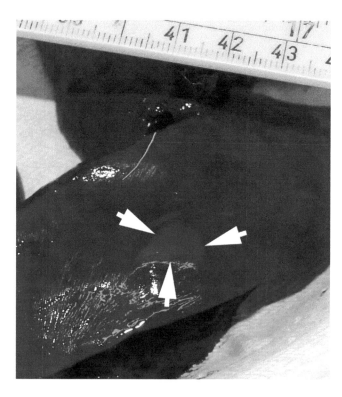

FIGURE 7.1   Part of the lung from a 5-month-old reindeer calf with septicemic pasteurellosis. Arrows point to the border of a trapezoid-shaped area of lighter colour. This area was firm and contained massive inflammatory changes with necrosis. This case is from an outbreak of septicemic pasteurellosis in northern Norway in October 2010. Only one out of eight examined calves had macroscopic inflammatory changes in the lungs. *Pasteurella multocida* was grown in pure culture from all organs of this and the other calves (Photo: Terje D. Josefsen).

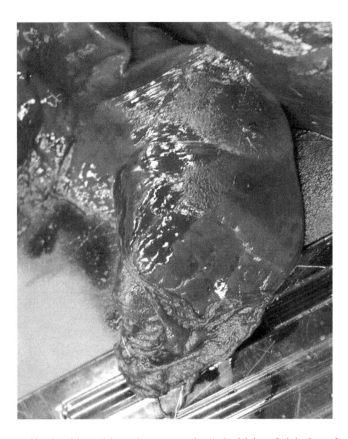

FIGURE 7.2   *Pasteurella* pleuritis and bronchopneumonia. Apical lobe of right lung from an adult female reindeer. The photo shows a thick layer of fibrin that is partly detached from the pleura, and brown discolouration of pleura where it has been covered by fibrin. Lung tissue in the area is otherwise dark red and firm. The pathological diagnoses were fibrinous pleuritis and fibrinopurulent bronchopneumonia. *Pasteurella multocida* was cultivated as the dominating bacteria from the affected parts of the lung. This case is from an outbreak of *Pasteurella* bronchopneumonia that occurred in northern Norway in February 2014. The affected reindeer were kept in a fenced area for winter-feeding. The number of reindeer that died is unknown. Lungs were sampled from five dead adult females, and four of them had fibrinopurulent bronchopneumonia with *P. multocida* as the dominant agent (Photo: Torill Mørk).

7.2.4   EPIDEMIOLOGY

Pasteurellosis in reindeer was first reported in northern Sweden and Norway, and was described as causing large epizootics in semi-domesticated reindeer herds in 1912–1914 (Magnusson 1913, Brandt 1914, Horne 1915). The outbreak started in Sweden in June 1912, in the Arjeplog municipality (Pite Lappmark) located just south of the Arctic Circle, and approached the Norwegian border. There were only scattered deaths at the beginning of the summer, but mortality increased along with increasing summer temperatures, reaching a peak in late July and the beginning of August, and ceasing in late September when the cold season started. A total of 1600 reindeer died, most of them (86%) being calves. The disease was entirely unknown to the reindeer herders in that area, which could indicate that it was newly introduced into the Scandinavian reindeer population at that time.

The disease also occurred on the Norwegian side of the border: In 1913, about 275 reindeer from five different herds were lost to the disease, equaling 10% of the animals. In 1914, only a single herd was affected, losing 25 calves. Samples were obtained during outbreaks in both countries, and the bacteriological diagnosis was verified via the identification of large numbers of *Pasteurella* in most samples.

Similar events have occurred in Jämtland, Sweden, in the summers of 1924 (1300 died) and 1959 (600 died) (Nordkvist and Karlsson 1962). In Norway, outbreaks of pasteurellosis were reported in 1912–14, 1956, 1966 and 1973 (Horne 1915, Skjenneberg 1957, Kummeneje 1976). In the outbreaks in northern Norway in 1956 and 1973, bronchopneumonia was the dominant form of the disease, still with high mortality among calves (Skjenneberg 1957, Kummeneje 1976). Further smaller outbreaks have occurred up to now; the last reported outbreak of septicemia in reindeer calves was in the corral of a slaughterhouse in October 2010, where 44 calves died and pasteurellosis was diagnosed in eight of 11 examined cases (Mørk et al. 2014).

In addition, sporadic cases and small outbreaks of pasteurellosis have been recorded in wild Eurasian tundra reindeer in Norway (Handeland 2014), though few details are published. Outbreaks of pasteurellosis have not been reported in caribou from North America, but there are some reports of pneumonia in caribou calves in Alaska, without further diagnostic specification (Sellers et al. 2003, Valkenburg 2003). From Canada, there is one report of *P. multocida* isolation from abscesses in caribou calves, as a result of bite wounds from a lynx (Bergerud 1971).

In Russia, a "diplococcal pneumonia" has been reported (Nikolaevskii 1961, Syroechkovskii 1995), affecting mainly reindeer calves and causing septicemia or bronchial pneumonia. This disease may resemble pasteurellosis, but *P. multocida* is not usually included among the diplococci.

The history of large die-offs and the presence of small disease outbreaks with high mortality makes pasteurellosis the most significant bacterial disease in reindeer in Scandinavia, and a disease that has the potential to cause large epizootics in reindeer in this area.

7.2.5 Diagnosis, Treatment and Management

Clinical history and pathological findings may be suggestive of pasteurellosis; however, confirmative diagnosis rests on detection of *P. multocida* in tissues or fluids from a sick or dead animal.

Pasteurella multocida is sensitive to antibiotics, and antibiotic treatment of calves has been tried with variable success (Kummeneje 1976, Korbi 1982). Treatment with antibiotics is likely to have a better effect if started early in the course of infection, before the animals are weakened. In reindeer husbandry, such treatment can often be challenging to conduct, and isolation of sick animals, as well as prevention of contact between flocks, is recommended to prevent further spread of the disease. Vaccination against pasteurellosis in calves was attempted during an outbreak in 1973, and seemed successful; however, the number of calves was too small to draw any conclusions (two vaccinated against two unvaccinated controls) (Kummeneje 1976).

7.3 BRUCELLOSIS

Ingebjørg H. Nymo

7.3.1 Aetiology

The genus *Brucella* resides within the family Brucellaceae, contains Gram-negative, non-motile, facultative intracellular, small coccobacillary bacteria and consists of 12 species with a wide variety of host preferences (Corbel and Banai 2005).

7.3.2 Pathogenesis, Pathology and Clinical Signs

When entering a host, *Brucella* spp. are engulfed by circulating polymorphonuclear cells and macrophages (Radostits et al. 2000). The hallmark of the pathogenic brucellae is their ability to replicate and survive within vacuoles derived from the endoplasmic reticulum, particularly in macrophages (reviewed by von Bargen et al. 2012).

The most common mode of entry for *Brucella* spp. is through ingestion of food and water contaminated with aborted material; however, infection can also occur through respiratory and conjunctival exposure, as well as entry through damaged skin. *Brucella* spp. can also be transmitted during breeding (i.e. semen) and lactation (i.e. milk), and by crossing the placenta from mother to offspring (Radostits et al. 2000, Corbel 2006). The bacteria are initially – during a bacteremic phase – transported in macrophages and neutrophils, and freely in the plasma, to regional lymph nodes, thereafter spreading to other lymphoid tissues, including the spleen. A second bacteremia results in a generalized infection affecting other target organs as well as their associated lymph nodes (Radostits et al. 2000).

Brucella suis biovar 4 is the causative agent of brucellosis in *Rangifer* (Corbel and Banai 2005) and may cause, for example, a retained placenta due to placentitis, metritis, abortion, stillbirth, mastitis, orchitis, epididymitis, synovitis, bursitis, abscesses, lymphadenitis and nephritis (Neiland et al. 1968, Rausch and Huntley 1978, Tessaro and Forbes 1986, Forbes 1991, Ferguson 1997). The clinical signs are, however, most often associated with the reproductive systems and joints (O'Reilly and Forbes 1995).

Brucella spp. localize in the pregnant uterus of ruminants due to the bacterium's predilection for cells producing the sugar alcohol erythritol, which is a constituent of normal ungulate foetal and placental tissue (Smith et al. 1962). This predilection was also shown in experimentally infected *R. t. tarandus* (Forbes and Tessaro 1993). Newly infected female *Rangifer* often abort in the first calving period after infection, 1 or 2 months prior to normal calving time (O'Reilly and Forbes 1995). Experimental infection of caribou (*R. t. granti*) 5 months into gestation confirmed this by yielding abortion during the last month of gestation. The foetuses had low weights, and *B. suis* bv. 4 was recovered from numerous organs. However, the mothers produced seemingly healthy fawns the next year (Rausch and Huntley 1978), as has also been reported in bovines (Radostits et al. 2000). Abortions are reported to occur frequently in reindeer populations in Siberia (Russia), where *B. suis* bv. 4 is regarded as enzootic (reviewed by Neiland et al. 1968). *Brucella suis* bv. 4 has been isolated in association with placentitis (Forbes 1991), metritis and retained placentas (Neiland et al. 1968) in *Rangifer* under natural conditions.

Brucella spp. has a predilection for the male reproductive organs in bovines (Radostits et al. 2000), and *B. suis* bv. 4 has been isolated from the testes of experimentally infected Eurasian tundra reindeer bulls (Forbes and Tessaro 1993). Under natural conditions, the bacterium has been isolated from purulent orchitis in *R. t. granti* (Neiland et al. 1968), from suppurative epididymitis in *R. t. groenlandicus* and from both suppurative and non-suppurative orchitis in *R. t. groenlandicus* and *R. t. tarandus* (Forbes 1991, Ferguson 1997).

A non-suppurative synovitis of the carpus may be observed in cattle infected with *B. abortus* (Radostits et al. 2000). *Brucella suis* bv. 4 has been isolated in association with both suppurative and non-suppurative carpal infections in *R. t. groenlandicus* and *R. t. granti*, and also from infections in other joints (Neiland et al. 1968, Forbes 1991). Serofibrinous carpal bursitis is reported as one of the typical conditions associated with brucellosis in *Rangifer* in Siberia (reviewed by Neiland et al. 1968). In North America, this condition is sometimes referred to as "caribou knees" (Figure 7.3), and affected animals may be seen to move slower than the rest of the herd (Neiland et al. 1968).

7.3.3 Epidemiology

Transmission of *Brucella* spp. between terrestrial animals usually takes place through contact with aborted, infected material. The risk posed to susceptible animals, following parturition of infected offspring, depends on the number of bacteria excreted. *Brucella* spp. achieves its greatest numbers in the pregnant uterus, in the foetus and in the foetal membranes in the first parturition after infection (Radostits et al. 2000). Counts of 10^3 to 10^5/g of *B. suis* bv. 4 have been reported from the uterus of experimentally infected *R. t. tarandus* (Forbes and Tessaro 1993), and *B. suis* bv. 4 has been isolated from aborted foetuses (Rausch and Huntley 1978). In the following parturitions, the bacterial numbers are usually lower in bovines (Radostits et al. 2000), as also seems to be the case for *Rangifer*, since mothers aborting one year were reported to produce healthy fawns the next year (Rausch and Huntley 1978).

The survival of the bacteria in the environment is of epidemiological importance (Radostits et al. 2000). *Brucella* spp. do not multiply outside the host, but may persist for years in frozen aborted

FIGURE 7.3   Characteristic "caribou knee" due to *Brucella suis* biovar 4 in the carpus of a barren ground caribou. The animal was from Yellowknife, Canada (Photo: Patricia Lacroix, Government of Northwest Territories, Environment and Natural Resources).

material and for months in moist conditions at 10–15°C (Crawford et al. 1990). They are, however, susceptible to sunlight, desiccation and standard disinfectants (Radostits et al. 2000).

Brucellosis is shown to be present in caribou in Alaska (Zarnke et al. 2006) and Canada (Curry et al. 2011). The source of *B. suis* bv. 4 introduction into North America has been debated (O'Reilly and Forbes 1995), and suggestions include the importation of reindeer from Siberia in 1891 (Lantis 1950, Dieterich 1981) or that brucellae were introduced with reindeer crossing the Bering Sea during the last ice age (Rausch and Huntley 1978). Brucellosis is also present in reindeer in Siberia (reviewed by Zheludkov and Tsirelson 2010). In other Arctic regions like Greenland (O'Reilly and Forbes 1995), mainland Norway (Åsbakk et al. 1999, Nymo et al. 2013) and Svalbard (Nymo, unpublished data), no indication of *Brucella* sp. infection has been detected in *Rangifer*, and there are no reports of the disease in reindeer in Finland and Sweden.

7.3.4   DIAGNOSIS AND MANAGEMENT

The gold standard in brucellosis diagnostics is isolation of the bacterium (Poester et al. 2010). Detection of antibodies (i.e. serology) is, however, presumptive evidence of infection, but there are considerable differences in the accuracy of the various serological tests (Poester et al. 2010). As with many infectious agents, serological cross-reaction is a major problem when detecting anti-*Brucella* antibodies (Corbel 1985), and serological methods must be validated for usage in *Rangifer*, preferably by comparison to a gold standard (Nymo et al. 2013).

Treatment of brucellosis in production animals is not feasible due to the intracellular presence of the bacteria and hence the necessity for long-lasting antibiotic treatment. Eradication of brucellosis in cattle is often based on culling of seropositive animals and vaccination (Radostits et al. 2000).

Vaccination of reindeer with killed *B. suis* bv. 4 yielded adequate protection under experimental settings until 43 months post-vaccination (Morton 1990). Vaccination of reindeer in Russia has been reported as partly successful (reviewed by O'Reilly and Forbes 1995). Some vaccine strains have been tested on *Rangifer* and found to be unsuitable, such as *B. melitensis* strain 1738 (reviewed by O'Reilly and Forbes, 1995), *B. melitensis* strain H-38 (Dieterich et al. 1979), *B. abortus* strain 19 (Dieterich et al. 1987, 1989) and *B. abortus* strain 45/20 (Dieterich et al. 1981).

7.3.5 ZOONOTIC IMPLICATIONS

Brucellosis is the most common zoonotic disease worldwide, with more than 500,000 new cases annually (reviewed by Pappas et al. 2006). Human brucellosis due to *B. suis* bv. 4 have been described in Alaska (Brody et al. 1966), Canada (Chan et al. 1989) and Siberia (reviewed by Zheludkov and Tsirelson 2010). *Brucella suis* bv. 4 has been isolated from bone marrow, lymph nodes, joints and blood of infected humans (Matas and Corrigan 1953, Corrigan and Hanson 1955, Toshach 1963, Brody et al. 1966, Chan et al. 1989, Forbes 1991), and the strains isolated from humans and *Rangifer* have been shown to be indistinguishable from each other (Meyer 1966), leaving no doubt about transmission from animals to man. The clinical signs in humans have been undulant fever, dehydration, increased pulse, lethargy, loss of appetite, malaise, headache, joint pain, hepatomegaly and splenomegaly (Matas and Corrigan 1953, Corrigan and Hanson 1955, Toshach 1963, Brody et al. 1966, Chan et al. 1989). The reported prevalence of *Brucella*-seropositives among people in close contact with *Rangifer* vary (Greenberg and Blake 1958, Brody et al. 1966), and probable risk factors are exposure to and ingestion of raw *Rangifer* products (reviewed by Toshach 1963, Chan et al. 1989, Greenstone 1993, Hueffer et al. 2013).

7.4 NECROBACILLOSIS

Terje D. Josefsen

Necrobacillosis is a common term for disease caused by the bacterium *Fusobacterium necrophorum*. The genus *Fusobacterium* consists of more than 20 species, of which many are able to cause disease in animals and man; however, *F. necrophorum* is the species that is most often recovered from cases of necrobacillosis (Jang and Hirsh 1994).

Fusobacterium necrophorum is an obligate anaerobic pleomorphic rod, often occurring in long filamentous forms. It is non-motile, non-spore-forming and part of the normal microbiota in the rumen of ruminants, numbering 10^5 to 10^6 per gram of ruminal content dry matter (Tan et al. 1994) in the bovine rumen. It can also be recovered from other parts of the alimentary tract, such as the oral cavity and the intestines (Tadepalli et al. 2009).

Necrobacillosis occurs worldwide, and *F. necrophorum* is a major pathogen in domestic ruminants. It is involved in different forms of digital disease in both cattle and sheep, either as a primary, synergistic or secondary pathogen (Hargis and Ginn 2012), and it is the etiologic agent in liver abscess formation in grain-fed cattle, known as "the rumenitis-liver abscess complex" in feedlot cattle (Tadepalli et al. 2009).

7.4.1 NECROBACILLOSIS IN REINDEER

Necrobacillosis in reindeer is primarily an infection of the digits and distal feet (digital necrobacillosis). Oral infections may occur together with foot lesions, and are in these cases suspected to be transferred from the foot by licking. However, oral lesions may also occur alone, probably due to ingestion of contaminated fodder. Suckling calves with oral lesions may transfer the infection to the udder of their mother. Cases with lesions in reproductive organs are reported during the rutting season (Skjenneberg and Slagsvold 1968, Rehbinder and Nikander 1999).

Fusobacterium necrophorum is not able to penetrate intact skin or mucous membranes; thus, necrobacillosis is principally a wound infection. Small skin or mucosal membrane abrasions or fissures are believed to be the port of entry. *Fusobacterium necrophorum* is regarded as the primary pathogenic agent of digital necrobacillosis in reindeer, but other opportunistic pathogens like *Trueperella pyogenes* and *Staphylococcus aureus* may contribute to the infection.

The first written report of digital necrobacillosis in Scandinavian reindeer husbandry is nearly 300 years old (Linné 1732). Horne (1898) and Bergman (1909) made classic reviews of the disease, while they and other sources are reviewed in Qvigstad (1941). The disease was called "reindeer foot disease" or "reindeer foot rot" in early reports. The Sami people, the indigenous reindeer-herding people of northern Fennoscandia, have named the disease "slubbo," meaning "club," referring to the leg being club-shaped (Figure 7.3) when swollen in distal parts (Horne 1898).

Reports of digital necrobacillosis in reindeer come from all areas inhabited by reindeer: Fennoscandia (reviewed in Skjenneberg and Slagsvold 1968), Russia and Siberia (Nikolaevskii 1961, Syroechkovskii 1995, Melnik et al. 2009), Alaska (Morton 1981, Woolington and Machida 2001) and Canada (Godkin 1986, Leighton 2001, Anonymous 2012). Since *Fusobacterium necrophorum* is part of the normal ruminal microbiota of reindeer (Aagnes et al. 1995), the potential for opportunistic infections will always be present.

Large disease outbreaks of digital necrobacillosis, affecting hundreds or thousands of domestic reindeer, are reported in literature, though often in an anecdotal form (Bergman 1909). Outbreaks of digital necrobacillosis in reindeer are often associated with crowding of animals on wet and muddy ground contaminated by faeces. When milking of reindeer hinds in summer was common practice, the animals were gathered in a corral several days each week, and for several hours at a time. These milking corals became muddy and heavily contaminated with faeces, and were probably a major contributing factor to outbreaks of necrobacillosis. Another contributing factor is weather conditions; the infection nearly always occurs in the warm season (late spring, summer, early autumn), and high temperatures have been associated with disease outbreaks. However, the amount of precipitation might be even more important, as moisture promotes *F. necrophorum* infections (Monrad et al. 1983).

Digital necrobacillosis more or less vanished from Fennoscandian reindeer husbandry after 1950. This is attributed to a change in reindeer husbandry, particularly the termination of reindeer milking, but also a general shift towards more extensive reindeer husbandry, with larger flocks ranging freely over wider areas. In Siberia, the disease is still present, causing considerable losses for reindeer husbandry (Melnik et al. 2009).

In Fennoscandia, new practices have arisen during the last few decades, such as gathering of reindeer for supplementary feeding in winter, or calving in fenced areas for predator protection. This has created new scenes for the disease and cases of oral necrobacillosis related to supplementary feeding have been reported (Nordkvist 1967, Rehbinder and Nikander 1999). An outbreak of oral necrobacillosis in suckling calves, born in a fenced area, occurred in both 2012 and 2013 (Åhman et al. 2013, Ågren et al. 2014), affecting at least 36 calves (5% of the calves) in the 2013 outbreak. The disease outbreaks were associated with high animal density and faecal contamination of the environment. The budding of new teeth in the young calves probably served as excellent ports of entry for infection of the oral mucosa.

Necrobacillosis is rare in wild reindeer, but scattered cases or small outbreaks have been recorded (Syroechkovskii 1995, Woolington and Machida 2001, Anonymous 2012). The causes of such outbreaks are largely unknown. Handeland and coworkers (2010) reported an outbreak of digital necrobacillosis in wild reindeer in southern parts of Norway in the summers of 2007 and 2008, affecting at least 80 reindeer altogether. When meteorological data from this outbreak was analyzed, it appeared that the number of warm days (daily maximum temperature above 15°C) was higher than normal for both summers. The amount of precipitation was also slightly above normal for both summers, but the largest deviation was that the number of days with precipitation was exceptionally high as compared to normal years. Thus, warm and humid summers were presumed to be a predisposing element for this disease outbreak (Handeland et al. 2010).

7.4.2 CLINICAL SIGNS, PATHOLOGICAL FINDINGS AND DIAGNOSIS

The principal clinical signs of digital necrobacillosis are lameness and swelling of distal parts of the foot (Figure 7.4). The swelling comprises the area from the fetlock joint to the coronary band (Handeland et al. 2010). After a few days, there may be ulceration of the skin over the swelling. The degree of lameness will vary with the severity of the lesions.

Usually only one foot is affected in each animal, but cases with two affected feet sometimes occur (Bergman 1909, Handeland et al. 2010). Handeland and coworkers (2010) reported slightly more hind limb lesions than fore limb lesions in their material (17 versus 10), whereas reindeer herders told Bergman (1909) that fore and hind limbs were equally affected.

Animals of all ages may be affected by digital necrobacillosis. Some have reported that young animals are more often diseased than older animals (Horne 1898, Bergman 1909, Skjenneberg and Slagsvold 1968), while Melnik et al. (2009) claimed that adults are most susceptible to infection, but that mortality is higher among calves.

Clinical signs of oral necrobacillosis include excessive salivation and problems chewing and swallowing feed, which predispose the animals to aspiration pneumonia (Handeland 2012).

FIGURE 7.4   Digital necrobacillosis. Adult male wild Eurasian tundra reindeer with digital necrobacillosis in right hind leg, Southern Norway, January 2015. Outbreaks (Handeland et al. 2010) and scattered cases of digital necrobacillosis have occurred in the Norwegian wild reindeer population since 2007. The reindeer in the photo was shot for animal welfare reasons and to reduce the risk of spreading the infection. The right hind leg was frozen and later sent to the Norwegian Veterinary Institute for examination. A diagnosis of digital necrobacillosis was made, based on pathological findings. *Fusobacterium necrophorum* could not be cultivated from the lesions, but cultivation is not a very sensitive method to verify the infection. Negative anaerobic culture may occur, and *F. necrophorum* may still be shown to be present by other methods (Handeland et al. 2010) (Photo: Erik M. Ydse, Norwegian Nature Inspectorate).

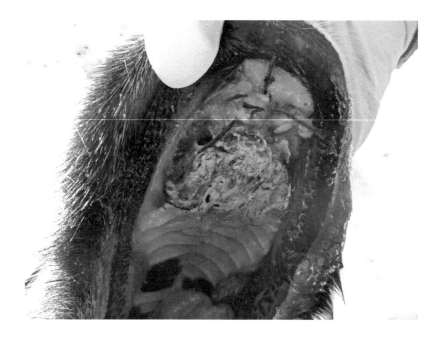

FIGURE 7.5   Oral necrobacillosis. Two-year-old male reindeer with severe necrotic lesion on dental pad and cranial part of the palate. The smell from the mouth was foul. A diagnosis of oral necrobacillosis was made, based on clinical sign. This case was one of several with oral necrobacillosis in a reindeer herd in Sweden in March/April 2016. The outbreak of disease came after a period of winter-feeding in a corral. A concurrent outbreak of orf-virus infection may have precipitated the oral necrobacillosis (Photo: Ingebjørg H. Nymo).

Fusobacterium necrophorum produce toxins that cause necrosis, and areas of tissue necrosis are the principal pathological finding in lesions. In digital necrobacillosis, necrotic areas may comprise the skin, subcutaneous tissue, tendon sheets, tendons, joint capsules and articular cartilage. Inflammation of bone (periostitis and osteomyelitis) may occur. Extensive fibrosis occurs in chronic cases, leading to increased swelling (Handeland et al. 2010). In oral necrobacillosis, various parts of the oral mucosa may be affected (Figure 7.5).

A tentative diagnosis may be based on clinical signs, but a confirmative diagnosis must be based on pathological examination and demonstration of the etiological agent (cultivation, PCR, histology, *in situ* hybridization) in typical lesions. Differential diagnoses include trauma and infections caused by bacteria other than *F. necrophorum*.

7.4.3   PROGNOSIS AND TREATMENT

Most cases have a fatal outcome, due to starvation, predation or, less commonly, spread to other parts of the body, with multiple organ involvement and septicemia. Without treatment, reported mortality has been up to 75–80% (Horne 1898, Bergman 1909, Syroechkovskii 1995). Immediate slaughter has been the preferred action when diseased animals are detected (Bergman 1909), since this action diminishes the spread of the disease, and at the same time saves the meat. Early attempts to treat individual animals by surgical debridement and antiseptics seemed successful (Bergman 1909), at least in view of survival, but animal welfare concerns were not discussed in these cases. Today, antibiotics are probably effective if treatment is initiated at an early stage of the disease, but treatment is not an option in wild reindeer, and may also be challenging in reindeer husbandry. Preliminary results of vaccination against necrobacillosis in Sakha (Russia) seem to have reduced both the number of infected animals and the mortality associated with the infection (Melnik et al. 2009).

7.5 CLOSTRIDIAL INFECTIONS AND INTOXICATIONS

Terje D. Josefsen

Clostridia are obligate anaerobic, spore-forming, Gram-positive, large rods that are straight or slightly curved. Most species are motile by peritrichous flagella. More than 100 *Clostridium* spp. are known, and they are widely distributed in soil, aquatic environments and decaying organic matter, and in the gastrointestinal tract of humans and animals.

Many species of clostridia are major pathogens, e.g. the neurotoxic species *Clostridium botulinum* and *C. tetani*, the histotoxic species *C. septicum*, *C. perfringens*, *C. chauvoei*, *C. novyi* and *C. sordellii*, and the enteropathogenic/enterotoxic species *C. perfringens* and *C. difficile*.

Very few documented cases of clostridial infections in reindeer are reported in the scientific literature, while the existing cases are divided into two categories: enterotoxemia or enteritis caused by *C. perfringens* (see the example in Figure 7.6), and clostridial infection and septicemia caused by histotoxic *Clostridium* spp.

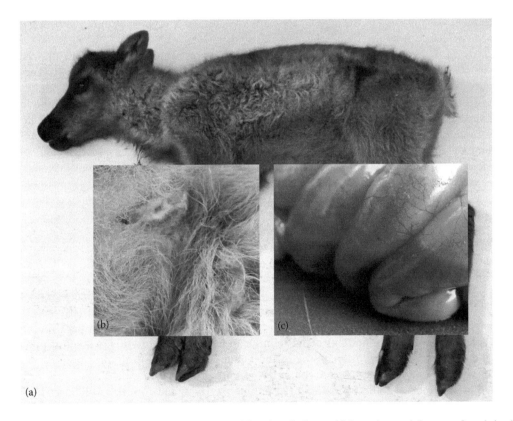

FIGURE 7.6 Clostridial enteritis. (a) A 17-day-old male reindeer calf, born in captivity, was found dead. Clinical signs of disease had not been observed. (b) A small amount of mucoid greenish faeces was apparent in the fur close to the anus. (c) At necropsy, the last third of the small intestine was hyperemic and dilated with ample watery bloodstained content. Rich growth of *Clostridium perfringens* and other *Clostridium* sp. was found in the intestinal contents. ELISA for *C. perfringens* epsilon toxin was negative (Photos: Terje D. Josefsen).

Clostridium perfringens enterotoxemia, involving *C. perfringens* type D, was diagnosed in three suckling calves (1–3 weeks of age) of semi-domestic Eurasian tundra reindeer (*R. t. tarandus*) in Norway (Josefsen et al. 2014). The calves were born in a fenced area to protect them against predators. Approximately ten calves died. Some had diarrhoea before they died; others were found dead without signs of disease. Four calves were submitted for necropsy. The three calves with enterotoxemia were all in good body condition, and had coagulated milk in their abomasums, indicating a short course of disease. The content in the small intestines was watery with gas bubbles, while cecal content was watery and colon and rectum content was mushy or liquid. Two of the calves had a very soft renal cortex ("pulpy kidneys"). No serosal hemorrhages or fluids in body cavities were observed. Cultivation from intestinal content (small and large intestines) revealed rich growth of *C. perfringens* in all calves. An ELISA kit for detection of *C. perfringens* enterotoxins demonstrated the presence of epsilon toxin at levels about eight times higher than the lower limit of positive samples from sheep and goat.

Clostridium perfringens enteritis has also been diagnosed in suckling reindeer calves in Sweden (Åhman et al. 2013). Like the case in Norway, these calves were born in a fenced area to protect them from predators. Thirty of 375 calves died within a few weeks, with signs of diarrhoea. Acute enteritis was diagnosed at necropsy. *Clostridium perfringens* was cultivated from the intestinal content and was regarded as the most likely cause of enteritis. No typing of the *C. perfringens* has yet been published from this outbreak.

Clostridium perfringens type A enterotoxemia was reported in four Eurasian tundra reindeer kept in a farm in Austria (Sipos et al. 2003). Clinical signs were acute and severe enteritis, and three reindeer died within 1–2 days, while the fourth recovered after intensive treatment.

A different clinical picture of *C. perfringens* enterotoxemia in reindeer was reported by Kummeneje and Bakken (1973). This disease occurred in adult animals, mostly females in good body condition, during the autumn translocation of animals from the coast to inland pastures in Finnmark County, northern Norway. Upon arrival at pastures rich in lichens, several cases of sudden death occurred. The authors witnessed that the reindeer fell down in convulsions and died. Three reindeer were necropsied. Major macroscopic findings were petechial and ecchymotic hemorrhages in muscles, subcutis, heart and kidneys. Lungs were congested and edematous and the ileum was hyperemic and dilated. Diarrhoea was not reported. Bacterial cultivation from organs (lung, liver, kidney, spleen) was negative, but *C. perfringens* was cultivated in large numbers from the ileum (3×10^7 to 3×10^8 per gram ileum content) and was identified as type A. Undiluted supernatant from centrifuged ileum content was injected intravenously into mice, to test if it contained any toxins. The mice all survived the 3-day observation period.

Interpreting the results of this report is challenging. The authors concluded that enterotoxemia was the cause of death based on four criteria: the change in diet to rich lichen pastures (rich in carbohydrates); the fact that adult females were most affected (anticipated to have the highest feed intake); the necropsy findings being "similar to that of enterotoxemia in small ruminants"; and the high number of *C. perfringens* in the ileum. With regard to the last point, it may be objected that *C. perfringens* type A commonly occurs in normal intestinal content in reindeer, also in high numbers, without causing disease. Embury-Hyatt and coworkers (2005), who investigated a syndrome of sudden death in farmed white-tailed deer, found no significant differences in numbers of *C. perfringens* type A in the intestines of affected compared to unaffected deer, and the numbers of bacteria in the ileum reached an order of magnitude similar to the reindeer case. The injection of ileal content into mice without detectable reaction, weaken the diagnosis. Still, no other plausible explanation can be suggested to explain this outbreak of peracute death in reindeer.

7.5.2   Histotoxic Clostridial Infections in General

Infection with histotoxic clostridia results in severe, often fatal, disease. A wide range of animals may contract clostridial infections; however, ruminants are the group most often affected by these diseases. Sheep are commonly vaccinated against a range of different clostridial infections: *C. perfringens* type B, C and D, *C. septicum*, *C. chauvoei*, *C. novyi*, *C. hemolyticum* and *C. tetani*.

Infection with histotoxic clostridia may occur through skin wounds, with the resulting lesions characterized by edema, hemorrhages and necrosis in muscles and adjacent connective tissue. If gas is produced within the lesions, the term "gas gangrene" is used; otherwise, "malignant edema" is the usual term. Gas gangrene is often associated with *C. perfringens* and malignant edema with *C. septicum*; however, different *Clostridium* spp. are able to cause both types of clinical presentations, and mixed infections also occur (Cooper and Valentine 2016).

Another way in which clostridial infections may arise is through activation of resident spores in the muscles or liver. Thus, both malignant edema and gas gangrene may appear without any penetrating skin lesions. The latent muscle spores are stimulated to germinate when a local event creates muscle damage or low oxygen tension. The disease terms "blackleg" (*C. chauvoei*) and "pseudo-blackleg" (*C. septicum*) are specifically used for gangrenous myositis arising from latent spores in muscles (Cooper and Valentine 2016), while "black disease" is a corresponding germination of latent *C. novyi* spores in liver, causing necrotic hepatitis (Cullen and Stalker 2016). The germination of muscle spores is initiated by bruises or other kind of muscular damage, while the germination of liver spores is initiated by damage to liver tissue caused by migrating liver flukes (Cooper and Valentine 2016, Cullen and Stalker 2016).

Braxy is a necrotizing abomasitis in sheep and sometimes cattle, caused by *C. septicum* (rarely *C. sordellii*). It occurs sporadically in young animals in temperate areas of the world. The factors that elicit the disease are largely unknown, but probably implicate local lesions in abomasal mucosa (Uzal et al. 2016).

7.5.3   Histotoxic Clostridial Infections in Reindeer

Infections with histotoxic clostridia are not considered to be of any importance in semi-domesticated or wild reindeer today. However, in Scandinavia at the end of the 19th century, several outbreaks of an apparently contagious disease caused high mortality in reindeer. The disease was called reindeer plague (*pestis tarandi*), and has subsequently been ascribed to *Clostridium septicum* (Skjenneberg and Slagsvold 1968). Rehbinder and Nikander (1999) suggest the reindeer plague was braxy (bradsot), caused by *C. septicum*. The perception that *C. septicum* was the etiological agent in the reindeer plague is widely quoted, but reading of the original sources (Lundgren 1898, Bergman 1901) provides no reason to believe that this is true. The organisms causing the disease are Gram-positive, spore-forming rods; however, they were reported to grow well aerobically, thereby excluding all species of clostridia.

Cases of wound infections caused by histotoxic clostridia probably occur in reindeer as well as in other species. A case of *C. perfringens* myositis after dart injection of sedative was reported by Herron and coworkers (1979).

A case of black disease (necrotizing hepatitis caused by *C. novyi*) was reported in a single reindeer in the UK (Voigt et al. 2009). It was a male yearling forest reindeer (*R. t. fennicus*) imported to Scotland 2 months before it died. Pathological findings included a rapidly decomposing carcass, fibrinous perihepatitis and multifocal liver necrosis. *Clostridium novyi* type B was cultured from the liver, and immunohistochemistry on liver sections showed *C. novyi* in close association with the liver necrosis. The pastures on which the reindeer had been grazing were known to cause infection with liver flukes in other deer species, but no infection of flukes could be shown in the reindeer.

Josefsen and coworkers (2014) provide a review of reindeer necropsies from northern Norway in 1998–2011. In their material of 337 whole carcass necropsies, three cases of clostridial septicemia were identified. All were adult reindeer found dead in late autumn or winter (November to March), without previous signs of disease. One of the animals was observed only a few hours earlier, grazing and behaving normally. Necropsy findings were a foul-smelling, inflated carcass, with widespread bloodstained subcutaneous edema, often also containing gas. The blood was poorly coagulated, swelling of the liver and spleen were variable, and the kidneys were pulpy, sometimes with sub-capsular bleedings. No skin lesions were observed. Cultivation of bacteria revealed pure culture of indole positive clostridia from all organs, identified as *C. sordellii*.

The diagnosis is challenging, as clostridia are common post-mortem invaders of carcasses, and the possibility of diagnosing agonal or post-mortem invaders as causative agents exists. For this reason, Josefsen and coworkers (2014) only diagnosed clostridial septicemia when carcasses were fresh and with pure culture of clostridia from organs. Another five cases with more or less the same history (adult reindeer, November to March, similar necropsy findings), but where the bacteriological results were more variable, were placed in the category "uncertain diagnosis."

Very few reports of similar cases are published. The Swedish wildlife pathologist Karl Borg briefly mention that generalized infections with clostridia occur in moose, with gross findings including swollen spleen, hemorrhages and edema in muscles and severe enteritis (Borg 1978). It may be that such cases occur sporadically in many deer species but that they are seldom reported due to the difficulties in distinguishing between a primary clostridial septicemia and post-mortem growth of the same agents.

7.6 ANTHRAX

Terje D. Josefsen and Torill Mørk

Anthrax is the name of the disease caused by *Bacillus anthracis*, a large, Gram-positive aerobe or facultative anaerobe, spore-forming rod belonging to the family Bacillaceae. Anthrax is a severe disease with worldwide distribution. The disease may affect virtually all mammalian species, including humans; however, the susceptibility to disease varies between species. Ruminants, both domestic and wild, usually develop peracute or acute fatal septicemia, while pigs, horses and humans develop subacute disease (Quinn et al. 2011). Dogs, foxes and carrion feeders like hyenas and jackals are relatively resistant, though not considered immune, to anthrax (Hugh-Jones and de Vos 2002).

7.6.1 ANTHRAX IN REINDEER

Anthrax in reindeer is primarily known from the Siberian Taiga, where huge outbreaks have occurred. "When reindeer were moved into the Taiga during the summer, the disease was such a problem that it was called Siberian or 'Yamal' disease. In some years, over 100,000 animals would die of it." (Gainer and Oksanen 2012, referring to Kolonin 1971).

Vaccination of reindeer against anthrax in Russia was started in 1928, and no epizootics were later reported in areas where vaccination was performed (Nikolaevskii 1961). According to Cherkasskiy (1999), there was a large epizootic in reindeer in 1931, followed by additional cases in 1932 and 1934. Syroechkovskii (1995) claims that there was an additional outbreak of anthrax in reindeer in Russia in 1969, in the areas of Yakutia, Taimyr and Evenkia. Recently, in July 2016, another large outbreak was reported in Yamal–Nenets Autonomous Okrug (Textbox 7.1). The previous known outbreak in this area was in 1941 (Anonymous 2016).

Reports of anthrax in reindeer outside Russia are scarce. There may have been some outbreaks in Scandinavia before 1900 but these are poorly documented. There are no reports concerning anthrax in caribou in North America.

7.6.2   Clinical Signs, Pathology and Diagnosis

The clinical course of anthrax in reindeer is reported by Nikolaevskii (1961) and does not seem to differ from the disease seen in domestic ruminants. A peracute form – or "lightning form," as Nikolaevskii expressed it – was common during large outbreaks. Clinical signs were labored respiration, dark blue mucous membranes, and sometimes bloodstained foam from the mouth and blood from anus. Death occurred within 1–2 hours of the initial clinical signs.

Nikolaevskii (1961) also reported an acute form, with a clinical course that could last for up to 24 hours. The clinical signs were more or less the same; however, the development was slower. Fever and swellings in the skin were additional signs in the more prolonged courses.

The pathological findings in reindeer that succumbed to anthrax resembled the findings in domestic ruminants (Nikolaevskii 1961): The blood was dark and failed to clot, and areas of yellowish subcutaneous edema were often present. Petechial hemorrhages were present on the pleura and on the kidneys beneath the capsule. The liver was congested with blood. A notable finding in reindeer was that, according to Nikolaevskii, the spleen was seldom enlarged. This differs from cattle, in which an enlarged spleen (splenomegaly) is the most significant finding (Valli et al. 2016). Splenomegaly is also reported in white-tailed deer with anthrax (Kellogg and Prestwood 1970), but is less prominent in sheep (Valli et al. 2016), and impala rarely develop splenomegaly (Hugh-Jones and de Vos 2002).

The pathological findings associated with anthrax are not specific, and the diagnosis relies on the finding of numerous typical bacilli with capsules in a stained blood smear, or cultivation of *B. anthracis* from blood.

7.6.3   Epidemiology and Pathogenesis

The key factor in the epidemiology of anthrax is the ability of *B. anthracis* to form spores that are highly resistant to environmental factors. The spores are resistant to extremes in temperature, pH and pressure; they can resist prolonged exposure to chemical disinfectants, ultraviolet light and ionizing radiation, and desiccation (Dragon and Rennie 1995). The spores can survive for many decades in water and soil, and, at the extreme, have been recovered from 200-year-old bones in Kruger National Park in South Africa (de Vos 1998).

The first victim in an outbreak of anthrax is believed to be infected by spores that are either ingested, inhaled or inoculated into wounds. Whether anthrax develops or not depends on the number of infecting spores, the susceptibility of the species and the immunological status of the host. Wild cervids are considered highly susceptible to anthrax (Fasanella 2012), and the large epizootics in Siberia indicate that reindeer are no exception. The spores germinate within the body, the bacteria multiply rapidly and septicemia develops within a short time. After one case has occurred, other animals may be infected with vegetative bacteria from the diseased or dead animal, as well as from spores formed around the carcass. The opening of a carcass by scavengers promotes sporulation, and spores may be found in the soil around an anthrax carcass a long time after the carcass has disappeared (Dragon et al. 2005).

In reindeer, all the large epizootics have occurred in summer. This has been explained by anticipating that the ground of the summer pastures has been heavily contaminated by spores, at least in some spots. Similarly, heavily contaminated spots have been identified in repeated outbreaks of anthrax in bison in Canada (Dragon et al. 1999). Whether reduced resistance (immunosuppression) also contributes to outbreaks of anthrax is not known. In bison, anthrax preferentially occurs in males, and outbreaks coincide with the rut. In reindeer, heat and insect harassment place a considerable stress on them, and this may contribute to reduced resistance to anthrax spores (Gainer and Oksanen 2012). When one or a few reindeer have contracted anthrax from spores in the environment, biting insects, numerous in the Taiga in summer, are anticipated to play a major role in the rapid spread of the disease in the herds, infecting thousands of reindeer within 2–3 weeks (Nikolaevskii 1961).

TEXTBOX 7.1   OUTBREAK OF ANTHRAX IN REINDEER IN SIBERIA, 2016

During July and August 2016, a large outbreak of anthrax took place in Yamal–Nenets Autonomous Okrug in northern Siberia (Russia). The number of dead reindeer was estimated at about 2500 (Anonymous 2016, Goudarzi 2016). Several hundred herders were evacuated from the area (Anonymous 2016). One 12-year-old boy died from the infection, and about 100 more people were hospitalized with the suspicion of anthrax, of which at least 20 cases were confirmed (Goudarzi 2016).

No scientific reports are yet available from this outbreak, but it is suggested that the cause of the outbreak could be thawing of permafrost after an unusually warm summer, which may have led to anthrax spores being shifted to upper soil layers, infecting reindeer during grazing (Goudarzi 2016, Vasilieva 2016). However, another factor is also important: In the outbreak area of Yamal–Nenets, the annual reindeer vaccination program against anthrax was stopped in 2007 (Vasilieva 2016). This decision now appears to have had devastating consequences for the reindeer communities.

During the outbreak, a buffer zone of 110 km was created, with mandatory vaccination and health control of animals and people within the zone. By the middle of August 2016 more than 200,000 reindeer and about 12,000 people were vaccinated, and as many as 700,000 reindeer will be included in a new vaccination program (Anonymous 2016).

7.6.4   TREATMENT AND MANAGEMENT

Vaccination has proved to be very efficient in controlling anthrax in reindeer (Nikolaevskii 1961, Cherkasskiy 1999). *Bacillus anthracis* is sensitive to antibiotics, but treatment is not possible to implement in reindeer husbandry, due both to the usually short clinical course in reindeer, and the practical problems with treating free-ranging animals. Anthrax carcasses should be burned or buried to prevent spread of spores. The carcasses should not be opened for necropsy, as this promotes spore formation. The diagnosis of a suspected carcass must rely on the careful extraction of a small amount of blood that can be used for smear preparation or cultivation.

7.7   MYCOBACTERIAL INFECTIONS

Terje D. Josefsen

7.7.1   TUBERCULOSIS

Tuberculosis is a disease caused by infection with mycobacteria in the *Mycobacterium tuberculosis* complex, comprising species like *Mycobacterium tuberculosis*, *M. bovis*, *M. caprae* and others. By far the most common cause of tuberculosis in cervids is infection with *M. bovis* (Gavier-Widén et al. 2012).

Natural infection with *M. bovis* in reindeer is exceedingly rare compared to other deer species (Palmer et al. 2006), and cases referred to in literature are poorly documented (Banfield 1954, Syroechkovskii 1995). However, experimental infection has shown that reindeer are susceptible to infection with *M. bovis*, developing caseonecrotic granulomas in both local and more distant lymph nodes after tonsillar inoculation (Palmer et al. 2006). The susceptibility to infection was comparable to similar experiments in red deer (*Cervus elaphus*); however, lesions were less severe, fewer and less widely disseminated than seen in white-tailed deer (*Odocoileus virginianus*) (Palmer et al. 2006). Thus, even though the disease is rare, *Rangifer tarandus* ssp. may contract tuberculosis in the same manner as other deer species, with similar development of the disease.

7.7.2 Paratuberculosis

Paratuberculosis (Johne's disease) is a disease caused by infection with *Mycobacterium avium* subsp. *paratuberculosis* (MAP). MAP is an acid-fast, aerobic, non-spore-forming, non-motile rod and is represented by three different strains. Type I and III (the sheep types) are isolated mainly from sheep and goats, though rarely from wildlife. Type II (the cattle type) is primarily isolated from cattle, but has a wider host range, and is the predominant type in wild cervids (Gavier-Widén et al. 2012).

7.7.2.1 Paratuberculosis in General

Paratuberculosis is a chronic granulomatous enteritis, primarily affecting ruminants. The disease occurs worldwide in both domestic and wild ruminants, and it is always fatal. Infection usually occurs orally in neonates or young animals; however, clinical signs do not develop until after 2 years of age in cattle, sheep and goats (Uzal et al. 2016). The onset of clinical disease occurs at a younger age in farmed deer, sometimes as early as 8 months of age (Mackintosh et al. 2004). Infected animals start to shed MAP while they are subclinically affected, and the number of bacteria in faeces increases as the disease progress to the clinical stage.

The main clinical sign is chronic emaciation. Diarrhoea or soft faeces may be present, but this sign shows variation between species and individual animals. Cattle often show intermittent or persistent diarrhoea throughout the clinical stages. Sheep and goat most often have normal faecal pellets until late stages of the disease (Uzal et al. 2016). In farmed red deer (*Cervus elaphus scoticus*), two clinical pictures are seen: a sporadic form with low incidence (1–3% per year) affecting animals in all ages and an "outbreak-form" affecting several young animals (up to 20% of a group) at the same time. Clinical signs are diarrhoea and weight loss in both forms, but while the sporadic cases may last for months, the clinical course during outbreaks may be as short as a few weeks, with profuse diarrhoea (Mackintosh et al. 2004).

7.7.2.2 Paratuberculosis in Reindeer: Occurrence, Clinical Signs and Pathological Findings

The only reported outbreak of paratuberculosis in reindeer husbandry occurred in Yakutia (now Sakha Republic) in Siberia, Russia, from 1954 on, and is widely referred to in scientific literature (Poddoubski 1957, Nikolaevskii 1961, Katic 1961, Skjenneberg and Slagsvold 1968, Syroechkovskii 1995). The disease was believed to be transmitted to reindeer from diseased cattle in the area. Polikarpov (1966) estimated that about 1–1.5% of the herd showed clinical signs and died each year. He also quoted old reindeer breeders telling him that the disease had been known in the area for a long time, and had special local names.

Nikolaevskii (1961) has described clinical signs and pathological findings associated with paratuberculosis in reindeer as follows: The disease typically occurred in reindeer older than 1.5 years, seldom as young as 1 year. Clinical signs were emaciation and diarrhoea. The diarrhoea could be persistent, in which case the disease progressed to death within 2–3 months, or it could disappear for shorter or longer periods and thereby prolong survival. Reindeer with paratuberculosis had normal body temperature but were lethargic, with anemic and sometimes even icteric mucous membranes. In reindeer that died from paratuberculosis, gross pathological changes in the intestines were rarely encountered. Only in cases where the disease had lasted for a prolonged time with recurring bouts of diarrhoea were gross changes in the distal part of the jejunum and ileum observed. These changes consisted of thickening of the intestinal wall and folds in the mucosa, so that the inner surface of the distal jejunum and ileum resembled the cerebral hemispheres in appearance. Mesenteric lymph nodes in young animals were often enlarged, yellowish and dry on the cut surface. Generalized swelling of other lymph nodes, oozing with fluid on the cut surface, was also noted. Some reindeer had edema of the mesentery and an increased amount of fluid in the thoracic cavity.

Besides the Siberian outbreak, paratuberculosis in reindeer is rarely reported. The disease has never been diagnosed in Fennoscandia (Norway, Sweden, Finland), where the total number of reindeer is about 700,000. This is probably due to the low occurrence of paratuberculosis in domestic species in Fennoscandia, particularly in the northern regions which are the main areas for reindeer husbandry. In Iceland, where paratuberculosis is enzootic in domestic ruminants, one single case of paratuberculosis in reindeer was suspected in 1973 (quoted in Fridriksdottir et al. 2000) but was never confirmed. As a paradox, the few reported European cases of reindeer paratuberculosis come from the United Kingdom, with a total stock of about 1500 reindeer. UK veterinary diagnostic laboratories filed two cases in 2004–2014 (Anonymous 2013). Del-Pozo and coworkers (2013) reported an additional case, which is summarized in Textbox 7.2.

TEXTBOX 7.2   PARATUBERCULOSIS IN REINDEER: A CASE REPORT

Del-Pozo and coworkers (2013) reported the following case:

A 4-year-old captive female tundra reindeer (*Rangifer tarandus tarandus*) developed weight loss and diarrhoea that, after some temporary response to treatment, progressively worsened, and the reindeer was euthanized about 5 months after the first veterinary consultation.

Main gross findings at necropsy were faecal soiling of perianal skin, poor body condition, serous exudate in thoracic cavity, and a well demarcated 3.5 cm diameter subpleural nodule in the apical lobe of the right lung. The intestines and mesenteric lymph nodes appeared normal.

Microscopical examination revealed granulomatous enteritis in the distal jejunum and ileum, and a granulomatous lymphadenitis in the mesenteric lymph nodes. Macrophages containing moderate to large numbers of slender acid-fast rods were frequently observed in both the intestines and lymph nodes. In the liver, there were multifocal granulomas with very rare intracellular acid-fast rods.

The subpleural node in the right lung consisted of multifocal to coalescing granulomas with caseous necrosis, resembling tuberculous granulomas caused by *Mycobacterium bovis*. However, PCR examination of tissue from the ileum and lung showed that *Mycobacterium avium* ssp. *paratuberculosis* was the only *Mycobacterium* sp. present in both tissues.

The descriptions given by Nikolaevskii (1961) and Del-Pozo and coworkers (2013) confirm that clinical signs of paratuberculosis in reindeer resemble those of farmed deer, with chronic weight loss and intermittent or persistent diarrhoea as the most prominent signs. Further, Nikolaevskii (1961) report the same two clinical pictures as Mackintosh and coworkers (2004) in farmed red deer: one with profuse diarrhoea and short course, and one with more intermittent diarrhoea and a slower progression of the disease. Both reports confirm that lesions in the intestines and mesenteric lymph nodes are most often not grossly visible, and that a confident diagnosis at necropsy relies on histologic examination combined with methods for detection of the bacteria.

7.7.2.3   Screenings for Paratuberculosis in Reindeer

There are no reports of paratuberculosis as a clinical disease in reindeer in Fennoscandia or in caribou in North America. Thus, prevalence of antibodies against MAP in reindeer in these areas could be expected to be low. In Canada, 76 female boreal caribou (*Rangifer tarandus caribou*) were tested, and one (1.3%) was seropositive (Johnson et al. 2010). In Norway, 325 semi-domesticated reindeer (*Rangifer tarandus tarandus*) were tested, of which 11 (3.4%) were seropositive, while all of 91 wild reindeer were negative (Tryland et al. 2004). The results of these studies could easily be disregarded as false positives, possibly reflecting serological cross reactions between closely related bacteria in the *M. avium* complex. However, a Canadian study (Forde et al. 2012) has provided new insights with respect to MAP in wild caribou. Faecal samples of 561 caribou from all over Arctic Canada, including West Greenland, were examined by PCR and cultivation. Fecal samples from 31 animals (5.5%) were PCR positive. Only one of these was also culture positive, where the bacteria were identified as MAP type II (cattle type). Most of the sampled caribou herds had no contact with domestic ruminants, and paratuberculosis had never been diagnosed in these herds or their surroundings. Finding MAP in the fecal samples of these animals raises new questions about the ecology and epidemiology of MAP and paratuberculosis, and the relationship between the agent and clinical disease.

7.7.2.4   Prevention, Management and Treatment

Most reindeer are free of the disease, and the main strategy must be to prevent introduction of the pathogen into reindeer grazing areas. In infected herds, various combinations of strategies may be used: biosecurity and hygiene to prevent new infections, test-and-cull programs, vaccination and genetic selection (Sweeney et al. 2012). There is no cure for paratuberculosis; however, lifelong medical treatment can alleviate clinical signs (Sweeney et al. 2012). Treatment in reindeer is not reported and is probably rarely indicated.

7.8   TICK-BORNE BACTERIAL INFECTIONS

Terje D. Josefsen

Ticks are vectors for many different pathogens, including bacteria, viruses and protozoa. *Borrelia burgdorferi* s.l., *Rickettsia* sp. and *Anaplasma* sp. are the most common pathogenic bacteria transmitted by ticks. *Borellia burgdorferi* s.l. and *Rickettsia* sp. are not reported to cause disease in reindeer. *Anaplasma* sp. are thus the only bacteria to be considered here.

7.8.1   *ANAPLASMA PHAGOCYTOPHILUM*

Tick-borne fever caused by *Anaplasma phagocytophilum* is the most widespread tick-borne disease in animals in Europe. The infection has been well known in domestic ruminants for decades but is now recognized as infecting a wider range of animals, including humans (Stuen et al. 2013). General clinical signs are high fever, anorexia and dullness. *Anaplasma phagocytophilum* infect phagocytic cells, especially neutrophilic granulocytes, leading to severe neutropenia that predisposes the animal to opportunistic bacterial infections.

Anaplasma phagocytophilum has a worldwide distribution. The major vector is ticks in the genus *Ixodes*; however, the *Ixodes* species differ in different parts of the world. Other tick genera may occasionally harbor *A. phagocytophilum* (Stuen et al. 2013).

A range of different wild cervid species are commonly infected by *A. phagocytophilum*, as shown by a number of screening surveys for *A. phagocytophilum* DNA in blood and tissue (reviewed in Stuen et al. 2013). However, no survey for this agent is reported in reindeer. Furthermore, there are no reports of natural infection with *A. phagocytophilum* in reindeer. However, an experimental infection of reindeer with blood from sheep infected with *A. phagocytophilum* showed that reindeer are susceptible to the infection (Stuen 1996). Details from this report are summarized in Textbox 7.3.

Though not documented, the incidence of infection by *A. phagocytophilum* is probably low in reindeer, since the preferred reindeer habitats are commonly at higher latitudes and in more alpine areas than the tick habitats. Global warming may change this, and tick-borne fever may occur more frequently in reindeer in the future (see also Chapter 15).

TEXTBOX 7.3   *ANAPLASMA PHAGOCYTOPHILUM*: EXPERIMENTAL INFECTION IN REINDEER

Stuen (1996) made the following experimental infection: Two adult female reindeer and one suckling calf (7 weeks old) were inoculated with tick-borne fever infected blood from sheep. Two additional female reindeer served as controls. The inoculated reindeer developed fever 5 days post infection. Maximum body temperature registered was 40.5°C and 40.9°C in the adults and 41.0°C in the calf. Fever (T. >40°C) lasted for 3 and 6 days in the adults, and 12 days in the calf. The infected reindeer were dull and lacked appetite during the febrile period. They developed neutropenia (less than 0.7×10^9 cells/litre) from day 7, reached a minimum of 0.3×10^9 cells/litre on days 8–10, and increased above 0.7×10^9 cells/litre on days 13 to 14. *Anaplasma phagocytophilum* appeared in neutrophils from day 5 post infection, together with the fever, and disappeared on day 21 (last sampling day) in the adult, but still persisted in the calf. The percentage of neutrophils infected reached its peak at days 6 or 7, and this peak ranged from 40 to 84% infected neutrophils.

One of the adult females died suddenly and unexpectedly on day 16 post infection. Two days earlier, this reindeer had developed black diarrhoea (melena) that persisted on day 16. Red urine was observed from this animal the evening before it died. Necropsy findings were severe anaemia, with an enlarged spleen. The content in the lower jejunum was bloodstained, while black in the large intestine. Blood samples taken on days ten, 14 and 16 showed haematocrit dropping from 45 to 10%. The cause of death was concluded to be circulatory failure related to anaemia. The exact mechanism that had caused the anaemia, and possible relation to the experimental infection, was discussed, but could not be firmly established.

7.8.2   ANAPLASMA OVIS

Anaplasma ovis is another tick-borne agent that may cause disease in reindeer. *Anaplasma ovis* is primarily an infection of sheep and goats, and it is highly prevalent in sheep in many Eurasian and southern European countries. It is usually not associated with clinical disease in otherwise healthy animals, though different kinds of stress or simultaneous infections with other agents may cause severe disease (Renneker et al. 2013). *Anaplasma ovis* infects erythrocytes, and infected erythrocytes are removed by the monocyte-macrophage system. This may cause anaemia and icterus (Valli et al. 2016).

Anaplasma ovis was diagnosed as the cause of disease in reindeer in Mongolia in 2004–2006 (Haigh et al. 2008). The disease occurred in the summer of each year. In 2004, 17 reindeer died suddenly at one location. During three field seasons, seven reindeer were observed to be sick, showing one or more of the following clinical signs: fever up to 39.7°C, pale mucous membranes, tachycardia and weight loss. By examination of blood smears, small inclusions resembling *Anaplasma* sp. were observed in erythrocytes. The inclusions were later identified as *A. ovis* by PCR.

7.9   MISCELLANEOUS BACTERIAL INFECTIONS

Terje D. Josefsen and Torill Mørk

7.9.1   COMMON OPPORTUNISTIC INFECTIONS: STAPHYLOCOCCI, STREPTOCOCCI, ESCHERICHIA COLI AND TRUEPERELLA PYOGENES

Some common opportunistic bacteria reside on the skin, in the mucous membranes or in the gastrointestinal tract as bacterial symbionts or commensals. These bacteria may cause infection when normal barriers are damaged either because of mechanical trauma, chemical substances, other pathogens or immunosuppression. The most important bacterial genera/species within this category are *Staphylococcus* spp., *Streptococcus* spp., *Escherichia coli*, and *Trueperella pyogenes* (formerly *Arcanobacterium pyogenes*, *Actinomyces pyogenes*, *Corynebacterium pyogenes*). Examples of such infections are given in Figures 7.7 and 7.8. These opportunistic agents are all common causes of infections in a wide variety of species, and the severity of the infection may vary from a single, well demarcated, clinically insignificant inflammatory focus, to severe generalized pyemia and septicemia. All of these agents are represented in the necropsy material reviewed by Josefsen and coworkers (2014).

Wild or free-ranging reindeer may have a lower incidence of opportunistic infections than domestic ruminants, due to the comparatively clean environment of huge forest or mountain pastures, with a low density of animals. However, crowding may occur in wild reindeer as well. Clausen and coworkers (1980) report infection with *E. coli* in caribou calves (*Rangifer tarandus groenlandicus*) on West Greenland. The disease occurred over several years in the same area, where between 10 and 20% of the calves became ill each year. Calves at the age of 3–8 weeks were affected. The disease started with heavy diarrhoea that disappeared without mortality, after which polyarthritis developed and the calves had problems following their mothers, ending up staggering around on their own. Death occurred 8–10 days after the onset of arthritis. A terminal acute peritonitis was diagnosed in some cases. *Escherichia coli* O55 was isolated as the etiological agent. The infection occurred in an area with particularly lush green pastures. The reindeer gathered in this area and grazed it repeatedly, resulting in heavy faecal contamination of the ground. This was anticipated to be the predisposing factor to the *E. coli* infections in the calves.

7.9.2 SALMONELLA

Subspecies and serotypes of *Salmonella enterica* are well-known pathogens in animals and man. After infection, salmonellae often establish a subclinical infection, localized in the ileum, cecum and colon, with shedding of small numbers of bacteria via the faeces. However, such infections

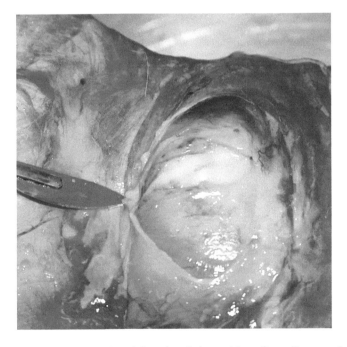

FIGURE 7.7 Staphylococcal abscess. An adult male reindeer with an 8 cm diameter abscess in the dorsal part of the neck. The reindeer died from emaciation in April, and the abscess was revealed during necropsy. *Staphylococcus aureus* was grown in pure culture from the pus. The cause of the abscess is unknown (Photo: Terje D. Josefsen).

FIGURE 7.8   Incidental opportunistic bacterial infection. Reindeer calves kept inside a fence during winter, for protection against predators. One calf in April with swelling of left front leg, dorsal to the fetlock joint. The swelling contained pus. Cultivation revealed mixed flora, dominated by streptococci (Photo: Terje D. Josefsen).

seem to be very rare in reindeer. All attempts to cultivate *Salmonella* from reindeer faeces have turned out negative (Table 7.1). Screening of serum samples for antibodies against *Salmonella* sp. in free-ranging reindeer has most often found a low prevalence (<1%), though a higher seroprevalence has been detected in a research herd that was kept in a restricted area (Table 7.1)

Clinical salmonellosis with enteritis and septicemia has occurred in reindeer kept in zoos (Eulenberger et al. 1985, Schröder 1985) but is not reported in wild or free-ranging reindeer. Kuronen and coworkers (1998) reported *Salmonella* Infantis in two dead 1-week-old reindeer calves from two separate herds in Finland, but they concluded that the *Salmonella* was an incidental finding, unrelated to the cause of death in the calves. *Salmonella* has not since been detected in reindeer in Finland, despite extensive testing of slaughter animals.

7.9.3   *YERSINIA*

Bacteria in the genus *Yersinia* are facultative anaerobe Gram-negative rods, belonging to the family Enterobacteriaceae. Different *Yersinia* spp. are widely distributed in the environment and a common part of the gut microbiota in a wide variety of mammals and birds.

Yersiniosis is caused by pathogenic strains of *Y. pseudotuberculosis* or *Y. enterocolitica*, and may take the form of enterocolitis, mesenteric lymphadenitis and sometimes septicemia. Yersiniosis caused by *Y. pseudotuberculosis* has been reported in cervids (Sanford 1995, Zhang et al. 2008) and has been diagnosed as a cause of enteritis in at least one reindeer in the UK (Foster 2010, Anonymous 2013). Details about the disease in reindeer have not been reported.

Yersina sp. has been detected in 108 (4.8%) of 2243 faecal samples from semi-domesticated reindeer in Norway and Finland (Kemper et al. 2006). *Yersinia enterocolitica* made up 29 (26.9%)

TABLE 7.1

Screening for *Salmonella* in Reindeer (*Rangifer tarandus tarandus*) in Fennoscandia and Iceland

Reindeer, Sampling Circumstances and Country	Sample	Method	Number of Animals	Positive	Reference
Semi-domestic reindeer, dead animals for necropsy, Norway	Faeces	Cultivation after non-selective pre-enrichment and selective enrichment	35	0	Aschfalk et al. 2003
Wild reindeer, sampled during hunting, Norway	Faeces	Cultivation after non-selective pre-enrichment and selective enrichment	153	0	Lillehaug et al. 2005
Semi-domestic reindeer, healthy animals for slaughter, Norway and Finland	Faeces	Cultivation after selective enrichment	2243	0	Kemper et al. 2006
Semi-domestic reindeer, healthy adult animals for slaughter, Norway	Serum	Indirect ELISA	2000	12 (0.6%)	Aschfalk et al. 2000
Semi-domestic reindeer, healthy animals for slaughter, Finland	Serum	Indirect ELISA	802	7 (0.9%)	Aschfalk and Müller 2003
Semi-domestic reindeer, research herd, kept in fenced area	Serum	Indirect ELISA	230	22 (9.6%)	Aschfalk and Müller 2003
Wild reindeer, sampled during hunting, Iceland	Serum	Indirect ELISA	59	2 (3.4%)	Aschfalk and Thórisson 2004

of the isolates; however, they all belonged to biogroup 1A, which is considered non-pathogenic. *Yersinia pseudotuberculosis* was not detected.

7.9.4 *Moraxella*

Moraxella spp. are aerobic, Gram-negative short rods, cocci or diplococci. Most *Moraxella* spp. are commensals of mucous surfaces, which sometimes give rise to opportunistic infection. However, *Moraxella bovis* is regarded as the causative agent of infectious bovine keratoconjunctivitis (IBK). The disease is a common ocular disease in cattle and is manifested by conjunctivitis, lacrimation, edema and corneal ulcer (Angelos et al. 2007, O'Connor et al. 2011).

Infectious keratoconjunctivitis (IKC) is a well-known disease in reindeer as well as in sheep, goats and other deer species (Tryland et al. 2009). The symptoms resemble those described for cattle, and in reindeer mostly calves are affected, though with different disease severities (Figure 7.9). In reindeer, the causative agent of IKC is most likely Cervid herpesvirus 2 (CvHV2) (Tryland et al. 2009); however, *Moraxella* sp. have also been associated with the disease (Romano et al. 2018). In Finland, *Moraxella* sp. was isolated from an outbreak of IKC (Oksanen 1995), as well as from seven of 20 animals with clinical symptoms of the disease in Norway in 2009. Two of these isolates were identified as *Moraxella bovoculi* (Tryland et al. 2009). *M. bovoculi* has also been isolated from the conjunctiva of reindeer without symptoms of eye disease, as well as from the nasal cavity (Mørk, unpublished data).

Moraxella bovoculi has been associated with IBK in cattle; however, the clinical significance seems unclear (Angelos et al. 2007, Angelos 2010, Gould et al. 2013). In an experimental study, reindeer did not develop clinical symptoms of IKC after inoculation with *M. bovoculi* in conjunctiva (Tryland et al. 2017).

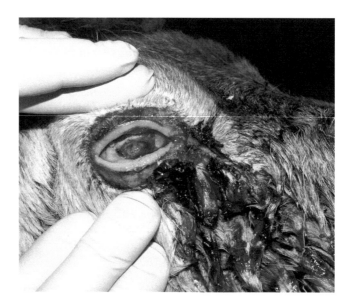

FIGURE 7.9   Eye infection with destruction of the eye. A 1.5-year-old reindeer with destruction of the right eye. Visible changes are swelling of both eyelids, abundant bloody exudate, chronic corneal ulcer and collapsed eye bulb. This and two other similar cases arose during an outbreak of infectious keratoconjunctivitis (IKC) in northern Norway in March 2009. The etiological agent in IKC is believed to be Cervid herpesvirus 2 (Tryland et al. 2009, 2017). Severe eye lesions like those in the photo are probably infected with opportunistic bacteria, such as *Moraxella bovoculi*, *Trueperella pyogenes* and *Pasteurella multocida*. Reindeer with such severe eye lesions are euthanized (Photo: Torill Mørk).

7.9.5   *LISTERIA*

Listeriosis is a disease caused by bacteria in the genus *Listeria*. *Listeria monocytogenes* is by far the major pathogenic species in this genus, though *Listeria ivanovii* has also been associated with disease in animals (Low and Donachie 1997).

Listeria monocytogenes is a Gram-positive, facultative anaerobic rod that is able to grow at temperatures ranging from 3 to 45°C (Low and Donachie 1997). It is widespread in the environment, and resides in soil as a saprophytic bacterium.

Listeria monocytogenes is able to cause disease in many different species, including man. The disease is most often seen in sheep and goats, and is strongly associated with the use of silage feed, due to the occurrence of vast amounts of *Listeria* sp. in silage with high pH (Low and Donachie 1997).

Three major forms of the disease are seen: encephalitis, septicemia and abortion. In reindeer, only the septicemic form has been reported and the disease seems to be rare, as very few cases are reported. Evans and Watson (1987) reported one case in a 2-day-old reindeer calf in a zoo in Michigan, USA, and the authors discuss the possibility that the calf was infected in utero before birth, due to the chronic nature of some lesions. Nyyssönen and coworkers (2006) reported four cases in reindeer calves 5 to 11 days old, all born in the same corral, where silage had been used as feed. Josefsen and coworkers (2014) reported a single case in a reindeer calf, aged between 1 and 3 weeks, born in a corral were the dams were fed pelleted reindeer feed. Whether silage also was used in this corral is unclear, as anamnestic information was scarce.

The reindeer calves with listeriosis were all found dead, and therefore specific clinical signs are not reported. The diagnosis is based on rich growth of *L. monocytogenes* in pure or nearly pure culture from several organs. Histological examination revealed multifocal necrosis, most frequent in the liver but also in lungs and spleen (Evans and Watson 1987, Nyyssönen et al. 2006, Josefsen et al. 2014).

In conclusion, listeriosis is a rare disease in reindeer, only reported in septicemic form in calves below 3 weeks of age, and often in relation to silage feeding.

7.9.6 ERYSIPELOTHRIX RHUSIOPATHIAE

Erysipelothrix rhusiopathiae is a non-motile, facultative anaerobic, Gram-positive rod. It is ubiquitous in nature, commonly present in decaying organic matter, and able to survive for long periods in the environment, including marine habitats (Wang et al. 2010). It is also a common commensal of a wide variety of mammals, birds and fish. The domestic pig is amongst the most important reservoirs of *E. rhusiopathiae*, which is carried in the tonsils of 30–50% of healthy pigs (Wang et al. 2010).

Erysipelothrix rhusiopathiae is a well-known pathogen in mammals and birds, and the cause of major diseases in farm animals, including erysipelas in swine, chronic polyarthritis in lambs and septicemia in poultry, turkeys and ducks. *Erysipelothrix rhusiopathiae* may gain access to the body via the gastrointestinal tract, through skin lesions or via bloodsucking or biting arthropods.

Infections with *E. rhusiopathiae* seem to be rare in cervids and only a few cases are reported in literature: one case of multifocal necrotic hepatitis in a 3-week-old calf of white-tailed deer (*Odocoileus virginianus*) in Iowa, USA (Bruner et al. 1984); one case of thrombotic endocarditis and septicemia in a single roe deer (*Capreolus capreolus*) in Europe (Eskens and Zschöck 1991) and septicemia in three moose (*Alces alces*) in Canada (Campbell et al. 1994).

Infections with *E. rhusiopathiae* are also rare in reindeer. The first report of *E. rhusiopathiae* in reindeer came from Russia in 1939, and this report is later widely used as a reference for *E. rhusiopathiae* infection in reindeer. Stepaykin (1939) examined 28 samples of necrotic material from reindeer with foot lesions, believed to be digital necrobacillosis, with the primary aim of isolating *Bacillus necrophorus* (now *Fusobacterium necrophorum*). However, there was overgrowth of a mixed flora with *E. coli*, *Proteus*, staphylococci and streptococci, and so mice were used as "biological filter": The mixed samples were injected into mice, so that any pathogenic bacteria present in the samples could multiply and kill the mice. In this way, the author happened to isolate *Bacillus erysipelatis suis* (now *E. rhusiopathiae*) from two samples. Stepaykin concluded that *E. rhusiopathiae* contributes to digital necrobacillosis in reindeer. This conclusion was rapidly contradicted by Revnivykh (1939), who pointed out that there was nothing in Stepaykin's study to substantiate a pathogenic role of *E. rhusiopathiae* in digital necrobacillosis. Present knowledge tends to support Revnivykh in that *F. necrophorum* are considered the primary pathogen in digital necrobacillosis. However, other pathogens readily participate in the infection and *E. rhusiopathiae* may be one of those.

Lately, more reports of *E. rhusiopathiae* in reindeer have been published. Josefsen and coworkers (2014) reported one case of severe subcutaneous and intermuscular phlegmonous inflammation of the right gluteal and thigh region of a 10-month-old reindeer calf, kept in a corral for winter-feeding. The cause of death was emaciation, probably due to reduced ability to move. *Erysipelothrix rhusiopathiae* was isolated in pure culture from the phlegmonous inflammation. Schwantje and coworkers (2014) gave a preliminary report of a case of *E. rhusiopathiae* infection in a yearling boreal caribou (*R. t. caribou*) that died without any obvious predisposing factors. Further, *E. rhusiopathiae* has been cultured from several dead caribou, and it is suggested that *E. rhusiopathiae* may be a more common cause of death in caribou and other wildlife species than previously thought. This gained support when *E. rhusiopathiae* was found to be the cause of considerable summer mortality in muskoxen in the years 2010–2013, in the Canadian Arctic (Kutz et al. 2015). Thus, *E. rhusiopathiae* may be an emerging wildlife pathogen in that area (Forde et al. 2016).

7.9.7 ACTINOMYCES BOVIS

Actinomyces bovis is a gram-positive, branching filamentous rod, and the etiological agent of classic "lumpy jaw" in cattle. Lumpy jaw is a severe chronic osteomyelitis, usually affecting the mandible.

Since the term "lumpy jaw" is sometimes used for any condition causing swelling of the mandible, we will here use the term "mandibular actinomycosis."

Actinomyces bovis is part of the normal oral flora of cattle and gains access to the jaw through lesions in the oral mucosa. Mandibular actinomycosis is most common in cattle, but it also occurs in sheep and goats, and occasionally in other animals (Craig et al. 2016). Other bacteria, like *Actinobacillus ligneriesi* and *Trueperella pyogenes*, may also cause mandibular osteomyelitis, representing differential diagnoses of actinomycosis.

So far, actinomycosis has not been diagnosed in reindeer. Mandibular lesions have been observed in barren ground caribou in Canada (Miller et al. 1975) and in the Western Arctic caribou herd of Alaska (Doerr and Dietrich 1979), with frequencies ranging from 0.4 to 7.0%. However, both authors believe that dental abscesses cause most mandibular lesions, and that osteomyelitis due to *Actinomyces* sp. has a very low frequency, if it occurs at all.

Leader-Williams (1980) described mandibular swellings in reindeer (*R. t. tarandus*) on the Antarctic island of South Georgia, where whalers introduced reindeer from Norway in 1911 and 1925. Frequencies ranged from 12 to 28% in reindeer more than 1-year-old. In at least one case, he was able to demonstrate "sulphur granules" (a characteristic sign of actinomycosis) in osteomyelitic lesions, whilst in Gram-stained smears he could demonstrate Gram-positive organisms. However, he was not able to cultivate the bacterium, thus lacking the final evidence for mandibular actinomycosis in reindeer. Leader-Williams later (1982) demonstrated that mandibular swellings occurred more often in reindeer carcasses than in apparently healthy reindeer shot for examination, and therefore he concluded that the disease contributes to mortality in this reindeer population.

In conclusion, it seems probable that mandibular actinomycosis may occur in reindeer, even though no cases are yet documented.

7.9.8   NOCARDIA

Bacteria in the genus *Nocardia* are aerobe, Gram-positive and partially acid-fast rods or branching filaments. *Nocardia* spp. are saprophytes found in soil or decaying vegetation, and are able to infect a wide variety of animals, including man. *Nocardia asteroides* is the main pathogenic species; however, other species may also cause infection.

A case of systemic nocardiosis due to *N. asteroides* has been reported in a 1.5-year-old farmed reindeer (Vemireddi et al. 2007), showing that reindeer may also contract this infection. The clinical course lasted for ten days and included intermittent fever and anaemia. Necropsy showed pyogranulomatous lesions in multiple organs, and indicated a chronic subclinical disease lasting several weeks.

7.9.9   MYCOPLASMA

Hemotropic mycoplasmas (hemoplasmas) are small bacteria without a cell wall, previously known as *Eperythrozoon-* and *Hemobartonella* spp. Hemotropic mycoplasmas parasitize red blood cells of several mammalian species, including man. Infections are usually chronic and subclinical; however, the organisms may cause anaemia, especially during periods of stress or immunosuppression.

Hemotropic mycoplasmas have been characterized in different cervid species all over the world (e.g. Grazziotin et al. 2011, Maggi et al. 2013, Tagawa et al. 2014), and have also been reported from a small reindeer herd kept at a research facility in Iowa, USA (Stoffregen et al. 2006). This report includes one positive sample from a separate reindeer herd in Tennessee, indicating that the finding in Iowa was not a single, sporadic event. The work of Stoffregen and coworkers (2006) was initiated due to recurring episodes of moderate to severe anaemia, for which the isolated *Mycoplasma*, closely related to other ruminant *Mycoplasma* species, was believed to be the main cause. No further studies in North American caribou or Eurasian tundra reindeer have yet been conducted to see if this infection exists in wild or semi-domesticated reindeer, ranging in their natural habitat.

In Norway, outbreaks of severe, often fatal, pneumonia in muskoxen (*Ovibos moschatus*) occurred in 2006 and 2012. The disease was originally diagnosed as pasteurellosis; however, subsequent studies identified *Mycoplasma ovipneumoniae* as the primary cause (Handeland et al. 2014). The organism was introduced to the muskoxen by subclinically infected sheep grazing in the same areas in summer. During this investigation, serum samples from 35 wild reindeer sharing the same pastures were examined, yet all were negative for antibodies against *M. ovipneumoniae*. So far, no evidence exists for mycoplasmas being involved in outbreaks of pneumonia caused by *Pasteurella multocida* in reindeer.

Mycoplasma conjunctivae has been identified as the cause of infectious keratoconjuctivitis in sheep in Norway (Åkerstedt and Hofshagen 2004) as well as in alpine chamois and ibex in Switzerland (Mayer et al. 1996, Tschopp et al. 2005). Infectious keratoconjunctivitis is a recurring problem in reindeer husbandry as well, but studies have indicated that *Mycoplasma* spp. have no role in the etiology of keratoconjunctivitis in reindeer (Evans et al. 2008, Romano et al. 2018), and that the primary and transmissible agent probably is Cervid herpesvirus 2 (Tryland et al. 2009, 2017).

7.9.10 LEPTOSPIRA

Leptospira spp. are Gram-negative, motile, helical bacteria that exploit a wide range of mammals as reservoir hosts, where they colonize the renal tubules (Birtles 2012). The infection is usually chronic and asymptomatic. Leptospiral infection has been reported in red deer (*Cervus elaphus*), fallow deer (*Dama dama*) and roe deer (*Capreolus capreolus*) (Birtles 2012).

Leptospirosis has been detected in reindeer in Russia, according to Syroechkovskii (1995); however, no scientific report about this is available. In Fennoscandia and North America, leptospiral infection or leptospirosis has not been reported in reindeer; however, antibodies to *Leptospira* sp. were found in one of 61 caribou (1.6%) in Alaska (Zarnke 1983).

REFERENCES

Aagnes, T. H., W. Sørmo, and S. D. Mathiesen. 1995. Ruminal microbial digestion in free-living, in captive lichen-fed, and in starved reindeer (*Rangifer tarandus tarandus*) in winter. *Appl. Environ. Microbiol.* 61(2):583–91.

Aalbæk, B., L. Eriksen, R. B. Rimler, P. S. Leifsson, A. Basse, T. Christiansen, and E. Eriksen. 1999. Typing of *Pasteurella multocida* from haemorrhagic septicaemia in Danish fallow deer (*Dama dama*). *APMIS* 107:913–20.

Ågren, E. O., E. Wikström, U. Rockström, M. Westman, and B. Åhman. 2014. Oral necrobacillosis in reindeer calves. Conference poster, available at www.slu.se/Global/externwebben/vh-fak/husdjurens-utfodring -och-vard/Posters/2014/Oral-necrobacillosis-in-reindeer-calves-small.pdf.

Åhman, B., E. Ågren, and E. Wikström. 2013. Health risks associated with corralled calving. Conference poster, available at www.slu.se/Global/externwebben/vh-fak/husdjurens-utfodring-och-vard/Posters /2013/Health_risks_associated_with_corralled_calving_Ahman_Agren_Wikstrom_2013.pdf.

Åkerstedt, J., and M. Hofshagen. 2004. Bacteriological investigation of infectious keratoconjunctivitis in Norwegian sheep. *Acta Vet. Scand.* 45:19–26.

Angelos, J. A. 2010. *Moraxella bovoculi* and infectious bovine keratoconjunctivitis; Cause or coincidence? *Vet. Clin. North Am. Food Anim. Pract.* 26:73–8.

Angelos, J. A., P. Q. Spinks, L. M. Ball, and L. W. George. 2007. *Moraxella bovoculi* sp. nov., isolated from calves with infectious keratoconjunctivitis. *Int. J. Syst. Evol. Microbiol.* 57:789–95.

Anonymous. 2012. Unusual lesions seen in Rankin Inlet caribou herd. Available from: Blog.healthywildlife .ca/unusual-lesions-seen-in-rankin-inlet-caribou-herd/ (accessed 27.01.2016).

Anonymous. 2013. Health and welfare of reindeer. The Veterinary Deer Society (www.vetdeersociety.com/), published 23.08.2013 (accessed 14.04.2016).

Anonymous. 2016. ProMED-mail 26.07–10.10.2016 (collection of different sources, available at www .promedmail.org/) (accessed 13.10.2016).

Åsbakk, K., S. Stuen, H. Hansen, and L. B. Forbes. 1999. A serological survey for brucellosis in reindeer in Finnmark county, northern Norway. *Rangifer* 19:19–24.

Aschfalk, A., N. Kemper, and C. Höller. 2003. Bacteria of pathogenic importance in faeces from cadavers of free-ranging or corralled semi-domestic reindeer in northern Norway. *Vet. Res. Com.* 27:93–100.

Aschfalk, A., S. Laude, and N. Denzin. 2000. Seroprevalence of antibodies to *Salmonella* spp in semidomesticated reindeer in Norway, determined by enzyme-linked immunosorbent assay. *Berl. Münch. Tierärztl. Wschr.* 115:351–4.

Aschfalk, A., and W. Müller. 2003. Unterschiedliche Antikörperprävalenzen für Salmonella spp. in freilebenden und in Gehegen gehaltenen, semidomestizierten Rentieren (*Rangifer tarandus tarandus*) in Finland. *Dtsch. Tierärztl. Wschr.* 110:498–502. (In German, with English abstract).

Aschfalk, A., and S. G. Thórisson. 2004. Seroprevalence of *Salmonella* spp. in wild reindeer (*Rangifer tarandus tarandus*) in Iceland. *Vet. Res. Com.* 28:191–5.

Banfield A. W. F. 1954. Preliminary investigations of the barren-ground caribou. Part II: Life history, ecology and utilization. Canadian Wildlife Service: *Wildl. Manag. Bull. Series 1.* 10B, 1–112.

Bergerud, A. T. 1971. The population dynamics of Newfoundland caribou. *Wildlife Monogr.* No. 25:1–55.

Bergman, A. M. 1901. Rennthierpest und Rennthierpestbacillen. *Zeitschrift für Tiermedizin* 5:241–83, 326–37. (In German).

Bergman, A. M. 1909. Om klöfröta och andra med progressive nekros förlöpande sjukdomar hos ren. *Meddelanden från Kungl. Medicinalstyrelsen*, No. 12:1–40. (In Swedish).

Birtles, R. 2012. Chapter 33: Leptospira infections, in *Infectious Diseases of Wild Mammals and Birds in Europe*, eds. D. Gavier-Widén, J. P. Duff, and A. Meredith. 402–408. Chichester: Wiley-Blackwell.

Borg, K. 1978. *Viltsykdommer* [Wildlife diseases]. Oslo: Landbruksforlaget. (In Norwegian).

Brandt, O. 1914. Pasteurellos hos ren [Pasteurellosis in reindeer]. *Svensk Vet. Tidskr.* 19:379–97, 434–43. (In Swedish).

Brody, J. A., B. Huntley, T. M. Overfiel, and J. Maynard. 1966. Studies of human brucellosis in Alaska. *J. Infect. Dis.* 116:263–9.

Bruner, J. A., R. W. Griffith, J. H. Greve, and R. L. Wood. 1984. *Erysipelothrix rhusiopathiae* serotype 5 isolated from a white-tailed deer in Iowa. *J. Wildl. Dis.* 20(3):235–6.

Campbell, G. D., E. M. Addison, I. K. Barker, and S. Rosendal. 1994. *Erysipelothrix rhusiopathiae*, serotype 17, septicemia in moose (*Alces alces*) from Algonquin Park, Ontario. *J. Wildl. Dis.* 30(3):436–8.

Chan, J., C. Baxter, and W. M. Wenman. 1989. Brucellosis in an inuit child, probably related to caribou meat consumption. *Scand. J. Infect. Dis.* 21:337–8.

Cherkasskiy, B. L. 1999. A national register of historic and contemporary anthrax foci. *J. Appl. Microbiol.* 87:192–5.

Clausen, B., A. Dam, K. Elvestad, H. V. Krogh, and H. Thing. 1980. Summer mortality among caribou calves in West Greenland. *Nord. Vet. Med.* 32:291–300.

Cooper, B. J., and B. A. Valentine. 2016. Muscle and tendon, in *Jubb, Kennedy and Palmer's Pathology of Domestic Animals*, ed. M. Grant Maxie. Vol. 1: 164–249. 6th edition. St. Louis: Elsevier.

Corbel, M. J. 1985. Recent advances in the study of *Brucella* antigens and their serological cross-reactions. *Vet. Bull.* 55:927–42.

Corbel, M. J. 2006. *Brucellosis in Humans and Animals.* eds. M. J. Corbel, S. S. Elberg, and O. Cosivi. Geneva: WHO Press.

Corbel, M. J., and M. Banai. 2005. Genus *Brucella* Meyer and Shaw 1920, in *Bergey's Manual of Systematic Bacteriology*, eds. G. M. Garrity, N. R. Krieg, J. T. Staley, and T. James. Volume Two: The Proteobacteria (Part C). 370–85. New York: Springer.

Corrigan, C., and S. Hanson. 1955. Brucellosis and miliary tuberculosis in an Eskimo woman. *Can. Med. Assoc. J.* 72:217–18.

Craig, L. E, K. E. Dittmer, and K. G. Thompson. 2016. Bones and joints, in *Jubb, Kennedy and Palmer's Pathology of Domestic Animals*, ed. M. Grant Maxie. Vol. 1: 16–163. 6th edition. St. Louis: Elsevier.

Crawford, R. P., J. D. Huber, and B. S. Adams. 1990. Epidemiology and surveilance, in *Animal Brucellosis*, K. Nielsen, and J. R. Duncan (eds.) 131–52. Boca Raton, Florida: CRC Press.

Cullen, J. M., and M. J. Stalker. 2016. Liver and biliary system, in *Jubb, Kennedy and Palmer's Pathology of Domestic Animals*, ed. M. Grant Maxie. Vol. 2: 258–352. 6th edition. St. Louis: Elsevier.

Curry, P. S., B. T. Elkin, M. Campbell et al. 2011. Filter-paper blood samples for ELISA detection of *Brucella* antibodies in caribou. *J. Wildl. Dis.* 47:12–20.

De Alwis, M. C. L., T. G. Wijewardana, A. I. U. Gomis, and A. A. Vipulasiri. 1990. Persistence of the carrier status in hemorrhagic septicaemia (*Pasteurella multocida* serotype 6:B infection) in buffaloes. *Trop. Anim. Health Prod.* 22:185–94.

Del-Pozo, J., S. Girling, J. McLuckie, E. Abbondati, and K. Stevenson. 2013. An unusual presentation of *Mycobacterium avium* ssp. *paratuberculosis* infection in a captive tundra reindeer (*Rangifer tarandus tarandus*). *J. Comp. Pathol.* 149:126–31.

de Vos, V. 1998. The isolation of viable and pathogenic *Bacillus anthracis* organisms from 200-year-old bone fragments from the Kruger National Park. In *Proc. ARC-Onderstepoort OIE International Congress with WHO-Cosponsorship on anthrax, brucellosis, CBPP, clostridial and mycobacterial diseases*, 9–15 August, Berg-en Dal, Kruger National Park. 22–4. Pretoria: Sigma Press.

Dieterich, R. A. 1981. Brucellosis, in *Alaskan Wildlife Diseases*, ed. R. A. Dieterich. 53–8. Fairbanks, Alaska: Institute of Arctic Biology.

Dieterich, R. A., B. L. Deyoe, and J. K. Morton. 1981. Effects of killed *Brucella abortus* strain 45/20 vaccine on reindeer later challenge exposed with *Brucella suis* type 4. *Amer. J. Vet. Res.* 42:131–4.

Dieterich, R. A., J. K. Morton-Dieterich, and B. L. Deyoe. 1979. Observations on reindeer vaccinated with *Brucella melitensis* strain H-38 vaccine and challenged with *Brucella suis* type 4, in *International Reindeer/Caribou Symposium*, 442–8.

Dieterich, R. A., A. Robert, and J. K. Morton. 1987. Effects of live *Brucella abortus* strain 19 vaccine on reindeer later challenge exposed with *Brucella suis* type 4. *Rangifer* 7:33–6.

Dieterich, R. A., A. Robert, and J. K. Morton. 1989. Effects of live *Brucella abortus* strain 19 vaccine on reindeer. *Rangifer* 9:47–50.

Doerr, J. G., and R. A. Dietrich. 1979. Mandibular lesions in the western arctic caribou herd of Alaska. *J. Wildl. Dis.* 15(2):309–18.

Dragon, D. C., D. E. Bader, J. Mitchell, and N. Woollen. 2005. Natural dissemination of *Bacillus anthracis* spores in northern Canada. *Appl. Environ. Microbiol.* 71(3):1610–5.

Dragon, D. C., B. T. Elkin, J. S. Nishi, and T. R. Ellsworth. 1999. A review of anthrax in Canada and implications for research on the disease in northern bison. *J. Appl. Microbiol.* 87:208–13.

Dragon, D. C. and R. P. Rennie. 1995. The ecology of anthrax spores: Tough but not invincible. *Can. Vet. J.* 36:295–301.

Embury-Hyatt, C. K., G. Wobeser, E. Simko, and M. R. Woodbury. 2005. Investigation of a syndrome of sudden death, splenomegaly, and small intestinal hemorrhage in farmed deer. *Can. Vet. J.* 46:702–8.

Eskens, U., and M. Zschöck. 1991. [Eryipelas in the roe deer – a case report]. *Tierarztl. Prax.* 19(1):52–3. (In German, English abstract).

Eulenberger, K., K. F. Schüppel, G. Krische et al. 1985. Beitrag zum Krankheitsgeschehen und zur Narkose bei Cerviden. *Verhandlunsbericht des 27. Internationalen Symposiums über die Erkrankungen der Zootiere*. Akademie-verlag, Berlin: 37–50. [In German].

Evans, A. L., R. F. Bey, J. V. Schoster, J. E. Gaarder, and G. L. Finstad. 2008. Preliminary studies on the etiology of keratoconjunctivitis in reindeer (*Rangifer tarandus tarandus*) calves in Alaska. *J. Wildl. Dis.* 44(4):1051–5.

Evans, M. G., and G. L. Watson. Septicemic listeriosis in a reindeer calf. 1987. *J. Wildl. Dis.* 23:314–7.

Fasanella, A. 2012. Chapter 25: Anthrax, in *Infectious Diseases of Wild Mammals and Birds in Europe*, eds. D. Gavier-Widén, J. P. Duff, and A. Meredith. 265–292. Chichester: Wiley-Blackwell.

Ferguson, M. A. D. 1997. *Rangiferine brucellosis* on Baffin Island. *J. Wildl. Dis.* 33:536–43.

Ferroglio, E. 2012. Chapter 23: Pasteurella infections, in *Infectious Diseases of Wild Mammals and Birds in Europe*, eds. D. Gavier-Widén, J. P. Duff, and A. Meredith. 310–317. Chichester: Wiley-Blackwell.

Forbes, L. B. 1991. Isolates of *Brucella suis* biovar 4 from animals and humans in Canada, 1982–1990. *Can. Vet. J.* 32:686–8.

Forbes, L. B., and S. V. Tessaro. 1993. Transmission of brucellosis from reindeer to cattle. *J. Am. Vet. Med. Assoc.* 203:289–94.

Forde, T., K. Orsel, J. De Buck et al. 2012. Detection of *Mycobacterium avium* subspecies *paratuberculosis* in several herds of arctic caribou (*Rangifer tarandus* ssp.). *J. Wild. Dis.* 48(4):918–24.

Forde, T., K. Orsel, R. N. Zadoks et al. 2016. Bacterial genomics reveal the complex epidemiology of an emerging pathogen in arctic and boreal ungulates. *Front. Microbiol.* doi: 10.3389/fmicb.2016.01759.

Foster, A. 2010. Common conditions of reindeer. 2010. *In Practice* 32:462–7.

Fridriksdottir, V., E. Gunnarsson, S. Sigurdarson, and K. B. Gudmundsdottir. 2000. Paratuberculosis in Iceland: Epidemiology and control measures, past and present. *Vet. Microbiol.* 77:263–7.

Gainer, R., and A. Oksanen. 2012. Anthrax and the Taiga. *Can. Vet. J.* 53:1123–5.

Gavier-Widén, D., M. Chambers, C. Gortázar et al. 2012. Chapter 20: Mycobacteria infections, in *Infectious Diseases of Wild Mammals and Birds in Europe*, eds. D. Gavier-Widén, J. P. Duff, and A. Meredith. 265–92. Chichester: Wiley-Blackwell.

Godkin G. F. 1986. The reindeer industry in Canada. *Can. Vet. J.* 27(12):488–90.

Goudarzi S. 2016. What lies beneath. *Sci. Am.*, 315:11–12 doi: 10.1038/scientificamerican1116-11.

Gould S., R. Dewell, K. Toffelmire et al. 2013. Randomized blinded challenge study to assess association between *Moraxella bovoculi* and infectious bovine keratoconjunctivitis in dairy calves. *Vet. Microbiol.* 164:108–15.

Greenberg, L., and J. D. Blake. 1958. An immunological study of the Canadian Eskimo. *Can. Med. Assoc. J.* 78:27–31.

Greenstone, G. 1993. Brucellosis – a medical rarity that used to be common in Canada. *Can. Med. Assoc. J.* 148:1612–3.

Grazziotin, A. L., A. P. Santos, A. M. S. Guimaraes et al. 2011. *Mycoplasma ovis* in captive cervids: Prevalence molecular characterization and phylogeny. *Vet. Microbiol.* 152:415–19.

Haigh J. C., V. Gerwing, J. Erdenebaatar, and J. E. Hill. 2008. A novel clinical syndrome and detection of *Anaplasma ovis* in Mongolian reindeer (*Rangifer tarandus*). *J. Wildl. Dis.* 44(3):569–77.

Handeland, K. 2012. Fusobacterium necrophorum infection, in *Infectious Diseases of Wild Mammals and Birds in Europe*, eds. D. Gavier-Widén, J. P. Duff, and A. Meredith. 428–30. Chichester: Wiley-Blackwell.

Handeland, K. 2014. Europas siste villrein – forvaltning, helsestatus og fremtidige utfordringer [Wild reindeer in Europa – management, health and future challenges]. *Norsk Vet. Tidskr.* 126(2):250–7. (In Norwegian).

Handeland, K., M. Boye, B. Bergsjø et al. 2010. Digital necrobacillosis in Norwegian wild tundra reindeer (*Rangifer tarandus tarandus*). *J. Comp. Path.* 143:29–38.

Handeland, K., T. Tengs, B. Kokotovic et al. 2014. *Mycoplasma ovipneumoniae* – a primary cause of severe pneumonia epizootics in the Norwegian muskox (*Ovibos moschatus*) population. *Plos One.* 9(9): e106116.

Hargis A. M., and P. E. Ginn. 2012. Ch. 17: The integument. In *Pathologic Basis of Veterinary Disease*, eds. J. F. Zachary and M. D. McGavin. 972–1084. 5th edition. St. Louis: Elsevier Mosby.

Herron, A. J., R. H. Garman, R. Baitchman, and A. L. Kraus. 1979. Clostridial myositis in a reindeer (*Rangifer tarandus*): A case report. *J. Zoo Anim. Med.* 10(1):31–4.

Horne H. 1898. Renens klovsyge [Reindeer foot disease]. *Norsk Vet. Tidskr.* 10:97–110. (In Norwegian).

Horne, H. 1915. Rensyke (Hæmorrhagisk septikæmi hos ren). *Norsk Vet. Tidskr.* 27:41–7, 73–84, 105–19. (In Norwegian).

Hotchkiss E. J., M. P. Gadleish, K. Willoughby et al. 2010. Prevalence of *Pasteurella multocida* and other respiratory pathogens in the nasal tract of Scottish calves. *Vet. Rec.* 167:555–60.

Hueffer, K., Parkinson, A. J., Gerlach, R., Berner, J., 2013. Zoonotic infections in Alaska: Disease prevalence, potential impact of climate change and recommended actions for earlier disease detection, research, prevention and control. *Int. J. Circumpolar Health* doi: dx.doi.org/10.3402/ijch.v72i0.19562.

Hugh-Jones, M. E., and V. de Vos. 2002. Anthrax and wildlife. *Rev. Sci. Tech. Off. Int. Epiz.* 21(2):359–83.

Jang S. S. and D. C. Hirsh. 1994. Characterization, distribution and microbiological associations of *Fusobacterium* spp. in clinical specimens of animal origin. *J. Clin. Microbiol.* 32(2):384–7.

Johnson, D., N. J. Harms, N. C. Larter et al. 2010. Serum biochemistry, serology, and parasitology of boreal caribou (*Rangifer tarandus caribou*) in the Northwest Territories, Canada. *J. Wildl. Dis.* 46(4):1096–107.

Josefsen, T. D., T. Mørk, K. K. Sørensen et al. 2014. Funn ved obduksjon og undersøkelse av organer fra rein 1998–2011 [Findings at necropsy or by examination of organs from reindeer in Norway 1998–2011]. *Norsk Vet. Tidskr.* 126:174–83. (In Norwegian with English abstract).

Katic, I. 1961. Paratuberculosis (Johne's disease) with special reference to captive wild animals. *Nord. Vet. Med.* 13:205–14.

Kellogg, F. E., and A. K. Prestwood. 1970. Anthrax epizootic in white-tailed deer. *J. Wildl. Dis.* 6:226–8.

Kemper, N., A. Aschfalk, and C. Höller. 2006. *Campylobacter* spp., *Enterococcus* spp., *Escherichia coli*, *Salmonella* spp., *Yersinia* spp., and *Cryptosporidium* oocysts in semi-domesticated reindeer (*Rangifer tarandus tarandus*) in northern Finland and Norway. *Acta Vet. Scand.* 48:7 doi: 10.1186/1751-0147-48-7.

Kolonin, G. V. 1971. [Evolution of anthrax. II. History of the spread of the disease and formation of the nosoareal]. *Zh. Mikrobiol. Epidemiol. Immunobiol.* 48(1):118–22. (In Russian).

Korbi, K. A. 1982. Sykdomsregistrering i reinkjøttkontrollen [Disease registration during reindeer postmortem inspection at slaughter]. *Norsk Vet. Tidskr.* 94:45–9. (In Norwegian).

Kummeneje, K. 1976. Pasteurellosis in reindeer in northern Norway. *Acta Vet. Scand.* 17:488–94.

Kummeneje K., and G. Bakken. 1973. Enterotoxaemia in reindeer. *Nord. Vet. Med.* 25:196–202.

Kuronen, H., V. Hirvelä-Koski, and M. Nylund. 1998. Salmonellaa poronvasoissa [salmonella in reindeer calves]. *Poromies.* 4–5:54 (In Finnish).

Kutz, S., T. Bollinger, M. Branigan et al. 2015. *Erysipelothrix rhusiopathiae* associated with recent widespread muskox mortalities in the Canadian Arctic. *Can. Vet. J.* 56:560–3.

Lantis, M. 1950. The reindeer industry in Alaska. *Arctic* 3:27–44.

Leader-Williams, N. 1980. Dental abnormalities and mandibular swellings in South Georgia reindeer. *J. Comp. Pathol.* 90:315–30.

Leader-Williams, N. 1982. Relationship between a disease, host density and mortality in a free-living deer population. *J. Anim. Ecol.* 51(1):235–40.

Leighton, A. F. 2001. *Fusobacterium necrophorum* infection, in *Infectious Diseases of Wild Mammals*, eds. E. S. Williams, and I. K. Barker. 493–496. 3rd edition. Ames, Iowa: Iowa State University Press.

Lillehaug, A., B. Bergsjø, J. Schau et al. 2005. *Campylobacter* spp., *Salmonella* spp., verocytotoxic *Escherichia coli*, and antibiotic resistance in indicator organisms in wild cervids. *Acta Vet. Scand.* 46:23–32.

Linné, C. 1732. *Iter Laponicum.*

Low, J. C., and W. Donachie. 1997. A review of *Listeria monocytogenes* and listeriosis. *The Vet. J.* 153:9–29

Lundgren, J. 1898. Die Rennthierpest. *Zeitschrift für Thiermedicin* 2:401–17. (In German).

Mackintosh, C. G., G. W. de Lisle, D. M. Collins, and J. F. T. Griffin. 2004. Mycobacterial diseases of deer. *New Z. Vet. J.* 52(4):163–74.

Maggi, R. G., M. C. Chitwood, S. Kennedy-Stoskopf, C. S. DePerno. 2013. Novel hemotropic *Mycoplasma* species in white-tailed deer (*Odocoileus virginianus*). *Comp. Immunol. Microbiol. Inf. Dis.* 36:607–11.

Magnusson, H. 1913. Om pasteurellos hos ren jämte ett bidrag till kännedomen om pasteurellans biologiska egenskaper. [Pasteurellosis in reindeer, with a contribution on the biological characteristics of Pasteurella]. *Skand. Vet. Tidsskr.* 4:127–34, 159–84.

Matas, M., and C. Corrigan. 1953. Brucellosis in an Eskimo boy. *Can. Med. Assoc. J.* 69:531.

Mayer, D., J. Nicolet, M. Giacometti, M. Schmitt, T. Wahli, and W. Meier. 1996. Isolation of *Mycoplasma conjunctivae* from conjunctival swabs of alpine ibex (*Capra ibex ibex*) affected with infectious keratoconjunctivitis. *Zentralbl. Veterinärmed.* B 43(3):155–61.

Melnik N. V., S. V. Kriukov, P. P. Rakhmanin, B. V. Solovev, A. A. Plokhova, I. D. Karavaev, and I. N. Semenova. 2009. An effective drug developed against necrobacillosis. *Rangifer*, Special Issue No. 18:55 (Conference abstract).

Meyer, M. E. 1966. Identification and virulence studies of *Brucella* strains isolated from Eskimos and reindeer in Alaska, Canada, and Russia. *Amer. J. Vet. Res.* 27:353–8.

Miller, F. L., A. J. Cawley, L. P. E. Choquette, and E. Broughton. 1975. Radiographic examination of mandibular lesions in barren-ground caribou. *J. Wildl. Dis.* 11(4):465–70.

Monrad, J., A. A. Kassuku, P. Nansen, and P. Willeberg. 1983. An epidemiological study of foot rot in pastured cattle. *Acta Vet. Scand.* 24:403–17.

Morton, J. K. 1981. Necrobacillosis, in *Alaskan Wildlife Diseases*, ed. R. A. Dieterich. Fairbanks, Alaska: University of Alaska.

Morton, J. K. 1990. Laboratory and field trials of killed *Brucella suis* type 4 vaccine in reindeer. *Rangifer* 10 (Special Issue No. 3):351 (Conference abstract).

Mørk, T., M. Sunde, and T. D. Josefsen. 2014. Bakterieinfeksjoner hos rein [Bacterial infections in reindeer]. *Norsk Vet. Tidsskr.* 126(2):222–8. (In Norwegian, with English abstract).

Neiland, K. A., J. A. King, B. E. Huntley, and R. O. Skoog. 1968. The diseases and parasites of Alaskan wildlife populations. Part I: Some observations on brucellosis in caribou. *J. Wildl. Dis.* 4:27–36.

Nikolaevskii, L. D. 1961. Chapter 8: Diseases in reindeer, in *Reindeer Husbandry*, ed. P. S. Zhigunov. 230–93. 2nd edition. Moscow. Translated from Russian by Israel Program for Scientific Translations, Jerusalem, 1968.

Nordkvist, M. 1967. Nekrobacillos hos renar [Necrobacillosis in reindeer]. *Svensk Vetr. Tidsskr.* 19:303 (In Swedish).

Nordkvist, M., and K. A. Karlsson. 1962. Epizootisk förlöpande infektion med *Pasteurella multocida* hos ren. [*Pasteurella multocida* epizooty in reindeer]. *Nord. Vet. Med.* 14:1–15. (In Swedish).

Nymo, I. H., J. Godfroid, K. Åsbakk et al. 2013. A protein A/G indirect enzyme-linked immunosorbent assay for the detection of anti-*Brucella* antibodies in Arctic wildlife. *J. Vet. Diagn. Invest.* 25:369–75.

Nyyssönen, T., V. Hirvelä-Koski, H. Norberg, and M. Nieminen. 2006. Septicaemic listeriosis in reindeer calves – a case report. *Rangifer* 26:25–8

O'Connor, A. M., H. G. Shen, C. Wang, and T. Opriessnig. 2011. Descriptive epidemiology of *Moraxella bovis*, *Moraxella bovoculi* and *Moraxella ovis* in beef calves with naturally occurring infectious bovine keratoconjunctivitis (pinkeye). *Vet. Microbiol.* 155:374–80.

Oksanen, A. 1995. Keratoconjunctivitis in corralled reindeer. Poster abstract from the 7th Nordic Workshop on Reindeer Research, Tromsø 22–23 Sept. 1993. *Rangifer* Report No. 1:50.

O'Reilly, K., and L. B. Forbes. 1995. *Brucellosis in the Circumpolar Arctic.* Vol. 1, 1st edition. Ottawa: Canadian Polar Commission.

Palmer, M. V., W. R. Waters, T. C. Thacker, W. C. Stoffregen, and B. V. Thomsen. 2006. Experimentally induced infection of reindeer (*Rangifer tarandus*) with *Mycobacterium bovis. J. Vet. Diagn. Invest.* 18:52–60.

Pappas, G., P. Papadimitriou, N. Akritidis, L. Christou, and E. V. Tsianos. 2006. The new global map of human brucellosis. *Lancet Infect. Dis.* 6:91–9.

Poddoubski, I. V. 1957. La paratuberculose. *Bull. Off. Int. Epizoot.* 48:469–76. (In French).

Poester, P., K. Nielsen, L. E. Samartino, and W. L. Yu. 2010. Diagnosis of brucellosis. *The Open Vet. Sci. J.* 4:46–60.

Polikarpov, V. A. 1966. [Diagnosis of paratuberculosis in reindeer]. *Veterinariia* 43:29–32. (In Russian).

Quinn, P. J., B. K. Markey, F. C. Leonard et al. 2011. *Veterinary Microbiology and Microbial Disease.* 2nd edition. Oxford: Wiley-Blackwell.

Qvigstad, J. 1941. Den tamme rens sykdommer [Diseases of domestic reindeer]. *Tromsø Museums Årshefter. Naturhistorisk avdeling* 59(1):1–56. (In Norwegian).

Radostits, O. M., C. C. Gay, D. C. Blood, and K. W. Hinchcliff. 2000. *Veterinary Medicine: A Textbook of the Diseases of Cattle, Sheep, Pigs, Goats and Horses.* 867–882. London: Elsevier Limited.

Rausch, R. L., and B. E. Huntley. 1978. Brucellosis in reindeer, *Rangifer tarandus* L., inoculated experimentally with *Brucella suis*, type 4. *Can. J. Microbiol.* 24:129–35.

Rehbinder C. and S. Nikander. 1999. *Ren och rensjukdomar.* Uppsala: Studentlitteratur. (In Swedish).

Renneker, S., J. Abdo, D. E. A. Salih et al. 2013. Can *Anaplasma ovis* in small ruminants be neglected any longer? *Transbound. Emerg. Dis.* 60 (Suppl. 2):105–12.

Revnivykh, A. G. 1939. [Occurrence of *Bac. erysipelatis suis* in reindeer]. *Sovyetsk. Vet.* 16(8):76–7 (In Russian).

Romano, J. S., T. Mørk, S. Laaksonen et al. 2018. Infectious keratoconjunctivitis in semi-domesticated Eurasian tundra reindeer (*Rangifer tarandus tarandus*): Microbiological study of clinically affected and unaffected animals with special reference to cervid herpesvirus 2. *BMC Vet. Res.* 14(1):15 doi: 10.1186 /s12917-018-1338-y.

Sanford, S. E. 1995. Outbreak of yersiniosis caused by *Yersinia pseudotuberculosis* in farmed cervids. *J. Vet. Diagn. Invest.* 7:78–81.

Schröder, H. D. 1985. Beitrag zu den Infektionskrankheiten der Cerviden. *Verhandlunsbericht des 27. Internationalen Symposiums über die Erkrankungen der Zootiere.* 51–54. Akademie-verlag, Berlin: (In German).

Schwantje, H., B. J. Macbeth, S. Kutz, and B. Elkin. 2014. British Colombia Boreal Caribou Health Program. Progress report year 1. (www.bcogris.ca/sites/default/files/bcip-2014-05-bc-boreal-caribou-health-program -year-end-report-28201429-ver-1c.pdf) (accessed 17.05.2016).

Sellers, R. A., P. Valkenburg, R. C. Squibb, B. W. Dale, and R. L. Zarnke. 2003. Natality and calf mortality of the Northern Alaska Peninsula and Southern Alaska Peninsula caribou herds. *Rangifer*, Special Issue No. 14:161–6.

Shivachandra, S. B., K. N. Viswas, and A. A. Kumar. 2011. A review of hemorrhagic septicemia in cattle and buffalo. *Anim. Health Res. Rev.* 12(1):67–82.

Sipos, W., L. Fischer, M. Schindler, and F. Schmoll. 2003. Genotyping of *Clostridium perfringens* isolated from domestic and exotic ruminants and swine. *J. Vet. Med. B*, 50:360–2.

Skjenneberg, S. 1957. Sykdom på reinkalver i Porsangerdistriktet våren 1956 [Disease in reindeer calves in the Porsanger district spring 1956]. *Medlemsblad Nor. Vet. Foren.* 9:153–9. (In Norwegian).

Skjenneberg, S. and L. Slagsvold. 1968. *Reindriften og dens naturgrunnlag.* Oslo: Universitetsforlaget. (In Norwegian. English translation: Eds. C. M. Anderson and J. R. Luick. 1979. *Reindeer Husbandry and its Ecological Principles.* Juneau, Alaska: U.S. Department of the Interior Bureau of Indian Affairs.)

Smith, H., A. E. Williams, J. H. Pearce et al. 1962. Foetal erythritol: A cause of the localization of *Brucella abortus* in bovine contagious abortion. *Nature* 193:47–9.

Stepaykin, P. P. 1939. [Occurrence of *B. erysipelatis suis* in reindeer]. *Sovyetsk. Vet.* 16(5):52–3. (In Russian).

Stoffregen, W. C., D. P. Alt, M. V. Palmer et al. 2006. Identification of a haemomycoplasma species in anemic reindeer (*Rangifer tarandus*). *J. Wildl. Dis.* 42(2):249–58.

Stuen, S. 1996. Experimental tick-borne fever infection in reindeer (*Rangifer tarandus tarandus*). *Vet. Rec.* 138:595–6.

Stuen, S., E. G. Granquist, and C. Silaghi. 2013. *Anaplasma phagocytophilum* – a widespread multi-host pathogen with highly adaptive strategies. *Front. Cell. Infect. Microbiol.* doi: 10.3389/fcimb.2013.00031.

Sweeney, R. W., M. T. Collins, A. P. Koets, S. M. McGuirk, and A. J. Roussel. 2012. Paratuberculosis (Johne's disease) in cattle and other susceptible species. *J. Vet. Intern. Med.* 26:1239–50.

Syroechkovskii, E. E. 1995. *Wild Reindeer.* Lebanon: Science Publishers Inc.

Tadepalli, S., S. K. Narayanan, G. C. Stewart, M. M. Chengappa, and T. G. Nagaraja. 2009. *Fusobacterium necrophorum*: A ruminal bacterium that invades liver to cause abscesses in cattle. *Anaerobe* 15:36–43.

Tagawa, M., K. Matsumoto, N. Yokoyama, and H. Inokuma. 2014. Prevalence and molecular analyses of hemotropic *Mycoplasma* spp. (hemoplasmas) detected in sika deer (*Cervus nippon yesoensis*) in Japan. *J. Vet. Med. Sci.* 76(3):401–7.

Tan, Z. L., T. G. Nagaraja, M. M. Chengappa. 1994. Selective enumeration of *Fusobacterium necrophorum* from the bovine rumen. *Appl. Environ. Microbiol.* 60(4):1387–9.

Tessaro, S. V., and L. B. Forbes. 1986. *Brucella suis* biotype 4: A case of granulomatous nephritis in a barren ground caribou (*Rangifer tarandus groenlandicus* L.) with a review of the distribution of *Rangiferine brucellosis* in Canada. *J. Wildl. Dis.* 22:479–83.

Toshach, S. 1963. Brucellosis in the Canadian Arctic. *Can. J. Public Health* 54:271–5.

Tryland, M., C. G. das Neves, M. Sunde, and T. Mørk. 2009. Cervid herpesvirus 2, the primary agent in an outbreak of infectious keratoconjunctivitis in semidomesticated reindeer. *J. Clin. Microbiol.* 47(11):3707–13.

Tryland, M., I. Olsen, T. Vikøren et al. 2004. Serologic survey for antibodies against *Mycobacterium avium* subsp. *paratuberculosis* in free-ranging cervids from Norway. *J. Wildl. Dis.* 40(1):32–41.

Tryland, M., J. S. Romano, N. Marcin et al. 2017. Cervid herpesvirus 2 and not *Moraxella bovoculi* caused keratoconjunctivitis in experimentally inoculated semi-domesticated Eurasian tundra reindeer. *Acta Vet. Scand.* 59:23 doi: 10.1186/s13028-017-0291-2.

Tschopp, R., J. Frey, L. Zimmermann, and M. Giacometti. 2005. Outbreaks of infectious keratoconjunctivitis in alpine chamois and ibex in Switzerland between 2001 and 2003. *Vet. Rec.* 157:13–8.

Uzal, F. A., B. L. Plattner, and J. M. Hostetter. 2016. Alimentary system, in *Jubb, Kennedy and Palmer's Pathology of Domestic Animals*, ed. M. Grant Maxie. Vol. 2: 1–257. 6th edition. St. Louis: Elsevier.

Valkenburg, P., R. A. Sellers, R. C. Squibb et al. 2003. Population dynamics of caribou herds in southwestern Alaska. *Rangifer*, Special Issue No. 14:131–42.

Valli, V. E. O., M. Kiupel, D. Bienzle, and R. D. Wood. 2016. Chapter 2: Hematopoietic system, in *Jubb, Kennedy and Palmer's Pathology of Domestic Animals*, ed. M. Grant Maxie. Vol. 3: 102–268. 6th edition. St. Louis: Elsevier.

Vasilieva, T. 2016. If you're left without reindeer, there is nothing else. Greenpeace International, blogpost available at www.greenpeace.org/international/en/news/Blogs/makingwaves/russia-anthrax-reindeer -indigenous/blog/57511/ (accessed 13.10.2016).

Vemireddi, V., A. Sharma, C. C. Wu, and T. L. Lin. 2007. Systemic nocardiosis in a reindeer (*Rangifer tarandus tarandus*). *J. Vet. Diagn. Invest.* 19:326–9.

Voigt, K., M. P. Dagleish, J. Finlayson, G. Beresford, and G. Foster. 2009. Black disease in a forest reindeer (*Rangifer tarandus fennicus*). *Vet. Rec.* 165:352–3.

von Bargen, K., J. P. Gorvel, and S. P. Salcedo. 2012. Internal affairs: Investigating the *Brucella* intracellular lifestyle. *Fems Microbiol. Rev.* 36:533–62.

Wang, Q., B. J. Chang, and T. V. Riley. 2010. *Erysipelothrix rhusiopathiae*. *Vet. Microbiol.* 140:405–17.

Wilson, B. A., and M. Ho. 2013. *Pasteurella multocida*: From zoonosis to cellular microbiology. *Clin. Microbiol. Rev.* 26:631–55.

Woolington J. D., and S. Machida. 2001. Caribou management report, units 9B, 17, 18 south, 19A and 19B. In *Caribou Management Report of Survey-Inventory Activities 1 July 1998–30 June 2000*, ed. C. Healy. 23–32. Project 3.0. Juneau, Alaska: Alaska Department of Fish and Game.

Zarnke, R. L. 1983. Serologic survey for selected microbial pathogens in Alaskan wildlife. *J. Wildl. Dis.* 19(4):324–9.

Zarnke, R. L., J. M. ver Hoef, and R. A. deLong. 2006. Geographic pattern of serum antibody prevalence for *Brucella* spp. in caribou, grizzly bears, and wolves from Alaska, 1975–1998. *J. Wildl. Dis.* 42:570–7.

Zhang, S., Z. Zhang, S. Liu, W. Bingham, and F. Wilson. 2008. Fatal yersiniosis in farmed deer caused by *Yersinia pseudotuberculosis* serotype O:3 encoding mannosyltransferase-like protein WbyK. *J. Vet. Diagn. Invest.* 20:356–9.

Zheludkov, M. M., and L. E. Tsirelson. 2010. Reservoirs of *Brucella* infection in nature. *Biol. Bull.* 37:709–15.

8 Viral Infections and Diseases

Morten Tryland, Carlos G. das Neves, Jörn Klein,
Torill Mørk, Maria Hautaniemi and Jonas J. Wensman

CONTENTS

What is a virus and what is a viral infection? Whether or not viruses represent a life form may be a question of a philosophical nature, but viruses in general have a tremendous replication potential when the conditions are right. In contrast to bacteria, they are not cells and they do not generate energy, and thus they are dependent on being inside a host cell to replicate. Their ability to survive, or stay infective, in the environment outside host cells over time is thus, in general, limited, although some viruses are more robust than others. Every virus family has its specific strategy for replication. When they replicate, they basically take over the metabolic machinery of the host cell, and reprogram it to produce thousands of copies of the virus. Most viral infections are asymptomatic, meaning that in spite of the fact that the virus may replicate and produce many progeny (virus particles), the host copes with the infection and manages to balance and compensate, and can often clear the viral infection. However, depending on the type of virus, the infected host and cell type, immune response and many other factors, some viral infections results in disease: the virus impacts the host cells to such a degree that normal functions cannot be sustained. If many cells are infected, or the whole organ, such as the lungs, the liver or the kidneys, the infection may introduce a deficiency of the normal function of the organ, producing disease symptoms, such as difficult breathing, or liver or kidney deficiency. Viruses also have different strategies for transmission, and whereas some viruses establish chronic infections, latency and integration of their genome in the genome of their host, such as HIV and herpesviruses, other viruses have a "hit and run" strategy, with a quick spread to new hosts, even before the onset of clinical symptoms, as is the case for foot and mouth disease virus and others. Based on the nature of the virus and the host, different strategies are needed to be able to avoid, detect, monitor and control viral infections. Viruses are not susceptible to antibiotics used against bacterial diseases, and there are only a few anti-viral drugs against a very restricted repertoire of viruses on the market, such as HIV and herpesviruses. These drugs are mostly used in human medicine, although some anti-viral treatments also have been investigated for domestic animals.

We will go through the most relevant viral infections in semi-domesticated reindeer and caribou and review their epidemiology, including transmission to other animals and man, diagnosis, clinical signs, treatment, and significance for individuals, herds and populations. We will also mention a few viral infections that might affect reindeer in the future.

8.1   *PARAPOXVIRUS* INFECTIONS AND CONTAGIOUS ECTHYMA

Morten Tryland and Maria Hautaniemi

Parapoxviruses cause proliferative processes in the skin and mucosal membranes (nose and mouth) of reindeer, called contagious ecthyma (contagious pustular dermatitis, contagious stomatitis, "scabby mouth").

8.1.1   AETIOLOGY

Genus *Parapoxvirus* is one of nine genera in the subfamily *Chordopoxvirinae*, of the *Poxviridae* family (Skinner et al., 2012). Members of the same genus have similar morphology and host range, and are genetically and antigenically related (Figure 8.1).

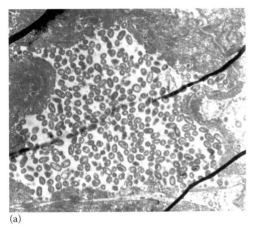
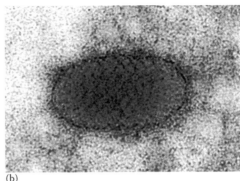

(a)   (b)

FIGURE 8.1   (a) Electronmicroscopy (transmission) of *parapoxvirus* particles in a section of a contagious ecthyma lesion from the lip of a semi-domesticated reindeer. (b) Electronmicroscopy (negative stain) showing the typical surface tubules covering an Orf virus particle, the type virus of genus *Parapoxvirus* (Photo: Morten Tryland).

Like other poxviruses, parapoxviruses are large (140–170 nm wide, 220–300 nm long), double-stranded DNA viruses, but in contrast to all other DNA viruses, poxviruses have an entirely cytoplasmic life cycle. Recognized members of genus *Parapoxvirus* currently consist of *Orf virus* (ORFV), *Bovine papular stomatitis virus* (BPSV), *Pseudocowpoxvirus* (PCPV) and *Parapoxvirus of red deer in New Zealand* (PVNZ) (Skinner et al., 2012). In addition, several tentative members of the genus have been suggested, such as Sealpox virus and a parapoxvirus of California sea lions (Sea Lion Poxvirus-1), Auzduk disease virus, Camel contagious ecthyma virus and Chamois contagious ecthyma virus. Traditionally, the classification of parapoxviruses has been based on natural host range and pathology, but recently also on molecular methods, such as restriction enzyme analyses, hybridization, DNA sequencing and phylogenetic analysis.

The disease contagious ecthyma has been diagnosed in semi-domesticated reindeer in Sweden (Nordkvist, 1973) and Finland (Büttner et al., 1995). In Norway, contagious ecthyma was first diagnosed in reindeer during an animal experiment (Kummeneje and Krogsrud, 1979), and later under natural herding conditions (Tryland et al., 2001). Virus characterizations have shown that the outbreaks of contagious ecthyma in reindeer in Fennoscandia were caused by ORFV, which also causes this disease in sheep and goats (Klein and Tryland, 2005), whereas more recent outbreaks in Finland have been due to PCPV, another parapoxvirus, associated with cattle (Tikkanen et al., 2004) (see Textbox 8.1).

8.1.2   PATHOGENESIS, PATHOLOGY AND CLINICAL SIGNS

The virus usually enters the skin or the oral mucosal membrane through abrasions. Small lesions of the skin, muzzle and oral mucosa from ice crusts or from coarse feed during supplementary feeding may predispose for the infection. The incubation period in reindeer is 3–5 days (Tryland et al., 2013). The lesions usually start as a papule that develops to a pustule and further to cauliflower-like proliferations, typically on the muco-cutaneous junction around the mouth, both on the skin (Figure 8.2a) and on the oral mucosa (Figure 8.2b). Skin lesions are vulnerable to abrasions and bleeding, and may sometimes develop into thick, black crusts.

Histological examination will reveal a proliferative dermatitis, with epidermal hyperplasia and the formation of elongated rete ridges, and hyperkeratosis can be present (Tryland et al., 2013). In experimentally inoculated animals, accumulation of inflammatory cells, especially lymphocytes and macrophages, were found at the dermal/epidermal junction and in the dermal papillae (Tryland et al., 2013). The first indication of contagious ecthyma in reindeer can be animals standing with

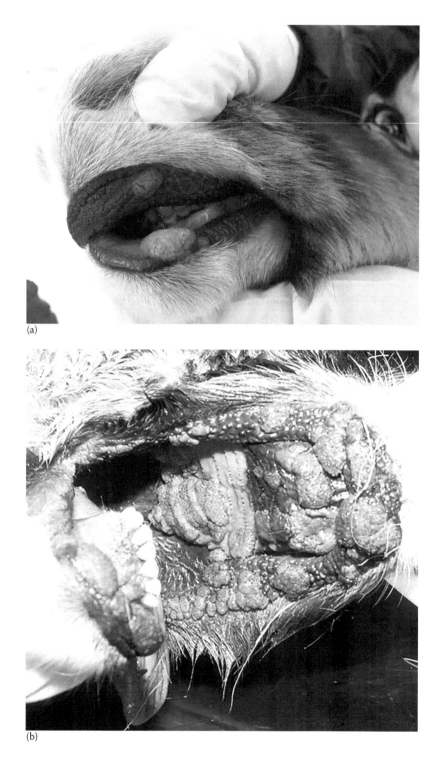

(a)

(b)

FIGURE 8.2 *Parapoxvirus* infection and contagious ecthyma in semi-domesticated reindeer. Proliferative lesions often develop at the muco-cutaneous junction around the mouth and can be visible both on the skin (a) and on the mucosa (b) (Photo: Ingebjørg H. Nymo and Morten Tryland).

fodder sticking out of their mouths, but being unable to chew. Food and regurgitated matter can accumulate in their mouths, and they can have greenish saliva coming out of the corner of their mouths. At closer range, a bad smell, usually from secondary bacterial infections of proliferative processes, can be prominent. At inspection, one (primary) or more (secondary) proliferative processes can be seen, on the skin of the muzzle and lips, and/or in the oral mucosa (chin, gingiva, tongue). The lesions can be isolated or coalescing, and sometimes cover larger parts of the skin or the mucosa. Experimental inoculation of otherwise healthy animals in good condition showed that some animals had contagious ecthyma lesions in a healing stage 4 weeks post inoculation (Tryland et al., 2013). Complications may occur due to problems with eating and secondary bacterial infections.

8.1.3 EPIDEMIOLOGY

ORFV is distributed worldwide in sheep and goats. ORFV is presumably transferred from sheep or goats to reindeer, either via direct contact, or indirectly through sharing pastures containing infected organic matter (scabs), or corrals, transport vehicles, food troughs, etc. (Tryland et al., 2001). Contagious ecthyma can appear on one or a few animals only, or as regular outbreaks, affecting hundreds of animals and having a high mortality (Tryland et al., 2001; Tikkanen et al., 2004). Finnmark County, northern Norway, accounts for approximately 73% of Norwegian reindeer, but the disease contagious ecthyma has never been reported in reindeer from this region. Nevertheless, parapoxvirus-specific DNA was detected in various tissues of reindeer carcasses from Finnmark, with no clinical symptoms that could be associated to contagious ecthyma (Tryland, 2002). This indicates that, although the virus has to be present to cause the disease, other factors also determine whether a disease outbreak occurs.

Parapoxvirus with DNA sequences (*B2L*, *GIF* and *IL10* PCR amplicons) indistinguishable from ORFV has been found in cutaneous lesions on the upper lip and the plantar side of the hooves/coronary bands of a caribou (*R. t. granti*) from Admiralty Bay, Northern Alaska, USA (Figure 8.3) (Tryland et al., 2018), but the prevalence of such infections in wild populations of *Rangifer* is unknown.

FIGURE 8.3 *Parapoxvirus* infection in caribou (*R. t. granti*, Alaska, USA), showing contagious ecthyma lesions of a less proliferative type at the muco-cutaneous junction (Photo: Jason Herreman, Department of Wildlife Management, North Slope Borough, Alaska, USA).

TEXTBOX 8.1   CONTAGIOUS ECTHYMA IN FINNISH REINDEER: TWO DIFFERENT VIRUSES INVOLVED

In Finnish reindeer, contagious ecthyma has been observed since the early 1990s. It is commonly seen in the southern parts of the reindeer herding districts through the winter. The disease is characterized by ulcerative skin lesions and erosions and papules in the mouth, but proliferative lesions and necrotic/purulent erosions have also been observed, with secondary bacterial infections. It is assumed that supplementary feeding, which increased over the 1990s and is nowadays a common practice in Finland during the winter months, contributes to increased infection pressure, impacting both the morbidity and mortality of the disease (Hirvelä-Koski, 2001). The most prominent epizootic in Finland took place during the winter of 1992–1993, when about 400 reindeer died and at least 2800 were affected. Sporadic outbreaks have been common, but severe outbreaks were also reported in 1999–2000 and in the late winter of 2007 (Hautaniemi et al., 2011).

In contrast to contagious ecthyma in reindeer elsewhere, two different parapoxvirus species, ORFV and PCPV, have been identified as the causative agents of the disease in Finnish reindeer. Whereas the early outbreaks, caused by ORFV, were severe, with proliferative lesions, later outbreaks from 1999–2000 have been associated with PCPV, with mostly mild inflammatory spots and ulcers in the mouth of the reindeer. As with ORFV, PCPV also occurs worldwide, but is associated with cattle rather than small ruminants, and causes chronic infections and lesions on the teats of cows and on the muzzle and in the oral mucosa of nursing calves.

Phylogenetic analyses have revealed that the outbreak in 1992–1993 was due to ORFV but that the outbreak in 1999–2000 was caused by PCPV (Tikkanen et al., 2004), which was subsequently confirmed by genome sequencing (Hautaniemi et al., 2010 and 2011). In samples from reindeer suffering from the outbreak in 2007, both ORFV and PCPV were identified. Further phylogenetic analyses of PPV-isolates from Finnish reindeer, cattle and sheep indicated that the viruses causing disease in reindeer are very closely related to the viruses infecting Finnish cattle (PCPV) and sheep (ORFV), suggesting that these viruses have been circulating among Finnish reindeer, cattle and sheep for the last decade or more (Hautaniemi et al., 2011). The transmission pattern of PCPV to reindeer is not well known. Although there have been a few PCPV epizootics in cattle in Finland (1971, 1974 and 1999) these were geographically remote from the region with affected reindeer (Tikkanen et al., 2004).

8.1.4   DIAGNOSIS AND MANAGEMENT

Several different viruses can produce lesions and processes of the skin and mucosa in ruminants. Although it has not been possible to clinically distinguish contagious ecthyma in reindeer caused by ORFV and PCPV, it is usually easy to distinguish contagious ecthyma from lesions produced by other viruses, such as papillomavirus and foot and mouth disease virus (FMDV), since contagious ecthyma is characterized by lesions of a proliferative and cauliflower-like nature and appears on the muzzle, lips and oral mucosa. The diagnosis should be further verified, by histology (tissue fixed in 10% buffered formalin), by the finding of characteristic virus particles by electron microscopy (negative staining or transmission), or by extracting DNA from tissue from a scab (sampled and kept dry or frozen) and running PCR with primers specific to the GIF gene (Klein and Tryland, 2005), or alternatively to the *B2L* (Inoshima et al., 2000) or the *IL10* genes (Klein and Tryland, 2005).

There is no specific treatment against ORFV, but supportive therapy and antibiotic treatment against secondary bacterial infections may be indicated. The humoral immune response is thought to be short lived in reindeer, as in sheep (Damon, 2007). In an experimental infection study, in which semi-domesticated reindeer were inoculated with ORFV obtained from a reindeer with contagious ecthyma, five of six inoculated animals developed an increase in the antibody titer around

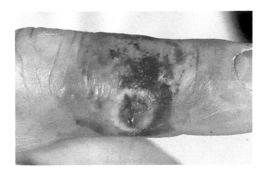

FIGURE 8.4   Parapoxvirus infection in a person that had been handling reindeer heads during slaughter from animals with no obvious clinical symptoms of contagious ecthyma (Photo: Riitta-Liisa Palatsi).

3–4 weeks post inoculation (Tryland et al., 2013). In the same study, two reindeer vaccinated with a commercial ORFV vaccine (Scabivax®, Schering Plough Animal Health Corp.) 28 days prior to the inoculation of ORFV (challenge) showed a similar humoral response as the non-vaccinated animals, but with an earlier onset and a sharper rise of the titer. However, the vaccine did not prevent infection and clinical signs, since the two vaccinated animals also developed lesions similar to the ones in the non-vaccinated animals, although less severe. It should, however, be kept in mind that a full vaccination and challenge trial with an ORFV vaccine in reindeer never has been conducted or validated (Tryland et al., 2013). Vaccination is not employed during natural herding conditions.

ORFV and PCPV, as well as another bovine pathogen, bovine papular stomatitis virus (BPSV), can also infect man, i.e. they are zoonotic. ORFV is usually transmitted to people from sheep and goats, but it has been shown that ORFV has also been transmitted from reindeer to people handling reindeer heads during slaughter (Palatsi et al., 1993) (Figure 8.4).

8.2   PAPILLOMAVIRUS

Morten Tryland

8.2.1   AETIOLOGY

Papillomaviruses (virus family Papillomaviridae) cause skin processes called papillomas or fibropapillomas in many animal species, including reindeer. Papillomaviruses were previously grouped together with polyomaviruses as Papovaviridae, but certain characteristics different from polyomaviruses led to papillomaviruses being designated as a separate family. Papillomaviruses consist of a group of non-enveloped viruses, 50–55 nm in size, containing a circular, double-stranded DNA genome of 7.4–8.6 kb.

8.2.2   PATHOGENESIS, PATHOLOGY AND CLINICAL SIGNS

Papillomavirus processes (i.e. papillomas, fibropapillomas or warts) develop as a result of introduction of the virus via skin abrasions. Papillomaviruses are epitheliotrophic, and infect and transform basal layer cells in squamous epithelia or mucous membranes, resulting in proliferation and benign processes of the skin (warts) and mucous membranes (condylomas). Some papillomaviruses are also associated with malignant conditions (cancer) in humans, in the uterine cervix as well as other tumours of the urogenital tract (Howley and Lowy, 2007).

Papillomavirus warts often appear as a single process, varying from a small firm nodule to larger processes, usually black or grayish in colour. In some cases, papillomaviruses may cause generalized infections and produce large and sometimes coalescing warts distributed around the body (Figure 8.5). In moose, the European elk papillomavirus (EEPV) induces pulmonary fibromatosis, in addition to warts (Moreno-Lopéz et al., 1986), a clinical feature that has not been reported for reindeer or caribou.

FIGURE 8.5   Papillomavirus infection in a wild Eurasian tundra reindeer (*R. t. tarandus*) at Hardangervidda, Norway. Papillomavirus infections are usually easy to distinguish clinically from parapoxvirus infections (Photo: Sondre Myrvang).

8.2.3   EPIDEMIOLOGY

Papillomaviruses have been isolated from many host species, including mammals, birds, snakes and turtles, and they are regarded as species specific, i.e. they can infect only one host species. On the other hand, one host species can be infected by several or many different papillomaviruses, and in humans, more than 100 different papillomaviruses have been reported, of which types 16 and 18 are associated with cervical cancer (Kim et al., 2014). Phylogenetic studies have indicated that the variability among the papillomavirus genomes is a result of point mutations rather than recombinations, suggesting that papillomaviruses have accompanied their host species during evolution (Howley and Lowy, 2007).

Although papillomaviruses are known to cause warts or condylomas, most papillomavirus infections are regarded as asymptomatic. A study of swab samples obtained from healthy skin of 19 different animal species revealed papillomavirus-specific DNA in eight species, representing 53 new papillomavirus species (Antonsson and Hansson, 2002).

Papillomaviruses have been isolated from reindeer (Moreno-Lopéz et al., 1987), white tailed deer (*Odocoileus virginianus*) and red deer (*Cervus elaphus*) (Groff et al., 1983), roe deer (*Capreolus capreolus*) (Erdeélyi et al., 2009) and moose (*Alces alces*) (Moreno-Lopéz et al., 1986). Papillomatosis, similar to conditions seen in other ungulates, has been observed in both semi-domesticated reindeer and in wild reindeer in Norway.

During an outbreak of infectious keratoconjunctivitis (IKC) in semi-domesticated reindeer, eye swab samples were obtained from both healthy and diseased animals. From a clinically affected reindeer, a full-length papillomavirus genome sequence was amplified, which was different from a previous reindeer isolate (RtPV1), and thus considered as a new species (RtPV2), belonging to a new genus of papillomaviruses (Smits et al., 2013). From another 1-year-old, a different papillomavirus was detected, called RtPV3. It was, however, not possible to conclude that the presence of the papillomaviruses in these two animals had any relevance to the clinical findings (Smits et al., 2013).

During a virus screening of free-ranging caribou (*Rangifer tarandus granti*) of the Western Arctic Caribou Herd, Alaska, USA, a complete sequence of the L1 gene, used for classification of papillomaviruses, was obtained. It was revealed that this virus should be designated as a new species

(RtPV4), belonging to a new genus. Smaller L1 fragments were found in many other samples, with similarities to papillomaviruses found in humans, cattle and goats (Schürch et al., 2014). Together, these findings indicate that there might be a huge variability of papillomaviruses in *Rangifer*.

Little is known about the prevalence of papillomavirus in reindeer and caribou. A recent epidemiological study on the distribution of roe deer papillomavirus (CcPV1) in roe deer (*Capreolus capreolus*) in Hungary revealed an endemic status, and showed that the likelihood of infection increased with animal density and the presence of waterways (Erdeélyi et al., 2009).

8.2.4 DIAGNOSIS AND MANAGEMENT

Most papillomavirus infections are asymptomatic and have little impact on herd and population levels. In some individuals, however, it is clear that the number and masses of papillomas may impact health and the general condition of the animal. It is also important to keep in mind that papillomas clinically may resemble Orf-virus infections (contagious ecthyma), and samples should be obtained for a viral diagnosis. Tissue samples are obtained and kept dry or frozen for DNA extraction and PCR with specific primers, or fixed in 10% buffered formalin for histology. Small pieces (1.0 × 1.0 mm) of tissue may also be fixed in McDowell's fixative for negative staining with 2% phosphotungstic acid and transmission electron microscopy, to identify the viral particles. Since papillomaviruses are very host specific, they are not zoonotic.

8.3 HERPESVIRUS INFECTIONS

Morten Tryland

Herpesviruses are highly disseminated in nature. In fact, most investigated animal species are host to one, or often several, herpesvirus species, and more than 200 different herpesviruses have been identified so far (Pellett and Roizman, 2007). Herpesviruses have a linear double-stranded DNA genome (124–241 kb), surrounded by an icosahedral capsid, a tegument and an envelope, and these characteristics have been the criteria for inclusion of a virus in the family Herpesviridae. The virus family is divided into three subfamilies, Alpha-, Beta- and Gammaherpesvirinae. A betaherpesvirus was recently identified in eye swab samples from three semi-domesticated reindeer in Norway, all having clinical symptoms of infectious keratoconjunctivitis (IKC) (Smits et al., 2013). The virus was designated cervid herpesvirus 3 (CvHV3), and shown to be different from CvHV1 and CvHV2, associated to red deer (*Cervus elaphus*) and reindeer, respectively (Thiry et al., 2006), but it is not known if it has any clinical impact on IKC or reindeer in general. In *Rangifer*, herpesviruses belonging to the Alpha- and Gammaherpesvirinae subfamilies are important pathogens.

8.3.1 ALPHAHERPESVIRUS

Morten Tryland

8.3.1.1 Aetiology

Alphaherpesviruses are enveloped with a spherical/pleomorphic shape (Figure 8.6), having a diameter of approximately 150–200 nm and a double-stranded genome consisting of 120–180 kb.

Serological surveys conducted from the early 1980s on indicated antibodies in reindeer and caribou in Alaska, Canada and Greenland, as well as in Norway, Sweden and Finland, reacting against antigens of bovine herpesvirus 1 (BoHV1) (Evans et al., 2012; das Neves et al., 2010). The isolation of herpesvirus from reindeer, designated cervid herpesvirus 2 (CvHV2), was reported from Finland (Ek-Kommonen et al., 1986) and Sweden (Rockborn et al., 1990), in both cases from animals treated with drugs (corticosteroids) to induce immunosuppression and reactivation of latent herpesvirus infections. In Norway, CvHV2 was isolated from reindeer from Troms County,

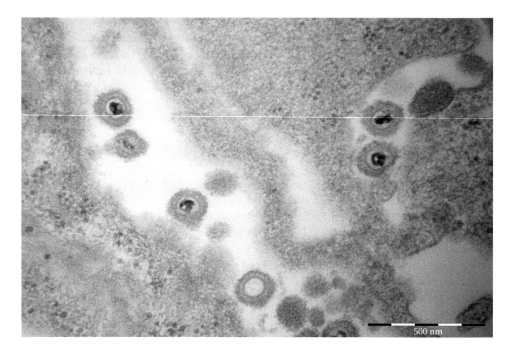

FIGURE 8.6   Alphaherpesvirus (CvHV2) particles in tissue sections of a reindeer lip with viral lesions (Photo: Carlos das Neves).

associated with infectious keratoconjunctivitis, IKC (Tryland et al., 2009), and from reindeer from Finnmark County, treated with dexametasone to reactivate latent herpesvirus infection (das Neves et al., 2009a).

CvHV2 has also been detected by PCR in eye and vaginal swabs as well as in trigeminal ganglia and foetuses from reindeer, representing many pasture districts in Finnmark County, Norway (das Neves et al., 2009b). Phylogenetic studies, based on sequencing of PCR amplicons from the genes encoding glycoproteins B, C and D have revealed a generally high degree of sequence homology between different CvHV2 isolates from Finland, Sweden and Norway, suggesting a common distribution of CvHV2 in these reindeer populations (das Neves et al., 2010), and also a high homology with similar gene regions in other alphaherpesviruses, such as bovine herpesvirus 1 (BoHV1) (Ros and Belák, 2002), whereas sequences obtained from glycoprotein E, revealing less homology, have been used to distinguish between alphaherpesviruses from different hosts (Thiry et al., 2007).

8.3.1.2   Pathogenesis, Pathology and Clinical Signs

The reindeer herpesvirus (CvHV2) enters and infects cells of the mucosal membranes. As for herpes simplex virus (HSV) in humans, the virus is believed to attach to the cell by one or more viral glypoproteins (gC, gB and maybe others) to cellular receptors, such as heparan sulfate proteoglycans (HSPGs) (Akhtar and Shukla, 2009). Once attached, the viral envelope fuses to the plasmamembrane of the cell, and the nucleocapsid enters the cytoplasm. The nucleocapsid is transported to a nuclear pore through which the viral DNA is transferred to the nucleus of the cell and can start to express its genes. Replication and production of progeny virus particles takes place within hours after infection. This normally causes damage to the cell, expressed as cytopathic effects (CPE), before the cell dies (lysis).

As with other herpesviruses, CvHV2 establishes a lifelong infection. Although the clinical signs may pass acute and chronic stages and disappear, the virus retracts in a latent form in nervous tissues, from where it can be reactivated (during stress, immune suppression or concomitant infections), travel along nerves back to the port of entry, and produce clinical symptoms (Engels and Ackermann, 1996). CvHV2 has been found in nerve ganglia (Ganglion trigeminale) in reindeer (das Neves et al., 2009b), but attempts to isolate the virus from Ganglion sacrale failed (Rockborn et al., 1990).

Although commonly distributed in most reindeer and caribou herds that have been investigated (das Neves et al., 2010), CvHV2 has previously been described as having little clinical impact, except for being able to make mucosal lesions, i.e. a port of entry for secondary bacterial infections, such as *Fusobacterium necrophorum*, and thus contributing to the disease alimentary necrobacillosis (Rehbinder and Nikander, 1999; Rockborn et al., 1990). However, the fact that CvHV2 is enzootic in the reindeer populations, that it may be transferred across the placenta to the offspring, that the virus shows latency and may be reactivated from nerve ganglia, and that it has been associated with outbreaks of transmissible keratoconjunctivitis, indicates that the pathogenic potential and impact of this virus in the reindeer herds has been underestimated (das Neves et al., 2010).

In an experimental infection study, pregnant reindeer were inoculated vaginally with CvHV2 and showed only mild clinical signs of vulvovaginitis. One of the animals gave birth to a calf that died the day after, having no colostrum in its stomach, which probably was the direct cause of death, but interestingly, CvHV2 was found in multiple organs of the calf (das Neves et al., 2009c). During another experiment, when a pregnant reindeer that was seropositive to CvHV2 but had no virus shedding from eye and nose (PCR) was treated with glucocorticosteroids, the virus was reactivated and produced blisters at the muco-cutaneous junction of the mouth, and the female aborted (Figure 8.7). The abortion may have been initiated by the treatment, but interestingly, CvHV2 was detected in multiple tissues of the foetus (das Neves et al., 2009b). Together, these findings strengthen the hypothesis that CvHV2 may contribute to abortions and weak borne calves in reindeer. In cattle, both natural infections and experimental studies have shown that alphaherpesvirus (BoHV1) infections can cause infertility, abortions and stillbirths, with a varying pathogenicity between different isolates (Graham, 2013).

In cattle, BoHV1 is not only associated with reproduction disorders and systemic infections of neonates, but is also a major pathogen associated to respiratory disease in the upper respiratory tract

FIGURE 8.7   Glucocorticosteroids were administered to a reindeer that had no clinical symptoms, but had antibodies against alphaherpesvirus. The animal developed lesions on the lip characteristic of alphaherpesvirus infections (CvHV2) (Photo: Carlos das Neves).

(infectious bovine rhinotracheitis [IBR]), often referred to as the "respiratory disease complex," in which BoHV1 seems to play an important role as a primary infection, followed by secondary bacterial infections and the progression of the disease to bronchopneumonia (Muylkens et al., 2007). A similar infection biology may be valid for CvHV2 in reindeer, since CvHV2 has been detected in macrophages of the lungs, associated with acute hemorrhagic and necrotizing pneumonia (das Neves et al., 2009a). CvHV2 may thus be important for paving the way for bacterial infections such as pasteurellosis (*Mannheimia hemolytica* [previously *P. hemolytica*] and *Pasteurella multocida*), which is known to cause pneumonia in reindeer (Kummeneje, 1976).

CvHV2 has also been identified as the primary agent during an outbreak of infectious keratoconjunctivitis (IKC) in reindeer. IKC is a contagious ocular infection involving both the cornea and the mucosa of the conjunctiva. The disease has been described since the 19th century, with a more thorough report presented in 1912 by the Swedish biologist Arvid M. Bergman (1872–1923) (Bergman, 1912), and has been called *Ĉalbmevikke* in Saami (Skjenneberg and Slagsvold, 1968). IKC can appear in one or a few animals at a time, it can affect one eye only or both eyes, and from early stages of the disease, it may heal spontaneously. However, IKC can also appear as large outbreaks, affecting hundreds of animals in a herd, predominantly calves and young animals, and progress to a devastating disease, which destroys the eye(s) and causes permanent blindness and severe pain for the animals, and economic loss for the herders (Rehbinder and Nilsson, 1995; Oksanen et al., 1996; Tryland et al., 2009).

The very early sign of disease is increased lacrimation, with an untidy, discoloured haircoat beneath the affected eye (Figure 8.8a), followed by increasing periorbital and corneal oedema, giving the eye a whitish or bluish appearance (Figure 8.8b). If not healed spontaneously, the infection progresses with increasing oedema, corneal ulcer, panophthalmitis and loss of the lens, followed by damage of the eye structures and permanent blindness (Figure 8.8c and d). As with infectious bovine keratoconjunctivitis (IBK), a similar disease in cattle, IKC is regarded as a multifactorial disease, but the reindeer alphaherpesvirus, CvHV2, has been identified as the primary pathogen (Tryland et al., 2009; Tryland et al., 2017), usually followed by secondary

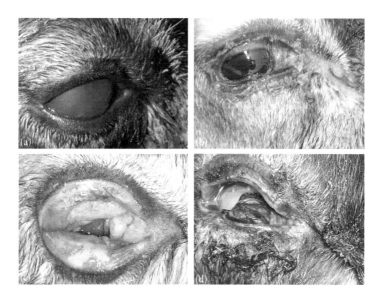

FIGURE 8.8   Stages of infectious keratoconjunctivitis (IKC). Alphaherpesvirus (CvHV2) has been shown to be the transmissible component when IKC affects a reindeer herd, where several dozen animals, mostly calves, are affected. (a) Reindeer with increased lacrimation and corneal oedema, giving the eye a bluish or whitish appearance. (b) Severe periorbital oedema, conjunctivitis and keratitis and shedding of pus. (c) Increased severity of IKC, with bleeding conjunctivitis. (d) Panophthalmitis – the function and the structure of the eye is destroyed (Photos: Sauli Laaksonen, Javier Sánchez Romano and Morten Tryland).

bacterial infections, such as *Moraxella bovoculi* and *Moraxella ovis*, *Pasteurella multocida* and *Trueperella pyogenes* and others (Kummeneje, 1976; Oksanen et al., 1996; Rehbinder and Nilsson, 1995; Aschfalk et al., 2003; Tryland et al., 2009; Sánchez Romano et al., 2018), which dominate during later stages of the disease. In an experimental study, it was shown that CvHV2 (but not *Moraxella bovoculi*, which was also isolated from reindeer with IKC), was able to cause clinical symptoms indistinguishable from IKC in semi-domesticated reindeer (Tryland et al., 2017). An interesting finding during this experiment was that the excretions from the eye, resembling whitish clots of pus, did not contain bacteria that could be cultivated under traditional bacteriological conditions (i.e. blood agar and incubation at 37°C, aerobic and anaerobic conditions). Thus, bacterial infections did not seem to be a prerequisite to cause any of the symptoms that characterize IKC in reindeer.

8.3.1.3 Epidemiology

Based on serological screenings, it has been shown that, except for the subspecies Svalbard reindeer, inhabiting the high arctic Svalbard archipelago, other subspecies and herds of reindeer and caribou that have been investigated are exposed to alphaherpesvirus infections, presumably CvHV2 (das Neves et al., 2010). In most investigated herds, a majority of the adult reindeer have been exposed, seroprevalence being typically around 50% and increasing with age, whereas much lower seroprevalenc has normally been detected for calves and young animals (das Neves et al., 2009d). Thus, it is believed that CvHV2 is normally latent in older animals, but becomes reactivated during stressful conditions for the animals, such as gathering, corralling, handling and transport. The virus is then shed on mucosal membranes and is again transmitted to other animals through direct contact and possibly also aerosols. They acquire a primary infection and have to build an immune response to the virus, and it has been shown that it takes more than 10 days to seroconvert after experimental tracheal and vaginal inoculation of reindeer (das Neves et al., 2009c). Meanwhile, the virus establishes infections of the mucosal membranes and causes lesions. Subsequently, bacteria may establish as secondary infections, which progressively drive the pathology, whether it is infections in the mouth (necrobacillosis) or in the eye (conjunctivitis and keratitis). Larger outbreaks of IKC usually affect calves and animals corralled for supplementary feeding.

8.3.1.4 Diagnosis and Management

As for the symptoms of IKC in reindeer, reindeer herders that have experienced such outbreaks recognize the very early indications that something is wrong, and they report increased lacrimation and observe that the fur coat below one or both eyes is wet, untidy and discoloured, and sometimes, a pus clot can be seen in the medial eye angle. At later stages, the bluish coloured cornea and the periorbital oedema, together with the excretions from the eye, are easily recognized, along with the fact that it is spreading and affecting more and more animals. CvHV2 may also produce mucosal lesions in the mouth, which may resemble other viral infections in ruminants, such as foot and mouth disease (FMD). A specific diagnosis is therefore often important.

Antibodies against CvHV2 can be detected with a commercial ELISA designed for alphaherpesvirus (i.e. BoHV1) in cattle (das Neves et al., 2009d), based on the immunological cross-reaction between similar alphaherpesviruses against the highly conserved viral antigen, glycoprotein B (gB). An alternative, or supplementary diagnostic assay, is the virus neutralization test (VNT), which detects antibodies that actually interfere with the virus and hinders infection of cells in culture. Detecting antibodies against CvHV2 means the animal has been exposed to the virus, but also, since it is a lifelong infection, that the animal carries the virus, either as a latent or as an active infection. An active infection implies shedding of virus on mucosal membranes, which normally makes it possible to detect CvHV2-specific DNA in swab samples obtained from the mucosal membranes (oral, ocular or vaginal), by a nested polymerase chain reaction (PCR) (Ros and Belák, 1999) or by real time PCR (qPCR), which also is a quantification assay (Wang et al., 2007). Detecting a latent infection is possible by testing nerve ganglia, such as the trigeminal, for CvHV2-specific DNA with PCR (das Neves et al., 2010). However, it should be kept in mind that CvHV2 may be found on mucosal membranes and in nerve ganglia of healthy animals without any clinical signs, or with clinical signs from other diseases and infections, and that the presence of the virus thus does not always reveal a causative relationship with the clinical condition.

There are no vaccines against alphaherpesvirus in reindeer. Although there are herds/populations with more and with less animals exposed (i.e. high and low seroprevalence), the virus is so commonly distributed among reindeer and caribou herds that the issue of being exposed on a herd level is of little value. Thus, the finding of CvHV2 exposure of a reindeer herd should not hamper contact or cooperation with other herders. Management of CvHV2 is therefore more focused on how to avoid severe clinical outbreaks, such as IKC. Any stressful situation for the animals is thought to potentially impact how the animal and the herd manage to cope with their CvHV2 status, i.e. whether a clinical outbreak occurs or not. Herding, gathering, handling and transport of animals will represent stress for the animals to various degrees, and may thus contribute to immunosuppression, reactivation of latent infections and disease outbreaks, such as IKC.

8.3.2   Gammaherpesvirus

Morten Tryland and Torill Mørk

8.3.2.1   Aetiology

The subfamily Gammaherpesvirinae consists of viruses affecting humans and other vertebrates, and one of the genera of this subfamily, genus *Macavirus*, contains several virus species that are closely related, being associated to the disease malignant catarrhal fever (MCF) in several ruminant species (Davison et al., 2009; Li et al., 2011). The most relevant viruses of this group, causing MCF in domestic and wild ruminants, are Ovine herpesvirus 2 (OvHV2) and Caprine herpesvirus 2 (CpHV2). OvHV2, which is assumed to be commonly distributed in domestic sheep as subclinical infections, may cause fatal disease in other ruminant species of the order Artiodactyla, such as Bovidae, Cervidae and Giraffidae, as well as Suidae (Li et al., 2014). The virus itself has never been isolated, but sequences obtained from tissues of MCF-affected animals have been identified (Li et al., 2014).

8.3.2.2   Pathogenesis, Pathology and Clinical Signs

MCF is a generalized lymphoproliferative disease, primarily affecting lymphoid tissues and epithelial cells of the gastrointestinal and respiratory tract, and is frequently fatal in cattle and other species (Li et al., 2014). No thorough descriptions of this disease in reindeer are available, but for MCF in cattle, caused by alcelaphine herpesvirus 1, the incubation time is 3 weeks, and the disease is characterized by fever, depression, leukopenia, severe inflammation of the eyes (bilateral ophthalmia), nasal and ocular discharges, swollen lymph nodes (lymphadenopathy), mucosal erosions and symptoms of the central nervous system (CNS) (Murphy et al., 1999).

During an outbreak of MCF in a zoo in Germany, several species were affected, including two reindeer. The reindeer displayed reduced uptake of feed and water, labored breathing, tympanic condition, and keratitis and conjunctivitis, and one of the reindeer died 3 days after the disease onset (Altmann et al., 1973).

Clinical symptoms of MCF in moose and roe deer have included apathy, incoordination and staggered gait, convulsions, impaired vision, corneal opacity, erosions in mucosal membranes, and purulent exudates from the respiratory tract and vagina, and histology has revealed corneal lesions, keratitis and uveitis (Vikøren et al., 2006).

8.3.2.3 Epidemiology

Domestic sheep (*Ovis aries*) are healthy carriers of OvHV2 and the virus is transmitted to domestic and wild ruminants, as well as pigs (Syrjälä et al., 2006). CpHV2, which is hosted by goats (*Capra aegagrus hircus*), can cause MCF in moose (*Alces alces*), Sika deer (*Cervus nippon*) and white tailed deer (*Odocoileus virginianus*) (Crawford et al., 2002; Li et al., 2003; Keel et al., 2003). The transmission of virus from carriers to susceptible ruminants may take place via nasal secretions (Li et al., 2004).

In a PCR screening of tissue samples from wild ruminants in Norway showing pathological symptoms of MCF, OvHV2 and/or CpHV2 specific DNA was detected in moose, red deer (*Cervus elaphus*) and roe deer (*Capreolus capreolus*) (Vikøren et al., 2006).

MCF has been diagnosed in one semi-domesticated reindeer under regular herding conditions (see Textbox 8.2) and also in reindeer in zoos (Altmann et al., 1973; Li et al., 1999; Kiupel et al., 2004). Nevertheless, serological screenings of wild (n = 44) and semi-domesticated (n = 3339) reindeer (*Rangifer tarandus tarandus*) in Norway have revealed a seroprevalence of 4% and 3.5%, respectively (Vikøren et al., 2006; das Neves et al., 2013). An antibody prevalence of 4% was found when screening 232 caribou (*Rangifer tarandus*) against a MCF virus in Alaska, using ELISA (Zarnke et al., 2002).

In ruminant wildlife species sharing pastures with reindeer and caribou, MCF seems to be enzootic in muskox (*Ovibos moschatus*) and Dall sheep (*Ovis dalli*) in Alaska, and in muskox in Norway (Zarnke et al., 2002; Vikøren et al., 2013), whereas low prevalence and, possibly, introduction of the virus from time to time may be the case in moose populations (Zarnke et al., 2002; Vikøren et al., 2015). Reindeer in zoos seem to have been infected through contact with other species, such as mouflon (*Ovis musimon*) and pygmy goats (*Capra hircus*) (Li et al., 1999).

Moose and roe deer can suffer from acute forms of MCF (Vikøren et al., 2006). Latent or subclinical gammaherpesvirus infections have been reported in several ruminant species (Powers et al., 2005), and this may also be the situation in reindeer, based on the serological screenings and the lack of clinical disease in semi-domesticated reindeer (das Neves et al., 2013). Preliminary sequence data of PCR amplicons obtained from extracted DNA from lymphocytes from ten seropositive reindeer representing four different herding districts in Finnmark County, Norway, indicated consistent homology between the amplicons from the reindeer investigated, but low homology with previously deposited gene sequences from gammaherpesvirus (GenBank; OvHV2, CpHV2, rupricapra herpesvirus, fallow deer herpesvirus, equine herpesvirus 2 and bovine herpesvirus 4) (das Neves et al., unpublished data), suggesting that the gammaherpesvirus circulating in reindeer in Norway is a hitherto undescribed virus, which may be transmitted from other reservoirs than sheep (OvHV2) and goat (CpHV2). Alternatively, this is a reindeer-specific gammaherpesvirus, possibly with restricted pathogenicity for reindeer, based on the fact that antibodies were fairly common in the same reindeer herds, whereas MCF has been reported in only one reindeer under regular herding conditions.

8.3.2.4 Diagnosis and Management

Antibodies against gammaherpesviruses can be detected with a direct competitive-inhibition enzyme-linked immunosorbent assay (cELISA) (Li et al., 2001; das Neves et al., 2013). Samples from affected tissues can be fixed in 10% buffered formalin, embedded in paraffin, sectioned and stained by haematoxylin and eosin (H&E) for histologic examination. Gammaherpesvirus-specific DNA sequences may be detected by PCR (Li et al., 1999; Vikøren et al., 2006). PCR can be conducted on extracted DNA from fresh, frozen or paraffin-embedded tissues, or from the lymphocyte fraction of the blood (peripheral blood lymphocytes [PBL]). Also, real time PCR assays have been developed to detect and quantify MCFV in clinical samples (Traul et al., 2007; Cunha et al., 2009).

TEXTBOX 8.2   GAMMAHERPESVIRUS INFECTION
AND MALIGNANT CATARRHAL FEVER IN REINDEER

An adult male semi-domesticated reindeer with symptoms of eye disease and reduced vision had been fed and taken care of by the owner, who suspected keratoconjunctivitis (IKC). Since the condition did not improve, the animal was euthanized and delivered for necropsy. The macroscopic findings were mainly from the skin and eyes (Figure 8.9). There was hair loss and thickening of the skin with crusts around the muzzle, in the axillary regions of both sides, and on distal parts of the legs. In some of the crusts, thick, yellow pus was present, from which the bacterium *Trueperella pyogenes* was cultivated. Both eyes showed swollen eyelids, corneal opacity and fibrinoporulent eyeflood. In addition, superficial lymph nodes were swollen, the spleen was moderately enlarged and both kidneys had multiple, light brown foci in the cortex.

The histopathological findings revealed non-purulent dermatitis and a non-purulent interstitial renitis. In the brain tissue (Figure 8.10), mononuclear cells were present around the blood vessels and necrotizing vasculitis was evident, both being typical findings for malignant catarrhal fever (MCF). PCR (Baxter et al., 1993) was conducted from samples of brain tissue and identified as ovine herpesvirus (OvHV-2).

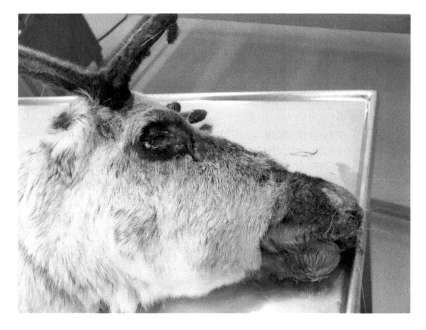

FIGURE 8.9   A semi-domesticated reindeer with gammaherpesvirus infection showed hair loss and thick crust in the skin around the muzzle. Both eyes displayed corneal opacity, hyperemia of conjunctiva and fibrinopurulent eye flood (Photo: Torill Mørk).

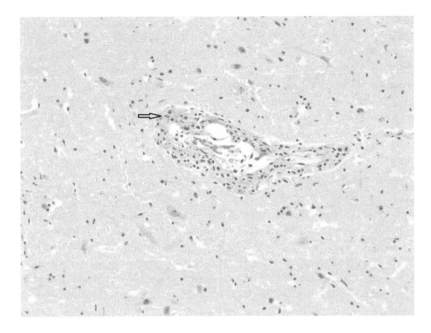

FIGURE 8.10   Necrotizing vasculitis in brain tissue (arrow pointing at the wall of a blood vessel) (Photo: Torill Mørk).

There are no thorough records on treatment of MCF in reindeer, but during the MCF outbreak in a zoo in Germany, the supportive treatment consisted of intravenous nutrition and fluid therapy and antibiotics against secondary bacterial infections (Altmann et al., 1973).

8.4   *PESTIVIRUS* INFECTIONS

Carlos G. das Neves and Jonas J. Wensman

8.4.1   AETIOLOGY

Pestiviruses are enveloped viruses, 40–60 nm in diameter, with a single-stranded positive sense RNA genome of about 12.5 kb consisting of one large open reading frame flanked by 5' and 3' noncoding regions (Schweizer and Peterhans, 2014). *Pestivirus* is a genus within the Flaviviridae family that comprises four well-defined species – bovine viral diarrhoea virus 1 and 2 (BVDV-1, BVDV-2), border disease virus (BDV) and classical swine fever virus (CSFV) – which cause severe and economically important diseases in cattle, sheep and pig production worldwide. In addition, a group of atypical pestiviruses, tentatively designated BVDV-3 (Liu et al., 2009), is causing disease in cattle. As far as it is known, pestiviruses do not exhibit zoonotic potential.

BDV is currently further divided into eight genotypes (Peletto et al., 2015). Although the natural hosts of pestiviruses are cattle, sheep and pigs, they are known to cross species barriers, infecting a wide range of animals in the order Artiodactyla (Nettleton, 1990), and several pestivirus species have been detected, both in domestic and wild animals (Ridpath, 2015; Vilcek and Nettleton, 2006). While a number of pestiviruses have been identified in ungulates, there are no reports on the isolation of pestivirus from wild or semi-domesticated reindeer. However, a pestivirus (V60) was isolated from a reindeer (*R. t. tarandus*) that died with signs of severe diarrhoea and anorexia in a German zoo in 1996 (Becher et al., 1999). Phylogenetic studies have shown this virus to be most closely related to BDV-2 (Avalos-Ramirez et al., 2001; Becher et al., 2003; Giangaspero et al., 2006).

8.4.2 PATHOGENESIS, PATHOLOGY AND CLINICAL SIGNS

Pestiviruses have never been isolated from wild or semi-domesticated reindeer, and therefore there is little information on routes of transmission, pathogenesis, pathology or clinical signs. One single experimental infection trial performed in the 1990s investigated the infection of reindeer with BVDV (Morton et al., 1990). Clinical signs included loose stools with blood and mucus, transient laminitis and coronitis, and serous to mucopurulent nasal discharge, as well as a relative leukopenia. Some lesions in the nasal mucosa were also observed. Viremia was detected 4 days post-infection and antibodies against BVDV were identified from day 21 post-infection onwards.

The pathogenesis of ruminant pestiviruses is complex with a wide variety of clinical signs. BVDV and BDV can infect their hosts either transiently or persistently. During primary infections, ruminant pestiviruses are believed to enter the host via the oropharynx. In infected cattle and sheep, viremia is observed for 1 to 2 weeks but may last longer in infections caused by more virulent strains (Schweizer and Peterhans, 2014).

In most cattle infected transiently with BVDV, clinical signs are mild, characterized by low-grade fever, diarrhoea and coughing. Transient infection of sheep with BDV is mostly unapparent; however, BDV infection in chamois (*Rupricapra rupricapra*) may lead to serious clinical disease (Marco, 2012; Marco et al., 2015).

Hosts persistently infected with BVDV/BDV are immunotolerant to the persisting viral strain. Tolerance includes both the cellular and humoral immune systems. Persistent infection by non-cytopathic strains is initiated in the early stages of foetal growth. Persistently infected (PI) foetuses may be aborted, or may develop and be born normally but continue to shed large quantities of virus for life. Persistent infection with BVDV has also been demonstrated in other ungulate species, as is the case with white tailed deer (Passler et al., 2010). The clinical signs of PI animals are quite variable. They may show poor growth and frequent infections due to severe virus-induced immunosuppression, but some may show no signs at all (Schweizer and Peterhans, 2014).

Mucosal disease (MD) is a lethal disease of cattle that is caused by super-infection with a pathogenic strain or the emergence of a cytopathogenic BVDV in an animal persistently infected with a non-cytopathogenic strain (usually by genetic mutations). This disease is usually characterized by high fever, anorexia, salivation, and erosions and ulcers in the gastrointestinal tract (Brownlie et al., 1984).

Border disease (BD) is an endemic disease of small ruminants, and clinical signs include abortion, stillbirths and birth of weak lambs showing tremor, abnormal body conformation and hairy fleece. In addition to small ruminants, BDV, the aetiological agent of BD, can infect pigs (Roehe et al., 1992), cattle (Becher et al., 1997), chamois (Arnal et al., 2004), bison and reindeer (Becher et al., 1999). BVDV can also be responsible for the development of BD in goats and sheep, as is the case in Norway (Løken, 1990, 1992) and Sweden (Carlsson, 1991), where BDVs have never been identified.

Whether reindeer can be persistently infected with pestivirus is unknown, and there is no information on outcomes of foetal infection in terms of abortion or PI shedders, but one can assume that similar methods of persistence, immune evasion and establishment of disease might also be present for reindeer.

8.4.3 Epidemiology

Along with isolation of pestiviruses from wildlife, serological screenings have indicated that many wildlife species have been exposed to pestiviruses and also indicated transmission of pestivirus between domestic and wild animal species (for a review, see Vilcek and Nettleton, 2006). Regarding reindeer and caribou (*R. tarandus*), several studies across the Arctic confirm the circulation of pestiviruses in these populations, although at various levels (for a review, see Larska, 2015). In Canada, the seroprevalence varied from 0 to 70% in caribou in the beginning of the 1980s (Elazhary et al., 1981), while a more recent study revealed a prevalence of up to 72% (Curry et al., 2014). In Alaska, a study in 1983 in Porcupine caribou (*R. t. granti*) identified a low prevalence of 3% (Zarnke, 1983), while in a recent study in two different Alaskan caribou populations, covering a sampling period of over 30 years (1981–2010), prevalence ranged from 0 to 60% (das Neves et al., unpublished data).

In wild reindeer (*R. t. tarandus*) in Norway (n = 810), a seroprevalence of 4.2% was found, varying from 0 to 51% between populations (Lillehaug et al., 2003). In semi-domesticated reindeer (*R. t. tarandus*), there is abundant evidence that pestiviruses are persistent in reindeer in Finnmark County, Norway: a seroprevalence of 17% in the 1990s (n = 326) (Stuen et al., 1993); 33% in reindeer carcasses found dead on the winter pasture in 2000 (n = 48) (Tryland et al., 2005); and 12.5% in a screening (2004–06) covering 16 reindeer herding districts in Finnmark (n = 3339) (das Neves et al., unpublished). As part of the latter study in Norway, serum neutralization tests (SNT) suggested that the virus circulating in the reindeer herds was indeed the same as, or most similar to, the isolate (V60; BDV-2) obtained from reindeer in a zoo (Becher et al., 2003), as compared to BVDV or BDV-1. A screening of semi-domesticated reindeer (*R. t. tarandus*) in 2001–2002 in Sweden (n = 1158) revealed a seroprevalence of 32%, including both seronegative herds and herds with higher seroprevalence, thus indicating the presence of PI animals (Kautto et al., 2012). The SNT in that study showed the highest neutralization titers against BDV-1, as compared to BVDV-1 and BVDV-2. Thus, due to low prevalence of BVDV in Scandinavia, it is most probably a BDV-like strain circulating in the Scandinavian reindeer population.

From all screened *Rangifer* subspecies so far, Svalbard reindeer (*R. t. platyrhynchus*), the subspecies existing only in the high arctic archipelago of Svalbard (Norway), remains the only *Rangifer* subspecies in which pestivirus seems to be absent (Stuen et al., 1993, das Neves et al., unpublished).

8.4.4 Diagnosis and Management

There are no reindeer-specific diagnostic tools available for pestivirus. The lack of isolation and characterization of circulating viruses in these populations and the lack of knowledge on possible clinical signs and pathology makes it more complicated to develop good diagnostic tools. Reindeer are nonetheless experimentally susceptible to BVDV (Morton et al., 1990) and most likely also to BDV, and existing diagnostic methods for these viruses may also be suitable for reindeer. Likewise, due to the genetic and antigenic similarity between reindeer pestivirus (V60) and BDV-2, some diagnostic solutions used for BDV can also be used for reindeer. For population screenings, there are several ELISA antibody and antigen diagnostic tests for both BVDV and BDV based on highly conserved viral proteins p80/p125 (Deregt et al., 1990), which therefore can detect a broad range of closely related viruses. Since these tests are designed for domestic animal species and not wildlife or reindeer, the choice of a competition or blocking design is preferred (Kautto et al., 2012). The SNT continues to be the gold standard test and several published protocols against different pestiviruses have been used for reindeer and other wild ungulates (Martin et al., 2011; Kautto et al., 2012). Young animals should be sampled to know if pestivirus is circulating in the flock, because of persisting antibody titers in older animals (Kautto et al., 2012).

Detection of viral RNA is possible by reverse transcription PCR (RT-PCR), and to increase the chances of identifying unknown pestiviruses in reindeer, assays should be designed to detect a broad range of pestiviruses (Elvander et al., 1998). From animals suspected of being infected (acutely or persistently), buffy coat cells, whole blood, leucocytes or serum are suitable sampling materials for RT-PCR. From dead animals, several tissue samples (such as lung and spleen) can also be processed.

BVDV and BDV are not strictly species specific. Hence, these viruses might be transmitted back to cattle/sheep from species not considered as primary hosts, a fact to be paid attention to when implementing management and control measures in bovine populations (Lindberg and Alenius, 1999). Control and eradication measures for pestivirus in domestic ruminants are mostly based on detection and removal of PI animals and protection of pestivirus-free herds. Similar control measures could be used in reindeer herds, even though there is no evidence of persistence of pestivirus in reindeer.

8.5 FOOT AND MOUTH DISEASE

Jörn Klein

8.5.1 AETIOLOGY

Foot and mouth disease (FMD) is a highly contagious viral disease, with a great potential for causing severe economic losses in susceptible cloven-hoofed animals. Reindeer and caribou, as well as other wildlife species, including several species of deer, bison, musk ox, Dall's sheep and moose, are susceptible to FMD. FMD is listed in the World Organization for Animal Health (OIE) *Terrestrial Animal Health Code* and must be reported to the OIE (Anonymous, 2011).

The causative agent of FMD is the foot and mouth disease virus (FMDV). The genome consists of positive sense single-stranded RNA of approximately 8.3 kb, and the virus belongs to the *Aphthovirus* genus of the family Picornaviridae. The high genetic and phenotypic variability of FMDV is reflected in the existence of seven serotypes: O, A, C, Asia 1, SAT 1, SAT 2 and SAT 3, and further numerous variants and lineages, described as topotypes (Samuel and Knowles, 2001). The antigenic diversity within those serotypes is crucial to consider when selecting vaccine strains, as cross-reactivity may be variable (Kitching et al., 2005). The virus capsid is composed of 60 copies of each of the four structural polypeptides, designated VP1 to VP4. A region in the G-H loop (around residues 140–160) of the VP1 protein exposed on the surface of the viral capsid was determined to be the main antigenic site recognized by neutralizing antibodies (Fowler et al., 2011).

8.5.2 PATHOGENESIS, PATHOLOGY AND CLINICAL SIGNS

The mechanism of spread of FMDV is primarily in the form of aerosols or saliva, or through indirect contact via people or contaminated surfaces, and then subsequently back to the respiratory tract (Alexandersen and Mowat, 2005). However, infection can also occur through lesions of the skin or mucous membranes, but this is a very inefficient entrance route, unless abrasions or cuts are present. Initial infection occurs predominantly at the epithelial surface of the soft palate and adjacent nasopharynx, with subsequent spread of virus through regional lymph nodes, followed by viraemia. The virus can be detected in the oral cavity by real time RT-PCR 1–3 days before the onset of viraemia (Alexandersen et al., 2003).

Clinical signs of FMD in cervidae vary; with red deer (*Cervus elaphus*) showing milder symptoms compared to white tailed deer (*Odocoileus virginianus*), roe deer (*Capreolus capreolus*) or muntjac deer (genus *Muntiacus*) (Moniwa et al., 2012). Clinical signs in white tailed deer appear to be similar to those in cattle, characterized by fever, often above 40°C, excessive salivation, lameness and depression. Lesions on the feet of white tailed deer seems to be more severe than those on the mucosa of the lips, dorsum of the tongue, or the dental plate (Moniwa et al., 2012). In red deer, clinical signs of FMD are either very mild or inapparent; however, indications for a possible carrier state in red deer are given (Kittelberger et al., 2015).

Little is published about pathogenesis and clinical signs in *Rangifer* species. A Swedish study from 1927 (Magnusson, 1927) described experimental infection of reindeer with FMDV. Clinical signs were described as an increase in body temperature, as well as lesions in the oral mucosa and on the hooves, but the report also stated that the FMDV infections in reindeer can be rather benign, probably reflecting the isolate of the FMDV in question. This is in line with the report from Kvitkin (1961), who described the clinical signs in reindeer after experimental infection as comparable to white tailed deer – as mild, with no or little salivation. On the contrary, Ogryzkov and coworkers (Ogryzkov, 1964) reported from an outbreak in Russia in 1955 a more severe disease, involving 12 reindeer herds and particularly affecting the calves.

Since all these published studies are relatively old, and a typing of the disease-causing FMDV could not be performed, it is difficult to determine the severity of FMD in reindeer. Modern studies have generally shown that the FMDV serotype A is often more virulent and causes more severe clinical signs than serotype O or C (Klein et al., 2006).

8.5.3   Subclinical and Persistent Infections

Inapparent infections of FMDV can be divided into two types: first, those animals that become infected and spread the virus without showing clinical signs, and second, those animals in which the virus persists in the host after it is recovering from FMD with clinical signs (Sutmoller and Casas, 2002).

Persistent infection by FMDV is defined as animals being virus positive for a minimum of 28 days, which are usually referred to as virus carriers (McVicar and Sutmoller, 1969; Sutmoller et al., 1968).

Unlike persistently infected animals, subclinically infected animals (i.e. those animals not developing clinical signs) may be highly contagious (Kitching, 2002). Besides the immune status of the host, the involved FMDV lineage also plays a role in the subclinical course of infection (Klein, 2009).

8.5.4   Epidemiology

Even though there are some common features in the spread of FMDV, each of the seven serotypes, and even their variants (topotypes), have different ways of transmission, clinical appearance and species tropism (Kitching, 2005).

The pandemic strain of FMDV serotype O is the most prevalent serotype worldwide, although there is no precise genetic explanation for this higher prevalence (Mason et al., 2003). Serotype A shows a great antigenic diversity between the variants of this serotype, and there is also often no immunological cross-protection between them (Klein et al., 2006; Wekesa et al., 2014; Zheng et al., 2015).

FMD has not been reported in North America for more than 60 years. The last U.S. outbreak occurred in 1929, while Canada and Mexico have been FMD-free since 1952–1953 (Mohler, 1952). In Scandinavia the last FMD outbreaks were in Norway in 1952, Finland in 1959, Sweden in 1966 (Westergaard et al., 2008) and Denmark in 1983 (Christensen et al., 2005). Outbreaks among semi-domesticated reindeer have been described in Russia up to 1955–1956 (Syroechkovskii, 1995). However, in the Trans-Baikal region of Russia, FMD outbreaks occur frequently (Anonymous, 2015a; Anonymous, 2015b), and it can be assumed that outbreaks in semi-domesticated reindeer herds also occur. A large and costly FMD outbreak, which also spread to France and the Netherlands, occurred in the UK in 2001 (Thompson et al., 2002).

8.5.5   Diagnosis and Management

Diagnosis of FMD is mainly done by virus isolation or by the detection of viral antigen or nucleic acid in tissues or fluid samples (blood, saliva, milk). Detection of virus-specific antibodies in blood can also be used for diagnosis, and antibodies to viral non-structural proteins (NSPs) can be used as indicators of infection, irrespective of vaccination status (Alexandersen et al., 2003). The preferred tissue for the diagnosis of FMD is epithelium from unruptured or freshly ruptured vesicles or vesicular fluid. Alternatively, oesophageal–pharyngeal fluid samples taken by probang cup in ruminants (Alexandersen et al., 2003) or throat swabs for qRT-PCR analysis can be used (Klein et al., 2008).

FMD is a genuine trans-boundary animal disease (TAD), severely affecting regional and international trade with animals and animal products. The World Organization for Animal Health (OIE) has therefore developed a Global FMD Control Strategy to control the disease, which divides countries into one of three disease states with regard to FMD: FMD present with or without vaccination, FMD-free with vaccination, and FMD-free without vaccination (Anonymous, 2011). North America and most of Europe, including Scandinavia, are FMD-free without vaccination, and outbreaks there would result in a stamping out procedure and destruction of animals, as well as targeted ring vaccination, although control measures are dependent on the country in question.

8.6 MISCELLANEOUS VIRAL INFECTIONS IN REINDEER AND CARIBOU

Morten Tryland

8.6.1 RABIES

Rabies is an acute viral infection of the brain in humans and most other warm-blooded animals. Rabies is caused by rabies virus, an RNA virus of genus *Lyssavirus* in the Rhabdoviridae family. Rabies virus is an enveloped, bullet-shaped virus particle, approximately 180 nm long and 80 nm wide. Whereas other lyssaviruses have bats as their main reservoirs, lyssavirus 1, known as the rabies virus in other mammal species, has its main reservoirs in foxes (*Vulpes vulpes* and others), raccoon dogs (*Nyctereutes procyonoides*), raccoons (*Procyon lotor*) and skunks (family Mephitidae) in the sub-arctic regions; the arctic fox (*Vulpes lagopus*) is the main reservoir throughout the arctic region (Mørk and Prestrud, 2004). Rabies is transmitted to reindeer and other hosts mainly by bites from a rabies-infected animal, shedding virus in the saliva. The virus enters nervous tissues at the inoculation site, and travels along nerves to the central nervous system (CNS), and may cause lameness. After reaching the brain, the virus proliferates and cause encephalomyelitis, before the virus is shed in saliva and the animal becomes infectious. During this phase, clinical symptoms are usually apparent in reindeer, consisting of weakness, incoordination, posterior paralysis, head weaving, confusion and aggression (Dieterich and Ritter, 1982). The animal will often lie flat on the ground, unable to stand or lie on its chest. Reindeer often show lesions and damage on their head and legs, caused by seizures and structures on the ground, and the disease will end fatally within a short time (Figure 8.11). Although reindeer and caribou are not reservoir hosts for rabies, these animals are affected from time to time during epizootics in foxes and other species. Rabies has been recorded in semi-domesticated reindeer (*R. t. tarandus*) and caribou (*R. t. granti*) in Canada (Dieterich and Ritter, 1982; Ritter, 1981), as well as in reindeer (*R. t. tarandus*) in Russia, in the northwest region (Mørk et al., 2011) and in the region of the Taymyr Peninsula further east (Revich et al., 2012).

Rabies was diagnosed for the first time in Svalbard reindeer (*R. t. platyrhynchus*) in 1981 (Ødegaard and Krogsrud, 1981). Investigations based on trapped arctic foxes has revealed a low prevalence of rabies, indicating that the virus is not endemic in the arctic fox population on the Svalbard archipelago, but is introduced from time to time by foxes migrating on sea ice, causing outbreaks (Prestrud et al., 1992, Mørk et al., 2011). The most recent outbreak, in which ten reindeer were found with rabies, occurred on the archipelago in 2011 (MacDonald et al., 2011). Since the outbreak occurred during the reindeer hunting season, prophylactic vaccination was conducted on 280 people. Hunters were advised to treat the meat so that contact with the CNS of the animal was avoided, to not eat the CNS, and to prepare the meat with thorough heat treatment (MacDonald et al., 2011).

In a vaccination trial using the vaccine Rabisin®, reindeer showed a good humoral response with more than 1.5 IU/ml 38 days post vaccination, but it was also revealed that 22 of 39 vaccinated animals had low antibody titers (<0.5 IU/ml) 1 year post-vaccination (Sihvonen et al., 1993).

FIGURE 8.11   A rabies outbreak occurred among arctic foxes on the Svalbard archipelago during 2011. Several Svalbard reindeer (*R. t. platyrhynchus*) were found with clinical rabies. This animal had seizures and the rocks on the ground have caused bleeding and damage (Photo: Per Andreassen).

8.6.2 PARAINFLUENZA VIRUS

Parainfluenzavirus 3 (PIV3) is an enveloped virus with a genome consisting of single-stranded and negative sense RNA, which belongs to genus *Respirovirus*, subfamily Paramyxovirinae in the virus family Paramyxoviridae. In cattle, PIV3 usually causes subclinical infections or only mild symptoms from the upper respiratory tract, but the virus may also pave the way for other potential pathogens, both viruses and bacteria. Serological screenings of cattle indicated that PIV3 is a very common pathogen. In Norway, 50.2% of 1348 calves > 150 days of age and representing 135 dairy herds had antibodies against PIV3 (Gulliksen et al., 2009). Antibodies against PIV3 have been detected in a wide range of wild ruminants (Frölich, 2000). In a serological study on semi-domesticated reindeer (*R. t. tarandus*) in Sweden (n = 50), PIV3 antibodies were detected in all the five herds investigated and in 54% of the animals (Rehbinder et al., 1991), whereas a similar screening of semi-domesticated reindeer (*R. t. tarandus*) in Finnmark County in Norway (n = 326) and Svalbard reindeer (*R. t. platyrhynchus*) (n = 40), no antibodies were detected against PIV3 (Stuen et al., 1993). Similarly, a serological screening of caribou in northern Quebec (n = 58; 1978–1979) revealed no antibodies against PIV3 (Elazhary et al., 1981). The role of PIV3 as a potential pathogen in reindeer and caribou is not known, but it is possible, when enzootic in the herds, that it may cause symptoms of the upper respiratory tract, especially in calves, and also induce secondary infections involving other pathogens such as *Pasteurella*, a syndrome often referred to as bovine respiratory disease complex (BRDC) in cattle (Grissett et al., 2015).

8.6.3 POLYOMAVIRUS

Polyomaviruses (family Polyomaviridae) are small virus particles with a genome consisting of a circular double-stranded DNA of approximately 5000 bp. Polyomaviruses are highly host specific and typically cause subclinical infections with lifelong persistence. In immunocompromized individuals, however, they may cause disease. There are no reports on serosurveys or isolation of polyomavirus from reindeer or caribou, but from eye and nose swab samples from a caribou (*R. t. granti*) from Alaska, evidence of a new polyomavirus was obtained by random amplification combined with a metagenomic analysis approach, although little is known about the possible clinical impact of such infections in reindeer and caribou (Schürch et al., 2014).

8.6.4   WEST NILE VIRUS

West Nile virus (WNV) is a zoonotic arthropodeborne virus (arbovirus) belonging to genus *Flavivirus* in the virus family Flaviviridae. It was first identified in the West Nile region of Uganda (1937), but after registered outbreaks in Algeria and Romania (1994–1996), and in New York City (1999) and elsewhere, it is now regarded as enzootic in parts of Europe, Africa, Asia, Australia, North America and Latin America (Nash et al., 2001). WNV is mainly transmitted by mosquitos (*Culex* spp. and others) and is maintained in enzootic cycles between the vector and avian hosts, with spread to humans and horses (Gray and Webb, 2014) and a wide range of wildlife species, including cervids (Santaella et al., 2005; for a review, see Jeffrey Root, 2013).

During the westward spread of WNV from New York, it was assumed that reindeer were not susceptible. Reindeer with clinical symptoms from the CNS were, as a matter of routine, tested for chronic wasting disease (CWD), a prion disease affecting the CNS, but not tested for WNV. In a tuberculosis experiment in reindeer in Iowa, USA, four reindeer (2–4 years old) were held as controls (Palmer et al., 2004). One day, two of the animals were found with fever (40°C), exhibiting paddling of all four limbs, and not being able to rise in sternal recumbency, and they were later euthanized. A third animal was observed febrile (40°C) and depressed, with tilted head, a flaccid tongue and dysphagia. The fourth animal was clinically normal and grazing, but was found dead the next day (Palmer et al., 2004). A diagnosis of WNV was established, which probably represents the first report of confirmed cases of WNV in reindeer. Since WNV may cause CNS symptoms in reindeer, WNV infection is, in enzootic regions, a differential diagnosis to CWD and also to brain worm (*Elaphostrongylus tarandi*) infections.

WNV surveillance is challenging, as the virus transmission cycle is complex and most human and animal cases remain asymptomatic (Gossner et al., 2017). The new distribution of WNV has demonstrated that mosquito-borne viruses are not restricted to tropical regions, which is relevant for *Rangifer* populations from a climatic change perspective.

8.6.5   VIRAL INFECTIONS AND CLIMATE CHANGE

With climatic changes, such as increases in precipitation and temperature that are predicted for the arctic region (Pachauri and Meyer, 2014), ecosystems will change over time. Conditions for all life will be affected, favoring some species but causing disadvantages for others. With altered distribution of species, introduction of new species and new insect vectors, the host-virus interactions will also change, and we may face viral infections that are new to the *Rangifer* species. Some examples, such as bluetongue virus (BTV) and Schmallenberg virus (SBV), are mentioned briefly in Chapter 16, Climate Change – Potential Impacts on Pasture Resources, and Health and Diseases of Reindeer and Caribou, when discussing climate change and vectors.

REFERENCES

Akhtar, J. and D. Shukla D. 2009. Viral entry mechanisms: Cellular and viral mediators of herpes simplex virus entry. *FEBS J.* 276(24):7228–36.
Alexandersen, S., M. Quan, C. Murphy, J. Knight and Z. Zhang. 2003. Studies of quantitative parameters of virus excretion and transmission in pigs and cattle experimentally infected with foot-and-mouth disease virus. *J Comp Pathol.* 129(4):268–82.
Alexandersen, S. and N. Mowat. 2005. Foot-and-mouth disease: Host range and pathogenesis. *Curr Top Microbiol Immunol.* 288:9–42.

Altmann, D., H. Kronberger and K. F. Schüppel. 1973. Bösartiges Katarrhalfieber (Coryza Gang- raenosa) bei zwei Elchen, zwei Rentieren und einer Hausziege in Thüringer Zoopark Erfurt. *Verhandlungsbericht des Internationalen Symposium* über *die Erkrankungen der Zootiere* 15:41–9.

Anonymous. 2011. Terrestrial animal health code: OIE World Organisation for Animal Health Paris. www.oie.int/international-standard-setting/terrestrial-code/access-online/ (25 January 2016).

Anonymous. 2015a. World Organisation for Animal Health. World animal health information database (wahid) interface. www.oie.int/wahis_2/public/wahid.php/Wahidhome/Home (25 January 2016).

Anonymous. 2015b. Surveillance, Federal Service for Veterinary and Phytosanitary. Epizoonotic Situation. www.fsvps.ru/fsvps/ya/chronology/ (25 January 2016).

Antonsson, A. and B. G. Hansson. 2002. Healthy skin of many animal species harbors papillomaviruses which are closely related to their human counterparts. *J Virol.* 76:12537–42.

Arnal, M., D. Fernandez-de-Luco, L. Riba, M. Maley, J. Gilray, K. Willoughby, S. Vilcek and P. F. Nettleton. 2004. A novel pestivirus associated with deaths in Pyrenean chamois (*Rupicapra pyrenaica pyrenaica*). *J Gen Virol.* 85:3653–7.

Aschfalk, A., T. D. Josefsen, H. Steingass, W. Müller and R. Goethe. 2003. Crowding and winter emergency feeding as predisposing factors for kerato-conjunctivitis in semi-domesticated reindeer in Norway. *Dtsch Tierarztl Wochenschr.* 110:295–8.

Avalos-Ramirez, R., M. Orlich, H. J. Thiel and P. Becher. 2001. Evidence for the presence of two novel pestivirus species. *Virology.* 286:456–65.

Baxter, S. I. F., I. Pow, A. Bridgen and H. W. Reid. 1993. PCR detection of the sheep-associated agent of malignant catarrhal fever. *Arch Virol.* 132:145–59.

Becher, P., M. Orlich, A. D. Shannon, G. Horner, M. Konig and H. J. Thiel. 1997. Phylogenetic analysis of pestiviruses from domestic and wild ruminants. *J Gen Virol.* 78:1357–66.

Becher, P., M. Orlich, A. Kosmidou, M. Konig, M. Baroth and H. J. Thiel. 1999. Genetic diversity of pestiviruses: Identification of novel groups and implications for classification. *Virology.* 262(1):64–71.

Becher, P., R. Ramirez, M. O. Avalos, S. C. Rosales, M. Konig, M. Schweizer, H. Stalder, H. Schirrmeier and H. J. Thiel. 2003. Genetic and antigenic characterization of novel pestivirus genotypes: Implications for classification. *Virology.* 311(1):96–104.

Bergman, A. 1912. Contagious keratitis in reindeer. *Scand Vet J.* 2:145–77.

Brownlie, J., M. C. Clarke and C. J. Howard. 1984. Experimental production of fatal mucosal disease in cattle. *Vet Rec.* 114(22):535–6.

Büttner, M., von Einem C., McInnes C., Oksanen A. 1995. Clinical findings and diagnosis of a severe parapoxvirus epidemic in Finnish reindeer. *Tierarztl Prax.* 23(6):614–8. (German).

Carlsson, U. 1991. Border disease in sheep caused by transmission of virus from cattle persistently infected with bovine virus diarrhoea virus. *Vet Rec.* 128(7):145–7.

Christensen, L. S., P. Normann, S. Thykier-Nielsen, J. H. Sorensen, K. de Stricker and S. Rosenorn, S. 2005. Analysis of the epidemiological dynamics during the 1982–1983 epidemic of foot-and-mouth disease in Denmark based on molecular high-resolution strain identification. *J Gen Virol.* 86:2577–84.

Crawford, T. B., H. Li, S. R. Rosenberg, R. W. Norhausen and M. M. Garner. 2002. Mural folliculitis and alopecia caused by infection with goat-associated malignant catarrhal fever virus in two sika deer. *J Am Vet Med Assoc.* 221(6):843–7.

Cunha, C. W., L. Otto, N. S. Taus, D. P. Knowles and H. Li. 2009. Development of a multiplex real-time PCR for detection and differentiation of malignant catarrhal fever viruses in clinical samples. *J Clin Microbiol.* 47(8):2586–9.

Curry, P. S., C. Ribble, W. C. Sears, W. Hutchins, K. Orsel, D. Godson, R. Lindsay, A. Dibernardo and S. J. Kutz. 2014. Blood collected on filter paper for wildlife serology: Detecting antibodies to *Neospora caninum*, West Nile virus, and five bovine viruses in reindeer. *J Wildl Dis.* 50(2):297–307. doi: 10.7589/2012-02-047.

Damon, I. 2007. Poxviruses. In *Fields Virology*, D. M. Knipe and P. M. Howley (eds.), 5th edition, Lippincott, Williams & Wilkins, London, UK, pp. 2947–75.

das Neves, C. G., T. Mørk, J. Thiry, J. Godfroid, E. Rimstad, E. Thiry and M. Tryland. 2009a. Cervid herpesvirus 2 experimentally reactivated in reindeer can produce generalized viremia and abortion. *Virus Res.* 145(2):321–8.

das Neves, C. G., E. Rimstad and M. Tryland. 2009b. Cervid herpesvirus 2 causes respiratory and fetal infections in semidomesticated reindeer. *J Clin Microbiol.* 47(5):1309–13. doi: 10.1128/JCM.02416-08.

das Neves, C. G., T. Mørk, J. Godfroid, K. K. Sørensen, E. Breines, E. Hareide, J. Thiry, E. Rimstad, E. Thiry and M. Tryland. 2009c. Experimental infection of reindeer with Cervid herpesvirus 2. *Clin Vacc Immunol.* 16(12):1758–65.

das Neves, C. G., J. Thiry, E. Skjerve, N. G. Yoccoz, E. Rimstad, E. Thiry and M. Tryland. 2009d. Alphaherpesvirus infections in semidomesticated reindeer: A cross-sectional serological study. *Vet Microbiol.* 139(3–4):262–9.

das Neves, C. G., S. Roth, E. Rimstad, E. Thiry and M. Tryland. 2010. Cervid herpesvirus 2 infection in reindeer: A review. *Vet Microbiol.* 143(1):70–80.

das Neves, C. G., H. I. Ihlebæk, E. Skjerve, W. Hemmingsen, H. Li and M. Tryland. 2013. Gammaherpesvirus infection in semidomesticated reindeer (*Rangifer tarandus tarandus*): A cross-sectional, serologic study in Northern Norway. *J Wildl Dis.* 49(2):261–9. doi: 10.7589/2012-07-185.

Davison, A. J., R. Eberle, B. Ehlers, G. S. Hayward, D. J. McGeoch, A. C. Minson, P. E. Pellett, B. Roizman, M. J. Studdert and E. Thiry. 2009. The order Herpesvirales. *Arch Virol.* 154:171–7.

Deregt, D., S. A. Masri, H. J. Cho and H. Bielefeldt Ohmann. 1990. Monoclonal antibodies to the p80/125 gp53 proteins of bovine viral diarrhea virus: Their potential use as diagnostic reagents. *Can J Vet Res.* 54(3):343–8.

Dieterich, R. A. and D. G. Ritter. 1982. Rabies in Alaskan reindeer. *J Am Vet Med Assoc.* 181:1416.

Ek-Kommonen, C., S. Pelkonen and P. F. Nettleton. 1986. Isolation of a herpesvirus serologically related to bovine herpesvirus 1 from a reindeer (*Rangifer tarandus*). *Acta Vet Scand.* 27:299–301.

Elazhary, M. A. S. Y., J. L. Frechette, A. Silim and R. S. Roy. 1981. Serological evidence of some bovine viruses in the caribou (*Rangifer tarandus caribou*) in Quebec. *J Wildl Dis.* 17(4):609–12.

Elvander, M., C. Baule, M. Persson, L. Egyed, A. Ballagi-Pordany, S. Belák and S. Alenius. 1998. An experimental study of a concurrent primary infection with bovine respiratory syncytial virus (BRSV) and bovine viral diarrhoea virus (BVDV) in calves. *Acta Vet Scand.* 39(2):251–64.

Engels, M. and M. Ackermann. 1996. Pathogenesis of ruminant herpesvirus infections. *Vet Microbiol.* 53:3–15.

Erdeélyi, K., L. Dencso, R. Lehoczki, M. Heltai, K. Sonkoly, S. Csányi and N. Solymosi, N. 2009. Endemic papillomavirus infection of roe deer (*Capreolus capreolus*). *Vet Microbiol.* 138:20–6.

Evans, A. L., C. G. das Neves, G. F. Finstad, K. B. Beckmen, E. Skjerve, I. H. Nymo and M. Tryland. 2012. Evidence of alphaherpesvirus infections in Alaskan caribou and reindeer. *BMC Vet Res.* 8:5. doi: 10.1186/1746-6148-8-5.

Fowler, V. L., J. B. Bashiruddin, F. F. Maree, P. Mutowembwa, B. Bankowski, D. Gibson, S. Cox, N. Knowles and P. V. Barnett. 2011. Foot-and-mouth disease marker vaccine: Cattle protection with a partial vp1 g-h loop deleted virus antigen. *Vaccine.* 29(46):8405–11. doi: 10.1016/j.vaccine.2011.08.035.

Frölich, K. 2000. Viral diseases of northern ungulates. *Rangifer.* 20:83–97.

Giangaspero, M., R. Harasawa, K. Muschko and M. Buttner. 2006. Characteristics of the 5' untranslated region of wisent (*Bison bonasus*) and reindeer (*Rangifer tarandus*) Pestivirus isolates. *Vet Ital.* 42(3):165–72.

Gossner, C. M., L. Marrama, M. Carson, F. Allerberger, P. Calistri, D. Dilaveris, S. Lecollinet, D. Morgan, N. Nowotny, M. C. Paty, D. Pervanidou, C. Rizzo, H. Roberts, F. Schmoll, W. Van Bortel and A. Gervelmeyer. 2017. West Nile virus surveillance in Europe: Moving towards an integrated animal-human-vector approach. *Euro Surveill.* 22(18). pii: 30526. doi: 10.2807/1560-7917.ES.2017.22.18.30526.

Graham, D. A. 2013. Bovine herpes virus-1 (BoHV-1) in cattle – A review with emphasis on reproductive impacts and the emergence of infection in Ireland and the United Kingdom. *Ir Vet J.* 66(1):15. doi: 10.1186/2046-0481-66-15.

Gray, T. J. and C. E. Webb. 2014. A review of the epidemiological and clinical aspects of West Nile Virus. *Int J Gen Med.* 7:193–203.

Grissett, G. P., B. J. White and R. L. Larson. 2015. Structured literature review of responses of cattle to viral and bacterial pathogens causing bovine respiratory disease complex. *J Vet Intern Med.* 29(3):770–80.

Groff, D. E., J. P. Sundberg and W. D. Lancaster. 1983. Extrachromosomal deer fibromavirus DNA in deer fibromas and virus-transformed mouse cells. *Virology.* 131(2):546–50.

Gulliksen, S. M., E. Jor, K. I. Lie, T. Løken, J. Åkerstedt, and O. Østerås. 2009. Respiratory infections in Norwegian diary calves. *J Diary Sci.* 92:5139–46.

Hautaniemi, M., N. Ueda, J. Tuimala, A. A. Mercer, J. Lahdenperä and C. J. McInnes. 2010. The genome of pseudocowpoxvirus: Comparison of a reindeer isolate and a reference strain. *J Gen Virol.* 91:1560–76. doi: 10.1099/vir.0.018374-0.

Hautaniemi, M., F. Vaccari, A. Scacliarini, S. Laaksonen, A. Huovilainen and C. J. McInnes. 2011. Analysis of deletion within the reindeer pseudocowpoxvirus genome. *Virus Res.* 160:326–32.

Hirvelä-Koski, V. 2001. Porojen suutauti: Kyselytutkimus taudin esiintymisestä ja merkityksestä porotaloudelle. *Abstract in English.* In Anonymous. ISBN 952-453-052-X, Oulun yliopistopaino, Oulu.

Howley, P. M. and D. R. Lowy. 2007. Papillomaviruses. In *Fields Virology*, D. M. Knipe and P. M. Howley (eds.), 5th edition, vol. 2, Lippincott Williams & Wilkins, London, UK, pp. 2299–354.

Inoshima, Y., A. Morooka and H. Sentsui. 2000. Detection and diagnosis of parapoxvirus by the polymerase chain reaction. *J Virol Meth.* 84:201–8.

Jeffrey Root, J. 2013. West Nile virus associations in wild mammals: A synthesis. *Arch Virol.* 158(4):735–52.

Kautto, A. H., S. Alenius, T. Mossing, P. Becher, S. Belák and M. Larska. 2012. Pestivirus and alphaherpesvirus infections in Swedish reindeer (*Rangifer tarandus* L.). *Vet Microbiol.* 156(1–2):64–71.

Keel, M. K., J. G. Patterson, T. H. Noon, G. A. Bradley and J. K. Collins. 2003. Caprine herpesvirus-2 in association with naturally occurring malignant catarrhal fever in captive sika deer (Cervus nippon). *J Vet Diagn Invest.* 15(2):179–83.

Kim, K. S., S. A. Park, K. N. Ko, S. Yi and Y. J. Cho. 2014. Current status of human papillomavirus vaccines. *Clin Exp Vaccine Res.* 3(2):168–75.

Kitching, R. P. 2002. Identification of foot and mouth disease virus carrier and subclinically infected animals and differentiation from vaccinated animals. *Rev Sci Tech.* 21(3):531–8.

Kitching, R. P. 2005. Global epidemiology and prospects for control of foot-and-mouth disease. *Curr Top Microbiol Immunol.* 288:133–48.

Kitching, R. P., A. M. Hutber and M. V. Thrusfield. 2005. A review of foot-and-mouth disease with special consideration for the clinical and epidemiological factors relevant to predictive modeling of the disease. *Vet J.* 169:197–209.

Kittelberger, R., C. Nfon, K. Swekla, Z. Zhang, K. Hole, H. Bittner, T. Salo, M. Goolia, C. Embury-Hyatt, R. Bueno, M. Hannah, R. Swainsbury, C. O'Sullivan, R. Spence, R. Clough, A. McFadden, T. Rawdon and S. Alexandersen. 2015. Foot-and-mouth disease in red deer – Experimental infection and test methods performance. *Transbound Emerg Dis.* doi: 10.1111/tbed.12363.

Kiupel, M., A. Wise, S. Bolin, P. Walker, T. Marshall and R. Maes. 2004. Malignant catarrhal fever in reindeer (*Rangifer tarandus*). In *Proceedings of the European Association of Zoo and Wildlife Veterinarians 5th Scientific Meeting.* European Association of Zoo and Wildlife Veterinarians, 19–23 May, Ebeltoft, Denmark, pp. 73–8.

Klein, J. and M. Tryland. 2005. Characterisation of parapoviruses isolated from Norwegian semi-domesticated reindeer (*Rangifer tarandus tarandus*). *Virol J.* 2:79. doi: 10.1186/1743-422X-2-79.

Klein, J., U. Parlak, F. Ozyoruk and L. S. Christensen. 2006. The molecular epidemiology of foot-and-mouth disease virus serotypes A and O from 1998 to 2004 in Turkey. *BMC Vet Res.* 2:35.

Klein, J., M. Hussain, M. Ahmad, M. Afzal and S. Alexandersen. 2008. Epidemiology of foot-and-mouth disease in Landhi dairy colony, Pakistan, the world largest buffalo colony. *Virol J.* 5:53.

Klein, J. 2009. Understanding the molecular epidemiology of foot-and-mouth-disease virus. *Infect Genet Evol.* 9(2):153–61.

Kummeneje, K. 1976. Pasteurellosis in reindeer in Northern Norway: A contribution to its epidemiology. *Acta Vet Scand.* 17:488–94.

Kummeneje, K. and J. Krogsrud. 1979. Contagious ecthyma (orf) in reindeer (*Rangifer t. tarandus*). *Vet Rec.* 105:60–1.

Kvitkin, J. P. 1961. Blood picture in the reindeer in normal and pathological cases. *Vet Res St Saratov Sci Works.* pp. 25–48. (In Russian).

Larska, M. 2015. Pestivirus infection in reindeer. *Front Microbiol.* 6:1187. doi: 10.3389/fmicb.2015.01187.

Li, H., W. C. Westover and T. B. Crawford. 1999. Sheep associated malignant catarrhal fever in a petting zoo. *J Zoo Wildl Med.* 30(3):408–12.

Li, H., T. C. McGuire, U. Muller-Doblies and T. B. Crawford. 2001. A simpler, more sensitive competitive inhibition enzyme-linked immunosorbent assay for detection of antibody to malignant catarrhal fever virus. *J Vet Diagn Invest.* 13:361–4.

Li, H., A. Wunschmann, J. Keller, D. G. Hall and T. B. Crawford. 2003. Caprine herpesvirus-2-associated malignant catarrhal fever in white-tailed deer (*Odocoileus virginianus*). *J Vet Diagn Invest.* 15(1):46–9.

Li, H., N. S. Taus, G. S. Lewis, K. Okjin, D. L. Traul and T. B. Crawford. 2004. Shedding of ovine herpesvirus 2 in sheep nasal secretions: The predominant mode for transmission. *J Clin Microbiol.* 42(12):5558–64.

Li, H., C. W. Cunha and N. S. Taus. 2011. Malignant catarrhal fever: Understanding molecular diagnostics in context of epidemiology. *Int J Mol Sci.* 12:6881–93.

Li, H, C. W. Cunha, N. S. Taus and D. P. Knowles. 2014. Malignant catarrhal fever: Inching toward understanding. *Annu Rev Anim Biosci.* 2:209–33. doi: 10.1146/annurev-animal-022513-114156.

Lillehaug, A., T. Vikøren, I. L. Larsen, J. Akerstedt, J. Tharaldsen and K. Handeland. 2003. Antibodies to ruminant alpha-herpesviruses and pestiviruses in Norwegian cervids. *J Wildl Dis.* 39(4):779–86.

Lindberg, A. L. and S. Alenius. 1999. Principles for eradication of bovine viral diarrhoea virus (BVDV) infections in cattle populations. *Vet Microbiol.* 64(2–3):197–222.

Liu, L., H. Xia, N. Wahlberg, S. Belák and C. Baule. 2009. Phylogeny, classification and evolutionary insights into pestiviruses. *Virology*. 385(2):351–7. doi: 10.1016/j.virol.2008.12.004.

Løken, T. 1990. Pestivirus infections in Norway. Epidemiological studies in goats. *J Comp Pathol*. 103(1):1–10.

Løken, T. 1992. Pestivirus infections in ruminants in Norway. *Rev Sci Tech*. 11(3):895–9.

MacDonald, E., K. Handeland, H. Blystad, M. Bergsaker, M. Fladberg, B. Gjerset, O. Nilsen, Ø. Os, S. Sandbu, E. Stokke, L. Vold, I. Ørpetveit, H. A. Gaup and O. Tveiten. 2011. Public health implications of an outbreak of rabies in arctic foxes and reindeer in the Svalbard archipelago, Norway, September 2011. *Euro Surveill*. 16(40).

Magnusson, H. 1927. Är renen mottaglig för muloch klövsjuka? *Skand Vet I*. 17:49–62.

Marco, I. 2012. Pestivirus of chamois and border disease. In *Infectious Diseases of Wild Mammals and Birds in Europe*, D. Gavier-Widen, J. P. Duff and A. Meredith (eds.), Wiley-Blackwell, p. 568.

Marco, I., O. Cabezón, R. Velarde, L. Fernández-Sirera, A. Colom-Cadena, E. Serrano, R. Rosell, E. Casas-Díaz and S. Lavín. 2015. The two sides of border disease in Pyrenean chamois (*Rupicapra pyrenaica*): Silent persistence and population collapse. *Anim Health Res Rev*. 16(01):70–7. doi: 10.1017/S1466252315000055.

Martin, C., C. Letellier, B. Caij, D. Gauthier, N. Jean, A. Shaffii and C. Saegerman. 2011. Epidemiology of Pestivirus infection in wild ungulates of the French South Alps. *Vet Microbiol*. 147(3–4):320–8. doi: 10.1016/j.vetmic.2010.07.010.

Mason, P. W., J. M. Pacheco, Q. S. Zhao and N. J. Knowles. 2003. Comparisons of the complete genomes of Asian, African and European isolates of a recent foot-and-mouth disease virus type O pandemic strain (PanAsia). *J Gen Virol*. 84:1583–93.

McVicar, J. W. and P. Sutmoller. 1969. The epizootiological importance of foot-and-mouth disease carriers. II. The carrier status of cattle exposed to foot-and-mouth disease following vaccination with an oil adjuvant inactivated virus vaccine. *Arch Gesamte Virusforsch*. 26(3):217–24.

Mohler, J. R. 1952. Foot-and-mouth disease. US Dept. of Agriculture. Series Volume 666.

Moniwa, M., C. Embury-Hyatt, Z. Zhang, K. Hole, A. Clavijo, J. Copps and S. Alexandersen. 2012. Experimental foot-and-mouth disease virus infection in white tailed deer. *J Comp Pathol*. 147(2–3):330–42.

Moreno-Lopéz, J., T. Morner and U. Pettersson. 1986. Papillomavirus DNA associated with pulmonary fibromatosis in European elks. *J Virol*. 57(3):1173–6.

Moreno-Lopéz, J., H. Ahola, A. Eriksson, P. Bergmann and U. Pettersson. 1987. Reindeer papillomavirus transforming properties correlate with a highly conserved E5 region. *J Virol*. 61(11):3394–400.

Mørk, T. and P. Prestrud. 2004. Arctic rabies – A review. *Acta Vet Scand*. 45(1–2):1–9.

Mørk, T., J. Bohlin, E. Fuglei, K. Åsbakk and M. Tryland. 2011. Rabies in the arctic fox *population*, Svalbard, Norway. *J Wildl Dis*. 47:945–57.

Morton, J., J. F. Evermann and R. A. Dieterich. 1990. Experimental infection of reindeer with bovine viral diarrhea virus. *Rangifer*. 10(2):75–7.

Murphy, F. A., E. P. J. Gibbs, M. C. Horzinek and M. J. Studdert. 1999. *Herpesvirirdae*. In *Veterinary Virology*, 3rd edition, Academic Press, San Diego, California, pp. 301–25.

Muylkens, B. J. Thiry, P. Kirten, F. Schynts and E. Thiry. 2007. Bovine herpesvirus 1 infection and infectious bovine rhinotracheitis. *Vet Res*. 38(2):181–209.

Nash, D., F. Mostashari, A. Fine, J. Miller, D. O'Leary, K. Murray, A. Huang, A. Rosenberg, A. Greenberg, M. Sherman, S. Wong and M. Layton. 2001. The outbreak of West Nile virus infection in the New York City area in 1999. *N Engl J Med*. 344(24):1807–14.

Nettleton, P. F. 1990. Pestivirus infections in ruminants other than cattle. *Rev Sci Tech*. 9(1):131–50.

Nordkvist, M. 1973. Munvårtsjuka - en ny rensjukdom? *Rennäringsnytt*. 8–9:6–8.

Ødegaard, Ø. A. and J. Krogsrud. 1981. Rabies in Svalbard: Infection diagnosed in arctic fox, reindeer and seal. *Vet Rec*. 109:141–2.

Ogryzkov, S. E. 1964. The pathology of foot and mouth disease in reindeer. Paper presented at the Trudy II vses. Konf. patol. Anat. Zhivotnykh, Mosk. Vet Akad., 1964.

Oksanen, A., S. Laaksonen and V. Hirvelä-Koski. 1996. Pink-eye in a winter-corralled reindeer herd. *Suomen eläinlääkärilehti*. 102:3.

Pachauri, R. K. and L. A. Meyer. 2014. IPCC. Climate Change 2014: Synthesis Report. Contribution of Working Groups I, II and III to the Fifth Assessment Report of the Intergovernmental Panel on Climate Change. IPCC, Geneva, Switzerland, 151 pp.

Palatsi, R., A. Oksanen, R. Sormunen and J. Karvonen. 1993. The first Orf virus epidemic diagnosed in man and reindeer in 1992–1993 in Finland. *Duodecim*. 109(21):1945–50. (In Finnish).

Palmer, M. V., W. C. Stoffregen, D. G. Rogers, A. N. Hamir, J. A. Richt, D. D. Pedersen and W. R. Waters. 2004. West Nile virus infection in reindeer (*Rangifer tarandus*). *J Vet Diagn Invest*. 16:219–22.

Passler, T., S. S. Ditchkoff, M. D. Givens, K. V. Brock, R. W. DeYoung and P. H. Walz. 2010. Transmission of bovine viral diarrhea virus among white-tailed deer (*Odocoileus virginianus*). *Vet Res.* 41(2):20. doi: 10.1051/vetres/2009068.

Peletto, S., C. Caruso, F. Cerutti, P. Modesto, S. Zoppi, A. Dondo, P. L. Acutis and L. Masoero. 2015. A new genotype of border disease virus with implications for molecular diagnostics. *Arch Virol.* doi: 10.1007/s00705-015-2696-4.

Pellett, P. E. and B. Roizman. 2007. The family *Herpesviridae*: A brief introduction. In *Fields Virology*, D. M. Knipe and P. M. Howley (eds.), 5th edition, Lippincott, Williams & Wilkins, London, UK, pp. 2479–99.

Powers, J. G., D. C. VanMetre, J. K. Collins, R. P. Dinsmore, J. Carman, G. Patterson, D. Brahmbhatt and R. J. Callan. 2005. Evaluation of ovine herpesvirus type 2 infections, as detected by competitive inhibition ELISA and polymerase chain reaction assay, in dairy cattle without clinical signs of malignant catarrhal fever. *J Am Vet Med Assoc.* 227(4):606–11.

Prestrud, P., J. Krogsrud and I. Gjertz. 1992. The occurrence of rabies in the Svalbard Islands of Norway. *J Wildl Dis.* 28:57–63.

Rehbinder, C., S. Belák and M. Nordkvist. 1991. A serological, retrospective study in reindeer on five different viruses. *Rangifer.* 12(3):191–5.

Rehbinder, C. and A. Nilsson. 1995. An outbreak of kerato-conjunctivitis among corralled, supplementary fed, semi-domesticated reindeer calves. *Rangifer.* 15:9–14.

Rehbinder, C. and S. Nikander. 1999. Ren och rensjukdomar. Studentlitteratur, Lund, Sweden. 247 pp. ISBN 91-44-01138-5.

Revich, B., N. Tokarevich and A. J. Parkinson. 2012. Climate change and zoonotic infections in the Russian Arctic. *Int J Circumpolar Health.* 71:18792.

Ridpath, J. F. 2015. Emerging pestiviruses infecting domestic and wildlife hosts. *Anim Health Res Rev.* 16(01):55–9. doi: 10.1017/S1466252315000067.

Ritter, D. 1981. Rabies. In *Alaskan Wildlife Diseases*, R. A. Dieterich (ed.), University of Alaska, Fairbanks, Alaska.

Rockborn, G., C. Rehbinder, B. Klingeborn, M. Lefler, K. Klintevall, T. Nikkilä, A. Landén and M. Nordkvist. 1990. The demonstration of a herpesvirus, related to bovine herpesvirus 1, in reindeer with ulcerative and necrotizing lesions of the upper alimentary tract and nose. *Rangifer.* Special issue No. 3:373–84.

Roehe, P. M., M. J. Woodward and S. Edwards. 1992. Characterisation of p20 gene sequences from a border disease-like pestivirus isolated from pigs. *Vet Microbiol.* 33(1–4):231–8.

Ros, C. and S. Belák. 1999. Studies of genetic relationships between bovine, caprine, cervine, and rangiferine alphahepresviruses and improved molecular methods for virus detection and identification. *J Clin Microbiol.* 37(5):1247–53.

Ros, C. and S. Belák. 2002. Characterization of the glycoprotein B gene from ruminant alphaherpesviruses. *Virus Genes.* 24(2):99–105.

Samuel, A. R. and N. J. Knowles. 2001. Foot-and-mouth disease type O viruses exhibit genetically and geographically distinct evolutionary lineages (topotypes). *J Gen Virol.* 82:609–21.

Sánchez Romano, J., T. Mørk, S. Laaksonen, E. Ågren, I. H. Nymo, M. Sunde, and M. Tryland. 2018. Infectious keratoconjunctivitis in semi-domesticated Eurasian tundra reindeer (*Rangifer tarandus tarandus*): Microbiological study of clinically affected and unaffected animals with special reference to cervid herpesvirus 2. *BMC Vet Res.* 14(1):15. doi: 10.1186/s12917-018-1338-y.

Santaella, J., R. McLean, J. S. Hall, J. S. Gill, R. A. Bowen, H. H. Hadow and L. Clark. 2005. West Nile virus serosurveillance in Iowa white-tailed deer (1999–2003). *Am J Trop Med Hyg.* 73:1038–42.

Schürch, A. C., D. Schipper, M. A. Bijl, J. Dau, K. B. Beckmen, C. M. Schapendonk, V. S. Raj, A. D. Osterhaus, B. L. Haagmans, M. Tryland and S. L. Smits. 2014. Metagenomic survey for viruses in Western Arctic Caribou, Alaska, through iterative assembly of taxonomic units. *PLoS One.* 9(8):e105227. doi: 10.1371/journal.pone.0105227.

Schweizer, M. and E. Peterhans. 2014. Pestiviruses. *Annu Rev Anim Biosci.* 2:141–63. doi: 10.1146/annurev-animal-022513-114209.

Sihvonen, L., K. Kulonen, T. Soveri and M. Nieminen. 1993. Rabies antibody titers in vaccinated reindeer. *Acta Vet Scand.* 34:199–202.

Skinner, M. A., R. M. Buller, I. K. Damon, E. J. Lefkowitz, G. McFadden, C. J. Mc Innes, A. A. Mercer, R. W. Moyerand C. Upton. 2012. *Poxviridae*. In *Virus Taxonomy: Classification and Nomenclature of Viruses: Ninth Report of the International Committee on Taxonomy of Viruses*, A. M. Q. King, M. J. Adams, E. B. Carstens and E. J. Lefkowitz (eds.), Elsevier Academic Press, San Diego, California pp. 291–309.

Skjenneberg, S., and L. Slagsvold. 1968. Reindriften og dens naturgrunnlag. Universitetsforlaget, Oslo, Norway.

Smits, S. L., C. M. Schapendonk, M. van Leeuwen, T. Kuiken, R. Bodewes, V. Stalin Raj, B. L. Haagmans, C. G. das Neves, M. Tryland and A. D. Osterhaus. 2013. Identification and characterization of two novel viruses in ocular infections in reindeer. *PLoS One* 8(7):e69711. doi: 10.1371/journal.pone.0069711.

Stuen, S., J. Krogsrud, B. Hyllseth and N. J. C. Tyler. 1993. Serosurvey of three virus infections in reindeer in northern Norway and Svalbard. *Rangifer.* 13(4):215–9.

Sutmoller, P., J. W. McVicar and G. E. Cottral. 1968. The epizootiological importance of foot-and-mouth disease carriers. I. Experimentally produced foot-and-mouth disease carriers in susceptible and immune cattle. *Arch Gesamte Virusforsch.* 23:227–35.

Sutmoller, P. and O. R. Casas. 2002. Unapparent foot and mouth disease infection (sub-clinical infections and carriers): Implications for control. *Rev Sci Tech.* 21:519–29.

Syrjälä, P., H. Saarinen, T. Laine, T. Kokkonen and P. Veijalainen. 2006. Malignant catarrhal fever in pigs and a genetic comparison of porcine and ruminant virus isolates in Finland. *Vet Rec.* 159(13):406–9.

Syroechkovskii, E. E. 1995. Diseases and parasites. In *Wild Reindeer.* E. Syroechkovskii and D. R. Klein (eds.). Science Publishers, Enfield, New Hampshire, pp. 152–67.

Thiry, J., V. Keuser, B. Muylkens, F. Meurens, S. Gogev, A. Vanderplasschen and E. Thiry. 2006. Ruminant alphaherpesviruses related to bovine herpesvirus 1. *Vet Res.* 37:169–90.

Thiry, J., F. Widen, F. Gregoire, A. Linden, S. Belák and E. Thiry. 2007. Isolation and characterisation of a ruminant alphaherpesvirus closely related to bovine herpesvirus 1 in a free-ranging red deer. *BMC Vet Res.* 3:26. doi: 10.1186/1746-6148-3-26.

Thompson, D., P. Muriel, D. Russell, P. Osborne, A. Bromley, M. Rowland, S. Creigh-Tyte and C. Brown. 2002. Economic costs of the foot and mouth disease outbreak in the United Kingdom in 2001. *Rev Sci Tech.* 21(3):675–87.

Tikkanen, M. K., C. J. McInnes, A. A. Mercer, M. Büttner, J. Tuimala, V. Hirvelä-Koski, E. Neuvonen and A. Huovilainen. 2004. Recent isolates of parapoxvirus of Finnish reindeer (*Rangifer tarandus tarandus*) are closely related to bovine pseudocowpox virus. *J Gen Virol.* 85:1413–18.

Traul, D. L., N. S. Taus, J. Lindsay Oaks, D. O'Toole, F. R. Rurangirwa, T. V. Baszler and H. Li. 2007. Validation of nonnested and real-time PCR for diagnosis of sheep-associated malignant catarrhal fever in clinical samples. *J Vet Diagn Invest.* 19(4):405–8.

Tryland, M., T. D. Josefsen, A. Oksanen and A. Aschfalk. 2001. Parapoxvirus infection in Norwegian semi-domesticated reindeer (*Rangifer tarandus tarandus*). *Vet Rec.* 149(13):394–5.

Tryland, M. 2002. Asymptomatic parapoxvirus infections in semi-domesticated reindeer (*Rangifer tarandus tarandus*). In *Proceedings of the XIVth International Poxvirus & Iridovirus Workshop*, 20–25 September, Lake Placid, New York, P75, p. 169.

Tryland, M., T. Mørk, K. Ryeng and K. K. Sørensen, K. 2005. Evidence of parapox-, alphaherpes- and pestivirus infections in carcasses of semi-domesticated reindeer (*Rangifer tarandus tarandus*) from Finnmark, Norway. *Rangifer.* 25(2):75–83.

Tryland, M., C. G. das Neves, M. Sunde and T. Mørk T. 2009. Cervid herpesvirus 2, the primary agent in an outbreak of infectious keratoconjunctivitis in semidomesticated reindeer. *J Clin Microbiol.* 47(11):3707–13. doi: 10.1128/JCM.01198-09.

Tryland, M., J. Klein, T. Berger, T. D. Josefsen, C. G. das Neves, A. Oksanen and K. Åsbakk. 2013. Experimental parapoxvirus infection (contagious ecthyma) in semi-domesticated reindeer (*Rangifer tarandus tarandus*). *Vet Microbiol.* 162(2–4):499–506. doi: 10.1016/j.vetmic.2012.10.039.

Tryland, M., J. S. Romano, N. Marcin, I. H. Nymo, T. D. Josefsen, K. K. Sørensen, and T. Mørk. 2017. Cervid herpesvirus 2 and not *Moraxella bovoculi* caused infectious keratoconjunctivitis in experimentally inoculated semi-domesticated Eurasian tundra reindeer. *Acta Vet Scand.* 59(1):23. doi: 10.1186/s13028-017-0291-2.

Tryland, M., K. B. Beckmen, K. A. Burek-Huntington, E. M. Breines and J. Klein. 2018. Orf virus infection in Alaskan mountain goats, Dall's sheep, muskoxen, caribou and Sitka black-tailed deer. *Acta Vet Scand.* 60(1):12. doi: 10.1186/s13028-018-0366-8.

Vikøren, T., H. Li, A. Lillehaug, C. M. Jonassen, I. Böckerman and K. Handeland. 2006. Malignant catarrhal fever in free-ranging cervids associated with OvHV-2 and CpHV-2 DNA. *J Wildl Dis.* 42(4):797–807.

Vikøren, T., S. Klevar, H. Li and A. G. Hauge. 2013. Malignant catarrhal fever virus identified in free-ranging musk ox (*Ovibos moschatus*) in Norway. *J Wildl Dis.* 49(2):447–50.

Vikøren, T., S. Klevar, H. Li and A. G. Hauge. 2015. A geographic cluster of malignant catarrhal fever in moose (*Alces alces*) in Norway. *J Wildl Dis.* 51(2):471–4.

Vilcek, S. and P. F. Nettleton. 2006. Pestiviruses in wild animals. *Vet Microbiol.* 116(1–3):1–12.

Wang, J., J. O'Keefe, D. Orr, L. Loth, M. Banks, P. Wakeley, D. West, R. Card, G. Ibata, K. Van Maanen, P. Thoren, M. Isaksson and P. Kerkhofs. 2007. Validation of a real-time PCR assay for the detection of bovine herpesvirus 1 in bovine semen. *J Virol Meth*. 144(1–2):103–8.

Wekesa, S. N., A. K. Sangula, G. J. Belsham, V. B. Muwanika, R. Heller, S. N. Balinda, C. Masembe and H. R. Siegismund. 2014. Genetic diversity of serotype A foot-and-mouth disease viruses in Kenya from 1964 to 2013; implications for control strategies in eastern Africa. *Infect Genet Evol*. 21:408–17. doi: 10.1016/j.meegid.2013.12.006.

Westergaard, J. M., C. B. Andersen and S. Mortensen. 2008. A foot and mouth disease simulation exercise involving the five Nordic countries. *Rev Sci Tech*. 27:751–8.

Zarnke, R. L. 1983. Serologic survey for selected microbial pathogens in Alaskan wildlife. *J Wildl Dis*. 19(4):324–9.

Zarnke, R. L., H. Li and T. B. Crawford. 2002. Serum antibody prevalence of malignant catarrhal fever viruses in seven wildlife species from Alaska. *J Wildl Dis*. 38(3):500–4.

Zheng, H., K. Lian, F. Yang, Y. Jin, Z. Zhu, J. Guo, J. Cao, H. Liu, J. He, K. Zhang, D. Li and X. Liu. 2015. Cross-protective efficacy of engineering serotype a foot-and-mouth disease virus vaccine against the two pandemic strains in swine. *Vaccine*. 33:5772–8.

9 Prions and Chronic Wasting Disease (CWD)

Morten Tryland

CONTENTS

9.1 CHRONIC WASTING DISEASE (CWD)

Chronic wasting disease (CWD) is a transmissible spongiform encephalopathy (TSE). TSEs, commonly known as prion diseases, are characterized as fatal, progressive neurodegenerative disorders which can affect man, domestic animals and wildlife. Of several TSEs in humans, Creutzfeldt-Jakob disease (CJD) and a new variant of CJD (vCJD), the latter being transmitted to man from cattle with Bovine spongiform encephalopathy (BSE), are the most well known. Examples of TSEs in animals are scrapie (sheep and goats; classical and atypical forms), Feline Spongiform Encephalopathy (FSE, cats), and CWD, affecting free-ranging and captive cervids. CWD is the only known TSE that affects free-ranging, non-domestic animals (Haley and Hoover, 2015). These diseases are all caused by prions.

TEXTBOX 9.1   THE APPEARANCE OF CHRONIC WASTING DISEASE (CWD) IN REINDEER IN NORWAY

The first CWD case in free-ranging cervids outside North America, and the first natural case in a reindeer (Figure 9.1), appeared unexpectedly in a wild Eurasian tundra reindeer (*R. t. tarandus*) in the Nordfjella mountain region, Norway (Figure 9.2) on March 15, 2016, during field research conducted by Norwegian Institute for Nature Research (NINA) (Benestad et al., 2016; Vikøren et al., 2016). While a helicopter was being used to approach reindeer to track the herd, a sick animal that was unable to follow the flock was observed. The animal, a 3- to 4-year-old female in fairly good to suboptimal condition (43 kg), died shortly after (Figure 9.1) and was routinely necropsied. Muscular hemorrhages were evident, but with no other macroscopic pathological findings.

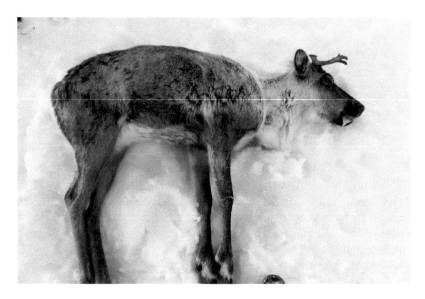

FIGURE 9.1   The first documented case of CWD in reindeer. A herd of wild Eurasian tundra reindeer (*R. t. tarandus*) in Nordfjella, Norway, in 2016, was approached by a helicopter. The animal left the herd and was soon after found in a moribund stage (Photo: Lars Nesse).

In May 2016, two moose (*Alces alces*) were diagnosed with a prion disease in the Selbu municipality, approximately 300 kilometers north–northeast from the first reindeer case, and during fall 2017, a red deer (*Cervus elaphus*) was diagnosed with CWD (Figure 9.2). In the Nordfjella region, where the first CWD reindeer case was diagnosed in 2016, several more CWD cases were detected, and a decision was made to cull all the wild reindeer in the northern management zone of Nordfjella in an attempt to eradicate the disease. During a regular hunt, followed by culling of the population by professional hunters (winter 2017–2018), a total of 18 CWD positive reindeer have been detected, totaling 12 out of approximately 1000 animals tested (December 2017).

The appearance of CWD in wild cervids in Norway represents a turning point in wildlife disease surveillance and management in Norway. A large-scale screening is being conducted to try to identify the prevalence and the geographical distribution of CWD in Norway, testing fallen stock and road kills, hunted animals and slaughtered semi-domesticated reindeer, as well as farmed red deer. These screenings will form baseline knowledge on how CWD can be managed in Norway, and represent guidelines on how to avoid spread to other cervid populations in neighboring countries, including semi-domesticated reindeer. Preliminary laboratory investigations and characterization of the prions indicates that the wild reindeer cases are similar to the CWD present in North America, whereas some results suggest the cases in moose and red deer might represent atypical CWD cases in older animals, being presumably less contagious.

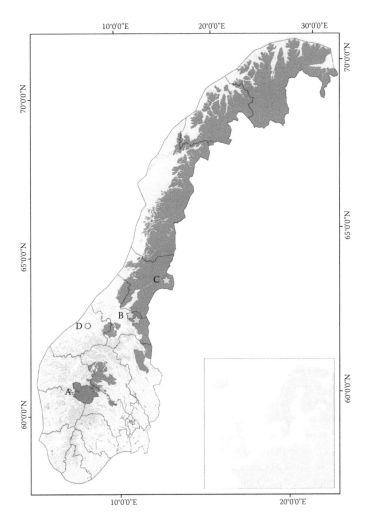

FIGURE 9.2   The geographical locations of the first documented cases of CWD in Norway (April 2018): (A) 18 wild Eurasian tundra reindeer (*R. t. tarandus*) in Nordfjella, (B) two moose (*Alces alces*) in Selbu, (C) one moose in Lierne and (D) one red deer (*Cervus elaphus*) in Gjemnes, Norway, 2016–2017 (Map: Bernt Johansen and Javier Sanchez Romano).

9.2   WHAT IS A PRION?

The diseases that are known today as prion diseases were for several decades called "slow virus infections," due to the assumption that they were caused by viruses having long incubation periods (Gajdusek, 1967). However, high doses of ultraviolet radiation, enough to inactivate most viruses, did not inactivate prions, whereas treatment with protein denaturation procedures did (Alper et al., 1978). The somewhat challenging conclusion was that the transmissible agent consisted of a partly proteinase resistant protein, named a "proteinaceous infectious particle," or a prion (Prusiner, 1982). A conformational misfolding (i.e. altered secondary structure) of the normal cellular prion protein (PrP^C) generates a prion, termed PrP^{Sc} (a term first used for scrapie). Thus, the amino acid sequence of PrP^C and PrP^{Sc} is identical, but the secondary protein structure has changed, from predominantly α-helices to one dominated by β-sheets. A prion consists of approximately 250 amino acids, but they tend to form aggregates (Saunders et al., 2012), and studies have shown that aggregates containing about 12 PrP^{Sc} molecules are the most infectious fractions, with a size of 12 nm, which is still only about half the size of the smallest known viruses, such as parvovirus (18–26 nm) (Silveira et al., 2005). Prions exhibit unusual resistance to physical and chemical procedures commonly used for microbial inactivation and disinfection, such as radiation, autoclaving (121°C for 1 hour) and a range of alcohols and detergents, but may be inactivated by autoclaving at 134°C (at 3 atm pressure, 1 hour) and with chlorine, sodium hydroxide and a few other chemicals with a strong protein denaturing effect (Hörnlimann et al., 2006).

9.3   PATHOGENESIS, PATHOLOGY AND CLINICAL SIGNS

A conformational misfolding of the normal cellular prion protein (PrP^C) into disease-provoking prions may occur spontaneously, or as a result of introduction of prions from an external source, since the pathogenic prion (PrP^{Sc}) can template further misfolding of PrP^C, thereby propagating the prion. The misfolding of the normal prion protein to a prion changes the overall secondary and tertiary structure of the protein, which makes them more resistant to endogenous proteases and breakdown. Hence, they aggregate and accumulate in the brain tissue, causing damage to neurons and causing the "spongiform" appearance of brain tissue, which is reflected in the common pathological term for these diseases, transmissible spongiform encephalopathies (TSEs). This progressive neurodegeneration leads to loss of neuronal functions with associated clinical signs, such as CNS symptoms, and the condition is always fatal.

TSEs generally have a long incubation period. Based on experimental oral infections, the incubation period for the onset of clinical signs was estimated to be about 15 months in mule deer (*Odocoileus hemionus hemionus*) and 12 and 34 months in Rocky Mountain elk (*Cervus elaphus nelsoni*), with most of the clinical cases observed in animals between 2 and 7 years of age (Williams and Miller, 2002). Infectious CWD prions have been detected in saliva, urine and blood, suggesting these body fluids are important in the horizontal transmission and dissemination of the disease. CWD has been transmitted experimentally to white tailed deer (*Odocoileus virginianus*) via saliva and blood (Mathiason et al., 2006).

In North American deer species, such as mule deer, elk (Wapiti, *Cervus canadensis*) and moose (*Alces shirasi*), the body condition in subclinical or early stages of CWD can be normal (Haley and Hoover, 2015). In later stages of the disease, animals are typically emaciated and may have aspiration pneumonia and watery rumen contents. Other clinical symptoms are altered behaviour and loss of fear of humans, depression, increased salivation, ruminal atony and a gradual loss of body condition (wasting).

Histopathology of lesions in clinically affected cervids are similar to those seen in other ruminant TSEs. Specific lesions are found in the gray matter of the CNS, and the lesions are bilaterally symmetrical, with neuronal degeneration and spongiform changes (vacuolization), and gliosis (i.e. a nonspecific reactive change of glial cells), but inflammatory cell response is not apparent (Williams, 2005). Experimentally, CWD in reindeer has appeared in a similar manner as in other susceptible cervid species, with weight loss, uneven hair coat, ataxia, excessive salivation and grinding of the teeth, before entering terminal stages of CWD (Mitchell et al., 2012).

The first reindeer diagnosed with CWD in Norway (see Textbox 9.1) showed abnormal behaviour and ataxia, and died upon stress when approached by a helicopter (Benestad et al., 2016). In contrast, most of the other wild reindeer from Nordfjella that have tested positive for CWD did not show clinical symptoms.

9.4   EPIDEMIOLOGY

CWD was first observed in captive mule deer in Colorado, USA, in 1967 (Williams and Young, 1980). From being diagnosed in two to three US states up until the mid-1990s, CWD has, during the past two decades, had an apparently rapid spread, now being present in 23 US states and in the provinces of Alberta and Saskatchewan in Canada. This increased detection over time may, however, also partly reflect increased surveillance efforts (Miller and Fischer, 2016). CWD has also been detected in two outbreaks in farmed elk in South Korea, after import of CWD positive animals from Canada (Sohn et al., 2002).

In captive herds with a recent introduction of CWD, the prevalence may be low (1%) but may reach 100% in endemic research facilities (EFSA, 2017). In free-ranging populations in enzootic regions of the USA, the prevalence has been estimated to be 0.5% for Rocky Mountain elk, 2.1% for white tailed deer and 4.9% for mule deer (Miller et al., 2000), but may reach levels of up to 30% (Saunders et al., 2012; EFSA, 2017). CWD has also been diagnosed in moose (*Alces alces shirasi*) in North America (Sigurdson, 2008).

In spite of potential habitat overlap with CWD-infected deer and elk herds in the enzootic regions of the USA and Canada, CWD has not been reported in any *Rangifer* subspecies of this region (i.e. *R. t. tarandus*, *R. t. caribou* and *R. t. granti*) (Mitchell et al., 2012). Further, an investigation of 100 caribou in Quebec, Canada, did not detect CWD in any of the animals (Lapointe et al., 2002).

Experimentally, CWD was successfully transmitted from white tailed deer (CWD^{WTD}) to reindeer (Mitchell et al., 2012). Two groups of three 6-month-old reindeer calves were orally inoculated with 5 grams each of brain tissue. Two of the three animals receiving CWD^{WTD} developed clinical symptoms 17–18 months post inoculation (pi), whereas the third animal, as well as three reindeer that had been inoculated with prions from Rocky Mountain elk (CWD^{ELK}), remained healthy; the latter group was euthanized after 23–61 months (Mitchell et al., 2012). The symptoms of the two affected reindeer consisted of weight loss and uneven hair coat, and after displaying ataxia, excessive salivation and grinding of the teeth, both individuals developed terminal CWD at 18.5 and 20 months pi, respectively (Mitchell et al., 2012). These experiments showed that reindeer were susceptible to PrP^{WTD}, and that clinical cases among reindeer and caribou may be expected in regions where CWD is enzootic in other cervids (Mitchell et al., 2012; Haley and Hoover, 2015).

In another experiment, reindeer were challenged by intracranial inoculation with brain material from CWD-affected elk, mule deer or white tailed deer. In this experiment, reindeer were susceptible to CWD from all of the three sources, and horizontal transmission, via contaminated urine, faeces and saliva, occurred. Clinical signs were observed 20.9 months after inoculation. Animals were found dead without clinical signs recorded, or showed loss of body condition and were found in a lethargic state on the ground (Moore et al., 2016).

The source of CWD prions in reindeer, moose and red deer in Norway 2016 is currently not known (VKM, 2016). Since import of live cervids to Norway is prohibited, importation of infected cervids is not likely to be the source of CWD in Norway. The use of urine lures to attract cervids during hunting is one possible source. These products were available but are now banned.

Another speculation is that the appearance of CWD in the USA, and Norway, may have arisen from scrapie in sheep (Williams and Young, 1980; Benestad et al., 2016). As for Norway, very few cases of classical scrapie have been diagnosed in sheep, the last case being in 2009. This is also the situation for atypical scrapie (Nor98), with only five to 13 cases annually the past 10 years. This, together with the fact that the distinctive molecular signature of Nor98 is a multiband western blot pattern, which was not seen for the reindeer prion, makes an association between CWD and scrapie in sheep less likely (Benestad et al., 2016).

During the period 2003–2004, semi-domesticated reindeer from Finnmark County, Norway (n = 771) (S. Sigurdarson, Iceland, personal communication) and from Finland (n = 1300) (H. Tapiovaara, Finnish Food Safety Authority Evira, personal communication) all tested negative for CWD, and during subsequent testing of a limited number of cervids each year for 12 years, no indications were found that this disease has occurred (VKM, 2016).

Due to the fact that the screenings that have been conducted have been somewhat separated in time and space and that a very low proportion of the cervid populations has been investigated, the possibility that CWD may already have existed in Norway for a decade or more cannot be excluded (EFSA, 2017).

9.5 ZOONOTIC POTENTIAL OF CWD

There is strong evidence that BSE prions crossed the species barrier and was transmitted to humans, causing the new variant of Creutzfeldt-Jakob disease (vCJD), predominantly in young people (Béringue et al., 2008). Since CWD prions are present in almost any tissue, such as muscle, fat, glands and organs, antler velvet and nervous tissues, including CNS, the potential for human exposure through handling and consumption of an infected cervid is almost unavoidable, and the degree of human exposure will increase with increasing CWD prevalence in the cervid populations (Saunders et al., 2012). There are no indications that CWD has caused disease in humans in the USA and Canada, but since CWD recently was detected in a new species in Norway (i.e. reindeer), and since the CWD prions detected in Norway in reindeer and moose may be different from each other and have not yet been fully characterized, these are good reasons for caution with respect to interspecies CWD transmission, including the zoonotic aspect (VKM, 2016).

9.6 DIAGNOSIS AND MANAGEMENT

CWD is usually diagnosed based on detection of accumulations of PrP^{CWD} in tissue sections of medulla oblongata of the brain, at the level of the obex (dorsal motor nucleus of the vagus nerve). This can be conducted by ELISA methods (Enzyme Linked Immunosorbent Assay) or western blot tests, often supported by immunohistochemistry (IHC) and pathological evaluations (Figure 9.3). A recent study presented a method for rapid ante-mortem detection of CWD proteins in saliva from exposed deer, with and without clinical symptoms of CWD, called real time quaking conversion (RT-QuIC), which may increase the sensitivity of detection (Henderson et al., 2013).

Bioassay studies are based on inoculation of experimental animals, usually rodents, and have revealed infectious prions in saliva, blood and faeces of infected deer (Mathiason et al., 2006) and white tailed deer (Haley et al., 2009), but the detection may have restricted sensitivity. PrP^{CWD} can be detected in live animals (i.e. during the incubation phase) by investigating biopsies from lymphatic tissues, such as the tonsils or lymphatic tissues of the rectum. This technique may be applicable for screening smaller herds (e.g. farmed cervids), but may seem laborious and less applicable for wild cervids or semi-domesticated reindeer.

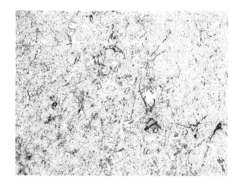

FIGURE 9.3 Immunohistochemistry of CNS (ventral midbrain) of a reindeer, showing accumulation of abnormal prion protein stained red brown (×100) (Photo: Sylvie Benestad).

Screenings of free-ranging cervids are necessary to monitor presence and prevalence of CWD. In the USA and Canada, surveillance programs are established to screen brain tissues from farmed and wild cervids. In the EU, a screening of brain tissues from hunted deer was conducted in 2006–2007 with negative results. In Norway, during 2004–2015, a total of 2163 cervids (i.e. wild and semi-domesticated reindeer, moose, red deer, roe deer and fallow deer) were screened with negative results (VKM, 2016).

Following the detection of CWD in reindeer, moose and red deer in Norway, a large-scale screening program is being conducted on fallen stock and slaughtered and hunted cervids of all species, with a certain focus around the geographical regions in which the CWD clinical cases have been identified (Figure 9.2). So far (March 2018) more than 40 000 cervids of different species and populations have been tested in Norway, revealing a total of 22 CWD cases. Eighteen of these cases are wild reindeer of the now culled Nordfjella population, in which all the CWD cases in reindeer so far have been found.

One of the most common flaws in CWD control efforts, based on the experience from North America, has been the underestimation of the affected area, and of how long the disease has been present. Based on the findings so far, it is possible that CWD was present for as long as a decade before it was detected in 2016 (Miller and Fischer, 2016; EFSA, 2017).

9.7   POTENTIAL CONSEQUENCES FOR CERVID POPULATIONS AND REINDEER HERDING

In the USA, eradication of CWD from farmed cervids is the goal of state, federal and industry programs, whereas eradication in free-ranging populations is regarded as unlikely (Williams, 2005).

Should CWD become enzootic in wild and semi-domesticated reindeer in Norway and maybe in Fennoscandia, it will represent a dramatic threat to the wild reindeer and other cervid populations, but will also presumably be devastating to the reindeer herding industry. Wherever space and natural resources allows, reindeer herding is based on a semi-nomadic production, with seasonal translocation of animals between summer and winter pastures, sometimes over large geographical distances. This allows calf production and growth on rich spring and summer pastures, and the use of more vulnerable winter pastures as maintenance food through winter, often with a substantial element of lichens. Live reindeer are also commonly traded between reindeer herders – again, often transported over vast geographical distances.

Translocation of CWD-infected animals has been the most important mode of spread of CWD in the USA and Canada (EFSA, 2017). Thus, restrictions on animal movements may be considered as one of several measures to try to restrict the spread of CWD (EFSA, 2017). It should, however, be kept in mind that reindeer herding, as we know it today, is based on and often totally dependent on translocation of live animals between different pastures. Strict regulations on animal movements may thus represent a major obstacle in offering summer and winter pastures for the animals. These aspects have to be taken into account when defining geographical units (i.e. infected and CWD free), should CWD enter the reindeer herding industry or become enzootic in the wild cervid populations. Further, the contact herds for each reindeer herd (i.e. which herds have contact through sharing pastures, corrals and transport vehicles) have to be identified, to be able to define effective management measures.

Another major impact on reindeer herding, should CWD be introduced and become enzootic, is that the disease, if it would appear in reindeer as in other cervid species, will affect animals between 2 and 7 years of age (EFSA, 2017). The number of animals developing clinical CWD, which will indicate the potential economic impact of the disease, can probably be partly reduced by increasing the frequency of slaughter of calves and young animals. This would, however, affect the age structure of the herd, with consequences that are not so easy to predict. Older animals are important to the herd, since they are leading animals, know the migration routes, have strength to dig in deep snow for food in winter and are experienced in supporting a calf through its first winter.

In designing control strategies, the mode of disease transmission is critical (Uehlinger et al., 2016). The effects of non-selective culling to decrease the population density and restrict the spread of CWD has been investigated, with the results depending on the assumed mode of transmission of CWD. If infection rate is determined by host density (density dependent), culling may contribute to a reduction in transmission of CWD. However, if the transmission is frequency dependent, transmission is independent of host density, but is rather associated to close contact within social groups (McCallum et al., 2001; Potapov et al., 2012), which probably applies to gregarious cervids like reindeer (Figures 9.4–9.6). Population control (i.e. reduced herd size) is thus less likely to be effective for semi-domesticated reindeer herds or for wild reindeer populations in reducing the transmission rate.

FIGURE 9.4   Semi-domesticated reindeer are gathered in Fram Tamreinlag, one of the herds bordering the Nordfjella wild reindeer herd where CWD has been detected in three animals. Stress, aerosols, dust and high animal density are factors facilitating transmission of infectious agents, including prions (Photo: Kjell Bitustøyl).

FIGURE 9.5   Wild reindeer gathering at Hardangervidda, forming groups with high animal density (Photo: Autocamera NINA/Olav Strand).

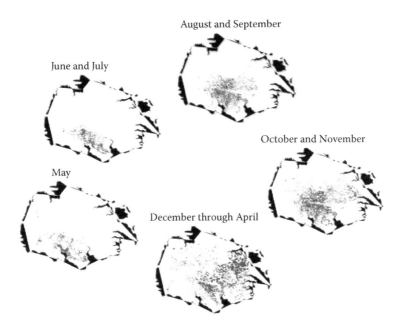

FIGURE 9.6   Map of the Hardangervidda mountain region (8 200 km²), south of the Nordfjella wild reindeer population. The Hardangervidda wild reindeer population constitutes approximately 12–14 000 animals in the summer herd. Signals from 101 GPS radio-collared reindeer (purple spots) during the period 2001–2016 show that the animals appear as a rather dense herd in every season of the year. In contrast to other cervid populations, culling of a wild reindeer population may thus not reduce transmission pressure for infectious diseases (Map: Olav Strand).

REFERENCES

Alper T., D. A. Haig, and M. C. Clark. 1978. The scrapie agent: Evidence against its dependence for replication on intrinsic nucleic acid. *J. Gen. Virol.* 41:503–16. doi: 10.1099/0022-1317-41-3-503.

Benestad, S. L., G. Mitchell, M. Simmons, B. Ytrehus and T. Vikøren. 2016. First case of chronic wasting disease in Europe in a Norwegian free-ranging reindeer. *Vet. Res.* 47(1):88. doi: 10.1186/s13567-016-0375-4.

Béringue, V., J. L. Vilotte and H. Laude. 2008. Prion agent diversity and species barrier. *Vet. Res.* 39(4):47. doi: 10.1051/vetres:2008024.

EFSA - Panel on Biological Hazards. 2017. Authors: S. Benestad, D. Gavier-Widen, M. W. Miller, G. Ru, G. C. Telling, M. Tryland, A. O. Pelaez and M. Simmons. Scientific opinion on chronic wasting disease (CWD) in cervids. *EFSA Journal* 15(1):4667, 72 pp. doi: 10.2903/j.efsa.2017.4667.

Gajdusek, D. C. 1967. Slow-virus infections of the nervous system. *N. Engl. J. Med.* 276(7):392–400.

Haley, N. J., D. M. Seelig, M. D. Zabel, G. C. Telling and E. A. Hoover. 2009. Detection of CWD prions in urine and saliva of deer by transgenic mouse bioassay. *PLoS One* 4:e4848. 10.1371/journal.pone.0004848.

Haley, N. J. and E. A. Hoover. 2015. Chronic wasting disease of cervids: Current knowledge and future perspectives. *Annu. Rev. Anim. Biosci.* 3:305–25. doi: 10.1146/annurev-animal-022114-111001.

Henderson, D. M., M. Manca, N. J. Haley et al. 2013. Rapid antemortem detection of CWD prions in deer saliva. *PLoS One.* 8(9):e74377. doi: 10.1371/journal.pone.0074377.

Hörnlimann B., D. Riesner and H. Kretzschamar. 2006. *Prions in Humans and Animals.* De Gruyter Berlin, New York, 683 pp.

Lapointe, J. M., D. Leclair, C. Mesher and A. Balachandran. 2002. Screening for chronic wasting disease in caribou in Northern Quebec. *Can. Vet. J.* 43:886–7.

Mathiason, C. K., J. G. Powers, S. J. Dahmes et al. 2006. Infectious prions in the saliva and blood of deer with chronic wasting disease. *Science* 314:133–6. doi: 10.1126/science.1132661.

McCallum, H., N. Barlow and J. Hone. 2001. How should pathogen transmission be modelled? *Trends Ecol. Evol.* 16(6):295–300.

Miller, M. W., E. S. Williams, C. W. McCarty et al. 2000. Epizootiology of chronic wasting disease in free-ranging cervids in Colorado and Wyoming. *J. Wildl. Dis.* 36:676–90.

Miller, M. W. and J. R. Fischer. 2016. The first five (or more) decades of chronic wasting disease: Lessons for the five decades to come. Transactions of the North American Wildlife and Natural Resources Conference 81. cpw.state.co.us/Documents/Research/CWD/Miller-Fischer_CWDlessons.pdf.

Mitchell, G. B., C. J. Sigurdson, K. I. O'Rourke et al. 2012. Experimental oral transmission of chronic wasting disease to reindeer (*Rangifer tarandus tarandus*). *PLoS One* 7(6):e39055:1–8.

Moore, S. J., R. Kunkle, M. H. W. Greenlee et al. 2016. Horizontal transmission of chronic wasting disease in reindeer. *Emerg. Infect. Dis.* 22(12):2142–5. doi: 10.3201/eid2212.160635.

Potapov, A., E. Merrill and M. A. Lewis. 2012. Wildlife disease elimination and density dependence. *Proc. Biol. Sci.* 279(1741):3139–45.

Prusiner, S. 1982. Novel proteinaceous infectious particles cause scrapie. *Science* 216(4542):136–44.

Saunders, S. E., S. L. Bartelt-Hunt and J. C. Bartz. 2012. Occurrence, transmission and zoonotic potential of chronic wasting disease. *Emerg. Infect. Dis.* 18(3):369–76.

Sigurdson, C. J. 2008. A prion disease of cervids: Chronic wasting disease. *Vet. Res.* 39:41.

Silveira, J. R., G. J. Raymond, A. G. Hughson et al. 2005. The most infectious prion protein particles. *Nature* 437:257–61. doi: 10.1038/nature03989.

Sohn, H. J., J. H. Kim, K. S. Choi et al. 2002. A case of chronic wasting disease in an elk imported to Korea from Canada. *J. Vet. Med. Sci.* 64(9):855–8.

Uehlinger, F. D., A. C. Johnston, T. K. Bollinger and C. L. Waldner. 2016. Systematic review of management strategies to control chronic wasting disease in wild deer populations in North America. *BMC Vet. Res.* 12(1):173.

Vikøren, T., S. Benestad, M. Haugum et al. 2016. Prionsjukdommen Chronic Wasting Disease (CWD) påvist på hjortedyr i Noreg. *Norsk veterinærtidsskrift* 128:405–6 (In Norwegian).

VKM. 2016. CWD in Norway. The Norwegian Scientific Committee for Food Safety. Authors: M. A. Tranulis, M. Tryland, G. Kapperud, E. Skjerve, R. Gudding, D. Grahek-Ogden. ISBN: 978-82-8259-216-1, Oslo, Norway.

Williams, E. S. and S. Young. 1980. Chronic wasting disease of captive mule deer: A spongiform encephalopathy. *J. Wildl. Dis.* 16:89–98.

Williams, E. S. and M. W. Miller. 2002. Chronic wasting disease in deer and elk in North America. *Rev. Sci. Tech. Off. Int. Epiz.* 21(2):305–16.

Williams, E. S. 2005. Chronic wasting disease. *Vet. Pathol.* 42(5):530–49.

The Impact of Infectious
Agents on *Rangifer* Populations

Anja M. Carlsson, Andy Dobson and Susan J. Kutz

CONTENTS

10.1 INTRODUCTION

Rangifer populations globally are host to a variety of infectious agents including viruses, bacteria, helminths (worms), protozoa and arthropods (see Chapters 6–9 for more details). Infectious agents are not only incredibly common, they are also an integral part of ecosystems with population and community level effects. They can interact with other ecological drivers, resulting in both detrimental and beneficial effects on ecosystem health and biodiversity (Gómez and Nichols 2013). Highly pathogenic agents, or those invading naïve populations, can have obvious and dramatic consequences for the dynamics and abundance of their host species. For example, the fungal disease white-nose syndrome has killed more than five million bats in North America, and chytrid fungus has wiped out amphibians in regions around the world. They can also have other and further reaching consequences, such as the recent 2016 anthrax outbreak in reindeer in Russia. The outbreak had social, economic and ecological consequences; 2,300 reindeer died, with significant economic loss for reindeer herders, who were also moved out of the area due to the zoonotic potential of the disease, which one boy succumbed to. Impacts of infectious agents may also be more subtle, where they negatively affect hosts without causing overt disease. Yet, they may still influence birth and death rates of individuals and, thereby, influence population dynamics. Subtle impacts may only become noticeable in certain contexts, such as in conjunction with other stressors, such as severe weather, poor habitat, predation and other infectious agents. Through these interacting and subtle effects, pathogens that initially appear less virulent may have equally important impacts on host fitness (reproduction and/or survival) and population dynamics as more virulent pathogens.

Despite the accumulating evidence demonstrating the importance of infectious agents for host population dynamics and ecosystem functioning, they are still often overlooked in wildlife biology and management (Joseph et al. 2013). While it is accepted that individual animals that are heavily infected show signs of disease, poor productivity and sometimes death, the role that infectious agents play at the population level, particularly in wild populations, has been vastly under-recognized, with the exception of a few key studies (Albon et al. 2002; Hudson et al. 1998). (See also Chapters 6–9 on parasites, bacteria, viruses and prions and their impacts on *Rangifer*.)

In this chapter we (i) present a number of case studies that illustrate the different mechanisms through which infectious agents can influence wildlife populations and (ii) outline a mathematical framework that allows us to examine how they can influence the population dynamics of *Rangifer*. Our intent is not to review all the literature pertaining to the impact of infectious agents in *Rangifer*, but to give an overview of how and when infections may have consequences for the dynamics and viability of wild *Rangifer* populations. To do this, we present case studies from a variety of different species and systems, using *Rangifer* examples where available. We also discuss some of the challenges involved and opportunities available for understanding the role of infectious agents in free-ranging wildlife.

10.1.1 Glossary

Important words and concepts discussed in the text are defined in the glossary below. These words are *italicized* the first time they appear in the text.

GLOSSARY	
Anthelmintics	A group of drugs that act against helminths (parasitic worms including nematodes, trematodes and cestodes).
Arrested development (Hypobiosis)	A temporary pause in the development of a parasite at specific point in its lifecycle.
Carrying capacity	The maximum abundance of a species that the environment can support in any particular location.

Cross-sectional study	A study where observations of individuals occur at a single time point, or over a short, defined, time period, providing a snapshot in time.
Definitive host	The host species in which sexual reproduction of a parasite takes place.
Demography	The study of the birth and death rates of a population and how they vary with size, growth, density, distribution, and available resources.
Density-dependent transmission	The phenomenon where the transmission rate of an infectious agent increases with host density.
Epidemiology	The branch of medicine that deals with the understanding of the incidence, distribution and control of diseases.
Epizootic	An outbreak of disease affecting many animals at the same time.
Frequency-dependent transmission	The phenomenon where the transmission rate of an infectious agent is associated with the frequency of interactions between hosts. This may be linked to host density, but not necessarily.
Intermediate host	An obligate host that is used by a parasite during its lifecycle. The intermediate host is required for the parasite to develop to the next stage. Depending on the parasite species, it may multiply asexually in the intermediate host, but never sexually.
Longitudinal study	A study that involves repeated observations of the same variables and individuals over time.
Macroparasites	Macroscopic parasites that do not multiply within their definitive hosts and usually have medium-short generation times: worms, ticks, bot flies (see Section 10.5.7).
Microparasites	Microscopic infectious agents that multiply within their *definitive* hosts, short generation times: viruses, bacteria, protozoa, fungi (see Section 10.5.4).
Prevalence of infection	Percentage of infected hosts in a given sample.
Recrudescence	Reappearance of symptoms and pathology of a disease after a period of recovery.
Reservoir host	A host species that may maintain an infectious agent and serves as a source from which other animals can be infected.
Sympatric host	A host species that occurs within the same or overlapping geographical area of another host species.
Sylvatic host	A wild animal host.
Tolerance	The host's ability to survive in the face of infection.
Population regulation	The tendency of a population to decrease in size when it is above a particular level but increase when below that level, as governed by density-dependent processes which return the population to its equilibrium density. Processes that influence birth and death rates (and immigration and emigration rates) in a density-dependent manner have the potential to regulate a population (see Section 10.7.2).
Vector	An organism, usually an arthropod, which may carry a parasite from one host to another. In some cases, obligate development of the parasite in the vector may be required, in which case this vector is either an intermediate or definitive host.

TEXTBOX 10.1 DEFINING 'PARASITE'

There are different definitions of 'parasite' used in the literature that can vary between fields of study and contexts. Here we focus on describing the 'parasite' terms commonly used in ecology, specifically in the context of modeling population dynamics. However, note that these broad classifications of 'parasites' are not always appropriate for other fields of study, such as medicine. One of the most widely accepted definitions is that a parasite is an organism that uses another organism (the host) as habitat and a source of nourishment (Park and Allaby 2017). Some definitions also state that parasites exert some level of harm on the host, exhibit adaptive structural modifications and/or have the ability to evade the immune system of the host (Zelmer 1998). A more definitive and explicit definition can also be useful, such as the generally well accepted list of attributes that define parasitic relationships provided by

Crofton (1971) when describing helminths, nematodes, trematodes, cestodes and arthropods: parasites are physiologically or metabolically dependent on their hosts, heavily infected hosts will be harmed by the parasite, the reproductive potential of the parasite exceeds that of the host, and the parasite population is not evenly distributed, but aggregated among the hosts, with a large proportion of the parasite population in a small proportion of the host population (Crofton 1971).

When studying population dynamics of different types of infectious agents, different model structures are used (see Section 10.5). For this purpose, infectious agents are generally divided into two broad classes: 'microparasites' and 'macroparasites' (Anderson and May 1979; May and Anderson 1979). Microparasites include viruses, bacteria and usually also protozoa. They multiply directly in the host and in large numbers. They reproduce rapidly, have short generation times and their hosts typically develop protective immunity of variable length. Models for microparasites divide the hosts into classes that represent different states of infection: susceptible, exposed, infectious and recovered (see Textbox 10.3). In contrast, macroparasites include most parasitic helminths, nematodes, trematodes and cestodes, as well as arthropods. Macroparasites do not multiply directly inside the definitive host; instead, each individual parasite infects the next host through transmission stages that are passed out into the external environment, where they either develop into stages that can reinfect hosts, or they infect intermediate hosts. Infections with these types of parasites are often chronic, leading to morbidity instead of mortality. Complete protective immunity to macroparasites does not usually occur.

10.2   QUANTIFYING IMPACTS OF INFECTIOUS AGENTS ON WILDLIFE POPULATION DYNAMICS

A key barrier to understanding the role of infectious agents in wildlife populations has been quantifying the impacts. This is, in large part, due to the challenges involved with performing controlled studies on free-ranging wildlife. Challenges include: study design type, logistics, difficulties in measuring the key population parameters of reproduction/recruitment and survival, imperfect 'tests' for pathogen presence and abundance, and the inability to control for the complexities of the system. This section, and Table 10.1, gives a general overview of some of the main points to consider when attempting to quantify the impacts of infectious agents on wild animal populations.

In general, studies on pathogen impacts tend to be either *cross-sectional* or *longitudinal* and may or may not involve experimental manipulation. Cross-sectional studies are typically done at a single point in time/over a short period of time and without any experimental manipulations. They are usually cheaper, faster, and useful for exploring associations, documenting current status and baselines. However, they cannot infer cause and effect relationships. Conversely, longitudinal studies, those that occur over a longer time period and include resampling of individuals, are considerably better for testing causal relationships, but require extended logistical and financial commitments. The studies best suited for inferring casual relationships are experimental. For example, augmentation of infectious agents in individuals or populations permits direct measurement of the impact of the agents (Pedersen and Fenton 2015). By doing this, through individual-based long-term studies, between-year and individual variation in fitness parameters can be controlled for (Stien, Irvine, Ropstad et al. 2002). The feasibility of this approach is often limited when studying large free-ranging animals such as *Rangifer*, partly due to the difficulties in finding matched controls and appropriate and approved pharmaceuticals that achieve the manipulation goal (e.g. parasite removal) (Pedersen and Fenton 2015). Experimental research on the gastrointestinal nematodes of Svalbard reindeer *Rangifer tarandus platyrchyncus* provides a good example of a well-designed long-term capture-recapture parasite removal study in a wild *Rangifer* population (see Section 10.3.1.1).

TABLE 10.1

Overview of the Key Points to Consider in Order to Quantify the Impacts of Infectious Agents on Host Populations

Element	Key Considerations
Study Design[1,2,3]	
Cross-sectional studies Observations of individuals occur at a single point in time or during a short, defined time period	• Provide individual status at a single time point • Not very resource intensive • Short time frame • Cannot infer cause and effect
Longitudinal studies Observations occur on the same individuals over a period of time	• Can track changes in individuals over time • Resource intense • Longer time commitment • Potentially invasive/disruptive
Experimental studies Using veterinary pharmaceuticals (anthelmintics, vaccines, antibiotics, etc.) or other mechanisms to disrupt the infectious agent-host system experimentally. These types of studies can be cross-sectional or longitudinal.	• Allow for comparison to a control group • Can infer cause and effect relationships • Pharmaceuticals usually not validated for wildlife and efficacy may not be known • Broad-spectrum drugs will remove multiple parasite species • Invasive
Mathematical models Portray infectious agent-host system as a set of equations	• Useful in determining what type of study/data to perform/ collect in the field and the sensitivity of the system to different factors • Can be specific and/or general and predictive • Provide insights into population change over long time scales • Can help scale up results to population level impacts • Allow exploration of management options
Measuring Fitness[4,5,6,7]	
Body condition, reproduction, recruitment, survival: The key parameters that influence birth and death rates	• Can be difficult to obtain data from free-ranging animals • Often involves live-captures or terminal sampling • Many hormonal assays need to be validated • Dispersal events may confound estimations of survival rates • Resource intensive
Steroid hormones Such as: estrogen, androgen, progestogen and glucocorticoid metabolites extracted from faeces and hair are used as health indicators	• Sampling can be non-invasive • Methods still need to be validated • Relationship between steroid levels and fitness not always clear
Traditional and local knowledge These are knowledge systems belonging to indigenous and local communities, comprising wisdom and practices developed over several generations. Fitness assessments are usually obtained from observations made of animals harvested by indigenous hunters.	• Inexpensive • Collaborative with indigenous stakeholders • Incorporates a different knowledge system • Potential for capturing large sample size encompassing a large geographical area • Data biased by hunter preference • Data can be influenced by individual variability in fitness assessments

(Continued)

TABLE 10.1 (CONTINUED)

Overview of the Key Points to Consider in Order to Quantify the Impacts of Infectious Agents on Host Populations

Element	Key Considerations
Detecting the Pathogen[8,9,10,11]	
Microscopy Isolation and visualization of the infectious agent in the host or of eggs/larvae/oocysts in faeces/blood/tissues: usually used for macroparasites	• No advanced technology needed • Sampling protocol/sampling size is critical due to aggregated distribution of parasites (see Section 10.5.7) • Sampling can be non-invasive; however, to count adult parasite burdens, the host requires terminal sampling • The number of 'offspring' detected is context-dependent (season, host condition, parasite and host species, adult parasite fecundity)
Identification of genetic material Isolation of the infectious agent from tissue or swab samples and cultivation/isolation/identification of genetic material: usually used for viruses and bacteria	• Identifies specific species/strains • Can be highly sensitive • Can be used for phylogenetic studies • Can provide data critical for epidemiological studies • Need molecular tools (e.g., DNA/RNA extraction, PCR, sequencing) • Samples can be difficult to obtain/store/ship

Exposure
Host exposure to infectious agent determined by measuring host antibody/cellular response to infection using serological tests
	• Blood samples sufficient
	• Relatively low financial and human resource cost for analyses
	• Dependent on host ability to develop an immune response
	• Results are context-dependent (e.g., varies between species, age, genetics, nutritional status, previous exposure)
	• Most tests developed for domestic livestock and not validated for *Rangifer*
	   • Impacts test interpretation (cutoff values, sensitivity and specificity)
	   • Detected antibody response may reflect immunological cross-reactions with *Rangifer* specific agents

[1] Stien, A., R. Irvine, E. Ropstad et al. 2002. The impact of gastrointestinal nematodes on wild reindeer: Experimental and cross-sectional studies. *J. Anim. Ecol.* 71 (6):937–45.

[2] Pedersen, A.B., and A. Fenton. 2015. The role of antiparasite treatment experiments in assessing the impact of parasites on wildlife. *Trends Parasitol.* 31 (5):200–11.

[3] Albon, S.D., A. Stien, R.J. Irvine et al. 2002. The role of parasites in the dynamics of a reindeer population. *Proc. Biol. Sci.* 269 (1500):1625–32.

[4] Kutz, S., J. Ducrocq, C. Cuyler et al. 2013. Standardized monitoring of *Rangifer* health during International Polar Year. *Rangifer* 33 (Sp. Iss. 21):91–114.

[5] Lyver, P.O.B., and A. Gunn. 2004. Calibration of hunters' impressions with female caribou body condition indices to predict probability of pregnancy. *Arctic* 57 (3).

[6] Carlsson, A.M., G. Mastromonaco, E. Vandervalk, and S. Kutz. 2016. Parasites, stress and reindeer: Infection with abomasal nematodes is not associated with elevated glucocorticoid levels in hair or faeces. *Conserv. Physiol.* 4 (1):cow058.

[7] Kofinas, G., P. Lynn, D. Russell, R. White, A. Nelson, and N. Flanders. 2003. Towards a protocol for community monitoring of caribou body condition. *Rangifer* (14):43–53.

[8] Hoar, B., M. Oakley, R. Farnell, and S. Kutz. 2009. Biodiversity and springtime patterns of egg production and development for parasites of the Chisana Caribou herd, Yukon Territory, Canada. *Rangifer* 29 (1):25–37.

[9] Forde, T.L., K. Orsel, R.N. Zadoks et al. 2016. Bacterial genomics reveal the complex epidemiology of anemerging pathogen in Arctic and boreal ungulates. *Front. Microbiol.* 7 (1759).

[10] Tryland, M., C.G.D. Neves, M. Sunde, and T. Mørk. 2009. Cervid herpesvirus 2, the primary agent in an outbreak of infectious keratoconjunctivitis in semidomesticated reindeer. *J. Clin. Microbiol.* 47 (11):3707–13.

[11] Curry, P.S., C. Ribble, W.C. Sears et al. 2014. Blood collected on filter paper for wildlife serology: Detecting antibodies to *Neospora caninum*, West Nile Virus, and five bovine viruses in *Rangifer tarandus* subspecies. *J. Wildl. Dis.* 50:297–307.

In the absence of population level experiments, computer simulations, using models populated with parameters defined in smaller-scale studies, can help us understand what the population level impacts might be under different scenarios. Mathematical modeling can also provide important insights into the study design for field experiments and predict responses to management interventions aimed at reducing impacts of pathogens (for more details see Section 10.5).

To demonstrate impact on a host population, one must demonstrate that infection has a direct or indirect impact on host fitness through impact on host birth and/or death rates. Fitness parameters typically measured include body condition, reproduction and survival, as these are key factors influencing birth and death rates (see Section 10.5.2). Obtaining these data often involves live-capture of the animal, or terminal sampling. Remote surveillance through telemetry and aerial surveys can be used to assess survival, and in some cases, reproductive success, yet are extremely costly (CircumArctic *Rangifer* Monitoring Assessment Network 2008; Kutz et al. 2013). The value of traditional knowledge and hunter observations to assess fitness of wildlife populations is becoming increasingly recognized (Lyver and Gunn 2004), yet remains to be critically evaluated as a mechanism for evaluating impacts of infectious agents at the population level (Carlsson, Veitch et al. 2016).

Non-invasive methods to gather and analyze samples that can provide information on the health of individuals, and potentially populations, is also a rapidly evolving field (Madliger et al. 2016). For example, *Rangifer* faecal samples have been used to obtain information on individuals' sex, age class and pregnancy status (Flasko et al. 2017; Joly et al. 2015; Madliger et al. 2016). Physiological stress, as measured by steroids in hair and faecal samples, is increasingly being applied as an indicator of health for *Rangifer*; however, further validation is needed (Ashley et al. 2011; Carlsson, Mastromonaco et al. 2016; Macbeth 2013; Wasser et al. 2011) (see Chapter 2 for an in-depth discussion).

Another challenge with assessing impacts of infectious agents is accurately detecting, identifying and quantifying the agent. The presence of, or exposure to, an infectious agent can be determined directly, by identifying the infectious agent through direct visualization and morphology or using DNA-based methods, or indirectly, by detecting the host response (e.g., by detecting an antibody or cellular response). Infection with macroparasites is usually measured directly by isolating and visualizing adult parasites or their offspring. For many macroparasites the clinical outcome of infection is dose-dependent and accurate estimates of parasite burdens are important to understand potential impacts (Hughes et al. 2009). Infection intensity is ideally estimated by isolating adult parasites from the host; however, this isn't always possible, and indirect measures include counting the juvenile stages (eggs, larvae or oocysts shed in the faeces, or present in the blood stream). However, production and shedding of these juvenile stages can vary by host species, parasite species, season, host condition (Hoar et al. 2009), and adult parasite burden and fecundity (Irvine et al. 2001). Furthermore, macroparasite distribution is often aggregated, where a few hosts are heavily infected while most are lightly infected (for more details on parasite aggregation, see Section 10.5.7). This has important implications for estimations of infection intensity in a population and for study design and data analysis (Wilson et al. 1996).

Bacteria and viruses may be detected directly through direct culture or by identifying the presence of genetic material (DNA and/or RNA) from tissue or swab samples (Forde et al. 2012; Tryland et al. 2009). However, tissue and swab samples from wildlife are often challenging to obtain, transport and store. Thus, the presence of bacteria and viruses in a population is often addressed indirectly, by measuring the host antibody response to infection by *serology*. In general, most serological tests used are developed for domestic livestock and their specific infectious agents, and have not been validated for use in *Rangifer* (Curry et al. 2014). As such, interpretation of results may be challenging, since the tests are not optimized for *Rangifer* (or other wildlife species). More often than not, the antibody response detected is likely not directed against the identified infectious agent expected to circulate in domestic animals and to which the test is designed, but rather reflects an immunological cross-reaction with a closely related *Rangifer* specific agent (das Neves et al. 2010).

Serological tests indicate exposure to the infectious agent. However, antibodies to infectious agents may persist for variable periods of time, making it difficult to infer infectious status from seropositivity and, by extension, to infer associations with infectious status to changes in host *demography*. But unlike methods that rely on direct isolation of an infectious agent during active infection, serological tests can give information about previous exposure. Serology can aid our understanding of prevalence, distribution and host interactions with a variety of infectious agents (Gilbert et al. 2013).

Despite the numerous challenges and complexities involved, there are a number of detailed studies that help us better understand how, and when, infectious agents may have an impact on population dynamics of their hosts. Section 10.3 summarizes the main findings from a few of these studies.

10.3 IMPACTS OF INFECTIOUS AGENTS ON *RANGIFER* POPULATIONS: EVIDENCE FROM THE FIELD

To influence population dynamics, infectious agents must ultimately alter the birth or death rates of their hosts. They can do this through a wide variety of direct and indirect mechanisms, including impacts on body condition, pregnancy rates, survival and behaviour (Figure 10.1). Additionally, how infectious agents impact their hosts will also be influenced by interactions with climate, predators and co-infecting infectious agents. We use case studies in the following sections to explore these issues.

10.3.1 IMPACTS ON BODY CONDITION

In domestic animals it is well established that parasites can have negative impacts on body condition (Fox 1997). In *Rangifer*, body condition is a key determinant for growth, fecundity and survival; thus, any impacts on body condition can have other, far-reaching, consequences for individual and population health (Parker et al. 2009; White et al. 2013). In this section, we will review one case that demonstrates the impact of parasites on body condition in *Rangifer* and provide a brief overview of evidence from other case studies.

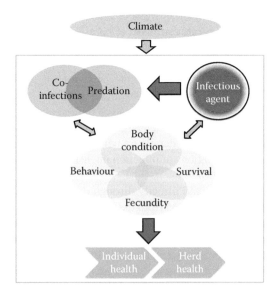

FIGURE 10.1 Conceptual overview of the direct and indirect impacts of infectious agents on individuals and populations. Infectious agents can directly affect survival, fecundity, body condition and behaviour. These factors can interact and can also interact with climate, predators and disturbance as well as co-infections, to lead to an outcome of growing, declining or stable populations.

10.3.1.1 *Ostertagia gruehneri* and Svalbard Reindeer

Perhaps the most convincing evidence of the impact of nematodes on *Rangifer* is the work done on Svalbard reindeer and their gastrointestinal nematodes. In a landmark study, researchers were able to quantify the impacts of nematodes on host fitness and population dynamics through an experimental longitudinal individual-based capture-recapture study. Using a combination of field research and modeling, researchers demonstrated that the abomasal nematode *Ostertagia gruehneri* had a negative effect on the body condition of reindeer. This effect was density-dependent, affected fecundity of the reindeer and was sufficient to regulate the population dynamics (Albon et al. 2002; Stien, Irvine, Ropstad et al. 2002).

To quantify the impact of nematodes on Svalbard reindeer fitness, researchers first used a combination of in-host worm burden counts and faecal egg counts from individually marked animals with known age to determine the population (and probable transmission) dynamics of the two dominant abomasal parasites: *O. gruehneri* and *Marshallagia marshalli*. They demonstrated that worm burdens and egg output of *O. gruehneri* peaked in the summer and those of *M. marshalli* peaked in winter (Carlsson et al. 2012; Carlsson et al. 2013; Irvine et al. 2001; Irvine et al. 2000; Stien, Irvine, Langvatn et al. 2002). They then administered a slow-release anthelmintic 'bolus' to half of the individually marked reindeer captured in April to reduce summer parasite burdens (primarily *O. gruehneri*) and evaluate the impact on reindeer condition and fecundity. Reindeer were captured 12 months later to determine their body mass, back fat and pregnancy status. On average, treated animals were 1.9 kg heavier, had 3.3 mm more back fat and had an 11% higher probability of being pregnant, compared to non-treated animals (Stien, Irvine, Ropstad et al. 2002) (Figure 10.1). The parasite's impact on host fecundity was *density-dependent* and increased with the annual mean estimate of *O. gruehneri* in the host population. Furthermore, the abundance of *O. gruehneri* was also density-dependent but time-delayed, being related to the host population density two years earlier. Thus, increases in host abundances led, after two years, to an increase in parasite abundance, which in turn led to poorer body condition of heavily infected individuals, and a decrease in calf production the following summer (Albon et al. 2002; Stien, Irvine, Ropstad et al. 2002). The time delay is most likely a result of ingested larvae entering a stage of 'arrested development', delaying the time when they develop into adults and start reproducing (Hoar et al. 2012; Irvine et al. 2000). A model parameterized with these specific host-parasite interactions showed that the parasite-induced reduction in host fecundity, mediated through negative impacts on body condition, was sufficient to cause a negative population growth rate, and regulate the population dynamics of Svalbard reindeer (Albon et al. 2002).

Establishing the role of *O. gruehneri* in regulating the population dynamics of Svalbard reindeer was possible because (i) the life-history strategy and population dynamics of the parasite were well documented, (ii) there was an available treatment which removed parasites during their peak abundance and (iii) it was possible to perform a longitudinal experiment, repeated over several years, allowing authors to control for inter-individual variability in host fitness. A cross-sectional study, conducted in the same area and time frame, did not detect an effect of parasitism on body condition, highlighting the difficulties for cross-sectional studies to detect subtle effects of parasitism (Stien, Irvine, Ropstad et al. 2002). The Svalbard reindeer provide an excellent study system for parasite host interactions because there are few confounding variables; there is no competition for food, there are no predators and human impact is minimal. However, these conditions are unique; studying the impact of infectious agents in other *Rangifer* populations likely involves complexities that are difficult to control for.

10.3.1.2 Evidence from Other Case Studies

Another experimental study has also demonstrated a link between parasite infection and a reduction in *Rangifer* body condition (Ballesteros et al. 2012). In this case, warble larvae (*Hypoderma tarandi*) were removed in autumn/winter by treating semi-domesticated reindeer with anthelmintics. Treated female reindeer had higher body mass the following summer. Treatment also tended to have a positive effect on the early summer body mass of their calves. These results suggest that development of warble flies in the host might have direct impacts on body condition that may compound the effects of disrupted activity patterns during insect oviposition (Ballesteros et al. 2012) (see Section 10.3.4.1). Unfortunately, the treatment used was a broad-spectrum drug (Ivermectin) that is effective against intestinal, pulmonary and tissue nematodes in addition to warbles. Thus, positive effects of treatments could be the consequence of removing other parasites instead of, or in addition to, warbles (Ballesteros et al. 2012). Although, the warble study did not take the next step to investigate how the observed reduction in body condition might influence the population dynamics of the herd, it is generally recognized that females in poor body condition are less likely to get pregnant and calves in poor body condition have lower survival rates (Cameron et al. 1993; Parker et al. 2009).

A few cross-sectional studies also provide compelling evidence that parasites negatively impact *Rangifer* body condition. For example, in northern Canada, Hughes et al. (2009) found a negative correlation between nematode burdens and body weight, and between warble abundance and back fat and pregnancy, in adult female caribou from the Dolphin and Union herd (Hughes et al. 2009). In two herds of West Greenland caribou, Kangerlussuaq-Sisimiut and Akia-Maniitsoq, Steele (2013) found that the most abundant gastrointestinal nematode in each herd, *Marshallagia marshalli* and *Ostertagia gruehneri*, respectively, was negatively associated with body condition indices. However, impacts of parasites may vary between species, host populations and seasons. For example, a longitudinal experimental study found no evidence to suggest that the winter transmitted gastrointestinal nematode *Marshallagia marshalli* had a negative impact on body condition or pregnancy of Svalbard reindeer (Carlsson et al. 2018; Carlsson et al. 2012), although the opposite was true for the summer transmitted nematode *O. gruehneri* (Stien, Irvine, Ropstad et al. 2002). Thus, time of sampling, parasite species and their life history, and characteristics of the host population are all important factors to consider.

10.3.2 IMPACTS ON FECUNDITY (PREGNANCY)

Many infectious agents have direct impacts on reproduction that are not mediated through impacts on other fitness parameters. These infectious agents may affect the productivity of a population directly (by causing testicular disease, abortions, foetal resorption or mummification, or sterility) or indirectly (by reducing the survival of newborn and young) influencing fertility. In a sense, infectious agents with these types of reproductive impacts are 'silent killers'; impacts are subtle and very difficult to detect in the field, as there will be poor calving success but no visible carcasses. Alternatively, there will be weak calves that are taken by predators. This frequently leads to predation being considered the main driver of population reduction, rather than a secondary consequence of parasitism (see Section 10.4). The protozoal parasites *Neospora caninum* and *Toxoplasma gondii* can act in this manner. These are transmitted by canids and felids, respectively, and are both associated with abortions/mummification/stillbirths, foetal abnormalities, weak calves and neonatal mortality in domestic livestock (Dubey et al. 2007; Innes 2010), with evidence of similar impacts in *Rangifer* (Kutz et al. 2012) (for more detail see Chapter 9).

10.3.2.1 Brucellosis in Caribou on Southampton Island

Brucella spp. are bacteria known as major causes of infertility in a broad range of domestic and wild mammals (Lopes et al. 2010). *Brucella suis* biovar 4, the causative agent of brucellosis in *Rangifer*, is reported from *Rangifer* across most of their range (Forbes 1991; Meyer 1966) and can cause abortions, retained placenta, sterility, bursitis and nephritis (Rhyan 2013) (see Chapter 7 for more details). In at least one population, the caribou on Southampton Island, it appears to have affected population trends.

Southampton Island, in northern Hudson Bay, Nunavut, Canada, is approximately 41,200 km². Historically, caribou and wolves were abundant on the island, but by 1953 both species were

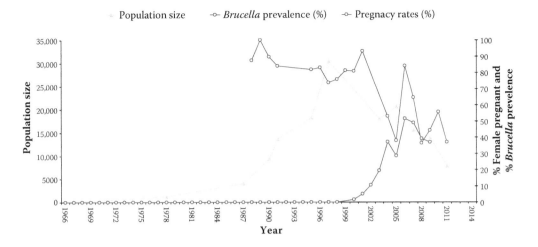

FIGURE 10.2 Infectious agents can have direct impacts on fecundity of the host with major consequences for population dynamics, but such effects are often difficult to demonstrate. In the case of caribou on Southampton Island, the increase in *Brucella* seroprevalence (red lines and circles) mirrors a decrease in pregnancy rates (blue lines and circles) and population abundance (gray lines and triangles); it seems likely that this bacterium has influenced the population dynamics (Campbell 2013).

extirpated. In 1967, barren-ground caribou were reintroduced to the island and subsequently flourished. The conditions on the island were favorable, with high forage quality, few predators and initial low harvesting pressure (Ouellet et al. 1997). From the 48 introduced caribou (28 females), the numbers increased rapidly, with a growth of about 27% per year, peaking around 30,000 animals in 1997. A population decline was first observed in 2003 and the downward trend continued, with an estimated population size of around 7,000 animals in 2013, but a slight recovery to around 12,000 animals in 2015 (Campbell 2006; Gunn et al. 2011).

During federally inspected commercial harvests in the 1990s and early 2000s, caribou were routinely tested for brucellosis. In 2000, *Brucella suis* biovar 4 was detected for the first time in 1.5% of the sampled carcasses (Campbell 2013); this increased to 58.8% in 2011. Pregnancy rates decreased from 93% in 2001 to 24% in 2005, and increased to 37% in 2011 (Campbell 2013). The significant drop in pregnancy rates coincides with appearance and emergence of *Brucella* (Figure 10.2) and strongly suggests a role of this bacterium in the population dynamics of the Southampton caribou; icing events, increasing predation, hunting pressure and low genetic variability may also have contributed to this decline. Although the role of *B. suis* biovar 4 in the caribou decline is compelling, a better understanding of the pathophysiology of infection, and analyses of the relationship between pregnancy rates and distribution, prevalence and demographics of infection are required to determine a clearer cause and effect relationship. Seropositivity to *B. suis* biovar 4 has been reported in several caribou herds (Ferguson 1997; Forbes 1991; Gunn et al. 1991), and further studies investigating its direct impact on fecundity, and the role in population declines and population dynamics, are needed.

10.3.2.2 Evidence from Other Case Studies

One compelling case demonstrating infections having direct impacts on fecundity comes from the studies on wood bison (*Bison bison athabascae* L.) in Wood Buffalo National Park in Canada. The decline of this population from the 1960s to 1990s perplexed managers. It was hypothesized that infections with two pathogens, *Mycobacterium bovis* (the cause of tuberculosis) and *Brucella abortus*, which had historically spilled over from domesticated cattle, were the cause of the decline (Joly and Messier 2004b; Joly and Messier 2005). Both these pathogens can have direct impacts on reproduction by causing foetal infection and loss. Tuberculosis can also impact reproduction indirectly

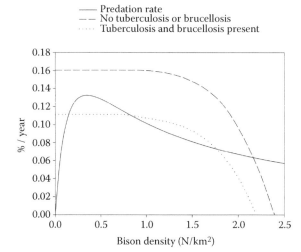

FIGURE 10.3   Infectious agents can have direct impacts on fecundity, but these may be context-dependent and difficult to detect. A model simulating population growth of the wood bison herd in Wood Buffalo National Park in Canada showed that herds that experienced predation that were also infected with *Brucella abortus* and *Mycobacterium bovis* (i.e. tuberculosis) would stabilize at low densities. Uninfected herds, experiencing only predation or food limitation, would increase and stabilize at higher than observed densities. These results indicate that the observed decline in the 1990s was due to a combination of infectious agents and predation (Joly and Messier 2004b).

by reducing pregnancy rates, mediated through impacts on body condition. By live-capturing bison, determining pregnancy status and screening for the presence of antibodies to tuberculosis and brucellosis, Joly and Messier (2005) found that bison that were positive for both diseases were less likely to be pregnant compared to bison that were positive for one, or neither, disease. Furthermore, infection with both pathogens was also related to lower over-winter survival (Joly and Messier 2005). Computer modeling simulation showed that infected herds experiencing predation would stabilize at low densities, while uninfected herds experiencing only predation or food limitation would increase and stabilize at higher than the observed densities (Figure 10.3) (Joly and Messier 2004b).

Another example, where impacts were more difficult to detect, comes from a long-term study of alpine chamois (*Rupicapra rupicapra*, L.) in France. In this seven-year study, seropositivity to three bacteria, *Salmonella abortus ovis*, *Chlamydophila abortus* and *Coxiella burnetii*, which are all causative agents of abortions in domestic livestock, explained 35% of the annual variation in chamois reproductive success (Pioz et al. 2008). Clinical signs of disease were not observed. The role of these bacteria in reproductive success was only detected after accounting for the impacts of weather and density, and this was only possible because of the long-term nature of the study.

10.3.3   IMPACTS ON SURVIVAL

Some infectious agents, particularly viruses and bacteria, can cause direct mortality of their hosts, whereas others may be less virulent (e.g. macroparasites) and only cause mortality through interaction with other biotic and abiotic elements of the ecosystem. Here we provide examples of infectious agents causing direct mortality and then briefly explore how infection, in combination with climate and host density, can impact survival and may make hosts more vulnerable to predation.

10.3.3.1   Direct Mortality

Rangifer are susceptible to a number of infectious agents that can cause direct mortality (for more information on specific impacts of viruses, bacteria and parasites on individual *Rangifer*, see Chapters 6–9). For example, anthrax, West Nile Virus and the brainworm *Parelaphostrongylus tenuis* are almost invariably fatal in *Rangifer* (see Chapters 6–9). While animals dying from anthrax can contaminate the environment with infectious bacterial spores for decades, for the latter two agents, *Rangifer* hosts normally die before transmission can occur, and neither circulates successfully in *Rangifer* populations in the absence of other hosts. Thus, these infectious agents will circulate independently of *Rangifer* population density, and the impacts on *Rangifer* will depend on what other competent hosts are *sympatric* and maintaining the parasites. If the competent host density and ecology are such that the infectious agents spill over regularly to the more susceptible *Rangifer*, then populations may be eliminated through 'parasite-mediated competition'. For example, the 'disappearance' of woodland caribou from Minnesota, USA, is thought to be due in part to mortality caused by *P. tenuis*, which is carried by white-tailed deer in the area (Jordan et al. 1998). Similarly, reintroduction efforts of caribou into Wisconsin, USA, were thought to have failed because of the high mortality caused by *P. tenuis*, which was maintained in the abundant sympatric white-tailed deer population (Trainer 1973).

10.3.3.2 Indirect Mortality

Impacts of infectious agents on host survival are often context-dependent and only become problematic when hosts are exposed to additional stressors. For example, parasites can interact with climatic conditions to induce population level effects. This has been demonstrated for gastrointestinal parasites in free-ranging Soay sheep (*Ovis aries* L.) on the island of Soay in the St. Kilda archipelago in Scotland. Gastrointestinal nematodes in these sheep can induce anorexia and reduce gut absorption of nutrients. In years with good climatic conditions, parasites seem to be of minor importance for sheep survival (Wilson et al. 2004). However, in years with bad winters and high population density, these parasites contribute to population crashes – those sheep with high parasite intensities have the highest mortality rates.

'Subclinical' infections can also make hosts more vulnerable to predation. This has been demonstrated in several different systems, including nematodes in snowshoe hares (*Lepus americanus*), *Trichostrongylus* in red grouse (*Lagopus lagopus scotius*), prions in mule deer (*Odocoileus hemionus*) and *Echinococcus* in moose (*Alces alces*) (Hudson et al. 1992; Joly and Messier 2004a; Krumm et al. 2010; Murray, Cary, and Keith 1997). In some systems this effect may be especially pronounced when hosts are nutritionally stressed (Murray et al. 1997). Interactions between infectious agents, predators, weather and other factors are explored further in Sections 10.3.5 and 10.4.

10.3.4 IMPACTS ON BEHAVIOUR

Some infectious agents can manipulate their hosts and change their behaviour in order to improve their chances of transmission (Heil 2016). For example, horsehair worms (*Paragordius tricuspidatus*) infect crickets, but they need an aquatic habitat to reproduce. To make sure they reach this habitat, the parasite alters the behaviour of its host, causing them to move towards and immerse themselves in water. Once in the water, the parasite emerges from the cricket and is able to reproduce (Sanchez et al. 2008). Most of these 'extreme' and spectacular 'mind-controlling' manipulations are described in parasitized insects, although *Toxoplasma gondii* has been associated with increased risky behaviour in rats and people (Webster 2001). In *Rangifer*, the best documented behavioural impacts are those caused by parasitic arthropods.

10.3.4.1 Impacts of Botflies and Other Arthropods on *Rangifer*

Rangifer reduce their exposure to biting flies by changes in activity and their migratory and social behaviour, with potential population-level consequences (Fauchald et al. 2007; Witter et al. 2012). Observational behavioural studies from different *Rangifer* herds and ecological settings have shown that insect harassment leads to decreased time lying and feeding, and increased time standing and engaging in insect avoidance behaviour (Hagemoen and Reimers 2002; Morschel and Klein 1997;

Witter et al. 2012). Leg-stomping, nose-dropping and running are stereotypical behaviours in response to oesterids, which are usually more severe than responses to mosquitoes. At the herd level, animals that are not directly affected, but in the vicinity of flies, also show disrupted activity patterns (Hagemoen and Reimers 2002; Witter et al. 2012). This type of movement is not only directly energetically costly, but also decreases time spent feeding.

Insect harassment peaks during the summer, a crucial period for *Rangifer* since this is when they restore their fat reserves lost during winter and fatten up for the coming rut. Pregnancy rates, growth rates and over-winter survival are all closely linked to summer grazing conditions (Reimers 1997). Any factors that prevent *Rangifer* from feeding optimally during this period could have a negative impact on body condition and productivity, and thus herd health. Indeed, a study on Norwegian reindeer found that reindeer carcass weights following a warm summer of intense insect harassment were lower when compared to a summer with fewer insects (Colman et al. 2003). Warble abundance (*Hypoderma tarandi*) has been negatively correlated to back-fat depth and pregnancy rates in the Dolphin and Union caribou herd (Hughes et al. 2009) and pregnancy rates in Greenland caribou (Cuyler et al. 2012). It is, however, not clear to what extent fitness costs are a consequence of energetic expenditure due modified behaviour in response to heavy insect harassment during the summer, or due to metabolic costs of hosting larvae during the winter (White et al. 2013).

In addition to impacts on fitness, there is some evidence that insects can influence populations by modifying their behaviour at a larger scale. For example, there is some support for the 'migratory escape' hypothesis (Avgar et al. 2014); where reindeer with post-calving migrations had significantly lower warble fly larval abundances compared to herds remaining on or near the calving grounds during the summer (Folstad et al. 1991). Parasite infection may also impact grazing strategies (Gunn and Irvine 2003). Svalbard reindeer (*Rangifer tarandus platyrhunchus*) avoided pastures where faecal contamination was increased, potentially reducing the risk of becoming infected with nematodes (van der Wal et al. 2000).

10.3.5 INTERACTIONS

Hosts and infectious agents do not live in a static environment, but are constantly influenced by other factors. For the infectious agent, it can be the immune system of the host or other co-infections; for the host, it can be a range of biotic and abiotic factors, such as predation, habitat, weather and climate. The nature of this interaction can change in certain contexts, which may also change the impact the infectious agent has on the host. Indeed, the infection outcome depends on the host, the 'agent', and the environmental conditions both experience. In this section, we review a few case studies that demonstrate how the impacts of infectious agents can be context-dependent by illustrating: (i) how stressful conditions may lead to the renewal (or *recrudescence*) of clinical symptoms, (ii) how resource availability, weather and parasite interactions can impact host survival, (iii) how exposure history and co-evolution is important for infection outcome, (iv) how co-infections with other parasites can alter virulence and disease outcomes, and finally, (v) how parasites may increase host mortality by making them more susceptible to predation.

We highlight the importance of considering the presence of other sympatric/*reservoir hosts*. For example, pathogens can have radically different impacts in different hosts; introduction or movement of infected sympatric/reservoir hosts may lead to disease outbreaks and, from a monitoring perspective, sympatric/reservoir hosts could be monitored to predict/counteract future outbreaks in the 'target' populations. The case of the brainworm *Parelaphostrongylus tenuis* illustrates some of these points (see Section 10.3.3.1 and Chapter 9). *P. tenuis* is asymptomatic in the primary host, white-tailed deer, but is lethal to caribou. With climate change, white-tailed deer are predicted to expand their range northward, bringing *P. tenuis* with them as a significant threat to woodland caribou (Chapter 9). The case of bovine tuberculosis (*Mycobacterium bovis*) in wood bison (*Bison bison athabascae*) in Wood Buffalo National Park also illustrates how *sylvatic hosts* can be a threat to domestic livestock, and vice versa. Original exposure and infection of the bison was likely a result of spill-over from contact with diseased cattle before bison were translocated to the park. Bovine tuberculosis has since been eradicated from domestic livestock in Canada but persists in the bison population, posing an infection risk to the domestic livestock industry and threatening the conservation of bison (Shury et al. 2015).

10.3.5.1 Stress and Recrudescence: Alphaherpesvirus in Reindeer

Alphaherpesviruses are part of the genus *Varicellovirus*, which replicate rapidly and can easily be transmitted in animal populations at high densities. Cervid herpes virus 2 (CvHv2) is the alpha-herpesvirus endemic in reindeer populations in Eurasia, and this same virus is thought to circulate in *Rangifer* in North America (das Neves, Rimstad et al. 2009; das Neves et al. 2010). Previously thought to be a relatively benign pathogen in *Rangifer*, CvHv2 is now recognized to have important impacts on this host, including causing respiratory infections, keratoconjunctivitis and abortions (das Neves, Rimstad et al. 2009; Tryland et al. 2009; Tryland et al. 2017) (see Chapter 8, Viral Infections and Diseases, for details). Those impacts are often visible in conjunction with other stressors, when the infection recrudesces.

The initial infection with alphaherpesvirus is often associated with mild to moderate disease. However, following 'recovery', the virus is not eliminated – rather, it remains in a latent state in the host and may be reactivated at a later time (recrudesce). Experimental reactivation of latent infections with CvHV2, or recrudescence, has been demonstrated in reindeer (das Neves, Mørk et al. 2009), and it is possible that reactivation induced by stressful conditions associated with corralling and high animal densities led to the keratoconjuctivitis outbreak observed in Norway in 2009 (das Neves et al. 2010). Similar mechanisms exist with chicken pox in humans (which is also caused by a herpesvirus), when recrudescence can occur many years after recovery from infection. Stress, or aging, leads to the appearance of shingles and the pathogen can be transmitted to younger susceptible hosts who have not been exposed to chicken pox (Steiner et al. 2007). In light of the additional stress that populations may face with rapid environmental change and anthropogenic disturbance in the Arctic, alphaherpesvirus is likely an important pathogen to consider in future studies on *Rangifer* health and population dynamics.

10.3.5.2 Parasites, Food and Weather: The Cost of Parasites in Soay Sheep

Soay sheep (*Ovis aries* L.) were introduced to the archipelago of St. Kilda, UK, during the bronze age (3000 to 5000 ybp). The sheep population fluctuates by 60–70% every three to four years. Population crashes are a result of complex interactions between the sheep, their food supply and climate, and occur in late winter, a vulnerable time for all northern ungulates. In years with bad winters and high population density, sheep may starve to death, leading to high rates of mortality. Mortality events may still occur during low density years, but at a much lower rate. Parasites contribute to these population crashes by inducing anorexia and reducing gut absorption capabilities (Pemberton 2004; Wilson et al. 2004).

In 1989, there was a large mortality event on St. Kilda, with two-thirds of the Soay sheep population dying over 2 weeks. Post-mortems found high nematode burdens and the likely cause of death was attributed to protein-energy malnutrition. The most common parasite in this system is the directly transmitted gastrointestinal strongyle, *Teladorsagia circumcincta*. Prior to this population crash, Gulland (1992) administered a slow-release anthelmintic bolus, a device that releases anthelmintics over a prolonged period of time (at least 100 days), to remove worm burdens from a subset of the population. Although there was no difference in mortality rates between treated and control animals, the daily survival rate was significantly higher in the treated group (Gulland 1992). A second, similar, experiment conducted prior to the crash in the winter of 1991–1992 demonstrated that mortality rates for all treated animals were lower compared to controls. Young males in particular benefited from treatment; 47% (n = 17) of control males died whereas all of the treated males survived (n = 18) (Gulland et al. 1993). The population crash in 1989 was severe (70% mortality) while the population crash in 1991–1992 was intermediate (44% mortality). Treatment only increased survival in

1991–1992, suggesting that the cost of parasitism may be context-dependent, where parasite removal only enhances ultimate survival in years of moderate food limitation (Wilson et al. 2004).

10.3.5.3 Parasites and Climate: *Setaria tundra* in Reindeer in Finland

In 2003, in the southern herd range in Finland, a sudden outbreak of peritonitis and mass appearance of nematodes in the abdominal cavity at the beginning of the slaughter season perplexed reindeer herders, veterinarians and scientists. The outbreak was associated with poor body condition, especially in calves. In 2004, the focus of the outbreak moved 100 km north and by 2005 only the northernmost slaughterhouses were free of the disease. The outbreak led to substantial economic losses and promoted research into the causative agent, the nematode *Setaria tundra* (Laaksonen et al. 2007). The study of this host-parasite system briefly outlined below illustrates the complex interactions between climate, parasites, and their arthropod and mammalian hosts that need to be understood in order to predict and manage disease outbreaks.

Setaria tundra is a filarioid nematode that, in adult form, lives in the peritoneal cavity of the *definitive host* (such as *Rangifer*). Adults produce microfilaria that circulate in the bloodstream. Arthropods take up the microfilaria during a blood meal, they develop into the infective stages and are transmitted when the insect feeds again (Laaksonen, Solismaa, Kortet et al. 2009). By collecting insects, and examining them for the presence of microfilaria, the main *vectors* in Finland were determined to be *Aedes* spp. and *Anopheles* spp. mosquitoes. Experimental studies of larval development in the insect host demonstrated that the development to infective stages is temperature dependent. Transmission is also dependent on the life span of female mosquitoes, which, in turn, is also dependent on temperature (Laaksonen et al. 2009).

The behavioural response of reindeer to climatic factors was also identified as a factor in promoting outbreaks. In warm summers, reindeer were reported to congregate around wet, mosquito-rich areas, likely due to the availability of drinking water, fresh food plants and improved thermoregulation. However, due to the microclimate also being favorable for mosquitoes, this behaviour likely increases infection risk (Laaksonen et al. 2009).

Laboratory experiments, field studies and historical weather data provided the background needed to model the temperature conditions necessary to cause outbreaks. Models suggest that temperatures above 14°C increase infection rates, but disease only emerges the following summer, if the weather conditions in that summer are still favorable. This coincides with observations made during the outbreaks, which all occurred in warmer-than-usual years, and explains why no peritonitis outbreaks were observed in the northernmost slaughterhouses (Laaksonen et al. 2010). These models could potentially be used to predict the timing of future outbreaks and allow proactive management of the situation.

10.3.5.4 Parasites and Exposure History: *Elaphostrongylus rangiferi* in Newfoundland Caribou

Elaphostrongylus rangiferi is a nematode common in reindeer in Fennoscandia and Russia. First-stage larvae are passed in the faeces of infected *Rangifer* and then invade a gastropod (slug/snail), the intermediate host, where they develop into infective third-stage larvae. Reindeer become infected by accidentally eating gastropods. Third-stage larvae released in the abomasum then migrate through the central nervous system and skeletal muscles (Lankester 2001). The parasite releases eggs that travel via the blood stream or lymphatics to the lungs and cause verminous pneumonia. The migration and maturation of the parasite can cause damage to the central nervous system and result in debilitating neurologic disease. In Fennoscandia, periodic epizootics of neurologic disease occur, primarily in young *Rangifer*, in late winter. Epizootics tend to be associated with favorable climatic conditions that increase parasite development rates in the gastropods and result in higher transmission rates (Halvorsen 2012). Severity of disease is dose-dependent, and adults only show clinical signs after high rates of exposure. The impact of the parasite on semi-domesticated reindeer has been reduced by the widespread use of ivermectin (Josefsen et al. 2014).

In 1908, *E. rangiferi* was introduced with infected reindeer from Norway to St. Anthony, at the northern tip of Newfoundland, Canada. The parasite subsequently invaded the native caribou population and, after 80 years, arrived in the Avalon Peninsula, in the far south of Newfoundland. Two epizootics of severe neurological disease caused by *E. rangiferi* have been reported in Newfoundland caribou: the first in the mid-1980s in central Newfoundland and the second in 1996 in the Avalon Peninsula, shortly after the parasite arrived in this area. The epizootic in the Avalon herd was unusual. Notably, adults as well as calves showed clinical signs of neurological disease. In contrast, only young animals appear affected in herds in Fennoscandia where *E. rangiferi* has been established for almost a century. Furthermore, the size of the Avalon herd also dropped significantly during the time of outbreak (from an estimated 7,000 animals to <2,500 in a three-year period (Mahoney and Weir 2009)), possibly partly as a result of infected animals of all age classes dying. This is consistent with the idea that naïve herds would be more vulnerable to new parasites. In this case, disease was seen across all age classes, likely because of a lack of previous exposure and, therefore, no acquired immunity in older animals (Ball et al. 2001).

10.3.5.5 Co-Infections

Most wildlife studies investigating the impact of infectious agents on host fitness and population dynamics have, to date, focused on 'one-host-one-pathogen' systems (Hellard et al. 2015). However, in nature, most hosts are infected by multiple infectious agents at once (Rigaud et al. 2010). At the within host individual level, infectious agents may interact either directly by competing for resources, or indirectly, via the host immune system (Pedersen and Fenton 2007). But interactions can also occur at the population level, where, for example, one infectious agent may impact the chance of transmission of another infectious agent (Hellard et al. 2015). Evidence from field studies and models increasingly suggests that these interactions are critical for understanding the community ecology and infection dynamics of infectious agents (Pedersen and Fenton 2007; Telfer et al. 2010). For example, Telfer et al. (2010) tracked infections of four different microparasites in individual wild voles over five years and found both positive and negative interactions. In all cases but one, infection with other parasite species explained more variation in infection risk than any other factors related to exposure risk (such as age, season and year). They found that one parasite could have contrasting effects on susceptibility to other parasites. For example, infection with the tick transmitted bacterium, *Anaplasma phagocytophilium*, increased likelihood of infection with cowpox virus but substantially decreased infection rates with the flea transmitted *Bartonella* spp.

Co-infections not only impact future infection risk but can also have important consequences for host fitness. For instance, a recent study in African Buffalo (*Syncerus caffer*) showed that buffalo infected with *Mycobacterium bovis*, the causative agent of bovine tuberculosis (BTB), were not only more likely to acquire the viral disease Rift Valley fever (RVF), but also up to six times more likely to abort due to RVF, compared to buffalo that were not infected with BTB (Beechler et al. 2015). Thus, although BTB may have limited population level effects on its own, it may exacerbate the effects of other diseases and, through that interaction, impact host population dynamics.

Experimental studies, such as those discussed in Sections 10.2.2 and 10.3.1.1, are crucial for studying within host interactions of parasites. Perturbation studies, where a specific group of infectious agents is removed using drugs, have shown that a successful reduction in abundance of the target species can lead to changes in the abundance of non-target species (Pedersen and Fenton 2015). These changes can negatively or positively affect host health, depending on the type of interaction. Thus, studies that only consider infectious agents in isolation may misinterpret results or overlook important interactions (Pedersen and Antonovics 2013). Understanding interactions between infectious agents, and how they shape disease dynamics, is crucial for optimizing control or conservation programs. For example, treating a population to remove a parasite might have unintended consequences and increase or exacerbate a disease caused by a co-infecting parasite. Alternatively, if it is easier to control a co-infecting parasite that facilitates the disease progression of the target parasite, then targeting the 'facilitator' may be a more efficient way to manage/prevent disease outbreaks (Lafferty 2010).

10.3.5.6   Parasites and Increased Susceptibility to Predation

Infectious agents can interact with predation to influence population dynamics – as, for example, in the case of snowshoe hares (*Lepus americanus*), nematodes and lynx (*Lynx canadensis*). Snowshoe hares have spectacular synchronized ten-year population cycles. These cycles, with a slight time lag, are mirrored by lynx, a specialist predator of snowshoe hares across the boreal region. The main drivers of the cycles are predation and food availability, where the latter is felt mainly in the winter and is indirect and mediated by predation: hares in poor body condition are more likely to be predated (Krebs et al. 2001). In an experimental study, snowshoe hares that had been treated to remove nematodes were more likely to survive than untreated controls. This effect was especially pronounced if food was limited. Authors concluded that parasitized hares were more vulnerable to predation, especially in years of nutritional stress, such as years with bad winters or high population density (Murray et al. 1997). Similar effects were experimentally quantified in the studies of red grouse (*Lagopus lagopus scoticus*) and their parasitic nematodes (*Trichostrongylus tenuis*) in northern Britain: birds taken by predators were observed to have higher worm burdens compared to those shot at random in the annual sports harvest. Experimental studies using game dogs to detect birds showed that the dogs found control birds with high parasite burdens at four times the rate of treated birds with reduced parasite burdens (Dobson and Hudson 1994; Hudson et al. 1992).

10.4   PREDATORS MAKE FOR HEALTHY HERDS?

It is a common management strategy to cull large carnivores in attempts to protect endangered wildlife and livestock (Ripple et al. 2014). However, the efficiency and outcomes of such practices are debatable (Walsh et al. 2012). Large carnivores have considerable effects on the structure and function of ecosystems and removing them can lead to 'unintentional' consequences, leading to cascading trophic interactions with impacts on other 'non-target' species (Hollings et al. 2013, 2014; Levi et al. 2012; Ripple et al. 2014).

Infectious agents, their hosts and predators interact with many different ecological communities. Changes in their interactions can influence other ecosystem processes and alter patterns of infection and disease (Orlofske et al. 2012; Ostfeld and Holt 2004; Packer et al. 2003). In fact, removing predators may increase infection pressure on prey, potentially exacerbating the problem managers are trying to solve (Packer et al. 2003). In some cases, predators may even be beneficial to the health of populations (Packer et al. 2003).

The 'healthy herd' hypothesis suggests that predators can keep the herds healthy in two main ways. First, predators can reduce host density and thereby reduce opportunities for spread of density-dependent diseases. Second, assuming that predators are not part of the parasite lifecycle, selective predation on infected individuals will remove a disproportionately large numbers of parasites from the host population, thus reducing infection pressure (Packer et al. 2003). Although evidence for the healthy herd hypothesis has not been demonstrated in *Rangifer*, there is supporting data from other systems and mathematical models.

In Section 10.3.5.6 we discussed how grouse and snowshoe hares infected with parasites are more vulnerable to predation. There is plenty of additional empirical data from different systems that support this. For example, Krumm et al. (2010) found that deer (*Odocoileus hemionus*) killed by mountain lions (*Puma concolor*) were more likely to be infected by chronic wasting disease compared to those killed by hunters, suggesting that mountain lions selectively prey on infected individuals (Krumm et al. 2010). Similarly, heavy infection with the cestode *Echinococcus canadensis* (previously *E. granulosus*) predisposes moose (*Alces alces*) to predation by wolves (Joly and Messier 2004a).

The increased vulnerability to predation of heavily infected animals can alter the abundance of infectious agents in the host population. Mathematical models demonstrate that selective predation on infected prey can decrease the number of diseased individuals and 'keep herds healthy', whereas predator removal can increase disease prevalence (Packer et al. 2003; Wild et al. 2011). There is also population-scale empirical evidence consistent with these models. The classic study on red grouse by Hudson et al. (1992) is a great example. In Scotland, game-keepers shoot predators in order to keep grouse populations at high densities. Analyses of the parasite burdens of grouse, where the number of game-keepers/unit area varied, demonstrated that the proportion of grouse with high nematode burdens was higher in areas with more game-keepers (thus fewer predators). These data support the hypothesis that predators play a role in reducing parasite abundance at a population level (Hudson et al. 1992).

Removal or loss of top predators can also influence trophic cascades and thereby community structure. Lafferty (2004) used data from a long-term kelp-forest monitoring program in the Channel Island National Park (USA) and found that in areas where lobsters (*Panulirus interruptus*), the main predators of urchins (*Strongylocentroutus purpuratus*), were fished, urchin populations increased to well above the host-density threshold for epidemics. As a consequence, urchins in fished areas were more likely to suffer high mortality from an urchin-specific bacterial disease (Lafferty 2004).

However, the relationship between predator populations and disease in prey populations is not unidirectional. The outcome of interactions depends on many different factors, such as predation selectivity, presence of alternative prey species, infection dynamics and community diversity and complexity (Johnson et al. 2006; Orlofske et al. 2012; Rohr et al. 2015). In some cases predators may enhance disease spread (Duffy et al. 2011). For example, in some cases, predators are obligate, definitive or intermediate hosts of parasites that are shared with their prey. The protozoan parasites *Toxoplasma gondii* and *Neospora caninum* are transmitted to caribou primarily through the faeces of infected felid and canid (such as lynx domestic cats, dogs, coyotes, arctic fox and wolves) definitive hosts, respectively. Although there is likely some trans-placental transmission, in the absence of these definitive hosts the parasites would likely disappear from *Rangifer* populations (Dubey et al. 2011; Innes 2010). The role that predators play in Rangifer population dynamics is discussed in Chapter 5.

10.5 MODELING IMPACTS OF PARASITES USING MATHEMATICAL MODELS

10.5.1 Why Model?

Mathematical models are now as synonymous to the study of wildlife disease and population management as microscopes and binoculars; when developed and used properly, models can provide vital insights into processes that cause population change on time scales that far exceed normal field studies. They also allow exploration of ranges of management options that would simply be neither possible, ethical nor economically viable in most situations. Their strength is that simple models can quickly be developed from a basic understanding of the natural history of the system; their shortcoming is that people assume they need to be parameterized from really detailed field studies and will then provide exact future population projections. Models should mainly be used for exploring ideas about how a population interacts with its food resources and natural enemies: predators and parasites. This should be a central exercise in determining which data to collect during field studies and how to develop and refine statistical methods for data analysis. Building a model is never the end goal of any research project! They are much more useful if developed as the research protocol is designed – they can then be used to optimize data collection in ways that address specific hypotheses whose insight can initially be developed using the model.

10.5.2 Birth, Death, Sex and Aging

The *demography* of all populations is defined by the interaction of the birth and death processes; this is as true for parasites and pathogens as it is for their hosts. The demography of *Rangifer* is defined by the rate at which females give birth to male and female calves, that then have to survive, develop and grow into adults that have acquired enough food resources to mate and, if female, conceive and produce the next cohort of calves. Parasites can change the rate at which all these demographic processes operate and thereby potentially influence the host (*Rangifer*) population structure and dynamics. They do this as fundamental components of their own demography with the additional twist that transmission of the parasite from host to host is an additional component of the parasite's birth rate. Furthermore, the acquisition of resources from the host for their own reproduction has the potential to increase both the mortality of the host and their own mortality rate; this additional type of mortality is driven by the parasite's intimate dependence on the host, and defines the virulence of the interaction between host and parasite.

In the simplest framework, *Rangifer* demography, annual rates of recruitment and survival of individuals in the population depend on survival, reproduction and age of first reproduction (for full details see Textbox 10.2). One key parameter that can be derived from these models is, *Ro*, the 'basic reproductive number' of the population. Ro can help us estimate if the population is growing or not; if Ro is greater than 1, the population will grow; if less than 1, it will decline. This applies to both pathogens and hosts and is fundamental for predicting if a pathogen can create a disease outbreak or not (Section 10.5.6). Individuals with traits that give them higher Ro will outcompete and replace individuals with smaller intrinsic Ro. We will return to this fundamental concept later in this section.

TEXTBOX 10.2   DEMOGRAPHIC FRAMEWORK FOR *RANGIFER*

The simplest demographic framework we can use to examine *Rangifer* demography divides time into years and simply considers the annual rates of recruitment and survival of individuals in the population. Models for this type of population can be defined by a handful of parameters: adult survival, s; calves produced each year by each female, c; and survival rates of calves to age one. Thus, if Rt is the size of the population in year t, changes in population numbers between two successive years are given by the simple formula

$$R_{t+1} = (s + s_1 c) R_t \qquad (10.1)$$

If *Rangifer* do not breed for the first time until age 2, then this framework requires slight modification:

$$R_{t+1} = s(R_t + s_1 c R_{t-1}) \qquad (10.2)$$

Acknowledging that *Rangifer* may take two years to reach sexual maturity leads us to create a time delay in the demography that slows and reduces the rates at which calves enter the breeding population. Time delays always reduce the net rate at which the population can grow and recover from disturbance or over-exploitation; they also create the potential for the population to illustrate less stable, and occasionally cyclic, dynamics by overshooting the resources available to them.

We can derive a simple expression for the annual rate of increase of the population by simply rearranging these first two equations to obtain the ratio of population sizes in two successive years.

$$\lambda_1 = \frac{R_{t+1}}{R_t} = s + s_1 c \qquad (10.3)$$

As the average annual survival of the adults is always less than one, then the population can only grow if the females produce calves that survive through to age 1 at an annual rate greater than (1-s). The annual rate of increase can be used to derive an expression for Ro, the 'basic reproductive number' of the population. This is an estimate of the average number of female offspring that a female will produce in her lifetime; if the population is constant, then Ro will also equal 1; if it is growing, females will produce more than one surviving female calf per lifetime and the population will grow. We will return to this concept when we discuss parasites, as Ro is fundamental for whether or not a parasite or pathogen creates a disease outbreak.

10.5.3 POPULATION REGULATION: FOOD, CLIMATE AND RECRUITMENT

If birth and death rates were always constant, populations would either grow exponentially (if net births are larger than the death rate) or decline to extinction. The reindeer of South Georgia provide an example of a population illustrating nearly exponential growth (Leader-Williams 1988). Several special conditions allowed this to occur: the reindeer had been introduced in small numbers to an area with unexploited food supply, no predators and very few parasites, whence births greatly exceeded deaths and the population grew rapidly for over 40 years (due to reindeer overgrazing of tussock and burnet, which had negative consequences for native burrowing birds; these introduced reindeer have now been eradicated from South Georgia). Other reindeer and caribou have exhibited similar patterns of growth when first introduced to new habitats. No population can grow exponentially forever; eventually food resources become limiting, and reduce birth rates, or predators discover the population and increase death rates, or, more frequently, parasites establish and either increase death rates or reduce birth rates.

In the absence of natural enemies, parasites and predators, resource limitation is the key mechanism that leads to changes in birth and death rates. The consequences of this have perhaps been best explored for *Rangifer* by Caughley and Gunn (1993) in a paper that eloquently illustrates how variation in rainfall in desert landscapes causes variation in forage available for caribou (and kangaroos in their comparison!). They use a relatively straightforward function that links forage availability to birth rates to show that while rainfall variation between years can cause annual variation in birth and death rates, the population is ultimately regulated by the average amount of forage available to them. When food resources become limiting, animals are unable to meet their daily metabolic needs and start to lose weight. This is likely to impact adult and juvenile survival, particularly during periods when energy demands are high: in winter, or when migrating. When resources are limiting, or overexploited, lactating females and those entering their breeding phase, may stop lactating and lose their calf, or fail to conceive altogether

All of these processes operate in ways that are likely to regulate the population: death rates will increase and birth rates will decrease when per capita resources are too low to meet metabolic requirements for growth, birth and survival. The population will then decline to a level where more resources are available (as less animals are eating them) and the population will start to increase again. In some cases, the shortage of resources is matched by a particularly severe winter or an epidemic disease outbreak and the population crashes to a low abundance, where birth rates are again high and death rates low; the low numbers of individuals means that it will take the population some time to regrow to former levels of abundance.

Using a simple model framework, Figure 10.4 (a and b) illustrates how changes in birth and death rates at different levels of abundance cause the population to settle to an equilibrium size, which is roughly analogous to the traditional wildlife management concept of *carrying capacity* (Horn 1968). The core of this figure assumes that birth and death rates are relatively constant at different densities and that we can easily measure these rates. In reality, the curves are likely to move

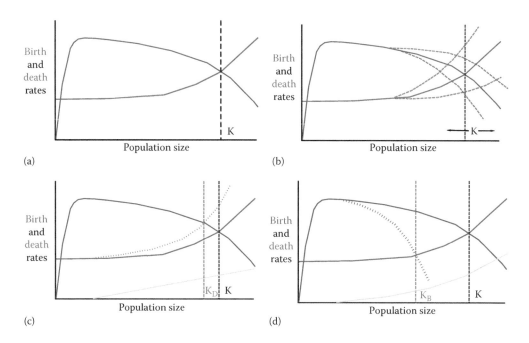

FIGURE 10.4   This figure illustrates how density-dependent changes in fecundity (birth rates) and/or mortality (death rates) as a population changes in abundance cause it to settle to an equilibrium abundance that is essentially analogous to the traditional concept of carrying capacity. In each figure the x-axis depicts population size, and birth rates are plotted in blue and death rates in red. In (a), birth and death rates change with density, with mortality increasing as the density gets large. Birth rates initially increase as it becomes easier for potential mates to find each other; they are then relatively constant until high densities, when they decline. The population will stabilize when birth rates equal death rates; this defines an abundance commonly called the carrying capacity, K, the equilibrium size of the population. Notice that birth rates can equal death rates at very low levels of abundance. This creates an Allee effect: a population density below which the population collapses to local extinction. (b) is directly analogous to (a) except we have now included variability in birth and death rates, which is likely driven by annual variations in climate that determine forage availability in different years and harsh weather conditions (such as 'rain on snow'/icing events) that lead to unusually high levels of mortality. This variability causes the carrying capacity to vary between years; it does not alter the fundamental feature of regulation that requires birth rates to decrease and/or death rates to increase when the population becomes large. Population dynamic models describe these different curves as functions that can be parameterized from empirical data. In (c) and (d) we add the influence of infectious agents into these figures. Initially we assume that the 'agents' increase mortality effects (green) and this effect is additive to other sources of mortality; similarly, we assume the mortality due to infectious agents increases as the host population size gets bigger. In (d) we assume the infectious agents reduce host birth rates (in green) and this reduces the rate of recruitment into the population. In both cases the effect of 'agents' is to reduce the host population below the abundance it would settle to in the absence of the parasite: K_D and K_B instead of K.

up and down depending on local climate conditions that determine forage quality and availability. This means that there will be no fixed carrying capacity; instead there will be a range of densities at which the population will settle to some range of abundances. These will be higher following a series of good years, and lower following a sequence of poor years. If the population is managed in ways that minimize fluctuations in abundance, then we have very limited amounts of information about the way in which birth and death respond to changes in density. Only when data is collected following perturbations can we identify the strength of the mechanisms regulating abundance. Predators or pathogens will add additional lines to these curves that further alter the position of the point where births balance deaths (Figure 10.4 c and d). Thus, the density that the population settles to will usually be lower than the abundance determined purely by resource limitation. An important consequence of this is known as the 'paradox of enrichment' (Berryman 1992; Rosenzweig 1971). This occurs when attempts to improve the forage quality for a herbivore species produce no increase in their abundance, but instead an increase in the abundance of predators and parasites (but also see Toyokawa [2017]). Wildlife managers have not had a happy time grasping the importance of the paradox of enrichment, nor of dealing with its implications for management (for example, see Sharp and Pastor 2011).

10.5.3.1 Energetic Models for *Rangifer*

A number of studies have developed energetic models of *Rangifer* dynamics that couple resources consumed as food with energy needed for fecundity and body maintenance (survival) (Boulanger et al. 2011; White, Daniel, and Russell 2013). These models provide important insights into the nutritional stresses that *Rangifer* species face at different ages and at different times of the year. These can be very different for females than for males and provide important insights into foraging and migratory strategies (Gunn et al. 2012). The next major step in their development will be to include information on the energetic cost of parasitism (Gunn and Irvine 2003).

10.5.4 PREDATION: A ONCE-IN-A-LIFETIME MOMENT

Adult *Rangifer* have a range of predators: wolves, bears, wolverine, cougars and even lynx prey on adults. Calves are also preyed upon by these predators, as well as by eagles and coyotes. How effective are these predators at regulating the abundance of their hosts? This remains a central question of wildlife management and many predator culling programs are based on the assumption that predators have a major impact on the abundance of their prey. There are significant problems with the assumption that predators can control their prey abundance – most significant is that predators have to catch their prey and then spend time eating and digesting it; this means that the rate at which prey are consumed as their abundance increases will always saturate and is always a decreasing proportion of population size (Lafferty et al. 2015). This makes it very hard for predators to regulate the size of their prey population; instead, the abundance of the prey population (and their food resources) usually determines the abundance of the predator population. Certainly predators reduce the abundance of their prey, as they cause an increase in mortality, but it is only very rarely that predators act in a purely regulatory fashion (Lafferty et al. 2015). While predators can only affect birth and death rates once, when they kill the prey, parasites can infect hosts on multiple occasions and for prolonged periods of time; some infections are even lifelong – potentially exerting impacts on hosts on several occasions. Trauma caused by different kinds of predators and their impact on *Rangifer* are further discussed in Chapter 5 ("Non-infectious Diseases and Trauma").

10.5.5 MICROPARASITES

Models for infectious agents ('microparasites'), such as herpesvirus and the bacteria *Brucella* sp. and *Erysipelothrix* sp., divide the host population into different stages of infection (Keeling and Rohani 2008). In the simplest case, where infections are chronic and hosts rarely recover without treatment, the host population can be divided into susceptible and infected classes (usually designated S and I) (Figure 10.5). More detail can be included by adding an exposed class, such that hosts move sequentially from susceptible, S, to exposed, E (through contact with infected hosts or free-living infectious stages), and then slowly develop symptoms that indicate they are infectious; they then remain chronically infected until death. Where hosts recover from infection, we can add a recovered class of hosts, R, to the model; these may have immunological resistance (also R) to reinfection. This immunity may, or may not, decrease through time and returns the recovered hosts to the susceptible class (Anderson and May 1991). A host moves between different classes of infection at rates that are dependent upon the development times of different stages of infection. The classic model for microparasite dynamics is hence termed the SEIR model, as it describes the flow of hosts

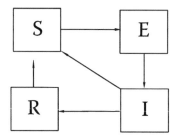

FIGURE 10.5 SEIR box and arrow diagram. The population is divided into susceptible hosts, S, who contact infected hosts, I. If transmission is successful this converts susceptible hosts to exposed hosts, E, who then develop into infected hosts. When they recover from infection, they enter the recovered class, R, who may eventually lose immunity and again become susceptible to infection. Models for this type of pathogen describe the rates of conversion between these classes of infection.

between susceptible, exposed, infectious and recovered states. The models used to describe these dynamics are outlined in Textbox 10.3. Here, two expressions are outlined: one that describes what happens with host abundance in the absence of the pathogen, and one that describes what happens when the pathogen is present, which helps us understand the impact of pathogens under different scenarios.

The crucial feature of the SEIR model is the function that defines transmission of the pathogen from Infected hosts to Susceptible hosts; the rate at which this occurs is driven by the transmission rate. It is important to note that there are two different types of transmission: *density-dependent* transmission and *frequency-dependent transmission* (Textbox 10.3). In density-dependent transmission, transmission rates increase with host density since the contact rate between susceptible and infected individual depends on their population density (i.e. the number of individuals in the population divided by the area). In frequency-dependent transmission, the contact rate between S and I individuals does not depend on population density, but on the frequency of contacts between infected and susceptible hosts (Begon et al. 2002). Thus, transmission is often a function of the duration of time for which hosts are infectious. Frequency-dependent transmissions can usually persist even in very small populations, since transmission is not dependent on population size, only on contacts that lead to transmission of the pathogen (Begon 2009).

TEXTBOX 10.3 MICROPARASITE MODELS: SEIR

The models described in Textbox 10.2 for healthy populations can fairly readily be modified to subdivide the host population into S, E, I and R classes by the inclusion of a function that describes the rates at which infected hosts contact susceptible hosts in ways that lead to successful transmission of the parasite. The host then moves between different classes of infection at rates that are dependent upon the development times of different stages of infection. This creates a problem for our current model framework (in Textbox 10.2), in that discrete models for the host's population assume everything happens on an annual cycle and time is neatly divided into annual units. This works less well for models that include parasites and pathogens, as transmission, infection, disease and recovery are all likely to operate on different time scales, some of which are much shorter than a year. We can deal with this by changing our discrete time model into a continuous time model. We can recapture critical aspects of the annual dynamics of the population if we set functions such as host and parasite birth rates as seasonal functions of temperature (or day length). The papers by Molnár et al. (Molnár, Dobson, and Kutz 2013; Molnár et al. 2013) provide a comprehensive description of how to do this, as does the book by Keeling and Rohani (Keeling and Rohani 2008).

We will ignore the complexities of seasonality initially, so let us rewrite equation (3) for discrete growth in a population with minimal age-structure as a continuous time equation where the birth rate is dependent upon available resources and thus population growth begins to slow when $N = b/\Delta$.

$$dN/dt = N\frac{m(b-\Delta N)}{(m+d)} - dN \tag{10.4}$$

Here N is the size of the adult population (formerly R, but we're going to use that now for recovered and resistant hosts), b and d are the birth and death rates of the host (1/b is life expectancy in years) and 1/m is the age of first reproduction. Notice we have included a small amount of age-structure into the population by only allowing a proportion of calves (m/(m + d)) to survive to reproductive age. In the absence of pathogens and predators, the population will settle to an abundance that we could traditionally think of as carrying capacity, K.

$$K\frac{mb - d(m+d)}{m} \tag{10.5}$$

This still retains all of the features of our simpler model and thus variations in b, m, d and resource availability, Δ, will all cause carrying capacity to be a distribution of abundances, rather than some fixed point. It is important to have an expression for host abundance in the absence of the pathogen as we can compare this abundance with the one the population settles to in the presence of the pathogen and thus determine the characteristics of pathogens that cause the largest impact on host abundance.

We can now add a pathogen to the model and subdivide N into susceptible, exposed, infectious and recovered hosts (note S + E + I + R = N).

$$dS/dT = N\frac{m(b-\Delta N)}{(m+d)} - dS - \beta\frac{SI}{N} \tag{10.6}$$

$$dE/dt = \beta\frac{SI}{N} - (d+\sigma)E \tag{10.7}$$

$$dI/dT = \sigma E - (d+\beta+\alpha)I \tag{10.8}$$

$$dR/dt = \beta I - (d)R \tag{10.9}$$

The crucial feature of this model is the function that defines transmission of the pathogen from I, infected hosts, to S, susceptible hosts. The rate at which this occurs is driven by the transmission rate, β; but it is possible for this to enter into the equations in two different ways. We have written this above as βSI/N, which is termed frequency-dependent transmission. Inherently we are assuming that contacts between infected and susceptible hosts are buffered by the presence of recovered and exposed hosts; this will slow rates of transmission when many hosts in the population have been exposed and recovered, but cause it to be much higher when all hosts are susceptible. An alternative form of transmission is termed density-dependent transmission; here the term in N is not included in the denominator (but set to 1) and transmission is determined by the density of susceptible and infected hosts. As *Rangifer* tend to live in large herds, it is likely that frequency-dependent transmission is the more accurate description of transmission within groups; in contrast, transmission between groups is likely to be density-dependent (where density is defined by the density of groups), although there are very little data to test this assumption!

10.5.6   The Basic Reproductive Number, Ro Revisited

The equations for SEIR (Textbox 10.3) can be rearranged to provide an expression for Ro, the basic reproductive number, of the parasite. As mentioned above, this is directly equivalent to the number of offspring that a female *Rangifer* produces in her lifetime. From the pathogen's perspective, it is the average number of new infected hosts produced by the first infected individual in the herd (Diekmann et al. 1990; Roberts and Heesterbeek 2013). In this case, the expression for Ro can be seen to be the product of the transmission rate, βN, and the average time the hosts spends in each class of infection, $(d + \sigma)$ and $(d + \delta + \alpha)$ (note that the time spent incubating is modified by the proportion that survive incubating to become infectious $\sigma/(d + \sigma)$).

$$R_0 = \frac{\beta N \sigma}{(d + \sigma)(d + \delta + \alpha)}.$$
(10.10)

If an infection is to spread in a population it must cross the transmission threshold, which occurs under conditions when Ro = 1. Thus, if Ro is < 1 the infection will eventually die out, since each current infection leads to fewer than one infection in the future. The infection will spread in the population if Ro is > 1. For the SEIR model described above, Ro will increase as the size of the host population, N (or the size of the social groups), increases. If the pathogen has a long incubation period $(1/\sigma)$, then Ro is likely to decrease, as only a small proportion of hosts will survive incubation to go on and transmit the disease.

The power of Ro lies in providing a concise algebraic expression for what are the important stages of the parasite's lifecycle that determine the rate at which it spreads through the host population. Ro can help us understand which stage of a pathogen's lifecycle to target in order to have the best chance of reducing transmission, and also what proportion of the population needs to be treated to reduce transmission or eliminate infection. For example, if we have a pathogen such as *Brucella* in a herd of reindeer, then we would need to treat a proportion greater than (1−1/Ro) of the herd if we want to eradicate the pathogen or prevent initial establishment.

Estimating transmission rates is always the hardest part of any ecological study of microparasite infections. The infective stages are too small to be seen and it often takes a newly infected host several days to show symptoms of infection. If data are available from serology of the numbers of individuals of different ages exposed to infection, then we can estimate transmission rates from the different proportions of individuals infected in different age classes. To a rough approximation Ro = L/A, where L is average life expectancy (1/d) and A is average age of first infection; thus if a reindeer lives for 8 years and average age for infection with *Brucella* is 4, then Ro for *Brucella* in reindeer is 8/4 = 2.

10.5.7   MACROPARASITES: NEMATODES WITH DIRECT LIFE CYCLES

Hosts can accumulate an increasing community of parasitic worms throughout their lives. The population dynamics of the worm parasites are different from those of the viruses and bacteria because we can no longer simply divide the host population into susceptible and infected hosts. Instead we have to acknowledge that the impact of the worms on the host is usually a function of the number of worms infecting the host; more worms lead to more damage and either decreased host survival or reduced fecundity. The statistical distribution of worms also complicates the dynamics of the interaction. The abundance of worms of any population is usually aggregated, with a few hosts harboring a high proportion of the adult worm population and many hosts harboring low worm burdens or remaining uninfected; this type of distribution is best described by the negative binomial distribution (Anderson and May 1978; May and Anderson 1978). The distribution arises mainly because the free-living infective stages are often aggregated in their distribution, so feeding in one location can expose individuals to many more infective stages than other individuals not feeding in that location. Even if the distribution of infective stages were random, small differences in feeding rate (due to body size, dominance status) and differences in immunological status, behaviour or genetic susceptibility interact with the spatial distribution of infective stages to amplify the degree of aggregation observed in the host population. This degree of aggregation tends to increase with age; thus, the youngest age class sampled may have a distribution that is almost random (or Poisson), but as the individuals age, small differences between them amplify the differences in susceptibility to give more aggregated distributions. In some cases, individuals with very high worm burdens will die, and this may cause the level of aggregation to be reduced in older age-classes, although the mean may continue to increase. If estimates of worm burden are based purely on counts of parasite eggs in faeces, they will tend to underestimate the degree of aggregation. This primarily occurs if the fecundity of female worms decreases in heavily infected hosts (intraspecific competition), as this may lead to heavily infected hosts producing similar numbers of eggs in the hosts' faeces as hosts with lower worm burdens.

Many nematodes, such as abomasal nematodes like *Ostertagia* and *Marshallagia*, have direct lifecycles where eggs that are shed in the faeces hatch and undergo several molts to develop into the infective third-stage larvae (L3s). Hosts are infected when they ingest these L3s while grazing. Inside the host, larvae molt twice and develop into egg-laying adults. From a modeling perspective (Textbox 10.4), the lifecycle can initially be divided into two stages: the population of worms living in the hosts, P, and the free-living stages that live in the environment, W (Figure 10.6). The free-living stages could be subdivided into eggs, and three sequential larval stages, but it is simpler to condense this to a single stage that has a development rate from egg to infective L3, and a mortality rate. Key here is that both the development and the mortality rates are dependent upon temperature, where both increase with increases in temperature. After L3s have been ingested and infected the host, they can develop into adults, or into encysted larvae in a state of *'arrested development'* or *'hypobiosis'*, where they rest for weeks to months before developing into reproductive adult worms (Michel 1974). The arrested larvae create a time delay in the lifecycle. Arrested development has likely evolved to reduce reproduction at times of the year when the free-living stages would have a very low chance of survival, or to avoid times when hosts are migrating and passing quickly through areas where the hosts are no longer present when the infective larvae emerge (Halvorsen et al. 1999). For an overview of the models describing the dynamics of macroparasites with direct lifecycles, see Textbox 10.4. Critically, these models can be used to predict how parasites with direct lifecycles may respond to changes in temperature, and how that may affect the lifecycle, transmission windows and potential range expansion.

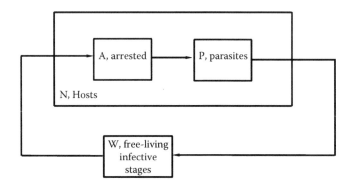

FIGURE 10.6   Macroparasite lifecycle/model box and arrow diagram. The hosts are assumed to have birth and death rates that are modified by the presence of parasites. The parasites have a free-living population of infective stages that are produced by adult worms living in the hosts. When ingested, the infective larvae may enter an arrested, hypobiotic stage when they are inert and cause little harm to the hosts. After some time delay they develop into adults where their abundance has detrimental impact on host fecundity and mortality. Not all hosts carry the same number of parasites; in general, hosts are born uninfected and gain a population of parasites through feeding on infected pasture. Because different individuals feed in different places, the distribution of parasites across the host population becomes aggregated with a large number of parasites in only a small proportion of the host population; these unfortunate hosts suffer most from the presence of the parasites. Models for these parasites again focus on rate of flow between different stages of the host and parasite population, but also need to consider the statistical distribution of parasites per host.

TEXTBOX 10.4   MODELS FOR MACROPARASITES WITH DIRECT LIFECYCLES

For macroparasites with direct lifecycles, we can again adapt our basic single species model (Textbox 10.2) and add in a population of adult and arrested larval worms, P, that live inside these hosts, and a population of free-living larvae, W, that live outside the hosts.

$$dN\!/\!dt = N\,\frac{m(b-\Delta N)}{(m+d)} - dN - \alpha P \tag{10.11}$$

$$dW\!/\!dt = \lambda P - \delta W - \beta WN \tag{10.12}$$

$$dA\!/\!dt = \beta WN - (\mu+d+h)A - \alpha\,\frac{PA}{N}\left(\frac{k+1}{k}\right) \tag{10.13}$$

$$dP\!/\!dt = hA - (\mu+d+\alpha)P - \alpha\,\frac{P^2}{N}\left(\frac{k+1}{k}\right) \tag{10.14}$$

Here μ is the death rate of the parasites in the hosts, β is the rate at which free-living larvae are ingested by the hosts, h is the rate at which arrested larvae develop into adult worms, α is the rate at which the each parasite increases the host's mortality and k is the parameter of the negative binomial distribution that reflects the degree of aggregation of the worms in the host population. If k is greater than 5, the worms have a random distribution.

More characteristically, the worms have values of k less than 0.1 – this considerably increases the rate at which worms kill heavily infected hosts and, of course, the worms in those hosts. These equations can be collapsed down to two equations if we assume that the dynamics of the free-living stages are relatively fast $(dW / dt = 0)$. The equation for hosts is identical to the above and for the adult parasites becomes

$$dP\!\!\Big/\!\!dt = \frac{\lambda PN\mu}{(N + H_0)(\mu + h)} - (\mu + d + \alpha)P - \alpha \frac{P}{N}\left(\frac{k+1}{k}\right)$$

Two things have happened here: the fast time scale of the larvae is reflected in the Ho term $(=\delta/\beta)$, the death rate of the free-living larvae divided by the rate at which they are ingested by the hosts. Similarly, only a proportion $(\mu/(\mu + h))$ of the arrested larvae survive hypobiosis to mature as adults. Additionally, we can use the machinery developed by Molnár et al. (2013) to examine the influence of temperature on the survival and development rate of the free-living larval stages. We can then again derive an expression for the basic reproductive number of the parasites:

$$R_0 = \frac{\lambda\beta Nh}{(N + H_0)(\mu + d + h)(\upsilon + d + \alpha)}$$

Although this is at first sight intimidating, closer inspection reveals that Ro is again the product of the birth, development, and transmission rates, and the mortality stages at each stage of the lifecycle. The proportion of worms that survive arrestment h/(μ + d + h) enters in a similar fashion to the proportion of hosts that survive incubation in the SEIR model. The huge fecundity of parasitic worms, λ, is offset by their very low probability of infecting a host. βN/(H + Ho) – remembering that Ho = δ/β, the mortality rate of free-living stages divided by the rate at which they are ingested by hosts. Molnár et al. (2013) have used this model to examine the dependence of Ro on seasonal changes in temperature; they do this using a model parameterized with *Ostertagia gruehneri*. The model provides an important insight into the likely response of this parasite to increasing temperatures in the Arctic; this leads not only to a longer transmission season, but also to a split in the original single peak transmission window into a bimodal, seasonal pattern of transmission, with transmission peaking in spring and summer and declining in the now hotter summer months. The model's projections match the observed data extremely well.

10.6   UNDERSTANDING THE ROLE OF INFECTIOUS AGENTS IN WILDLIFE POPULATIONS: CHALLENGES AND OPPORTUNITIES

Today, the role infectious agents have in shaping ecosystems, including wildlife behaviour and ecology, is recognized (Gómez and Nichols 2013; Hudson, Dobson, and Lafferty 2006), with researchers, managers and conservation biologists more frequently integrating disease-focused studies into their programs. With habitat fragmentation, industrial activity and climate change creating novel opportunities for infectious agents to spread and infect their hosts (Acevedo-Whitehouse and Duffus 2009; Altizer et al. 2013; Gottdenker et al. 2014), continued monitoring and modeling of *Rangifer* health and changes in infectious agent-host communities and interactions is imperative. Although we have made impressive advances in our understanding of these interactions, there are plentiful opportunities for further research that will help us overcome some of the challenges we face when trying to understand the role of infectious agents in population dynamics of *Rangifer*. We will not attempt to outline all the factors that can influence infectious agent-host interaction and wildlife health research, or the full potential scope of future research directions, as this has been discussed elsewhere (Deem et al. 2001; Hudson et al. 2002; Tompkins et al. 2011), but some important areas to consider in future programs are briefly discussed below.

First, we need comprehensive baseline data on the diversity and distribution of infectious agents in all populations to establish 'normal' or expected values in different regions. Changes in baseline parameters before or after an event can inform hypotheses that certain 'agents' had an impact on host health. Regular monitoring of 'agent' prevalence and diversity will also allow for early detection of changes in disease status and provide data to inform timely implementation of management actions and control measures. Second, standardization of methods and assays would allow comparisons across herds and regions, making large-scale patterns easier to identify (Kutz et al. 2013). Today, comparisons between studies are often complicated or hindered by the variability in the type of assays/methods used, since their sensitivity, specificity and cutoff values will differ between studies. Third, pathological examinations should be conducted whenever possible to determine the cause of death. Although this is particularly difficult for wild *Rangifer*, it is also particularly important in these populations, since our baseline knowledge of the abundance, diversity, and impacts of infectious agents still remains poor. Fourth, molecular epidemiology studies can help us identify the source of outbreaks and track their progression, potentially helping us prevent future epidemics. Isolating and identifying different strains of infectious agents will help us understand if *Rangifer* pathogens are also cycling in other wildlife and domestic animal populations (Forde et al. 2016) and can help us control disease outbreaks by preventing 'spill-over' events. Fifth, continued archiving of data and samples (and analysis of these) is crucial since they allow retrospective epidemiological analysis. A sample that may seem insignificant today may become pivotal tomorrow (Cook et al. 2016). Sixth, and critically, there is a need for studies specifically designed to investigate the prevalence and diversity of infectious agents and risk factors for disease. Today, many studies rely on samples collected for other purposes and are not stratified adequately to allow comparisons between seasons, sexes, regions, ecotypes and other risk factors. To better understand the impact and role of infectious agents in *Rangifer* population dynamics, we need studies and controlled experiments specifically designed for that purpose. Finally, modeling is a critical component of understanding observed trends, predicting new patterns, and developing and refining hypotheses and research programs on *Rangifer* health and disease dynamics.

The intention of this chapter was to illustrate that infections with different infectious agents can have important consequences for wild animal populations, and are not only important because of disease they may cause in individual animals. In fact, infectious agents can have significant population level impacts even in the absence of observable clinical disease. Scaling relationships between disease impacts at the individual and population level are not straightforward. Epidemics of highly virulent infectious agents that kill individual hosts can have large impacts on populations if many hosts are affected, leading to die-offs. This is especially problematic for small isolated threatened species, as these populations may not recover, particularly if reservoir hosts are present that perpetuate infections. However, the impacts of highly virulent pathogens are usually short-lived, since the host often dies before transmitting the pathogen. Thus, infectious agents with the greatest population impact may be those with intermediate virulence that have some, but not massive, effects at the individual level. As they do not kill their host, they can transmit further to many hosts, potentially leading to major effects at the population level. Ironically, those types of infectious agents are rarely monitored, slipping 'under the radar', since they do not cause overt disease and mortality.

As discussed, the experimental population level experiments needed to detect subtle context-dependent impacts of infectious agents are difficult to perform in most wildlife species, including *Rangifer*, due to logistical, ethical and methodological restrictions. In these cases, theoretical models can provide significant and important insight into the potential role/impact of infectious agents in a population. Interdisciplinary collaboration needs to be an integral component if long-term research, management and conservation programs are to be successful. A recent review evaluating the extent to which disease ecology theory is integrated into wildlife disease management found that academics and government agencies do collaborate to manage wildlife diseases, but do not consistently integrate theory with management (Joseph et al. 2013). Although some theoretical concepts, such as density-dependence in transmission, were frequently applied, others, such as the role of predators in regulating disease and pathogen evolution, were barely acknowledged. The authors outline ways to facilitate a predictive framework and suggest that in efficient collaborations there is great potential for natural field experiments, where theoretical predictions can be tested using large-scale manipulation in real systems, which is often logistically unfeasible for academics working in isolation. Management interventions can be evaluated critically in the context of current disease ecology theory, and adaptive management approaches allow testing of competing hypothesis. One of the roadblocks identified for successful use of predictive models in management was that theoreticians often prefer general models applicable at a broader scale, while managers need specific and detailed models, but system-specific detail necessary to populate these models is often missing. If researchers and managers collaborate closely it provides opportunities to test disease ecology theory in a management perspective while building system-specific understanding, which will lead to more effective control strategies (Joseph et al. 2013).

Finally, programs can also benefit by capitalizing on the increasing public interest in science and increased support and acceptance for 'citizen science' and community-based monitoring approaches in research and management (Lawson et al. 2015). People have lived together with *Rangifer* for millennia; hunters and reindeer herders are the 'eyes on the land' and depend on *Rangifer* for their livelihood. They hold traditional knowledge that can contribute to an integrated view of individual and population health and can help identify infectious agents present in herds in the past, or new agents that are entering the herd today (Carlsson, Veitch et al. 2016; Russell et al. 2013). Including community members as integral parts of research and management programs builds trust between researchers, managers and users and can improve the speed of decision making and the support, uptake and implementation of new management directives (Danielsen et al. 2005).

REFERENCES

Acevedo-Whitehouse, K., and A.L.J. Duffus. 2009. Effects of environmental change on wildlife health. *Philos. Trans. R. Lond. B. Biol. Sci.* 364 (1534):3429–38.

Albon, S.D., A. Stien, R.J. Irvine et al. 2002. The role of parasites in the dynamics of a reindeer population. *Proc. Biol. Sci.* 269 (1500):1625–32.

Altizer, S., R.S. Ostfeld, P.T.J. Johnson, S. Kutz, and C.D. Harvell. 2013. Climate change and infectious diseases: From evidence to a predictive framework. *Science* 341 (6145):514–19.

Anderson, A.M., and R.M. May. 1991. *Infectious Diseases of Humans: Dynamics and Control.* Oxford: Oxford Scientific Press.

Anderson, R.M., and R.M. May. 1978. Regulation and stability of host-parasite population interactions. I. Regulatory processes. *J. Animal Ecol.* 47:219–47.

Anderson, R.M., and R.M. May. 1979. Population biology of infectious diseases: Part I. *Nature* 280:361–7.

Ashley, N.T., P.S. Barboza, B.J. Macbeth et al. 2011. Glucocorticosteroid concentrations in feces and hair of captive caribou and reindeer following adrenocorticotropic hormone challenge. *Gen. Comp. Endocrinol.* 172 (3):382–91.

Avgar, T., G. Street, and J.M. Fryxell. 2014. On the adaptive benefits of mammal migration. *Can. J. Zool.* 92 (6):481–90.

Ball, M.C., M.W. Lankester, and S.P. Mahoney. 2001. Factors affecting the distribution and transmission of *Elaphostrongylus rangiferi* (Protostrongylidae) in caribou (*Rangifer tarandus caribou*) of Newfoundland, Canada. *Can. J. Zool.* 79 (7):1265–77.

Ballesteros, M., B.J. Bardsen, K. Langeland et al. 2012. The effect of warble flies on reindeer fitness: A parasite removal experiment. *J. Zool.* 287 (1):34–40.

Beechler, B., C. Manore, B. Reininghaus et al. 2015. Enemies and turncoats: Bovine tuberculosis exposes pathogenic potential of Rift Valley fever virus in a common host, African buffalo (*Syncerus caffer*). *Proc. Biol. Sci.* 282 (1805):20142942.

Begon, M. 2009. Ecological epidemiology. In *The Princeton Guide to Ecology*, edited by S.A. Levin, 220–6. Princeton, New Jersey: Princeton University Press.

Begon, M., M. Bennett, R.G. Bowers et al. 2002. A clarification of transmission terms in host-microparasite models: Numbers, densities and areas. *Epidemiol. Infect.* 129 (1):147–53.

Berryman, A.A. 1992. The orgins and evolution of predator-prey theory. *Ecology* 73 (5):1530–5.

Boulanger, J., A. Gunn, J. Adamczewski, and B. Croft. 2011. A data-driven demographic model to explore the decline of the Bathurst caribou herd. *J. Wild. Manage.* 75 (4):883–96.

Cameron, R.D., W.T. Smith, S.G. Fancy, K.L. Gerhart, and R.G. White. 1993. Calving success of female caribou in relation to body weight. *Can. J. Zool.* 71 (3):480–6.

Campbell, M. 2006. Monitoring condition, feeding habits and demographic parameters of island bound barrenground caribou (*Rangifer tarandus groenlandicus*) Southampton Island, Nunavut. Final status report: 3. Iqaluit: Government of Nunavut Department of Environment.

Campbell, M. 2013. Research update to the Department of Environment: Population estimate of a declining population of island bound barren-ground caribou (*Rangifer tarandus groenlandicus*), Southampton Island, NU. Department of Environment, Kivalliq Region.

Carlsson, A.M., K.J. Irvine, K. Wilson et al. 2012. Disease transmission in an extreme environment: Nematode parasites infect reindeer during the Arctic winter. *Int. J. Parasitol.* 42 (8):789–95.

Carlsson, A.M., R.J. Irvine, K. Wilson, and S.J. Coulson. 2013. Adaptations to the Arctic: Low-temperature development and cold tolerance in the free-living stages of a parasitic nematode from Svalbard. *Polar Biol.* 36 (7):997–1005.

Carlsson, A.M., G. Mastromonaco, E. Vandervalk, and S. Kutz. 2016. Parasites, stress and reindeer: Infection with abomasal nematodes is not associated with elevated glucocorticoid levels in hair or faeces. *Conserv. Physiol.* 4 (1):cow058.

Carlsson, A.M., A. Veitch, R. Popko, S. Behrens, and S. Kutz. 2016. Monitoring wildlife health for conservation and food security in the Canadian Arctic – A case study from the Sahtu settlment area in the Northwest Territories. In *One Health Case Studies: Addressing Complex Problems in a Changing World*, edited by S.C. Cork et al., In review. Sheffield, UK: 5m Publishing.

Carlsson A.M., A.B. Albon, S.J. Coulson, E. Ropstad, A. Stein, K. Wilson, L.E. Vebjorn, R.J. Irvine. 2018. Little impact of over-winter parasitism on a free-ranging ungulate in the high Arctic. *Func. Ecol.* 32 (4):1046–1056.

Caughley, G., and A. Gunn. 1993. Dynamics of large herbivores in deserts: Kangaroos and caribou. *Oikos* 67 (1):47–55.

CircumArctic *Rangifer* Monitoring Assessment Network. 2008. *Rangifer* health and body condition monitoring protocols level 1 and 2. Circumarctic *Rangifer* Monitoring and Assessment Network. Accessed 31/3/2014. www.caff.is/resources/field-protocols.

Colman, J.E., C. Pedersen, D.O. Hjermann et al. 2003. Do wild reindeer exhibit grazing compensation during insect harassment? *J. Wildl. Manage.* 67 (1):11–19.

Cook, J.A., K.E. Galbreath, K.C. Bell et al. 2016. The Beringian Coevolution Project: Holistic collections of mammals and associated parasites reveal novel perspectives on evolutionary and environmental change in the North. *Arctic Science* 3 (3):585–617.

Crofton, H. 1971. A quantitative approach to parasitism. *Parasitology* 62 (02):179–93.

Curry, P.S., C. Ribble, W.C. Sears et al. 2014. Blood collected on filter paper for wildlife serology: Detecting antibodies to *Neospora caninum*, West nile virus, and five bovine viruses in *Rangifer tarandus* subspecies. *J. Wildl. Dis.*:297–307.

Cuyler, C., R.R. White, K. Lewis et al. 2012. Are warbles and bots related to reproductive status in West Greenland caribou? *Rangifer* 32 (2):243–57.

Danielsen, F., N. Burgess, and A. Balmford. 2005. Monitoring matters: Examining the potential of locally-based approaches. *Biodivers. Conserv.* 14 (11):2507–42.

das Neves, C.G., T. Mørk, J. Thiry et al. 2009. Cervid herpesvirus 2 experimentally reactivated in reindeer can produce generalized viremia and abortion. *Virus Res.* 145 (2):321–8.

das Neves, C.G., E. Rimstad, and M. Tryland. 2009. Cervid herpesvirus 2 causes respiratory and fetal infections in semidomesticated reindeer. *J. Clin. Microbiol.* 47 (5):1309–13.

das Neves, C.G., S. Roth, E. Rimstad, E. Thiry, and M. Tryland. 2010. Cervid herpesvirus 2 infection in reindeer: A review. *Vet. Microbiol.* 143 (1):70–80.

Deem, S.L., W.B. Karesh, and W. Weisman. 2001. Putting theory into practice: Wildlife health in conservation. *Conserv. Biol.* 15 (5):1224–33.

Diekmann, O., J.A.P. Heesterbeek, and J.A.J. Metz. 1990. On the definition and the computation of the basic reproduction ratio R_0 in models for infectious-diseases in heterogeneous populations. *J. Math. Biol.* 28 (4):365–82.

Dobson, A., and P. Hudson. 1994. The interaction between the parasites and predators of red grouse *Lagopus lagopus scoticus*. *Ibis* 137:S87–96.

Dubey, J.P., M.C. Jenkins, C. Rajendran et al. 2011. Gray wolf (*Canis lupus*) is a natural definitive host for *Neospora caninum*. *Vet. Parasitol.* 181 (2-4):382–7.

Dubey, J.P., G. Schares, and L.M. Ortega-Mora. 2007. Epidemiology and control of neosporosis and *Neospora caninum*. *Clin. Microbiol. Rev.* 20 (2):323–67.

Duffy, M.A., J.M. Housley, R.M. Penczykowski, C.E. Caceres, and S.R. Hall. 2011. Unhealthy herds: Indirect effects of predators enhance two drivers of disease spread. *Funct. Ecol.* 25 (5):945–53.

Fauchald, P., R. Rødven, B.J. Bårdsen et al. 2007. Escaping parasitism in the selfish herd: Age, size and density-dependent warble fly infestation in reindeer. *Oikos* 116 (3):491–9.

Ferguson, M.A. 1997. *Rangiferine brucellosis* on Baffin Island. *J. Wildl. Dis.* 33 (3):536–43.

Flasko, A., M. Manseau, G. Mastromonaco et al. 2017. Fecal DNA, hormones, and pellet morphometrics as a noninvasive method to estimate age class: An application to wild populations of Central Mountain and Boreal woodland caribou (*Rangifer tarandus caribou*). *Can. J. Zool.* 95 (5):311–21.

Folstad, I., A.C. Nilssen, O. Halvorsen, and J. Andersen. 1991. Parasite avoidance – the cause of post-calving migrations in *Rangifer*. *Can. J. Zool.* 69 (9):2423–9.

Forbes, L.B. 1991. Isolates of *Brucella suis* biovar 4 from animals and humans in Canada, 1982–90. *Can. Vet. J.* 32 (11):686–8.

Forde, T., K. Orsel, J. De Buck et al. 2012. Detection of *Mycobacterium avium* subspecies paratuberculosis in several herds of arctic caribou (*Rangifer tarandus* ssp.). *J. Wildl. Dis.* 48 (4):918–24.

Forde, T.L., K. Orsel, R.N. Zadoks et al. 2016. Bacterial genomics reveal the complex epidemiology of an emerging pathogen in Arctic and boreal ungulates. *Front. Microbiol.* 7 (1759).

Fox, M. 1997. Pathophysiology of infection with gastrointestinal nematodes in domestic ruminants: Recent developments. *Vet. Parasitol.* 72 (3):285–308.

Gilbert, A.T., A.R. Fooks, D.T. Hayman et al. 2013. Deciphering serology to understand the ecology of infectious diseases in wildlife. *EcoHealth* 10 (3):298–313.

Gómez, A., and E. Nichols. 2013. Neglected wildlife: parasitic biodiversity as a conservation target. *Int. J. Parasitol. Parasites Wildl.* 2:222–7.

Gottdenker, N.L., D.G. Streicker, C.L. Faust, and C. Carroll. 2014. Anthropogenic land use change and infectious diseases: A review of the evidence. *EcoHealth* 11 (4):619–32.

Gulland, F. 1992. The role of nematode parasites in Soay sheep (*Ovis aries L.*) mortality during a population crash. *Parasitology* 105 (03):493–503.

Gulland, F., S. Albon, J. Pemberton, P. Moorcroft, and T. Clutton-Brock. 1993. Parasite-associated polymorphism in a cyclic ungulate population. *Proceedings of the Royal Society of London, Series B: Biological Sciences* 254 (1339):7–13.

Gunn, A., and R.J. Irvine. 2003. Subclinical parasitism and ruminant foraging strategies – A review. *Wildl. Soc.* 31 (1):117–26.

Gunn, A., T. Leighton, and G. Wobeser. 1991. Wildlife diseases and parasites in the Kitkmeot region, 1984–90. Department of Renewable Resources, Government of the Northwest Territories.

Gunn, A., K.G. Poole, and J.S. Nishi. 2012. A conceptual model for migratory tundra caribou to explain and predict why shifts in spatial fidelity of breeding cows to their calving grounds are infrequent. *Rangifer* 32 (2):259–67.

Gunn, A., D. Russell, and J. Eamer. 2011. Northern caribou population trends in Canada. In *Canadian Biodiversity: Ecosystem Status and Trends 2010, Technical Thematic Report No. 10*, iv + 71 p. Ottawa, ON: Canadian Councils of Resource Ministers.

Hagemoen, R.I.M., and E. Reimers. 2002. Reindeer summer activity pattern in relation to weather and insect harassment. *J. Anim. Ecol.* 71 (5):883–92.

Halvorsen, O. 2012. Reindeer parasites, weather and warming of the Arctic. *Polar Biol.* 35 (11):1749–52.

Halvorsen, O., A. Stien, J. Irvine, R. Langvatn, and S. Albon. 1999. Evidence for continued transmission of parasitic nematodes in reindeer during the Arctic winter. *Int. J. Parasitol.* 29 (4):567–79.

Heil, M. 2016. Host manipulation by parasites: Cases, patterns, and remaining doubts. *Front. Ecol. Evol.* 4:80.

Hellard, E., D. Fouchet, F. Vavre, and D. Pontier. 2015. Parasite-parasite interactions in the wild: How to detect them? *Trends Parasitol.* 31 (12):640–52.

Hoar, B., M. Oakley, R. Farnell, and S. Kutz. 2009. Biodiversity and springtime patterns of egg production and development for parasites of the Chisana Caribou herd, Yukon Territory, Canada. *Rangifer* 29 (1):25–37.

Hoar, B.M., A.G. Eberhardt, and S.J. Kutz. 2012. Obligate larval inhibition of *Ostertagia gruehneri* in *Rangifer tarandus*? Causes and consequences in an Arctic system. *Parasitology* 139 (10):1339–45.

Hollings, T., M. Jones, N. Mooney, and H. McCallum. 2013. Wildlife disease ecology in changing landscapes: Mesopredator release and toxoplasmosis. *Int. J. Parasitol. Parasites Wildl.* 2:110–18.

Hollings, T., M. Jones, N. Mooney, and H. McCallum. 2014. Trophic cascades following the disease-induced decline of an apex predator, the Tasmanian Devil. *Conserv. Biol.* 28 (1):63–75.

Horn, H.S. 1968. Regulation of animal numbers: A model counter-example. *Ecology* 776–8.

Hudson, P.J., A.P. Dobson, and K.D. Lafferty. 2006. Is a healthy ecosystem one that is rich in parasites? *Trends Ecol. Evol.* 21 (7):381–5.

Hudson, P.J., A.P. Dobson, and D. Newborn. 1992. Do parasites make prey vulnerable to predation? Red grouse and parasites. *J. Anim. Ecol.* 61:681–92.

Hudson, P.J., A.P. Dobson, and D. Newborn. 1998. Prevention of population cycles by parasite removal. *Science* 282 (5397):2256–8.

Hudson, P.J., A. Rizzoli, B.T. Grenfell, H. Heesterbeek, and A.P. Dobson. 2002. *The Ecology of Wildlife Diseases*. Oxford: Oxford University Press.

Hughes, J., S.D. Albon, R.J. Irvine, and S. Woodin. 2009. Is there a cost of parasites to caribou? *Parasitology* 136 (2):253–65.

Innes, E.A. 2010. A brief history and overview of *Toxoplasma gondii. Zoonoses Public Health* 57 (1):1–7.

Irvine, R., A. Stien, J. Dallas et al. 2001. Contrasting regulation of fecundity in two abomasal nematodes of Svalbard reindeer (*Rangifer tarandus platyrhynchus*). *Parasitology* 122 (06):673–81.

Irvine, R., A. Stien, O. Halvorsen, R. Langvatn, and S. Albon. 2000. Life-history strategies and population dynamics of abomasal nematodes in Svalbard reindeer (*Rangifer tarandus platyrhynchus*). *Parasitology* 120 (03):297–311.

Johnson, P.T., D.E. Stanton, E.R. Preu, K.J. Forshay, and S.R. Carpenter. 2006. Dining on disease: How interactions between infection and environment affect predation risk. *Ecology* 87 (8):1973–80.

Joly, D.O., and F. Messier. 2004a. The distribution of *Echinococcus granulosus* in moose: Evidence for parasite-induced vulnerability to predation by wolves? *Oecologia* 140 (4):586–90.

Joly, D.O., and F. Messier. 2004b. Testing hypotheses of bison population decline (1970–1999) in Wood Buffalo National Park: Synergism between exotic disease and predation. *Can. J. Zool.* 82 (7):1165–.

Joly, D.O., and F. Messier. 2005. The effect of bovine tuberculosis and brucellosis on reproduction and survival of wood bison in Wood Buffalo National Park. *J. Anim. Ecol.* 74 (3):543–51.

Joly, K., S.K. Wasser, and R. Booth. 2015. Non-invasive assessment of the interrelationships of diet, pregnancy rate, group composition, and physiological and nutritional stress of barren-ground caribou in late winter. *PLoS One* 10 (6).

Jordan, P.A., J.L. Nelson, and J. Pastor. 1998. Progress towards the experimental reintroduction of woodland caribou to Minnesota and adjacent Ontario. *Rangifer* 18 (5):169–81.

Josefsen, T.D., A. Oksanen, and B. Gjerde. 2014. Parasitter hos rein i Fennoskandia, en oversikt. *Norsk Veterinærtidskrift* 126:185–201.

Joseph, M.B., J.R. Mihaljevic, A.L. Arellano et al. 2013. Taming wildlife disease: Bridging the gap between science and management. *J. Appl. Ecol.* 50 (3):702–12.

Keeling, M.J., and P. Rohani. 2008. *Modeling Infectious Diseases in Humans and Animals*. Princeton, New Jersey: Princeton University Press.

Krebs, C.J., R. Boonstra, S. Boutin, and A.R. Sinclair. 2001. What drives the 10-year cycle of Snowshoe hares? *Bioscience* 51 (1):25–35.

Krumm, C.E., M.M. Conner, N.T. Hobbs, D.O. Hunter, and M.W. Miller. 2010. Mountain lions prey selectively on prion-infected mule deer. *Biol. Lett.* 6 (2):209–11.

Kutz, S., J. Ducrocq, C. Cuyler et al. 2013. Standardized monitoring of *Rangifer* health during International Polar Year. *Rangifer* 33 (Sp. Iss. 21):91–114.

Kutz, S.J., J. Ducrocq, G.G. Verocai et al. 2012. Parasites in ungulates of Arctic North America and Greenland: A view of contemporary diversity, ecology, and impact in a world under change. *Adv. Parasitol.* 79:99–252.

Laaksonen, S., J. Kuusela, S. Nikander, M. Nylund, and A. Oksanen. 2007. Outbreak of parasitic peritonitis in reindeer in Finland. *Vet. Rec.* 160 (24):835–41.

Laaksonen, S., J. Pusenius, J. Kumpula et al. 2010. Climate change promotes the emergence of serious disease outbreaks of filarioid nematodes. *Ecohealth* 7 (1):7–13.

Laaksonen, S., M. Solismaa, R. Kortet, J. Kuusela, and A. Oksanen. 2009. Vectors and transmission dynamics for *Setaria tundra* (Filarioidea; Onchocercidae), a parasite of reindeer in Finland. *Parasit. Vectors.* 2 (1):3.

Laaksonen, S., M. Solismaa, T. Orro et al. 2009. *Setaria tundra* microfilariae in reindeer and other cervids in Finland. *Parasitol. Res.* 104 (2):257–65.

Lafferty, K.D. 2004. Fishing for lobsters indirectly increases epidemics in sea urchins. *Ecol. Appl.* 14 (5):1566–73.

Lafferty, K.D. 2010. Interacting parasites. *Science* 330 (6001):187–8.

Lafferty, K.D., G. DeLeo, C.J. Briggs et al. 2015. A general consumer-resource population model. *Science* 349 (6250):854–7.

Lankester, M.W. 2001. Extrapulmonary lungworms of cervids. In *Parasitic Diseases of Wild Mammals*, 2nd Edition, edited by W. Samuel et al., 228–78. Ames, Iowa: Iowa State University Press.

Lawson, B., S.O. Petrovan, and A.A. Cunningham. 2015. Citizen science and wildlife disease surveillance. *EcoHealth* 12 (4):693–702.

Leader-Williams, N. 1988. *Reindeer on South Georgia,* 1st Edition, edited by N. Leader-Williams. Cambridge: Cambridge University Press.

Levi, T., A.M. Kilpatrick, M. Mangel, and C.C. Wilmers. 2012. Deer, predators, and the emergence of Lyme disease. *PNAS* 109 (27):10942–7.

Lopes, L., R. Nicolino, and J. Haddad. 2010. Brucellosis – Risk factors and prevalence: A review. *Open Vet. Sci. J.* 4 (1):72–84.

Lyver, P.O.B., and A. Gunn. 2004. Calibration of hunters' impressions with female caribou body condition indices to predict probability of pregnancy. *Arctic* 57 (3).

Macbeth, B.J. 2013. An evaluation of hair cortisol concentration as a potential biomarker of long-term stress in free-ranging grizzly bears (*Ursus arctos*), polar bears (*Ursus maritimus*) and caribou (*Rangifer tarandus* sp.). PhD, Department of Veterinary Biomedical Sciences, University of Saskatoon.

Madliger, C.L., S.J. Cooke, E.J. Crespi et al. 2016. Success stories and emerging themes in conservation physiology. *Conserv. Physiol.* 4 (1):cov057.

Mahoney, S., and J. Weir. 2009. Caribou data synthesis – progress report. Overview of the status of woodland caribou in insular Newfoundland: Research methodology, results, interpretations and future projections. Sustainable Development and Strategic Science, Government of Newfoundland and Labrador, St. John's, Newfoundland and Labrador, Canada.

May, R.M., and R.M. Anderson. 1978. Regulation and stability of host-parasite population interactions. II. Destabilizing processes. *J. Animal Ecol.* 47:249–67.

May, R.M., and R.M. Anderson. 1979. Population biology of infectious diseases: Part II. *Nature* 280:455–61.

Meyer, M.E. 1966. Identification and virulence studies of Brucella strains isolated from Eskimos and reindeer in Alaska, Canada, and Russia. *Am. J. Vet. Res.* 27 (116):353–8.

Michel, J.F. 1974. Arrested development of nematodes and some related phenomena. *Ad. Parasitol.* 12:279–343.

Molnár, P.K., A.P. Dobson, and S.J. Kutz. 2013. Gimme shelter – The relative sensitivity of parasitic nematodes with direct and indirect life cycles to climate change. *Glob. Chang. Biol.* 19 (11):3291–305.

Molnár, P.K., S.J. Kutz, B.M. Hoar, and A.P. Dobson. 2013. Metabolic approaches to understanding climate change impacts on seasonal host-macroparasite dynamics. *Ecol. Lett.* 16 (1):9–21.

Morschel, F.H., and D.R. Klein. 1997. Effects of weather and parasitic insects on behavior and group dynamics of caribou of the Delta herd, Alaska. *Can. J. Zool.* 75 (10):1659–70.

Murray, D.L., J.R. Cary, and L.B. Keith. 1997. Interactive effects of sublethal nematodes and nutritional status on snowshoe hare vulnerability to predation. *J. Anim. Ecol.*:250–64.

Orlofske, S.A., R.C. Jadin, D.L. Preston, and P.T. Johnson. 2012. Parasite transmission in complex communities: Predators and alternative hosts alter pathogenic infections in amphibians. *Ecology* 93 (6):1247–53.

Ostfeld, R.S., and R.D. Holt. 2004. Are predators good for your health? Evaluating evidence for top-down regulation of zoonotic disease reservoirs. *Front. Ecol. Environ.* 2 (1):13–20.

Ouellet, J.-P., D.C. Heard, S. Boutin, and R. Mulders. 1997. A comparison of body condition and reproduction of caribou on two predator-free arctic islands. *Can. J. Zool.* 75 (1):11–17.

Packer, C., R.D. Holt, P.J. Hudson, K.D. Lafferty, and A.P. Dobson. 2003. Keeping the herds healthy and alert: Implications of predator control for infectious disease. *Ecol. Lett.* 6 (9):797–802.

Park, C., and M. Allaby. 2017. *Dictionary of Environment & Conservation.* Oxford University Press.

Parker, K.L., P.S. Barboza, and M.P. Gillingham. 2009. Nutrition integrates environmental responses of ungulates. *Funct. Ecol.* 23 (1):57–69.

Pedersen, A.B., and J. Antonovics. 2013. Anthelmintic treatment alters the parasite community in a wild mouse host. *Biol. Lett.* 9 (4).

Pedersen, A.B., and A. Fenton. 2007. Emphasizing the ecology in parasite community ecology. *Trends Ecol. Evol.* 22 (3):133–9.

Pedersen, A.B., and A. Fenton. 2015. The role of antiparasite treatment experiments in assessing the impact of parasites on wildlife. *Trends Parasitol.* 31 (5):200–11.

Pemberton, J. 2004. *Soay Sheep: Dynamics and Selection in an Island Population.* Cambridge University Press.

Pioz, M., A. Loison, D. Gauthier et al. 2008. Diseases and reproductive success in a wild mammal: Example in the alpine chamois. *Oecologia* 155 (4):691–704.

Reimers, E. 1997. *Rangifer* population ecology: A Scandinavian perspective. *Rangifer* 17 (3):105–18.

Rhyan, J.C. 2013. Pathogenesis and pathobiology of brucellosis in wildlife. *Rev. Sci. Tech.* 32 (1):127–36.

Rigaud, T., M.-J. Perrot-Minnot, and M.J. Brown. 2010. Parasite and host assemblages: Embracing the reality will improve our knowledge of parasite transmission and virulence. *Proc. Biol. Sci.* 277 (1701):3693–702.

Ripple, W.J., J.A. Estes, R.L. Beschta et al. 2014. Status and ecological effects of the world's largest carnivores. *Science* 343 (6167):1241484.

Roberts, M.G., and J.A.P. Heesterbeek. 2013. Characterizing the next-generation matrix and basic reproduction number in ecological epidemiology. *J. Math. Biol.* 66 (4–5):1045–64.

Rohr, J.R., D.J. Civitello, P.W. Crumrine et al. 2015. Predator diversity, intraguild predation, and indirect effects drive parasite transmission. *PNAS* 112 (10):3008–13.

Rosenzweig, M.L. 1971. Paradox of enrichment: Destabilization of exploitation ecosystems in ecological time. *Science* 171 (3969):385–7.

Russell, D., M.Y. Svoboda, J. Arokium, and D. Cooley. 2013. Arctic Borderlands Ecological Knowledge Cooperative: Can local knowledge inform caribou managment? *Rangifer* 33 (21):71–8.

Sanchez, M.I., F. Ponton, A. Schmidt-Rhaesa et al. 2008. Two steps to suicide in crickets harbouring hairworms. *Anim. Behav.* 76 (5):1621–4.

Sharp, A., and J. Pastor. 2011. Stable limit cycles and the paradox of enrichment in a model of chronic wasting disease. *Ecol. Appl.* 21 (4):1024–30.

Shury, T.K., J.S. Nishi, B.T. Elkin, and G.A. Wobeser. 2015. Tuberculosis and brucellosis in wood bison (*Bison bison athabascae*) in Northern Canada: A renewed need to develop options for future management. *J. Wildl. Dis.* 51 (3):543–54.

Steele, J.F. 2013. "The Devil's in the Diversity: Divergent Parasite Faunas and Their Impacts on Body Condition in Two Greenland Caribou Populations," MSc thesis, Department of Veterinary Medical Sciences, University of Calgary.

Steiner, I., P.G. Kennedy, and A.R. Pachner. 2007. The neurotropic herpes viruses: Herpes simplex and varicella-zoster. *Lancet Neurol.* 6 (11):1015–28.

Stien, A., R. Irvine, E. Ropstad et al. 2002. The impact of gastrointestinal nematodes on wild reindeer: Experimental and cross-sectional studies. *J. Anim. Ecol.* 71 (6):937–45.

Stien, A., R.J. Irvine, R. Langvatn, S.D. Albon, and O. Halvorsen. 2002. The population dynamics of *Ostertagia gruehneri* in reindeer: A model for the seasonal and intensity dependent variation in nematode fecundity. *Int. J. Parasitol.* 32 (8):991–6.

Telfer, S., X. Lambin, R. Birtles et al. 2010. Species interactions in a parasite community drive infection risk in a wildlife population. *Science* 330 (6001):243–46.

Tompkins, D.M., A.M. Dunn, M.J. Smith, and S. Telfer. 2011. Wildlife diseases: From individuals to ecosystems. *J. Anim. Ecol.* 80 (1):19–38.

Toyokawa, W. 2017. Scrounging by foragers can resolve the paradox of enrichment. *Open Science* 4 (3):160830.

Trainer, D.O. 1973. Caribou mortality due to the meningeal worm (*Parelapostrongylus tenuis*). *J. Wildl. Dis.* 9 (4):376–8.

Tryland, M., C.G.D. Neves, M. Sunde, and T. Mørk. 2009. Cervid herpesvirus 2, the primary agent in an outbreak of infectious keratoconjunctivitis in semidomesticated reindeer. *J. Clin. Microbiol.* 47 (11):3707–13.

Tryland, M., J.S. Romano, N. Marcin et al. 2017. Cervid herpesvirus 2 and not *Moraxella bovoculi* caused keratoconjunctivitis in experimentally inoculated semi-domesticated Eurasian tundra reindeer. *Acta Vet. Scand.* 59 (1):23.

van der Wal, R., J. Irvine, A. Stien, N. Shepherd, and S.D. Albon. 2000. Faecal avoidance and the risk of infection by nematodes in a natural population of reindeer. *Oecologia* 124 (1):19–25.

Walsh, J., K. Wilson, J. Benshemesh, and H. Possingham. 2012. Unexpected outcomes of invasive predator control: The importance of evaluating conservation management actions. *Anim. Conserv.* 15 (4):319–28.

Wasser, S.K., J.L. Keim, M.L. Taper, and S.R. Lele. 2011. The influences of wolf predation, habitat loss, and human activity on caribou and moose in the Alberta oil sands. *Front. Ecol. Environ.* 9 (10):546–51.

Webster, J.P. 2001. Rats, cats, people and parasites: The impact of latent toxoplasmosis on behaviour. *Microbes Infect.* 3 (12):1037–45.

White, R.G., C.J. Daniel, and D.E. Russell. 2013. CARMA's integrative modeling: Historical background of modeling caribou and reindeer biology relevant to development of an energy/protein model. *Rangifer* 33 (2):153–60.

Wild, M.A., N.T. Hobbs, M.S. Graham, and M.W. Miller. 2011. The role of predation in disease control: A comparison of selective and nonselective removal on prion disease dynamics in deer. *J. Wildl. Dis.* 47 (1):78–93.

Wilson, K., B. Grenfell, J. Pilkington et al. 2004. Parasites and their impact. In *Soay Sheep: Dynamics and Selection in an Island Population*, edited by T. Clutton-Brock and J. Pemberton, 113–65. Cambridge, United Kingdom: Cambridge University Press.

Wilson, K., B. Grenfell, and D. Shaw. 1996. Analysis of aggregated parasite distributions: A comparison of methods. *Funct. Ecol.*:592–601.

Witter, L.A., C.J. Johnson, B. Croft, A. Gunn, and M.P. Gillingham. 2012. Behavioural trade-offs in response to external stimuli: Time allocation of an Arctic ungulate during varying intensities of harassment by parasitic flies. *J. Anim. Ecol.* 81 (1):284–95.

Zelmer, D.A. 1998. An evolutionary definition of parasitism. *Int. J. Parasitol.* 28 (3):531–3.

11 Meat Quality and Meat Hygiene

Eva Wiklund, Gunnar Malmfors, Gregory L. Finstad,
Birgitta Åhman, Lavrans Skuterud, Jan Adamczewski
and Kerri Garner

CONTENTS

11.1 MEAT QUALITY

Eva Wiklund, Gunnar Malmfors and Gregory L. Finstad

11.1.1 INTRODUCTION

Consumer opinion is increasingly important to meat industries worldwide. Although the quality preferences (e.g. flavour, tenderness, nutrient content) of different consumer groups may vary, consistency is crucial: the quality of every purchase should be the same.

Production systems like reindeer husbandry, when the animals graze during most of the year, are usually considered more animal-friendly and ethical compared with the standard commercial production methods for beef, pork, or poultry. Reindeer meat is a high-quality product that is also attractive to the health-conscious consumer for its low fat content, favourable fat composition, and high mineral content.

11.1.2 REINDEER MEAT PRODUCTION AND STATISTICS

The examples in this chapter are taken from the traditional reindeer husbandry cultures in Fennoscandia (Sweden, Norway and Finland) and Alaska (with roots in Russia and Fennoscandia). These industries are mainly focused around pasture-based meat production systems where the reindeer are free ranging – not enclosed in fenced areas – in forests and on the mountain tundra. However, reindeer are occasionally fed supplements or replacements, particularly during winter when pastures cannot provide enough nutrition for maintenance and growth (Staaland and Sletten 1991; Åhman and Danell 2001). The reindeer (*Rangifer tarandus tarandus*) represents a meat producing species that exists both as wild and domestic (farmed) animals. Fennoscandia leads the world in commercial reindeer meat production from semi-domestic or farmed animals. The slaughter, boning, and processing of carcasses from these animals are done at facilities specifically designed for reindeer but include the technology developed for the commercial processing of other species of meat animals (Wiklund and Smulders 2011).

11.1.2.1 Fennoscandia

The Sami people inhabiting the region between Tromsø in northern Norway and the Kola Peninsula in northwest Russia have been reindeer herders for thousands of years. Traditionally, the reindeer were managed in an intensive manner where the herders tended their herds all year round. During the 20th century, reindeer herding gradually became more extensive. Sami families are not nomadic but live in communities and use all kinds of modern technology for reindeer herding, such as helicopters, airplanes, motorbikes, and snowmobiles (Bäck 1993). However, the traditional knowledge around reindeer is still very much alive in the Sami reindeer herders' everyday life.

The number of reindeer (winter stock) in Fennoscandia is approximately 700.000 [Sweden: 248.655 (Swedish Sami Parliament 2015); Finland: 203.700 (Muuttoranta 2014); Norway: 246.800 (Reindriftsforvaltningen 2013)] and varies according to changes in weather conditions, availability of pastures, and predation. Meat is the main product for the reindeer industry, and to maximize meat production it is crucial to align the numbers of reindeer to the available grazing resources. Herd composition and slaughter strategies are also important tools to influence meat production (Holand 2007). The highest possible proportion of reproductive females combined with a slaughter scheme based on calves is a common approach used in the Fennoscandian countries. The proportion of slaughtered calves per total number of slaughtered reindeer today exceeds 70% (Muuttoranta 2014; Swedish Sami Parliament 2015). About 2.000–3.000 tons of reindeer meat is produced annually in Finland (Muuttoranta 2014), 1.900 tons in Norway (Reindriftsforvaltningen 2013) and 1.200 tons in Sweden (Swedish Sami Parliament 2015). Due to the small production volumes, reindeer meat is a very exclusive gourmet product which is in high demand and often on the menu of high-end restaurants. Almost no reindeer meat is exported from Fennoscandia. The meat is consumed fresh but is also marketed as cold- or hot-smoked and dried meat products.

11.1.2.2 Alaska

Domestic reindeer were introduced to Alaska from Siberia, Russia in 1892 to establish a predict-able meat source and economic development for Alaskan Natives (Stern et al. 1980). Most of the imported animals were Chukchi stock, but reindeer from the Tungus of eastern Siberia were also purchased. At first, Chukchi herders were employed to train the Inupiat in handling and herding, but cultural differences created friction between the two groups. The Chukchi herders returned to Siberia and Sami herders from Norway and Finland were brought to the Seward Peninsula to teach reindeer husbandry to the Inupiat.

The introduced reindeer did well and over 500.000 reindeer could be found in Alaska, spread from Barrow to the Aleutian Islands, during the 1920s (Stern et al. 1980). The numbers of reindeer on the Seward Peninsula decreased drastically from 127.000 in 1927 to 25.000 in 1950, largely because of the influence of World War II, when the cessation of close herding, overgrazing brought about by the open grazing method, predation, and the presence of wild caribou led to large reindeer losses (Stern et al. 1980). After a revitalization program initiated by the Bureau of Indian Affairs (BIA) in 1941 to improve reindeer management, 17 new herds under private management were started on the Seward Peninsula. To be granted a grazing permit on government lands, the herder must develop a grazing management plan in cooperation with the Natural Resources Conservation Service (NRCS 1953). In 1971 the Reindeer Herder's Association was formed, which united the herders into a political organization to work with government agencies and to advocate efforts to further develop the reindeer industry (Bader and Finstad 2001).

Since the 1970s, reindeer herding has been a significant economic factor in villages on the Seward Peninsula (Schneider et al. 2005). Sales of velvet antler and meat generated over one million dollars in annual revenue for the rural communities of the Seward Peninsula during the early to mid 1990s (Carlson 2005). Most herders believe that meat sales provide the economic backbone for the industry and manage their herds accordingly. All present-day herders castrate excess males to reach a ratio of one intact male for every 15 to 20 females in their herds. Although velvet antler sales generated US\$10.3 million while reindeer meat sales generated US\$9.6 million from 1987 to 2003 (Alaska Agricultura Statistics 2006), herders believe development of the meat industry is key to their long-term economic success (Finstad et al. 2006).

Ground meat and steaks are preferred by US consumers. Thus, a slaughtered animal must be of sufficient size to produce a cut of meat that is large enough to satisfy typical American expectations of a pan-fried or grilled steak. Hence, the Alaska reindeer industry must rely on a larger framed animal with fully developed muscle groups rather than calves as in the Fennoscandian reindeer industry.

11.1.3 Pre-Slaughter Handling, Stress and Meat Quality

Reindeer handling before slaughter includes various methods: gathering and herding using such aids as snow machines, helicopters, and motor bikes; selecting animals for slaughter using a lasso or by hand in a system of corrals; road transport of animals on trucks; and holding animals in corrals outside the slaughter plant. Some of these handling routines have been discussed by veterinary authorities from the perspective of stress and animal welfare. In particular, long road transports and the use of helicopters when gathering and herding reindeer have been pointed out as factors that could cause severe stress to the animals (Rehbinder 1990).

The first published articles dealing with the effects of pre-slaughter handling on reindeer meat quality were two Norwegian studies (Skjenneberg et al. 1974; Hanssen et al. 1984). Because of new directives for reindeer slaughter in Sweden (Swedish National Food Agency 1993), an extensive project was started in 1991 to evaluate several handling routines for their effects on meat quality. This project was summarized in a PhD thesis (Wiklund 1996).

Because the pH value of meat gives useful information about shelf life, tenderness, colour, and water-holding properties, it is a good indicator of meat quality. All the mentioned quality attributes are important for both fresh meat and for meat used as raw material for further processing. Meat pH values of 5.5–5.7 measured approximately 24 hours post slaughter – so-called ultimate pH – is within the normal range, while values over 5.8 result in reduced shelf life, especially for vacuum packaged meat (Gill 2004). Meat with very high pH (> 6.2), so-called DFD (Dark, Firm, Dry) meat, is a quality defect found in all meat species.

As a part of the Swedish project, a comprehensive survey of meat pH for reindeer (n = 3.400) demonstrated pH values of 5.8 and higher for 29% of the measured carcasses, i.e. meat with an obvious risk of reduced shelf life (Wiklund et al. 1995). It was also concluded that reindeer shoulder meat often had very high pH values (Wiklund et al. 1995), something already known to reduce the quality of processed reindeer meat products (Niinivaara and Petäjä 1985). Road transport of reindeer by truck and herding animals by helicopter did not elevate the meat pH values; however, the use of a lasso when selecting slaughter animals seemed to be very stressful for the reindeer and resulted in significantly higher pH values (Wiklund et al. 1996a, 1996b). Increased meat pH values were recorded in reindeer that were slaughtered as early as 30 minutes post lasso selection/handling, and when the selection in the herd had continued for 6 hours, even more muscle energy had been consumed and higher pH values were measured (Wiklund et al. 1997a). One of the major conclusions from the Swedish project was that, of the handling routines investigated, lasso selection clearly had the most negative effect on reindeer meat quality (Figure 11.1).

In the early 1980s, some published studies focused on pre-slaughter handling and stress (Rehbinder et al. 1982; Essén-Gustavsson and Rehbinder 1984). This research showed that all types of handling cause elevated levels of stress metabolites (cortisol, ASAT, urea) in reindeer blood. The same metabolites were, therefore, measured in the Swedish project in the 1990s, where it was concluded that the lasso selection procedure resulted in the highest levels of all these stress metabolites (Wiklund et al. 1996b). The studies from the 1980s tried to correlate the increased values of the stress metabolites in the blood to chemical changes in the muscles that could explain a phenomenon called "stress-flavour" in the meat. No such correlation was found (Rogstadkjærnet and Hanssen 1985; Hanssen and Skei 1990).

Today, modern handling methods (truck transport and herding with helicopter) are used as frequently as the more traditional lasso selection. The reindeer have become more familiar with these

FIGURE 11.1   Pre-slaughter handling stress can result in high meat pH values and reduced shelf life of the meat. The use of a lasso when selecting reindeer for slaughter has been proven to have a significant negative effect on reindeer meat quality (Photo: E. Wiklund).

new techniques and today are not at all affected in the same way as they were when these handling methods were introduced about 35 years ago. In contrast, the use of a lasso seems to be a routine that is so unpleasant for the animals they never get used to this technique.

11.1.4 FEEDING, GRAZING AND BODY CONDITION

Meat pH values are directly correlated to the levels of muscle energy (glycogen) at slaughter. If the glycogen stores in the muscles are low, meat pH values will be elevated, which results in the meat quality problems previously described. Low muscle glycogen might be a consequence of poor physical condition, stress or intense physical activity. Therefore, meat ultimate pH value can also be regarded as an indirect measure of pre-slaughter handling stress. It has been demonstrated that reindeer in good physical condition produce meat with optimal pH values, whether they were fed a commercial feed mixture or grazed (Wiklund et al. 1996a).

Finnish results (Petäjä 1983) showed that almost all reindeer carcasses from animals slaughtered late in the season (midwinter or early spring) had extremely high pH values (pH > 6.2). The author suggested that this was related to poor grazing conditions, which significantly reduced animal body condition.

Body condition can be determined on either living animals or on the carcasses following slaughter. For living animals, weight, body measurements (length, circumference) and various body condition scoring systems that estimate body fat and muscle content are often used. At slaughter, carcass weight, grading scores and measurements of fat deposits in the carcass can provide body condition information.

11.1.5 CARCASS TREATMENTS FOR IMPROVED MEAT QUALITY

It is well known that the carcass handling conditions during development of rigor mortis are very important in controlling meat tenderization. Carcass suspension techniques have been studied for beef and shown to affect the tenderness of different muscles (Lundesjö Ahnström 2008). The most common method to hang a carcass is by the Achilles tendon, but the pelvic suspension technique is also used (where the carcass is hung by a hook through the pelvic bone). Pelvic suspension stretches different muscles on the carcass compared to Achilles tendon suspension. Generally, the more stretched the muscles are during rigor development the more tender the meat will be. Most of the valuable cuts from the carcass (from the hind-quarter region) are more stretched by pelvic suspension than if the carcass is hung by the Achilles tendon, although this is not true for the tenderloin (*M. psoas major*). The two suspension methods have been compared for reindeer and it was demonstrated that pelvic suspension improved tenderness in the striploin (*M. longissimus*) and the inside (*M. semimembranosus*) (Wiklund et al. 2012).

When carcasses are cooled too quickly, the muscles tend to contract substantially, which is called "cold-shortening" (Savell et al. 2005). The resulting meat is very tough and will not age or tenderize with time. To prevent cold-shortening, carcasses from domestic animals are frequently electrically stimulated. During electrical stimulation, a current is run through the carcass for a short time (0.5–1 minute), causing the muscles to contract rapidly and exhausting their energy (glycogen) stores, thereby accelerating rigor attainment, and consequently the natural enzymatic tenderizing process in the resulting meat is enhanced. Electrical stimulation is used at commercial slaughter abattoirs in many countries for species like cattle, sheep and goats. In New Zealand, electrical stimulation of deer carcasses is part of the normal slaughter routine at abattoirs where deer are slaughtered; however, the reindeer slaughter plants in Fennoscandia do not use electrical stimulation for reindeer carcasses.

Most reindeer in Alaska are slaughtered through a state-regulated field slaughter system. Field slaughtered reindeer meat can be marketed locally provided animals are slaughtered on snow when ambient temperature is below 0°C; carcasses are allowed to freeze in the field and the meat is not

FIGURE 11.2   The majority of reindeer in Alaska are slaughtered through a state-regulated field slaughter system. The meat can be marketed locally, provided animals are slaughtered on snow when ambient temperature is below 0°C (Photo: E. Wiklund).

thawed until in the hands of the consumer (Alaska Department of Environmental Conservation 2003). As ambient air temperatures during field slaughtering are usually below −10°C, instant chilling and freezing of the carcasses inevitably occurs and the risk of cold-shortening is obvious in field slaughtered reindeer (Figure 11.2). This risk could be minimized with electrical stimulation immediately post slaughter. Portable electrical stimulation equipment that can be connected to a generator or a battery is available and has been proven to function well in the harsh environment during the winter field slaughter of reindeer in the Seward Peninsula of Alaska (Wiklund et al. 2008).

11.1.6   MEAT CHEMICAL COMPOSITION

As already mentioned, feeding reindeer commercial feed mixtures or letting them graze on good pastures will positively affect the animal's body condition and increase muscle glycogen stores, and thereby give meat pH values in the optimal range for good meat quality. What the reindeer eat will also affect the chemical composition of the meat. Grazing reindeer produce meat higher in "good fats" (i.e. polyunsaturated fatty acids, or PUFA) while meat from animals fed grain-based feed mixtures have been found to have higher levels of saturated fatty acids (Wiklund et al. 2001a; Table 11.1). Reindeer meat has a low fat content, but the fat composition is still important for meat shelf life and for the quality of processed meat products. As a continuation of the Swedish reindeer study in the 1990s, a project started in 2001 focused on the chemical composition of reindeer meat. This project was concluded in a PhD thesis in 2005 (Sampels 2005) and is discussed below.

During storage, meat fats will get rancid (oxidation) and be broken down (lipolysis). Although some of the components formed during oxidation and lipolysis are important for the typical character of different meat products, too much of these components will deteriorate the quality. Two common methods of processing reindeer meat are smoking and drying. It has been demonstrated that the drying process speeds up oxidation and lipolysis in the meat, while smoking seemed to be a much more gentle process (Sampels et al. 2004). Compared with saturated fats, PUFA will oxidize more easily, which increases the importance of the difference in fat

TABLE 11.1

Mean Values for Fatty Acid Composition (% of Total Fatty Acids) in *M. longissimus dorsi* from Pasture and Pellet-Fed Reindeer (*Rangifer tarandus tarandus*)

Fatty Acid	Pasture (n = 6)	Pellets (n = 9)	Degree of sign.[1]
	Polar Lipids		
SFA[2]	25.4	26.3	n.s
MUFA	17.3	16.0	*
PUFA (n-6)	31.9	39.4	***
PUFA (n-3)	14.2	7.5	***
(n-6)/(n-3)	2.2	5.3	***
	Neutral Lipids		
SFA	53.0	54.6	n.s.
MUFA	37.6	39.2	*
PUFA (n-6)	2.6	2.3	n.s
PUFA (n-3)	1.4	0.3	***
(n-6)/(n-3)	1.9	8.1	***

Source: Data from Wiklund et al. (2001a).

[1] n.s. = $P > 0.05$; * = $P \le 0.05$; ** = $P \le 0.01$; *** = $P \le 0.001$.

[2] SFA: Saturated fatty acids; MUFA: Monounsaturated fatty acids; PUFA: Polyunsaturated fatty acids.

composition between grazing reindeer and animals fed grain-based feeds. This information can be used by slaughter and meat processing companies to sort their raw material in the best way for optimal processing.

An experimental reindeer feed mixture was manufactured in Sweden with the same nutrient content as the standard feed made by the same company. In the experimental feed, linseed cake replaced the fat source in the standard feed (Renfor Bas, Lantmännen, Sweden). Linseed contains high levels of the fatty acid 18:3 omega-3. Meat from reindeer fed the linseed feed mixture had a fat composition much more similar to that of grazing reindeer compared to the meat from the control group of animals eating the standard feed (Sampels et al. 2006). In Alaska, a reindeer feed mixture made from locally produced ingredients has been developed by the Reindeer Research Program at the University of Alaska Fairbanks (Finstad et al. 2007). In this feed mixture, fishmeal replaces soybean meal as a protein source. In a feeding experiment, the quality attributes of meat from three groups of reindeer were compared: animals grazed on the Seward Peninsula, animals fed the mixture with fishmeal and animals fed a mixture with soybean meal. When meat from the fishmeal and soybean meal groups was compared, only a slight difference in fat composition was found. Meat from grazing reindeer had significantly higher levels of PUFA than the other two groups (Finstad et al. 2007).

11.1.7 Sensory Quality of Reindeer Meat

Sensory evaluation of food products has become an increasingly important research field over the last 35 years. In early meat taste tests, randomly selected people (usually all members in a research group, for example) answered questions about meat tenderness or colour. Formally trained panel members can judge different quality attributes (tenderness, flavour, etc.) on a scale from, for example, zero to ten, while disregarding personal preferences for the products (Figure 11.3). However, in the 1990s in Sweden, no professionally trained taste panel had experience in evaluating reindeer or other game meats. Therefore, collaboration of Swedish scientists with the Norwegian Food Research

FIGURE 11.3   Sensory evaluation is an important complement to technological measurements of meat quality. The combined information from a trained expert panel and consumer preference tests will provide the best description of product quality. (Photo: courtesy of Reindeer Research Program, University of Alaska Fairbanks, USA.)

Institute was initiated. With the Institute's help, the Department of Domestic Sciences at Uppsala University in Sweden learned to select and train a taste panel for reindeer meat sensory work. This collaboration further resulted in experienced staff from Uppsala University assisting in selection and training of a taste panel to evaluate reindeer meat at the University of Alaska Fairbanks (UAF) Cooperative Extension Service Kitchen.

As a complement to the results of a trained panel, it is common to use consumer preference tests where untrained people are asked for their personal preference for a product. In most of the published reindeer studies a combination of these two sensory tests have been used. In the Swedish studies, using meat samples for which pH values and stress metabolites in the blood had already been analyzed, a trained panel evaluated samples from reindeer exposed to all the various pre-slaughter handling routines. The only meat samples that stood out as different to the taste panel were samples from reindeer fed commercial feed. In a comparison of grain-based diets versus grazing diets, meat from reindeer fed grain-based feed mixtures had the most "untypical", mild taste. Meat from grazing reindeer had a stronger and more typical reindeer flavour (Wiklund et al. 2003). In a consumer preference test of the same meat, participants identified the same differences between the two types of meat, and 50% preferred meat from the grazed animals, while the other half preferred meat from grain-fed reindeer (Wiklund et al. 2003).

In Alaska, the trained panel found no differences in sensory attributes between meat from grazed reindeer and animals fed the mixture based on fishmeal or the one based on soybean meal (Finstad et al. 2007). Comments from the consumer test regarding gamey and reindeer flavour were most common on the samples from grazing reindeer. For the fishmeal, no negative effect on any sensory attribute was reported (Finstad et al. 2007). With respect to the sensory evaluation of meat handled through different suspension techniques (Achilles tendon vs. pelvic suspension), the trained panel found that pelvic suspension improved tenderness in the striploin

(*M. longissimus*) and the inside (*M. semimembranosus*) compared with samples from Achilles tendon hung carcasses (Wiklund et al. 2012). The positive effect of electrical stimulation on meat tenderness was clearly demonstrated by consumers judging meat from field slaughtered and electrically stimulated reindeer carcasses more tender than meat from unstimulated animals (Wiklund et al. 2008).

Consumers often mention tenderness as the most important sensory attribute for meat. In the mid 1990s, research collaboration between the Swedish reindeer project and Utrecht University in the Netherlands was initiated to study tenderness in reindeer meat. Biochemical measurements of tenderizing enzymes and microscopical studies of the meat tenderizing process were carried out. Reindeer meat was demonstrated to be much more tender than beef, and post-slaughter ageing of the meat was not necessary to optimize tenderness; when carcasses were boned 1 day after slaughter, the meat was already tender (Barnier et al. 1999). This phenomenon was explained by high activity of the tenderizing enzymes (Wiklund et al. 1997b) and small muscle fibre size in reindeer meat (Taylor et al. 2002).

11.1.8   MEAT QUALITY: CONCLUSIONS

Although research on reindeer meat production and quality is limited, it is of great value to reindeer herders and their enterprises worldwide, as they are primarily engaged in meat production. The reindeer industries in Fennoscandia and Alaska are producing one of the most prestigious sources of animal proteins, which attract premiums in high value markets. Reindeer meat has earned this prestige due to its naturalness, provenance, and inherent leanness and tenderness, all qualities that endear it to some of the most discerning consumers of meat. For these consumers, it is of great importance that reindeer meat is different from beef, pork and chicken.

The image of reindeer meat has been frequently discussed, especially in the Fennoscandian countries. In some of these countries, new brand names and labelling have been of particular interest, and various criteria to measure and guarantee the desired quality of reindeer meat have been debated. In Sweden, a quality label for reindeer meat – Renlycka – was launched in 2011 (Wiklund 2014). The most exceptional feature of this label is that it acknowledges both the unique origin (Sami produced) and three measures (EUROP carcass grading score, meat pH value, and natural grazing) to represent optimal meat quality (Wiklund 2014). For marketing, reindeer industries in Fennoscandia emphasize such attributes as "natural", "exotic", "exclusive" and "healthy". The introduction of new production methods, handling routines, feeding strategies and slaughter techniques should be well balanced to fit into the image of reindeer meat as a desirable product for modern consumers concerned about health and the environment.

11.2   MEAT CONTROL

Eva Wiklund

11.2.1   INTRODUCTION

In Fennoscandia, the rules applied for animal transport, veterinary inspection of living animals and carcasses, stunning methods, slaughter hygiene, carcass grading, and chilling conditions for reindeer are similar to those applied for other domestic species. Most of the specialized reindeer slaughter plants are approved to EU standards for other meat species.

Control programs for reindeer meat were introduced in both Sweden and Norway after the Chernobyl accident. Monitoring of radioactive caesium was initially made by taking meat samples from each reindeer carcass at the slaughterhouse, followed by analysis at a laboratory. Control of radioactive caesium at reindeer slaughter is still carried out in some areas of Sweden according to a control program applied (and updated annually) by the National Food Agency.

Historically, reindeer were slaughtered at the selection site, i.e. at various locations in the forests or mountain tundra surrounding the reindeer herding districts of the Sami people. At these selection sites, permanent corrals for handling and sorting of animals are available and also some types of basic slaughter facilities. New directives regarding reindeer meat inspection were introduced starting in the 1990s (e.g. by the Swedish National Food Agency 1993). Consequently, many of the former outdoor slaughter sites were closed, the numbers of reindeer transported to slaughter increased, and new mobile slaughter facilities were developed (Wiklund 1996).

Today, reindeer in Sweden are slaughtered according to the same regulations that apply for other domestic animals:

- Veterinary inspection of live animals before slaughter, and of the carcass and organs after slaughter
- Stunning of the animals using a captive bolt before bleeding
- Similar requirements regarding hygiene in the slaughter facilities, among staff and for the equipment used, as for other domestic animals
- Compulsory carcass grading according to the EUROP system, i.e. the meat (lean) and fat content of the carcasses is estimated
- Chilling of carcasses following similar protocols to those of other domestic species

The reindeer slaughter plants are EU approved facilities and are in most cases completely specialized for reindeer.

In Sweden, carcasses from cattle, pigs, sheep, goats, horses, and reindeer that are marketed as food for human consumption must be graded according to the EUROP-scale. These regulations are established in EU directives, in Swedish law, and in the code of statutes from the Swedish Board of Agriculture. The Board of Agriculture supervises the grading system so that carcasses are evaluated in a similar manner all over the country to ensure that all rules regarding carcass grading are applied properly. The purpose of the grading system is to describe – as accurately as possible – the usability of each carcass, along with its content of meat (lean), fat, and bones. The grading system also facilitates communication between producer and consumer, so that the producer gets increased opportunities for feedback on quality of production and better guidance to meet market demands.

Swedish reindeer carcasses have been graded according to the EUROP system since 1994: the fat and lean content of the carcass are estimated visually by specially trained staff at the slaughter plant (Swedish Board of Agriculture 2004). When judging carcass conformation, the following sub-classes are used; E, U, R+, R, R−, O+, O, O−, P+, P, and P−. On the EUROP-scale, E stands for the best body condition (extremely well developed/muscled carcass) and P for the poorest (poorly muscled/very lean carcass). A normal variation in grading scores for reindeer carcasses is typically between the groups R and P on the scale. According to statistics from the Swedish Sami Parliament (2015), 99% of the carcasses from reindeer slaughtered during 2013–2014 (n = 54 104) were graded in the EUROP groups R to P: 3% in sub-classes P−, P, and P+; 86% in sub-classes O−, O, and O+; and 10% in sub-classes R−, R, and R+.

In a Swedish study of carcass composition for different reindeer categories (bulls, cows, and calves) content (percentages) of meat, fat, bones, valuable meat cuts, and EUROP grading scores were reported (Table 11.2; Wiklund et al. 2000). In Table 11.2, the different EUROP sub-classes for body condition are translated to numbers and explained in the footnotes to the table.

TABLE 11.2

Carcass Characteristics, Yield and Composition (Least-Squares Means ± Standard Errors) in Reindeer Bulls, Cows and Calves

Trait	Bulls (n = 21)	Cows (n = 15)	Calves (n = 12)	Degree of sign.[1]
Carcass weight (Cw), kg	42.2[a] ± 1.8	27.6[b] ± 2.1	19.6[c] ± 2.2	***
EUROP conformation[2]	6.4[a] ± 0.4	3.4[b] ± 0.4	4.4[b] ± 0.5	***
Leg, kg	14.9[a] ± 0.6	9.4[b] ± 0.7	7.7[b] ± 0.8	***
Leg, % of Cw	32.1[a] ± 1.8	34.5[ab] ± 2.1	39.4[b] ± 2.2	*
Saddle, kg	8.2[a] ± 0.4	5.3[b] ± 0.4	4.0[c] ± 0.4	***
Saddle, % of Cw	16.8 ± 1.2	19.2 ± 1.3	20.3 ± 1.4	n.s
Striploin, kg	1.5 ± 0.2	1.4 ± 0.2	0.9 ± 0.2	n.s
Striploin, % of Cw	9.4 ± 2.2	5.0 ± 2.6	4.8 ± 2.8	n.s
Topside, kg	3.2[a] ± 0.4	1.5[b] ± 0.5	1.3[b] ± 0.5	**
Topside, % of Cw	5.8[ab] ± 0.2	5.4[a] ± 0.3	6.6[b] ± 0.3	*
Shoulder, kg	10.9[a] ± 1.3	5.5[b] ± 1.5	3.8[b] ± 1.6	**
Shoulder, % of Cw	17.3 ± 0.9	19.9 ± 1.0	19.5 ± 1.0	n.s.
Bone, kg	8.1[a] ± 0.7	7.8[a] ± 0.8	4.9[b] ± 0.9	*
Bone, % of Cw	18.7[a] ± 1.1	23.2[b] ± 1.3	25.1[b] ± 1.4	**
Fat, kg	4.7 ± 1.9	2.6 ± 2.1	0.01 ± 2.3	n.s
Fat, % of Cw	6.5 ± 2.0	1.9 ± 2.3	0.06 ± 2.5	n.s

Source: Data from Wiklund, E. et al., Composition and quality of reindeer (*Rangifer tarandus tarandus* L) carcasses, in *Proceedings: 46th International Congress of Meat Science and Technology*, Buenos Aires, Argentina, 2000.

[1] n.s. = p > 0.05; * = p ≤ 0.05; ** = p ≤ 0.01; *** = p ≤ 0.001. Means in the same row having the same superscript are not significantly different (p > 0.05).

[2] The EUROP system used in Sweden:

E	U	R+	R	R–	O+	O	O–	P+	P	P–
14	11	9	8	7	6	5	4	3	2	1

11.2.4 RADIOACTIVE CAESIUM

Birgitta Åhman and Lavrans Skuterud

11.2.4.1 Background

Early in the 1960s it was discovered that the food chain lichen–reindeer–man was particularly efficient in transferring radioactive fallout from nuclear weapons tests to humans (Lidén 1961; Lidén and Gustafsson 1967). Among the various radioactive elements, isotopes of the alkaline metal caesium (^{134}Cs and ^{137}Cs) pose a special problem because they behave similarly to potassium in biological systems and, therefore, are found in various edible tissues.

Reindeer are typically free ranging and eat naturally growing vegetation. Lichens are a main food source in winter, while grass, herbs and other vascular plants dominate the diet during the snow-free periods. Fungi are also eaten when available. Lichens are slow growing, have no roots and absorb nutrients from air and precipitation (Mattsson 1975). A dense carpet of ground lichens will absorb most of the radionuclides in a fallout situation (Papastefanou et al. 1989). Many fungi effectively absorb caesium from the soil (Barnett et al. 1999), while the transfer to vascular plants is generally considerably lower, but highly variable, depending on soil properties and plant species (Ehlken and Kirchner 2002).

Although the contamination levels in reindeer were relatively high during the nuclear weapons test era in the 1960s, those levels were exceeded by more than a factor of ten after the 1986 Chernobyl accident in Ukraine. That accident caused significant and inhomogeneous fallout in

Fennoscandia. Much of the areas in Sweden and Norway receiving fallout were ranges for reindeer, while reindeer ranges in Finland were less affected. Early monitoring of reindeer in Sweden, Norway, and Finland confirmed that reindeer would be considerably more contaminated than most other livestock (Skogland 1986; Åhman 1986; Rissanen et al. 1987). Based on the experience from the nuclear weapons tests fallout (e.g. Westerlund et al. 1987) it was already recognized in July 1986 that the Chernobyl fallout would probably affect reindeer herding for decades. The highest activity concentrations in reindeer meat were recorded during the winter after the accident. In Norway, maximum levels reached 150.000 Bq/kg fresh weight (Strand et al. 1992). In Sweden, the highest values were around 80.000 Bq/kg (Åhman and Åhman 1994). The variation in diet, together with seasonal migrations resulted in pronounced seasonal variations of ^{137}Cs in reindeer (Åhman and Åhman 1994), in some areas with differences up to 20 times between summer and winter.

In 1986, all reindeer meat from central and southern Norway was banned due to the contamination (Tveten et al. 1998). The situation in Sweden was similar, with 78% of the total reindeer slaughter being banned during autumn 1986 and the following winter (Åhman 1999). The high contamination levels and the expected long-term duration of the problem triggered an intensive search in Sweden and Norway for countermeasures that could reduce contamination levels and consequences for the herders. The most commonly implemented countermeasures were change of slaughter time (taking advantage of the seasonal variation described above), clean feeding, and use of caesium binders (e.g. Hove 1993; Åhman 1996, 1999; Tveten et al. 1998).

Fortunately, the initial yearly decline in ^{137}Cs was more rapid than what had been expected based on experiences from the nuclear weapon test fallout. Effective halftimes (T_{eff}) around 2.5–3 years were observed in both Sweden and Norway (Åhman and Åhman 1994; Skuterud et al. 2005), whereas the halftimes were 6–7 years from the 1960s onwards (Westerlund et al. 1987). This may be explained by a corresponding rapid decline in lichens (Rissanen et al. 2005). However, some years after the Chernobyl fallout, the decline became slower and was also shown to be associated with remaining contamination from the nuclear bomb tests (Åhman et al. 2001). In sum, the high Chernobyl fallout and the slow decline will for many years necessitate countermeasures, control, and elevated permissible levels for radioactive caesium in reindeer meat in Sweden and Norway, well beyond the 30th anniversary of the accident.

11.2.4.2   Permissible Levels of Radioactive Caesium in Reindeer

At the time of the Chernobyl accident, there were no internationally agreed recommendations on permissible levels for radioactive contamination in foods. Due to the consequences of the accident, various countries adopted different permissible levels reflecting the contamination problems in the respective countries. Some international bodies (e.g. the EC and the UN's Codex Alimentarius) later developed regulations and recommendations, but generally these are applicable only to relatively short periods following potential new contamination situations.

A common basis for derivation of permissible levels of radioactive contamination in foods is often the criteria that no member of the population shall receive an extra radiation dose from the contamination exceeding 1 mSv (millisievert) per year, although there are no definite dose limits in the case of accidents (ICRP 2007). Using the ingestion dose conversion factor for ^{137}Cs, 1 mSv/year corresponds to an intake of about 77.000 Bq/year (in adults). Permissible levels can then be derived from this "limit", depending on contamination levels and dietary habits, and may result in different permissible levels between countries.

Following the Chernobyl fallout, the authorities in Norway initially chose 300 Bq/kg as the intervention level for radioactive caesium in food. For foodstuffs other than milk and baby food, this was raised to 600 Bq/kg in June 1986. Following the finding of the high contamination levels in reindeer meat during summer and autumn 1986, the intervention level for trade of this product was increased further to 6.000 Bq/kg in November 1986 (and also made applicable to game, freshwater fish, and wild fungi in 1987). The intervention level was reduced from 6.000 to 3.000 Bq/kg in 1994. In Sweden, the initial intervention limit was set to 300 Bq ^{137}Cs per kilogram of fresh food for all foodstuffs except milk and baby food, the limit for which was lower. After a year, this limit was raised to 1.500 Bq/kg for reindeer meat, game, freshwater fish, wild berries and mushrooms.

Both in Norway and Sweden the justification for the elevated permissible levels for radioactive caesium in reindeer was the generally low intake of reindeer meat by the general consumer. Further, it is important to note that the permissible levels did not apply to the reindeer herders and others with high consumption: These consumers were recommended to adopt the general limit for basic foodstuffs (i.e. 600 Bq/kg in Norway and 300 Bq/kg in Sweden). Both in Norway and Sweden, health authorities also developed dietary advice (booklets) for consumers of reindeer and game. Surveys of the reindeer herders have shown that the advice, coupled with other measures taken, has significantly reduced their radiation doses (Strand et al. 1992; Skuterud and Thørring 2012).

11.2.4.3   Control Programs in Norway and Sweden

Control programs for reindeer meat were introduced in both Sweden and Norway after the Chernobyl accident. Because of low initial intervention limits in both countries, only a small part of the reindeer meat produced during the first year after the accident was approved for human consumption. Raising the intervention levels and introducing counter measures enabled continued reindeer meat production in these countries.

Monitoring of radioactive caesium was initially made by taking meat samples from each reindeer carcass at the slaughterhouse for analysis at a laboratory. Control of radioactive caesium at reindeer slaughter is still carried out in some areas of Sweden according to a control program applied (and updated annually) by the National Food Agency. Laboratory analyses have, however, gradually been replaced by direct measurement on the reindeer carcasses (Figure 11.4). This is possible since radioactive caesium emits gamma radiation that can be detected outside the organism (Brynildsen and Strand 1994).

Both Sweden and Norway developed methods and programs for external monitoring of animals with NaI detectors before slaughter – "live monitoring" – in the years after the Chernobyl accident (Figure 11.5). These methods have the advantage that slaughtering can be avoided for animals whose meat would be condemned because of too much radioactive caesium. Instead, animals with excess contamination levels can be subjected to countermeasures, or they can be replaced by less contaminated individuals. Although animal owners may receive compensation for their condemned animals, condemning their products is highly unsatisfactory for producers in the long term.

Live monitoring is generally considered a relative rough procedure. It is quick (the animals are monitored for 10 seconds), resulting in fewer counts and poorer analytical precision than can generally be achieved in laboratory conditions (due the short counting time, the variable sizes of animals, variability in background radiation, etc.). However, although caesium is relatively evenly distributed in muscle, differences between muscle groups exist (Rissanen et al. 1990; Åhman 1994; Skuterud et al. 2004), and these differences may vary if the animals are in periods of increasing or decreasing caesium intake. Thus, analysis of a sample of meat in the laboratory may have lower analytical error, but may still not give a more correct estimate of the concentrations of radioactive caesium in the body.

Monitoring of live reindeer is required in areas where animals may be over the limit. In Norway, after a herd has been gathered, some animals in the herd are monitored (the size of the sample is equal to the square root of the number of animals going to be slaughtered). If none of the animals in this sample are above the limit (3.000 Bq/kg), the herd is approved for slaughtering and trade. If one animal is above the limit, then all the animals (those planned to be slaughtered) must be monitored in order to decide whether the animals can be sent for slaughter and if clean feeding is required. In Sweden, a minimum of 30 reindeer from the herd are usually monitored. If the mean activity concentration is over 1.200 Bq/kg, usually all reindeer in the herd are fed for 45–80 days (depending on contamination level) prior to slaughter. Herders are then compensated for the costs of feeding. Clean feeding is also applied for reindeer for own consumption, to ensure that these are below the recommended limit 300 Bq/kg for those that consume large amounts of reindeer meat.

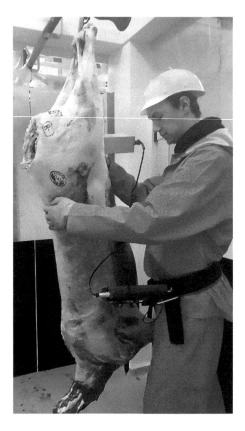

FIGURE 11.4   Direct monitoring of radioactive caesium in a reindeer carcass (Photo: B. Åhman).

FIGURE 11.5   Monitoring radioactive caesium in a live reindeer (Photo: T.A. Østmo).

Slaughtered reindeer from contaminated areas of Sweden (as defined in the control program from the National Food Agency) should be controlled at the slaughterhouse as part of the regular meat inspection. This control presently (2015) affects 15 of the 51 reindeer herding districts in Sweden, although, because of the seasonal variation, reindeer do not have to be controlled at all times of the year. The main control is usually done by direct monitoring with the NaI detector. A minimum of 50 carcasses are monitored, and if the mean value of this sample is below 800 Bq/kg and the highest value is below 1.500 Bq/kg, all reindeer from the herd are approved. In case the contamination level is higher, or it cannot be guaranteed that all reindeer in the group have been gathered from the same area, all carcasses are monitored. If the measured value is within a certain range around the Swedish intervention limit, 1.500 Bq/kg (between 1.300 and 3.000 Bq/kg), a meat sample is taken for analysis and sent to the laboratory. Based on the laboratory analyses, carcasses below 1.500 Bq/kg are accepted for human consumption and those over this limit are condemned. Very few reindeer carcasses (less than 0.1% of all slaughtered reindeer) have exceeded the allowed limit during the last decade. The highest recorded activity concentration in a reindeer carcass in Sweden during the last five years (2009–2014) was around 5.500 Bq/kg, and very few have actually been over 2.000 Bq/kg.

In Norway, animals in six reindeer herding districts may still (by 2015) have contamination levels above the limit of 3.000 Bq/kg. However, clean feeding of reindeer in corrals has not been necessary in Norway for several years, because the number of animals with contamination levels above 3.000 Bq/kg generally is low (less contaminated individuals are slaughtered instead). Nevertheless, following a rich mushroom season, as late as in September 2014, more than 800 reindeer in the Vågå herd could not be slaughtered as planned because of elevated contamination levels – reaching a maximum of 8.200 Bq/kg. These animals were released onto the ranges, and instead slaughtered during winter (an option called "clean feeding on less contaminated pastures").

Few reindeer with high levels of radioactive caesium are presently found, which is a combined effect of the rapid initial decline of radioactive caesium in reindeer, as mentioned above, and the applied counter measures (clean feeding and adapted timing of slaughter; Figure 11.6). The effects of sale limits and counter measures on the radiation dose from reindeer meat to humans were large

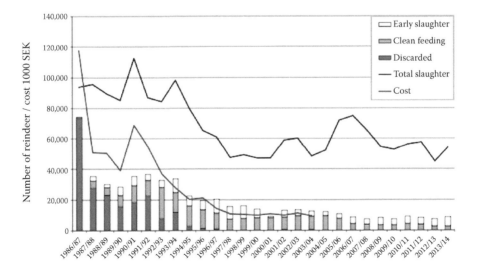

FIGURE 11.6 Counter measures applied in Sweden after the Chernobyl accident to reduce the radiation dose to humans from reindeer meat. The low intervention level (300 Bq/kg), which was applied the first year after the accident (later raised to 1 500 Bq/kg), explains the high number of discarded reindeer carcasses in this year. Counter measures like timing of slaughter ("early slaughter") and clean feeding were gradually introduced and dramatically reduced the number of reindeer with too-high levels of radioactive caesium at slaughter. Based on data from the Swedish Sami Parliament. Total costs are not available after 2004.

in the first years after the Chernobyl accident. However, the radiation doses have gradually been reduced as radioactive caesium levels in reindeer declined (Åhman 1999).

11.3 MEAT HYGIENE, MICROBIOLOGY AND STORAGE

Eva Wiklund

11.3.1 INTRODUCTION

Reindeer meat is traditionally sold as a frozen product in Scandinavia and Alaska; however, the demand for fresh chilled meat is slowly increasing. There is very limited knowledge available about the properties of chilled reindeer meat in relation to handling, packaging and storage. On the contrary, New Zealand has extensive experience in marketing deer meat (venison) as fresh chilled meat to Europe and the U.S. The New Zealand Deer Industry, together with the venison processing companies, have developed good protocols for processing, hygiene, handling, packaging and storage (Wiklund et al. 2010).

The knowledge about reindeer meat quality has increased substantially over the last 30 years as a result of investigations of a variety of aspects like pre-slaughter handling, stress, feeding strategies, carcass composition, chemical composition of the meat and carcass handling techniques (see Section 11.1). Two of the most important meat quality attributes – tenderness and flavour – valued by consumers as crucial when judging the eating quality of meat have been demonstrated to be unique for reindeer meat compared with other meat types. Reindeer meat is significantly more tender compared with beef (Barnier et al. 1999), and the typical flavour of reindeer meat is a direct reflection of the natural pastures in the forest and on the mountain tundra consumed by the reindeer (Wiklund et al. 2003). It is, therefore, not unlikely that other aspects of reindeer meat quality, like the properties of fresh chilled meat in relation to packaging and storage, will differ from those of other meats.

11.3.2 MICROBIOLOGICAL/HYGIENIC MEAT QUALITY

Shelf life of fresh meat is often determined by microbiological activity, e.g. the total amount and types of microorganisms present on the meat. A critical limit often used to judge microbiological/ hygienic quality of meat is 7 $\log_{10}$ CFU (Colony Forming Units)/g of aerobic microorganisms. Values of 7 $\log_{10}$ CFU/g and above indicate that the meat is not fit for human consumption (Wiklund et al. 2001b; 2010). Factors of importance for the microbial growth on chilled meat are pH, slaughter hygiene, and chilling conditions/temperatures.

Finnish studies of microbiological quality on reindeer meat indicated problems with high levels of aerobic microorganisms (6–7 $\log_{10}$ CFU/g) in shredded meat, but also unacceptable levels of faecal Streptococci and *Staphylococcus aureus* in cold-smoked vacuum packaged reindeer meat products (Hatakka et al. 1990). The authors compared several different reindeer processing plants and the mentioned hygienic problems were mainly related to two of the investigated plants. A list was compiled of the possible reasons for the deteriorated meat quality, where factors related to slaughter plant operation, building structures, and working methods were identified to be of inferior standards for optimal food quality and security (Hatakka et al. 1990). In addition, Finnish researchers reported that the hygienic quality of reindeer carcasses from controlled stationary reindeer slaughter plants were not superior to that of the reindeer carcasses from field slaughter plants (Vaarala and Korkeala 1999). However, during bad weather conditions (rain and consequently mud on the ground, causing dirty animals), field slaughtered reindeer carcasses had the highest microbiological contamination (Vaarala and Korkeala 1999).

Microbiological data reported for New Zealand venison demonstrated values of 2 $\log_{10}$ CFU/g and 4 $\log_{10}$ CFU/g for venison stored at −1.5°C for 3 and 9 weeks, respectively (Wiklund et al. 2010). The low storage temperature combined with good hygienic quality of the meat make it possible for the New Zealand venison industry to guarantee long shelf life (up to 12–14 weeks) for these chilled

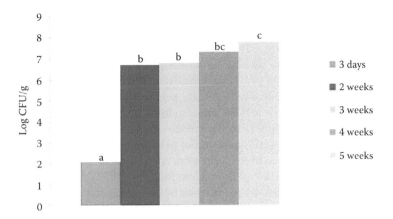

FIGURE 11.7 Microbiological quality ($\log_{10}$ CFU/g aerobic microorganisms) in reindeer samples (*M. longissimus dorsi*, n = 6 for each treatment) stored in vacuum bags at +4°C for 3 days, 2, 3, 4 and 5 weeks post-slaughter. Means with different letters are significantly different (p ≤ 0.05), based on data from Wiklund (2011).

venison cuts. In contrast, Swedish reindeer meat had poorer hygienic quality and shelf life; microbiological data of 5 $\log_{10}$ CFU/g for reindeer meat (*M. semimembranosus*) stored for 10 days at +4°C and 6.8 $\log_{10}$ CFU/g for reindeer loins (*M. longissimus*) stored 3 weeks at +4°C have been reported (Wiklund et al. 2001b; Wiklund 2011; Figure 11.7). One factor of critical importance explaining the higher levels of contamination of the Swedish reindeer carcasses is that the reindeer carcasses were transported from the slaughter facility to a processing facility before boning and packaging, which means extra manual handling – and higher risk of contamination – of every carcass. In New Zealand, all deer are slaughtered and processed/boned at the same facility. It is very common for reindeer carcasses in Sweden to be handled as described in the cited study above (Wiklund 2011). Therefore, it is essential that the significance of hygiene during slaughter and handling of carcasses is stressed by the slaughter and processing companies. The increase in demand for fresh, chilled reindeer meat will require a completely different perspective on the relationship between microbiology and shelf life compared with the current situation, where most of the meat is still sold frozen.

From Norway, it has been reported that new hygiene and animal welfare regulations have resulted in stricter requirements for the reindeer slaughter plant control systems (Hagen and Gaarden 2014). According to the Norwegian experience, the traditional Sami slaughter technique has generally adapted well to the new EU standards, although close monitoring and enforcement by the supervisory authority is essential, particularly during a start-up period. In a Norwegian study of microbiological quality and shelf life of fresh chilled reindeer meat (*M. semimembranosus*) packaged in vacuum or in modified atmosphere (MAP), it was concluded that MAP was the most optimal packaging method (Kvalvåg Pettersen et al. 2014). Longer shelf life and better quality was obtained using MAP; however, after storage for 3 weeks at +4°C, microbiological data for the different treatments varied between 5.5 and 6.0 $\log_{10}$ CFU/g, i.e. in the same range as the values reported for reindeer meat in the mentioned Swedish studies (Wiklund et al. 2001b; Wiklund 2011) and much higher than the data reported for New Zealand venison (Wiklund et al. 2010).

11.3.3 Meat pH

In countries that export large quantities of chilled meat (South America, Australia, and New Zealand), meat pH determines if the meat is to be exported as frozen or chilled. A pH value of 5.5–5.7 is the optimal range, while values over 5.8 result in reduced shelf life, especially for vacuum packaged meat (see also Sections 11.1.3 and 11.1.4). Meat pH values are directly correlated to the levels of muscle energy (glycogen) at the time of slaughter (Gill 2004). If glycogen stores in the muscles are low, meat pH will be elevated. Low muscle glycogen stores might result from poor physical condition, intense physical activity, or stress during pre-slaughter handling. In 1995, 29% of measured reindeer carcasses in Sweden (n = 3.400) had pH values of 5.8 and higher, i.e. they produced meat with an obvious risk of reduced shelf life (Wiklund et al. 1995).

The microbiological effects of distinct ultimate pH values on shelf life and safety of reindeer meat have also been studied. While at 3 days post mortem, microbial numbers tended to be higher in high pH meat (p < 0.10), all samples reached values around 6 $\log_{10}$ CFU/g after 2 weeks of refrigerated vacuum storage (+4°C), regardless of initial pH. However, at that time Enterobacteriaceae Colony Counts (ECC) of samples with high pH were approximately one log unit higher than those with intermediate and low pH values (Wiklund and Smulders 1997). Such observations have been explained by Gill and Newton (1981) to be the result of low glycogen stores, favouring the predominance of proteolytic microbial flora rather than the general positive effect high pH values have on microbial growth (Wiklund and Smulders 1997; Figures 11.8 and 11.9).

Increased demand for and handling of vacuum packed fresh, chilled reindeer meat will immediately highlight any problems with pH values that are currently "invisible" in the frozen products. As for other sectors in the meat industry, the quality and shelf life of the fresh, chilled meat must be guaranteed. Routine pH measurements of all reindeer carcasses intended for production of vacuum packaged fresh, chilled meat would contribute to optimizing reindeer meat quality and shelf life.

11.3.4 CARCASS CHILLING CONDITIONS

Except for storage time, temperature is the most important extrinsic factor influencing the storage life of fresh meat (O'Keeffe and Hood 1981; Jeyamkondan et al. 2000). Maximum storage life is achieved when meat is held at −1.5°C, which is the lowest temperature that can be maintained indefinitely without the muscle freezing (Gill and Jones 1992). Hence, to obtain maximum shelf life of chilled beef, lamb and venison from New Zealand, Australia and South America, the meat should be equilibrated to −1.5°C before load-out from the processing plant and kept at this temperature during storage and the extended transport period (Rosenvold and Wiklund 2011).

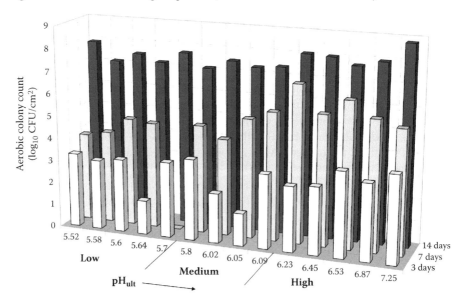

FIGURE 11.8   Aerobic colony count ($\log_{10}$ CFU/cm²) of individual reindeer longissimus samples of various ultimate pH, as assessed after 3, 7 and 14 days of refrigerated (+4°C) vacuum storage. Based on data from Wiklund and Smulders (1997).

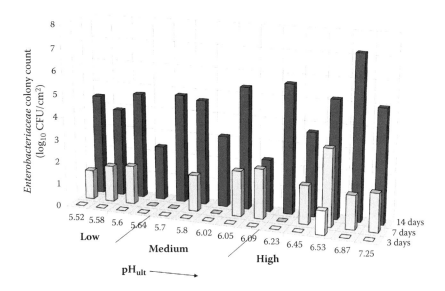

FIGURE 11.9 Enterobacteriacea colony count ($\log_{10}$ CFU/cm^2) of reindeer longissimus samples of various ultimate pH, as assessed after 3, 7 and 14 days of refrigerated (+4°C) vacuum storage. Based on data from Wiklund and Smulders (1997).

There is a substantial difference between the storage temperature (+4°C) used in the Swedish studies mentioned earlier (Wiklund and Smulders 1997; Wiklund et al. 2001b; Wiklund 2011) and the optimal storage temperature for fresh, chilled meat described above. However, the meat industry in Sweden – including the reindeer slaughter and processing companies – is not set up to operate according to, for example, New Zealand handling and transport standards, which means keeping the whole handling chain of fresh, chilled meat to a temperature of −1.5°C. Nevertheless, it would be essential for the reindeer meat companies that now are starting to place fresh, chilled reindeer meat on the market in other EU countries to standardize and possibly adjust/lower the storage temperatures throughout the handling chain. Improved chilling standards, increased awareness around the issues of hygiene and manual handling of carcasses, and routine control of meat pH provide significant opportunities to extend the shelf life of fresh, chilled reindeer meat.

11.4 SUBSISTENCE HARVEST OF CARIBOU IN NORTHERN CANADA

Jan Adamczewski and Kerri Garner

11.4.1 INTRODUCTION

Aboriginal people in northern Canada subsisted for thousands of years on hunting, trapping, and fishing, and developed an intimate knowledge of the landscapes, waters, fish and wildlife in their regions (Kendrick et al. 2005). Few wildlife resources have been as central to Aboriginal cultures and survival in northern Canada as caribou (Gordon 2005). In the northern boreal forest and tundra of Canada's mainland, migratory tundra caribou herds have sometimes reached numbers of tens or hundreds of thousands where other food sources were scarce. Caribou were the main source of food, clothing, tools and shelter for many Dene and Inuit peoples. A brief overview is provided in this section of the importance and history of subsistence harvest of caribou in northern Canada.

11.4.2 History and Cultural Value

Caribou have been a vital resource for Dene and Inuit peoples in northern Canada for countless generations. In these hunting cultures, caribou have been much more than a key subsistence food; caribou and caribou hunting have become part of people's cultural, physical, and spiritual well-being (Gordon 1996; Dogrib Treaty 11 Council 2001; Kendrick et al. 2005; Beaulieu 2012). Caribou had long been known by Aboriginal people to vary greatly in abundance on a time-scale of decades (Zalatan et al. 2006; Beaulieu 2012). Sometimes when caribou were scarce and there were few alternatives like moose or fish, times were hard and people starved (Beaulieu 2012). The key to living with the caribou properly was respect for caribou and the land.

> Respect is shown by taking only what is needed, using all parts of the harvested animals, and discarding any unused parts in respectful ways. Respect is also shown by having and sharing knowledge of the caribou. A lack of knowledge, and therefore respect, will result in the caribou migrating elsewhere and a population decline.
>
> **Dogrib Treaty 11 Council 2001**

Longstanding traditions of hunting at water crossings and on traditional trails have been known across the range of migratory caribou herds in northern Canada (Gordon 1996; Bergerud et al. 2008). In the ranges of the Beverly and Qamanirjuaq herds on the north-central Canadian mainland, biologists mapped and described key water crossings (Williams and Gunn 1982). Aboriginal hunters knew of these crossings and traditional caribou trails much earlier and relied on them to hunt caribou. In the range of the Beverly herd, more than a thousand archaeological sites in the Thelon River and nearby watersheds showed evidence of up to 8,000 years of use of water crossings by as many as nine successive caribou-hunting Aboriginal cultures (Gordon 1977, 1996, 2005; Table 11.3; Figure 11.10). These findings suggest that some migratory tundra caribou herds have been on the landscape for very long periods, and that the relationship between hunters and caribou is ancient. Bergerud et al. (2008) wrote, "We believe that the George River herd has traditionally summered northeast of Indian House Lake (in Labrador) for the past 7,500–4,000 years," based in part on archaeological evidence of repeated use of traditional water crossings by Aboriginal hunters.

For some Dene cultures, hunting in traditional territories over time led to strong associations with individual caribou herds. The Tłıchǫ (also called Dogrib) people and the Yellowknives Dene had a long history of hunting and knowledge of the Bathurst herd, while the Athabasca Denésǫłıné (also called Chipewyan) of northern Saskatchewan had a similar history and association with the Beverly herd (Figure 11.11; Gordon 1996). The Tłıchǫ Agreement, a land claim and self-governance agreement that came into effect in 2005, refers frequently to the Bathurst herd and the need to manage this herd and its range soundly for future generations. Further west, a number of Aboriginal peoples, including the Gwich'in and Inuvialuit in the Northwest Territories and several First Nations in northern Yukon, have a similar ancient association with the Porcupine caribou herd. These associations continue in the present day. The Porcupine Caribou Management Board (PCMB) and the Beverly and Qamanirjuaq Caribou Management Board (BQCMB) were both formed in the 1980s to allow Aboriginal communities dependent on these herds to work for conservation collaboratively with governments and other agencies.

11.4.3 Scale and Value of Caribou Harvest

Migratory tundra caribou herds have sometimes been estimated at peak numbers of several hundred thousand. Examples are the Bathurst herd in 1986 at 470,000 (Boulanger et al. 2011), the Qamanirjuaq herd in 1994 at 500,000 (Campbell et al. 2010), and the George River herd in the late 1980s at 600,000 or more (Bergerud et al. 2008). At these herd sizes, annual Aboriginal harvest levels have sometimes numbered in the thousands. In some regions, harvest by non-Aboriginal hunters has also been substantial. In the early 1990s, estimates of annual caribou harvest from the George

TABLE 11.3

Water Crossings in Beverly Caribou Summer Range with Associated Archaeological Sites and Aboriginal Hunting Cultures

Beverly water crossing number	17	18	19	20	21	22	23	24	25	26	27
Water crossing name (from Williams and Gunn 1982)	East Aberdeen Lake	E-Central Aberdeen Lake	Central Aberdeen Lake	West Aberdeen Lake	Dubawnt River	East Thelon Sanctuary	Central Thelon Sanctuary	West Thelon Sanctuary	La Du Bois	Mary Frances Lake	Lockhart River
Number of archaeological sites near crossing	8	None	1	8	18	5	6	64	2	None	8
Years ago — Evidence of aboriginal culture	Yes	No	Yes	Yes	Yes	Yes	Yes	Yes	Yes (Undetermined)	No	Yes
Present–300 Inuit	X		X	X	X	X	X	X			
Present–200 Dene Chipewyan					X		X	X			X
200–1300 Late Taltheilei				X	X	X	X	X			
1300–1800 Middle Taltheilei	X		X	X	X	X	X	X			X
1800–450 Early Taltheilei			X	X	X	X		X			X
2450–2600 Earliest Taltheilei					X			X			
2650–3540 Pre-Dorset ASTt	X		X	X	X		X	X			X
3500–6500 Shield Archaic			X	X	X	X	X	X			
7000–8000 Northern Plano				X	X			X			

Note: Water crossing numbers and names are from Williams and Gunn (1982). Archaeological sites and evidence of Aboriginal cultures over time near the water crossings are primarily from Gordon (1996) and were identified with the assistance of B. Gordon, L. Johanis and F. Krist, March 2012.

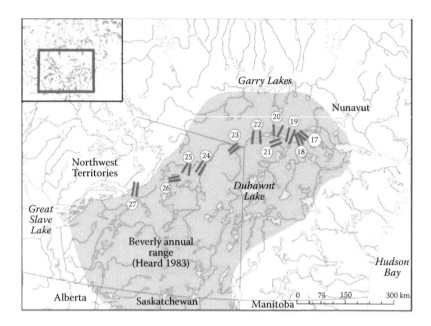

FIGURE 11.10   Traditional Caribou Water Crossings in Beverly Caribou Summer Range in northern Canada (adapted from Williams and Gunn 1982), with annual range of Beverly herd based on Heard (1983).

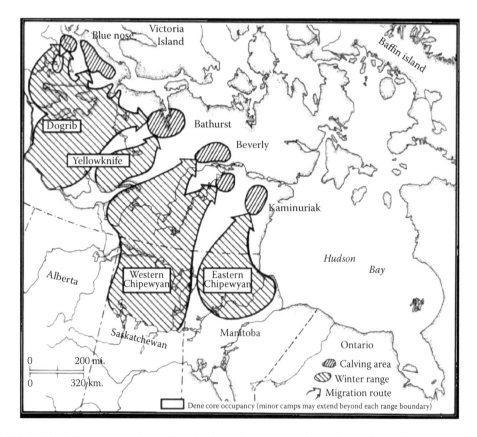

FIGURE 11.11   Migratory tundra caribou herds and associated Aboriginal cultures in northern Canada (from Gordon 1996; earlier version in Kelsall 1968).

River and Leaf River herds included 8.000–15.000 in Labrador by Aboriginal and non-Aboriginal hunters, 9.000–15.000 by sports hunters in northern Québec, and 11.000 by Aboriginal hunters in Québec (Jean and Lamontagne 2004). For 2005–2006, the BQCMB provided estimates of 10.300 Qamanirjuaq caribou and 3.800 Beverly caribou (BQCMB 2006) harvested, with much of this hunt from Aboriginal hunters and a small proportion from resident and guided outfitter hunters. Annual harvest of Bathurst caribou was estimated at 7.000–21.000 per year in the late 1980s and early 1990s and at about 4.000–6.000 per year in 2007–2009 (Boulanger et al. 2011), with a high percentage of this harvest being Aboriginal. At times of low caribou numbers, winter ranges used by herds shrank substantially and hunters had to either travel long distances to find caribou or shift to other sources of meat (Beaulieu 2012).

InterGroup Consultants (2013) assessed the economic value of the 2005–2006 Qamanirjuaq and Beverly caribou harvest (BQCMB 2006) at $20 million, much of this value being meat replacement. This represents a substantial annual part of the traditional economy in small communities that often have few other economic opportunities, and where groceries are often expensive. The cultural and spiritual benefits of maintaining a long-established way of life as caribou hunters is less easily measured, but remain of immense value to northern communities and Aboriginal cultures in Canada.

11.4.4 Subsistence Harvest of Caribou: An Uncertain Future

Abundance of migratory tundra caribou varies by individual herd, but there is generally synchrony on a regional scale (Gunn 2003). Most migratory caribou herds in Canada reached peak numbers in the 1980s or early 1990s and declined in the late 1990s and 2000s (Vors and Boyce 2009; Festa-Bianchet et al. 2011). The Bathurst herd had fallen from an estimated 470.000 in 1986 to an estimated 32.000 in 2009, with a declining trend (Boulanger et al. 2011). The George River herd that may have exceeded 600.000 in the late 1980s (Bergerud et al. 2008) was estimated in 2014 at 14.200, with a steeply declining trend. Under these circumstances, management actions have by necessity been taken to reduce harvest, and sometimes in a very short time-frame. For the Bathurst herd, an annual harvest of 4.000–6.000 was reduced in one winter (2009–2010) to 300 or less (Boulanger et al. 2011).

Under land claims and treaties across Canada, Aboriginal subsistence harvest of game takes priority over harvest by non-Aboriginal hunters and commercial harvest. Decision-making about managing caribou in the Canadian North is no longer made solely by government agencies, with co-management boards and Aboriginal governments now having a greater voice in wildlife management. Trust between different ways of thinking about caribou takes time to develop and in some cases, extensive work remains to be done in building relationships. Shared wildlife management has not made decisions about managing caribou any less difficult, and decisions about limiting harvest have been the most difficult of all. For Aboriginal hunting cultures long dependent on caribou, the sometimes sudden loss of harvesting opportunities from caribou herds at low numbers has meant serious hardships to hunters, their cultures, and their communities.

Hardships resulting from loss of caribou hunting are multi-faceted and go beyond the lack of availability of this critical food source, particularly in isolated communities where store bought food is expensive and often unhealthy. This shift in diet results in higher costs of living, a less healthy diet, and less time spent on the land. It also means less caribou hide, bone and antler available for clothing and crafts. A critically important impact of limitation of Aboriginal harvest is the potential cessation of cultural and spiritual practices associated with the caribou hunt and all of the activities that take place with it for men, women, and youth (Figure 11.12). If hunters are limited from hunting for many years, particularly in today's fast changing world, there is potential for these cultural practices and traditional knowledge to be lost, and no longer transferred from one generation to the next (Figures 11.13 and 11.14). This knowledge and these practices cannot be transferred to the same degree in a home or a classroom and will realistically only be shared when "on the land". The overall impact of the limitation on hunting caribou to Aboriginal people begins at the individual level, but will also be felt at the family and community level.

FIGURE 11.12   Cutting up caribou meat in camps is often a social activity with multiple generations participating. (Photo: S. Cavanagh, courtesy of Beverly and Qamanirjuaq Caribou Management Board.)

Some factors affecting caribou abundance can be managed (harvest, industrial development, and
predators), but other factors, including weather in all seasons, will continue to affect caribou and may
be key to the overall long-term changes in caribou abundance (Gunn 2003). The caribou declines
have been widespread in northern Canada and recovery will take years, if not decades. Sometimes
caribou herds have declined to low numbers and remained there for decades (Valkenburg et al. 1994).
As the future of the migratory caribou herds is uncertain, so too is the ability of caribou-dependent
hunting cultures to maintain a way of life that has existed for thousands of years. Management plans
like those developed collaboratively for the Cape Bathurst, Bluenose-West and Bluenose-East caribou herds (ACCWM 2014), the BQCMB for the Beverly and Qamanirjuaq caribou herds (BQCMB
2014) and the PCMB for the Porcupine herd (PCMB 2010) suggest there is hope of sound caribou
management that sustains caribou and caribou-hunting cultures in northern Canada.

FIGURE 11.13   In earlier times, hunters sometimes bundled much of the meat of a caribou into its hide
and carried the meat back to camp. (Photo: J. Stephenson, courtesy of Beverly and Qamanirjuaq Caribou
Management Board.)

FIGURE 11.14   Drying thin strips of meat continues to be a way of preserving the meat as well as substantially reducing its weight. (Photo: P. Bernier, courtesy of Beverly and Qamanirjuaq Caribou Management Board.)

REFERENCES

Advisory Committee for Cooperation on Wildlife Management (ACCWM). 2014. Taking Care of Caribou – The Cape Bathurst, Bluenose-West, and Bluenose-East Barren Ground Caribou Herds Management Plan (Final). C/O Wek'èezhìi Renewable Resources Board, 102A, 4504 – 49 Avenue, Yellowknife, Northwest Territories, Canada. Available at: www.grrb.nt.ca/pdf/wildlife/caribou/CB_BNW_BNE _Mgmt_Plan_FINAL.pdf.

Åhman, B. 1994. Body burden and distribution of ^{137}Cs in reindeer. *Rangifer* 14 (1):23–28.

Åhman, B. 1996. Effect of bentonite and ammonium-ferric(III)-hexacyanoferrate(II) on uptake and elimination of radiocaesium in reindeer. *Journal of Environmental Radioactivity* 31 (1):29–50.

Åhman, B. 1999. Transfer of radiocaesium via reindeer meat to man – Eeffects of countermeasures applied in Sweden following the Chernobyl accident. *Journal of Environmental Radioactivity* 46 (1):113–120.

Åhman, B. and G. Åhman. 1994. Radiocesium in Swedish reindeer after the Chernobyl fallout: Seasonal variations and long-term decline. *Health Physics* 66 (5):503–512.

Åhman, B. and Ö. Danell. 2001. Reindeer feeding – Possibilities, effects and economy. In *Programme and Abstracts, 11th Nordic Conference on Reindeer Research, Kaamanen, Finland*. R.E. Haugerud (Ed.), 14.

Åhman, B., S.M. Wright and B.J. Howard. 2001. Effect of origin of radiocaesium on the transfer from fallout to reindeer meat. *The Science of the Total Environment* 278 (1–3):171–181.

Åhman, G. 1986. Studier av radioaktivt cesium i svenska renar. Översikt över pågående undersökningar 1986 (Studies of radiocesium in Swedish reindeer: Investigations during 1986). *Rangifer* 6 (1, Appendix):53–64.

Alaska Agriculture Statistics. 2006. U.S. Department of Agriculture: Palmer, Alaska, USA.

Alaska Department of Environmental Conservation. 2003. Regulations for reindeer slaughtering and processing (18 AAC 32.600) and regulations for reindeer for retail sale to or at a market (18 AAC 31.820). State of Alaska, USA.

Bäck, L. 1993. Reindeer management in conflict and co-operation. A geographic land use and simulation study from northernmost Sweden. *Nomadic Peoples* 32:65–80.

Bader, H.R. and G.L. Finstad. 2001. Conflicts between livestock and wildlife: An analysis of legal liabilities arising from reindeer and caribou competition on the Seward Peninsula of Western Alaska. *Environmental Law* 31:549–580.

Barnett, C.L., N.A. Beresford, P.L. Self, B.J. Howard, J.C. Frankland, M.J. Fulker, B.A. Dodd and J.V.R. Marriott. 1999. Radiocaesium activity concentrations in the fruit-bodies of macrofungi in Great Britain and an assessment of dietary intake habits. *The Science of the Total Environment* 231:67–83.

Barnier, V.M.H., E. Wiklund, A. van Dijk, F.J.M. Smulders and G. Malmfors. 1999. Proteolytic enzyme and inhibitor levels in reindeer (*Rangifer tarandus* L) vs. bovine longissimus muscle, as they relate to ageing rate and response. *Rangifer* 19:13–18.

Beaulieu, D. 2012. Dene traditional knowledge about caribou cycles in the Northwest Territories. *Rangifer* Special Issue 20:59–67.

Bergerud, A.T., S.N. Luttich and L. Camps. 2008. *The Return of Caribou to Ungava.* McGill-Queen's University Press: Canada.

Beverly and Qamanirjuaq Caribou Management Board (BQCMB). 2006. Beverly and Qamanirjuaq Caribou Management Board 24th Annual Report 2005–2006. Available at: www.arctic-caribou.com/PDF /BQCMB_2005_2006_Annual_Report.pdf.

Beverly and Qamanirjuaq Caribou Management Board (BQCMB). 2014. Beverly and Qamanirjuaq Caribou Management Plan. Available at: www.arctic-caribou.com/PDF/bqcmb_managementplan_detailed 2014.pdf.

Boulanger, J., A. Gunn, J. Adamczewski and B. Croft. 2011. A data-driven demographic model to explore the decline of the Bathurst caribou herd. *Journal of Wildlife Management* 75:883–896.

Brynildsen, L.I. and P. Strand. 1994. A rapid method for the determination of radioactive caesium in live animals and carcasses, and its practical application in Norway after the Chernobyl reactor accident. *Acta Veterinaria Scandinavica* 35 (4):401–408.

Campbell, M., J. Nishi and J. Boulanger. 2010. A Calving Ground Photo Survey of the Qamanirjuaq Migratory Barren-Ground caribou (*Rangifer tarandus groenlandicus*) Population – June 2008. Technical Report Series 2010 No. 1–10. Government of Nunavut, Department of Environment. Available at: env.gov.nu.ca /sites/default/files/report_-qamanirjuaq_caribou_nov_2010.pdf.

Carlson, S.M. 2005. Economic impact of reindeer-caribou interactions on the Seward Peninsula. MSc thesis, University of Alaska Fairbanks, Fairbanks, Alaska, USA.

Dogrib Treaty 11 Council. 2001. Caribou Migration and the State of their Habitat, Final Report. Submitted to the West Kitikmeot Slave Society, Yellowknife, Northwest Territories, Canada.

Ehlken, S. and G. Kirchner. 2002. Environmental processes affecting plant root uptake of radioactive trace elements and variability of transfer factor data: A review. *Journal of Environmental Radioactivity* 58 (2–3):97–112.

Essén-Gustavsson, B. and C. Rehbinder. 1984. The influence of stress on substrate utilization in skeletal muscle fibres of reindeer (*Rangifer tarandus* L). *Rangifer* 4:2–8.

Festa-Bianchet, M., J.C. Ray, S. Boutin, S.D. Côté and A. Gunn. 2011. Conservation of caribou (*Rangifer tarandus*) in Canada: An uncertain future. *Canadian Journal of Zoology* 89:419–434.

Finstad, G.L., K.K. Kielland and W.S. Schneider. 2006. Reindeer herding in transition: Historical and modern day challenges for Alaskan reindeer herders. *Nomadic Peoples* 10:31–49.

Finstad, G., E. Wiklund, K. Long, P.J. Rincker, A.C.M. Oliveira and P.J. Bechtel. 2007. Feeding soy or fish meal to Alaskan reindeer (*Rangifer tarandus tarandus*) – Effects on animal performance and meat quality. *Rangifer* 27:59–75.

Gill, C.O. 2004. Spoilage, factors affecting – Microbiological. In *Encyclopedia of Meat Sciences*, W. Jensen (Ed.), 1324. Elsevier: Oxford, United Kingdom.

Gill, C.O. and S.D.M. Jones. 1992. Efficiency of a commercial process for the storage and distribution of vacuum packaged beef. *Journal of Food Protection* 55:880–887.

Gill, C.O. and K.G. Newton. 1981. Microbiology of DFD beef. In *The Problem of Dark-Cutting in Beef*, D.E. Hood and P.V. Tarrant (Eds.), 305–327. Martinus Nijhoff: Den Haag, the Netherlands.

Gordon, B.C. 1977. Prehistoric Chipewyan harvesting at a barrenland caribou water crossing. *Western Canadian Journal of Anthropology* 7:69–83.

Gordon, B.C. 1996. People of Sunlight, People of Starlight: Barrenland Archaeology in the Northwest Territories of Canada. Mercury Series, Archaeological Survey of Canada, Canadian Museum of Civilization: Hull, Québec, Canada.

Gordon, B.C. 2005. 8000 years of caribou and human seasonal migration in the Canadian Barrenlands. *Rangifer* Special Issue 16:155–162.

Gunn, A. 2003. Voles, lemmings and caribou – population cycles revisited? *Rangifer* Special Issue 14:105–112.

Hagen, A. and H.K. Gaarden. 2014. Offentlig kontroll og slaktting av rein (Official control and slaughter of reindeer). *Norsk Veterinær tidskrift* 126:134–138 (in Norwegian with English summary).

Hanssen, I., A. Kyrkjebø and P.K. Opstad. 1984. Physiological responses and effects on meat quality in reindeer (*Rangifer tarandus*) transported on lorries. *Acta Veterinaria Scandinavica* 25:128–138.

Hanssen, I. and T. Skei. 1990. Lack of correlation between ammonia-like taint and polyamine levels in reindeer meat. *Veterinary Record* 127:622–623.

Hatakka, M., H. Korkeala and K. Salminen. 1990. Poronlihatuotteiden hygieeninen laatu (The hygienic quality of reindeer meat products). *Suomen Eläinlääkärilehti* 6:275–282 (in Finnish with English table legends and summary).

Heard, D.C. 1983. Hunting patterns and the distribution of the Beverly, Bathurst and Kaminuriak caribou herds based on tag returns. *Acta Zoologica Fennica* 175:145–147.

Holand, Ø. 2007. Herd composition and slaughtering strategy in reindeer husbandry – revisited. *Rangifer Report* 12:21–33 (in Norwegian with English abstract).

Hove, K. 1993. Chemical methods for reduction of the transfer of radionuclides to farm animals in semi-natural environments. *The Science of the Total Environment* 137:235–248.

ICRP. 2007. The 2007 Recommendations of the International Commission on Radiological Protection. ICRP publication 103. *Annals of the ICRP* 37 (2–4):1–332.

InterGroup Consultants. 2013. Economic Valuation and Socio-Cultural Perspectives of the Estimated Harvest of the Beverly and Qamanirjuaq Caribou Herds (revised version Oct. 3, 2013). Prepared by: InterGroup Consultants Ltd., Winnipeg, Manitoba, Canada, for the Beverly and Qamanirjuaq Caribou Management Board, Stonewall, Manitoba, Canada.

Jean, D. and G. Lamontagne. 2004. Northern Québec Caribou (*Rangifer tarandus*) Management Plan 2004–2010. Ministère des Ressources naturelles et de la Faune – secteur Faune Québec, Direction de l'aménagement de la faune du Nord-du-Québec. Québec City, Québec, Canada. 80 pp.

Jeyamkondan, S., D.S. Jayas and R.A. Holley. 2000. Review of centralized packaging systems for distribution of retail-ready meat. *Journal of Food Protection* 63:796–804.

Kelsall, J.P. 1968. The Migratory Barren-Ground Caribou of Canada. Environment Canada, Canadian Wildlife Service, Vol. 3, 340 pp.

Kendrick, A., P.O'B. Lyver and Łutsel K'e Dene First Nation. 2005. Denésoliné (Chipewyan) knowledge of barren-ground caribou (*Rangifer tarandus groenlandicus*) movements. *Arctic* 58:175–191.

Kvalvåg Petterson, M., A. Ådland Hansen and M. Mielnik. 2014. Effect of different packaging methods on quality and shelf life of fresh reindeer meat. *Packaging Technology and Science* 27:987–997.

Lidén, K. 1961. Cesium137 burdens in Swedish Laplanders and reindeer. *Acta Radiol.* 56 (3):237–240.

Lidén, K. and M. Gustafsson. 1967. Relationships and seasonal variation of ^{137}Cs in lichen, reindeer and man in northern Sweden 1961–1965. In *Radioecological Concentration Processes. Proceedings of an International Symposium Held in Stockholm 25–29 April, 1966.* B. Åberg and F.P. Hungate (Eds.), 193–208. Oxford: Pergamon Press.

Lundesjö Ahnström, M. 2008. Influence of pelvic suspension on beef meat quality. PhD thesis No. 2008:61, Department of Food Science, Swedish University of Agricultural Sciences, Uppsala, Sweden. Available at: pub.epsilon.slu.se/1808/1/Maria_Lundej%C3%B6_Ahnstr%C3%B6m_Thesis.pdf.

Mattsson, L.J.S. 1975. [137]Cs in the reindeer lichen *Cladonia Alpestris*: Deposition, retention and internal distribution 1961–1970. *Health Physics* 28:233–248.

Muuttoranta, K. 2014. Current state of and prospects for selection in reindeer husbandry. PhD thesis No. 2014:32, Department of Agricultural Sciences, Faculty of Agriculture and Forestry, University of Helsinki, Finland. Available at: ethesis.helsinki.fi.

Niinivaara, F.P. and E. Petäjä. 1985. Problems in the production and processing of reindeer meat. In *Trends in Modern Meat Technology,* B. Krol, P.S. van Roon and J.H. Houben (Eds.), 115–120. Pudoc: Wageningen, the Netherlands.

NRCS (Natural Resource Conservation Service), United States Department of Agriculture. 1953. Amendment No. 4, Title 9, Administrative Regulations, May 17, 1954, and Comptroller General's Opinion B-115665 of October 1, 1953, 33CG:133.

O'Keeffe, M. and D.E. Hood. 1981. Anoxic storage of fresh beef. II. Colour stability and weight loss. *Meat Science* 5:267–281.

Papastefanou, C., M. Manolopoulou and T. Sawidis. 1989. Lichens and mosses: Biological monitors of radioactive fallout from the Chernobyl accident. *Journal of Environmental Radioactivity* 9:199–207.

Petäjä, E. 1983. DFD meat in reindeer meat. In *Proceedings 29th European Congress of Meat Researcher Workers,* 117–124. Salsomaggiore, Italy.

Porcupine Caribou Management Board (PCMB). 2010. Harvest Management Plan for the Porcupine Caribou Herd in Canada. Available at: www.pcmb.ca/documents/Harvest%20Management%20 Plan%202010.pdf.

Rehbinder, C. 1990. Management stress in reindeer. *Rangifer* Special Issue No. 3:267–288.

Rehbinder, C., L.-E. Edqvist, K. Lundström and F. Villafane. 1982. A field study of management stress in reindeer (*Rangifer tarandus* L). *Rangifer* 2:2–21.

Reindriftsforvaltningen. 2013. Totalregnskap for reindriftsnæringen. Økonomisk utvalg, Nov 2013. Statistics of Norwegian reindeer husbandry (in Norwegian), Reindriftsforvaltningen, Alta, Norway.

Rissanen, K., B. Åhman, J. Ylipieti and T. Nylén. 2005. Radiocaesium in reindeer pasture. In *Radiological Protection in Transition. Proceedings of the XIV Regular Meeting of the Nordic Society for Radiation Protection, NFSF, Rättvik, Sweden, 27–31 August 2005.* J. Valentin, T. Cederlund, P. Drake, I.E. Finne, A. Glansholm, A. Jaworska, W. Paile and T. Rahola (Eds.), 307–310. Swedish Radiation Protection Authority: Stockholm, Sweden.

Rissanen, K., T. Rahola and P. Aro. 1990. Distribution of cesium-137 in reindeer. *Rangifer* 10 (2):57–66.

Rissanen, K., T. Rahola, E. Illukka and A. Alfthan. 1987. Radioactivity of Reindeer, Game and Fish in Finnish Lapland After the Chernobyl Accident in 1986. Vol. STUK-A63, Annual Report STUK. Finnish Centre for Radiation and Nuclear Safety, Helsinki.

Rogstadkjærnet, M. and I. Hanssen. 1985. Ammonia-like taint and creatine, creatinine and dimethylamine contents in reindeer meat. *Acta Veterinaria Scandinavica* 26:143–144.

Rosenvold, K. and E. Wiklund. 2011. Retail colour display life of chilled lamb as affected by processing conditions and storage temperature. *Meat Science* 88:354–360.

Sampels, S. 2005. Fatty acids and antioxidants in reindeer and red deer – Emphasis on animal nutrition and consequent meat quality. PhD thesis No. 2005:31, Department of Food Science, Swedish University of Agricultural Sciences, Uppsala, Sweden. Available at: pub.epsilon.slu.se/800/1/Avhandling_nr _31_2005.pdf.

Sampels, S., J. Pickova and E. Wiklund. 2004. Fatty acids, antioxidants and oxidation stability of processed reindeer meat. *Meat Science* 67:523–532.

Sampels, S., E. Wiklund and J. Pickova. 2006. Influence of diet on fatty acids and tocopherols in *M. longissimus dorsi* from reindeer. *Lipids* 41:463–472.

Savell, J. W., S.L. Mueller and B.E. Baird. 2005. The chilling of carcasses. *Meat Science* 70:449–459.

Schneider W.S., K. Kielland and G. Finstad. 2005. Factors in the adaptation of reindeer herders to caribou on the Seward Peninsula, Alaska. *Arctic Anthropology* 42:36–49.

Skjenneberg, S., E. Jacobsen and H. Movinkel. 1974. pH-verdien i reinkjott etter forskjelling behandling av dyrene for slakt. *Nordisk Veterinärmedicin* 26:436–443 (in Norwegian).

Skogland, T. 1986. High radio-cesium contamination of wild reindeer from southern Norway following the Chernobyl accident. *Rangifer* 6(1, Appendix):72.

Skuterud, L., E. Gaare, I.M. Eikelmann, K. Hove and E. Steinnes. 2005. Chernobyl radioactivity persists in reindeer. *Journal of Environmental Radioactivity* 83:231–252.

Skuterud, L., Ø. Pedersen, H. Staaland, K.H. Røed, B. Salbu, A. Liken and K. Hove. 2004. Absorption, retention and tissue distribution of radiocaesium in reindeer: Effects of diet and radiocaesium source. *Radiation and Environmental Biophysics* 43:293–301.

Skuterud, L. and H. Thørring. 2012. Averted doses to Norwegian Sami reindeer herders after the Chernobyl accident. *Health Physics* 102 (2):208–216. doi 0.1097/HP.0b013e3182348e12.

Staaland, H. and H. Sletten. 1991. Feeding reindeer in Fennoscandia: The use of artificial food. In *Wildlife Production, Conservation and Sustainable Development*, L.A. Renecker and R.J. Hudson (Eds.), 227. Agricultural and Forestry Experiment Station, University of Alaska Fairbanks: Fairbanks, Alaska, USA.

Stern, R.O., E.L. Arobio, L.L. Naylor and W.C. Thomas. 1980. Eskimos, Reindeer and Land. Agricultural and Forestry Experimental Station, University of Alaska Fairbanks. Bulletin 59, 205 pp.

Strand, P., T.D. Selnæs, E. Bøe, O. Harbitz and A. Andersson-Sørlie. 1992. Chernobyl fallout: Internal doses to the Norwegian population and the effect of dietary advice. *Health Physics* 63 (4):385–392.

Swedish Board of Agriculture. 2004. Föreskrifter om ändring i Statens jordbruksverks föreskrifter (SJVFS 1998:127) om klassificering av slaktkroppar. Jönköping, Sweden (in Swedish).

Swedish National Food Agency. 1993. Regulations regarding meat inspection etc. at reindeer slaughter. SLV FS 1993:5, H 197:2 (in Swedish).

Swedish Sami Parliament. 2015. Statistics of Swedish reindeer husbandry (in Swedish) sametinget.se/statistik _rennaring. Accessed 3 March 2015.

Taylor, R.G., R. Labas, F.J.M. Smulders and E. Wiklund. 2002. Ultrastructural changes during ageing in *M. longissimus* from moose (*Alces alces*) and reindeer (*Rangifer tarandus tarandus*). *Meat Science* 60:321–326.

Tveten, U., L.I. Brynildsen, I. Amundsen and T.D.S. Bergan. 1998. Economic consequences of the Chernobyl accident in Norway in the decade 1986–1995. *Journal of Environmental Radioactivity* 41 (3):233–255.

Vaarala, A.M. and H.J. Korkeala. 1999. Microbiological contamination of reindeer carcasses in different reindeer slaughterhouses. *Journal of Food Protection* 62:152–155.

Valkenburg, P., D.G. Kelleyhouse, J.L. Davis, and J.M. Ver Hoef. 1994. Case history of the Fortymile Caribou Herd, 1920–1990. *Rangifer* 14:11–22.

Vors, L.S. and M.S. Boyce. 2009. Global declines of caribou and reindeer. *Global Change Biology* 15: 2626–2633.

Westerlund, E.A., T. Berthelsen and L. Berteig. 1987. Cesium-137 body burdens in Norwegian Lapps, 1965–1983. *Health Physics* 52 (2):171–177.

Wiklund, E. 1996. Pre-slaughter handling of reindeer (*Rangifer tarandus tarandus* L) – Effects on meat quality. Doctoral thesis, Department of Food Science, Swedish University of Agricultural Sciences, Uppsala, Sweden.

Wiklund, E. 2011. Microbiological shelf life of fresh, chilled reindeer meat (M. longissimus dorsi). *Rangifer* 31:85–90.

Wiklund, E. 2014. Experiences during implementation of a quality label for meat from reindeer. In *Trends in Game Meat Hygiene. From Forest to Fork*, Paulsen, P., Bauer, A. and F.J.M. Smulders (Eds.), 295–303. Wageningen Academic Publishers: the Netherlands.

Wiklund, E., A. Andersson, G. Malmfors and K. Lundström. 1996a. Muscle glycogen levels and blood metabolites in reindeer (*Rangifer tarandus tarandus* L) after transport and lairage. *Meat Science* 42:133–144.

Wiklund, E., A. Andersson, G. Malmfors, K. Lundström and Ö. Danell. 1995. Ultimate pH values in reindeer meat with particular regard to animal sex and age, muscle and transport distance. *Rangifer* 15:47–54.

Wiklund, E., V.M.H. Barnier, F.J.M. Smulders, K. Lundström and G. Malmfors. 1997b. Proteolysis and tenderisation in reindeer (*Rangifer tarandus tarandus* L) bull longissimus thoracis muscle of various ultimate pH. *Meat Science* 46:33–43.

Wiklund, E., G. Finstad, G. Aguiar and P.J. Bechtel. 2012. Does carcass suspension technique influence reindeer (*Rangifer tarandus tarandus*) meat quality attributes? *Animal Production Science* 52:731–734.

Wiklund, E., G. Finstad, L. Johansson, G. Aguiar and P.J. Bechtel. 2008. Carcass composition and yield of Alaskan reindeer (*Rangifer tarandus tarandus*) steers and effects of electrical stimulation applied during field slaughter on meat quality. *Meat Science* 78:185–193.

Wiklund, E., L. Johansson and G. Malmfors. 2003. Sensory meat quality, ultimate pH values, blood parameters and carcass characteristics in reindeer (*Rangifer tarandus tarandus* L) grazed on natural pastures or fed a commercial feed mixture. *Food Qual. Pref.* 14:573–581.

Wiklund, E., I. Hansson and G. Malmfors. 2000. Composition and quality of reindeer (*Rangifer tarandus tarandus* L) carcasses. In *Proceedings: 46th International Congress of Meat Science and Technology*. Buenos Aires, Argentina.

Wiklund, E., R. Kemp, G.J. le Roux, Y. Li and G. Wu. 2010. Spray chilling of deer carcasses–effects on carcass weight, meat moisture content, purge and microbiological quality. *Meat Science* 86:926–930.

Wiklund, E., G. Malmfors and K. Lundström. 1997a. The effects of pre-slaughter selection of reindeer bulls (*Rangifer tarandus tarandus* L) on technological and sensory meat quality, blood metabolites and abomasal lesions. *Rangifer* 17:65–72.

Wiklund, E., G. Malmfors, K. Lundström and C. Rehbinder. 1996b. Pre-slaughter handling of reindeer bulls (*Rangifer tarandus tarandus* L) – Effects on technological and sensory meat quality, blood metabolites and muscular and abomasal lesions. *Rangifer* 16:109–117.

Wiklund, E., J. Pickova, S. Sampels and K. Lundström. 2001a. Fatty acid composition in *M. longissimus lumborum*, ultimate muscle pH values and carcass parameters in reindeer (*Rangifer tarandus tarandus* L) grazed on natural pasture or fed a commercial feed mixture. *Meat Science* 58:293–298.

Wiklund, E., C. Rehbinder, G. Malmfors, I. Hansson and M.-L. Danielsson-Tham. 2001b. Ultimate pH values and bacteriological condition of meat and stress metabolites in blood of transported reindeer bulls. *Rangifer* 21:3–12.

Wiklund, E. and F.J.M. Smulders. 1997. Tenderness and microbiological condition of reindeer (*Rangifer tarandus tarandus* L.) longissimus muscle of various ultimate pH. In *Proceedings WAVFH World Congress on Food Hygiene*. 165. The Hague, the Netherlands.

Wiklund, E. and F.J.M. Smulders. 2011. Muscle biological and biochemical ramifications of farmed game husbandry with focus on deer and reindeer. In *Game Meat Hygiene in Focus. Microbiology, Epidemiology, Risk Analysis and Quality*, P. Paulsen, A. Bauer, M. Vodansky, R. Winkelmayer, and F.J.M. Smulders (Eds.), 297–311. Wageningen Academic Publishers: the Netherlands.

Williams, T.M. and A. Gunn. 1982. Descriptions of Water Crossings and Their Use by Migratory Barren-Ground Caribou in the Districts of Keewatin and Mackenzie, NWT. NWT Wildlife Service, Yellowknife, Northwest Territories, Canada. File Report 27.

Zalatan, R., A. Gunn and G.H.R. Hare. 2006. Long-term abundance patterns of barren-ground caribou using trampling scars on roots of *Picea mariana* in the Northwest Territories, Canada. *Arctic, Antarctic and Alpine Research* 38:624–630.

12 Assessment and Treatment of Reindeer Diseases

Sauli Laaksonen

CONTENTS

Diseases and parasites are natural parts of the healthy ecosystems of wild cervids. Their natural circulation occurs without the need for human interference.

Nature seeks balance through its own regulatory systems, and it has its own means of health care. The domestication of reindeer for human needs has led to disturbances in this balance. In the past, reindeer populations were regulated by predators, parasites and various diseases that cut part of the benefit of meat production from the industry (herders). Today, this role has been assumed by man. Man decides on the regulating and harvesting of the population so that the profits from reindeer herding are as high as possible. Diseases, parasites and predators are now competitors. In the history of reindeer herding, epizootics have occurred one after another and led to the death of hundreds of thousands of reindeer. Today, epizootics are rare, but outbreaks still occur, and new pathogens have emerged in the reindeer herding area. In addition, domestication, including increased population density, supplementary feeding, antiparasitic treatment, fencing and transportation, has entailed new challenges in reindeer health care and veterinary practice. Although reindeer that are sick or in poor condition must be taken care of and sometimes treated for their condition, very little scientific knowledge is published dealing with veterinary practice in reindeer management. Thus, treatments, including choice of drugs and dosages, are often based on the clinical experience of reindeer veterinarians, and are often modified from treatment of domesticated ruminants.

12.1 SICK REINDEER AND CLINICAL EVALUATION

Clinical examination of reindeer is complicated by their thick coats and the stress they feel from being held and from the presence of man. Further, suffering or pain is not easy to detect, and often, a reindeer will eat until the very end, even if it is seriously ill (Nikolaevskii, 1961a; Rehbinder and Nikander, 1999; Laaksonen, 2016).

12.1.1 THE CHARACTERISTICS OF HEALTHY REINDEER (FIGURE 12.1)

- The animal is attentive, alert, and aware of its surroundings
- The eyes are bright and clear
- Ears are erect and mobile
- Hair coat is vibrant and clean, and the winter coat is well developed
- Antlers are symmetrical and normally developed
- Nostrils are clean
- Visible mucous membranes in the eyes and nostrils, and in the vagina, are pure pink
- Behaviour is active; for example, the animal reacts continuously to insects
- Movement and anatomy are normal
- The animal is ruminating at rest and faeces and urine are typical for the species. (Nikolaevskii, 1961a; Laaksonen, 2016).

FIGURE 12.1   A healthy reindeer is alert, the eyes are bright and the fur is vibrant and clean; body condition score is near 3 (Photo: Sauli Laaksonen).

12.1.2 ANAMNESIS

The disease history, anamnesis, is important in order to achieve diagnosis. It includes information about morbidity, mortality, meat inspection findings, medication and antiparasitic treatment and feeding history, as well as the environment, including pasture conditions.

12.1.2.1 Habitat and Pastures

The condition of reindeer pastures provides information on possible health risks resulting from environmental conditions, toxins or foreign chemical pollutants that may have entered the dietary intake of reindeer. For example, industrial or mining areas may be sources of such risks. Then again, the wear of entire pasture areas and the use of abnormal food plants are indicative of populations that are far too dense. Abnormally long migrations may indicate worn pastures, or, for example, that grazing has not been possible due to human activities. The result can be increased feeding of reindeer on pastures used by domestic animals, which provides an increased opportunity for inter-species transmission of pathogens.

Exceptional weather conditions, heat, precipitation, etc., may predict disease outbreaks or even the appearance of new diseases. A large number of predators can cause increased stress and prevent the normal use of pastures. These situations can lead to weakened physical condition and increased susceptibility to diseases.

12.1.3 BEHAVIOUR

Reindeer are normally relatively sensitive and tend to avoid human contact, and their behaviour is best observed from a distance. Disappearance of shyness towards humans is one of the most common and easily recognizable signs of disease. Aggression may be related to, for example, mating behaviour or calf defence, but it can also be a sign of disease or starvation. A sick reindeer may separate itself from the herd, and often it will be the last to lie down, or it will just remain standing all night (Nikolaevskii, 1961a; Laaksonen, 2016).

12.1.4 POSTURE AND MOVEMENT

Abnormal posture or hanging of the head or ears usually indicate a serious disease (Figure 12.2). Weakness of the hind legs may be associated with many diseases, such as elaphostrongylosis or muscular dystrophy, and injuries. Seriously ill reindeer lie on one side (Figure 12.3), often with the head and neck extended. This state may be associated with eye jitter, teeth grinding, and changes in respiratory frequency. These are often signs of impending death. Head dozing can be a sign of chemical poisoning, such as lead poisoning, or of a central nervous system disease. Itching and scratching, in turn, suggest external or internal parasites or central nervous system diseases.

12.1.5 RESPIRATION

In summer, the respiratory rate of reindeer at rest is about 20–50 times per minute, but only 8–16 times per minute in winter. Stress, such as capture, holding or heat may increase the respiratory frequency to 100–180 times per minute (Nikolaevskii, 1961b; Nieminen, 1994), and abdominal muscles are used actively for breathing. Tachypnea (increased respiration) is a sign of elevated body heat, fever, or stress, but it can also be a sign of a general disease condition or lung disease. Coughing occurs in the context of many respiratory diseases and is best detected in the morning (Nikolaevskii, 1961a; Laaksonen, 2016).

Lameness can suggest pain in the limbs or in the thoracic and abdominal cavity. Unsteadiness, disequilibrium, and circling suggest a most serious disease of the central nervous system or poisoning. Lameness is best detected immediately after the reindeer has got up from the ground (Nikolaevskii, 1961a; Laaksonen, 2016).

FIGURE 12.2 A sick reindeer may separate itself from the herd. A hanging head or hanging ears usually indicate a serious disease (Photo: Sauli Laaksonen).

FIGURE 12.3 The old saying that the snow does not melt from the back of the sick is true. This reindeer is suffering from chronic besnoitiosis ("sand reindeer") (Photo: Sauli Laaksonen).

12.1.6 Structure and Secretions

Swelling of the joints, jaws, udder, the abdomen or the groin is an easily recognizable condition. Swelling may be related to several different inflammatory or general diseases and undernourishment. Drooling or increased lacrimation suggests early symptoms of oral infections (Figure 12.4). Increased salivation can also occur during the change of teeth in young animals. Secretion or leaking from various body orifices is easily visible and may be caused by local or general diseases.

12.1.7 Skin and Fur

Various chronic diseases or dietary deficiencies can cause changes in the shine, vibrancy, and renewal of the coat. Curled and dry hair, often combined with paper-thin skin, is an indication of reduced well-being. Delayed change of winter or adult coat is a sign of illness or poor nutritional conditions during the summer (Figure 12.5). Wet skin and coat may indicate the reindeer wet belly syndrome. Inflammations (Figure 12.6) and ectoparasites cause skin changes. Biting of hair and skin may be caused by *Lappnema auris* or *Besnoitia* parasites or the microfilariae of filarioid parasites in the skin, such as *Setaria tundra*, *Rumenfilaria andersoni*, and *Onchocerca* spp. The skin in summer is highly pigmented during fur change, providing protection from UV radiation (Nikolaevskii, 1961b).

12.1.8 Antlers

Antlers are a good indicator of reindeer health. They give information about the well-being of reindeer during their growth: the nutritional status, diseases and dietary mineral content (about 60% of the antler is mineral), and injuries to vulnerable growing velvet antlers (Nikolaevskii, 1961b; Rehbinder and Nikander, 1999; Laaksonen, 2016) (Figure 12.7). Heavy parasite infestation has been shown to cause antler asymmetry (Markusson and Folstad, 1997). Delayed release of the velvet skin reflects a reduced state of health in the summer – for example, due to ageing, unfavourable weather conditions, parasite load or food shortages (Nikolaevskii, 1961b; Laaksonen, 2016).

FIGURE 12.4   Leaks from the mouth, difficulties in chewing, and feed drooping from the corners of the mouth indicate serious oral infections or dental defects (Photo: Sauli Laaksonen).

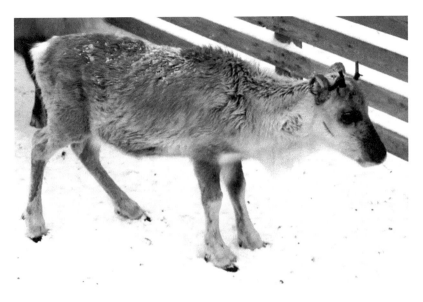

FIGURE 12.5   Poorly developed winter coat and small antlers indicate a chronic disease or deficiency. This calf is in poor condition (Photo: Sauli Laaksonen).

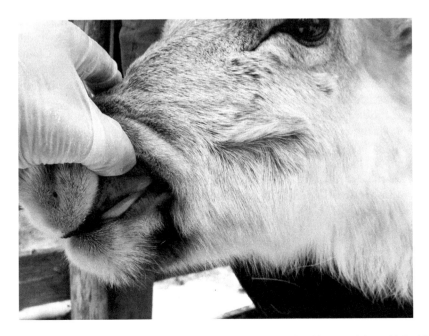

FIGURE 12.6   Paring of the fur on the cheek can be seen when the animal is scratching with its hind leg and often indicates problems in the mouth (Photo: Sauli Laaksonen).

FIGURE 12.7   Healthy antlers are well developed, the bone material is strong and opal-coloured, and the tips of the peaks are strong and sharp. In males, antlers are at their largest at 6–7 years of age. The number of peaks are indicative of the animal's condition (Photo: Sauli Laaksonen).

12.1.9   FUNCTION OF THE DIGESTIVE TRACT

The teeth, tongue, and oral mucous membranes of a sick reindeer must always be inspected. The contraction rate and fullness of rumen are detected by palpating the left abdominal surface. The accumulation of gas buildup at the top of the rumen is a sign of digestive disorder, pain, or stress. The pH of the normal reindeer rumen is about 6.7, and the rumen contraction rate is about 1–2 times per minute (Nieminen, 1994). Decreased appetite may indicate the start of a disease, or inferior feed. The ending of rumination is a sign of illness. Changes in faecal appearance or composition, such as diarrhoea (Figure 12.8), can be signs of poisoning, general diseases, intestinal infections, parasites, or improper nutrition and ruminal disorders (e.g. associated to supplementary feeding). Staining of the hair around the anus is characteristic (Nikolaevskii, 1961a; Laaksonen, 2016). Blood in the stools is a sign of a more serious condition. A sample of the ruminal content can be taken by a veterinarian through the abdominal wall or via a tube through the esophagus. Reindeer often have decreased appetite even if they also have stomatitis or rumen fermentation disorders, but they may be able to eat almost normally during serious infections, e.g. peritonitis or pleuritis. Also, during starvation caused by lack of energy or overly fiber-rich food, reindeer may eat until death (Rehbinder and Nikander, 1999; Laaksonen, 2016).

12.1.10   URINE

In summer, a healthy reindeer's urine is clear and almost colourless. In winter, it is slightly yellowish (Nikolaevskii, 1961b). Physical overload or babesiosis can turn it reddish because of increased myoglobin or haemoglobin caused by the destruction of muscle cells or red blood cells (erythrocytes), respectively (Rehbinder and Nikander, 1999).

FIGURE 12.8   Diarrhoea refers to incorrect feeding, a ruminal disorder, or an intestinal infection (Photo: Sauli Laaksonen).

12.1.11   BODY CONDITION CLASSIFICATION (SCORING)

Body condition classification is a simple way to assess the health and well-being of reindeer. In this method, the thickness of the subcutaneous adipose tissue and muscle mass is estimated by visual inspection and sometimes by palpation. For the inspection, the back (fillet) and pelvic areas are suitable (Figure 12.9). The method can be used to assess the state of individual reindeer or entire herds. The method is particularly suitable for assessing the reindeers' winter condition, to detect overfeeding or the possible need for emergency feeding (Laaksonen and Nieminen, 2005).

Category 1, very thin, emaciated (Figure 12.10)

Spinal column and the gaps between vertebrae are easily detected. The muscles are withered, the skin attached on the bones. The edge of the transverse process is sharp. Pelvic bones are clearly displayed, with a deep hollow between the spinous processes. The tuber coxae is evident, with deep gaps between the tuber coxae and spinous processes.

Category 2, thin (Figure 12.11)

The spinal column and vertebrae gaps are still palpable. There is a hollow between the spinous processes and on the pelvic bones. The spinous processes and gaps between vertebrae are still palpable. There is an evident gap between spinous and transverse processes. The edges of transverse processes are evident. The gaps between tuber coxae and spinous processes are indented.

Category 3, normal (Figure 12.12)

The spinal column is not, or is hardly, palpable because of fat and muscle tissue. Muscles on the back and pelvis are flat or only slightly sunken. The spinous processes are not, or are slightly, palpable because of fat and muscle tissue cover. The gap between the spinous and transverse processes is even or only mildly sunken. The edge of the transverse processes is only a minor shelf. The gaps between the tuber coxae and spinous processes are mellow.

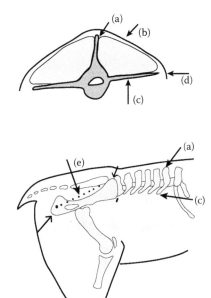

FIGURE 12.9   In body condition scoring, thickness of the fatty tissue and muscle mass over the back (b) and pelvic area (e) is assessed. (a) Spinous process, (b) back muscle area, (c) transverse process, (d) edge of transverse process, (e) pelvic muscle area (Illustration: Sauli Laaksonen).

FIGURE 12.10   Category 1, very thin and emaciated reindeer (Illustration: Sauli Laaksonen).

FIGURE 12.11   Category 2, thin reindeer (Illustration: Sauli Laaksonen).

FIGURE 12.12   Category 3, normal reindeer (Illustration: Sauli Laaksonen).

FIGURE 12.13   Category 4, fat reindeer (Illustration: Sauli Laaksonen).

Category 4, fat (Figure 12.13)

The spinal column is not palpable under the fat tissue; the back line is flat. The line between the spinous processes is convex. A midline groove on the back constituted by back muscles is evident. All bone edges are rounded. The spinous processes are hardly palpable under the fatty tissue. The gaps between the spinous and transverse processes are convex; there is a groove on the backbone (spine) with banks on both sides formed by muscles. Transverse processes form a smooth edge. The gaps between the tuber coxae and spinous processes are rounded (Laaksonen and Nieminen, 2005).

12.1.12   BODY TEMPERATURE

Reindeer body temperature varies greatly with age and physiological state. Capturing and holding an animal represents sources of stress, followed by a rise in body temperature, especially if the animal is not accustomed to handling. Normal rectal body temperature in adults is 38–40°C, and in calves, 39–41°C, with slightly lower values in winter and higher values in summer (Nikolaevskii, 1961a). Body temperature is measured from the rectum. A lubricant such as soap is used, and the thermometer is inserted gently into the anus to a depth of around 5 cm, and turned slightly sideways so that the thermometer touches the wall of the intestine.

12.1.13   MUCOUS MEMBRANES

The mucous membranes of the eyes, the mouth and the vagina are inspected. In healthy reindeer, they display a pure pink colour. Dark red or cyanotic mucous membranes (Figure 12.14) may indicate general inflammation and lack of oxygenation. Pale or yellowish mucosa suggest chronic diseases (Nikolaevskii, 1961a; Laaksonen, 2016).

12.1.14   BLOOD SAMPLING AND BLOOD VALUES

In reindeer, blood samples are taken from the jugular vein, at about one-third of the length of the neck downwards from the chin (Figure 12.15). Blood parameters can be used to evaluate

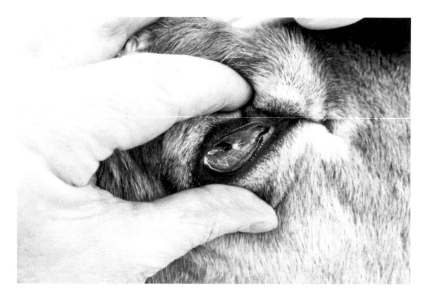

FIGURE 12.14   Redness of mucous membranes is caused by increased blood circulation and is usually a sign of an inflammatory condition (Photo: Sauli Laaksonen).

FIGURE 12.15   In reindeer, blood samples are usually taken from the jugular vein, at about one-third of the length of the neck downwards from the chin (Photo: Sauli Laaksonen).

the health condition of the reindeer (e.g. haematology and blood biochemistry), as well as the presence of antibodies against infectious agents (i.e. serology) and to detect the presence of pathogens in the blood.

Further details on haematology and serum chemistry, as well as reference values for some subspecies, are dealt with in Chapter 13, Haematology and Blood Biochemistry Reference Values for *Rangifer*.

12.1.15   AGE ASSESSMENT

A trained eye can tell the basic age class of a reindeer from the teeth present, as well as their appearance and wear. Females reach their full body measurements at about 5 and males at 6 years of age, followed by the phenomena associated with aging (Nikolaevskii, 1961b; Nieminen and Petersson, 1990; Nieminen and Helle, 1980; Laaksonen, 2016). The antlers of the calves are spikes, but two- or three-pointed antlers also occur. With age, the number of points will increase. Antlers are at their largest in males at 6–7 years of age, and in females at 4–5 years of age. After that the antler size begins to decline (Nieminen, 1994).

Changes in the hair coat also occur. In an aging reindeer, the long hair shortens and the hair on different body parts has equal length, the colour becomes lighter and it may begin to look yellow, and the hair on the crown becomes curly (Nikolaevskii, 1961b; Laaksonen, 2016).

Tooth change is the most reliable method of age evaluation in young animals. The deciduous incisors of a calf are usually replaced by permanent teeth after the first year of life (from 10 to 16 months). The first permanent molars erupt at 4 to 5 months of age, the second at 12 months, and the third at 25–30 months of age. Premolars will be replaced by permanent teeth at the age of about 20 months.

Reindeer teeth do not grow in length during life, but wear does cause changes. The sharp chisel-shaped incisors wear out, and their root canal becomes visible with age. The wear and tear in the medial incisors begin as early as at 3 years of age, in the second at 4, and in the third incisors at 5 years of age. Wear in the canine teeth can be detected at the age of 6 years, and from then on, the decline will continue until there are only small nodules left at the age of 12–14 years (Figure 12.16) (Nikolaevskii, 1961b; Nieminen, 1994, Laaksonen, 2016).

As the premolars and molars wear off, the sharp edges composed of enamel become shallower and the teeth become aligned (Figures 12.17 and 12.18). The changes caused by wear are affected by the quality of food and regional differences. Tooth cement accumulates at the root of the tooth in accordance with the seasons; in winter the cement is dense and in summer it is thinner. In a microscopic examination of a cross section of a root, usually conducted on the root of the first incisor, the zones are distinguishable as dark and light zones (Nieminen, 1994; Laaksonen, 2016).

12.1.16   VIABILITY ASSESSMENT OF AGING FEMALE REINDEER

It is important to recognize when it is time to remove aging female reindeer from breeding. The external distinguishing features of declined vitality are the reduced size of the animal, the lightening of the hair colour, and the rounding of the peaks of the antlers. The removal of velvet skin is delayed in autumn, and shreds of velvet skin often remain. The fur becomes lighter and curls at the sides and at the forehead and the vibrancy of the hair disappears (Figure 12.19).

The inspection of teeth is relevant: The incisors are the easiest to examine. The middle front teeth wear out first and an estimation of the animal's ability to utilize nutrition can be done from their wear. "When the middle incisors are at the level of gums, it is time to remove the reindeer from breeding," is an often-used evaluation. It must be remembered, however, that the incisors may still be in good condition, but the molars can be seriously damaged, or vice versa. In such case, other signs of vitality must be observed (Nikolaevskii, 1961a; Laaksonen, 2016).

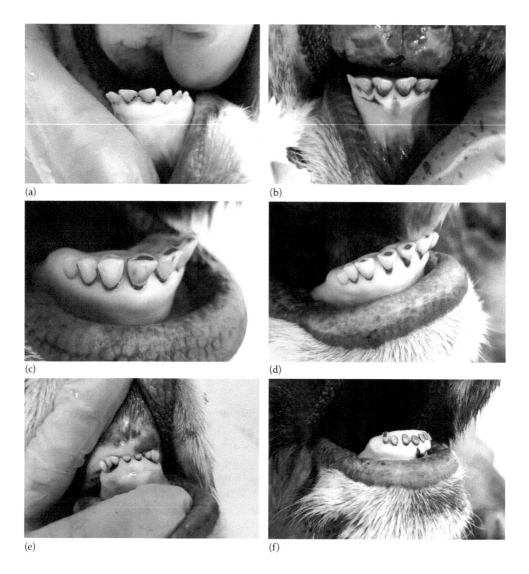

(a)   (b)   (c)   (d)   (e)   (f)

FIGURE 12.16   The incisors begin to wear already after the third year of life. The picture shows the wear of teeth at different ages: (a) 6 months; (b) 12 months; (c) 5 years; (d) 8 years; (e) and (f) 13 years (Photos: Sauli Laaksonen).

FIGURE 12.17   At the age of 18 months, the enamel edges of the molars are high (Photo: Sauli Laaksonen).

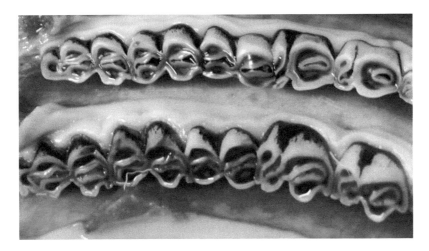

FIGURE 12.18   With age, the enamel edges become shallower and the teeth become aligned (Photo: Sauli Laaksonen).

FIGURE 12.19   The external distinguishing features of declined vitality are the reduced size of the animal, the lightening of the hair colour, and the rounding of the peaks of the antlers. The fur shortens, becomes lighter, and curls at the sides and at the forehead, and the vibrancy of the hair disappears (Photo: Sauli Laaksonen).

12.2   DISEASES OF REINDEER

12.2.1   STRESS

Understanding the importance of stress is the key to understanding the diseases of reindeer. It is the most common and most important cause of reindeer illness. Reindeer adapt quite easily to new conditions, but they are not tame, and therefore they are sensitive to stress (Rehbinder et al., 1982; Rehbinder, 1990a; Wiklund, 1996; Rehbinder and Nikander, 1999; Stubsjøen and Moe, 2014; Laaksonen, 2016).

Many infectious agents that are normally harmless or cause disease to a small extent in reindeer may cause disease and even disease outbreaks during stressful conditions. The Arctic habitat is increasingly changing. Climate change and the changes resulting from it in the natural habitat of the reindeer, motorized reindeer herding, the competing of agriculture and forestry, and the fragmentation of pastures, toxic substances, overdense populations, contact with farm animals, and supplementary feeding may be mentally or physically stressful factors for reindeer physiology. High infection pressure, often combined with a weakened immune system due to stress, may enable the pathogen to overcome the body's defence mechanisms, and a disease breaks out (Wiklund, 1996; Rehbinder and Nikander, 1999; Laaksonen, 2016).

Stress is caused by various risk factors such as threat, harassment, fear, tension and the presence of predators and humans. Significant stress factors are the gatherings, round-up and handling of reindeer and especially capturing by lasso (Figure 12.20) (Rehbinder, 1990a; Wiklund, 1996; Bengtson, 2004; Stubsjøen and Moe, 2014). Conditions in winter enclosures, especially in animal-dense conditions, and improper feeding may cause stress. Long-term stress is caused by exceptional climatic conditions such as heat, cold and wetness, as well as the harassment caused by an abundance of blood-sucking insects. Lack of food, wear or fragmentation of pastures, and overdense populations also cause long-term stress. Calves are more sensitive to stress, and separation from their mothers may cause severe stress. Reindeer which are used to handling are less sensitive to stress than untamed reindeer (Rehbinder, 1990a; Wiklund, 1996). Weakened reindeer become stressed more easily than those in good condition. Stress is characterized by its ability to accumulate (Rehbinder, 1990a). This has been seen in pet reindeer that have been moved away from their natural habitats to township territories. They have been found to suffer from a prolonged stress-like symptom that will often eventually result in indisposition and wasting (Laaksonen, 2016). See also Chapter 2: *Rangifer* Health – A Holistic Perspective.

Stress also causes systemic changes in reindeer. These changes are linked to the duration of stress. The concentration of urea in the urine will increase. As little as 4 hours of strenuous chasing may cause abomasal ulcers (Rehbinder and Nikander, 1999). Degenerative changes in musculature and in the heart muscle generally occur. Stress experienced by reindeer before slaughter decreases the quality of the meat (Rehbinder, 1990a; Wiklund, 1996; Laaksonen et al., 2017a).

Good nutritional status, an interference-free environment, and calm reindeer handling will prevent the onset of stress. In supplementary feeding and corralling, good principles of hygiene are followed. In all situations, the handling of reindeer must be calm, and no unnecessary people should be present. For instance, during treatment, reindeer are held only for the time required for

FIGURE 12.20   Manual handling and restraint have been found to be one of the major stress factors for reindeer (Photo: Sauli Laaksonen).

giving treatment. Reindeer are not to be dragged by the antlers or limbs. A cover over the eyes, e.g. a towel or a t-shirt, for the period of treatment can help to soothe sensitive animals. Repeated treatment should be avoided (Wiklund, 1996; Rehbinder and Nikander, 1999; Laaksonen, 2016). When gatherings and round-ups are planned, the focus should be on the well-being of the reindeer. During gathering, the reindeer should be herded and directed by taking advantage of their natural herd instinct. Motorized gathering facilitates the work of reindeer herders, but it may be harmful to the reindeer if their handling time is extended (Wiklund, 1996; Rehbinder and Nikander, 1999; Laaksonen, 2016). If a reindeer or the entire herd is suffering from stress, they should be provided an adequate period of rest, with fresh water and food provided. If a reindeer has become overheated during handling, it may be cooled by pouring water over it (Rehbinder, 1990a; Laaksonen, 2016).

Intensive herding with shepherd dogs can also cause stress. Cold, heat and snow cover, as well as round-ups held very late in the autumn, will increase the effects of stress. In corrals, manual handling, forcing or the use of a lasso or vimba (i.e. a long stick with a lasso at the end) should be avoided as much as possible. The fences and loading passages to vehicles must be designed in such a way that passing through them falls naturally to the reindeer, and that the reindeer have to be forced as little as possible. Slaughter should be planned in such a way that transfer distances are kept as short as possible and the transportation vehicles must be designed for the purpose (Rehbinder, 1990a; Wiklund, 1996; Rehbinder and Nikander, 1999; Stubsjøen and Moe, 2014; Laaksonen, 2016; Laaksonen et al., 2017a). Round-ups and transport of reindeer must be avoided in very cold weather, such as in temperatures of –30°C or below, or when the temperature is higher than +18°C (Laaksonen, 2016).

12.2.2 Diseases of the Head and Eyes

12.2.2.1 Teeth

12.2.2.1.1 Damage and Wear

The inspection of teeth, mouth and tongue is included in the basic clinical examination of a sick reindeer (Nikolaevskii, 1961a; Laaksonen, 2016).

Wear of and serious damage to teeth are common causes of inviability in reindeer. The first maxillary molars appear to be particularly sensitive to wear and damage (Figure 12.21), and the absence of one molar alone may cause serious problems. The factors that have a significant effect on the wear of the teeth are dietary composition and soil quality, and the condition of the winter pastures. For example, teeth wear out more quickly in reindeer that feed in sandy areas. Reindeer have adapted to soft plant diets, and coarse hay or pellets may cause teeth damage (Nieminen, 1994; Laaksonen, 2016). Sometimes in young reindeer, the changing milk molar tooth is retained, and left to ride on the permanent tooth due to the compression of two adjacent teeth.

A reindeer with dental defects has difficulty in chewing, and increased salivation occurs. Feed may accumulate inside the cheeks and cause swelling, sometimes on both sides (Figure 12.22). Bacterial infections may strike the damaged mucosa. Heavily worn teeth will gradually lead to a decline in the reindeer's general condition (Laaksonen, 2016).

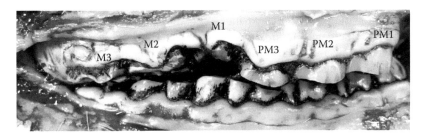

FIGURE 12.21 The first maxillary molars appear to be particularly sensitive to wear and damage (Photo: Sauli Laaksonen).

FIGURE 12.22   Poor teeth cause feed to accumulate inside the cheeks and cause swelling (Photo: Sauli Laaksonen).

12.2.2.2   Mouth and Tongue Diseases

Coarse fodder and wooden, worn-out feeding containers can cause damage to the tongue and oral mucosa (Figure 12.23). Metal containers in freezing temperatures can damage the tongue. Pox and herpes viruses are common causes of infections in the oral region, producing mucosal lesions, which often are prone to secondary bacterial infections, e.g. *Fusobacterium necrophorum* and the disease necrobacillosis (see Chapters 7 and 8 on bacterial and viral infections and diseases).

The throat bot (*Cephenemyia trompe*) and the tongue worm (*Linguatula arctica*) may cause irritation and inflammation of the nasal cavities. A condition called wood tongue (actinobacillosis), caused by bacteria of the genus *Actinobacillus*, is characterized by the growth of hard connective tissue in the tongue or jaw and is the consequence of a deep puncture wound, e.g. a stick jab, in the mouth or tongue (Rehbinder and Nikander, 1999; Laaksonen, 2016). The first symptoms are difficulty in eating and chewing, accumulation of feed inside the cheeks, and increased salivation and purulent discharge from the mouth and nostrils are common. Necrosis and abscesses, leading to loss of appetite and mortalities, occur in advanced cases.

Bacterial infections can be treated with antibiotics. Abscesses are incised for pus removal and rinsed by antiseptic solutions. If treatment is started right after the appearance of the first symptoms, it is often effective. Supportive therapy is important to a poorly eating reindeer. As a prophylactic

FIGURE 12.23 A rough or icy feed may cause damage to mucous membranes and expose them to *Fusobacterium necrophorum* infections (necrobacillosis) (Photo: Sauli Laaksonen).

measure against infectious diseases, attention must be focused on feeding hygiene. If the cause of the disease is contagious, prompt isolation of the sick animal is important to avoid transmission.

12.2.2.3 Eye Diseases

12.2.2.3.1 Occurrence

Eye disorders occur very commonly in reindeer for various reasons (Figures 12.24 through 12.27). Eye damage may be caused by a stab by an antler or a stick, a throat bot fly larva mistakenly shot into the eye, particles of feed and vegetation, or dust. Blackflies may bite the eyelids until they bleed, and clear the way for inflammation and infections, often by staphylococcal and streptococcal bacteria (Nikolaevskii, 1961c; Rehbinder and Nikander, 1999). *Lappnema auris* and *Besnoitia* parasite can sometimes cause changes, hairlessness and inflammation of the eyelids (Laaksonen, 2016). Strong ultraviolet radiation may cause damage to the eye and expose it to infection (Rehbinder and Nikander, 1999; Laaksonen, 2016).

Significant outbreaks of contagious eye infections have occurred in history, and smaller outbreaks still occur every year. The infections are most common in calves, and they often start after round-ups (Nikolaevskii, 1961a; Rehbinder and Nikander, 1999; Laaksonen, 2016).

The most significant contagious eye infections in reindeer are caused by the reindeer herpes virus (Cervid herpesvirus 2, CvHV2). Contagious eye infections can also be caused by *Listeria monocytogenes* (Figure 12.24), *Moraxella*, and *Pasteurella* bacteria (Tryland et al., 2009; Laaksonen, 2016;

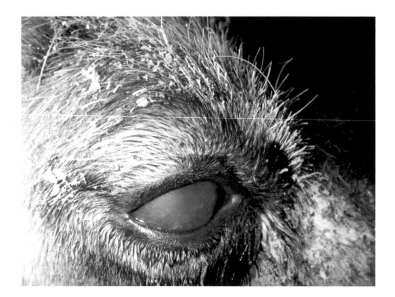

FIGURE 12.24   Eye infections occur very commonly in reindeer, often appearing as disease outbreaks. This reindeer had keratoconjunctivitis, from which the bacterium *Listeria monocytogenes* was isolated (Photo: Sauli Laaksonen).

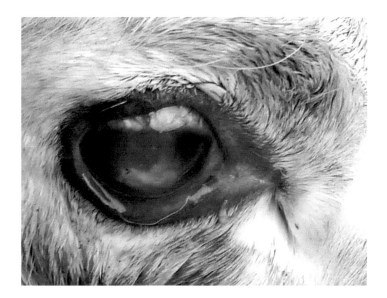

FIGURE 12.25   Acute corneal injury can be caused by external trauma or the larvae of warble fly. Local suitable antibiotic treatment is recommended against bacterial infections. Left untreated, such conditions leads to blindness (Photo: Sauli Laaksonen).

Romano et al., 2018; Tryland et al., 2017). Infectious agents can spread via close contact, dust, or insects, and also *Listeria* can spread via poor-quality silage.

The early symptoms of contagious eye infections are irritable eyes and lacrimation. Conjunctival inflammation occurs as congestion of blood and oedema. The cornea turns opaque (grey or bluish) due to oedema and loses its ability to transmit light, and the reindeer may become temporarily blind. If the inflammation is severe or pierces the cornea, the result is inflammation of the whole eye, and, without treatment, ultimately blindness (Rehbinder and Nikander, 1999; Laaksonen, 2016; Tryland et al., 2017).

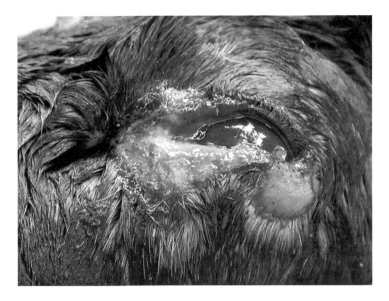

FIGURE 12.26   Blepharitis caused by unknown agent (Photo: Sauli Laaksonen).

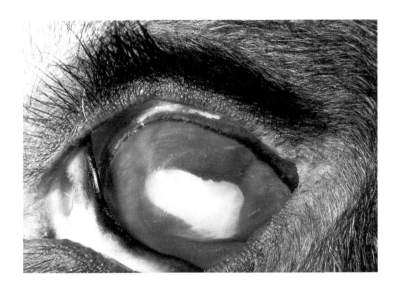

FIGURE 12.27   Cataracts occur randomly in reindeer, the reason being unknown (Photo: Sauli Laaksonen).

Mild eye infections usually heal spontaneously. If infection is treated in the early stages, before the cornea has burst or the infection has spread, the prognosis is good. Antibiotic treatment after cleaning the eye with sterile salt water and, in mild cases, antiseptic eye rinse can be used (Figure 12.28). In severe cases, the removal of the entire eye has been used as a successful treatment for breeding reindeer (Laaksonen, 2016).

In preventing summertime eye infections of calves, the avoidance of stress is essential – round-ups should not be held in hot weather or in dusty places, and drinking water should always be available. Under winter conditions, feeding hygiene is important: the quality of the feed is good, there are many feeding points, and compartmentation is used to prevent stress caused by excessive contacts and injuries (Rehbinder and Nikander, 1999; Laaksonen, 2016).

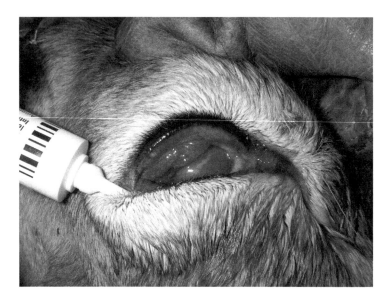

FIGURE 12.28   Eye infections can be treated locally by applying a suitable antibiotic preparation. If the tube is in contact with the eye, as shown, it should not be used on another animal (or should be sterilized first) (Photo: Sauli Laaksonen).

12.2.2.4   Ear Diseases

Ear diseases are rare in reindeer, but they may occur – for example, as a result of unsuccessfully performed ear marking with knife. Poor circulation caused by overly extensive ear notches during ear marking may cause the rest of the ear to become frostbitten, leading to necrosis. The notch wound can sometimes become infected. In very cold temperatures, bleeding may be profuse (Figure 12.29). Eartags may cause tearing in the earlobe. The ear mite (*Chorioptes* sp.) is common in reindeer, but it rarely causes symptoms.

Lappnema auris may manifest itself as a thickened and hairless earlobe, a so-called hot ear (Figure 12.30). The *Besnoitia* parasite may also cause thickening and hairlessness of the earlobe. Hanging ears and pus are signs of an ear infection. If the inflammation progresses into the inner ear, balance disorders may occur.

Inflammations are treated with local and general antibiotic preparations. The antiparasitic drug ivermectin is effective against hot ear and the ear mite. Ear notches should not be made in very cold temperatures. The notching knife and the handler's hands should be kept clean during ear marking (Laaksonen, 2016).

12.2.2.5   Antler Diseases

12.2.2.5.1   Antler Accidents

Growing velvet antlers are very susceptible to injury.

Dragging reindeer by the antlers, or seizing a running reindeer too vigorously by the antlers with hands or a lasso, can result in breaking the antlers, and in some cases, in fracturing the skull or the pedicles (Figure 12.31). Antler accidents can happen during power struggles between adult reindeer, and sometimes the antlers can get locked together, resulting in the death of both parties (Rehbinder and Nikander, 1999; Laaksonen, 2016). *Besnoitia tarandi* infection has been associated with broken antlers and velvet retention (Rehbinder et al., 1981).

A broken growing antler is usually left hanging by the antler skin. Bleeding can be severe, and the condition is painful to the reindeer because the growing antler is highly vascularized and

FIGURE 12.29 If ear marking notches are made in very cold temperatures, bleeding may be profuse (Photo: Sauli Laaksonen).

FIGURE 12.30 "Hot ear" may be caused by *Lappnema auris* parasites or the microfilaria of *Onchocerca* nematodes (Photo: Sauli Laaksonen).

innervated. The antler may also ossify in a wrong position and later, as it grows, apply pressure to the skull or the eye.

The fracture of an ossified antler does not require treatment. A broken growing antler must be cut below the fracture site. The operation demands local anaesthesia. To prevent excessive bleeding, a compressive tourniquet is placed below the fracture site. If the growing site of the antler (pedicle) is broken, the antler can be removed and the skin sutured over the site for protection. In future, the new antler will be a spike of varying size. If the skull is fractured, the only treatment option is euthanasia (Nikolaevskii, 1961a; Rehbinder and Nikander, 1999; Laaksonen, 2016).

FIGURE 12.31   Rough handling can cause a broken or partly broken antler or even a skull fracture. Such conditions require veterinary treatment (Photo: Sauli Laaksonen).

In summer, the handling of reindeer must be extremely gentle. During autumn round-ups the facilities must be planned in such a way that there is no need for the lasso or vimba, or for dragging the animals by the antlers. Netted materials should not be used in enclosures where reindeer are handled (Rehbinder and Nikander, 1999; Laaksonen, 2016).

12.2.2.5.2   Antler Freeze-Up

In unfavourably icy conditions, snow and ice may accumulate on the antlers. Snow or ice lumps can become so big that they prevent the reindeer from eating. This may lead to the starvation of the animal. Ice lumps at the base of the antlers can cause the brain to freeze (Nikolaevskii, 1961c).

12.2.2.5.3   Cactus or "Perruque" Antlers

Cactus antlers is a phenomenon of uncontrollable horn tissue growth to a cactus-shaped mass, and the antler will not be shed. This phenomenon is sometimes called "frostbite" by herders. The cause of cactus antlers in cervid species can be a disorder in the production of testosterone, inflammation or contamination by environmental toxins and chemicals which damage the testicles and subsequently reduces the production of testosterone or estrogen-like compounds which interfere with the normal development of testicles (Fox et al., 2015). In reindeer, the condition occurs most commonly in castrated males (Laaksonen, 2016). The growing tissue mass may press on the eye or the skull, and the structure is susceptible to inflammation. Flies often lay eggs in the folds of the antler mass or between the antler and skin.

12.2.2.6   Removal of Antlers

The removal of an antler may be carried out if it has particularly dangerous spikes, or if the animal is aggressive, to avoid damage to other reindeer or the reindeer handlers, such as during transportation of reindeer in motor vehicles (Rehbinder and Nikander, 1999; Laaksonen, 2016). Antlers can also be removed in case of trauma or due to cactus antlers.

Antlers can be painlessly removed after the velvet period by sawing them about 5 cm above the pedicle (Figure 12.32). The antler actively grows for about 3 months, after which it becomes ossified, and its innervation and vasculature disappear. This usually happens by mid-September at the

FIGURE 12.32   The antlers can be painlessly cut during bone antler period. When handling a strong male, anaesthesia and analgesia is recommended (and in some countries is mandatory by law) to relieve pain and avoid the development of captive myopathy (Photo: Sauli Laaksonen).

latest, but sometimes later in fall for calves and castrated males. The circulatory activity in the antler can be deduced by feeling its surface: an active antler does not freeze in cold weather and snow melts on it. The removal of an antler in growth or a cactus antler has to be done by a veterinarian. It requires analgesia, local anaesthesia and suppression of bleeding (Rehbinder and Nikander, 1999; Laaksonen, 2016).

12.2.3   Skin and Fur Diseases

Skin and coat disorders occur most frequently in autumn and winter in corralled animals. Normal annual moulting begins in May–July, and by November the winter coat is again fully formed (Nieminen, 1994). Parasites are the most common causes of changes in reindeer skin and coat. In addition, poor nutritional status, lack of protein and chronic diseases may cause disorders of the coat and the skin. The *Besnoitia* parasite may manifest as skin changes causing the so-called "hot leg" syndrome (Rehbinder et al., 1981b; Laaksonen, 2016). Some fungi, such as the ringworm and the *Candida* species, may cause skin inflammations, ring-shaped areas of flaking skin that spread at the perimeters, under unhygienic conditions, especially in fenced animals. Accidents, old barbed wire, and metal objects as well as predators may cause cuts, abrasions and bruises to the skin. Sometimes a maggot fly lays its eggs in such lesions and the larvae will develop. The papilloma virus sometimes causes proliferative formations on the reindeer skin (Rehbinder and Nikander, 1999; Laaksonen, 2016).

In spring, the warble fly larvae appear as swellings on the skin. Chewing lice (*Bovicola tarandi*) cause hairlessness and itching, mainly in late winter. The deer ked (*Lipoptena cervi*) begins to bother reindeer in the early autumn. Even a small number of deer keds make the reindeer itch and shake, causing scratching with antlers or rear legs, leading to hair loss and skin irritation (Figure 12.33) and sometimes secondary bacterial infections (Kynkäänniemi et al., 2014; Laaksonen, 2016).

FIGURE 12.33   Deer ked is the most common cause of skin and coat changes in the southern reindeer herding area in Finland. Symptoms (itching and hair loss caused by scratching) appear in early autumn (Photo: Sauli Laaksonen).

Antiparasitic treatments are effective for most of the above-mentioned ectoparasites. There is no effective treatment against the *Besnoitia* parasite. Small wounds can be cleaned and in most cases they will heal spontaneously. Larger wounds can be stitched up by a veterinarian. Bite wounds will often require antibiotic treatment, and also drainage into the wound for pus removal (Nikolaevskii, 1961a; Laaksonen, 2016).

12.2.4   SPINE, BONE AND JOINT DISEASES

The most common diseases occurring in the spine and bones of reindeer start with a trauma. Fractures, dislocations and tears may occur in the neck and spine, chest and limbs and sometimes also in the skull due to major damage to the antlers.

Accidents result from collisions with motor vehicles, or from the combination of reindeer harassment and rocky or drained terrain or from wire fences. Accidents may also happen during reindeer handling, transportation and loading. Unskilled capturing of reindeer with lassos or vimbas can cause damage to the cervical spine or injuries to extremities. Fences made of steel mesh in reindeer handling corrals often cause severe cervical spine injuries. The poor physical condition of reindeer in early spring makes them extremely susceptible to injuries. Wounds or general infections may cause arthritis, and in some countries, Brucellosis is a specific cause of arthritis (Laaksonen, 2016).

In case of spine injury, the rear end of a reindeer is partially or completely paralyzed. In less severe cervical lesions, the reindeer's neck is stiff and it will not be able to lift its head. The walk can be shaky and wobbly. The condition is similar to symptoms caused by the meningeal worm (*Elaphostrongylus rangiferi*). Ligament injuries or joint dislocations can also cause lameness in varying degrees, depending on the location and severity of the injury.

The ability to move well is vital for reindeer. Thus, if a reindeer is severely lame, euthanasia is the best option. Mild spine lesions may heal spontaneously, or with support treatment combined with analgesic drugs (e.g. NSAIDs). In case of more serious lesions, such as a limb fracture, the only option is to euthanize the reindeer. Fractures in the limbs of calves have been treated with support bandages and splints, but this requires special arrangements and close follow-up by the owner. Rib fractures are common, and they will heal spontaneously. It is possible to correct hip and elbow dislocation if the injury is very recent, but the prognosis is poor. Bacterial arthritis is treated with general antibiotics and analgesic drugs. In the case of single toe fracture, hoof amputation is possible (Nikolaevskii, 1961a; Laaksonen, 2016).

12.2.5 Hoof Diseases

Hoof injuries caused by the environment are common. They can lead to dermatitis in the injured area and the accumulation of pus under the sole. The best-known cause of reindeer hoof inflammation is necrobacillosis caused by the bacterium *Fusobacterium necrophorum*. The historical, Sami name of the disease is "slubbo", indicating the club-shaped hooves and tarsal/carpal joints (Figure 12.34). Slubbo used to appear as devastatingly large-scale epizootics, but this is not common anymore. The virus infection causing foot-and-mouth disease is also characterized by blisters and lesions in the hoof.

Unfavourable environmental and weather conditions, such as rocky or muddy terrain, or cold, wet, dry or hot weather, predispose the hoof to wear and inflammatory diseases. Drought during the summer will harden the otherwise soft pads and expose them to damage (Nikolaevskii, 1961c). One of the first symptoms is swelling just above the hoof. Inflammation in all hooves leads to immobility and starvation. If the hoof wall comes loose, the condition is extremely painful to the animal.

Treatment with antibiotics is efficient in the early stage of hoof infection, but if it has progressed deep into the bones and tissues, it is a justification for euthanasia. In case of a sole abscess, the abscess is opened and treated with local antiseptic substances. No treatment exists for the loosening of the hoof wall. In this case, the reindeer must be euthanized (Nikolaevskii, 1961a; Laaksonen, 2016). Good hygiene in feeding and farming, as well as the quarantine of diseased animals, are important measures for prevention. The ground of corrals and round-up fences must be dry and absorbent enough, and the area large enough, so that it will not be saturated with faeces and urine.

FIGURE 12.34 Digital necrobacillosis, or "slubbo", can cause devastatingly large-scale epizootics (Photo: Herdis Gaup Aamot).

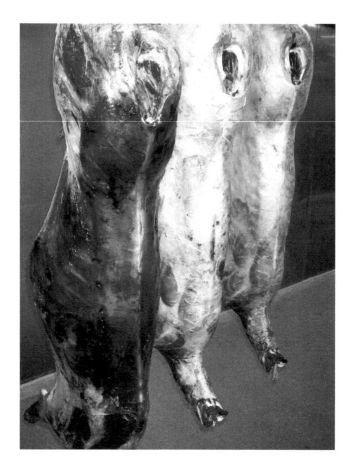

FIGURE 12.35 Contusions and bruises in tissues are often caused by social interactions, handling, or transport of reindeer (Photo: Sauli Laaksonen).

12.2.6 MUSCLE AND TENDON DISEASES

Muscle and tendon diseases are generally linked to trauma and bacterial and parasitic infections. External injuries cause contusions and haemorrhage in tissues, and stress may cause degenerative myopathia (Figure 12.35) (Rehbinder and Nikander, 1999). Such injuries are often caused by social interactions, handling, or transport of reindeer. Bacterial inflammations, especially necrobacillosis, may cause muscular abscesses. The meningeal worm (*E. rangiferi*) lives on the surfaces of muscle membranes, the legworm (*Onchocerca* spp.) on the surfaces of tissues and tendons. Toxoplasmosis and sarcocystosis cause muscle cysts, as do also the infectious larvae of muscle cysticerci (*T. krabbei*).

Selenium deficiency likely causes dystrophia (Rehbinder and Nikander, 1999; Laaksonen, 2016). Muscles turn pale and powerless, and the reindeer shows various symptoms of paralysis and weakness. Lichen is very rich in selenium, and it is unlikely that reindeer living on natural feed would suffer from selenium deficiency. However, it is possible that corralling or heavy additional feeding may cause selenium deficiency (Rehbinder and Nikander, 1999; Flueck et al., 2012; Vikøren et al., 2011; Laaksonen, 2016).

12.2.6.1 Paralysis and Myodegeneration

Many conditions cause symptoms of paralysis in reindeer. The best-known cause of paralysis is the meningeal worm. However, it is likely not the most common cause. All general illnesses that lead to the weakening of a reindeer's condition may lead to a paralysis-type condition, when the animal is weak and powerless or unable to get up. Traumas, especially those that affect the spine or the cervical spine, may cause paralysis, and so do fractures or violent contusions in muscles. In case of stress or great exertion, the reindeer may develop a syndrome much like exertional rhabdomyolysis, where muscles turn pale or dark. The external symptoms of overheating are similar to paralysis.

Muscular dystrophy cause changes in single muscles and muscle groups, and pale or dark areas in muscle tissue are seen. Exertional rhabdomyolysis manifests as weakness of the rear end, and a stooped position or inability to get up. The symptoms appear quickly, and the animal's urine is dark due to degradation of muscle cells. The muscles are soft and bloody. The treatment of paralysis symptoms is possible only after a diagnosis has been made. Selenium and vitamin B products are used for muscular weakness. Measures of supportive care, rubbing, turning and well-insulated and soft bedding are important factors in the treatment.

12.2.6.2 DFD Meat

After slaughter, most of the glycogen in the reindeer muscles is broken down into lactic acid, which causes a drop of pH from about 7.2 to about 5.5 in slightly more than 24 hours. If the animal has had fever, has a generalized infection or a chronic illness or the reindeer has been excessively strained during gathering, corralling, transport or slaughtering, the glycogen level in its muscles is lower than that of a healthy or unstressed animal, and the final pH of the meat will be higher and the meat will become dark, firm and dry (DFD). Late slaughter during the winter months sets the scene for DFD meat, as reindeer have already started to use up their energy reserves, and slaughtering should be completed in December. The formation of DFD will greatly decrease the value of the meat, affecting bacterial growth, water-holding capacity, colour and flavour. DFD meat cannot be hung for maturing. It has to be processed immediately, and preferably used for products that will be heated. If the reindeer meat is even mildly DFD, it is unsuitable for cold-smoked products (Petäjä, 1982; Rehbinder, 1990a; Wiklund et al., 1995; Wiklund, 1996). See also Chapter 11, Meat Quality and Meat Hygiene.

12.2.7 Respiratory Diseases

Respiratory diseases, some of which can be extremely acute and serious, often occur in reindeer.

There are several external factors that predispose reindeer to respiratory infections: unfavourable weather, cold or heat, combined with exertion. Galloping in cold weather can make the lungs to cool too much and cause damage. On the other hand, hot weather can dry the lungs. Sometimes dry feed can be inhaled and cause inflammation or aspiration pneumonia. This can also occur easily when medical treatment is administered orally. A heavy lungworm or meningeal worm infection may damage the lung tissue and expose the tissues to secondary bacterial infections. Throat bot larvae may cause irritation in the upper respiratory tract, or even end up in the lungs. The tongue worm (*Linguatula arctica*) can irritate the nasal cavity. The bacterium *Pasteurella multocida* causes severe pneumonia in reindeer, and sometimes also outbreaks with mass mortality, especially in calves, so that the disease has sometimes been called reindeer plague. Disease outbreaks and deaths occur especially in hot and humid summers, or when other factors lower the resistance of reindeer, e.g. after the summer ear marking of calves. *Mycoplasma* spp. may also cause disease and even mass deaths in reindeer (Stoffregen, 2006).

Fusobacterium necrophorum infections, manifested as oral necrobacillosis, may become generalized and spread to the lungs or pleurae, and be fatal (Figure 12.36). Sometimes cysts caused by *Echinococcus canadensis* are so large that they cause respiratory distress. Various fungi, such as *Aspergillus fumigatus*, may cause pneumonia to a weakened reindeer, especially if the ground was not frozen before permanent snow fell (Rehbinder et al., 1994).

General respiratory symptoms include nasal discharge, cough, quickened breathing and fever. In chronic cases, the animal loses weight and eventually dies. Incipient cough can best be noticed in cold mornings. The progression of serious diseases caused by *Pasteurella* bacteria can be so quick that symptoms do not occur before sepsis and death (Figure 12.37).

FIGURE 12.36   Pleuritis from respiratory infections is a common finding in reindeer and is often caused by *Fusobacterium necrophorum* (Photo: Sauli Laaksonen).

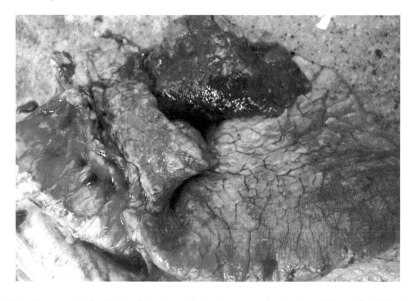

FIGURE 12.37   Pneumonia is distinguished as a dark dense area in the lung (Photo: Sauli Laaksonen).

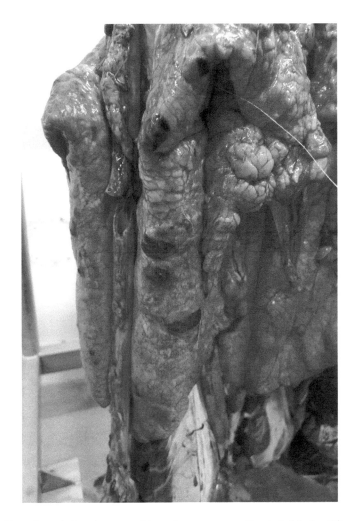

FIGURE 12.38 Stressing reindeer in very cold weather can cause direct visible frostbite and emphysema in the lungs (Photo: Sauli Laaksonen).

Emphysema may be caused by overexertion of reindeer in cold or hot weather (Figure 12.38) (Nikolaevskii, 1961a; Laaksonen, 2016). Bacterial respiratory inflammations can be treated with antibiotics. If treatment is initiated in an early stage, it may be efficient even against *Pasteurella* infections. Necrobacillosis usually causes respiratory symptoms so late that treatment is not efficient.

12.2.8 VASCULAR SYSTEM DISEASES

Heart diseases are occasionally diagnosed in reindeer during slaughter. The most common is pericarditis, and sometimes endocarditis is also diagnosed. Pericarditis and endocarditis are serious diseases that diminish the vitality of the reindeer. Infection occurs when bacteria reach the heart via blood circulation. In pericarditis, there are inflammatory adhesions and pus between the pericardium and the heart muscle. Endocarditis presents typically as proliferative formations in the cardiac valves. The symptoms are dyspnea and cyanosis of mucous membranes, which is due to the weakened pumping ability of the heart. The animal is wasting, and will eventually die.

The larval forms (microfilaria) of the *Setaria tundra* and *Rumenfilaria andersoni* worms occur in the bloodstream, with densities as large as thousands in 1 mL of blood (Laaksonen et al., 2017b). The effect of these larval forms on the health of reindeer is not known, although they can block up small capillaries. Redwater or piroplasmosis, caused by the *Babesia* parasite, is spread by ticks and occurs in Asian reindeer. It parasitizes in red blood cells and breaks them, causing the urine to become red. The parasite is the cause of great losses in Siberia (Rehbinder, 1990b). *Trypanosoma* protozoa are also found in reindeer's bloodstream, but their significance is not known (Rehbinder and Nikander, 1999). There are other parasites that use blood circulation as the route of their migration during their development, such as the meningeal worms.

12.2.9 Digestive Tract Diseases

Diseases relating to the function, nourishment supply and inflammation of the digestive tract are extremely common in supplementary-fed reindeer. Common causes are unphysiological nourishment, changes in feeding, bad or mouldy feed or lack of nourishment. Unsanitary conditions and contaminated drinking water or snow add the risk of bacterial infection (Rehbinder and Nikander, 1999; Laaksonen, 2016).

12.2.9.1 Ruminal Indigestion

A general disease, spoiled or unsuitable feed, contaminated drinking water, and quick changes in feeding cause changes in the function of the rumen microbial community, leading to malfermentation and poor digestion of feed. Ruminal indigestion can be divided into two groups: acid and alkaline. Diagnosis can be reached by taking a ruminal sample through the left paralumbar fossa, but in most cases external symptoms and anamnestic information is sufficient to diagnose the condition.

12.2.9.2 Alkaline Indigestion

Spoiled, bacteria-rich feed or drinking water, or feed high in protein or urea content are common causes of alkaline indigestion. Heavy fertilization of forage with highly nitrogenous fertilizers adds to the risk. Alkaline rumen is detrimental to the ruminal microbial community and harmful bacteria will take over (Nieminen et al., 1998; Rehbinder and Nikander, 1999; Laaksonen, 2016).

Mild cases can be almost asymptomatic, but mild diarrhoea and appetite fluctuations are common. In more severe cases, the ruminal motion stops entirely, the animal loses its appetite, and diarrhoea and tympany may occur. The rumen pH will rise to over 7. If untreated, the condition will lead to the gradual exhaustion of the animal.

In mild cases when typically the whole herd is affected, corrective measures of feeding are often sufficient treatment. In more severe cases, treatment may be life-saving. The aim is to restore the rumen function by making the ruminal environment favourable for the revival of the microbes. The acidity of the rumen can be lowered by making the animal drink about 1 L of a solution that has 5% acetic acid (vinegar), and by using rumen recovery products intended for domestic animals. Drinking cold water will also advance its recovery. Oral administration of ruminal fluid from a healthy reindeer is an efficient way of giving it new ruminal microbes. This can be done by robbing cud balls from a ruminanting reindeer's mouth, or taking the rumen content of a slaughtered animal. Water used for soaking lichen and good quality hay, with yeast added, can be administered to make the rumen recover. The reindeer is offered lichen and good, fine hay. It is also possible to force-feed the animal, by rolling hay into small balls that are then pushed between the reindeer's teeth. Bovine propylene glycol preparations can be used to ensure sufficient energy. Paste preparations are safe to administer.

12.2.9.2.1 Ammonia Poisoning

Ammonia poisoning may occur if reindeer graze in heavily fertilized cultivated areas or if the feed is inferior. If the feed is rich in highly soluble NPN compounds (non-protein-nitrogen), e.g. feed from fertilized cultivated areas, ammonia is produced in the rumen so quickly that the ruminal microbes cannot make use of it and the pH level of the rumen will increase. In acute poisoning, commonly occurring symptoms are signs of central nervous system dysfunction, anxiety, compulsive movements, convulsions and death. A chronic form also occurs, which presents with appetite loss and diarrhoea and deteriorated general condition of the animal.

Treatment is essentially the same as in alkaline ruminal indigestion. The further absorption of ammonia is prevented by making the rumen content acidic by administering, e.g. acidic acid or lactic acid (10% acetic acid in 2–3 L of water). In the chronic form of the disease, vitamin B supplements are necessary.

12.2.9.2.2   Nitrate and Nitrite Poisoning

Nitrate and nitrite poisoning may occur in farmed reindeer, or if reindeer eat green grains or heavily fertilized grass. Many cabbage plants as well as green grains may contain a high level of nitrate. Due to disturbances in fermentation, the nitrate and nitrite levels of poor-quality silage may be dangerously high. Poisoning may also result from contaminated drinking water. In the rumen, nitrate is quickly changed into nitrite, which is toxic.

The symptoms of nitrite poisoning appear in a few hours. As a result of the poisoning, the animal's blood pressure will drop quickly and methaemoglobin is produced. Methaemoglobin cannot bind oxygen; the animal will suffer from lack of oxygen and it will suffocate. The blood of an animal suffering from nitrite poisoning is dark and has poor clotting ability. Treatment is mainly the same as in ammonia poisoning.

12.2.9.3   Acid Indigestion, Carbohydrate Poisoning and Ruminal Acidosis

These are the most common forms of indigestion in reindeer that are given additional feed or supplementary fed.

If the reindeer diet is suddenly changed to contain mainly easily digestible carbohydrates, the result is very often acidosis. It may occur also if the feed contains too much grain, bread, root vegetables or fruits, or if the feed does not contain enough fiber. Fast- digesting carbohydrates are broken down into fatty acids, especially into lactic acid. The pH value of the rumen will decrease from normal to below 5.5. The ruminal microflora will die, and the rumen function will stop. The reindeer is tired and cannot stand up, and it may have difficulties in breathing (Figure 12.39). Despite treatment, the condition is often fatal.

Starving reindeer, and reindeer whose digestion has already adapted to the use of winter nourishment, are particularly susceptible to carbohydrate poisoning. The condition may also occur in situations of additional feeding, if there are not many feeding sites and dominant animals get to devour too much food. The acute form of the disorder has been linked to supplementary feeding, especially in the early days of that activity.

Nowadays, the chronic form of acid indigestion is likely more common, including symptoms like watery diarrhoea and a deteriorated general physical condition, often resulting in inflammation of the forestomachs, damage to mucous membranes, atrophy and darkening of rumen papillae, flabby ruminal wall and laminitis. Chronic acidity leads to the formation of minuscule abscesses in the ruminal wall, from which they can spread to the liver (Josefsen et al., 1997; Rehbinder and Nikander, 1999; Nieminen et al., 1998; Laaksonen, 2016). Acidic rumen content favours *Clostridium* bacteria and can cause a condition called enterotoxaemia.

The most important treatment of the acute acidosis is hydration (Figure 12.40). This can be done by giving water per os, by an intra ruminal tube or by intravenous administration. In an acute condition, the reindeer can be given 50 g per day of sodium bicarbonate (i.e. baking soda) mixed in water, to decrease the acidity of the rumen. In chronic cases, decreasing the acid level with yeast is generally sufficient. A suitable dose is half a sachet of dry yeast (5–7 g) or half a packet of fresh yeast (25 g). Vitamin B supplements revive the rumen microflora. Energy deficit can be treated with propylene glycol, about 100 g per day – for example, in paste form. Feeding cud taken from a healthy animal to the sick reindeer is a good way to build up new rumen microflora. Rumen revival products meant

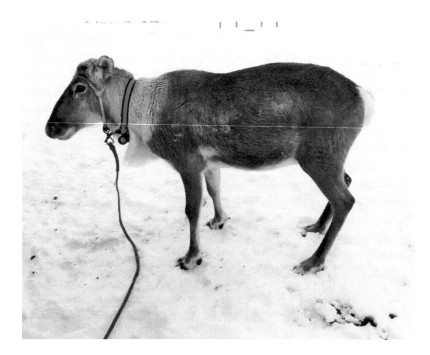

FIGURE 12.39   A reindeer suffering from acute acidosis. The rumen is paralyzed and swollen, and the stool is loose (diarrhoea). The body posture signals abdominal pain (Photo: Sauli Laaksonen).

FIGURE 12.40   Mastering the correct and safe infusion technique is key to the efficient treatment of many reindeer disease conditions. Do not lift the head too high; pour the drink into the mouth in small doses and wait for the animal to swallow. You can encourage the animal to swallow by stroking its throat or by getting it to chew the hose of the bottle (Photo: Sauli Laaksonen).

for domestic animals can also be used as supplementary treatment. Rumen revival may be continued by offering the reindeer good, leafy hay. Feeding is continued with bunches of leafy twigs or lichen, and feeding is normalized over a period of 2 to 3 weeks (Rehbinder and Nikander, 1999; Laaksonen, 2016).

In the control of all ruminal indigestions, the quality of feed and water is important, and stools must not contaminate feed. New feeds or changes of feed batches should be introduced over a period of 2 to 3 weeks, so that the ruminal microflora will have time to adapt. Care should be taken when reindeer are given feed rich in fast-acting carbohydrates, such as grains. Reindeer feeding starts in autumn by offering small amounts of all types of feed that the animals will be given during the winter. There must be several feeding troughs so that the dominant animals will not get oversized portions. The feed can also be spread over an area of fresh clean snow (Rehbinder and Nikander, 1999; Laaksonen, 2016).

12.2.9.4   Tympany

Tympany is a common condition that forms if microbial fermentation gases are trapped in the rumen and cannot be released by eructation. The gases accumulate in the rumen, ruminal contractions cease, and the rumen fills up and distends. It can be caused by various conditions of pain, indigestion, and malfermentation, or accumulation of foam caused by, for example, clover-rich forage or grazing on fresh pastures. Forage poor in roughage, or cold or wet forage, can also cause tympany. A hungry reindeer, when it gets to devour feed, is prone to bloating. Tympany can also be caused by esophageal obstruction. If a reindeer must be anaesthetised for medical operations, there is a risk of tympany when the rumen function is weakened, and the animal cannot belch. Tympany is also caused by many severe illnesses. It occurs in tetanus and in cases of severe abdominal infections, adhesions, or abscesses, when contractions are prevented by anatomical changes or pain.

A tympanic reindeer is soon in agony, and often kicks its belly with its hind legs. The left flank in front of the pelvis, at the paralumbar fossa, is distended. Due to the pressure of the enlarged rumen, the reindeer's breathing is quickened and strained. As the condition advances, the visible surfaces of mucous membranes begin to turn cyanotic and the animal will lie down. Untreated, the condition leads to painful death.

As first aid, the animal's front is lifted up to relieve the pressure on the lungs. A rope can be placed in the reindeer's mouth. Chewing the rope increases salivation, while the area of the rumen is carefully massaged. Drinking vegetable oil lowers the surface tension of water and helps to clear out the foam; it also aids belching. The animal's thorax must always be upright. If the condition is severe and the reindeer seems to be choking, a round, greased tube can be pushed into the mouth and through the esophagus into the rumen. If the tube goes into the trachea, the animal's breathing will be felt in the tube. Rumen odour coming from the tube indicates that the tube has reached the correct place. In extreme cases, a so-called cannulation of the rumen may be necessary. A thick (4–8 mm) cannula is pushed into the highest spot on the left side of the rumen so that it will pierce through the abdominal cavity and the ruminal wall, allowing the gas to be released through the cannula (Figure 12.41). This measure requires further treatment in order to prevent inflammation. As follow-up treatment, the same methods are used as in ruminal indigestion. If the condition is caused by esophageal obstruction, it is possible to try to gently move the obstruction by massaging the blockage or by pushing it into the rumen with a greased tube. In most cases, tympany requires treatment by a veterinarian, but evacuation of the gas by the owner may be life-saving (Rehbinder and Nikander, 1999; Laaksonen, 2016).

12.2.9.5   Straw Feed and Hay Belly

The reindeer has adapted to the use of soft and very fresh vegetation. Its teeth and digestion and the mucous membranes in the gastrointestinal tract are not adapted to the use of coarse fodder such as straw or hay. The digestibility and nutritional values of these are poor, and their volume is large. Coarse feed causes the rumen to dilate, which causes the ruminal wall to stretch and become

FIGURE 12.41   In life-threatening cases, a so-called cannulation of the rumen may be necessary. A thick (4–8 mm) cannula is pushed into the highest spot on the left side of the rumen so that it will pierce through the abdominal cavity and the ruminal wall. Gas is released through the cannula (Photo: Sauli Laaksonen).

thinner, weakening the ruminal motility. Coarse feed may cause ulceration and bacterial infections in the ruminal walls. A dilated rumen is visible on the left side of the abdominal area, and it can be partly filled with air. The animal will lose weight and its condition will weaken. Usually, the reindeer continues to eat well until it perishes from starvation. Correcting feeding and offering the animal an energy supplement, such as propylene glycol, is advised (Rehbinder and Nikander, 1999; Laaksonen, 2016).

12.2.9.6   Water Belly

Water belly is a condition in which a reindeer's abdominal cavity is dilated, and a splashing sound is heard when the animal moves or when its abdomen is pressed.

It is assumed that this condition is caused by the ingestion of overly coarse feed that the reindeer cannot digest well, combined with protein-rich feeding. The reindeer forestomachs dilate when they are filled with hay, and at the same time, the need for water increases due to the protein in the feed. The forestomach walls become tired, stretched and flabby, and their contraction is disrupted. Water belly may also result from foreign bodies or obstructions in the omasum or by chronic acidity of the rumen. Other factors may be a chronic peritonitis that has produced adhesions, as well as abscesses or tumours. Long-lasting stress has also been suggested as a cause. Treatment is similar to that of hay belly, but the prognosis is poor (Rehbinder and Nikander, 1999; Laaksonen, 2016).

12.2.9.7   Foreign Bodies, Blockages and Omasum Stenosis

Common causes of stenosis are various plastic particles, plastic clamps, wrappers, and bale strings. Open junkyards or plastic bags discarded in the wilderness are hazardous to reindeer, as they are sometimes prone to eating waste. Hair biting and eating, which can be induced by ectoparasites, insect-borne nematodes or the *Besnoitia* parasite, may sometimes cause hairball formation. Hairballs may block the omasum or travel all the way to the intestine and cause a blockage. Abscesses and inflammations may damage the vagus nerve, causing a functional blockage. Esophagus blockages are preventable (Figure 12.42). The early symptoms of stenosis are similar to those of other ruminal disorders, and diagnosis

FIGURE 12.42 Esophagus blockages can be prevented by offering dry pellets from a feeding bowl that hangs high (Photo: Sauli Laaksonen).

is difficult. The reindeer loses its appetite and eventually stops eating. The rumen content is watery and the abomasum is dry.

The prognosis for treatment is poor, but an attempt may be to make the reindeer drink 2–3 dL of paraffin oil. The careful removal of bale strings and clamp or fodder bale wrappers is a good preventive measure. Problems arise from unfenced or unauthorized landfills and littering. Parasite control is important in captive conditions (Rehbinder and Nikander, 1999; Laaksonen, 2016).

12.2.9.8 Wet Belly

Wet belly is a phenomenon that occurs from time to time and is associated with supplementary feeding and free grazing. The condition is often noticed in late winter, but can also occur in autumn when reindeer are rounded up. The cause or causes are not exactly known, but one theory is protein overfeeding or timothy-grass feeding (Åhman et al., 2002). Another theory is that mouldy feeds and toxins they contain would be the cause of the disease. Many observations suggest that various contagious infections, combined with stress, could evoke the disease: the reindeer's skin temperature increases and melts the snow, causing the insulating capability of its coat to deteriorate.

In mild cases, and in the early stage of the disease, the armpits or sides often get wet. In connection with stomatitis, salivation and drooling is common. Later, the wetness spreads under the abdomen, and reaches as deep as the skin (Figure 12.43) (Laaksonen, 2016). The skin becomes red and feels hot. Depending on the case, the animal may eat, but diseased animals often stop eating and mortality is common. The recommended treatment is often a change in feeding. If there is a slight possibility of an infectious cause, treatment with antibiotics and isolation of diseased animals have proven to be efficient measures.

12.2.9.9 Abomasal Inflammations, Ulcer and Abomasitis

This disease has been reported to be particularly associated with stress (Figure 12.43) (Rehbinder, 1990a; Wiklund, 1996). Lengthy motorized reindeer round-ups, transport and pens without compartments may cause stress. The disease may also be caused by abomasal parasites (e.g. *Ostertagia gruehneri*), especially in calves.

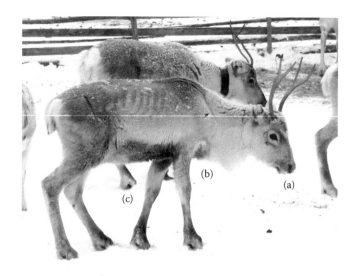

FIGURE 12.43 Viruses that cause oral infections often set off vigorous salivation and drooling (a), and, thereby, symptoms *resembling* wet belly syndrome. As the reindeer is lying down, its head is either in its armpit (b) or on its side (c), and the areas get wet from its saliva (Photo: Sauli Laaksonen).

The symptoms are non-specific, varying from unnoticeable to an animal not eating and death. Pain and bloody stools, as well as anaemia, may occur. Sudden death occurs if the abdominal wall is perforated by an ulcer. Severe abomasal parasitic inflammation may cause starvation, anaemia and calf deaths, especially at farms or in dense populations.

Supportive treatment and correct feeding is important. Linseed mucus, with the seeds sieved off, or other products for mucous membrane protection can be used for protecting the abomasal mucosa. Anthelmintic medication is effective against abomasal worms.

12.2.9.10 Peritonitis

Peritonitis is a common condition, and perhaps the most common fatal disease in semi-domesticated reindeer (Laaksonen, 2016). The reindeer peritoneum reacts easily to various factors that cause inflammation. Bacteria may gain access to the peritoneum through a damaged ruminal or abomasal wall, or, for example, through a puncture wound resulting from an antler stab. Generalized necrobacillosis often spreads to peritoneal membranes (Figure 12.44). Hay feed and acid indigestion are predisposing factors to a damaged wall. Peritonitis and peritoneal reactions caused by parasites are extremely common, especially those caused by the peritoneal worm (*S. tundra*) and legworm (*Onchocerca* spp.). *Taenia hydatigena* tapeworm causes fluid-filled cysts on the surfaces of organs or in the peritoneum, and the cervid lymphatic worm (*Rumenfilaria andersoni*) causes greenish yellow inflammatory changes on the surface (i.e. serosa) of the rumen.

The symptoms vary greatly depending on the type and extent of the inflammation, and it is hard to reach a diagnosis. The reindeer may often continue eating until it dies. The reindeer peritoneum is very capable of forming fibrin in order to keep the inflammation localized. In acute peritonitis, the congestion, accumulation of pus and secretion of fluid in the abdominal cavity are increased (Figure 12.45). As the local inflammation becomes chronic, the fibrin turns into connective tissue. The result is adhesions or abscesses. Adhesions develop in the abdominal walls and between organs, and they may impede the function of the digestive tract and cause blockages. These are very common findings in older slaughter reindeer (Rehbinder et al., 1975; Rehbinder and Nikander, 1999; Laaksonen, 2016). The treatment of peritonitis is usually based on antibiotics.

FIGURE 12.44   An inflamed and ulcerated abomasum of a stressed reindeer (Photo: Sauli Laaksonen).

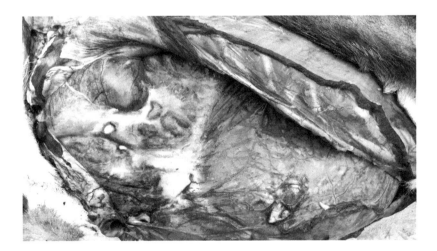

FIGURE 12.45   Peritonitis caused by necrobacillosis is the most common cause of death in reindeer in Finland. Typically, findings in acute inflammation are congestion and accumulation of pus, and in chronic cases, adhesions and abscesses (Photo: Sauli Laaksonen).

12.2.9.11   Enteritis

The most common cause of diarrhoea is nutritional indigestion. Severe enteritis may be caused by bacterial diseases such as clostridiosis, yersiniosis or colibacillosis. Parasitic infections such as coccidiosis, cryptosporidiosis, giardiosis, abomasal parasites and rumen flukes may cause diarrhoea in calves. Intestinal roundworms do not commonly cause problems under normal conditions in nature, but they may cause disease in weakened animals, and in reindeer at farms. In calves, a severe tapeworm infection may cause diarrhoea. Where found, tuberculosis and paratuberculosis are causes of chronic diarrhoea (Figure 12.46).

FIGURE 12.46   Calving in corrals in unhygienic conditions is a risk for faeces contamination, enteritis and mortality caused by *E. coli* bacteria (Photo: Sauli Laaksonen).

The most common symptom is diarrhoea, i.e. watery stools. Severe diarrhoea makes the animal lose up to 10% of the weight of the calf per day due to water loss, and also electrolytes are lost, especially sodium. This leads to rapid dehydration and death. A dehydrated calf has sunken eyes, and its skin has lost its elasticity. *Clostridium* bacteria may cause severe bloody enteritis and watery diarrhoea, usually when the animal has been fed heavy doses of concentrate feed. The disease often results in the quick death of the animal. In yersiniosis, the diarrhoea is greenish and watery, and may turn bloody, and sudden deaths also occur. Depending on the strain, colibacillosis and coccidian parasites can cause watery or bloody diarrhoea and calf mortalities, especially at calving corrals (Oksanen et al., 1990).

Mild cases of diarrhoea usually heal by balancing the diet. In severe cases, it is important to take care of the animal's fluid balance. Fluids can be administered by giving oral rehydration solutions that contain glucose and electrolytes. Severe cases may need intravenous or subcutaneous rehydration. In cases of suspected bacterial diarrhoea, treatment with antibiotics may be necessary. Lactic acid bacteria can be used in recovery. Anthelmintic drugs are used when the diarrhoea is caused by parasites. The animals can be put on a diet of hay, leafy twigs, lichen, and, if necessary, energy supply, such as propylene glycol pastes. In preventing enteritis, a balanced diet, water, and good feeding hygiene are of primary importance. Feed must not be contaminated by stools. Systematic control of parasites is essential.

12.2.10   Deficiency Diseases

Nutrition-dependent vitamin and micronutrient deficiency diseases in reindeer are not well known. It is assumed that they do not occur in animals grazing in nature if the nutritional status is good, or reindeer are fed with commercial pelleted feed for reindeer. Based on experiential knowledge, selenium deficiency may occur in farmed reindeer in selenium-poor areas. Stress is known to promote selenium-deficiency-based muscular dystrophy. Micronutrient deficiencies can be prevented by providing mineral licks and by giving the reindeer selenium supplements in selenium-poor areas. (See also Chapter 5, Non-Infectious Diseases and Trauma.)

FIGURE 12.47   A starved reindeer is apathetic and weak, and its mobility is limited (Photo: Herdis Gaup Aamot).

12.2.10.1   Starvation and Cachexia (Prostration)

Starvation was previously common among free-ranging reindeer over winter when the weather conditions were harsh and pastures were locked by ice or thick snow. Thousands of head could be lost. Starvation caused by lack of nourishment may still occasionally occur during unusual winter weather conditions in herds that are not being fed. However, lack of summer pastures may become an even bigger problem, or if grazing on summer pastures is interrupted for some reason, e.g. if the reindeer are continuously disturbed (Laaksonen, 2016). In addition, starvation is caused by inappropriate quality feed resulting in indigestion, as described above. It may also occur in reindeer that are fed, if there are too few feeding spots, or if all animals cannot get to eat due to discrimination or conflicts. Wasting is also connected with many chronic conditions, such as chronic wasting disease (CWD), tuberculosis, paratuberculosis, tumours, bad teeth and severe parasite loads.

Fat reserves melt away, first from under the skin, then from the abdominal cavity and around organs, and lastly, if starvation goes on for a long time, from the heart and the bone marrow. The colour of the fat starts to change from white to yellowish, and later disappears, replaced by gelatinous tissue. The animals' body condition score gets closer to one. Little by little, the animals become anaemic, fur gets worse, skin thins out and subcutaneous swelling occurs. Wasting causes gradual reduction of organs and muscle tissue. The reindeer's resistance deteriorates, and the animal may also be affected by other diseases. Behavioural changes also occur, the animal's reactivity diminishes and aggression occurs. At the end, the animal is unable to stand (Figure 12.47) (Rehbinder and Nikander, 1999; Josefsen et al., 2007).

12.2.10.2   Emergency Feeding

Emergency feeding must be started in situations when the reindeer are in danger of starvation. The situation may develop slowly, over several weeks. The earlier the danger is noticed and measures are taken to help the reindeer, the better the outcome. Emergency feeding must be started slowly, as their digestive systems have adapted to winter nourishment. The declined ruminal microflora must be given time to recover and adapt to the offered feed. This will take a few weeks. The best start-up feeds in emergency feeding are lichens. Lichen, especially, has been noted to promote the recovery of the ruminal papillae. It should be offered three to four times a day or more and in small quantities. Hungry reindeer may be aggressive towards each other. Little by little, the diet is extended with some dry hay, gathered when young, and, if possible, with dried leafy twigs. After a few days, the feeding of commercial, energy-rich feed can be started, again in small quantities. Good candidates are compound feeds, high in propylene glycol, especially meant for starving reindeer. If there is access to appropriate, good quality, unfrozen, and pre-dried silage, it can also be offered. Starving reindeer have a reduced tolerance to various pathogens, and therefore, all stress must be avoided.

A starved reindeer requires individual care. The animal must be isolated from other reindeer, and extra heat must be provided, e.g. by placing the animal in an appropriate shelter on a bed of dry litter. It is urgent that its fluid balance be corrected. To ensure the energy demands are met, propylene glycol pastes can be used. The rumen revival process is begun by making the reindeer drink water from poaching (stewing) hay and lichen. A cake of fresh yeast and commercial cattle supplements for rumen revival may also be used. Feeding a cud ball taken from a healthy reindeer is an efficient remedy. Salivation can be increased by giving the reindeer something to chew, e.g. a bundle of twisted hay. Lichen and beard moss is also offered immediately. The animal can be tempted to drink by mixing lichen or moss into water. Otherwise, feeding is done following the same principles as in emergency feeding, i.e. little by little. If starving reindeer are brought to farm feeding, it is recommended to put the animals in quarantine. Several disease outbreaks have been caused by late gatherings of reindeer, when supplementary feeding comes too late (Laaksonen, 2016).

12.2.11 LIVER DISEASES

Generally, changes in the liver are noticed during reindeer slaughter. The liver is an important part of the defence system of the organism. It gathers nutritional toxins and bacteria – in reindeer, in particular, harmful substances from lichens that are carried by the bloodstream. Therefore, the heavy metal and environmental toxin content of the liver is high.

Many disease conditions are seen in the liver. Parasitic changes in the liver or on its surface are extremely common. The peritoneal worm (*Setaria tundra*) causes granulomatous accumulation of fibrin on the liver surface and legworm (*Onchocerca* spp.) causes pale scars and calcifications in the liver tissue (Figure 12.48). It is seldom that echinococcus (*E. canadensis*) forms cysts in the liver, as

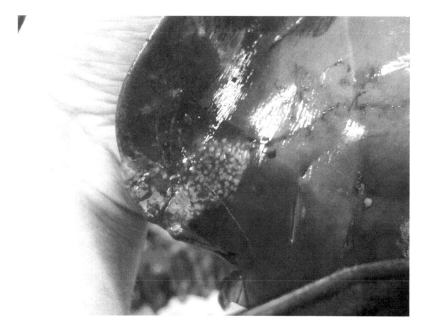

FIGURE 12.48 The migrating larvae of *Onchocerca* spp. cause nodules in the liver, which disappear during the winter (Photo: Sauli Laaksonen).

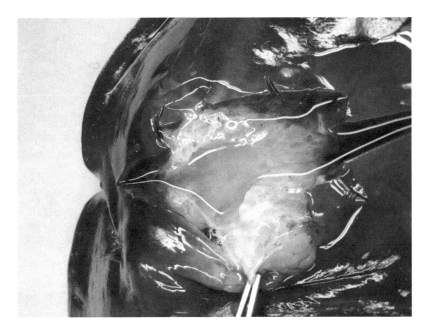

FIGURE 12.49   Benign tumour in biliary tract is harmless, but a relatively common finding (Photo: Sauli Laaksonen).

in reindeer it favours the lungs. Heavy formation of cysts from a parasitic infection caused by *Taenia hydatigena* (Figure 12.49) or the lancet liver fluke (*Dicrocoelium dendriticum*) parasites may cause cirrhosis of the liver (Rehbinder and Nikander, 1999; Laaksonen, 2016).

Liver abscesses are also common findings. Occasionally, fatty degeneration and deterioration of the liver occur. General infections cause liver atrophy. Many gastric disorders, such as acid indigestion, and coarse hay cause ruminal ulceration and inflammation that spread and form abscesses in the liver (Figure 12.50).

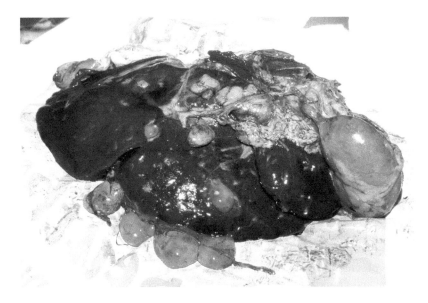

FIGURE 12.50   Cysts of *Teaenia hydatgena* can sometimes be so numerous that they interfere with liver function (Photo: Herdis Gaup Aamot).

FIGURE 12.51   A fatty liver is pale or yellowish, enlarged, and it has rounded edges (Photo: Sauli Laaksonen).

12.2.11.1   Fatty Liver

Fatty liver (Figure 12.51) may be caused by increased fat content of the blood. The fat may come from nutrition or from the quick retrieval of stored fat to fill the needs of the organism, e.g. due to oestrus or pregnancy, or during starvation. Fatty liver is also commonly found in male cervids that are in rut. In calves, fatty degeneration is frequently seen during slaughter and may be caused by stress (Laaksonen, 2016).

12.2.11.2   Icterus

Icterus is a condition in which biliary pigments (bilirubin) accumulate in tissues. Yellow discolouration occurs in the sclera of the eyes, internal organs, tendons, cartilage and, for example, surfaces of joints. Yellow discolouration is a sign of liver dysfunction or a blockage in the bile duct, which allows the bile pigments to get into the bloodstream. Abscesses, tumours, or liver flukes may cause blockages in the bile ducts. Icterus may also be caused by rapid deterioration of blood cells, when haemoglobin is broken down and excess bilirubin is carried around the body in the bloodstream. This condition may occur, for example, in situations with redwater caused by the *Babesia* parasites. Yellowness is commonly found in healthy cervids, evidently coming from what they eat (Laaksonen and Paulsen, 2015).

12.2.12   Urinal Diseases

In addition to the liver, the kidneys are important cleaning organs in the organism. They also filter foreign matter and chemicals, such as heavy metals, from the bloodstream. Nephritis, kidney deformations, tumours and mineral accumulation in the kidneys are occasionally found in reindeer. In addition, blockages of the urinary tract sometimes occur, especially in young males (Laaksonen, 2016). Various bacteria may spread to the kidneys via the bloodstream, or via the urinary tract. The urinary tract of a reindeer that was castrated young may be left smaller due to insufficient secretion of male hormones, which makes it prone to blockages (Laaksonen, 2016).

   The symptoms of renal failure are undefined and, in general, the animal is tired and loses weight. The smell of urine and ammonia may be detected on its breath. Normal reindeer urine is pale and bright, and it has a distinct tangy smell. In diseases of the kidney and urinary tract, urine may turn dark or cloudy and slimy. Bleeding may also occur. Sudden, heavy exertion may colour the urine

FIGURE 12.52 When a bull has been castrated at young age, urinary tract blockage may develop and lead to the formation of a permanent fistula to the urethra, from which urine is constantly running onto the animal's legs. The urine causes brown discolouration of the fur under the belly and on the insides of the thighs (Photo: Sauli Laaksonen).

brown, due to the breakdown of muscle cells. In case of urinary tract blockage, the reindeer tries to urinate repeatedly, with legs apart. The condition may lead to rupturing of the urinary tract. The first sign of a ruptured urinary tract is underbelly swelling, as urine accumulates in subcutaneous tissues. The condition may lead to the formation of a permanent fistula, from which urine is constantly running onto the animal's legs and hair (Figure 12.52).

If a diagnosis is reached, bacterial urinary tract infections can be treated with antibiotics in the early stage. In sheep and goats, uroliths (urinary stones) are dissolved by acidizing the urine. To achieve this, 7–10 g of ammonium chloride or 5 g of ascorbic acid (vitamin C) per 30 kg of weight is used. In prevention, fresh drinking water must be provided if the reindeer's diet is rich in protein or grain. Offering salt, e.g. licks, is beneficial, as salt makes the animals drink more.

12.2.13 CENTRAL NERVOUS SYSTEM DISEASES

Central nervous symptoms are common and always alarming in reindeer. It is difficult to distinguish the symptoms caused by advanced disease, general weakness, decreased consciousness and behavioural changes from actual CNS symptoms. Dysfunctions in ruminal fermentation may lead to the absorption of substances that will cause central nervous symptoms. The *Listeria* bacterium may cause meningitis, and generalized abscesses may spread to the brain (Nyyssönen et al., 2006). The meningeal worm may cause an inflammatory reaction on the meninx or spinal membranes that will result in rear-end paralysis. Mouldy or otherwise contaminated feed may contain toxic compounds. Vestibular organ function is disturbed by middle ear infections. Various injuries may damage the cervical spine or spinal column. Rabies and CWD cause severe central nervous symptoms (Nikolaevskii, 1961a; Rehbinder and Nikander, 1999; MacDonald et al., 2011; Mitchell et al., 2012). Treatment is in accordance with the cause. Bacterial infections are treated with antibiotics. Supportive treatment and injections of vitamin B are used to improve the general condition.

12.2.14   REPRODUCTIVE SYSTEM DISEASES

12.2.14.1   Fertility

Many factors may disturb the reproductive functions of reindeer. Unfavourable conditions in summer, lack of nutrition, or heavy infestation by insects may cause the animals stress and weakening of body condition and oestrus. Other causes are unfavourable farming practices and lack of compartments, or unsuccessful additional feeding. Many other factors may also cause similar stress, such as a high number of predators, or disturbances caused by untrained hunting dogs during the autumn oestrus. See also Chapters 1.6, *Rangifer* Reproduction Physiology, and 1.7 *Rangifer* Mating Strategies.

12.2.14.2   Foetal Death and Abortions

12.2.14.2.1   Abortion

An embryo or foetus that dies in the uterus in early pregnancy is disintegrated and absorbed, sometimes indicated by some vaginal discharge. If the foetus dies later during the pregnancy, it is usually born prematurely, or, rarely, remains in the uterus in a mummified form. Stress, general illnesses, and starvation are considered common causes of abortion. *Listeria* bacterium infection, the source of which often is poor-quality silage, causes general infection and also abortions. The parasite *Toxoplasma gondii* occurs in reindeer, and can be transferred from mother to calf through the placenta and cause calf death and abortion (Rehbinder and Nikander, 1999; Laaksonen, 2016). *Neospora caninum* infections have also been reported to be associated with abortions in reindeer (Kutz et al., 2012) (Figure 12.53).

The opening and pushing stage of the abortion process may take days, during which time the animal hunches as if it was trying to urinate (Figure 12.52). The reindeer can be helped, but only after the birth canal is fully open. Treatment is often necessary, especially as it is common following abortions that the afterbirth is retained, or that the uterus is prolapsed.

FIGURE 12.53   Abortion during the last trimester of the pregnancy is often a slow and difficult process (Photo: Herdis Gaup Aamot).

12.2.14.3   Pregnancy Determination

Using an ultrasound device, it is possible to determine pregnancy as early as 3 weeks after fertilization. When the doe is 4 months pregnant, the right side of her abdomen will grow. After 5 months, the calf's heartbeat can be heard through the right side of the groin. One sign of an "empty" (not pregnant) female is that it has already shed its antlers in winter, before calving.

12.2.14.4   Calf Mortality

By the time of calf ear marking, some of the calves of the year will have disappeared. Regional and annual differences for calf loss are huge and are linked to weather conditions, disease outbreaks and the number of predators in the area. Poor hygiene in calving paddocks may cause umbilical infections and also contagious intestinal infections in calves, especially colibacillosis. Poor-quality feed may cause listeriosis.

12.2.14.5   Dystocia

When calves are born, reindeer are usually grazing freely, to ensure a natural and successful birth. Then, however, any possible, though rare, difficulty or obstruction of birth will go unnoticed. In calving paddocks, birth difficulties occur occasionally, especially in calf pregnancies, i.e. when 1-year-old females are calving.

Abnormal presentation or position of the calf is probably the most common cause of delivery difficulties. Before birth, the calf lies on its back in the uterus, facing either forwards or backwards. In the beginning of the delivery, the calf will rotate until it lies with its back upwards. In normal birth, the calf comes out either front legs first, with head on or between them, or with the hind legs first. The positions are called anterior presentation and posterior presentation, respectively (Figure 12.54a, b), and the delivery is normally successful in either case (see also Chapter 1.6, *Rangifer* Reproductive Physiology, Figures 1.6.2 and 1.6.3).

The most common cause of malposition is the death of the foetus, e.g. in premature deliveries, when the foetus does not actively correct its position. A pelvis that is too small may also be the cause, especially in unwanted pregnancies, i.e. calves getting pregnant and calving as 1-year-olds.

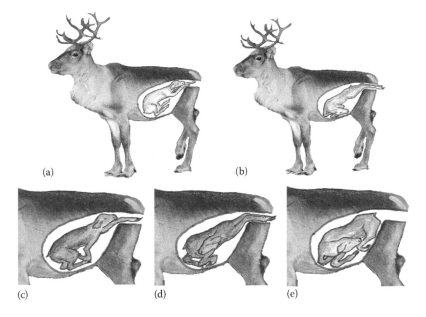

FIGURE 12.54   Normal anterior (a) and posterior (b) presentations and the most common malpositions of a calf. (c) Front leg malposition. (d) Head bent sideways or down. (e) Posterior presentation, one or both legs of the calf are presented, in this case with both legs in malposition. (Photo: Sauli Laaksonen.)

12.2.14.6   Obstetrics

Delivery assistance that is given in time may, at best, save both the doe and the calf. The calf is usually born within half an hour from the beginning of strong contractions. If this does not happen, the doe must be examined after 1 hour at the latest. If the foetus is visible in the birth canal, but is not born within 30 minutes, assistance must be given. It is best to observe the delivery at a distance, as the doe might be delaying the birth due to the presence of humans.

Giving delivery assistance is a demanding task. The doe must be held gently but firmly by an assistant. A high level of cleanliness is maintained. Small hands are an advantage. Hands and arms are carefully washed with warm water and soap. The doe's vulva and anus and the area around them is likewise washed. Hands and arms are lubricated with a suitable lubricant, or with liquid soap. In case of abortion, it is recommended to wear disposable sleeves.

With the fingers clamped, the hand is carefully pushed into the vagina and uterus, little by little, always between the pushing contractions. Using your hand, feel the foetus to determine its position. To do this, first you need to find the calf's head or tail. After this, ensure the position of the feet. If the foetus is in anterior presentation, the bottoms of the hooves face downwards. In posterior presentation, the bottoms of the hind hooves are facing up.

When the malposition has been identified, the calf is gently pushed back into the uterus, in the rhythm of the contractions. This procedure allows room to correct the malposition. When the limbs have been turned, cup the sharp hooves inside your palm, to keep them from puncturing the uterine wall. Similarly, when turning the head, the sharp lower teeth must be protected. If necessary, use more lubricant, which also makes the delivery of the calf much easier.

The extraction of the calf is assisted by pulling the legs. Traction is directed downwards at an angle of 45 degrees, using as little force as possible. The traction must be continuous, sustaining a small amount of pressure all the time, to prevent the calf from reverting into the uterus between contractions. After the birth, slime is quickly cleaned from the calf's nostrils and mouth, to help it breathe.

When the calf is born, people must move swiftly away to observe if the mother is strong enough to take care of the calf and lick it dry. This is beneficial, as it will attach the mother to the calf, preventing her from abandoning it. This may happen if the calf is handled too much by people. If the mother is too weak to take care of the calf, it must be dried with a towel and massaged gently at the same time. The umbilical cord, if still attached, must be tied and cut near the abdomen. The mother and her calf should now be left in peace, and the situation is observed from afar (Dieterich et al., 1990; Rehbinder and Nikander, 1999; Laaksonen, 2016).

12.2.14.6.1   Typical Malposition and Obstruction

If the front leg or legs is malpositioned, the calf is gently pushed back and the leg is straightened while protecting the hoof. The calf is pulled by the legs, and the head is directed to the birth canal (Figure 12.54c).

If the head is bent sideways or down, the calf is pushed back and the neck is followed up to the head, which is directed to the birth canal while protecting the teeth, avoiding damage to the uterus. The front legs are straightened and pulled into the birth canal while making sure that the head stays in the correct position (Figure 12.54d).

In posterior presentation, one or both hind legs of the calf can be malpositioned. The calf is pushed back, the leg is felt down to the hock, and the leg is pulled up so that the hoof is reached. Sheltering the hooves to avoid damage to the uterus, the legs are straightened and the calf is pulled out (Figure 12.54e).

12.2.14.6.3   Dead Calf

Often when delivery assistance is begun, the calf has already died in the uterus. The calf may already be stiff, which makes its correction more difficult, and there is a greater risk for the uterus wall to be damaged. In such cases, it may be necessary to have a veterinarian to section the calf and take one piece out after another.

12.2.14.6.4 Putrid Calf

If a difficult birth has been prolonged and the foetus has died, or in case of sloughing a dead foetus, the calf may already have turned so soft in the high temperature inside the uterus that it cannot be pulled out intact. In such case, the general condition of the mother is usually poor. The calf's hooves and even limbs may come loose during assisted delivery. A case like this requires treatment of the mother by a veterinarian.

12.2.14.6.5 Mummy

A foetus may be mummified inside the uterus. This may occur if the foetus dies in utero in late pregnancy, and, for some reason, its extraction will not begin. The same may also occur due to difficult birth, if the foetus was not able to enter the birth canal and the cervix has closed again (Figure 12.55). This condition is often noticed only during slaughtering.

12.2.14.6.6 Uterine Prolapse

Uterine prolapse is most common after a late abortion but may also occur after a normal birth. The uterus is extruded from the birth canal, and, without treatment, will quickly become necrotic. As first aid, the uterus is washed or bathed in cold water or snow, to prevent swelling. After washing and lubrication, starting from the base, attempt to gently push the uterus back inside. If the uterus is successfully replaced inside, the animal is kept with its rear end elevated for 1 hour. Sometimes, if the uterus is heavily swollen, it cannot be pushed back inside, and the uterus must be surgically removed (Figure 12.56). Uterine prolapses that have occurred after abortions are usually noticed too late. When the uterus is infected, this can be fatal for the mother. Sometimes the uterus has become necrotic and dried up, and the mother has survived, and the case is detected during slaughter in the autumn.

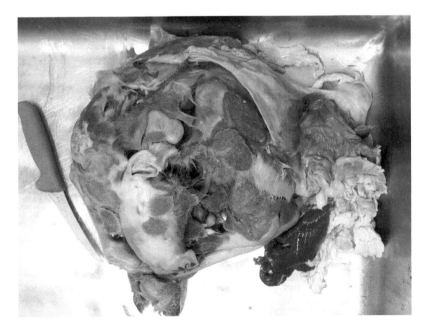

FIGURE 12.55 A mummified foetus is often not noticed until slaughter, when the uterus is opened (or abdomen is opened) (Photo: Sauli Laaksonen).

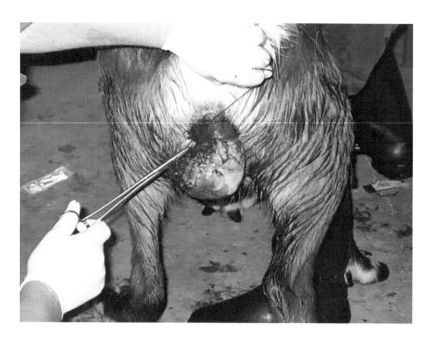

FIGURE 12.56   Uterine prolapse demands urgent medical treatment; in chronic cases the uterus must be surgically removed (Photo: Sauli Laaksonen).

12.2.14.6.7   Placental Retention

The placenta is generally delivered within 1–2 hours after birth. Retention occurs randomly in normal deliveries and is common with dystocia and abortion (Figure 12.57). If there is infection, local antibiotics and rinses are used for therapy.

12.2.14.6.8   Orphaned Calf

Sometimes a baby calf loses its mother due to accident, disease or disturbance. An orphaned calf may be noticed during summer markings, or may be just found somewhere in the pasture. The situation calls for immediate action to save the calf. The first thing to do is to take the calf to a warm place where it will be sheltered from unfavourable elements.

Colostrum is the first milk the mother produces for a few days after giving birth. It contains many important proteins (i.e. immunoglobulines/antibodies) that will protect the calf against pathogens. The antibodies are absorbed from the intestine of the calf during the first 24 hours, but after that, absorption no longer occurs. If the calf did not receive colostrum, its chances of survival are thus reduced.

The content of reindeer milk is: dry matter 29.8–35.7%, fat 14.5–19.7%, raw protein 10.3%, lactose 2.6–3%, ash 1.4–1.5% and energy content 6.7–8.4 (10) MJ/l (reviewed by Nieminen, 1994). A suitable reindeer milk replacer is bovine colostrum, which is found in the freezer of almost every dairy cattle farmer, and can even be bought from shops. The milk replacer is warmed up and fed to the calf as soon as possible, about 500 ml in total, divided into several smaller portions and fed to the calf during the first 24 hours of its life. The milk is offered from a nursing bottle meant for children or sheep. The hole in the teat can be enlarged a little with a heated needle. However, the hole must not be too big, as the calf needs to suck. Sucking will close the esophageal groove, and the milk will go directly to the abomasum. If the hole is too big, the milk may go to the lungs and cause a serious infection. If the calf is too weak to suck, the milk can be squirted straight into its mouth with a syringe that has a small tube at the tip. Force-feeding must be done very carefully so that the milk does not enter the calf's airways. Squirt only small amounts at a time and wait for the calf to swallow, before giving it another dose. The head must be kept horizontal but not any higher. Swallowing can be helped by stroking the calf under the jaw towards its throat.

FIGURE 12.57   Placenta retention occurs randomly in normal deliveries and is commonly associated with dystocia and abortion (Photo: Sauli Laaksonen).

When the calf is fed, it is important to stroke the area of the anus and underbelly, e.g. with a towel dipped in warm water. This procedure imitates the mother's licking that stimulates the functioning of the calf's urinary organs and intestine. After the colostrum stage, the calf is fed mother's milk replacer. This can be made from cow's milk or from milk replacer meant for bovine calves by adding vegetable oil until the fat content has increased to about 20%. Light cream (19% fat content) has also been used successfully as milk replacer. Baby vitamin drops (vitamins A, D, and E) are added to the drink, using the same dosage as for babies. The drink is offered lukewarm from the bottle, cupping the teat and the calf's nose in one hand. The hand supports the calf's head and the teat stays in its mouth if it moves its head, keeping the calf from swallowing air.

If the calf will not suck, which can happen if it is already a few weeks old when treatment begins, a special bottle can be used. A hose about 10 cm long and 1 cm in diameter is attached to the bottle. The drink is offered to the calf in several portions, during the first 2 weeks 5 to 7 times a day, and from then on gradually reducing the feeding times down to three to five a day. The dosage per day is about 1 L of milk replacer, or one-tenth of the calf's live weight. Overfeeding must be avoided, as then the milk will go into the rumen and will be spoiled. A sign of this is pale, yellowish diarrhoea. Only 1 day's dose of the milk replacer is prepared at the time, and it is kept refrigerated between feedings. The bottle or other feeding dish must also be kept clean.

There must always be high-quality hay, grass, lichen and silage available to the calf. Usually, it will already start tasting them on the very first days of its life. The food can be introduced to the calf by picking it up in small amounts and placing it into its mouth. There must also always be fresh water for the calf to drink. Continue feeding the milk replacer, gradually reducing the times and quantities, until the calf's forestomach function has developed so much that it will get along eating vegetable food. This stage will last from 1 to 2 months. It must be noted that the younger the orphaned calf is when it is taken into care, the more it will imprint on the handlers (Dieterich et al., 1990; Laaksonen, 2016).

12.2.14.6.9   Castration

When a male reindeer is no longer used for breeding, it is castrated, usually at about 5–7 years of age. Reindeer that are used for pulling sleighs or in tourism are often castrated earlier. It is not wise to

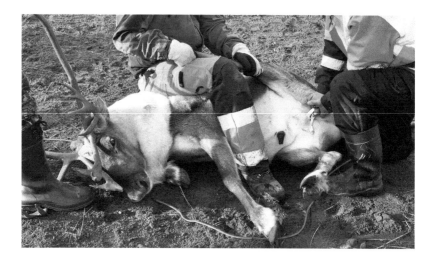

FIGURE 12.58   Castration is painful and must not be conducted without local anaesthesia. Otherwise, the reindeer may resist it vigorously, which may lead to a capture myopathy (Photo: Sauli Laaksonen).

castrate males during or right after the rut, as then their testicles are large and the blood circulation in them is still heavy, thus making the animals more vulnerable to complications. It is wiser to wait for the males to recover and get into shape again. The only permitted castration method is the use of anaesthetic drugs and a special "burdizzo" pliers (laws and regulations may vary between countries). The strong pressure from the pliers will crush the ducts that transport the semen from the epidydimis (vasa deferentia), nerves and blood vessels to the testicles, after which the testicles will gradually wither away (Laaksonen, 2016). The testes may also be surgically removed, but after this the animal will grow small, and have underdeveloped antlers which will remain covered with skin (Nikolaevskii, 1961a).

For a successful castration, the reindeer is forced to lie down. One hind leg is pulled forward to make room and give better visibility. It is well advised, and in some countries also compulsory, to perform castrations with the help of general and local anaesthesia, followed by analgesic treatment after the castration (Nikolaevskii, 1961a; Laaksonen, 2016) (Figure 12.58). The vasa deferentia can be felt clearly between the testes and the groin. The vasa deferentia are placed between the pliers, which are quickly snapped shut. The pressure is held on for five seconds. The result may be secured with another crushing made at about a 1 cm distance from the first. Every precaution should be taken to avoid breaking the skin, as this could lead to a severe infection. The skin may break if the scrotum is caught and squeezed between the angle in the jaw of the pliers, or if the handling is clumsy. As a result of unsuccessful castration, the reindeer may develop deformed cactus antlers. To a reindeer, castration is a stressful and painful procedure, and, even after some time, it may cause capture myopathy and death. The use of alternative methods, such as autumn or spring slaughtering of bucks, could be worthwhile (Laaksonen, 2016).

12.3   MEDICATION OF REINDEER

12.3.1   GENERAL MEDICATION

Health care is intended to prevent the need for medical care. However, a sick reindeer must be helped in every way. Sometimes, the reindeer owner may decide to slaughter the animal before it becomes generally sick (i.e. fever, not eating) so that it can be used for consumption. If the condition is already serious, treatment should be considered, but if the treatment effect is likely to be poor and the animal will suffer, a better solution is to euthanize the animal.

Little research exists on the medical treatment of reindeer. In fact, it is based on veterinarians' empirical experience and adaptations of treatments used for other ruminants. It must be remembered that reindeer are cervids, and their metabolism is different from that of production animals. For instance, the metabolism of reindeer slows down considerably during the winter months (Nilssen et al., 1984; Säkkinen, 2005). This will influence drug metabolism and the determination of withdrawal periods for meat to avoid remaining drugs and exposure to humans. Since almost no drugs are registered for reindeer, veterinarians often use drugs approved for other ruminants, with at least 28 days of withdrawal for meat in EU countries. This is in accordance with the cascade principle (2001/82/EC), which allows the veterinarian to use his/her clinical judgement to treat an animal by deciding which drug to use when no registered drug is available. The basic prerequisite for treatment is to isolate a sick reindeer to a location where it can be captured quickly and where it cannot transmit a possible infective agent to other reindeer.

Anthelmintics and vaccinations are administered subcutaneously (Figure 12.59). The correct location is at the boundary of the middle and back thirds of the neck, in the area of loose skin. Intramuscular drugs are administered into the neck. If a reindeer is medically treated for several days in a row, the location of the injection must be varied from side to side, to avoid bleeding and severe local reactions to the drug. Administering medication into the cervical vein is slightly more demanding, but it is a good place for giving many drugs that might cause irritation to tissues, such as vitamin preparations and fluids. If the injection site is dirty, it must first be properly cleaned and disinfected. If drugs with withdrawal periods are used, the identification of the reindeer must be secured, e.g. with colourants.

In the use of antibiotics, basic antibiotics like penicillin (G) are sufficient. Broad-spectrum antibiotics should not be used, to avoid the development of antibiotic resistant bacteria. In Finland the most frequently used antibiotics are penicillin and tetracyclins, which are effective in most cases. If treatment is started, the whole course of antibiotic treatment must be given to the animal even though the health condition may improve after a few days. Short courses often lead to the emergence of strains that are resistant to antibiotics, and the condition may exacerbate and become severe. Non-steroidal anti-inflammatory drugs (NSAID) such as meloxicam, kaprofen and ketoprofen are commonly used by reindeer practitioners as analgesics for reindeer, e.g. in combination with general and local anaesthesia during castrations of bulls to prevent capture myopathy. Experiences are good even though there is no scientific documentation of the efficacy or adverse reactions in reindeer.

FIGURE 12.59   A good, secure holding technique is essential for the safety of both the animal and the handlers (Photo: Sauli Laaksonen).

For general anaesthesia of reindeer, see recommended drugs and doses in Chapter 15, Restraint and Immobilization. It must be noted that only xylazine and detomidine have been approved for production animals and can be used according to the cascade principle for reindeer intended for human consumption. According to empirical experiences, local anaesthetics such as lidocaine and procaine-hydrochloride are well tolerated and are commonly used for reindeer. For obstetrics, medication is very rarely used for semi-domesticated reindeer (see Chapter 1.6, *Rangifer* Reproductive Physiology).

For many disease conditions in reindeer, supportive treatment is the most essential and life-saving measure. A draught-free space that offers enough shelter and can be heated if necessary, plenty of bedding, and fresh drinking water are the essential requirements that will help with survival. Supportive medication, most often fluids or energy preparations, are administered orally. It is important to master a safe technique of giving drink to the animal (see also Chapter 4, Feeding and Associated Health Problems).

12.3.2 Control of Parasite Infestations in Semi-Domesticated Reindeer

12.3.2.1 Pasture Circulation and Hygiene

In parasite control, understanding their life cycle is a prerequisite. For most parasites, it is typical that their life cycle includes a phase when the parasite lives outside its host animal, either in the environment or inside an intermediate host. This phase of the life cycle is necessary so that the parasite will be able to infect yet another host animal.

Reindeer that graze in the same area for a long time are more likely to be infected from the soil and from intermediate hosts than are reindeer that migrate. Pasture circulation is the best natural way to prevent infections. In nature, parasitic eggs or larvae will become infectious in 2 days or in several months, and they will maintain their ability to infect hosts for about a year (except for the *Trichuris* roundworm parasite). The evolution of vector-transmitted parasites to an infectious stage inside an insect will take at least 2 weeks. Herding the reindeer to new pastures at intervals of a few days or weeks will reduce the parasites' chances of infecting new hosts. The herd should return to the old pasture after a year has passed, as then the parasitic eggs and larvae have been destroyed. The natural resistance of calves is at its lowest until they are about a month old. During this time, they are more susceptible to infection. Good condition and abundant lactation of the mother will protect the calf from infections. At calving paddocks, infection pressure may be high and, therefore, taking care of paddock hygiene is especially important. At worn pastures the risk of infection is higher, animals must eat from closer to the ground, and more parasites will end up inside the reindeer. The same happens if reindeer are fed directly from the ground in farm conditions.

In modern reindeer herding, especially in Finland, it is difficult to carry out pasture circulation that requires active herding. However, following the principles of pasture circulation is beneficial if, for some reason, the reindeer seem to stay for a long time in a small area, such as an area under cultivation. In winter farming and at feeding grounds, the spread of infections can be reduced by using many feeding places, compartmentalizing, paddock cleaning or circulation and offering the feed in such a way that it cannot be contaminated by stools or stool-contaminated soil. Correct, sufficient reindeer nutrition will reduce all possible harmful effects of all parasites.

Contacts with production animals, e.g. through feed at pastures or fields, increase the possibility of parasitic and other infections. Contacts are prevented by pasture fencing and feed hygiene. Transfers or migrations of various cervids increase the chances of spread of new parasite species. Leaving dead reindeer in the wilderness or making the carcasses available for predators will increase the chance of life cycle completion of tapeworms and many protozoa. Feeding raw reindeer meat or organs to dogs has the same effect. As cats spread toxoplasmosis, they must be kept away from reindeer farms or paddocks.

12.3.2.2 Antiparasitic Treatment

Due to heavy infection pressure or the insufficiency of natural control means, it is often necessary to use chemical antiparasitic treatment in reindeer herding. The use of antiparasitic drugs must be justifiable regarding the well-being of animals, and the acquired benefit must be profitable to the reindeer owners. Parasite treatment may be justified when certain climatic and weather elements are known to favour mass-infestations of parasites, e.g. after warm and moist summers, when control measures can be started well in advance (Nikolaevskii, 1961a, c; Oksanen, 2003; Laaksonen, 2016).

12.3.2.3 How to Identify Reindeer with Heavy Parasite Load

Identification of symptoms caused by parasites is difficult. The effect of infections caught in the previous summer are often visible in the critical months of spring. A reindeer carrying a heavy parasite load is more susceptible to various other diseases (Nikolaevskii, 1961a; reviewed by Carlson et al., 2016). The general symptoms of a heavily parasitized reindeer are malnutrition, weight loss, decline and dullness of fur and a slowing down of moult.

Warble fly larvae are clearly visible only in spring as subcutaneous elevations that have breathing holes in the middle. In spring, throat bots cause respiratory symptoms, coughing, hawking and spluttering, as the larvae begin their vigorous growth. Heavy *Setaria* infestation causes weight loss as early as autumn. Severe legworm infection in the hocks may cause lameness. Lungworm infection may occur in the form of cough in spring. A severe meningeal worm infection may cause respiratory symptoms, due to the lung migration of the larvae.

The *Besnoitia* parasite causes "hot leg", when the reindeer bites and chews the parasite-infested areas, which leads to hair loss and skin thickening. "Hot ear" causes typical skin changes. Reindeer infested by the deer ked stand out due to their behaviour and due to damage to their fur that already appears in early autumn. Other changes caused by ectoparasites are typically invisible until late winter. As early as during autumn slaughtering, it is already possible to make accurate observations on the occurrence of parasites and, if necessary, start extensive control measures. This is especially important during mass-infestations of the *Setaria* parasite, when slaughtering should be started earlier to avoid the most harmful effects produced by this parasite.

The presence of intestinal parasites, lungworms, and meningeal worms can sometimes be proven with eggs or larvae found in stools. The parasites do not reproduce regularly. For instance, abomasal and lungworms may spend the winter as dormant larvae and do not become active until late spring. Meningeal worms reproduce after midwinter has passed. Vector-transmitted roundworm infections can be shown from blood samples, but not before December, when the parasites caught in summer have become sexually mature (Oksanen, 2003; Laaksonen, 2016).

12.3.2.4 Antiparasitic Medicines

The antiparasitic treatment of reindeer was begun in Fennoscandia in the late 1970s with organic phosphorus compounds, which were effective but extremely toxic compounds. Ivermectin, synthesized from the fermentation products of the *Streptomyces avermitilis* bacterium, took over the antiparasitic medicine market of the early 1980s, including in reindeer parasite control. There are only two antiparasitic medicines registered for reindeer: ivermectin and doramectin. Both substances are effective against the warble fly and throat bot fly larvae. Due to their special features, ivermectin products are more commonly used for reindeer antiparasitic treatment. The product is safe for reindeer and humans, and there is practically no possibility of overdose. Ivermectin has a paralyzing effect on the nervous system of parasites, but it does not affect the host animal.

Ivermectin kills the already existing parasites and protects the reindeer against new infections for 2–4 weeks. In addition to warble fly and deer bot fly, ivermectin is effective against other ectoparasites, deer keds and lice. The drug is also effective against the *Setaria* parasite, adult intestinal roundworms and lungworms. However, its effect on the dormant forms of abomasal and lung worms is not well known. Ivermectin is effective against adult meningeal worms that reside on muscle membranes, but parasites that are on the meninges or in the spinal cord are safe from the medicine, as it cannot pass from the bloodstream to the spinal fluid. Ivermectin is partially effective against the reindeer tongue worm, but it is likely that it is not effective against legworms (*Onchocerca*) or lymphatic roundworms (*Rumenfilaria*).

Antiparasitic medicines registered for other animals can also be used for reindeer, but, as with ivermectin and doramectin, they all have a minimum withdrawal period of 28 days in meat use. Other endoparasitic medicines, such as in the benzimidazole group (e.g. fenbendazole), are occasionally used for reindeer. Pyrethrin derivatives, such as deltamethrin-containing solutions, can be used as ectoparasitic medicines (Oksanen, 2003; Laaksonen, 2016).

12.3.2.5   Antiparasitic Treatment in Practice

Ivermectin antiparasitic drug is administered to reindeer by subcutaneous injection, at a dose of 200 µg/kg live weight. Underdosing must never be used, as many parasites, especially intestinal roundworms, are quick to develop resistance against ivermectin. Underdosing happens easily, if ivermectin is dosed by using ivermectin products meant for other animals, and the drug is administered either by mouth or in solution applied to the skin. In either case, it is impossible to reach a sufficient drug content in the gastrointestinal tract against roundworms.

Ivermectin is injected subcutaneously in the rear third of the neck. The animal must be kept very still, as the substance causes pain and irritation if it gets into muscle. The reindeer must be clean, and the procedure cannot be done in muddy conditions. The needle is changed as often as is necessary, and always changed if blood or dirt is visible. In Finland, ivermectin injections are administered only by veterinarians, whereas in other countries, trained non-veterinarians may also conduct such treatment (Figure 12.60).

Drugs against intestinal parasites are given safely by mouth in paste form from ready-packed tubes. Pyrethroid insecticides are administered on the reindeer's skin as a "pour-on" preparation. The product must be administered on the skin at the hair roots and will, due to its oily nature, disseminate in the fat layer of the skin and cover the whole body surface.

A record has to be kept of antiparasitic treatment, and the reindeer that are treated must be well marked so that any treated reindeer are not accidentally included in those that are slaughtered. Fence structures and the execution of antiparasitic treatment must be planned well to put as little strain on the reindeer as possible during the procedure. If the reindeer must run, or ropes or lassos have to be used, the execution has been poor, and the harm it has caused may be greater than the benefit from the antiparasitic treatment.

FIGURE 12.60   About 80% of Finnish reindeer are routinely treated with ivermectin against endo and external parasites during autumn round-ups or later in winter collars (Photo: Sauli Laaksonen).

The recommended time for reindeer antiparasitic treatment is from the end of September to the end of December. Calves have been experimentally administered antiparasitic medicine during summer round-ups, but no benefits have been identified with regards to parasite load or additional growth. Apparently, infection pressure is so great at the pastures for free-ranging reindeer that reindeer will quickly catch new infections. In areas where the *Setaria* parasite or the deer ked reside, early administration of antiparasitic medicine is recommended to prevent permanent hair damage and winter-time energy loss. Although treatment administered as late as February has not been seen to lead to subcutaneous infections caused by dead warble fly larvae under the skin, it benefits the reindeer if the larvae are destroyed when small, and thus any skin damage is prevented. After all, one suspected risk factor for necrobacillosis is the skin damage caused by the warble fly.

The *Setaria* and *Rumenfilaria* parasites begin to produce larvae in the bloodstream of reindeer in December. In February, the quantity of larvae in the blood may already be quite high. These larvae are extremely sensitive to ivermectin, and the simultaneous death and breakdown of a large quantity of larvae in the bloodstream may cause problems in the kidneys and lymphatic nodes (Nelson, 1966). Symptoms like these in the context of antiparasitic treatment have been described in other animal species and in humans. This suspicion is supported by the practical experience of reindeer herders, who observe that reindeer suffer from lack of appetite for a few days after a late administration of antiparasitic medicine (Laaksonen, 2016). Another reason to avoid antiparasitic treatment of reindeer in February is that any excess strain on already-pregnant females should be avoided. Although ivermectin is not toxic to the foetus, stress during the last trimester of pregnancy may cause abortions in does.

With reindeer that are farmed all year round, systematic parasite control is essential, as infection pressure, especially that of intestinal and lung roundworms, may be very great. Pasture circulation is a good solution, when there are two pastures available that can be changed once a year. The reindeer are treated with fenbendazole products in the beginning of summer. In autumn, the reindeer are injected with ivermectin, which is effective against ectoparasites. It must be remembered that the larvae of reindeer gastrointestinal roundworms are infective even in winter, and sometimes even three antiparasitic treatments may be necessary – for example, of reindeer in parks and zoos. During mass-infestation of insects, farmed reindeer can be protected with insecticides, e.g. long-lasting deltamethrin solutions (Nikolaevskii, 1961c; Oksanen, 2003; Laaksonen, 2016).

Dogs that move freely around reindeer herding areas, such as shepherd dogs or hunting dogs, must be treated with antiparasitic medicines containing praziquantel or epsiprantel twice a year. These drugs are effective against *Echinococcus* tapeworms.

12.4   HEALTH MONITORING STRATEGIES FOR SEMI-DOMESTICATED REINDEER

Monitoring the health and well-being of semi-domesticated reindeer populations is an important part of the observation of the Arctic environment and its changes.

Increased morbidity or mortality are always alarming and call for a thorough examination, including sampling and laboratory research. In situations where a pathogen is threatening the health of reindeer in the surrounding areas, targeted sampling is the most important tool. The monitoring may be carried out by tissue and blood samples obtained from healthy reindeer during slaughter and also fallen stock and necropsies. Other methods include investigation of possible vector insects. The samples can be examined, for example, to detect antibodies against disease agents in the bloodstream, or the disease agents themselves.

Early notice of changes in well-being and health of semi-domesticated reindeer is important on the population level. It enables early, vital alarm and control actions – for example, if nutritional conditions are poor or diseases are emerging in the population.

Semi-domesticated reindeer, either herded or free-ranging, migrate freely on pastures for most of the year. The health status of herded flocks can be monitored throughout the year. They are usually gathered twice a year for round-ups: in midsummer for calf marking and in autumn/winter for slaughter selection, during which their health can be monitored. They may also be monitored throughout the year by regular follow-up on pastures; nowadays, this is done mostly by locating the herds using satellite collars. Collars which alarm when the reindeer does not move anymore, indicating death, are also widely used. Although large differences exist between countries and regions, the majority of semi-domesticated reindeer in Fennoscandia are, to some extent, supplementary fed in winter, in fences or on mountain pastures, and are in this way continuously looked after.

The greater the number of reindeer which get sick or die with similar symptoms, the more important further investigations are. It should be noted that in the case where a hazardous or infectious animal disease is suspected, it is always important to inform the official veterinarian for the area. Then the guidance and sampling is carried out by the authorities. It is desirable that all reindeer herding units have an anticipatory contingency plan for such situations. A single case of reindeer disease can be significant, especially if an apparently healthy reindeer in good condition is found dead without any obvious reason.

The reindeer slaughtering period, with ante- and post-mortem control, provides an excellent picture of reindeer health, and also allows efficient collection of samples. Mean slaughter weight of calves and body condition assessments reflect the health and nutritional conditions during summer on pastures. The data of ante- and post-mortem meat inspection, especially if continued for years, are valuable. Data on calf survival and loss of animals in general may also give valuable information about health status.

12.4.1   FIELD NECROPSY AND SAMPLE COLLECTION

In some situations, it is not possible or reasonable to send a whole reindeer carcass to a research laboratory, for example, due to geographic distances or weather conditions. In this case reindeer necropsy and sampling may have to be performed under field conditions to determine the cause of the disease (Figure 12.61). Smaller organ samples may be chilled and sent to the diagnostic lab for further investigations. If the reindeer has been decomposing for a long time, it is not advisable to start the necropsy. It is best performed by two people: one of them ("dirty") makes the actual opening and the other ("clean") takes care of protection, cleaning, sample packing, notes and taking photos. Good hygiene and personal protection from potentially infectious agents are most important in carrying out the operation. Hygiene is also essential for the researchability of samples.

- Protect yourself from infections; use protective gloves and clothing.
- Do not cut sick or pathologically changed tissues and organs.
- Wash your hands and clothing with hot water and soap after handling the carcass.
- Carefully wash and disinfect all equipment used.
- Do not give dogs or other animals access to any parts of sick reindeer.
- If you suspect dangerous animal diseases or zoonoses, or if a mass mortality has occurred, always contact an official veterinarian first.

If blood is excreted from one or more body openings, anthrax may be suspected. In this case, a necropsy should not be conducted, and specific samples and measures should be taken by professionals.

View the carcass before starting the necropsy and make a record of the disease history (i.e. anamnesis), any traces of agony, signs or footprints of possible predators, injuries, position of death and an estimate of the time of death. Photos of all the observations are always very useful. The condition of the reindeer, changes in hair, ectoparasites, leaks from body openings and possible swellings are checked and recorded. If the whole carcass is to be sent to the laboratory, it is useful to skin the abdomen and chest area to help cool the carcass and slow down decay. This is especially important in warm weather.

The reindeer is placed on its back against the ground. Teeth, mouth and tongue are checked first. The animal is cut between the shoulder blades and the chest and the legs bent to the side. Hindquarters are cut by sliding the joints up to the hip and bent to the side. This way the reindeer stays on its back during the remaining necropsy.

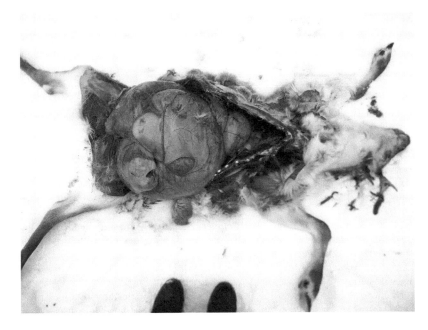

FIGURE 12.61 A reindeer has been necropsied in the field for sample collection, in an attempt to find out what caused its death (Photo: Sauli Laaksonen).

Then the reindeer is skinned on the abdomen and chest area to the neck to examine the possible ruptures and injuries under the skin. At the same time, the udder and the genitals are checked. The abdominal cavity is opened from the back of the breastbone to the front edge of the pelvis, guarding the gastrointestinal tract. If the reindeer has begun to deteriorate, the formation of gases has generally expanded the abdomen and the intestinal tract is particularly vulnerable. The sternum is removed. The best way to do this is by cutting the cartilage joints between the ribs and the sternum. A saw can also be used if needed.

Thereafter, the abdominal and thoracic organs can be examined and possible organ samples can be collected cleanly. Blood or serum samples are collected from the uncoagulated blood that has leaked into the abdominal or thoracic cavity or cardiac chambers. If there is no uncoagulated blood, blood clots can be collected. If an intestinal sample is taken, the bowel is bound at both ends before removing. Samples are packed in clean plastic bags for transport.

Always contact the laboratory that will receive the samples before the necropsy, and get instructions on which type of samples they would prefer and how to pack, store, and send them. Usually, all samples for diagnostic purposes are cooled to close to +0°C to keep the tissues as researchable as possible. The sample should not freeze, even though many of the animals found in winter have already been frozen. Samples are often transported to the laboratory by public means of transport. Attention must be paid to packaging the samples in order to minimize the vehicle and environmental contamination. After cooling, the animal or parts of the animal are packaged in airy and dry material such as newsprint. If it is an organ sample, it must first be packaged in a foil or baking paper so that the newspaper does not stick to the surface of the sample. Subsequently, the sample is transferred to a tight, fluid-tight plastic bag. Finally, the sample is packed for transport in a durable box, which can be, for example, Styrofoam, cardboard or wood. During the warm season, coolers are added to different parts of the sample.

A covering letter must always be attached to the shipment. Important information is: the contact details of the sender, the type of specimen, the time and place, and possible observations of the animal's symptoms and information about other dead animals. Photographs are a great help for the investigator.

In the case of extensive screening studies of diseases, the institution responsible for the study provides instructions for sampling and delivery. The sampling request includes specific instructions for samples of the body, organs, or blood.

REFERENCES

Åhman, B., A. Nilsson, E. Eloranta, and K. Olsson. 2002. Wet belly in reindeer (*Rangifer tarandus tarandus*) in relation to body condition, body temperature and blood constituents. *Acta Vet Scand* 43:85–97.

Bengtson, Y.A. 2004. Secondary effects of stress on domestic animals with specific reference to predator attacks. Thesis. Swedish University of Agricultural Sciences. The County Administrative Board of Gävleborg, Report 15.

Carlsson, A.M., G. Mastromonaco, E. Vandervalk, and S. Kutz. 2016. Parasites, stress and reindeer: Infection with abomasal nematodes is not associated with elevated glucocorticoid levels in hair or faeces. *Conserv Physiol* 4(1).

Dieterich, R.A., K. Jamie, and J.K. Morton. 1990. *Reindeer Health Aide Manual*. Second Edition. Agricultural and Forestry Experiment Station, Cooperative Extension Service, University of Alaska Fairbanks and U.S. Department of Agriculture cooperating.

Flueck, W.T., J.M. Smith-Flueck, J. Mionczynski, and B.J. Mincher. 2012. The implications of selenium deficiency for wild herbivore conservation: A review. *Eur J Wildl Res* 58:761.

Fox, K.A., B. Diamond, F. Sun et al. 2015. Testicular lesions and antler abnormalities in Colorado, USA mule deer (*Odocoileus hemionus*): A possible role for epizootic hemorrhagic disease virus. *J Wildl Dis* 51:166–76.

Josefsen, T.D., T.H. Aagnes, and S.D. Mathiesen. 1997. Influence of diet on the occurrence of intraepithelial microabcesses and foreign bodies in the ruminal mucosa of reindeer calves (*Rangifer tarandus tarandus*). *Zentralbl Veterinarmed A* 44:249–57.

Josefsen, T.D., K.K. Sørensen, T. Mørk, S.D. Mathiesen, and K.A. Ryeng. 2007. Fatal inanition in reindeer (*Rangifer tarandus tarandus*): Pathological findings in completely emaciated carcasses. *Acta Vet Scand* 49:27.

Kutz, S.J., J. Ducrocq, G.G. Verocai et al. 2012. Parasites in ungulates of Arctic North America and Greenland: A view of contemporary diversity, ecology, and impact in a world under change. *Adv Parasitol* 79:99–252.

Kynkäänniemi, S.M., M. Kettu, R. Kortet et al. 2014. Acute impacts of the deer ked (*Lipoptena cervi*) infestation on reindeer (*Rangifer tarandus tarandus*) behaviour. *Parasitol Res* 113(4):1489–97.

Laaksonen, S. 2016. *Tunne poro – poron sairaudet ja terveydenhuolto*. [*The diseases and health care of reindeer*]. Riga, Latvia: Livonia print.

Laaksonen, S., and M. Nieminen. Poron terveyden mittarit. 2005. [The health indicators of reindeer]. *Poromies* 2:42–5.

Laaksonen S., and P. Paulsen. 2015. *Hunting Hygiene*. Netherlands: Wageningen Academic Publishers.

Laaksonen, S., P. Jokelainen, J. Pusenius, and A. Oksanen. 2017a. Is transport distance correlated with animal welfare and carcass quality of reindeer (*Rangifer tarandus tarandus*)? *Acta Vet Scand* 59:17.

Laaksonen, S., A. Oksanen, S. Kutz, P. Jokelainen, A. Holma-Suutari, and E. Hoberg. 2017b. Filarioid nematodes, threat to arctic food safety and security. In *Game Meat Hygiene: Food Safety and Security*, P. Paulsen, A. Bauer, and F.J.M. Smulders (eds.). Netherlands: Wageningen Academic Publishers.

MacDonald, E., K. Handeland, H. Blystad et al. 2011. Public health implications of an outbreak of rabies in arctic foxes and reindeer in the Svalbard archipelago, Norway, September 2011. *Eurosurveillance* 16:2–5.

Markusson, E., and I. Folstad. 1997. Reindeer antlers: Visual indicators of individual quality? *Oecologia* 110:501–7.

Mitchell, C.B., C.J. Sigurdson, K.I. O'Rourke et al. 2012. Experimental oral transmission of chronic wasting disease to reindeer (*Rangifer tarandus tarandus*). *PLoS ONE* 7(6):e39055.

Nelson, G.S. 1966. The pathology of filarial infections. *Helminth Abstr* 35:311–36.

Nieminen, M. 1994. *Poron ruumiinrakenne ja elintoiminnat*. [*The anatomy and physiology of reindeer*]. Pohjolan Sanomat Oy, Kemi.

Nieminen, M., and T. Helle. 1980. Variations in body measurements of wild and semi-domestic reindeer (*Rangifer tarandus*) in Fennoscandia. *Ann Zool Fennici* 17:275–83.

Nieminen, M., and C.J. Petersson. 1990. Growth and relationship of live weight to body measurements in semi-domesticated reindeer (*Rangifer tarandus* L.) *Rangifer* 10(3):353–61.

Nieminen, M., V. Maijala, and T. Soveri. 1998. *Poron ruokintaa*. [*Feeding of reindeer*]. Riista- ja kalataloudentutkimuslaitos.

Nikolaevskii, L.D. 1961a. Diseases of reindeer. In *Reindeer Husbandry*, P.S. Zhigunov (ed.). Jerusalem: Israel Program for Scientific Translations 230–84.

Nikolaevskii, L.D. 1961b. General outline of the anatomy and physiology of reindeer. *In Reindeer Husbandry*, P.S. Zhigunov (ed.). Jerusalem: Israel Program for Scientific Translations 5–56.

Nikolaevskii, L.D. 1961c. Reindeer hygiene. *In Reindeer Husbandry*, P.S. Zhigunov (ed.). Jerusalem: Israel Program for Scientific Translations 57–77.

Nilssen, K.J., J.A. Sundsfjord, and A.S. Blix. 1984. Regulation of metabolic rate in Svalbard and Norwegian reindeer. *Am J Physiol* 247:837–41.

Nyyssönen, T., V. Hirvelä-Koski, H. Norberg, and M. Nieminen. 2006. Septicaemic listeriosis in reindeer calves – A case report. *Rangifer* 26:25–8.

Oksanen, A., M. Nieminen, T. Soveri et al. 1990. The establishment of parasites in reindeer calves. *Rangifer*, Special Issue 10(5):20–1.

Oksanen, A. *Parasitbehandling av renar*. 2003. Umeå: Sámiid Riikkasearvi/SSR, forskningsförmedlingen.

Petäjä, E. 1982. Poronlihan tervalihaisuus ja sen estäminen. *Helsingin yliopiston lihateknologian laitoksen julkaisuja*. Helsinki No. 132.

Rehbinder, C. 1990a. Management stress in reindeer. *Rangifer*, Special Issue No. 3.

Rehbinder, C. 1990b. Some vector borne parasites in Swedish reindeer (*Rangifer tarandus tarandus* L). *Rangifer*, Special Issue 10(3):67–73.

Rehbinder, C., D. Christensson, and V. Glatthard. 1975. Parasitic granulomas in reindeer. A histopathological, parasitological and bacteriological study. *Nord Vet Med* 27:499–507.

Rehbinder, C., M. Elvander, and M. Nordkvist. 1981b. Cutaneous besnoitiosis in a Swedish reindeer (*Rangifer tarandus* L.). *Nordisk Veterinaer* 33:270–2.

Rehbinder, C., L.-E. Edqvist, K. Lundstrom, and F. Villafane. 1982. En fåltstudie av stress hos ren i samband med olika hanteringsformer. *Rangifer* 2:2–21.

Rehbinder, C., R. Mattsson, and K. Belak. 1994. Aspergillosis in reindeer (*Rangifer tarandus tarandus* L). A case report. *Rangifer* 3:131–2.

Rehbinder, C., and S. Nikander. 1999. *Ren och rensjukdomar*. Studenlitteratur, Sveden.

Romano, J.S., T. Mørk, S. Laaksonen et al. 2018. Infectious keratoconjunctivitis in semi-domesticated Eurasian tundra reindeer (*Rangifer tarandus tarandus*): Microbiological study of clinically affected and unaffected animals with special reference to cervid herpesvirus 2. *BMC Vet Res* 14:15.

Säkkinen, H. 2005. Variation in the blood chemical constituents of reindeer. Significance of season, nutrition and other extrinsic and intrinsic factors. Acta Universitatis Ouluensis A, Scientiae Rerum Naturalium 440.

Stoffregen, W.C., D.P. Alt, M.V. Palmer et al. 2006. Identification of a haemomycoplasma species in anemic reindeer (*Rangifer tarandus*). *J Wildl Dis* 42:249–58.

Stubsjøen, S.M., and R.O. Moe. 2014. Stress og velfred hos rein – en oversikt. *Norsk Vet Tidskr* 2:118–22.

Tryland, M., C.G. das Neves, M. Sunde, and T. Mørk. 2009. Cervid herpesvirus 2, the primary agent in an outbreak of infectious keratoconjunctivitis in semidomesticated reindeer. *J Clin Microbiol* 47:3707–13.

Tryland, M., J.S. Romano, N. Marcin et al. 2017. Cervid herpesvirus 2 and not Moraxella bovoculi caused keratoconjunctivitis in experimentally inoculated semi-domesticated Eurasian tundra reindeer. *Acta Vet Scand* 59(1):23.

Vikøren, T., A.B. Kristoffersen, S. Lierhagen, and K. Handeland. 2011. A comparative study of hepatic trace element levels in wild moose, roe deer, and reindeer from Norway. *J Wildl Dis* 41:569–79.

Wiklund, E., A. Andersson, G. Malmfors, K. Lundstrom, and Ö. Danell. 1995. Ultimate pH values in reindeer meat with particular regard to animal sex and age, muscle and transport distance. *Rangifer* 15:47–54.

Wiklund, E. 1996. Pre-slaughter handling of reindeer (*Rangifer tarandus tarandus* L) effects on meat quality. Uppsala, Sweden: Department of Food Science, Swedish University of Agricultural Sciences.

APPENDIX: FINNISH MEAT INSPECTION AND DECISION FORM FOR DATA COLLECTION (TRANSLATED FROM FINNISH)

Aluehallintovirasto
Guovlohálddahusdoaimmahat

REGIONAL STATE ADMINISTRATIVE AGENCIES OF LAPLAND

Lappi

REINDEER MEAT INSPECTION DECISION AND FEEDBACK

Official veterinarian	Date	Recipient	
Address	Reindeer herding cooperative		
Postal number and postal address	Reindeer abattoir, name		Abattoir number

MEAT INSPECTION DECISION AND REASONINGS

Meat inspection finding	Slaughter identification	Findings/inspected (n)	Condemned, whole carcass (n)	Partly condemned (n)	
				Parts of carcasses	Organs
Ante mortem inspection (n)		*	*		
Contamination (dirty fur)					
Eye infection					
Diarrhoea					
Central nervous symptoms					
Wet belly					
Other reason					
Body condition class (1–4) of calves					
Abnormal fur					
Abnormal antler development (calf)					
Post mortem inspection (n)		*	*	*	*
Abnormal odor, colour or structure					
Bruises					
Fractures					
Cahexia					
Tumour					
Aspiration					
Contamination					
Besnoitia					
Echinococcosis					
Deer ked (*Lipoptena cervi*)					
Lung changes					
Taenia hydatigena (*cysticercus tenuicollis*)					
Muscle cyst worm (*Taenia krabbei*)					
Warble fly (*Hypoderma tarandi*)					
Nose bot (*Cephenemyia trompe*)					
Setaria tundra					
Onchocerca					
Hot ear (*Lappnema auris*)					
Liver scars (parasitic granuloma)					
Other parasite					
Generalized inflammation					
Skin inflammation					
Abomasitis					
Pneumonia					
Pleuritis					
Pericarditis					
Arthritis					
Enteritis					
Stomatitis					
Papillomatosis					
Abscess					
Peritonitis					
Other inflammation					
Other reason					

* mandatory information

13 Haematology and Blood Biochemistry Reference Values for *Rangifer*

Morten Tryland and Timo Soveri

CONTENTS

13.1 HAEMATOLOGY AND SERUM/PLASMA BIOCHEMISTRY

Measuring *Rangifer* health can be done using a variety of indicators. Haematology and serum/plasma biochemistry can provide insights into the physiological status of an individual, blood disorders and other abnormalities.

Haematology is the science concerned with the study of blood, blood producing organs and blood related diseases. Haematological values include quantification and differentiation of red and white blood cells, hemoglobin, packed cell volume, cell morphology and other measures (see Table 13.1). Haematology can also be used to detect and assess diseases in other organ systems, e.g. changes in the number and characteristics of white and red blood cells can infer infection and other abnormalities.

Clinical biochemistry is based upon chemical analyses of body fluids, usually serum or plasma, for their substances. Serum and plasma are the clear, acellular fluids that separate from the cells after clotted and nonclotted blood, respectively, is centrifuged. Plasma is essentially the serum plus the coagulation factors, such as fibrinogen. To obtain plasma, blood needs to be collected in tubes with an anticoagulant (e.g. EDTA, heparin and others). Serum and plasma contain proteins (without and with the coagulation proteins, respectively), enzymes, electrolytes, minerals and hormones, as well as exogenous substances if present, e.g. microorganisms (viruses, bacteria, etc.), drugs and toxins.

Many physiological conditions, like strenuous exercise and stress, may be indicated by certain changes in haematology and blood chemistry. These changes may also impact meat quality, an issue that is dealt with in Chapter 11, Meat Quality and Meat Hygiene. Blood analysis may be helpful to detect insufficiencies and disease of organs, mineral imbalances, poisoning, malnutrition or starvation and other types of diseases. It is, however, beyond the scope of this book to present details on how to interpret blood haematology and biochemistry, and the readers are referred to other sources (e.g. Desai and Samir, 2009; Kaneko et al., 2014).

Serology is the study of specific immunoglobulins such as IgG and IgM, usually referred to as antibodies, to trace a humoral immune response in an animal after exposure to foreign proteins (i.e. antigens, like parasites, bacteria or viruses). Serology is thus an important tool to detect previous exposure of an animal or a population to an infectious agent.

TABLE 13.1

Haematology Values for Norwegian Wild, Free-Ranging Eurasian Tundra Reindeer (*Rangifer tarandus tarandus*) after Immobilization with Medetomidine-Ketamine by Dart from a Helicopter

Parameter[a]	Unit	Range[b]	Mean ± SE	Median[b]
RBC	× 10⁹/L	9.24–11.96	10.70 ± 0.17	10.60
HGB	g/L	145–211	173 ± 2.58	173
HCT	L/L	0.39–0.57	0.47 ± 0.01	0.48
MCV	fL	40.3–48.0	44.4 ± 0.3	44.3
MCH	pg	15.0–17.5	16.3 ± 0.1	16.3
MCHC	g/L	351–384	366 ± 2	366
RDW	%	13.0–16.4	14.4 ± 0.2	14.2
HDW	g/L	19.9–40.3	24.9 ± 0.8	23.8
PLT	× 10⁹/L	74–603	253 ± 24	225
MPV	fL	3.4–6.9	5.2 ± 0.2	5.5
WBC	× 10⁹/L	0.96–5.19	2.96 ± 0.19	2.86
Neutrophils	× 10⁹/L	0.56–4.52	1.91 ± 0.18	1.75
Neutrophils	%	37.5–87.2	62.0 ± 2.8	62.5
Lymphocytes	× 10⁹/L	0.37–1.69	0.90 ± 0.06	0.87
Lymphocytes	%	8.9–53.3	31.4 ± 2.6	29.5
Monocytes	× 10⁹/L	0.0–0.03	0.01 ± 0.01	0.01
Monocytes	%	0.0–1.3	0.4 ± 0.1	0.4
Eosinophils	× 10⁹/L	0.00–0.01	0.00 ± 0.00	0.00
Eosinophils	%	0.0–0.4	0.1 ± 0.0	0.0
Basophils	× 10⁹/L	0.01–0.13	0.05 ± 0.01	0.04
Basophils	%	0.5–4.7	1.5 ± 0.3	1.1
LUC	× 10⁹/L	0.03–0.33	0.16 ± 0.01	0.15
LUC	%	2.4–7.5	5.26 ± 0.5	4.5

[a] RBC = red blood cell count; HGB = hemoglobin; HCT = haematocrit; MCV = mean corpuscular volume; MCH = mean corpuscular hemoglobin; MCHC = MCH concentration; RDW = red cell distribution width; HDW = hemoglobin cell distribution width; PLT = platelet count; MPV = mean platelet volume; WBC = white blood cell count, LUC = large unstained cells.

[b] Measurements are based on 18–29 individuals (Modified after Miller et al., 2013: see publication for details).

13.2   PRE-ANALYTICAL FACTORS THAT MAY IMPACT BLOOD PARAMETER MEASUREMENTS

There are numerous pre-analytical factors that might have impact on haematological and serum/plasma biochemistry measurements and results, and thus introduce bias to the dataset.

Stress, capture, immobilization and restraint methods (e.g. net-gunning, chemical immobilization, use of lasso) prior to blood sampling may impact blood constituents. Strenuous exercise and stress may change blood values, typically by quickly elevating cortisol concentrations and elevating the activity of lactate dehydrogenase (LDH), alkaline phosphatase (ALP) and creatine kinase (CK) (Rehbinder and Edqvist, 1981; Arnemo et al., 1994; Marco and Lavin, 1999; Säkkinen et al., 2004). Chemical immobilization may be less stressful than physical restraint, but a period of chasing, by foot, snowmobile or helicopter, is usually necessary to get close enough for darting. In addition, the drug or drug combination itself may introduce changes in blood parameters (Wolkers et al., 1994; Soveri et al., 1999; Arnemo and Ranheim, 1999).

Many factors may also impact haematology and serum/plasma biochemistry after the collection of the sample. Length of time between collection and analyses, and storage methods, can influence results. Haematology analyses should be conducted on fresh blood samples, preferably on the day of collection. This is not always possible in remote regions with restricted infrastructure, and sometimes a full haematology analysis is not possible. However, through the use of direct blood smears and portable analytic devices, valuable information, albeit a reduced number of variables, can still be obtained.

Storage of samples may also impact serum or plasma biochemistry results. Serum or plasma samples can be frozen, preferably at –80°C, but should nevertheless be analyzed as soon as possible. Storage at –20°C or warmer, repeated freezing and thawing, and storage for long periods (months to years) may alter the concentrations of several analytes and also reduce enzyme activity over time (Thoresen et al., 1995).

In addition to storage, there may be certain characteristics of the blood sample, such as hemolysis and lipemia, which may impact blood analysis. Hemolytic samples – for example, caused by freezing or difficulties in sampling – appear as slightly pink to dark red or almost black. Lipemic samples are slightly yellow to completely white and opaque, and are easily recognized by visual inspection. When analyzed by spectrophotometric methods, usually in automated analyzers, the degree of hemolysis and lipemia is recognized and, if possible, adjusted for, since hemolysis and lipemia may interfere with the measurements and introduce bias (i.e. over- and underestimation) of a wide range of serum chemistry analytes (Tryland and Brun, 2001; Lippi et al., 2006; Koseoglu et al., 2011). However, sometimes the grade of hemolysis or lipemia is so severe that the samples are not suitable for such analyses or it might be necessary to dilute the sample prior to analysis. One challenge with the interpretation of the results obtained from such samples is that there is no harmonization between different laboratories on how to evaluate the possible interference of hemolysis and lipemia on the outcome of the measurements (Farell and Carter, 2016).

Although pre-analytical factors, as exemplified above, may be the main sources of variation in clinical biochemistry analysis, variation may also be seen if different methods are used and even with the same methods performed in different laboratories (Lumsden, 1998). Thus, when reporting and comparing haematology and serum chemistry results, it is relevant to include the methods that were used and ensure that the units of measurement are the same when comparing different sample sets and reference values. Preferably, SI units (International System of Units; French: *Système international d'unités*, SI) should be used (Braun, 2009).

13.3   VARIABILITY AND NORMALITY

Haematology and serum chemistry values vary between different animal species. Within a species there will be a "normal range" or "reference values" that delineate the range of values that might be expected in healthy individuals. Such reference values are lacking for many wildlife species, and sometimes we have to refer to comparative knowledge for closely related species or subspecies.

Animals encounter different environments, defined by their owners (housing, food, exercise, etc.) and the ecological settings (ecotypes, pasture resources, food availability, etc.). These conditions are reflected in physiological and metabolic responses that influence haematology and serum/plasma biochemistry values. Gender, age and seasonal physiological changes also impact constituents of the blood, such as feeding and fasting, rut and breeding, pregnancy and strenuous activity such as long seasonal migrations, as well as change of hair coat (Nieminen 1980; Soveri et al., 1999; Hassan et al., 2012; Miller et al., 2013).

The number of observations (e.g. individuals represented) for each analyte should be > 40 in order to statistically calculate reference values for a species or subspecies (Lassen, 2006). Depending on the purpose, it may also be necessary to establish separate reference values for young and older animals, males and females, captive and wild, etc.

Reference values are often referred to as "normal blood parameters." The word "normal" is derived from the Latin word *norma*, which means guide or rule. Since normal is a challenging and often a subjective entity, we often use other words, such as "expected" or "within a certain range," when dealing with a set of reference values (Farver, 1997).

When facing variability, and sometimes a low number of observations, the task of establishing reference values may seem unattainable (Tryland, 2006). One pitfall when facing small numbers, suboptimal sample quality and long term storage of the samples is to think that whichever way the sampling and analyses are conducted, bias will be introduced, and thus, it is of no use trying to establish reference values. However, it is often better to obtain a set of reference values – recognizing, presenting and discussing the conditions and the factors that might have had impact on the analysis – rather than not conduct such efforts at all because the result will probably not be perfect. Such datasets, although they have built-in weaknesses and uncertainties, may still be valuable reference ranges for this species or population. The factors behind the variation, discussed above, including the laboratory methods and their analytical variation, should thus ideally be identical in the observed population and in the reference population.

When dealing with animals in captivity, and especially over a period of time (e.g. zoos and parks), blood and serum/plasma biochemical analyses are valuable monitoring tools, and it can be very helpful to establish reference values on an individual level, including measurements from different seasons, feeding regimes, pregnancy status, etc.

Haematology and clinical chemistry reference range data are often presented as the mean (average) value and the standard deviation (SD, or the Greek letter sigma in lower case, σ) as a measure of variation or dispersion of a set of values, generated from the analysis of a representative sample from the population. An animal will (by convention) be considered as normal for a particular analyte if its value for the analyte is within the mean ±2 SD, although the multiple of SD can be defined as smaller or greater than 2 (Farver, 2008). Using a Gaussian distribution of the analyte for a normal (healthy) population, 95% of the distribution will be located within 1.96 SD of the mean (i.e. close to 2), meaning that using the mean ±2 SD rule, approximately 2.5% of the healthy animals would have values for the particular analyte below the lower reference limit (mean –2 SD) and 2.5% of the animals would have values above the higher reference limit (mean +2 SD), giving a 5% chance that a true "normal animal" would be classified as abnormal for the analyte (Farver, 2008).

If the Gaussian distribution is not the best assumption for the analyte, meaning the analyte for the population has an asymmetric or skewed distribution, with a tail to the left or right, other calculations than mean ±2 SD can be used to find better reference intervals for that particular analyte. Either the analyte values can be transformed (logarithmic or square root transformation), rendering their distribution more Gaussian, or percentiles can be calculated as boundaries for the reference intervals, such as the 2.5th and the 97.5th percentiles, or the 5th and 95th percentiles, leaving extreme values outside the reference intervals since they may impact such intervals to a great extent (Farver, 2008). It is also common to present the range of all the measurements for the measured population, which are the minimum and the maximum values measured.

13.4 BLOOD REFERENCE VALUES FOR REINDEER AND CARIBOU

To be able to evaluate blood results of reindeer and caribou, it is important to have reference values, suggesting the expected ranges for the subspecies in question. Even better is to have established reference data for the same population, and also for both genders, for calves, for adult and old animals and for different seasons. For most *Rangifer* subspecies and populations, data on haematology and serum/plasma biochemistry is scarce or not available, but some reports exist for some subspecies (Catley et al., 1990; Timisjärvi et al., 1976; Soveri et al., 1992). However, many of these reports are rather old and the methods and equipment that were used for analyses have changed compared to the automated analyzers that are used today. Thus, we have chosen to present reference values that have been obtained more recently.

Table 13.1 (haematology) and Table 13.2 (serum biochemistry) are reference ranges for wild Eurasian tundra reindeer (*R. t. tarandus*) in Norway (Miller et al., 2013). The data is based on reindeer from the Nordfjella wild reindeer herd in southeast Norway, with 31 animals (nine males and 22 females) analyzed for haematology parameters (Table 13.1) and 29 animals (eight males and 29 females) for serum biochemical parameters (Table 13.2). The animals, all presumably healthy adults ≥ 1.5 years old, were immobilized with medetomidine-ketamine by dart from a helicopter prior to sampling (Arnemo et al., 2011).

Table 13.3 presents plasma biochemistry reference values for 127 semi-domesticated Eurasian tundra reindeer (*R. t. tarandus*) from herds from seven different regions of Norway (Tana, Lakselv, Tromsø, Hattfjelldal, Fosen, Røros and Valdres), thus representing animals exposed to

TABLE 13.2

Serum Biochemical Values for Norwegian Wild Eurasian Tundra Reindeer (*Rangifer tarandus tarandus*) after Immobilization with Medetomidine-Ketamine by Dart from a Helicopter

Parameter[a]	Unit	Range[b]	Mean ± SE	Median[b]
AST	U/L	57–211	104 ± 6	96
ALT	U/L	22–77	38 ± 2	34
ALP	U/L	100–483	257 ± 20	231
CK	U/L	130–928	356 ± 37	320
LD	U/L	644–1671	961 ± 41	923
GGT	U/L	10–49	24 ± 2	22
GD	U/L	0–10	6 ± 2	3
Amylase	U/L	12–76	45 ± 3	47
Lipase	U/L	5–30	14 ± 1	13
Protein (total)	g/L	52–71	62 ± 1	61
Urea	mmol/L	1.5–15	4.5 ± 0.6	3.1
Creatinine	μmol/L	148–229	178 ± 3	178
Uric acid	μmol/L	0–34	4 ± 2	0
Total bilirubin	μmol/L	1–5	2 ± 0	2
Cholesterol	mmol/L	1–2	1.5 ± 0	1.5
Triglycerides	mmol/L	0.1–0.5	0.2 ± 0	0.2
Free fatty acids	mmol/L	0.1–1.3	0.6 ± 0.1	0.5
B-HBA	mmol/L	0.4–1.5	0.8 ± 0.1	0.7
Glucose	mmol/L	0.9–14.8	6.7 ± 0.7	6.4
Phosphorus	mmol/L	0.7–2.2	1.4 ± 0.1	1.3
Calcium	mmol/L	2.2–2.7	2.4 ± 0	2.4
Magnesium	mmol/L	0.69–1.04	0.88 ± 0.02	0.9
Sodium	mmol/L	136–153	144 ± 1	144
Potassium	mmol/L	2.6–4.4	3.6 ± 0.1	3.5
Chloride	mmol/L	96–107	101 ± 0	101
Iron	μmol/L	16–39	27 ± 1	26
Copper	μmol/L	7–14	11 ± 0	10
Zinc	μmol/L	8–14	11 ± 0	11
Selenium	μg Se/g blood	0.17–0.27	0.21 ± 0.01	0.2
Cortisol	nmol/L	110–477	299 ± 15	300

[a] AST: aspartate aminotransferase; ALT: alanine transaminase; ALP: alkaline phosphatase; CK: creatine kinase; LD: lactate dehydrogenase; GGT: gamma-glutamyl transpeptidase; GD: glutamate dehydrogenase.

[b] Measurements are based on 17–31 individuals (Modified after Miller et al., 2013: see publication for details).

TABLE 13.3

Clinical Plasma Chemistry Values for 127 Semi-Domesticated Eurasian Tundra Reindeer (*Rangifer tarandus tarandus*) Presented for Adults and Calves as Mean, Range (Min-Max) and 2.5%–97.5% Percentiles. Blood was Collected in Vacutainer Blood Tubes from the Jugular Vein during Physical Restraint of the Animals

Parameter	Unit	Adults (>16 Months, n = 53)			Calves (4–8 Months, n = 74)		
		Mean	Range	2.5–97.5% Percentiles	Mean	Range	2.5–97.5% Percentiles
Enzymes[a]:							
AST	U/L	128	64–385	76–321	112	65–268	72–185
ALT	U/L	57	19–100	39–92	58	35–114	37–88
GGT	U/L	14	0–31	0.5–29	14	0–48	3.5–32
CK	U/L	882	125–9232	147–3938	595	87–5400	140–2419
GLDH	U/L	2	0–12	0.0–9.8	2	0–10	0.0–5.4
Proteins:							
Protein, total	g/L	70	53–91	55–89	70	56–87	59–84
Albumin	g/L	33	26–38	29–37	33	29–37	29–37
Metabolites:							
Urea	mmol/L	13	3–25	2.6–24.5	12	2–24	2.2–23.6
Creatinine	μmol/L	167	121–210	125–207	170	98–218	113–213
Cholesterol	mmol/L	2	1–3	1.1–2.1	2	1–3	1.1–2.5
Non-esterified fatty acids	mmol/L	2	1–3	1.0–2.1	2	0–3	0.6–2.7
Glucose	mmol/L	7	3–10	4.2–10.0	7	5–12	5.0–10.4
Bilirubin	μmol/L	6	4–13	3.9–11.7	6	2–11	3.1–9.8
Triglycerides	mmol/L	0	0–1	0.2–0.5	0	0–1	0.2–0.6
Glycerol	μmol/L	194	81–364	91–341	202	79–428	85–388
Ranbut (D3-Hydroxybutyrate)	mmol/L	0	0–1	0.1–0.6	0	0–1	0.1–0.5
Minerals:							
Phosphate, inorganic	mmol/L	2	1–3	1.0–2.8	2	1–3	1.0–3.0
Sodium	mmol/L	130	117–143	118–141	130	118–140	121–139
Chloride	mmol/L	105	98–111	99–111	105	99–111	100–110

[a] AST: aspartate aminotransferase; ALT: alanine transaminase; GGT: gamma-glutamyl transpeptidase; ALP: alkaline phosphatase; CK: creatine kinase; LD: lactate dehydrogenase; GGT: gamma-glutamyl transpeptidase; GD: glutamate dehydrogenase.

different climate, ecosystems and herding conditions (M. Tryland and T. Soveri, unpublished data) (Table 13.3). The animals were corralled and sampled under physical restraint by using a Vacutainer and a hypodermic needle in the jugular vein. Plasma was prepared and stored at −80°C until analysis.

In Table 13.4, we present serum biochemistry values for free-ranging adult female boreal woodland caribou (*R. t. caribou*) from northeast British Columbia (n = 81, 2012–2013) (Bondo et al., 2018) and Northwest Territories (n = 104, 2003–2009) (Johnson et al., 2010), Canada. All animals were caught by net-guns deployed from a helicopter.

TABLE 13.4

Serum Biochemistry Reference Values for Free-Ranging Live-Captured Adult Female Boreal Woodland Caribou (*Rangifer tarandus caribou*) from Northeast British Columbia and from Northwest Territories, Canada

Parameter[a]	Unit	Northeast British Columbia (n = 81)			Northwest Territories (n = 108)
		Mean or Median[b]	Range	Percentiles 2.5–97.5	Percentiles 2.5–97.5
Enzymes:					
AST	U/L	69[b]	23–152	40–119	42–163
ALT	U/L	55[b]	23–141	27–110	–
GGT	U/L	17[b]	0–90	1–64	
CK	U/L	206[b]	78–780	79–462	96–1614
GLDH	U/L	2[b]	0–18	1–11	–
Proteins:					
Protein, total	g/L	70	55–83	62–83	45–86
Albumin	g/L	43	34–49	36–47	28–46
Globulin	g/L	26[b]	16–42	20–41	17–45
Albumin-Globulin ratio		1.6	1.0–2.4	1.0–2.2	0.9–1.8
Metabolites:					
Beta-hydroxybutyrate	U/L	585	281–1060	338–844	
Urea	mmol/L	1.3[b]	0.8–4.2	1.0–2.5	1.0–4.3
Creatinine	μmol/L	210[b]	132–299	158–276	123–294
Cholesterol	mmol/L	1.2	0.63–1.7	0.86–1.56	–
Non-esterified fatty acids	mmol/L	0.6[b]	0.1–1.8	0.2–1.6	–
Glucose	mmol/L	6.7	1.6–10.8	3.2–10.3	4.1–13.0
Bilirubin total	μmol/L	1[b]	0–4	1.0–3.0	0.5–5.0
Bilirubin conjugated	μmol/L	1[b]	0–1	0–1	–
Bilirubin free	μmol/L	1[b]	0–4	0–3	–
Minerals:					
Calcium	mmol/L	2.49[b]	1.51–2.77	2.24–2.74	1.64–3.15
Phosphorus	mmol/L	2.0	0.80–2.86	1.29–2.75	0.73–2.53
Calcium : Phosphorus ratio		1.2[b]	0.77–3.3	0.88–1.81	–
Sodium	mmol/L	143[b]	122–156	125–150	93–170
Potassium	mmol/L	5.6[b]	3.4–28.8	3.7–28.1	3.0–8.0
Sodium : Potassium ratio		26[b]	4–44	4–40	–
Chloride	mmol/L	95[b]	86–104	87–99	55–106
Magnesium	mmol/L	1.1[b]	0.9–1.4	0.9–1.2	0.6–1.4

[a] Enzyme abbreviation: AST: aspartate aminotransferase; ALT: alanine transaminase; GGT: gamma-glutamyl transpeptidase; CK: creatine kinase; GLDH: glutamate dehydrogenase.

[b] Mean is presented for parameters that were normally distributed, median for parameters that were not normally distributed, marked with b.

The table is modified from Bondo and coworkers (2018) and Johnson and coworkers (2010) (see publications for details).

Table 13.5 displays serum chemistry reference range values for 16 Svalbard reindeer (*R. t. platyrhynchus*). The animals were chemically immobilized by darting from the ground, using medetomidine-ketamine (M. Tryland, unpublished data).

TABLE 13.5

Serum Biochemistry Reference Values for Svalbard Reindeer (*Rangifer tarandus platyrhynchus*) (n = 16), Chemically Immobilized with Medetomidine and Ketamine by Darting, Approaching the Animals by Foot

Parameter	Unit	Mean	Range	2.5–97.5% Percentiles
Enzymes[a]:				
AST	U/L	83	66–119	66–112
ALT	U/L	37	20–54	20–54
CK	U/L	210	116–395	121–385
AP	U/L	90	36–144	39–139
LD	U/L	495	356–603	368–602
Amylase	U/L	29	20–43	20–41
Lipase	U/L	16	12–20	12–19
Proteins:				
Protein (total)	g/L	58	54–63	54–63
Albumin[b]	g/L	41	39–43	39–43
Alpha-1 globulins[b]	g/L	3	2.9–3.8	3.0–3.7
Alpha-2 globulins[b]	g/L	3	2.7–4.5	3.0–4.1
Beta-1 globulins[b]	g/L	2	1.7–2.7	2.0–2.7
Beta-2 globulins[b]	g/L	1	0.9–1.6	1.0–1.6
Gammaglobulins[b]	g/L	7	5.5–9.5	6.0–9.4
Albumin/Globulin Ratio	–	2	2.0–2.9	2.0–2.9
Metabolites:				
Urea	mmol/L	6	3.1–8.8	3–8.5
Creatinine	μmol/L	146	117–201	117–196
Cholesterol	mmol/L	2	1.3–4.1	1–3.7
Non-esterified fatty acids	mmol/L	0	0.1–0.6	0–0.6
Glucose	mmol/L	13	9.7–14.8	10–14.8
Bilirubin (total)	μmol/L	1	1–2	1–1.6
Triglycerides	mmol/L	0	0.1–0.3	0–0.3
Glycerol	μmol/L			
Minerals:				
Phosphate, inorganic	mmol/L	2	1.2–2.1	1–2.1
Sodium	mmol/L	135	130–141	130–141
Chloride	mmol/L	100	97–107	97–106
Calcium	mmol/L	2	2.1–2.5	2–2.5
Magnesium	mmol/L	1	0.8–1.4	1–1.3
Potassium	mmol/L	4	3.6–5.5	4.0–5.2
Hormones:				
Cortisol	nmol/L	192	72–326	79–324
T4 total	nmol/L	132	103–160	104–159

[a] AST: aspartate aminotransferase; ALT: alanine transaminase; CK: creatine kinase; LD: lactate dehydrogenase.

[b] Quantified by protein electrophoresis.

It is important to look at most of these parameters as indicative and as a helpful diagnostic tool, to be used in addition to the disease history (anamnesis), the clinical symptoms present, and the conditions and the environment the animal is experiencing.

REFERENCES

Arnemo, J. M., T. Negard, and N. E. Søli. 1994. Chemical capture of free-ranging red deer (*Cervus elaphus*) with medetomidine-ketamine. *Rangifer* 14:123–7.

Arnemo, J. M., and B. Ranheim. 1999. Effects of medetomidine and atipamezole on serum glucose and cortisol levels in captive reindeer (*Rangifer tarandus tarandus*). *Rangifer* 19:85–9.

Arnemo, J. M., A. L. Evans, A. L. Miller, and Ø. Os. 2011. Effective immobilizing doses of medetomidine-ketamine in free-ranging, wild Norwegian reindeer (*Rangifer tarandus tarandus*). *J Wildl Dis* 47:755–8.

Bondo, K. J., B. Macbeth, H. Schwantje et al. 2018. Health status of live-captured boreal caribou in Northeastern British Columbia during a year of unusually high mortality (In Prep).

Braun, J.-P. 2009. Communicating with precision in veterinary clinical pathology: Definitions, units, and nomenclature. *Vet Clin Pathol* 38:416–7. DOI:10.1111/j.1939-165X.2009.00197.x

Catley, A., R. A. Kock, M. G. Hart, and C. M. Hawkey. 1990. Haematology of clinically normal and sick captive reindeer (*Rangifer tarandus*). *Vet Rec* 126:239–41.

Desai, M. D., and P. Samir. 2009. *Clinician's Guide to Laboratory Medicine: Pocket.* MD2B, Houston, TX, USA. pp. 1–255.

Farell, C. J., and A. C. Carter. 2016. Serum indices: Managing assay interference. *Ann Clin Biochem* 53:527–38.

Farver, T. B. 1997. Concepts of normality in clinical biochemistry. In *Clinical Biochemistry of Domestic Animals.* 5th edition. J. J. Kaneko, J. W. Harvey, M. L. Bruss, editors. Academic Press, London. pp. 1–19.

Farver, T. B. 2008. Concepts of normality in clinical biochemistry. In *Clinical Biochemistry of Domestic Animals.* 6th edition. J. J. Kaneko, J. W. Harvey, M. L. Bruss, editors. Elsevier, London. pp. 1–26.

Hassan, A. A., T. M. Sandanger, and M. Brustad. 2012. Selected vitamins and essential elements in meat from semi-domesticated reindeer (*Rangifer tarandus tarandus* L.) in mid- and northern Norway: Geographical variations and effect of animal population density. *Nutrients* 4:724–39.

Johnson, D., N. J. Harms, N. C. Larter et al. 2010. Serum biochemistry, serology, and parasitology of boreal caribou (*Rangifer tarandus caribou*) in the Northwest Territories, Canada. *J Wildl Dis* 46:1096–107.

Kaneko, J. J., J. J. Kaneko, and C. E. Cornelius. 2014. *Clinical Biochemistry of Domestic Animals.* 2nd edition. Elsevier, Academic Press, 454 pp.

Koseoglu, M., A. Hur, A. Atay, and S. Cuhadar. 2011. Effects of hemolysis interferences on routine biochemistry parameters. *Biochem Med* 21:79–85.

Lassen, E. D. 2006. Laboratory evaluation of the liver. In *Veterinary Hematology and Clinical Chemistry.* M. A. Thrall, D. C. Baker, T. W. Campbell, D. DeNicola, M. J. Fettman, E. D. Sassen, A. Rebar, G. Weiser, editors. Blackwell Publishing, Ames, Iowa, pp. 355–75.

Lippi, G., G. L. Salvagno, M. Montagnana, G. Brocco, and G. C. Guidi. 2006. Influence of hemolysis on routine clinical chemistry testing. *Clin Chem Lab Med* 44:311–6.

Lumsden, J. H. 1998. 'Normal' or reference values: Questions and comments. *Vet Clin Pathol* 27:102–6.

Marco, I., and S. Lavìn. 1999. Effect of the method of capture on the haematology and blood chemistry of red deer (*Cervus elaphus*). *Res Vet Sci* 66:81–4.

Miller, A. L., A. L. Evans, Ø. Os, and J. M. Arnemo. 2013. Biochemical and hematologic reference values for free-ranging, chemically immobilized wild Norwegian reindeer (*Rangifer tarandus tarandus*) during early winter. *J Wildl Dis* 49:221–8.

Nieminen, M. 1980. Nutritional and seasonal effects on the haematology and blood chemistry in reindeer (*Rangifer tarandus tarandus* L). *Comp Biochem Physiol, Part A: Mol Integr Physiol* 66A:399–413.

Rehbinder, C., and L.-E. Edqvist. 1981. Influence of stress on some blood constituents in reindeer (*Rangifer tarandus* L). *Acta Vet Scand* 22:480–92.

Säkkinen, H., J. Tornbeg, P. J. Goddard, E. Eloranta, E. Ropstad, and S. Saarela. 2004. The effect of blood sampling method on indicators of physiological stress in reindeer (*Rangifer tarandus tarandus*). *Domest Anim Endocrinol* 26:87–98.

Soveri, T., S. Sankari, and M. Nieminen. 1992. Blood chemistry of reindeer calves (*Rangifer tarandus*) during the winter season. *Comp Biochem Physiol Comp Physiol* 102(1):191–6.

Soveri, T., S. Sankari, J. S. Salonen, and M. Nieminen. 1999. Effects of immobilization with medetomidine and reversal with atipamezole on blood chemistry of semi-domesticated reindeer (*Rangifer tarandus tarandus* L.) in autumn and late winter. *Acta Vet Scand* 40:335–49.

Thoresen, S. I., A. Tverdal, G. Havre, and H. Morberg. 1995. Effects of storage time and freezing temperature on clinical chemical parameters from canine serum and heparinized plasma. *Vet Clin Pathol* 24:129–33.

Timisjärvi, J., M. Reinilä, and P. Järvensivu. 1976. Haematological values for the Finnish reindeer. *Blut* 32:439–42.

Tryland, M., and E. Brun. 2001. Serum chemistry of the minke whale from the northeastern Atlantic. *J Wildl Dis* 37:332–41.

Tryland, M. 2006. Normal serum chemistry values in wild animals. *Vet Rec* 158:211–2.

Wolkers, J., T. Wensing, and G. W. T. A. Bruinderink. 1994. Sedation of wild boar (*Sus scrofa*) and red deer (*Cervus elaphus*) with medetomidine and the influence on some haematological and serum biochemical variables. *Vet Q* 16:7–9.

14 Caribou and Reindeer in Parks and Zoos

Douglas P. Whiteside and Owen M. Slater

CONTENTS

Reindeer and caribou are popular cervids in zoos, particularly those in northern latitudes. Within the Species 360 (formerly International Species Inventory System) database, there are approximately 500 *Rangifer* spp. in 136 zoological institutions, comprising American woodland caribou, reindeer, tundra reindeer, Eurasia tundra reindeer and European forest reindeer.

14.1   HUSBANDRY REQUIREMENTS

Caribou and reindeer should be maintained on large natural ground paddocks that are bordered by secure fences that are at least 2.5 m (8 ft.) high to prevent escape and the entry of wild deer or predators. Some zoos will have a stand-off barrier from the fence to prevent bulls in rut from entangling their antlers. The ability to divide the pasture or have multiple pastures is of great benefit for pasture rotation and maintenance, and for separating herd members when needed (Figures 14.1 and 14.2). The exhibit should provide for adequate shade and windbreak through the use of natural plantings and artificial structures. The provision of rubbing structures, such as logs, industrial snow sweeping brushes and dead trees, is important when animals are losing their velvet. Off exhibit areas with well-developed handling facilities are ideal for herd management. Behavioural husbandry can be accomplished by training for medical or husbandry procedures and through environmental enrichment such as the addition of enrichment devices, changing the type and location of exhibit furnishing, the use of novel scents and a variety of low starch food items.

FIGURE 14.1   Caribou exhibit at the Calgary Zoo, Alberta, Canada. Note the ability to separate the natural pastures as needed (Photo: Doug Whiteside).

FIGURE 14.2   Reindeer exhibit at Wuppertal Zoological Gardens, Germany (Photo: Maya Kummrow).

14.2   NUTRITION

Rangifer species are browsers with opportunistic grazing on leafy plants and short leafy grasses, but have limited ability to digest long grass fibers rich in cellulose, such as mature timothy hay (*Phleum pratense*). Feeding grass hays or silage leads to a reduction in rumen papillae density and weight loss, and can lead to rumenal impactions (Josefsen et al., 1996; Norberg and Mathiesen, 1998; Olsen et al., 1995). Food intake is approximately 2–3% of body weight, depending on climate. In most zoos, feeding a combination of leafy forbs such as alfalfa (lucerne) hay and seasonal browse or commercially available dried moss, in combination with a pelleted feed formulated for browsing species with a beet pulp base as the pectin source rather than grains, has proven effective (Masters and Flach, 2014). Pellets make up 30–40% of the diet and hay and browse the remainder of the diet.

14.3 INFECTIOUS DISEASES

Captive *Rangifer* are susceptible to a number of infectious diseases in captivity. For more specific information on the pathogenesis and diagnosis of the diseases highlighted in this chapter, please refer to other chapters within this book. These disease lists are not exhaustive and each facility will have variations, especially between North America and Europe.

14.3.1 VIRAL AND PRION DISEASES

Although serological evidence of exposure to various viral diseases, especially bovine respiratory diseases, has been documented in captive reindeer and caribou, clinical disease is rare in zoos. *Rangifer* species are susceptible to malignant catarrhal fever (Ovine Herpesvirus-2), with clinical signs including sudden death, pyrexia, oculonasal discharge, corneal opacities, neurological signs and dysentery, so care not to house them in close proximity to sheep and goats is paramount (Flach 2003; Kiupel et al., 2004; Li et al., 1999; Masters and Flach, 2014). Cervine herpesvirus Type-2 has been associated with keratoconjunctivitis in reindeer, and is purported to be a factor in some types of pneumonia, and possibly abortion and neonatal mortality, in captive reindeer and caribou (Foster, 2010). Fibropapillomatosis associated with *Rangifer tarandus* Papillomavirus-1 (RPV) has been documented in zoos. Lesions are noted on haired skin and antlers (Figure 14.3) and may be severe in some cases, but can spontaneously regress (Rector and Van Ranst, 2013). Pseudocowpoxvirus and cervidpoxvirus causing crusting or chronic proliferative nodular lesions on the muzzle, eyelids, tongue, lips, gums and palate also have been described (Hautaniemi et al., 2010). West Nile virus has been associated with fatal lymphohistiocytic encephalomyelitis in captive *Rangifer* in North America (Palmer et al., 2004).

Chronic wasting disease has never been reported in captive *Rangifer* sp., but has recently been reported in a free ranging reindeer (Benestad et al., 2016), and they can be infected experimentally (Mitchell et al., 2012). This should be a careful consideration if zoos are planning to exhibit reindeer or caribou with other more susceptible cervids.

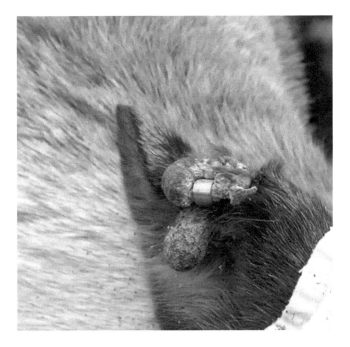

FIGURE 14.3   Papillomatous lesion on the ear of an American woodland caribou (Photo: Michelle Oakley).

14.3.2 BACTERIAL DISEASES

Tuberculosis (*Mycobacterium bovis*) and brucellosis (*Brucella suis* biovar 4) infections are important bacterial diseases of captive *Rangifer*, and strict quarantine, routine testing and biosecurity measures should be in place to maintain disease free herds.

Johne's disease (*Mycobacterium avium paratuberculosis*) is also of concern for captive animals and efforts should be directed at prevention. Similar to goats and camelids, clinical signs of weight loss without diarrhoea may be seen, and animals may be affected at a younger age (Vansnick, 2004). Screening by faecal PCR or faecal culture can be done, and deceased animals should have tissues collected for Johne's disease testing. Screening animals prior to importation is also recommended (Del-Pozo et al., 2013; Fowler and Boever, 1986). Other bacterial diseases of concern for captive animals include those associated with *Clostridium* species (tetanus, blackleg, bacillary hemoglobinuria, enterotoxemia, gas gangrene/malignant edema) (Herron et al., 1979, Mackintosh et al., 2002; Voight et al., 2009). Keeping stocking densities low, avoiding sudden changes in diet, reducing the incidents of trauma, and vaccination of animals are highly recommended. Neonatal diarrhoea due to *Salmonella* and *E. coli* are not uncommon (Mackintosh et al., 2002). Depending on the facility and its historical issues with neonatal diarrhoea, some facilities will handle newborn calves within the first 24 hours to apply a disinfectant to the umbilical cord and give additional medications as indicated (Dieterich and Morton, 1990). Other bacteria linked to disease in captive *Rangifer* include *Fusobacterium necrophorum* (foot rot, necrotic stomatitis), leptospirosis, listeriosis, pasteurellosis (pneumonia) and yersiniosis (Evans and Watson, 1987; Flach, 2003; Mackintosh et al., 2002).

14.3.3 FUNGAL DISEASES

Dermatophytosis associated with *Trichophyton verrucosum* has been reported in caribou and is the most likely pathogen in cases of ringworm in *Rangifer* species in zoological collections (Koroleva, 1976). Systemic mycoses (*Zygomycetes, Aspergillus* sp.) is rare and usually associated with immunosuppression in young animals (Rehbinder et al., 1994; Yokota et al., 2004).

14.3.4 PARASITIC DISEASES

Captive *Rangifer* seem particularly predisposed to acquiring and developing clinical signs associated with parasitic diseases. Attention to pasture maintenance, faecal removal and keeping densities of animals low will invariably help reduce the prevalence of these diseases. Many facilities have routine anti-parasitic programs to screen and treat for parasitic pathogens.

Northern facilities are prone to marked morbidity and mortality due to warble fly (*Hypoderma tarandi*) and nasal bot fly (*Cephenemyia trompe*) infections (Foster, 2010; Haigh et al., 2002). Myiasis can occur, especially if the antler velvet has been damaged. In Europe, mange during the winter months (*Chorioptes* sp.) has been noted at some zoos and is difficult to treat (M. Bertelsen, personal communication). Other external parasites of concern for captive animals include ticks (*Amblyomma, Ixodes* and *Dermacentor* sp.) and their associated pathogens (*Borrelia borgdorferri* and *Babesia*) (Haigh et al., 2002; Bartlett et al., 2009).

Internal parasites of importance include besnoitiosis (*Besnoitia tarandi*), lungworm (*Dictyocaulus viviparous*), muscle worm (*Parelophostrongylus andersoni*) and meningeal worms (*Parelophostrongylus tenuis, Elaphostrongylus rangiferi*) (Flach, 2003; Glover et al., 1990; Haigh et al., 2002). Gastrointestinal parasites are very common in captive facilities. These include nematodes [*Nematodirus tarandi, Nematodirella longissimespiculata, Capillari* sp., *Trichuris* sp., *Ostertagia* sp., pinworms (*Skrjabinema tarandi*)], tapeworms (*Taenia hydatigena, T. krabbei* and *Echinococcus granulosus*), *Cryptosporidium* and giardia. Toxoplasmosis (*Toxoplasma gondii*), *neospora*, *sarcocystis* and fascioloidiasis also should be considered potential pathogens (Flach, 2003; Haigh et al., 2002).

14.4 NON-INFECTIOUS DISEASES

Trauma is not uncommon in captive reindeer and caribou, and is more prevalent during the rut when seasonal fighting and aggression is heightened. To reduce aggression in male reindeer, anecdotal reports note that medroxyprogesterone acetate can be effective. Administration of 200 to 400 mg per animal in August and repeated in October have decreased aggression in non-breeding males or in males after breeding. Animals are noted to be very tractable, retain their antlers, and go on to breed normally in subsequent years. Further study is required to determine its impact on spermatogenesis, semen quality, and fertility (Blake et al., 2007).

Caribou and reindeer, like all ungulates, are prone to capture myopathy and efforts should be directed at prevention (Rehbinder, 1990; Spraker, 1993). This should include only working with animals during cooler weather when possible, and performing restraint and handling in a well-designed handling facility that minimizes processing times, handling stress and injury potential and is safe for both animals and staff. Experienced capture and processing teams are needed to reduce handling times; blindfolds should be used and all noise and talking kept to an absolute minimum. Determine and communicate chase, handling and restraint cut-off times beforehand and adhere to them throughout the procedures. Monitor closely for hyperthermia and hypoxemia while handling. Rectal body temperatures above 40°C and blood oxygen concentration (SpO_2) below 85% need to be acted on immediately.

During periods of warmer weather, shaded areas and misters to promote cooling on warm days is important, especially if animals are housed in warmer climates than are typical for the species.

Reindeer and caribou are sensitive to sudden dietary changes such as grain overload or sudden access to a lush pasture. Clinical signs are similar to those in other ruminants, including rumen stasis with bloating, weakness, depression, neurological signs and cardiovascular collapse associated with endotoxins. Treatment is based on clinical signs and response to therapy. Nutritional deficiencies have been reported in captive reindeer and caribou, and include copper deficiency, thiamine (vitamin B1) deficiency and vitamin E/selenium deficiencies. Other diet-related gastrointestinal complications from being fed inappropriate diets have led to abomasal ulceration, rumenal impactions, and severe intestinal hemorrhage, compounded by secondary opportunistic bacterial infections, including *Escherichia fergusonii* infection (Foster et al., 2010, Masters and Flach, 2014; McSloy, 2014). Sand impactions of the abomasum can be a significant cause of morbidity and mortality in calves (Bell and Dieterich, 2010), with some institutions keeping their calves off sandy paddocks for at least a month after birth (M. Bertelsen, personal communication).

Dental attrition is common in older caribou and reindeer. Clinical signs include dropping of feed from the mouth when masticating, accumulations of food material in the cheek areas and weight loss (Figure 14.4). It is important to provide good quality and easily digestible feed to geriatric animals.

Osteoarthritis is also not an uncommon clinical problem in aging *Rangifer*. Empirical use of various therapeutic agents, including non-steroidal anti-inflammatory drugs (NSAIDS), glucosamine, and methylsulfonylmethane (MSM) are used in its management (Whiteside, 2014). The authors have had good success with oral meloxicam at 0.3–0.5 mg/kg administered every 24–48 hours as needed.

Neoplasia can occur in captive caribou and reindeer but is not common. Lymphoma (Järplid and Rehbinder, 1995), conjunctival squamous cell carcinoma (Gonzalez-Alonso-Alegre et al., 2013) and cholangioma (Cunningham et al., 1993) have been reported in the literature.

FIGURE 14.4   Woodland caribou with food impaction associated with dental attrition (Photo: Michelle Oakley).

14.5   PREVENTIVE MEDICINE

A well-designed preventive medicine program for captive *Rangifer* is crucial to maintaining a healthy herd. The program should be tailored to the geographic area and meet the disease surveillance requirements of provincial or state and federal agencies.

Best practices in husbandry are of vital importance with captive reindeer and caribou. Given their nomadic life histories, they are particularly susceptible to diseases associated with high densities and lack of pasture rotation. At minimum, two separate pastures should be available for a herd so that animals can be rotated between them to reduce parasite burden, improve pasture and soil condition and allow for separation of animals for breeding, calving and if aggression occurs. Food and water troughs should be elevated and kept free from debris and excretions to limit disease transfer. Any faeces should be removed from these areas daily and animals monitored daily.

Although vaccine protocols are empirical and there are no veterinary licensed products for cervids, vaccination of *Rangifer* with a killed rabies and multivalent clostridium vaccine (7- or 8-way) is considered prudent (Mackintosh et al., 2002; Sihvonen et al., 1993). Booster vaccines should be administered as per label instructions. Clostridium annual booster vaccines should be given 1–2 months before the breeding season, when there is the highest risk of trauma. Other potential vaccines will be dependent on the location of the facility. West Nile virus vaccines may be prudent in endemic areas (Palmer et al., 2004). In some regions where *Brucella* is endemic, reindeer have been vaccinated (Davis and Elzer, 2002). Recent research shows that only the *Brucella suis* biovar 4 (killed) vaccine is effective in reindeer (Blake et al., 2007). Consultation with provincial/state and federal governments before administration of *Brucella* vaccines is recommended to ensure compliance with World Organization for Animal Health (OIE) recommendations.

At minimum, bi-annual screening of the herd for parasite identification and burden is recommended. Based on the results, clinical signs and disease risk analysis, prophylactic administration of anti-parasitic medications should be used where appropriate.

Depending on the ground cover, periodic hoof trimming may be required. Placing a concrete slab or abrasive gravel around feed and water troughs will help to promote hoof wear and reduce the need for trimming.

Having a well-functioning handling system will greatly reduce the need for chemical immobilization of captive caribou and reindeer for routine procedures such as sample collection (blood, hair, faeces), disease testing (serial TB testing), antler removal, medicating (oral, IM, IV) and hoof trimming. Acclimating the animals to the handling system, and the administration of a tranquilizer or sedative, can greatly reduce the stress of handling and restraint during these procedures and should be strongly considered.

Depending on federal regulations, herd testing for tuberculosis, brucellosis and chronic wasting disease may be required for any imports and exports and to maintain disease free herd status in a captive facility. All incoming animals should be quarantined for a period of time based on disease risk. During quarantine, it is recommended that each animal has a complete physical examination, with blood collected for routine analysis and disease screening, and remaining serum banked. Serial faecal samples should be tested for parasites.

Any mortality that occurs in the herd should be subject to a complete post mortem examination with appropriate tissue collection for further diagnostics, and samples submitted for chronic wasting disease, tuberculosis and any other regional diseases of concern.

14.6 PHYSICAL RESTRAINT AND CHEMICAL IMMOBILIZATION OF CAPTIVE *RANGIFER*

With proper planning and conditioning, captive caribou and reindeer can be handled and restrained for minor procedures (sample collection, injections) via a properly designed handling chute (e.g. a chute with a drop floor and squeeze). The design should have large enclosures leading to smaller separation pens where the animals can be selected out and funneled into raceways that ultimately lead to a manual or hydraulic squeeze. Interior walls need to be smooth and high enough to prevent animals from jumping up or seeing over them. Raceways leading to the squeeze area should be dark and should dampen noise transmission. Small, curtained punch out holes should be placed at eye level to allow for assessment of the animal and catwalks placed above the facility to encourage movement of individuals if needed. Regular feeding in the smaller pens and having the animals routinely go through the chute system will greatly increase their compliance when the time comes for them to be restrained (Flach, 2003).

In general, reindeer and caribou housed in captivity require lower doses of anesthetic drugs in comparison to their wild counterparts. These animals generally do not have long chase times, and darting or hand injections can be performed with minimal anxiety for the animal, leading to quicker and smoother inductions. Induction, maintenance and recovery can be much more controlled than in a free ranging environment, but attention to potential hazards and having pre-determined escape routes for staff planned out beforehand is recommended.

For chemical immobilization for short procedures, the authors have had good success with a combination of ketamine (2.5 mg/kg), medetomidine (0.05–0.1 mg/kg) and butorphanol (0.03–0.1 mg/kg) or midazolam (0.03–0.1 mg/kg) given intramuscularly. Use of intranasal medetomidine alone, or in combination with ketamine, in situations where the animal can be adequately restrained beforehand have been successfully used to sedate and lightly anesthetize free ranging caribou (Slater et al., 2014). Application of this technique in captive situations could have benefits, particularly for short procedures on animals that become uncooperative with physical restraint alone. Other intramuscular combinations include ketamine and medetomidine

(Caulkett and Arnemo, 2014), or butorphanol-azaperone-medetomidine (BAM). Tiletamine/
zolazepam (Telazol® or Zoletil®) in combination with xylazine or medetomidine also works well,
but combinations that are shorter acting with quicker recovery times are usually preferable in
captive situations.

Once anesthetized, animals can be intubated and maintained on a gaseous anesthetic if more
invasive or prolonged procedures are to be carried out. Hyperthermia is not uncommon in these
situations and body temperature should be monitored every 5 minutes in conjunction with pulse,
SpO_2 and respiratory rate, with follow-up as clinically indicated. Whenever possible animals should
be maintained in sternal recumbency with the head elevated and the oral cavity pointed towards the
ground to minimize complications associated with ruminal tympany, regurgitation and potential
aspiration. Limbs should be kept in normal anatomical positions and to the sides of the animal to
minimize complications associated with pressure neuropathies.

Due to their increased aggression and the physiological stressors these animals experience during breeding season, special precautions are advised with the use of physical or chemical restraint
methods in rutting *Rangifer*. If feasible, a hands-off approach during this time is considered to be the
best approach. Other limitations for *Rangifer* capture and handling occur during antler growth when
the animals are in velvet, from May to July in the northern hemisphere (Shury, 2014). Damage to
the highly vascularized velvet can lead to marked hemorrhage, pain and infestation of wounds with
fly larvae (myiasis). If handling of animals in velvet or rutting bulls is necessary, well-constructed
handling facilities are crucial.

Short and long acting tranquilizers in the form of intramuscular neuroleptic drugs such as per-
phenazine or zuclopenthixol acetate (Clopixol-Acuphase®) can greatly facilitate movement, handling and transport of both free ranging and captive reindeer and caribou. The latter works within
1–2 hours of administration and can last for up to 3–4 days. Animals tend to be much calmer during
movement, loading, transport and handling (Flach, 2003; Caulkett and Arnemo, 2014).

14.7   PHYSICAL EXAMINATION, DIAGNOSTICS AND SURGERY

A systematic approach to the physical examination should always be followed. Specimen collection
and handling is analogous to other ruminant species. Venipuncture can be accomplished most commonly from the jugular, cephalic or lateral saphenous veins. The interpretation of the haematological and serum biochemical values is similar to that for other ruminant species. Reference ranges for
Rangifer species held in captivity are available from Species 360 (www.species360.org).

Diagnostic modalities utilized in caribou and reindeer include radiography, ultrasonography and
endoscopy. Where available, computed tomography or magnetic resonance imaging also can be
very useful for evaluating morbidity (Figure 14.4).

Surgical management of traumatic injuries is the most frequent surgical problem encountered
in a captive setting. Other surgical indications include antler removal, contraceptive surgeries (e.g.
vasectomy), umbilical hernia repair and dystocias. Techniques utilized in other ruminant species
are applicable.

Multimodal analgesia is an important component of surgical management. Pharmacokinetic and
clinical efficacy studies of analgesics have not been published for *Rangifer* species, so extrapolation
is made from domestic ruminant species.

ACKNOWLEDGEMENTS

The authors thank Dr. Mads Bertelsen (Copenhagen Zoo), Dr. Maya Kummrow (Wuppertal
Zoological Gardens) and Dr. Michelle Oakley for their contributions to the chapter.

REFERENCES

Bartlett, S.L., Abou-Madi, N., Messick, J.B., Birkenheuer, A., and Kollias, G.V. 2009. Diagnosis and treatment of *Babesia odocoilei* in captive reindeer (*Rangifer tarandus tarandus*) and recognition of three novel host species. *J Zoo Wildl Med.* 40(1):152–159.

Bell, C., and Dieterich, R.A. 2010. Translocation of reindeer from South Georgia to the Falkland Islands. *Rangifer* 30(1):1–9.

Benestad, S.L., Mitchell, G., Simmons, M., Ytrehus, B., and Vikøren, T. 2016. First case of chronic wasting disease in Europe in a Norwegian free-ranging reindeer. *Vet Res.* 47(1):88–94.

Blake, J.E., Rowell, J.E., and Shipka, M.P. 2007. Reindeer reproductive management. In *Current Therapy in Large Animal Theriogenology*, 2nd Edition. Eds. R.S. Youngquist and W.R. Threlfall. Pp 970–974. St. Louis, Missouri: Saunders Elsevier Inc.

Caulkett, N., and Arnemo, J.M. 2014. Cervids (Deer). In *Zoo Animal and Wildlife Immobilization and Anesthesia*, 2nd Edition. Eds. G. West, D. Heard, N. Caulkett. Pp 823–829. Somerset, NJ, USA: Wiley.

Cunningham, A.A., Tyler, N.J., and Levene, A. 1993. Cholangioma in a Svalbard reindeer (*Rangifer tarandus platyrhynchus*). *Vet Rec.* 132(5):112–113.

Davis, D.S., and Elzer, P.H. 2002. Brucella vaccines in wildlife. *Vet Microbiol.* 90(1):533–544.

Del-Pozo, J., Girling, S., McLuckie, J., Abbondati, E., and Stevenson, K. 2013. An unusual presentation of *Mycobacterium avium* spp. *paratuberculosis* infection in a captive tundra reindeer (*Rangifer tarandus tarandus*). *J Comp Pathol.* 149:126–131.

Dieterich, R.A., and Morton, J.K. 1990. *Reindeer Health Aide Manual*, 2nd Edition. Agricultural and Forestry Experiment Station, Cooperative Extension Service and the University of Alaska Fairbanks. 77 pp.

Evans, M.G., and Watson, G.L. 1987. Septicemic listeriosis in a reindeer calf. *J Wildl Di.* 23(2):314–317.

Flach, E. 2003. Cervidae and Tragulidae. In *Zoo and Wild Animal Medicine*, 5th Edition. Eds. M.E. Fowler and E.R. Miller. Pp 634–649. Elsevier Science.

Foster, A. 2010. Common conditions of reindeer. *In Practice.* 32:462–467.

Foster, G., Evans, J., Tryland, M., Hollamby, S., MacArthur, I., Gordon, E., Hareley, J., and Voight, K. 2010. Use of citrate adonitol agar as a selective medium for the isolation of *Escherichia fergusonii* from a captive reindeer herd. *Vet Microbiol.* 144:484–486.

Fowler, M.E., and Boever, W.J. 1986. Artiodactylids: Cervidae. In *Zoo and Wild Animal Medicine*, 2nd Edition. Eds. M.E. Fowler. Pp 981–985. Philadelphia: W.B. Saunders.

Glover, G.J., Swendrowski, M., and Cawthorn, R.J. 1990. An epizootic of besnoitiosis in captive caribou (*Rangifer Tarandus Caribou*), reindeer (*Rangifer Tarandus Tarandus*) and mule deer (*Odocoileus Hemionus Hemionus*). *J Wildl Dis.* 26(2):186–195.

Gonzalez-Alonso-Alegre, E.M., Rodriguez-Alvaro, A., Martinez-Nevado, E., Martinez-de-Merlo E.M., and Sanchez-Maldonado, B. 2013. Conjunctival squamous cell carcinoma in a reindeer (*Rangifer tarandus tarandus*). *Vet Ophthalmol.* 16(1):113–116.

Haigh, J.C, Mackintosh, C., and Griffith, F. 2002. Viral, parasitic and prion diseases of farmed deer and bison. *Rev Sci Tech Off Int Epiz.* 21(2):219–248.

Hautaniemi, M., Ueda, N., Tuimala, J., Mercer, A.A., Lahdenperä, J., and McInnes, C.J. 2010. The genome of pseudocowpoxvirus: Comparison of a reindeer isolate and a reference strain. *J Gen Virol.* 91:1560–1576.

Herron, A.J., Garman, R.H., Baitchman, R., and Kraus, A.L. 1979. Clostridial myositis in a reindeer (*Rangifer tarandus*): A case report. *J Zoo Anim Me.* 10:31–34.

Järplid, B., and Rehbinder, C. 1995. Lymphoma in reindeer (*Rangifer tarandus tarandus L.*). *Rangifer* 15(1):37–38.

Josefsen, T.D., Tove, H., Aagnes, T.H., and Mathiesen, S.D. 1996. Influence of diet on the morphology of the ruminal papillae in reindeer calves (*Rangifer tarandus tarandus L.*) *Rangifer* 16(3):119–128.

Kiupel, M., Wise, A., Bolin, S., Walker, P., Marshall, T., and Maes, R. 2004. Malignant catarrhal fever in reindeer. 5th EAZWV meeting, Denmark. Pp 73–78.

Koroleva, V.P. 1976. Rasprostravennost vozbuditelei dermatomikozov Zhivotnykh v raznykh zonakh soyuza. *Byulleten Vsesoyuznogo Instituta Eksperimental'noi Veterinarii* 25:49–52.

Li, H., Westover, W.C., and Crawford, T.B. 1999. Sheep associated malignant catarrhal fever in a petting zoo. *J Zoo Wildl Med.* 30:408–412.

Mackintosh, C., Haigh, J.C., and Griffin, F. 2002. Bacterial diseases of farmed deer and bison. *Rev Sci Tech Off Int Epiz.* 21(2):249–263.

Masters, J.M., and Flach, E. 2014. Tragulidae, Mochidae, and Cervidae. In *Zoo and Wildlife Medicine*, 8th Edition. Eds. R.E. Miller, M.E. Fowler. Pp 611–625. Philadelphia: W.B. Saunders.

McSloy, A. 2014. Basic veterinary management of reindeer. *In Practice* 36:495–500.

Mitchell, G.B., Sigurdson, C.J., O'Rourke, K.I., Algire, J., Harrington, N.P., Walther, I., Spraker, T.R., and Balachandran, A. 2012. Experimental oral transmission of chronic wasting disease to reindeer (*Rangifer tarandus tarandus*). *PloS one* 7(6):e39055.

Norberg, H.J., and Mathiesen, S.D. 1998. Feed intake, gastrointestinal system and body composition in reindeer calves fed early harvested first cut timothy silage (*Phleum pratense*). *Rangifer* 18(2):65–72.

Olsen, M.A., Aagnes, T.H., and Mathiesen, S.D. 1995. Failure of cellulolysis in the rumen of reindeer fed timothy silage. *Rangifer* 15(2):79–86.

Palmer, M.V., Stoffregen, W.C., Rogers, D.G., Hamir, A.N., Richt, J.A., Pedersen, D.D., and Waters, W.R. 2004. West Nile virus infection in reindeer (*Rangifer tarandus*). *J Vet Diagn Invest*. 16(3):219–222.

Rector, A., and Van Ranst, M. 2013. Animal papillomaviruses. *Virol*. 445(1):213–223.

Rehbinder, C. 1990. Management stress in reindeer. *Rangifer* 10(3):267–288.

Rehbinder, C., Mattsson, R., and Belak, K. 1994. Aspergillosis in reindeer (*Rangifer tarandus tarandus* L). A case report. *Rangifer* 14(3):131–132.

Shury, T. 2014. Physical capture and restraint. In *Zoo Animal and Wildlife Immobilization and Anesthesia*, 2nd Edition. Eds. G. West, D. Heard, N. Caulkett. Pp 109–124. Somerset, NJ, USA: Wiley.

Sihvonen, L., Kulonen, K., Soveri, T., and Nieminen, M., 1993. Rabies antibody titres in vaccinated reindeer. *Acta Vet Scand*. 34(2):199–202.

Slater, O.M., Risling, T.E., Caulkett, N.A., Goldhawk, C., Schwartzkopf-Genswein, K., Pajor, E., Cook, J., Cook, R., Parker, K.L., and Schwantje, H. 2014. *Effects of Capture and Long Distance Transport on Mountain Caribou*. North American Caribou Workshop, Whitehorse, Canada.

Spraker, T.R. 1993. Stress and capture myopathy in artiodactylids. In *Zoo and Wildlife Animal Medicine – Current Therapy*, 3rd Edition. Ed. M.E. Fowler. Pp 481–488. Philadelphia: W.B. Saunders.

Vansnick, E. 2004. Johne's disease in zoo animals: Development of molecular tools for the detection and characterisation of *Mycobacterium avium* subspecies *paratuberculosis*. dspace.itg.be/handle/10390/757. Accessed October 13, 2014.

Voight, K., Dagleish, M., Finlayson, J., Beresford, G., and Foster, G. 2009. Black disease in a forest reindeer (*Rangifer tarandus fennicus*). *Vet Rec*. 165:352–353.

Whiteside, D.P. 2014. Analgesia. In *Zoo Animal and Wildlife Immobilization and Anesthesia*, 2nd Edition. Eds. G. West, D. Heard, N. Caulkett. Pp 83–108. Somerset, NJ, USA: Wiley.

Yokota, T., Shibahara, T., Ishikawa, Y., Kadota, K., Yamaguchi, M., and Jimma, K. 2004. Concurrent fatal listeriosis, zygomycosis and aspergillosis in a reindeer (*Rangifer tarandus*) calf. *Vet Rec*. 154(13):404–406.

15 Restraint and Immobilization

Marianne Lian, Alina L. Evans, Kimberlee B. Beckmen,
Nigel A. Caulkett and Jon M. Arnemo

CONTENTS

15.1 INTRODUCTION

Free-ranging reindeer and caribou (*Rangifer tarandus* sspp.) are handled for research, management or treatment of individual animals, whereas semi-domesticated reindeer are routinely handled as part of traditional husbandry. Pursuit, herding and restraint are stressful and may cause physiological and emotional distress. Reindeer and caribou are easily excitable animals, and all handling should be thoroughly planned and conducted cautiously to prevent morbidity or mortality. In general, planned herding, capture and restraint should be carried out when animals are in peak body condition, such as in late fall or winter, preferably with snow on the ground. Also, captures should be avoided in the last month of gestation, around calving and when antlers are in velvet. Due to medical or other reasons, physical restraint or chemical immobilization may have to be carried out at any time of the year, and people in charge of the operation should be prepared and trained to deal with difficult circumstances and possible complications and emergencies.

Although the biology and ecology of the subspecies of *Rangifer tarandus* differ, e.g. the body mass of adults range from 50 to more than 300 kg, their physiological response to stressors, including herding, physical restraint and anesthetic drugs, is assumed to be similar. Prevention, diagnosis and treatment of subsequent adverse effects to acute stress require a basic understanding of the neural and endocrine responses to stressors (Arnemo and Caulkett 2007). Stress may have consequences for animal welfare, meat quality and anesthetic risk, and may also bias physiological and ecological data. On the other hand, the use of anesthetic drugs also affects vital signs and other physiological variables used to assess the stress response. Stress physiology is reviewed in Chapter 2 (*Rangifer* Health – A Holistic Perspective). Here, the basic physiology of reindeer and caribou and important aspects of acute stress relevant to physical and chemical restraint are covered.

15.2   VITAL SIGNS AND PHYSIOLOGICAL RESPONSES TO STRESS

Rectal temperature and heart and respiratory rates (vital signs) of restrained animals should be frequently monitored throughout the handling period. Safe restraint and immobilization require knowledge of reference ("normal") values for vital signs and when corrective measures should be initiated due to critical levels. In addition to vital signs, blood oxygenation and capillary refill time (CRT) should be monitored in anesthetized animals. Basic principles of monitoring are covered by Ozeki and Caulkett (2014). Advanced monitoring techniques and equipment, e.g. blood gas analysis, electrocardiography, blood pressure measurement and capnography, are not routinely used in field anesthesia of reindeer and caribou and are therefore beyond the scope of this chapter. For these techniques, readers are referred to the standard textbooks in veterinary anesthesia (Grimm et al. 2015, West et al. 2014).

Reindeer and caribou are prey animals that are easily disturbed by human presence (Panzacchi et al. 2013). Naïve wild mammals are usually not stressed by aircraft or ground motor vehicles. Wild Norwegian reindeer and caribou, however, are easily frightened by helicopters. This response might be due to experiences with eagles as their natural predators. Helicopter approach might induce a panic flight response in an entire herd, which might include hundreds of animals. The speed and agility of reindeer and caribou can make them difficult targets for aerial netting or darting. Reindeer can achieve a maximum running speed of 60 km/h (Blix et al. 2011), and herding or pursuit with a helicopter or snowmobile for physical or chemical capture should be kept as short as possible (<1 min of strenuous running) in order to avoid distress, hyperthermia and risk of capture stress syndrome (Arnemo et al. 2011, Arnemo et al. 2014a, Cattet 2011, Caulkett and Arnemo 2014) (Figure 15.1).

FIGURE 15.1   Reindeer and caribou (*Rangifer tarandus* sspp.) can achieve a maximum running speed of 60 km/h, and aerial net-gunning or darting is a challenge. Pursuit time should be <1 min to avoid distress such as hyperthermia. Open-mouth breathing is a sign of increased body temperature (Photo: Jon M. Arnemo).

15.2.1 Thermoregulation

The body temperature of resting reindeer in winter is 38.1–38.6°C (Blix et al. 2011, Dieterich and Morton 1990, Mercer et al. 1985). Monitoring of body temperature is usually done with a rectal thermometer, which may give a reading that is lower than the core body temperature (Ozeki et al. 2014). The rectal temperature should be measured immediately after induction in the anesthetized animal. In addition to the actual recording, it is important to monitor the trend in body temperature during handling time, by repeating the measurement every 5–10 minutes. A rectal temperature of 39.0°C is slightly above normal and is not a concern. However, body temperatures in animals that have been running prior to physical or chemical capture can rapidly increase to life-threatening levels due to increased rates of cellular metabolism and energy consumption. Rectal temperatures of 41.0–42.0°C are not uncommon in reindeer and caribou chemically immobilized from a helicopter, even when the pursuit time is kept to a minimum (Arnemo et al. 2011, Lian et al. 2016). Helicopter-darted caribou with body temperatures above 43.0°C had high mortality rates (Fuller and Keith 1981). In Svalbard reindeer physically captured with a net from snowmobiles, the mean rectal temperature was 39.9°C, with the highest recorded value being 41.8°C (Omsjoe et al. 2009). In contrast, calm Svalbard reindeer stalked on foot and chemically immobilized by remote drug administration had mean rectal temperatures of 38.3–39.1°C in two different studies using the same drug combination (Arnemo and Aanes 2009, Evans et al. 2013). This reflects how heat-stress in *Rangifer* is closely related to the capture method, with additive effects from physiological and mental stressors.

Due to their extremely thick fur insulation in winter, reindeer and caribou are prone to a stress-induced increase in body temperature, even at low ambient temperatures, and capture and restraint of animals in winter coat is generally not recommended above 0°C (Arnemo and Aanes 2009, Arnemo et al. 2011, Cattet 2011, Evans et al. 2013). Causative factors of increased body temperature include fear, metabolic heat generated by excessive muscular exertion, high ambient temperature, sun exposure and interference with normal thermoregulatory mechanisms by anesthetic drugs. Signs of heat stress in reindeer start at a brain temperature of 39.0°C, when animals resort to open-mouth panting in order to dissipate heat from their highly vascularized tongue, with regular intermittent periods of closed-mouth breathing, probably due to the need for selective brain cooling (Blix et al. 2011) (Figure 15.2).

FIGURE 15.2 Radio collared caribou (*R. t. caribou*) recovering from carfentanil-xylazine immobilization. Open-mouth breathing indicates hyperthermia (Photo: Alaska Department of Fish and Game/Dominic Demma).

From a clinical point of view, hyperthermia is a body temperature ≥2°C greater than normal and is cause for concern (Arnemo et al. 2014a, Cattet 2011, Caulkett and Arnemo 2014, Caulkett and Arnemo 2015). Hyperthermic animals typically show snow-eating, increased respiratory rate and open-mouth panting. In anesthetized animals, signs can include rapid and shallow breathing. A rectal temperature in excess of 41.0°C is an emergency and should be treated aggressively. If the body temperature exceeds 42.0°C, cellular damage starts. Depending on the season and ambient temperature, treatment can include moving the animal into shade, spraying with cold water, cooling in a lake or stream, fanning, packing ice or snow in the inguinal and axillary regions, cool water enemas and intravenous fluids. Intravenous administration of 20°C fluids can help decrease body temperature even at low volumes (5–10 ml/kg). Hyperthermia may be prevented by avoiding immobilization or capture on warm days or limiting activities to the coolest part of the day. Avoid prolonged intensive pursuit, keep stress to a minimum and use the least stressful methods for physical restraint. Protect the animal from direct exposure to the sun. In a field situation, it is difficult to actively cool a large animal. The best option could be to release a physically restrained animal or to antagonize anesthetic agents to allow the animal's normal thermoregulatory mechanisms to recover. Hyperthermia greatly increases metabolic rate and thereby oxygen demand, and is a particularly serious complication in the face of hypoxemia (low oxygen level in the blood). In this condition, oxygen delivery simply cannot keep pace with the racing metabolic activity and increased oxygen consumption. Hyperthermic anesthetized animals should always receive supplemental oxygen (Figure 15.3).

Hypothermia is a body temperature ≥2°C lower than normal (Arnemo et al. 2014a, Caulkett and Arnemo 2015) and is usually a concern when animals are chemically immobilized during low ambient temperatures and in young or aged animals, animals with a small body mass and animals with loss of insulation due to poor body condition. Translocating animals, especially if wet, by sling loading under the helicopter increases risk of hypothermia and preventive measures should be taken. When dealing with a hypothermic animal, body temperature should be monitored frequently with special attention to trends and whether it continues to decrease, stabilize or increase.

FIGURE 15.3   Free-ranging Norwegian reindeer (*R. t. tarandus*) anesthetized with medetomidine-ketamine. Supplemental oxygen administered with a nasal line (Photo: Marianne Lian).

Hypothermia with body temperature below 34.0°C may result in prolonged recovery, acid-base imbalance and heart arrhythmia. Causes of hypothermia include inadequate circulation, low ambient temperature, evaporative cooling from wind-chill, wet fur, or precipitation and drugs that impair thermoregulation. Supportive procedures should begin when a declining trend in body temperature is first noted and include an immediate attempt to increase body temperature by drying wet animals, covering the animal and providing external heat sources such as hot water bottles. Protect the immobilized animal from low ambient temperatures and exposure to wind and precipitation. Keep it warm and dry by covering with blankets or a sleeping bag. Minimize conductive heat loss by maintaining the animal insulated from direct ground contact.

Restraint and capture should be avoided under extreme cold conditions because of the risk of hypothermia and frostbite. The recommended lower temperature limit for capture by net-gun is −30°C and for capture by remote drug delivery is −20°C (Cattet 2011).

15.2.2 CARDIOVASCULAR FUNCTION

Monitoring of cardiovascular function includes recording heart rate and variability, pulse rate and quality and CRT. In a field situation, heart or pulse rate and CRT are the easiest and most useful variables to assess.

Heart rate is measured by auscultation of the heart using a stethoscope or by palpation of the chest wall, usually on the left side under or directly behind the elbow. Pulse rate is recorded by palpating an accessible artery such as the auricular, palmar, femoral or saphenous arteries. Reference values for resting heart rate in *Rangifer* are seasonal and dependent on food intake, being 50–60 beats/min in summer and 35–45 beats/min in winter (Eloranta et al. 2002, Fancy and White 1985, Fancy and White 1986, Mesteig et al. 2000, Nilssen et al. 1984, Nilsson et al. 2006). Disturbance, foraging and physical activity increase the heart rate. The maximum heart rate of reindeer under experimental conditions is 250 beats/min (Timisjärvi et al. 1979). Reindeer running at 5–10 km/hr have heart rates of 60–120 beats/min in winter and 90–140 beats/min in summer (Nilssen et al. 1984). According to Cattet (2011), the heart rate of anesthetized caribou should remain between 50 and 130 beats/min. Reported heart rates in anesthetized animals, however, are typically 35–80 beats/min (Arnemo and Aanes 2009, Arnemo et al. 2011, Evans et al. 2013, Jalanka and Roeken 1990, Lian et al. 2016, Risling et al. 2011). Heat-stress and hypoxemia may cause tachycardia. In general, heart rates of chemically immobilized animals vary, and depend on the amount of running prior to capture, drugs, body temperature and blood oxygen levels.

The major neural response to acute stress is a generalized and immediate activation of the sympathetic nervous system, often called the "fight-or-flight" response. One of the most striking characteristics is the immediate effect on the cardiovascular system: Increased heart rate (doubling in 3–5 seconds); increased arterial blood pressure (doubling in 10–15 seconds); increased cardiac output (secondary to increased cardiac contractility and heart rate); increased blood flow (by as much as 400% from the resting state) to skeletal and cardiac muscles and the pulmonary circulation (due to vasodilatation), concurrent with decreased blood flow (by as much as 90% from the resting state) to organs such as the gastrointestinal tract and the kidneys (due to vasoconstriction); and contraction of the spleen with subsequent release of stored erythrocytes into the circulation (increasing the packed cell volume by as much as 50%).

CRT can be used to assess peripheral perfusion. Following digital compression to blanch an area of the gum, capillary refill should occur in less than 2 seconds. At the same time, the colour of the oral mucosa can be used to assess blood oxygenation (see p. 470). The oral mucosa of reindeer

```

and caribou tends to be pigmented and may render both methods less useful. As an alternative, the mucosa of the conjunctiva of the eye or the tongue can be used.

### 15.2.3 Respiratory Function

Monitoring of respiratory function includes recording of respiratory rate and assessment of depth and quality of breathing. Resting respiratory rate in reindeer may be as low as 7 breaths/min (Blix et al. 2011), but the reference values of lying and standing animals are more likely 23 and 36 breaths/min, respectively (White and Yousef 1978). Mental and physical stress will increase the respiratory frequency. Under experimental conditions, rates above 250 breaths/min have been recorded (Blix et al. 2011). Most drugs depress respiratory function, with signs of hypoventilation that can be recognized by a lowered respiratory frequency and shallow breathing. However, due to intra-pulmonary causes, including ventilation/perfusion mismatch leading to venous admixture, hypoventilation can occur even without any decreased respiratory rate. This is a potentially dangerous situation since the oxygen demand is increased due to stress-induced increase in metabolic rate and body temperature in chemically immobilized animals. Normal homeostatic regulation functions, including heat loss from panting, are impaired during anesthesia. Reported respiratory rates in chemically immobilized reindeer (Arnemo et al. 2011) and caribou (Lian et al. 2016) were 6–24 and 8–40 breaths/min, respectively. In Svalbard reindeer stalked on foot and chemically immobilized, the mean respiratory rate was 13 breaths/min (Arnemo and Aanes 2009). The respiratory rate of anesthetized animals should be 6 breaths/min or higher, and each breath should be quiet and characterized by full expansion and relaxation of the rib cage (Cattet 2011).

Hypoxemia is a common side effect in anesthetized reindeer (Arnemo and Aanes 2009, Arnemo et al. 2011, Evans et al. 2013, Lian et al. 2016, Risling et al. 2011), and blood oxygenation should be frequently monitored. A simple technique is to assess the mucous membrane colour, which should be pink in animals with normal blood oxygenation. A pale blue or dark blue colour indicates hypoxemia. This assessment is, however, complicated, as it is highly subjective and dependent on the observer's experience. Blood oxygenation can be indirectly measured by pulse oximetry (Fahlman 2014, Kreeger and Arnemo 2012, Ozeki and Caulkett 2014). This is done using a relatively cheap, easy-to-use, portable, battery-operated unit (a pulse oximeter) suitable for field use. A sensor is applied to the tongue of an anesthetized animal and the monitor measures the relative oxygen saturation of hemoglobin in arterial blood and the pulse rate. Normal saturation at sea level is >95% and values <90% are considered hypoxemic. However, the trend is more informative than the actual reading and decreasing values indicate that hypoxemia is developing (Kreeger and Arnemo 2012). Studies in reindeer (Evans et al. 2013) and caribou (Lian et al. 2016), using a blood gas analyzer, showed that nasal insufflation of supplemental oxygen (1–2 L/min) was effective for counteracting hypoxemia in anesthetized animals. It is strongly recommended that supplemental oxygen is used during chemical immobilization and that people handling anesthetized animals are able to perform cardiopulmonary resuscitation (CPR), including endotracheal intubation and manual ventilation, in case of emergencies like severe respiratory depression or arrest (Boysen 2014). If left untreated, hypoxemia will lead to hypoxia (low oxygen levels in tissues) with subsequent adverse effects on vital organs including brain, heart, kidneys and liver (McDonell and Kerr 2015). General aspects of airway management and oxygen therapy in non-domestic animals are found in Cracknell (2014) and Fahlman (2014) (Figure 15.4).
```

FIGURE 15.4 Semi-domestic Norwegian reindeer (*R. t. tarandus*) anesthetized with medetomidine-ketamine and monitored with a pulse oximeter. Relative arterial oxygen saturation is 92% and pulse rate is 39 beats/min (Photo: Jon M. Arnemo).

15.2.4 BLOOD PARAMETERS

Blood parameters can be used to assess the general health status of an animal as well as the physiological stress response elicited by physical or chemical capture and restraint. Reference values for haematology and/or serum biochemistry have been reported for both reindeer (Catley et al. 1990, Dieterich 1993, Miller et al. 2013, Milner et al. 2003, Nieminen and Timisjärvi 1981, Nieminen and Timisjärvi 1983, Soveri et al. 1992, Timisjärvi et al. 1981) and caribou (Johnson et al. 2010, McEwan 1967, Messier et al. 1987). There are several studies on the effects on blood parameters due to chemical, physical or mental stressors (Arnemo and Ranheim 1999, Essén-Gustavsson and Rehbinder 1984, Hyvärinen et al. 1976, Karns and Crichton 1978, Omsjoe et al. 2009, Pösö et al. 1994, Ranheim et al. 1997, Rehbinder et al. 1982, Rehbinder and Edqvist 1981, Säkkinen et al. 2004, Soveri et al. 1999, Timisjärvi et al. 1990, Wiklund et al. 1996).

The stress response is a series of complex and interrelated hormonal and neural events that is influenced by numerous agents and stimuli. Fear, strenuous exercise, starvation, infection, anesthesia and pain are stressors that can elicit identical activation of the hypothalamic-pituitary-adrenal (HPA) axis. Also, animals are known to secrete similar amounts of glucocorticoids during exercise, restraint, social interactions, pain and mating (Arnemo and Caulkett 2007, Colborn et al. 1991, Koolhaas et al. 2011). The baseline (resting) level of cortisol in reindeer is probably below 15 nmol/L (Säkkinen et al. 2004) and the maximum level found after administration of ACTH was approximately 250 nmol/L (Säkkinen et al. 2005).

Several studies in *Rangifer* show some of the inherent difficulties of using blood cortisol levels to assess stress. Light restraint and blood sampling by jugular venipuncture induced six times higher plasma cortisol levels compared to remote sampling using automatic blood sampling equipment

(Säkkinen et al. 2004, Sire et al. 1995, Arnemo and Ranheim 1999). In wild Norwegian reindeer kept in a corral and captured and restrained by means of a lariat in January, the mean (range) serum cortisol level in ten adult females was 441 (249–596) nmol/L, whereas the corresponding values for 48 calves were 238 (105–433) nmol/L (Arnemo, unpublished data). Semi-domestic reindeer had values ranging from 75–140 nmol/L after lariat captures compared with 100–200 nmol/L after emotional stress (Lund-Larsen et al. 1978, Rehbinder 1990). In addition to handling, physical restraint and body condition, anesthetic drugs and method of drug administration influence blood cortisol levels. The effects of anesthetic drugs on cortisol secretion are very complex. Alpha-2 adrenoceptor agonists such as xylazine and detomidine are known to reduce the stress response, measured as blood levels of cortisol. However, medetomidine, a more potent alpha-2 adrenoceptor agent, increases the secretion of cortisol in reindeer. Very little is known about specific effects on glucocorticoid secretion of other sedative or anesthetic agents in *Rangifer* or other non-domestic animals (Arnemo and Caulkett 2007).

In five semi-domestic reindeer kept in a research facility, the mean serum cortisol concentration was 350 nmol/L 30 minutes after administration of medetomidine, two times the pre-treatment level (Arnemo and Ranheim 1999). At 60 minutes post-medetomidine, when sedation was reversed by atipamezole, a decline in serum cortisol had started, but pre-treatment levels were not reached until 4 hours after administration of medetomidine. In captive, semi-domestic reindeer immobilized with medetomidine-ketamine (Ryeng et al. 2002), cortisol levels increased 2- to 3-fold during a 30-minute monitoring period, reaching mean levels of 220 to 300 nmol/L, in three different experiments using different doses and methods of drug administration. After hand syringe injection, a higher increment was seen when the doses were increased by 50%, and a further increase was seen when the drugs were administered by dart syringe injection. In free-ranging reindeer, the mean serum cortisol level was 299 (110–477) nmol/L in 30 adult animals immobilized with medetomidine-ketamine from a helicopter (Miller et al. 2013), with no correlation between cortisol levels and sex or induction time.

In caribou captured by net-gun from a helicopter, Oakley et al. (2004) found that the mean serum cortisol level in animals sedated with medetomidine in order to reduce stress (n = 18) was not significantly different from untreated controls (n = 18): 182 versus 192 nmol/L. Johnson et al. (2010) reported that the mean (central 95% range) in 103 adult, female caribou captured by net-gunning from a helicopter was 170 (63–362) nmol/L. In contrast, Svalbard reindeer captured in winter by net from snowmobiles and manually restrained had a mean cortisol level of 52 (3–127) nmol/L (Omsjoe et al. 2009). These levels are only slightly higher than serum cortisol levels in undisturbed Svalbard reindeer shot in winter (Nilssen et al. 1985) and are surprisingly low compared to the stress-induced cortisol response reported in several other *Rangifer tarandus* sspp. Obviously, measurement of cortisol alone cannot be used to differentiate between non-threatening stress and distress in reindeer and caribou.

Fear, chasing, physical restraint and/or chemical immobilization will elicit an acute stress response that can compromise the physiologic homeostasis of an animal and cause distress. The capture event, capture method and anesthetic drugs influence physiological parameters and the homeostasis of the animal. Muscular activity associated with excitement, chase and resistance to handling results in an increase in body temperature and lactic acid build-up once the anaerobic threshold is exceeded. Increased lactic acid levels can result in a subsequent metabolic acidosis (Haga et al. 2009).

Lactate is perhaps the easiest blood stress variable to measure directly in the field. A small, handheld unit (Lactate Pro 2, Arkray, Kyoto, Japan) measures the lactate level within a minute from one drop of whole blood (Beckmen and Lian 2014). The reference range for lactate in resting large domestic mammals is <2.2 mmol/L. Captive reindeer and caribou that are sedated after hand injection in a chute typically have a lactate level <4 mmol/L, which then resolves below detection within 15 minutes. In contrast, free-ranging caribou and wild Norwegian reindeer darted from a helicopter had lactate levels of 8.4 (2.1–15.5) and 8.7 (3.3–11.0), respectively (Evans et al. 2013, Lian et al. 2016).

On the other hand, Svalbard reindeer stalked and darted from the ground had undetectable lactate levels (Evans et al. 2013).

15.2.5 Other Considerations

Before any physical or chemical capture procedure, an appropriate plan of action must be devised. Preferably, a detailed protocol should be written as part of the planning and the selected method of capture should be based on recommendations from current literature and through consulting experts to ensure that the most suitable technique is used. General and in-depth coverage of wildlife capture are found in Caulkett and Arnemo (2015), Kock and Burroughs (2012), Kreeger and Arnemo (2012), Nielsen (1999) and West et al. (2014).

Helicopters or snowmobiles are often used for capture of reindeer and caribou but can be hazardous, and appropriate training and protective equipment will help to minimize the risk of injury to personnel. Detailed reviews of helicopter safety are found in Nielsen (1999), Kock and Burroughs (2012) and Caulkett and Shury (2014).

In cases of acute distress, immediate release of the animal should be considered unless effective treatment can be initiated. Emergency drugs including epinephrine, atropine, doxapram, local anesthetics and reversal agents should always be carried. Fluids for intravenous (IV) treatment of shock or hypo- or hyperthermia should be available. During the process of capture, physical trauma may be inflicted on the animal. Minor injuries can be treated and a surgical kit and antibiotics should be carried (see checklist in Kreeger and Arnemo 2012).

A thorough understanding of the inherent risks associated with animal capture is indispensable for prevention of capture-related injuries. Careful planning, close monitoring and early intervention are essential for preventing morbidity and mortality. Typically, the highest rates of morbidity and mortality are seen in the early stages of a project before the capture methods are refined, drug doses are adjusted and the capture team has gained experience and training. Moreover, an increased risk of mortality may also be seen when captures are carried out for specific purposes like health evaluation of animals under environmental or pathogenic stress. A capture-related mortality rate greater than 2% is unacceptable and requires that the anesthetic protocol should be re-evaluated (Kreeger and Arnemo 2012). By using established methods and techniques, a skilled and experienced team will minimize the risk of capture-related morbidity and mortality. In the event of mortalities, the situation should be evaluated to assess the cause of the mortality and to establish future preventive measures. A necropsy should be performed either on-site or by sending the cadaver to a diagnostic laboratory. The necropsy is the only way to determine if the animal was undergoing an unrelated pathological process or if a healthy animal died due to capture.

15.3 PHYSICAL CAPTURE

Techniques for the physical capture of reindeer and caribou include net-gunning, water capture and drive-nets. In traditional reindeer husbandry, and to a limited extent in wild reindeer, the animals are herded into corrals and manually restrained, often by means of a lariat, for tagging, weighing, slaughtering or other purposes. Most methods are based on the use of a helicopter, snowmobiles or all-terrain vehicles.

Physical restraint can induce greater stress than chemical restraint (Boesch et al. 2011, Cattet et al. 2003), with higher rates of trauma and mortality (DelGiudice et al. 2001, DelGiudice et al. 2005, Jacques et al. 2009, Kreeger and Arnemo 2012, Webb et al. 2008), and methods should be selected in order to reduce the stress response. Physical restraint should be of short duration and use appropriate restraint techniques, including hobbles and eye covers. Administration of sedatives or anesthetics are recommended to avoid injuries and to prevent or reduce distress (Arnemo et al. 2005, Cattet et al. 2004, Mentaberre et al. 2010, Oakley et al. 2004). Also, injury and even death of humans are not uncommon during physical capture of wildlife (Jessup et al. 1988; López-Olvera et al. 2009).

Jessup et al. (1988) reported that over a 12-year period in New Zealand, there were 127 helicopter crashes and 25 human fatalities during net-gun captures of red deer. Physical capture and restraint methods for wildlife were reviewed by Shury (2014), whereas techniques for caribou were covered by Haigh (1976), Patenaude (1982), International Wildlife Veterinary Services (1991) and Cattet (2011). General aspects of physical capture and restraint are found in Fowler (2008).

15.3.1 HELICOPTER NET-GUNNING

In Alaska and Canada, caribou are routinely captured by net-gunning from helicopters (Cattet 2011, Compton et al. 1995, Ferguson 2015, Smith et al. 2000, Stuart-Smith et al. 1997, Valkenburg et al. 1983). Typically, this is executed in areas where subsistence hunting is conducted throughout the year and residues of immobilizing drugs are a concern. Helicopter net-gunning requires experienced personnel and a close cooperation between pilot and net-gunner. It is often necessary to pass the group of caribou more than once, and usually a big herd has to be split into smaller groups. Inevitably, more animals than the animal targeted will be disturbed due to the herding nature of caribou. However, chase time should be kept to a minimum (<1 min of strenuous running) and pursuit should be aborted if the targeted animal shows signs of fatigue (Cattet 2011). After the helicopter closes in on the animal and the net-gun is fired off (a 5x5 m square nylon mesh with corner weights discharged from a gun), the helicopter lands within seconds, and the personnel run up to the animal to stop movement and hobble the feet (Shury 2014, Smith et al. 2000). Large bulls can get the net wound up in their antlers and might require more than one net. After the animal is hobbled and removed from the net, it is important to blind-fold the animal, both to reduce fear and to protect the eyes from snow, dirt and direct sunlight. Following restraint, handling procedures are fast, and the animal is usually collared, weighed and sampled in less than 10 min (Figure 15.5).

In spite of quick handling times, net-gunning is generally a very stressful capture method for caribou. A study from Alaska found significantly higher lactate levels in net-gunned animals

FIGURE 15.5 Manual restraint of a caribou (*R. t. caribou*) captured by net-gunning (Photo: Alaska Department of Fish and Game/Brian Pearson).

(mean lactate 21.7 mmol/L) when compared to chemically immobilized individuals (mean lactate 4.9 mmol/L) (Beckmen and Lian 2014). Two other studies showed that intranasal administration of sedatives reduces stress and facilitates handling in net-gunned cervids. Oakley et al. (2004) found significantly better relaxation scores and lower body temperatures, heart and respiratory rates, and muscle enzyme levels in net-gunned caribou given intranasal medetomidine compared to controls treated with saline. No difference was found in cortisol levels, most likely due to the fact that this drug increases blood cortisol. Net-gunned elk (*Cervus elaphus manitobensis*) that received intranasal xylazine after restraint showed increased relaxation and reduced blood levels of cortisol and muscle and liver enzymes compared to control animals given saline (Cattet et al. 2004).

In British Columbia, Canada, intranasal medetomidine is routinely used in net-gunned mountain caribou that require restraint for translocation flights of <1 hour to maternal calving pens. Medetomidine is administered intranasally at a dose of 10 mg (diluted to a volume of 3 ml) to adult female caribou at the capture site, the animal is blindfolded and restrained in a body bag for translocation. The medetomidine-induced sedation greatly facilitates procedures such as ultrasonography, blood sampling and collar fitting. At the termination of sampling procedures, the effects of medetomidine is antagonized with atipamezole at five times the medetomidine dose, resulting in a rapid, reliable recovery (Schwantje and Caulkett, unpublished data). This technique has been used safely up to 6 weeks prior to parturition.

In general, mortality rates are often higher with net-gunning than with chemical immobilizations (Jacques et al. 2009, Valkenburg et al. 1983, Webb et al. 2008). The main cause of mortality, or reason for euthanasia, is trauma during the capture event (Barrett et al. 1982, Compton et al. 1995, Shury 2014, Valkenburg et al. 1983). Caribou are relatively small and thin-skinned and are easily traumatized during net-gun captures (Valkenburg et al. 1983). Ferguson (2015) reported a mortality rate of 4% in net-gunned caribou in Canada and found that injuries occurred in 30% of the captures. No thorough studies on post capture morbidity and mortality in net-gunned caribou have been found, but evidence from other species shows a significant risk of capture myopathy (Jacques et al. 2009). As a general recommendation, net-gunning should only be conducted in areas where chemical immobilization is not an option, utilizing experienced personnel and pilots.

15.3.2 Drive-Nets

This technique was developed for caribou in the 1960s by DesMeules et al. (1971) and has been used in more recent studies (Compton et al. 1995, Smith et al. 2000). Caribou are hazed into upright nets and physically restrained by personnel. The method can be effective but requires at least one person per animal that gets entangled (Shury 2014). Sedative drugs will facilitate handling and reduce struggling and stress in the animals (Arnemo et al. 2005).

15.3.3 Netting from Snowmobiles

Netting from snowmobiles has been developed into the standard capture method for Svalbard reindeer, and hundreds of animals have been captured during the past 20 years (Milner et al. 2003, Omsjoe et al. 2009). A net is held between two snowmobiles and the target animal is chased until it gets entangled in the net. The animal is then manually restrained and hobbled before processing. No sedatives were administered in these studies and the use of head-covers was not reported. Although blood cortisol levels were low, half of the animals had rectal temperatures above 40°C, with some individuals developing extreme hyperthermia close to 42°C. Intranasal medetomidine at 8 mg (diluted to a volume of 2 ml) can be used to induce deep sedation (Evans, personal observation) (Figure 15.6).

FIGURE 15.6   Netting of a Svalbard reindeer (*R. t. platyrhynchus*) from snowmobiles (Photo: Erik Ropstad).

15.3.4   WATER CAPTURE

Water capture of caribou was developed in the late 1950s and was used to tag more than 2,400 animals over 7 years on a lake in Manitoba, Canada (Miller and Robertson 1967). In Alaska, caribou have been captured by boat on the Kobuk River over the last 30 years (James Dau, personal communication). The method uses a look-out point that has been used by caribou hunters for over 7,000 years, with a good view of the surrounding terrain and several kilometers of river on both sides. Personnel sit on the ridge and watch for caribou during the autumn migration. Three boats are used: a capture boat, helping boat and calf boat (Hicks 1997). The capture boat approaches the caribou from behind and one person reaches out from the front of the boat to grab the antlers. Another person throws the anchor upstream and, as a result, the boat and caribou swing into alignment with the river current. This person can then take hold of the caribou's tail. A third person does sampling and places the collar. For bull caribou, the helper boat comes alongside so that the caribou can be held from both sides. For females with calves, the calf boat lifts in the calf to weigh and mark it, and then releases it at the same time as the female. While this method results in a very low risk of hyperthermia and is extremely efficient (upwards of 20 animals per day), it requires a large team (10 people). Additionally, the physiological status of the animals is difficult to assess (Figure 15.7). A study from the Kobuk River found elevated lactate levels and creatine kinase (a muscle enzyme). When compared with caribou darted from helicopter, the river-captured caribou had three times higher lactate levels (Beckmen and Lian 2014). Karns and Crichton (1978) evaluated the effects of handling and physical restraint on 22 blood parameters in caribou captured by boat on a lake and concluded that the animals were not under a great deal of stress. Vital signs were not recorded. Another study using boat captures (Miller and Robertson 1967) reported that 11 animals (0.5%) were accidentally killed during the tagging operations.

15.3.5   CORRALS

In traditional reindeer husbandry, animals are routinely herded into corrals for tagging, transport to other areas, or slaughtering. Wild reindeer have also been herded into corrals for management purposes. In January–March of 1995 and 1996, almost 2,000 animals from a herd in south-central

FIGURE 15.7   Water capture of a caribou (*R. t. caribou*) (Photo: Alaska Department of Fish and Game/Jim Dau).

Norway were slowly driven into corrals by use of snowmobiles and funnel nets and processed by manual restraint for ear tagging, blood sampling and weighing. Although blood cortisol values were high in a subsample of individuals, the behaviour of wild reindeer was similar to what is considered normal for semi-domestic animals in a corral (Arnemo, personal observation). However, reindeer herders on the Seward Peninsula in Alaska report that reindeer will become extremely difficult to herd when just a few caribou are present in the herd. The herd fractionates and becomes agitated (Greg Finstad, personal communication).

15.4   CHEMICAL IMMOBILIZATION

Chemical immobilization of free-ranging reindeer and caribou is a form of veterinary anesthesia conducted under difficult circumstances. It is generally not possible to access the patient for a pre-operative physical examination or laboratory work. Induction of anesthesia in free-ranging animals can be extremely stressful and the risk of severe side effects, injuries and death can never be completely eliminated. In addition, all immobilizing drugs are toxic and some are potentially lethal to humans (Haymerle et al. 2010, Kock and Burroughs 2012, Kreeger and Arnemo 2012, Caulkett and Shury 2014). Chemical immobilization should only be considered when necessary for research or management. Captures should be carried out by a team of professionals with proper training, experience and expertise in wildlife capture, veterinary anesthesia, animal handling, and human and animal first aid and CPR techniques.

Drug administration is done by hand, pole, blow pipe or dart syringe injection. Animals may be stalked, approached in a vehicle or helicopter or netted. Contrary to widespread belief, chemical immobilization from a helicopter is perhaps the least stressful and safest capture method for wild animals. In most circumstances, the animal can be rapidly approached, darted, found, handled and treated, minimizing stress and the risks of the capture operation (Figure 15.8). A study in Alaska (Beckmen and Lian 2014) compared lactate and creatine kinase in caribou captured by three different methods: Chemical immobilization from helicopter, helicopter net-gunning and river captures. Physically restrained caribou had lactate levels four times higher than chemically immobilized caribou. This indicates that physical restraint causes more distress for caribou than helicopter darting and thereby may present a higher risk for capture myopathy (Paterson 2014). Creatine kinase

FIGURE 15.8   Aerial darting of caribou (*R. t. caribou*) (Photo: Alaska Department of Fish and Game/ Dominic Demma).

did not differ significantly between the groups, but showed a trend towards higher values in animals captured with physical restraint.

Caribou and reindeer can be difficult to anesthetize (Caulkett and Arnemo 2014). Effective doses are highly variable depending on capture technique, method of drug administration, degree of human conditioning, and subspecies. Doses required for immobilization of free-ranging animals are usually much higher than those required in captive semi-domesticated individuals (Kreeger and Arnemo 2012). In general, overdosing is probably safer than underdosing, because extended induction times may cause increased stress and risk of complications.

For remote drug administration with a dart, using a cartridge fired or CO_2-powered rifle, CO_2-powered pistol or blow pipe, the target animal has to be approached to within 5–40 m for accurate drug delivery. There have been major advances in the equipment available for remote drug delivery. These systems have the potential to produce serious injury and even death if they are used inappropriately (Cattet et al. 2006). The major sources of injury arise from dart-impact trauma, high-velocity injection of dart contents and inaccurate dart placement. Dart trauma results from dispersion of kinetic energy (E_k) (J) on impact, represented by the following equation: $E_k = \frac{1}{2} mv^2$, where m = mass of the dart (kg) and v = velocity of the dart (m/s) (Kreeger and Arnemo 2012). Obviously, high velocity is the major factor that will cause trauma. A good general rule is to use the lowest velocity that will provide an accurate trajectory at a given distance. Reindeer and caribou have thin skin which is easily penetrated, and dart velocity should be reduced compared to what is used in other cervids (Berntsen 1994, Ryeng et al. 2002, Valkenburg et al. 1999). The other major factor is the mass of the dart. Darts with a lower mass will have less impact energy at a given velocity. This should be a consideration in the choice of a darting system, particularly when dealing with calves or individuals in poor body condition that might be more prone to trauma.

Inaccurate dart placement can also cause severe injuries. This most frequently occurs if darts penetrate the abdomen, thorax or vital structures of the head and neck. The major factors that can lead to inaccurate dart placement include insufficient target practice with the darting system, an attempt to place a dart over an excessive range, and inherent inaccuracy of the darting system.

The final source of injury is related to high-velocity injection of dart contents. Systems that expel drugs via an explosive charge disrupt tissue and produce trauma. Injection volume should be minimized to decrease the degree of tissue trauma. When possible, the use of darts that deliver their contents via compressed air should be utilized to minimize trauma. The choice of system depends on the range required, the dart size and individual characteristics of the target animal. Reviews of remote drug delivery systems and manufacturer information can be found elsewhere (Isaza 2014, Kock and Burroughs 2012, Kreeger and Arnemo 2012).

Dart placement is the most important determinant of induction time, with the quickest absorption occurring after injection in large muscle masses, such as the neck, shoulder or hindquarters. One study in Norwegian reindeer (Arnemo et al. 2011) showed that the mean induction time in animals with optimal hits was 5.6 min compared to 11.1 min in individuals with suboptimal hits. Other factors that may influence induction time include the dose received, the animal's physical condition, age and sex, and its sensitivity to the immobilizing drugs. Animals that are excited or stressed can have induction times that are considerably longer than calm animals. As soon as the dart is placed, the time should be recorded and the animal must be carefully observed to ensure that it is not lost or injured during the induction period. If a helicopter is used in pursuit, it must retreat so that the animal can be observed with less stress for the animal. The helicopter may need to steer an animal away from potential hazards such as cliffs or open water. Observing the induction via the pilot or observer in a light fixed-wing is preferred to a hovering helicopter.

Careful initial approach to an immobilized animal is important. The animal should be observed from a distance to determine that there are no purposeful movements and that the position of the head and neck enable free airways. Respiratory movements can also be observed from a distance. If the animal is not at immediate risk, a set waiting period from the time of sternal recumbency to physical handling should be predetermined and maintained to prevent stimulating an arousal by premature approach. A set of vital signs, including rectal temperature, respiratory rate and heart rate, should be obtained immediately after approaching the animal. The eyes should be lubricated with an isotonic ophthalmic solution or gel, and a blindfold placed to decrease visual stimulation and to protect the cornea from dust and sunlight. Earplugs are useful to minimize auditory stimuli (Figure 15.9).

FIGURE 15.9 Eye cover and ear plugs in an anesthetized forest reindeer (*R. t. fennicus*) (Photo: Sauli Laaksonen).

Depth of anesthesia should be closely monitored throughout the procedure. Factors that increase the risk of sudden arousal include loud noises (especially distress vocalization of off-spring), movement of the animal or changes in the body position, or painful stimuli. Techniques for monitoring depth of anesthesia depend on the anesthetic agent. Degree of central nervous system depression can be classified as level I (mildly affected, voluntary movement and intact reflexes), level II (no voluntary movement and intact reflexes), level III (unconsciousness, depressed reflexes, muscular relaxation) or level IV (ceased respiration, dilated pupils) (Lian et al. 2016). The choice of drug depends on the length of the remaining procedures and depth of anesthesia needed. Top-up drugs can be given intramuscularly or intravenously, with intravenous drugs acting faster and with shorter duration than intramuscular drugs. When potent opioids are used, however, it is preferable to have the immobilized animal in a level I–II.

Immobilized reindeer and caribou should be positioned in sternal recumbency, with the head slightly higher than the body and the nose pointing downwards, permitting drainage of saliva and rumen contents. Regurgitation is an emergency in an immobilized animal. If inhalation of stomach contents occurs, aspiration pneumonia can result, requiring prolonged intensive care that usually is not possible in a field situation. A long-acting broad-spectrum antibiotic should be considered for animals showing signs of aspiration, such as observed aspiration events, rumen contents in the nostrils, coughing or harsh lung sounds.

Bloat (tympany) results from the inability to relieve gas from the rumen through eructation and is usually caused by lateral recumbency or by rumenal atony following administration of alpha-2 adrenergic agonists. Supportive treatment consists of placing the immobilized animal in sternal recumbency with the neck extended and the head forward, with the nose pointing downwards. Move the animal from side to side over the brisket and elevate the front quarters of smaller animals. If positioning does not relieve the bloat, insert a lubricated and properly sized tube via the esophagus into the rumen to relieve pressure. In animals immobilized with alpha-2 adrenergic agonists, bloat will resolve following the administration of a specific antagonist. In the face of severe bloat, procedures should be completed and the immobilization should be reversed. The last resort is trocharization of the rumen using a large bore (2.5 mm), long (60–80 mm) needle. A long-acting broad-spectrum antibiotic should be administered, unless contraindicated, if a trochar was used.

An uncommon cause of hypoxemia includes pneumothorax due to dart penetration of the thoracic cavity, causing air in the pleural space. It may be possible to treat pneumothorax with thoracocentesis and proper equipment should be carried to perform the procedure in the field.

Space limitations dictate what type of equipment can be carried in a field situation. However, capture should not take place unless all necessary equipment can be brought to the animal. A portable oxygen source should be part of the standard field equipment. Oxygen therapy is fundamental for prevention and treatment of hypoxemia during anesthesia. Lightweight aluminum cylinders, combined with a sturdy regulator, are handy for field use. Portable, battery-powered oxygen concentrators are also useful, especially in remote locations where logistics with oxygen cylinders can be difficult (Fahlman et al. 2012). See Fahlman (2014) for an in-depth coverage of oxygen therapy.

Equipment for airway support, including appropriately sized endotracheal tubes and a bag to enable mechanical ventilation, should be included in an emergency kit. Intubation can be difficult in reindeer and caribou. The best technique is to maintain the animal in sternal recumbency with the head and neck extended upwards. Use a laryngoscope with a long, flat blade and stiffen the endotracheal tube with a stylet. The epiglottis is long and mobile and the flat blade of the laryngoscope should be carefully placed on the dorsum of the epiglottis, depressing it ventrally. The opening to the glottis can then be visualized and intubation can proceed. To facilitate intubation, the depth of anesthesia may be increased with an additional dose of anesthetic agent unless the animal is already in respiratory distress.

15.4.1 IMMOBILIZING DRUGS

This section is an overview of pharmacological agents and recommended protocols for sedation, immobilization or anesthesia of *Rangifer*. In-depth coverage of the pharmacology of anesthetic drugs is beyond the scope of this chapter, and readers are referred to the standard textbooks in veterinary anesthesiology, Grimm et al. (2015) and West et al. (2014).

The major classes of sedative and anesthetic drugs used for chemical restraint and capture are alpha-2 adrenoceptor agents, opioids and cyclohexamines (Caulkett and Arnemo 2015, Kreeger and Arnemo 2012). Other anesthetic agents may be indicated for special purposes. In addition, various "tranquilizers" can be used to reduce stress during handling and transport. When drugs from different classes are combined, potentiating effects are achieved and the dose of each component can be reduced, and fewer side effects are seen.

Alpha-2 adrenoceptor agents include xylazine, romifidine, detomidine, medetomidine and dexmedetomidine. These drugs induce sedation, analgesia and muscle relaxation and can be used for immobilization of captive individuals at high doses. Although xylazine has been used as the sole immobilizing drug in caribou (Doherty and Tweedie 1989), alpha-2 adrenoceptor agents should always be combined with a cyclohexamine or an opioid for capture of free-ranging reindeer and caribou. The relative potencies of xylazine, romifidine, detomidine and medetomidine in sheep are 1:3:5:15 (Kreeger and Arnemo 2012). Dexmedetomidine, the active component of medetomidine, is twice as potent as medetomidine. The recommended antagonist (reversal agent) for all these drugs is atipamezole. It should be noted that atipamezole has a shorter elimination half-life than medetomidine in Norwegian reindeer. Ranheim et al. (1997) reported that light resedation, lasting for 3–5 hours, was seen 0.5–1 hour after intravenous administration of atipemezole in reindeer sedated with medetomidine. Other reversal agents, such as tolazoline and yohimbine, can be used, but they are less potent and less specific for the receptor binding sites.

Opioids used in *Rangifer* are etorphine, carfentanil and thiafentanil. These agents are potent analgesics that induce sedation and immobilization. They are extremely toxic to humans. In most cervids, carfentanil is approximately three times more potent than etorphine and thiafentanil (Kreeger and Arnemo 2012). Opioids are usually combined with an alpha-2 adrenoceptor agonist or a tranquilizer. Due to human abuse, carfentanil was taken off the market in 2016 and is no longer available. Opioid antagonists include naltrexone and diprenorphine. The latter is a partial agonist/antagonist and is only used to reverse etorphine when naltrexone is unavailable.

The cyclohexamines, ketamine and tiletamine, are dissociative anesthetics that induce loss of conscious perception. They cause muscle rigidity and are therefore always combined with an alpha-2 adrenoceptor agent or a benzodiazepine in order to produce adequate muscle relaxation, lower the dose requirements and avoid rough inductions and recoveries. Ketamine is the dissociative agent of choice in reindeer and caribou. Tiletamine is two to three times more potent than ketamine but causes extended recoveries due to its longer elimination half-life. There are no known antagonists to cyclohexamines, and reversals with alpha-2 adrenoceptor antagonists in animals immobilized with combinations of an alpha-2 adrenoceptor agent and a cyclohexamine should be delayed for 30 and 45 min, respectively, after administration of ketamine or tiletamine.

Blake et al. (2007) stated that rutting *Rangifer* might be extremely sensitive to alpha-2 adrenoceptor agents and cyclohexamines. Although unpublished, the increased sensitivity of rutting Alaskan caribou to xylazine has been recognized by veterinarians and field biologists, and even at extremely low doses, the mortality rate experienced in rutting bulls is unacceptably high. Thus, using extra caution or avoiding these drugs in rutting bulls whenever possible is warranted.

Anesthesia of pregnant animals has potential consequences for the foetus and there are concerns over the use of anesthetic drugs in *Rangifer* during late gestation. In general, the risks are categorized by gestational trimester (Raff 2015). Anesthesia during the first trimester is associated with possible foetal teratogenesis, spontaneous abortion and foetal death. Anesthetic exposure during the middle trimester is considered the safest period of gestation, although spontaneous abortion and foetal death have been reported. Anesthesia during the last trimester carries a risk of premature labor and foetal death. Alpha-2 adrenoceptor agonists are known to induce myometrial contractions and to reduce uterine blood flow and oxygen delivery. All these drugs are marketed with a caution against their use in pregnant domestic animals. Free-ranging reindeer and caribou are typically captured during midwinter, i.e. pregnant females are in their late second or early final trimester. Stress, such as chasing and physical restraint, may increase the anesthetic risk. There are no published studies on the effects of capture and anesthesia of pregnant *Rangifer*. A general recommendation is to limit the length of exposure to strenuous exercise and anesthesia. Care should be taken to closely monitor the animal in order to prevent or treat thermoregulatory, respiratory or cardiovascular distress.

Drugs and doses for zoo caribou and reindeer are found in Chapter 14 (Caribou and reindeer in Parks and Zoos). Inhalational anesthesia is recommended for prolonged or very invasive procedures in captive animals, but the method is not covered in this chapter focusing on field methods. Readers are referred to Grimm et al. (2015) and West et al. (2014).

Reindeer and caribou are food-producing animals and attention should be paid to possible drug residues in meat. Regulations differ between countries and drug withdrawal times may vary. Thus, most of the immobilizing drugs used in caribou and reindeer are administered "extra label," i.e. they are not formally approved for this species and no withdrawal times have been set (Kreeger and Arnemo 2012).

15.4.2 Caribou

Caribou have been chemically immobilized for various purposes since 1964. Etorphine, either alone or in combination with acepromazine or xylazine, replaced the obsolete neuromuscular blocking agents (Patenaude 1982). However, etorphine was not readily available in concentrations higher than 1 mg/ml, resulting in a volume problem for administration by dart (Fong 1982; Valkenburg et al. 1983).

When carfentanil was introduced in 1986 at a concentration of 3 mg/ml, it became the drug of choice for caribou immobilizations in Alaska. Since then, carfentanil-xylazine has been the primary drug combination for immobilization of free-ranging caribou (Adams et al. 1988; Boertje et al. 1996, Valkenburg et al. 1999). However, as of August 2016, carfentanil is off the market and is not expected to be available in the future. In a recent study, thiafentanil-azaperone-xylazine was evaluated for immobilization of caribou calves in Alaska (Lian et al. 2016).

The current dose recommendations in Alaska vary with age and body size. All of the following doses are for aerial darting. For calves captured at 5 months of age in fall (40–60 kg), carfentanil at 1.5 mg and xylazine at 20 mg per animal is used (Table 15.1). As an alternative a combination of thiafentanil (1.5 mg), azaperone (25 mg) and xylazine (20 mg) can be used. Naltrexone at 100 mg/mg carfentanil and 1 mg atipamezole per 10 mg of xylazine or tolazoline at 2 mg/kg are the commonly used antagonists. Thiafentanil is reversed by naltrexone at 10–30 mg/mg thiafentanil (Lian et al. 2016). For adult caribou in the spring season, typically weighing 50–140 kg, the recommended combination is carfentanil (3.75 mg) and xylazine (75 mg). For adult bulls >180 kg, the recommended protocol is carfentanil (4 mg) and xylazine (50–75 mg). The reversal protocols are the same as for calves immobilized with this combination, except that when tolazoline is used rather than atipamezole, it is administered at 1 mg/kg in animals >180 kg.

In Canada, potent opioids are not recommended due to year-round subsistence hunting. The current drug combinations of choice are xylazine (2–3 mg/kg) with ketamine (6–9 mg/kg), xylazine (2–3 mg/kg) with zolazepam-tiletamine (4–6 mg/kg), or medetomidine (0.15–0.2 mg/kg) with ketamine (2–2.7 mg/kg) (Cattet et al. 2005, Cattet 2011, Caulkett and Arnemo 2014). The recommended antagonist is atipamezole used at 1 mg atipamezole per 10 mg xylazine and 5 mg atipamezole per mg of medetomidine.

Antagonists should preferably be given intramuscularly, but in an emergency situation, slow intravenous administration is an option (Figure 15.10).

TABLE 15.1

Recommended Drug Protocols and Doses for Immobilization of Reindeer and Caribou (*Rangifer tarandus* sspp.)

Subspecies	Type	Drugs	Administration	Indication	References
Caribou (*R. t. caribou*)	Free-ranging	Medetomidine 0.15–0.2 mg/kg Ketamine 2–2.7 mg/kg	Aerial darting	Immobilization, anesthesia	Cattet 2011
Caribou (*R. t. caribou*)	Free-ranging	Carfentanil 1.5 mg/calf, 3.5 mg/adult, 4 mg/adult >180 kg Xylazine 20 mg/calf, 50–75 mg/adult	Aerial darting	Immobilization	Lian et al. 2016 Valkenburg et al. 1999
Caribou (*R. t. caribou*)	Free-ranging	Thiafentanil 1.5 mg/calf Azaperone 25 mg/calf Xylazine 20 mg/calf	Aerial darting	Immobilization	Lian et al. 2016
Caribou (*R. t. caribou*)	Free-ranging	Medetomidine 10–13 mg/adult	Hand, intranasal	Sedation in net-gunned animals	Oakley et al. 2004
Norwegian reindeer (*R. t. tarandus*)	Free-ranging	Medetomidine 10 mg/adult Ketamine 200 mg/adult	Aerial darting	Immobilization, anesthesia	Arnemo et al. 2011 Evans et al. 2013
Norwegian reindeer (*R. t. tarandus*)	Captive	Medetomidine 0.15 mg/kg Ketamine 0.75 mg/kg	Ground darting	Immobilization	Ryeng et al. 2002
Norwegian reindeer (*R. t. tarandus*)	Captive	Medetomidine 0.1 mg/kg Ketamine 0.5 mg/kg	Hand	Immobilization, anesthesia	Ryeng et al. 2002
Svalbard reindeer (*R. t. platyrhynchus*)	Free-ranging	Medetomidine 10 mg/adult Ketamine 200 mg/adult	Ground darting	Immobilization, anesthesia	Arnemo and Aanes. 2009 Evans et al. 2013
Forest reindeer (*R. t. fennicus*)	Captive	Medetomidine 0.06–0.08 mg/kg Ketamine 0.6–08 mg/kg	Ground darting	Immobilization, anesthesia	Jalanka and Roeken 1990

FIGURE 15.10   Weighing of a caribou (*R. t. caribou*) immobilized with carfentanil-xylazine (Photo: Alaska Department of Fish and Game/Torsten Bentzen).

15.4.3 Norwegian Reindeer

Ryeng et al. (2002) found that that the method of drug administration significantly influenced the effective immobilizing doses of medetomidine-ketamine in captive, semi-domestic Norwegian reindeer (Eurasian tundra reindeer; *R. t. tarandus*). Doses of 0.1 mg/kg medetomidine and 0.5 mg/kg ketamine given by hand syringe intramuscularly induced complete immobilization in 12 animals kept indoors. When these animals were administered the drugs by dart injection outdoors, the effective immobilizing doses were 50% higher (Table 15.1). The effects of medetomidine were reversed by 5 mg atipamezole per mg of medetomidine intramuscularly.

For free-ranging wild Norwegian reindeer, the same subspecies as semi-domestic reindeer, effective doses for aerial darting are much higher, presumably due to the effect of stress induced by the helicopter chase. Arnemo et al. (2011) used 12–14 mg medetomidine and 60–70 mg ketamine per adult and reported mean induction times of 5.6 min for optimal hits (n = 16) and 11.1 min for suboptimal hits (n = 13). Reversals were achieved by 5 mg atipamezole per mg medetomidine IM. In a follow-up study (Evans et al. 2013), doses were changed to 10 mg medetomidine and 200 mg ketamine per adult. The main side effect was hypoxemia, which was corrected with intranasal oxygen at 2 L/min (Evans et al. 2013) (Figure 15.11).

15.4.4 Svalbard Reindeer

The drug combination of choice for chemical capture of free-ranging Svalbard reindeer is medetomidine-ketamine administered by dart injection after stalking on foot (Arnemo and Aanes 2009, Evans et al. 2013) (Figure 15.12). Initially, a standard dose of 7–8 mg medetomidine and 140–160 mg ketamine per adult was used (Arnemo and Aanes 2009), but in a follow-up study, this dose was increased to 10 mg medetomidine and 200 mg ketamine per adult (Evans et al. 2013) see Table 15.1. The main side effect was hypoxemia, which was corrected with supplemental oxygen (1 L/min) (Evans et al. 2013). Atipamezole at 5 mg per mg of medetomidine intramuscularly reverses immobilization. Different doses of medetomidine-ketamine have been used in other studies (Berntsen 1994, Tyler et al. 1990), but sample sizes were either low or not reported, and data on physiological monitoring were limited or not recorded.

FIGURE 15.11   Radio collared Norwegian reindeer (*R. t. tarandus*) recovering from anesthesia with medetomidine-ketamine (Photo: Jon M. Arnemo).

FIGURE 15.12   Svalbard reindeer (*R. t. platyrhyncus*) are either captured by ground darting or by netting from snowmobiles (Photo: Ronny Aanes).

15.4.5   FOREST REINDEER

Jalanka and Roeken (1990) pioneered the use of medetomidine-ketamine and atipamezole in non-domestic mammals, and their recommended doses for ground darting of captive forest reindeer were 0.06–0.08 mg/kg medetomidine and 0.6–0.8 mg/kg ketamine, followed by 4–5 mg atipamezole per mg medetomidine for reversal. Although these authors divided the atipamezole half intravenously and half subcutaneously, the current consensus is that recoveries will be smoother if the antagonist is administered intramuscularly only. Nieminen et al. (1990) used detomidine alone at doses ranging from 0.04–0.1 mg/kg (hand syringe injection) to 0.1–0.3 mg/kg (ground darting) for sedation or immobilization of 126 semi-domesticated reindeer and 15 wild forest reindeer. From the report it is unclear whether the dose range included both subspecies, but the authors noted that induction times were slightly longer in forest reindeer (8–10 min) compared to semi-domesticated reindeer (8 min). The use of an antagonist was not mentioned.

15.4.6   SEDATION AND TRANQUILIZATION

In some situations, it may be desirable to sedate or tranquilize animals rather than subject them to immobilization or anesthesia. The use of intranasal medetomidine has been discussed for deep sedation. Medetomidine has the advantage of complete "reversibility of sedation" with atipamezole. On some occasions it may be desirable to tranquilize animals for repeated handling or transport. Captive semi-domestic reindeer can be sedated with 0.5 mg/kg xylazine intramuscularly, whereas higher doses (1.5–3 mg/kg) will induce immobilization in calm animals (Arnemo et al. 2014b). Medetomidine (0.05 mg/kg intravenously or 0.1–0.15 mg/kg intramuscularly) has similar effects in animals used to handling (Arnemo et al. 2014b, Arnemo and Ranheim 1999, Soveri et al. 1999). Nieminen et al. (1990) tested various doses of detomidine alone for hand syringe (0.04–0.3 mg/kg) and dart (0.1–0.3 mg/kg) in semi-domestic Norwegian reindeer and reported that initial effects were seen after 3–5 min and that the mean induction time was 8 min. Interestingly, the authors did not reverse the effects of detomidine and found that spontaneous recoveries started after 60–110 min. They concluded that detomidine was effective and safe but pointed out the high individual variation in response to the drug. Azaperone, administered at a dose of 0.2–0.4 mg/kg IM has been used to tranquilize reindeer and caribou for short translocations. The actual duration of this drug has not been determined in this species, but it may be effective for up to 6 hours. Zuclopethixol acetate has been shown to reduce handling stress in wapiti (*Cervus canadensis*) and white-tailed deer (*Odocoileus virginianus*) for up to 4 days (Read et al. 2000, Read and McCorkell 2002). Zuclopenthixol acetate has been successfully used at a dose of 1 mg/kg to reduce handling stress in reindeer during long distance translocations to new facilities (Caulkett, personal observation).

15.4.7 EUTHANASIA

Equipment and drugs for euthanasia or humane killing of irreversibly diseased or injured animals should always be available during captures. The method of euthanasia depends on the situation (Woodbury 2014). In a restrained animal, intravenous administration of a chemical agent such as pentobarbital is recommended. Under anesthesia, in a field situation, a saturated solution of potassium chloride can be administered intracardiacally to avoid the risks to scavengers that occur with barbiturates. A firearm or a bolt gun can also be used to stun the animal, followed by exsanguination. Carcasses of drugged animals should not be left in the field because of the risks posed to scavengers and humans. In some instances, however, the animal cannot be restrained and has to be killed with a hunting rifle. If firearms or bolt guns are used, personnel must be trained in the use of the firearms, the appropriate ammunition, and placement of the bullet/bolt to hit the heart-lung area or the brain.

REFERENCES

Adams, L.G., P. Valkenburg, and J.L. Davis. 1988. Efficacy of carfentanil citrate and naloxone for field immobilization of Alaskan caribou. *North American Caribou Workshop* 3: 167–168.
Arnemo, J.M., and R. Aanes. 2009. Reversible immobilization of in free-ranging Svalbard reindeer (*Rangifer tarandus platyrhynchus*) with medetomidine-ketamine and atipamezole. *Journal of Wildlife Diseases* 45: 877–880.
Arnemo, J.M., and N.A. Caulkett. 2007. Stress, in *Zoo Animal and Wildlife Anesthesia and Immobilization*, eds. G. West, D. Heard, and N.A. Caulkett. 103–109. Ames: Blackwell Publications.
Arnemo, J.M., A.L. Evans, Å. Fahlman, and N.A. Caulkett. 2014a. Field Emergencies and Complications, in *Zoo Animal and Wildlife Anesthesia and Immobilization*, eds. G. West, D. Heard, and N.A. Caulkett. 139–147. 2nd Ed. Ames: Wiley-Blackwell.
Arnemo, J.M., A.L. Evans, A.L. Miller, and Ø. Os. 2011. Effective immobilizing doses of medetomidine-ketamine in free-ranging, wild Norwegian reindeer (*Rangifer tarandus tarandus*). *Journal of Wildlife Diseases* 47: 755–758.
Arnemo, J.M., A.L. Evans, M. Lian, and Ø. Os. 2014b. Anaesthesia of reindeer [in Norwegian with English abstract]. *Norsk Veterinærtidsskrift* 126: 150–153.
Arnemo, J.M., and B. Ranheim. 1999. Effects of medetomidine and atipamezole on serum glucose and cortisol levels in captive reindeer (*Rangifer tarandus tarandus*). *Rangifer* 19: 85–89.
Arnemo, J.M., T. Storaas, C.B. Khadka, and P. Wegge. 2005. Use of medetomidine-ketamine and atipamezole for reversible immobilization of free-ranging hog deer (*Axis porcinus*) captured in drive nets. *Journal of Wildlife Diseases* 42: 467–470.
Barrett, M.W., J.W. Nolan, and L.D. Roy. 1982. Evaluation of a hand-held net-gun to capture large mammals. *Wildlife Society Bulletin* 10: 108–114.
Beckmen, K.B., and M. Lian. 2014. Evaluation of stress in caribou (*Rangifer tarandus granti*) attributed to different capture methods. Proceedings of the 63rd Wildlife Disease Association Annual International Conference, 27 July–1 August, Albuquerque, New Mexico: 42.
Berntsen, F. 1994. Medical immobilization of Svalbard reindeer. Abstracts of the 1st European Conference of the European Section, Wildlife Disease Association, 22–24 November, Paris, France: 15.
Blake, J.E., J.E. Rowell, and M.P. Shipka. 2007. Reindeer Reproductive Management, in *Current Therapy in Large Animal Theriogenology*, eds. R.S. Youngquist, and W.R. Threlfall. 970–974. 2nd Ed. Saunders: Elsevier.
Blix, A.S., L. Walløe, and L.P. Folkow. 2011. Regulation of brain temperature in winter-acclimatized reindeer under heat stress. *Journal of Experimental Biology* 214: 3850–3856.

Boertje, R.D., P. Valkenburg, and M.E. McNay. 1996. Increases in moose, caribou, and wolves following wolf control in Alaska. *Journal of Wildlife Management* 60: 474–489.

Boesch, J.M., J.R. Boulanger, P.D. Curtis, H.N. Erb, J.W. Ludders, M.S. Kraus, and R.D. Gleed. 2011. Biochemical variables in free-ranging white-tailed deer (*Odocoileus virginianus*) after chemical immobilization in clover traps or via ground-darting. *Journal of Zoo and Wildlife Medicine* 42: 18–28.

Boysen, S. 2014. Zoo and Wildlife CPR, in *Zoo Animal and Wildlife Anesthesia and Immobilization*, eds. G. West, D. Heard, and N.A. Caulkett. 125–138. 2nd Ed. Ames: Wiley-Blackwell.

Catley, A., R.A. Kock, M.G. Hart, and C.M. Hawkey. 1990. Haematology of clinically normal and sick captive reindeer (*Rangifer tarandus*). *Veterinary Record* 126: 239–241.

Cattet, M. 2011. *Standard Operating Procedure: Capture, Handling and Release of Caribou.* Northwest Territories Environment and Natural Resources. Wildlife Care Committee. Version 2. www.enr.gov .nt.ca/sites/default/files/guidelines/nwtwcc_sop_caribou.pdf.

Cattet, M., T. Shury, and R. Patenaude. 2005. *The Chemical Immobilization of Wildlife.* 2nd Ed. Saskatoon: Canadian Association of Zoo and Wildlife Veterinarians.

Cattet, M.R.L., A. Bourque, B.T. Elkin, K.D. Powley, D.B. Dahlstrom, and N.A. Caulkett. 2006. Evaluation of the potential for injury with remote drug-delivery systems. *Wildlife Society Bulletin* 34: 741–749.

Cattet, M.R.L., N.A. Caulkett, C. Wilson, T. Vandenbrink, and R.K. Brook. 2004. Intranasal administration of xylazine to reduce stress in elk captured by net gun. *Journal of Wildlife Diseases* 40: 562–565.

Cattet, M.R.L., K. Christison, N.A. Caulkett, and G.B. Stenhouse. 2003. Physiologic responses of grizzly bears to different methods of capture. *Journal of Wildlife Diseases* 39: 649–654.

Caulkett, N., and J.M. Arnemo. 2014. Cervids (Deer), in *Zoo Animal and Wildlife Anesthesia and Immobilization*, eds. G. West, D. Heard, and N.A. Caulkett. 823–829. 2nd Ed. Ames: Wiley-Blackwell.

Caulkett, N., and T. Shury. 2014. Human Safety During Wildlife Capture, in *Zoo Animal and Wildlife Anesthesia and Immobilization*, eds. G. West, D. Heard, and N.A. Caulkett. 181–187. 2nd Ed. Ames: Wiley-Blackwell.

Caulkett, N.A., and J.M. Arnemo. 2015. Comparative Anesthesia and Analgesia of Zoo Animals and Wildlife, in *Veterinary Anesthesia and Analgesia*, eds. K.A. Grimm, L.A. Lamont, W.J. Tranquilli, S.A. Green, and S.A. Robertson. 764–776. The Fifth Edition of Lumb and Jones. Ames: Wiley Blackwell.

Colborn, D.R., D.L. Thompson, T.L. Roth, J.S. Capehart, and K.L. White. 1991. Responses of cortisol and prolactin to sexual excitement and stress in stallions and geldings. *Journal of Animal Science* 69: 2556–2562.

Compton, B.B., P. Zager, and G. Servheen. 1995. Survival and mortality of translocated woodland caribou. *Wildlife Society Bulletin* 23: 490–496.

Cracknell, J. 2014. Airway Management, in *Zoo Animal and Wildlife Anesthesia and Immobilization*, eds. G. West, D. Heard, and N.A. Caulkett. 53–64. 2nd Ed. Ames: Wiley-Blackwell.

DelGiudice, G.D., B.A. Mangipane, B.A. Sampson, and C.O. Kochanny. 2001. Chemical immobilization, body temperature, and post-release mortality of white-tailed deer captured by Clover trap and net-gun. *Wildlife Society Bulletin* 29: 1147–1157.

DelGiudice, G.D., B.A. Sampson, D.W. Kuehn, M.C. Powell, and J. Fieberg. 2005. Understanding margins of safe capture, chemical immobilization, and handling of free-ranging white-tailed deer. *Wildlife Society Bulletin* 33: 677–687.

DesMeules, P., B.R. Simard, and J.M. Brassard. 1971. A technique for the capture of caribou, *Rangifer tarandus*, in winter. *Canadian Field Naturalist* 85: 221–229.

Dieterich, R.A. 1993. Medical Aspects of Reindeer Farming, in *Zoo and Wild Animal Medicine*, ed. M.E. Fowler. 123–127. Current therapy 3. Denver, Colorado: Saunders.

Dieterich, R.A., and J.K. Morton, eds. 1990. *Reindeer Health Aide Manual.* 2nd Ed. AFES Misc. Pub 90-4, CES 100H-00046. Fairbanks: Agricultural and Forestry Experiment Station, Cooperative Extension Service, University of Alaska Fairbanks.

Doherty, T.J., and P.R. Tweedie. 1989. Evaluation of xylazine hydrochloride as the sole immobilizing agent in moose and caribou – and its subsequent reversal with idazoxan. *Journal of Wildlife Diseases* 25: 95–98.

Eloranta, E., H. Norberg, A. Nilsson, T. Pudas, and H. Säkkinen. 2002. Individually coded telemetry: A tool for studying heart rate and behavior in reindeer calves. *Acta Veterinaria Scandinavica* 43: 135–144.

Essén-Gustavsson, B., and C. Rehbinder. 1984. The influence of stress on substrate utilization in skeletal muscle fibers of reindeer (*Rangifer tarandus* L). *Rangifer* 4: 2–8.

Evans, A.L., M. Lian, Ø. Os, R. Andersen, O. Strand, C. das Neves, M. Tryland, and J.M. Arnemo. 2013. Physiological evaluation of medetomidine-ketamine anesthesia in free-ranging Svalbard reindeer (*Rangifer tarandus platyrhynchus*) and wild Norwegian reindeer (*R. t. tarandus*). *Journal of Wildlife Diseases* 49: 1037–1041.

Fahlman, Å. 2014. Oxygen Therapy, in *Zoo Animal and Wildlife Anesthesia and Immobilization*, eds. G. West, D. Heard, and N.A. Caulkett. 69–81. 2nd Ed. Ames, Iowa: Wiley-Blackwell.

Fahlman, Å., N. Caulkett, J.M. Arnemo, P. Neuhaus, and K.E. Ruckstuhl. 2012. Efficacy of a portable oxygen concentrator with pulsed delivery for treatment of hypoxemia during anesthesia of wildlife. *Journal of Zoo and Wildlife Medicine* 43: 67–76.

Fancy, S.G., and R.G. White. 1985. Energy expenditures by caribou while cratering in snow. *Journal of Wildlife Management* 49: 987–993.

Fancy, S.G., and R.G. White. 1986. Predicting energy expenditures for activities of caribou from heart rates. *Rangifer Special Issue No. 1*: 123–130.

Ferguson, M.A.D. 2015. Gun-netting of Arctic tundra caribou: Inuit guided choices, and led to improvements and low impacts. 14th Arctic Ungulate Conference, 21–26 August, Røros, Norway.

Fong, D.W. 1982. Immobilization of caribou with etorphine plus acepromazine. *Journal of Wildlife Management* 46: 560–562.

Fowler, M.E. 2008. *Restraint and Handling of Wild and Domestic Animals*. 3rd Ed. Ames, Iowa: Wiley-Blackwell.

Fuller, T.K., and L.B. Keith. 1981. Immobilization of woodland caribou with etorphine. *Journal of Wildlife Management* 45: 745–748.

Grimm, K.A., L.A. Lamont, W.J. Tranquilli, S.A. Green, and S.A. Robertson, eds. 2015. *Veterinary Anesthesia and Analgesia*, The Fifth Edition of Lumb and Jones. Ames, Iowa: Wiley Blackwell.

Haga, H.A., S. Wenger, S. Hvarnes, Ø. Os, C.M. Rolandsen, and E.J. Solberg. 2009. Plasma lactate concentrations in free-ranging moose (*Alces alces*) immobilized with etorphine. *Veterinary Anaesthesia and Analgesia* 36: 555–561.

Haigh, J.C. 1976. Capture of woodland caribou in Canada. Proceedings of the Annual Conference of the American Association of Zoo Veterinarians, 110–115.

Haymerle, A., A. Fahlman, and C. Walzer. 2010. Human exposures to immobilising agents: Results of an online survey. *Veterinary Record* 167: 327–332.

Hicks, M.V., ed. 1997. Federal Aid in Wildlife Restoration Management Report of Survey-Inventory Activities 1 July 1994–30 June 1996: Caribou. 160. Fairbanks: Alaska Department of Fish and Game.

Hyvärinen, H., T. Helle, M. Bieminen, P. Väyrynen, and R. Väyrynen. 1976. Some effects of handling reindeer during gatherings on the composition of their blood. *Animal Production* 22: 105–114.

International Wildlife Veterinary Services. 1991. *Wildlife Restraint Series*. Salinas, California: International Wildlife Veterinary Services.

Isaza, R. 2014. Remote Drug Delivery, in *Zoo Animal and Wildlife Anesthesia and Immobilization*, eds. G. West, D. Heard, and N.A. Caulkett. 155–169. 2nd Ed. Ames, Iowa: Wiley-Blackwell.

Jacques, C.N., J.A. Jenks, C.S. Deperno, J.D. Sievers, T.W. Grovenburg, T.J. Brinkman, C.S. Swanson, and B.A. Stillings. 2009. Evaluating ungulate mortality associated with helicopter net-gun captures in the Northern Great Plains. *Journal of Wildlife Management* 73: 1282–1291.

Jalanka, H.H., and B.O Roeken. 1990. The use of medetomidine, medetomidine-ketamine combinations, and atipamezole in nondomestic animals: A review. *Journal of Zoo Wildlife Medicine* 21: 259–282.

Jessup, D.A., R.K. Clark, R.A. Weaver, and M.D. Kock. 1988. The safety and cost-effectiveness of net-gun capture of desert bighorn sheep (*Ovis canadensis nelsoni*). *Journal of Zoo Animal Medicine* 19: 208–213.

Johnson, D., N.J. Harms, N.C. Larter, B.T. Elkin, H. Tabel, and G. Wei. 2010. Serum biochemistry, serology, and parasitology of boreal caribou (*Rangifer tarandus caribou*) in the Northwest Territories, Canada. *Journal of Wildlife Diseases* 46: 1096–1107.

Karns, P.D., and V.F.J. Crichton. 1978. Effects of handling and physical restraint on blood parameters in woodland caribou. *Journal of Wildlife Management* 42: 904–908.

Kock, M.D., and R. Burroughs, eds. 2012. *Chemical and Physical Restraint of Wild Animals: A Training and Field Manual for African Species*. 2nd Ed. Greyton, South Africa: International Wildlife Veterinary Services.

Koolhaas, J.M., A. Bartolomucci, B. Buwalda, S.F. de Boer, G. Flügge, S.M. Korte, P. Merlo, R. Murison, B. Oliver, P. Palanza, G. Richter-Levin, A. Sgoifo, T. Steimer, O. Stiedl, G. van Dijk, M. Wöhr, and E. Fuchs. 2011. Stress revisited: A critical evaluation of the stress concept. *Neuroscience and Biobehavioral Reviews* 35:1291–1301.

Kreeger, T.J., and J.M. Arnemo. 2012. *Handbook of Wildlife Chemical Immobilization*, 4th Ed. Wheatland: TJ Kreeger.

Lian, M., K.B. Beckmen, T.W. Bentzen, D.J. Demma, and J.M. Arnemo. 2016. Thiafentanil-azaperone-xylazine and carfentanil-xylazine immobilizations of free-ranging caribou (*Rangifer tarandus granti*) in Alaska. *Journal of Wildlife Diseases* 52: 327–334.

López-Olvera, J.R., I. Marco, J. Montané, E. Casa-Díaz, G. Mentaberre, and S. Lavín. 2009. Comparative evaluation of effort, capture and handling effects of drive nets to capture of roe deer (*Capreolus capreolus*), southern chamois (*Rupicapra pyrenaica*) and Spanish ibex (*Capra pyrenaica*). *European Journal of Wildlife Research* 55: 193–202.

Lund-Larsen, T.R., J. Kofstad, and A. Aakvaag. 1978. Seasonal changes in serum levels of aldosterone, cortisol and inorganic ions in the reindeer (*Rangifer tarandus*). *Comparative Biochemistry and Physiology* 60A: 383–386.

McDonell, W.N., and C.L. Kerr. 2015. Physiology, Pathophysiology, and Anesthetic Management of Patients with Respiratory Disease, in *Veterinary Anesthesia and Analgesia*, eds. K.A. Grimm, L.A. Lamont, W.J. Tranquilli, S.A. Green, and S.A. Robertson. 513–555. The Fifth Edition of Lumb and Jones. Ames, Iowa: Wiley Blackwell.

McEwan, E.H. 1967. Hematological studies of barren-ground caribou. *Canadian Journal of Zoology* 46: 1031–1036.

Mentaberre, G., J.R. López-Olvera, E. Casa-Díaz, L. Fernández-Sirera, I. Marco, and S. Lavín. 2010. Effects of azaperone and haloperidol on the stress response of drive-net captured Iberian ibexes (*Capra pyrenaica*). *European Journal of Wildlife Research* 56: 757–764.

Mercer, J.B., H.K. Johnsen, S.D. Mathiesen, and A.S. Blix. 1985. An intra ruminal heat exchanger for use in large conscious animals. *Rangifer* 5: 10–14.

Messier, F., J. Huot, F. Goudreault, and A. Tremblay. 1987. Reliability of blood parameters to assess the nutritional status of caribou. *Canadian Journal of Zoology* 65: 2413–2416.

Mesteig, K., N.J.C. Tyler, and A.S. Blix. 2000. Seasonal changes in heart rate and food intake in reindeer (*Rangifer tarandus tarandus*). *Acta Physiologica Scandinavica* 170: 145–151.

Miller, A.L., A.L. Evans, Ø. Os, and J.M. Arnemo. 2013. Biochemical and hematological reference values for free-ranging wild Norwegian reindeer (*Rangifer tarandus tarandus*). *Journal of Wildlife Diseases* 49: 221–228.

Miller, D.R., and J.D. Robertson. 1967. Results of tagging of caribou at Little Duck Lake, Manitoba. *Journal of Wildlife Management* 31: 150–159.

Milner, J., A. Stien, R.J. Irvine, S.D. Albon, R. Langvatn, and E. Ropstad. 2003. Body condition in Svalbard reindeer and the use of blood parameters as indicators of condition and fitness. *Canadian Journal of Zoology* 81: 1566–1578.

Nielsen, L. 1999. *Chemical Immobilization of Wild and Exotic Animals*. Ames: Iowa State University Press.

Nieminen, M., E. Tanhuanpää, and T. Vähe-Vahe. 1990. Detomidine immobilization in wild and semi-domesticated reindeer. *Rangifer Special Issue No. 1*: 58.

Nieminen, M., and J. Timisjärvi. 1981. Blood composition of the reindeer. I. Haematology. *Rangifer* 1: 10–26.

Nieminen, M., and J. Timisjärvi. 1983. Blood composition of the reindeer. II. Blood chemistry. *Rangifer* 3: 16–32.

Nilssen, K.J., K. Bye, J.A. Sundsfjord, and A.S. Blix. 1985. Seasonal changes in T_3, FT_4, and cortisol in free-ranging Svalbard reindeer (*Rangifer tarandus platyrhynchus*). *General and Comparative Endocrinology* 59: 210–213.

Nilssen, K.J., H.K. Johnsen, A. Rognmo, and A.S. Blix. 1984. Heart rate and energy expenditure in resting and running Svalbard and Norwegian reindeer. *American Journal of Physiology* 246: R963–R967.

Nilsson, A., B. Åhman, H. Norberg, I. Redbo, E. Eloranta, and K. Olsson. 2006. Activity and heart rate in semi-domesticated reindeer during adaptation to emergency feeding. *Physiology and Behavior* 88: 116–123.

Oakley, M., T.S. Jung, M. Kienzler, R. Farnell, L. LaRocque, J. McLelland, and P. Merchant. 2004. Intranasal sedation of woodland caribou captured by net gun to reduce stress. 10th North American Caribou Conference, May 4–6, Girdwood Alaska.

Omsjoe, E.H., A. Stien, J. Irvine, S.D. Albon, E. Dahl, S.I. Thoresen, E. Rustad, and E. Ropstad. 2009. Evaluating capture stress and its effects on reproductive success in Svalbard reindeer. *Canadian Journal of Zoology* 87: 73–85.

Ozeki, L., and N. Caulkett. 2014. Monitoring, in *Zoo Animal and Wildlife Anesthesia and Immobilization*, eds. G. West, D. Heard, and N.A. Caulkett. 43–51. 2nd Ed. Ames, Iowa: Wiley-Blackwell.

Ozeki, L.M., Å. Fahlman, G. Stenhouse, J.M. Arnemo, and N. Caulkett. 2014. Evaluation of the accuracy of different methods of monitoring body temperature in anesthetized brown bears (*Ursus arctos*). *Journal of Zoo and Wildlife Medicine* 45: 819–824.

Panzacchi, M., B.V. Moorter, P. Jordhøy, and O. Strand. 2013. Learning from the past to predict the future: Using archaeological findings and GPS data to quantify reindeer sensitivity to anthropogenic disturbance in Norway. *Landscape Ecology* 28: 847–859.

Patenaude, R.P. 1982. Chemical Immobilization of North American Caribou, in *Chemical Immobilization of North American Wildlife*, eds. L. Nielsen, J.C. Haigh, and M.E. Fowler. 370–379. The Humane Society: Proceedings of the North American Symposium: Chemical Immobilization of Wildlife, Milwaukee, Wisconsin, 2–6 April.

Paterson, J. 2014. Capture Myopathy, in *Zoo Animal and Wildlife Anesthesia and Immobilization*, eds. G. West, D. Heard, and N.A. Caulkett. 171–179. 2nd Ed. Ames, Iowa: Wiley-Blackwell.

Pösö, A.R., M. Nieminen, S. Sankari, and T. Soveri. 1994. Exercise-induced changes in blood composition of racing reindeer (*Rangifer tarandus tarandus* L). *American Journal of Physiology* 267: R1209–R1216.

Raff, M.R. 2015. Anesthetic Considerations During Pregnancy and for the Newborn, in *Veterinary Anesthesia and Analgesia*, eds. K.A. Grimm, L.A. Lamont, W.J. Tranquilli, S.A. Green, and S.A. Robertson. 708–719. The Fifth Edition of Lumb and Jones. Ames, Iowa: Wiley Blackwell.

Ranheim, B., T.E. Horsberg, U. Nymoen, N.E. Søli, N.J.C. Tyler, and J.M. Arnemo. 1997. Reversal of medetomidine-induced sedation in reindeer (*Rangifer tarandus tarandus*) with atipamezole increases the medetomidine concentration in plasma. *Journal of Veterinary Pharmacology and Therapeutics* 20: 350–354.

Read, M., N.A. Caulkett, and M. McCallister. 2000. Evaluation of zuclopenthixol acetate to decrease handling stress in wapiti. *Journal of Wildlife Diseases* 36, 450–459.

Read, M.R., and R.B. McCorkell. 2002. Use of azaperone and zuclopenthixol acetate to facilitate translocation of white-tailed deer (*Odocoileus virginianus*). *Journal of Zoo and Wildlife Medicine* 33: 163–165.

Rehbinder, C. 1990. Management stress in reindeer. *Rangifer Special Issue No 3*: 267–288.

Rehbinder, C., and L.E. Edqvist. 1981. Influence of stress on some blood constituents in reindeer (*Rangifer tarandus* L). *Acta Veterinaria Scandinavica* 22: 480–492.

Rehbinder, C., L.E. Edqvist, K. Lundström, F. Villafañe. 1982. A field study of management stress in reindeer (*Rangifer tarandus* L). *Rangifer* 2: 2–21.

Risling, T.E., Å. Fahlman, N.A. Caulkett, and S. Kutz. 2011. Physiological and behavioral effects of hypoxemia in reindeer immobilized with xylazine-etorphine. *Animal Production Science* 51:1–4.

Ryeng, K.A., S. Larsen, and J.M. Arnemo. 2002. Medetomidine-ketamine in reindeer (*Rangifer tarandus tarandus*): Effective immobilizing doses by hand- and dart-administered injection. *Journal of Zoo and Wildlife Medicine* 33: 397–400.

Säkkinen, H., J. Tornberg, P.J. Goddard, E. Eloranta, E. Dahl, E. Ropstad, and S. Saarela. 2005. Adrenal responsiveness of reindeer (*Rangifer tarandus tarandus*) to intravenously administered ACTH. *Animal Science* 81: 399–402.

Säkkinen, H., J. Tornberg, P.J. Goddard, E. Eloranta, E. Ropstad, and S. Saarela. 2004. The effects of blood sampling methods on indicators of physiological stress in reindeer (*Rangifer tarandus tarandus*). *Domestic Animal Endocrinology* 26: 87–98.

Shury, T. 2014. Physical Capture and Restraint, in *Zoo Animal and Wildlife Anesthesia and Immobilization*, eds. G. West, D. Heard, and N.A. Caulkett. 109–124. 2nd Ed. Ames, Iowa: Wiley-Blackwell.

Sire, J.E., A. Blom, Ø.V. Sjaastad, E. Ropstad, T.A.B. Nilsen, Ø. Pedersen, and M. Forsberg. 1995. The effect of blood sampling on plasma cortisol in female reindeer (*Rangifer tarandus tarandus* L). *Acta Veterinaria Scandinavica* 36: 583–587.

Smith, K.G., J. Ficht, D. Hobson, T.C. Sorensen, and D. Hervieux. 2000. Winter distribution of woodland caribou in relation to clear-cut logging in west-central Alberta. *Canadian Journal of Zoology* 78: 1433–1440.

Stuart-Smith, A.K., C.J.A. Bradshaw, S. Boutin, D.M. Hebert, and A.B. Rippin. 1997. Woodland caribou relative to landscape patterns in northeast Alberta. *Journal of Wildlife Management* 61: 622–633.

Soveri, T., S. Sankari, and M. Nieminen. 1992. Blood chemistry of reindeer calves (*Rangifer tarandus*) during the winter season. *Comparative Biochemistry and Physiology* 102A: 191–196.

Soveri, T., S. Sankari, J.S. Salonen, and M. Nieminen. 1999. Effects of immobilization with medetomidine and reversal with atipamezole on blood chemistry of semi-domesticated reindeer (*Rangifer tarandus tarandus* L.) in autumn and late winter. *Acta Veterinaria Scandinavica* 40: 335–349.

Timisjärvi, J., L. Hirvonen, P. Järvensivu, and M. Nieminen. 1979. Electrocardiogram of the reindeer, *Rangifer tarandus tarandus*. *Laboratory Animals* 13: 183–186.

Timisjärvi, J., M. Nieminen, and H. Hyvärinen. 1981. Hematological values for reindeer. *Journal of Wildlife Management* 45: 976–981.

Timisjärvi, J., M. Nieminen, J. Leppäluoto, T. Lapinlampi, P. Saukko, E. Eloranta, and P. Soppela. 1990. Effects of running and immobilization on blood constituents in reindeer. *Rangifer Special Issue No. 3*: 419.

Tyler, N.J.C., R. Hotvedt, A.S. Blix, and D.R. Sørensen. 1990. Immobilization of Norwegian reindeer (*Rangifer tarandus tarandus*) and Svalbard reindeer (*R. t. platyrhynchus*) with medettomidine and medetoomidine-ketamine and reversal of immobilization with atipamezole. *Acta Veterinaria Scandinavica* 31: 479–488.

Valkenburg, P., R.D. Boertje, and J.L. Davis. 1983. Effects of darting and netting on caribou in Alaska. *Journal of Wildlife Management* 47: 1233–1237.

Valkenburg, P., R.W. Tobey, and D. Kirk. 1999. Velocity of tranquilizer darts and capture mortality of caribou calves. *Wildlife Society Bulletin* 27: 894–896.

Webb, S.L., J.S. Lewis, D.G. Hewitt, M.W. Hellickson, and F.C. Bryant. 2008. Assessing the helicopter and net gun as a capture technique for white-tailed deer. *Journal of Wildlife Management* 72: 310–314.

West, G., D. Heard, N.A. Caulkett. eds. 2014. *Zoo Animal and Wildlife Anesthesia and Immobilization*. 2nd Ed. Ames, Iowa: Wiley-Blackwell.

White, R.G., and M.K. Yousef. 1978. Energy expenditure of reindeer walking on roads and on tundra. *Canadian Journal of Zoology* 56: 215–223.

Wiklund, W., A. Andersson, G. Malmfors, and K. Lundström. 1996. Muscle glycogen and blood metabolites in reindeer (*Rangifer tarandus tarandus* L.) after transport and lairage. *Meat Science* 2: 133–144.

Woodbury, M. 2014. Euthanasia, in *Zoo Animal and Wildlife Anesthesia and Immobilization*, eds. G. West, D. Heard, and N.A. Caulkett. 149–153. 2nd Ed. Ames, Iowa: Wiley-Blackwell.

16 Climate Change
Potential Impacts on Pasture Resources, Health and Diseases of Reindeer and Caribou

Morten Tryland, Virve Ravolainen and Åshild Ønvik Pedersen

16.1 INTRODUCTION

Across their northern distribution range, *Rangifer* subspecies are living under a rapidly changing climate, and some of the populations are decreasing (Vors and Boyce, 2009). Reindeer and caribou are herbivorous animals and thus rely on plants as forage for their existence. Foraging resources of *Rangifer* are expected to change due to climate-driven changes in the physical environment, particularly because of the currently warming surface temperatures (Walsh et al., 2011), the longer snow-free season (Derksen and Brown, 2012) and the more frequent extreme weather events that the Arctic experiences (IPCC, 2014).

The changes in Arctic vegetation due to climate change are forecasted to be of great magnitude, with a general productivity increase, an increased shrub abundance, an advancing tree-line and a changing functional composition as the most prominent processes (Chapin et al., 2005; Myers-Smith et al., 2011; Elmendorf et al., 2012; Macias-Fauria et al., 2012; Xu et al., 2013). With a 2°C warming scenario, forest and tall shrubs encroaching on the open tundra have been predicted to cause a loss of up to 60% of the dwarf-shrub tundra habitat (Kaplan and New, 2006). However, vegetation is not changing everywhere, as the on-going vegetation responses to climate change are strongly modified by soil moisture and nutrient levels, current climate regimes and the traits of the plants (Elmendorf et al., 2012; Christie et al., 2015; Myers-Smith et al., 2015).

Are the predicted climate-driven vegetation changes likely to impact the quantity, quality or availability of forage at a magnitude that will affect *Rangifer* production or survival? Diet is an inherent part of this question, so we first briefly summarize *Rangifer* diet before discussing what implications climate-driven vegetation change may have for *Rangifer* (see also Chapter 3, *Rangifer* Diet and Nutritional Needs).

Winter pastures may, to a greater extent, become unavailable for reindeer and caribou due to increased winter rain precipitation and icing, potentially increasing supplementary feeding of semi-domesticated reindeer, and increased exposure to a contaminated environment. Further, climate change may directly affect the presence and prevalence of insects, including blood-sucking insects such as ticks, mosquitos and midges, which are potential vectors of pathogenic parasites, bacteria and viruses. A climate-driven gradual shift in the life conditions for these vectors may thus change the exposure patterns and infection pressure on *Rangifer* populations throughout their distribution range, and we may face new health challenges and diseases (Tryland, 2012).

In the following sections, we will briefly look at how the predicted climate change may impact pasture and foraging, as well as insect vectors and their distribution, and how these factors may directly and indirectly impact health and diseases of reindeer and caribou.

16.2   REINDEER AND CARIBOU DIET

The different *Rangifer* subspecies have adapted to many different types of forage throughout their distribution range (Figure 16.1). This encompasses a variety of vegetation types, from coniferous and deciduous forests to tall or low shrub tundra, and to the High Arctic polar deserts (see Chapter 3, *Rangifer* Diet and Nutritional Needs, for a detailed description of habitat choice and diet). In a brief literature review (Table 16.1), we found several hundred plant taxa mentioned as part of *Rangifer* diet, representing functionally different plants, such as evergreen and deciduous plant species, forbs, grasses, sedges, rushes, shrubs, ferns, mosses, lichens and macrofungi. Hence, the content

FIGURE 16.1   There is great spatial and seasonal variation in *Rangifer* forage resources. In summers, *Rangifer* can find very nutritious food plants, such as grasses and forbs (Photo: Lawrence Hislop).

TABLE 16.1

Overview of Published Information on *Rangifer* Diet

Geographic Region	*Rangifer* Type	Winter	Summer	Spring	Reference	Method
Boreal forest	Woodland caribou	Terrestrial lichens (*Cladina* sp.), graminoids (dead)	Terrestrial lichens (*Cladina* sp.), forbs (*Maianthemum*, *Menyanthes*), graminoids		Thompson et al., 2015	Observe natural diet selection with cameras
Russian Arctic, Wrangel	Reindeer	Lichens, grasses, willows	Graminoids, willows (also in autumn)	Willows, graminoids	Rozenfeld et al., 2012	Microhistology of pellets
Arctic Canada	Caribou		Willow (85%), sedge, legumes, Dryas		Larter and Nagy, 2004	Microhistology of pellets
Arctic Alaska	Caribou	Lichens, graminoids (changing from lichen to graminoid over time)			Joly et al., 2007	Microhistology of pellets
Arctic Alaska	Reindeer (captive)		Forbs, lichens, shrubs, graminoids		Trudell and White, 1981	Rumen samples
Arctic Alaska	Reindeer	Lichens, horsetails, forbs, mosses			Ihl and Klein, 2001	Microhistology of pellets
	Woodland caribou (captive, late autumn)		Arboreal lichen, forbs (*Pachistima myrsinites*)		Rominger and Oldemeyer, 1990; Rominger et al., 2000	Observing captive caribou
Arctic Alaska and alpine Norway			Willows, sedges, forbs, grasses		Skogland, 1980	Rumen sample microhistology
Arctic Alaska and Yukon	Caribou	Lichens, mosses, shrubs			Gustine et al., 2012	Microhistology and isotope analyses of pellets

(Continued)

TABLE 16.1 (CONTINUED)

Overview of Published Information on *Rangifer* Diet

Geographic Region	*Rangifer* Type	Winter	Summer	Spring	Reference	Method
High Arctic Svalbard	Svalbard reindeer	Graminoids, willow (*Salix polaris*), mosses	Graminoids, forbs (*Bistorta vivipara, Oxyria digyna*), willow (*Salix polaris*)		Bjorkvoll et al., 2009	Microscopy of rumen content
High Arctic Svalbard	Svalbard reindeer	Willow (*Salix polaris*), mosses	Forbs (*Oxyria digyna, Pedicularis* sp.)		Bjune, 2000	Pollen analysis of pellets
High Arctic Svalbard	Svalbard reindeer		Goose droppings		van der Wal and Loonen, 1998	Observation of pellet ingestion
High Arctic Svalbard	Svalbard reindeer		Lichens (*Cetraria delisei, Stereocaulon* sp.), willow (*Salix polaris*), forbs (*Saxifraga oppositifolia, Bistorta vivipara*)		Joo et al., 2014	Molecular analysis of pellet content
Low Arctic Fennoscandia	Semi-domesticated reindeer			Terrestrial lichens (*Cladonia* sp.), graminoids, dwarf shrubs (ericoids), bryophyte	Ophof et al., 2013	Microhistology of pellets

Note: A search was performed in ISI Web of Science with the words "*Rangifer*" and "Diet" (3.12.2015), which resulted in 154 matches. Articles on method development and feeding experiments using a few plants only, as well as some isotope-based studies that did not present other diet information, were excluded.

of nutrients, as well as that of phenols, tannins and other complex low-digestibility compounds, in the reindeer diet varies greatly, reflecting the plants available in specific regions and seasons (for an overview of plant nutrition traits, see Aerts and Chapin, 2000). *Rangifer* populations forage on everything from arboreal lichens and the lush forest field layer to the very sparse High Arctic gravel communities. Thus, reindeer and caribou can be described as "selective generalists," in the sense that they utilize many types of plants as food, but graze selectively on plant parts and species, changing their diet over the season as plants and plant parts of different quality become available.

16.3   CLIMATE-DRIVEN TRENDS: IMPACT ON VEGETATION PRODUCTIVITY AND SEASONALITY

Summer temperatures are increasing in the Arctic, which generally can cause a positive growth response in vegetation at northern latitudes (Myneni et al., 1997; Xu et al., 2013; Guay et al., 2014). However, such "Arctic greening" is not happening everywhere and there is strong regional variation in vegetation responses to climate. Large-scale remote-sensing analysis has shown that in some parts of the western and central Eurasian Arctic and in the western North American Arctic vegetation productivity was, rather, decreasing during the period of 1982–2008 (Beck and Goetz, 2011), and in most of the northern parts of the Arctic there was no greening trend from the early 1980s to 2008 (Guay et al., 2014). Hence, the *Rangifer* herds that undertake long migrations, such as the Bathurst and Porcupine caribou herds in Canada, may experience both a general greening and a "browning" trend (reduced productivity) within their range. Other *Rangifer* populations that move over shorter distances, such as the Southampton caribou population north of Hudson Bay, Canada, and the Svalbard reindeer population, may have their entire ranges in areas that have no documented increase in vegetation productivity or where greening decreased over the last decade (see CARMA [carma.caff.is] for range maps; Guay et al., 2014 for trends in vegetation; Vickers et al., 2016). Thus, the potential consequences that these large-scale "greening" or "browning" trends have for the *Rangifer* populations remain to be investigated.

16.3.1   *RANGIFER* FORAGING PATCHES

Rangifer typically forage in patches of vegetation. Plant species and plant groups are different in their growth rates and life history (e.g. generation time, age, etc.) as well as in reproductive and nutritional strategies (for an overview, see Aerts and Chapin, 2000; Bråthen and Ravolainen, 2015). Assessments of changes in forage resources in *Rangifer* pastures are important for our understanding of possible food-driven climate impacts on *Rangifer*. Warmer temperatures at northern latitudes should generally produce more plant biomass. The above-ground biomass of Arctic plants (van der Wal and Stien, 2014) and secondary growth (thickening of woody plant stems) (Macias-Fauria et al., 2012; Myers-Smith et al., 2015) increases when summer temperatures increase. Following this logic, experimentally, warming of plants should lead to an increase in total biomass of plants. This is, however, not always the case since species such as deciduous shrubs, graminoids (grasses and sedges) and forbs typically gain abundance at the expense of mosses and lichens in experiments where the temperature is artificially increased (Elmendorf et al., 2012). Different growth rates and competitive abilities are part of the explanation. Another clear modification to a simplified "warm, long summers = more plant biomass" equation is that the abundance of forbs, grasses and shrubs is found to increase in the southern parts of the Arctic, while it is not changing in the high north (Elmendorf et al., 2012). Some plant communities in the northernmost parts of the Arctic have resisted warming temperatures over decades. For instance, this applies to evergreen shrub and mountain aven communities that have the same community composition despite experimental warming or warmer ambient temperatures (Hudson and Henry, 2010; Prach and Rachlewicz, 2012). We can expect that plant communities dominated by deciduous and non-woody plants such as forbs are those that can be expected to increase fastest in abundance. At present, analysis of changes in vegetation coverage in response to climate warming has mainly been restricted to tall shrubs. They have increased in spatial distribution from 1950 to 2002 in Arctic Alaska, and particularly so along streams and on floodplains (Tape et al., 2012). The fact that only a few studies have focused on changes of plant community or vegetation type probably has its reasons in methodological constraints. Field registrations are typically made in small plots of 1 m² or even smaller, and high-resolution remote-sensing data is still limited, leading to very large unit sizes in remote-sensing studies. Thus, to fully understand plant (both woody plants and forbs and graminoids) changes relevant to *Rangifer* diet, we need novel combinations of field and remote-sensed data.

16.3.2   AVAILABILITY OF HIGH QUALITY FORAGE

The concentration of plant nutrients changes during the growing season, and reindeer follow the green-up of plants (Iversen et al., 2014). The highest nitrogen values are found in plant tissue in early spring (van der Wal et al., 2000). If climate change alters temperatures and the length of the growing season, this can be expected to reduce the nutrient content of *Rangifer* forage plants earlier in the season. Indeed, studies that experimentally expose plants to increased temperatures have found support for such quality decreases. For instance, increased temperatures over the growing season in experimental plots lead to lower nitrogen concentration in grasses and sedges compared to control plots growing under ambient conditions (Doiron et al., 2014). A longer growth season due to earlier snow-melt results in lower nutrient concentration in some species of reindeer forage plants (van der Wal et al., 2000).

In seasonal environments, which are characteristic of *Rangifer* habitats, calf production is timed to coincide with the annual peak of resources. The advancement of plant phenology in response to increased temperatures may cause a mismatch between resource demands by reproducing herbivores and resource availability. For example, calf mortality rates and offspring production dropped for caribou in Greenland as spring temperatures increased (Post and Forchhammer, 2008). However, in Fennoscandian semi-domesticated reindeer, both reproductive success and the body mass of the calf at birth were higher when spring came earlier (Tveraa et al., 2013). These studies used plant developmental state and large-scale productivity indices as proxies for plant forage quantity or quality, respectively. Thus, methodological differences are probable causes of the seemingly different impacts of the timing of forage availability and *Rangifer* reproduction, and the general relevance of these results to other *Rangifer* herds remains to be investigated. New methods (e.g. Near Infra Red Spectrometry) that can allow us to link the landscape scale nutrient levels to the *Rangifer* use of forage (Jean et al., 2015; Aikens et al., 2017) are promising tools to move us towards a better understanding of what kind of a net effect changed seasonality and forage availability have for the animals.

16.4   FORAGING IN WINTER LANDSCAPES

In the winter, *Rangifer* are dependent on reaching their forage by digging through the snow layers. Snow depth, snow hardness and ground ice are major determinants of forage access at this time of the year. Rain-on-snow (ROS) events, which cause ice layers in the snow and thick ground ice that blocks the pastures and hinders the reindeer from localizing the lichen through smell, are a well-known challenge to *Rangifer* in winter (Figure 16.2), when the plant quality and quantity are at their lowest (Putkonen and Roe, 2003). The milder, wetter winter climates, with increased frequency of ROS events (Hansen et al., 2014), affect survival and reproduction of reindeer (Putkonen and Roe, 2003).

The ROS phenomenon encases and destroys the vegetation (Bokhorst et al., 2012) and reduces herbivore food availability. For instance, in the High Arctic Svalbard reindeer populations, ROS events impact population growth rates (Hansen et al., 2011) and reproduction (Stien et al., 2010). Such extreme ROS events were reported from northwest Russia; they caused starvation of

FIGURE 16.2   In the winter time, food resources can be temporarily locked down due to excessive ice on the ground (Photo: Jennifer Forbey).

semi-domesticated reindeer on the Yamal Peninsula in November 2006 and 2013 (Forbes et al., 2016). These events caused severe ecological as well as long-lasting socio-economic impacts for the local reindeer herders because 61,000 reindeer out of a population of approximately 275,000 animals died from starvation. This is the largest recorded mass die-off ever recorded in this region. This study suggests a link between sea ice loss in the region (Barents Sea and Kara Sea) and more frequent and intense ROS events. Events like these will thus also have profound implications for the future of nomadism of the indigenous people depending on reindeer husbandry. Additionally, several behavioural adaptations are linked to ice-covered landscapes that lock the winter pastures. For instance, reduced forage availability for the Svalbard reindeer is coped with by switching the foraging niche through feeding on kelp and seaweed along the seashore and by moving to areas with better foraging conditions (Stien et al., 2010; Hansen and Aanes, 2012; Loe et al., 2016). The latter movement strategy also apparently resulted in lower over-winter loss of body mass, lower mortality rate and higher subsequent reproduction (Douhard et al., 2016; Loe et al., 2016).

Although population responses to extreme winter weather events have been studied (Tews et al., 2007), the question of whether greater plant biomass in summer or higher mortality rates due to starvation in winter are more important for reindeer and caribou populations remains unsolved. We do know, however, that *Rangifer* can move to better foraging areas in response to icing (Loe et al., 2016), and the mere fact that *Rangifer* have successfully maintained populations in coastal tundra regions with high variability in climate and weather over time suggest they are, as a species, relatively robust in response to variation in climate (see also Chapter 3, *Rangifer* Diet and Nutritional Needs).

16.4.1   HERBIVORE SEASONAL EFFECTS ON FORAGE AND COMPETITION BETWEEN SPECIES

Rangifer impact vegetation not only through foraging itself, but also by trampling and by fertilizing the ground with nutritious manure. A recent review systematically examined documented effects from *Rangifer* on the vegetation, and found many different effects and few general trends (Bernes et al., 2015). The authors suggest this may be due to the differing methods and focus of the different studies. Studies that address how populations of reindeer and caribou may impact the vegetation include cases of: decrease of preferred summer forage at high reindeer densities (Bråthen et al., 2007), decrease in shrub abundance (Ravolainen et al., 2014), lower winter forage (lichen) after years of high reindeer densities (Falldorf et al., 2015), decrease in mosses and lichens at the expense of graminoids (Hansen et al., 2007) and changing the vegetation from moss to graminoid dominated, leading to higher rates of nutrient cycling and warmer soil temperatures (Brooker and van der Wal, 2004). Further, forage abundance can also decrease due to the animals themselves, but so far there are few cases where this has been severe enough to represent a major part of the reason for a *Rangifer* population decline. An example was found on St. Matthew Island in the 1960s, where the *Rangifer* population increased to a level where they depleted their own food resource (Klein, 1968).

Vegetation in the Arctic is not only grazed by *Rangifer*; in some regions, the effects of goose grazing on vegetation can be severe. Geese changed the state of the whole ecosystem from productive tundra to non-productive, dead tundra in Hudson Bay (Jefferies et al., 2006). Similar situations may also happen elsewhere in the future where the goose populations are increasing (Pedersen et al., 2013). Whether geese deplete high quality forage resources to the degree where they compete with *Rangifer* is a question yet to be investigated. Also, small rodents can remove a substantial amount of biomass in reindeer grazing areas (Ravolainen et al., 2011). However, in the Arctic, many of the small rodents have large inter-annual population fluctuations, and their effect on plants with regard to reindeer forage availability would be dependent on the rodent population cycle. It is unlikely, although not quantified at scales relevant for reindeer foraging, that small rodents would cause a vegetation impact large enough to represent competition for food.

16.5 CHANGE IN PASTURE PREFERENCES AND MIGRATION ROUTES

Changes in climate and subsequent ecosystem alterations and adaptations may affect the reindeer pastures to such an extent that the *Rangifer* herds may find or prefer new regions for summer and winter pastures, and thus also change their distribution and migration routes (Sharma et al., 2009). In addition, other animal species may also change their distribution within the reindeer ecosystems and habitats. New animal species may thus be introduced into new regions and be exposed to reindeer and their environment for the first time. These mechanisms have the potential to introduce new pathogens to the reindeer and caribou herds. Since some pathogens are shared among cervid species, it is most likely that pathogens which in general have a wide host range, such as Orf virus and *Pasteurella* spp., may directly or indirectly be transmitted from one cervid species to another, via direct contact or through shared environment.

The expansion and increased distribution of moose (*Alces alces*) and red deer (*Cervus elaphus*) in Fennoscandia over the past few decades (Cederlund and Bergström, 1996) are examples of increased habitat overlap and new contact between semi-domesticated reindeer and other cervid species. However, not much is known about whether this has also introduced new infectious agents that might be pathogenic to *Rangifer*. In fact, it seems that several of the ruminant viral infections that are well known in livestock, such as alphaherpesvirus, gammaherpesvirus and pestivirus, once thought to be potentially transmissible between domestic animals and reindeer and other wild cervids, have shown to be rather species-specific, with genetic characteristics distinguishing them from each other (Becher et al., 2003; Thiry et al., 2007; das Neves et al., 2010; das Neves et al., 2013; Kautto et al., 2012). However, the recent (2016) detection of prions and chronic wasting disease (CWD) in reindeer and moose in Norway (see Chapter 9, Prions and Chronic Wasting Disease) is an example of an infectious agent that may affect multiple cervid species.

An example of possible impacts of a changed migration route for caribou is the changed migration of the Western Arctic Herd (WAH) of caribou (*R. t. granti*). This particular herd of caribou, counting approximately 200,000 individuals (2016; Alaska Dept. of Fish and Game), have their calving ground on the tundra plains north of Brooks Range, Alaska. In fall, they migrate south to winter pastures in boreal forest regions. On their way south, they used to pass Seward Peninsula, but since the mid-1990s, the caribou started to migrate westward and populated western parts of the peninsula. Here, they met the herds of semi-domesticated reindeer (*R. t. tarandus*) owned by Inuit communities of the region. Many of the reindeer herds mingled with and left with the caribou, and some herders lost most of their animals. Further, since the caribou also represented an important prey for brown bears and wolves, these predators followed the caribou westwards out on Seward Peninsula, and established permanent populations there, which has since been a challenge for reindeer herding in this region. Since 2003, 11 of 14 reindeer operations were no longer commercially viable, after about 100 years of history of reindeer herding in that area. Thus, an altered migration may lead to large local and regional changes in the life conditions of both *Rangifer* and the people dependent on them.

16.6 INCREASED FEEDING OF SEMI-DOMESTICATED REINDEER

Although the access to rich pastures for semi-domesticated reindeer may increase during summer due to climate change, the production limitations are usually associated with scarce or unavailable (i.e. ice-locked) winter pastures (Moen, 2008). Many reindeer herders are conducting supplementary feeding, either as an emergency measure in shorter periods to avoid critical situations, or as a more regular and long-term strategy. Regular winter feeding is especially common in Finland, where reindeer in many places are kept permanently in fences and corrals throughout the year, but it is also becoming more common in Sweden and Norway (Helle and Jaakkola, 2008; Staaland and Sletten, 1991; Turunen and Vuojala-Magga, 2014). Either reindeer feed is transported to the reindeer at their mountain pastures or the animals are kept in corrals for certain periods.

As discussed in Chapter 4 (Feeding and Associated Health Problems), successful feeding of reindeer is dependent on the right amounts of high quality food resources, as well as a period of adaptation to new types of feed so that the digestive tract can adjust. Feeding of reindeer also implies that the animals are held in a restricted area (corralled), or that they gather at designated sites on their pastures. At these sites, reindeer will have increased direct nose-to-nose contact, but will also be exposed to urine and faeces from each other to a much larger degree than in a vast pasture. In addition, feeding may also represent increased stress for the animals, both in terms of contact with people, but also through a tighter social setting, in which hierarchy and competition for feed has to be addressed continuously.

Thus, the animals will share infectious agents to a larger extent within the herd and also between herds if more herds are visiting the same feeding locations. This may cause stress that increases the production of the hormones glucocorticosteroids, which are known to have immunosuppressive effects (Coutinho and Chapman, 2011). In addition, keeping reindeer in the same corrals over time may create a dirty and unhealthy environment for the animals. Such conditions increase exposure to "environmental" and potentially pathogenic infectious agents that may cause disease outbreaks under certain circumstances. In the following, we will present a few examples of diseases that can be transmitted through supplementary feeding of animals.

The alphaherpesvirus Cervid herpesvirus 1 (CvHV2) has been identified as the causative agent of infectious keratoconjunctivitis (IKC), as indicated during an outbreak in Norway (Tryland et al., 2009) and experimentally (Tryland et al., 2017) (see Chapter 8, Viral Infections and Diseases). CvHV2 is enzootic in the reindeer populations in Fennoscandia. Serological surveys in Finnmark County, Norway, have indicated that around 50% of adult reindeer, up to 100% in some herds, and approximately 8% of the calves had been exposed to CvHV2 (das Neves et al., 2010). As do other herpesviruses, CvHV2 establishes lifelong infections and latency, from which it can be reactivated upon stimuli such as stress. Larger outbreaks of IKC, sometimes affecting dozens or hundreds of animals, have been reported from feeding-corrals. Outbreaks typically follow after a period of poor body condition, as well as stress due to gathering, handling and transport and close contact between animals.

Another pathogen that seems to be associated with feeding and corralling is the parapoxvirus Orf (and also pseudocowpoxvirus in Finland), causing the disease contagious ecthyma (see Chapter 8, Viral Infections and Diseases). The animals develop cauliflower-like processes in the oral mucosa and/or on the muco-cutaneous junction or in the skin, typically around the mouth, sometimes to such an extent that they stop eating and die from emaciation. This disease has typically appeared in corrals in which reindeer have been kept over weeks or months for supplementary feeding during winter, creating stress and close animal-to-animal contact. Sometimes outbreaks of contagious ecthyma have been associated with corrals previously used for sheep (Tryland et al., 2001).

The two viral infections mentioned above (CvHV2 and parapoxvirus) cause lesions in the skin and oral mucosa, which may pave the way for bacterial infections. Digital necrobacillosis ("slubbo" in Sami) is caused by infection with the bacterium *Fusobacterium necrophorum*, characterized by severe and necrotizing lesions in and around the inter-digital cleavage and distal parts of the legs (see Chapter 7, Bacterial Infections and Diseases). Digital necrobacillosis used to be particularly common in animals that were kept in corrals over time, such as in milking and draft animals. While milking of reindeer is still being practised in Southeastern Siberia (Russia), this practice more or less ceased in the Sami reindeer herding region during the 1960s (Holand et al., 2002), and snow scooters and quads have replaced the working animals. As a result, digital necrobacillosis is hardly seen anymore, but oral necrobacillosis, affecting the oral mucosa and the gastrointestinal tract, has become more common (Figure 16.3). Oral necrobacillosis often occurs in herds corralled for winter feeding, and may cause severe damage to the gum, gingiva and tongue, and also to other parts of the gastrointestinal tract (see Chapter 7, Bacterial Infections and Diseases).

Pasteurellosis in reindeer is a respiratory disease caused by the bacterium *Pasteurella multocida*. Acute pasteurella-associated pneumonia, especially in calves and young animals, has been reported associated with stress, transport and feeding situations. In some cases, calves have died in the slaughter fence after a per-acute disease, showing no clinical signs before dropping dead (see Chapter 7, Bacterial Infections and Diseases).

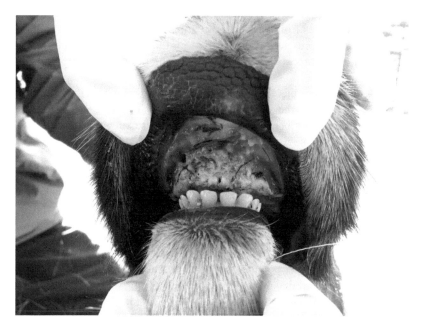

FIGURE 16.3   Feeding of semi-domesticated reindeer may be a valuable measure for the reindeer herder when natural pastures are scarce, but it may also contribute to severe disease outbreaks. In this disease outbreak, eye infections, contagious ecthyma and oral necrobacillosis, as illustrated here, occurred at the same time in a herd corralled for supplementary feeding in winter (Photo: Ingebjørg H. Nymo).

Sometimes two or more of the diseases mentioned above appear at the same time in a herd, ignited by the presence of the pathogens, but fuelled by the stress situation and the challenge these pathogens all pose for the immune system of the host (Figure 16.3). In some outbreaks, it seems that viral infections, such as CvHV2 and parapoxvirus, have been the initial infection(s), creating lesions in the mucosal membranes that are subsequently infected by *Fusobacterium necrophorum* and other opportunistic pathogens.

In addition to these specific diseases, there is reason to believe that the importance of parasites, such as liver flukes (*Dicrocoelium dendriticum*), lungworm and gastrointestinal parasites, may increase should regular feeding of reindeer throughout the winter become a common practice rather than just an emergency measure. Feeding reindeer in corrals during the winter months may thus necessitate a stronger focus on the early detection of disease symptoms, increased use of veterinary services and also increased efforts on prophylactic measurements (see Chapter 12, Assessment and Treatment of Reindeer Diseases). Increased feeding of semi-domesticated reindeer will thus probably be associated with more routine use of vaccines, anti-parasitic drugs and medical treatment, including antibiotics. This will represent one of several challenges for the high quality status reindeer meat has today among the general public as a natural, healthy, ecologically produced and "clean" product with a characteristic wildlife taste, as compared to meat from livestock (see Chapter 11, Meat Quality and Meat Hygiene).

16.7 CLIMATE CHANGE AND ARTHROPOD DISEASE VECTORS

In the phylum Arthropoda, we find invertebrate animals of the class Insecta, the insects. Many insects are important pollinators of plants and they represent important sources of nutrition for many mammals and birds. However, some insects also feed on animals, either as a more or less permanent parasite, such as lice, or as temporary blood-sucking parasites, such as ticks, mosquitos and midges. Through this blood-sucking activity, many insects inject saliva with anticoagulants into the new host, to hinder blood clotting in the small, narrow tube they are sucking blood with. Through this activity, they may also transmit infectious agents, such as parasites, bacteria and viruses, from one host to another, and act as important factors in the epidemiology of many infectious diseases (Capinera, 2010).

In addition to generally increased temperatures in northern regions, climate change may impact wind and precipitation patterns and humidity, and may also represent increased weather variability (IPCC, 2014). Since the development and activity of arthropods are particularly sensitive to weather and climate conditions, arthropod borne infectious diseases may be the category of infectious diseases that are most impacted by climate change (Wittmann and Baylis, 2000; Ogden and Lindsay, 2016).

Insects are poikilothermic creatures, unlike birds and mammals, and thus have the same temperature as their surroundings. In fact, the generally low temperatures in Arctic regions are close to the minimum requirements for insect locomotion, thus largely determining their activity pattern (Strathdee and Bale, 1998). But it is not only temperature that matters. Whereas the relationship between climate and periodic biological phenomena such as bird migration and plant flowering (i.e. phenology) at lower latitudes is primarily related to temperature, snow cover seems to be an even more important parameter in snow-dominated environments, impacting the seasonal development of insects to a large extent (Strathdee and Bale, 1998; Høye et al., 2007).

In the Arctic, studies of the impact of climate change on insect populations are limited (Kattsov et al., 2005; Boggs, 2016), but in a study on the activity of High Arctic flying arthropods and climate, it was shown that air temperature, advancement of phenology (i.e. earlier snow-melt and earlier onset of spring) and exposure to solar radiation were factors that influenced these insects and their ecological processes the most (Høye and Forchhammer, 2008).

Blood-sucking insects also have a short generation time. They may thus respond and adapt to climate change during short time periods, which also impacts their role as vectors for infectious diseases for *Rangifer* (Boggs, 2016).

Insecta constitutes a very diverse animal group, and even those that are temporary parasites feeding on *Rangifer*, such as ticks, mosquitos and midges, have fundamental differences in their life cycles, life history and behaviour, and may thus respond to altered weather and climate conditions in very different ways. In the following section, we will present more details on how climate change may impact the role of ticks, mosquitos and midges as vectors for infectious agents in Arctic and sub-Arctic regions.

16.7.1 Deer Ked (*Lipoptena cervi*)

The deer ked, or louse fly *Lipoptena cervi* (order Diptera, family Hippoboscidae), flies in the autumn, when most other insects of the summer have disappeared. Since the deer ked also can temporarily settle on humans, many people, such as hunters and berry pickers, have experienced being invaded by deer ked during the autumn. When the deer ked has found a host, it drops its wings and settles in the hair coat of the animal. Females lay their eggs on the hair of the host. The eggs encyst and drop to the ground during late autumn and winter, and adults emerge after many months, during the next summer and autumn (Kaunisto et al., 2011). Both males and females feed on blood from their host.

Although the deer ked has been present in southern parts of Fennoscandia for decades, its distribution northwards has increased (Välimäki et al., 2010; Kaunisto et al., 2011). Deer ked were reported in reindeer for the first time in 2004 in Finland, in wild forest reindeer (*Rangifer tarandus fennicus*). The parasite was associated with fur and skin changes and larger spots of alopecia, suggesting that the presence of the deer ked in the ecosystems of reindeer may cause energy loss and pose a challenge during hard winter conditions (see Chapter 6, Parasitic Infections and Diseases). In 2008, about 20% of semi-domesticated reindeer in Finland that were not treated with ivermectin as prophylaxis against ectoparasites had changes caused by deer ked. Deer ked infestation can also cause restlessness and changed behaviour among reindeer (Kynkäänniemi et al., 2014). The distribution of this parasite is still, to a large extent, south of the distribution areas for reindeer, but with climate change and milder winters, it is likely that the deer ked will continue its northbound migration, also reaching the major ecosystems inhabited by reindeer.

16.7.2 Ticks

Ticks are arthropods of the class Arachnida. Ticks can survive under different environmental conditions but are sensitive to climate. They require a relative humidity of 80% or more and generally thrive in areas with moderate to high rainfall and vegetation, especially deciduous woodland and mixed forest regions. There are several species of ticks, but in Fennoscandia, *Ixodes ricinus* is the most common species and a potential vector for infectious agents among semi-domesticated Eurasian tundra reindeer (*R. t. tarandus*). As ticks are becoming more common findings in the reindeer herding areas (Figure 16.4), tick-borne infections may also gain importance for the health and disease status of reindeer in the future.

Ixodes ricinus feeds on mammals such as rodents, hares and roe deer and other cervids. The ticks have brief feeding periods on three different hosts during their stages as larva, nymph and adult. The development from larva to adult normally takes two years. In Europe, *Ixodes ricinus* may be a vector for a range of potential pathogens, such as *Borrelia burgdorferi s.l.* (Lyme borreliosis), *Anaplasma phaocytophilum* (tick-borne fever), *Francisella tularensis* (tularemia), *Rickettsia* spp., *Babesia* spp., Louping ill virus, tick-borne encephalitis virus (TBE) and others, of which many are zoonotic (Medlock et al., 2013).

In Norway, the geographic distribution of *Ixodes ricinus* and semi-domesticated reindeer overlap only to a small extent, with the northernmost known established populations of the ticks being at the coast in Brønnøysund in Nordland County, south of the Arctic Circle (Hvidsten et al., 2015). These northern tick populations, however, which may affect local herds of semi-domesticated reindeer, may represent discontinuous foci only. It was recently concluded that ticks, investigated by

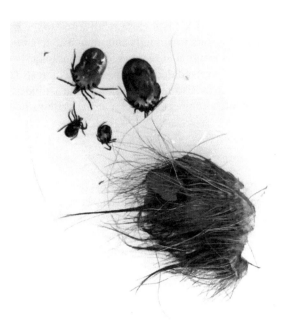

FIGURE 16.4   Ticks (*Ixodes ricinus*) are increasingly being reported on semi-domesticated reindeer in Norway, Sweden and Finland, and may be potential vectors for reindeer pathogens. These ticks, at different stages of the life cycle, were found on a reindeer in Nordland County, Norway, in 2016 (Photo: Morten Tryland).

examination of ears of hunted (August–December, 2001–2003) moose, red deer and roe deer in Norway, were enzootic only up to 63°30' N (i.e. the municipality of Hitra) and occurred more sporadically further north. This is in line with a previous study from the 1940s (Tambs-Lyche, 1943), thus not indicating a great northward expansion of the distribution of *I. ricinus* in Norway during the 60-year period between the two studies (Handeland et al., 2013).

Nevertheless, the distribution of *Ixodes ricinus* is changing in Europe. Key drivers for its geographical expansion to higher altitudes and latitudes include increased winter temperatures (increased winter survival), increased temperatures throughout the year (extended development period), snow cover (snow preventing ground freezing, increasing winter survival), expansion of deciduous woodland (increased survival and development), as well as favourable wildlife and forest management practices (Medlock et al., 2013). Thus, ground surface temperature, spring precipitation, duration of snow cover, bush encroachments and also abundance of red deer and farm animals were relevant factors when addressing altitudinal and latitudinal shifts in the distribution of *Ixodes ricinus* from 1978 to 2008 in three ecologically different districts of Western Norway (Jore et al., 2014).

In Sweden, two separate surveys (the first distributed in 1994 and the second in 2009) targeting dog owners, home owners and hunters, indicated that during this time period *Ixodes ricinus* became more abundant in south and central parts of Sweden and also expanded its range in northern Sweden. Ticks have previously been reported as occasional findings on reindeer in the southern reindeer herding ranges in the counties of Jämtland and Västerbotten, but have also recently been found on reindeer in the northern part in the county of Norrbotten. The main reasons for this expansion are suggested to be increased numbers and geographical distributions of roe deer (*Capreolus capreolus*) and red deer (*Cervus elaphus*), as well as a warmer climate, which has favoured both the cervid populations and the survival and proliferation of the ticks (Jaenson et al., 2012). There are also preliminary reports of ticks on reindeer in Finland.

Ixodes persulcatus, also called "taiga tick," is a close relative to *Ixodes ricinus* and is another important vector for many infectious agents in animals and man. The taiga tick has recently been detected in Sweden, has a scattered distribution in Finland and is continuously distributed in Latvia and Estonia and eastwards in Russia, in mixed deciduous-coniferous forests (Jaenson et al., 2016). If the distribution of the taiga tick expands to higher altitudes and latitudes, as has been shown for *Ixodes ricinus*, it may also become an important vector in ecosystems hosting reindeer in this region.

16.7.3 Mosquitos and Midges

Mosquitos and biting midges are economically important vectors of many arthropod borne pathogens and parasites to humans, livestock and wildlife. Midges are insects in the family Ceratopogonidae and genus *Cullicoides*, whereas mosquitos are members of the family Culicidae, with the subfamilies Anophelinae and Culicinae. As we have seen for ticks, biting midges and mosquitos are also sensitive to climatic conditions, and climate change may impact the distribution and range of these insects (Elbers et al., 2015).

In contrast to the situation for ticks, blood-sucking mosquitos and midges are usually numerous in the ecosystems inhabited by *Rangifer* (Figure 16.5). Climate change may, by increasing the length of the summer and the snow-free period in Arctic and sub-Arctic ecosystems, contribute to increased insect harassment for *Rangifer* (Moen, 2008). In a study of how mosquitos (Culicidae), black flies (Simulidae) and oestrid flies (Oestridae) – the latter group including bot flies and warble flies – impacted feeding behaviour of caribou, the time the animals spent feeding was reduced to the largest extent when all three insect types were present in combination (Witter et al., 2012). In fact, increased intensity and duration of insect harassment due to climate change has been hypothesized as one of several potential factors that may have caused declines of *Rangifer* populations in the circumpolar north during the last 20 years (Witter et al., 2012; Bastille-Rousseau et al., 2013).

FIGURE 16.5 Mosquitos and midges are, in contrast to ticks, often present in great abundance in *Rangifer* habitats throughout their circumpolar range, and represent potential disease vectors. When reindeer are corralled, they have little opportunities to get away from the insects (Photo: Sauli Laaksonen).

The presence and distribution of these potential vectors is largely unknown, and even fewer reports exist about which parasites, bacteria and viruses they may host, and if these are potential pathogens for reindeer and caribou. If the vectors experience altered geographical distribution, the distribution and the epidemiology of the pathogens will also change. Further, if one insect vector, such as a mosquito species, increases its distribution, it may at some point overlap with the distribution of a different mosquito species, with which it used to have no contact. If the expanding mosquito species is carrying a potential reindeer pathogen, it may thus, by infecting local fauna, introduce this pathogen to a new vector species with a different distribution, causing a more sudden and larger shift of the distribution of the potential pathogen.

A few pathogens are mentioned in the next section that are transmitted by midges and/or mosquitos, and that may achieve altered importance or may potentially be introduced to *Rangifer* populations in the future, due to different climatic change scenarios.

16.8 SOME EXAMPLES OF ARTHROPOD-BORNE INFECTIONS RELEVANT TO *RANGIFER*

16.8.1 SETARIA TUNDRA

The adult stage of the roundworm *Setaria tundra* appears as an approximately 10 cm long, white, lively worm in the abdomen of reindeer, although it may also be found in the pleural cavity and the pericardium. If present, it is visible during post mortem inspection, associated with fibrineous peritonitis and pleuritis, adherances and accumulation of fluids in the body cavities. In such cases, the carcasses may be discarded on the post mortem control, as they are not suitable for consumption. More common is the condemnation of the liver from animals with many worms. The larval stage, called microfilaria, is spread by mosquitos (genera *Aedes* and *Anopheles*), and an infected reindeer may have as many as 4000 microfilaria per millilitre of blood. A mean summer temperature of 14°C seems to drive the emergence of disease due to *Setaria tundra*, thus contributing to economic loss and negative impacts on free ranging and semi-domesticated *Rangifer* populations (Laaksonen et al., 2010).

See also Chapter 6 (Parasitic Infections and Diseases) and Chapter 10 (The Impact of Infectious Agents on *Rangifer* Populations) for further details.

16.8.2 *ANAPLASMA PHAGOCYTOPHILUM* AND TICK-BORNE FEVER

Tick-borne fever, caused by the bacterium *Anaplasma phagocytophilum*, is a world-wide-distributed disease affecting livestock and a wide range of wild mammals, as well as humans (Stuen et al., 2013). The bacterium is transmitted by ticks, and thus, the epidemiology of tick-borne fever is associated with the habitats and life conditions of the vector. With climate change, altered vegetation and an increased distribution of ticks, also affecting ecosystems inhabited by *Rangifer*, ticks may gain increased importance as vectors for pathogens, including *A. phagocytophilum*.

Eurasian tundra reindeer have experimentally been found susceptible to infections with *Anaplasma phagocytophilum* (Stuen, 2007; see Chapter 7, Bacterial Infections and Diseases). Antibodies against *A. phagocytophilum* were recently detected in two reindeer herds in Nordland County (Norway), but at a very low prevalence (M. Tryland, unpublished data). See also Chapter 7 (Bacterial Infections and Diseases) for further details on *A. phagocytophilum* and *Rangifer*.

Rangifer may theoretically also be susceptible to *Borrelia burgdorferi* sensu lato and possibly other tick-borne pathogens, but exposure of reindeer to such infections have scarcely been addressed. Among *Ixodes ricinus* ticks gathered from local fauna in Brønnøysund (Norway), one of the world's northernmost locations having a stable tick population, the prevalence of *Borrelia burgdorferi* sensu lato was recently shown to be 21% in nymphs and 46% in adults (Hvidsten et al., 2015).

16.8.3 Bluetongue Virus (BTV)

Bluetongue virus (BTV) belongs to genus *Orbivirus*, family Reoviridae, and is a small virus (70–80 nm) with a double stranded RNA genome. The virus is transmitted by biting midges (*Culicoides* spp.), but can also have oral and transplacental transmission (Backx et al., 2009). BTV causes non-contagious acute disease in naïve sheep, with fever, excessive salivation, oedema of the face and cyanosis of the tongue and lips (thus the name "bluetongue"), and may have a moderate to high morbidity and mortality. Cattle, which usually show less clinical symptoms upon infection than do sheep, is an important species for maintenance of the virus in endemic regions. BTV also infects a wide range of other domestic animals, and most species of wild ruminants and camelids are susceptible, although they are frequently asymptomatically infected (Fernández-Pacheco et al., 2008; García-Bocanegra et al., 2011; Corbière et al., 2012; Niedbalski, 2015). Thus, the epidemiology of BTV is associated with a multi-host, multi-vector and multi-pathogen (26 serotypes) transmission, in which wild ruminants are regarded as important (Ruiz-Fons et al., 2014).

The mechanisms of the spread of BTV are movement of viraemic hosts or animal products from endemic areas or from outbreaks, or by the spread of infected midges via air streams (Carpenter et al., 2009). The epidemiology of BTV is associated with climate, and thus climate change, in several ways. The longevity of the vector is associated with both temperature and humidity, and it has been shown that the transmission potential of the vector *Culicoides sonorensis* for BTV10 and BTV16 (serotypes) increased with higher temperatures. The vector survival, as such, decreased, but this was compensated for by a decrease in the extrinsic incubation period, i.e. the time of uptake of the agent to the time when the vector is infective (Wittmann et al., 2002). Vectors are reported to survive for months at 10°C, whereas at such low temperatures, the virus replication may cease (Mullens et al., 1995).

Since 2006, BTV expanded its distribution to northern Europe, possibly associated with climate change (Wilson and Mellor, 2008; Falconi et al., 2011). White-tailed deer (*Odocoileus hemionus*) seem to be one cervid species that commonly show clinical signs upon BTV infection, such as severe depression, fever, anorexia, hyperemia and ulcers of the oral mucosa, nasal discharge, respiratory distress, periorbital oedema, cyanotic tongue and increased salivation, among other symptoms (Falconi et al., 2011). Due to its abundance and distribution, red deer (*Cervus elaphus*) is believed to be involved in the epidemiology of BTV in Europe (Falconi et al., 2011).

Bluetongue appeared in Denmark in 2007, Sweden in 2008 and Norway in 2009, but has to date not been detected in Finland in spite of intensive surveillance. In Norway, it was distributed in the southern tip of the country, but did not reach the core regions of the wild reindeer herds in the south. Due to effective control regimes in animal husbandry, such as restrictions on animal movement and also, in some countries, vaccination, the disease in domestic animals is now mostly limited to the Mediterranean region, possibly indicating that local wildlife populations were not able to maintain BTV (Ruiz-Fons et al., 2014). There are no indications that wild or semi-domesticated reindeer were exposed to BTV during this epizootic, and there are no reports on transmission to or disease in reindeer under natural conditions. A recent serological screening (2013–2015) of 450 semi-domesticated reindeer in Norway and 635 semi-domesticated reindeer in Finland (*R. t. tarandus*) revealed no antibodies against BTV (M. Tryland et al., unpublished data; S. Laaksonen et al., unpublished data).

16.8.4 Schmallenberg Virus (SBV)

In late summer 2011, a disease with non-specific symptoms, such as fever, diarrhoea and reduced milk production, was reported in dairy cattle in Germany and the Netherlands. The first virus isolate, a negative sense single stranded RNA virus (genus *Orthobunyavirus*, family Bunyaviridae), was obtained from a cow in the city of Schmallenberg, Germany, and the virus was called Schmallenberg virus (SBV). The virus spread rapidly in Europe, and in September 2013, SBV had been reported in 27 European countries (Gibbens, 2012; Wernike et al., 2014). The virus is transmitted by biting midges (*Culicoides* spp.) or mosquitos and causes disease mainly in domestic and wild ruminants, but is not pathogenic for humans.

Adult ruminants may have a subclinical infection or show mild and non-specific symptoms, whereas the more severe symptoms are congenital malformations in newborn lambs, goat kids and calves, as well as premature birth or stillbirth and the birth of mummified foetuses (Doceul et al., 2013; Wernike et al., 2014). Since autumn 2011, several serological screenings have demonstrated the appearance of antibodies against SBV in many wild ruminants in Europe, such as red deer (*Cervus elaphus*), roe deer (*Capreolus capreolus*), fallow deer (*Dama dama*), sika deer (*Cervus Nippon*), European bison (*Bison bonasus*) and mouflon (*Ovis orientalis*), and also in wild boar (*Sus scrofa*) (Linden et al., 2012; Larska et al., 2014; Laloy et al., 2014; Díaz et al., 2015; Mouchantat et al., 2015). In Norway, a cow was found infected, carrying a malformed calf. Further, antibodies against SBV were found in cattle, and the virus was detected in biting midges in the southern part of the country (Wisløff et al., 2014).

No reports exist on SBV in reindeer or in other host species in regions inhabited by wild or semi-domesticated reindeer, but based on the findings in other deer species, it is to be assumed that reindeer are also susceptible to SBV. A serological screening of 187 wild (2010–2013) and 450 semi-domesticated reindeer (2013–2015) in Norway, and 635 semi-domesticated reindeer in Finland, all *R. t. tarandus*, revealed no antibodies against SBV (das Neves et al., unpublished data; M. Tryland et al., unpublished data; S. Laaksonen et al., unpublished data).

16.9 CONCLUDING REMARKS

Climate change will, beyond doubt, affect *Rangifer* throughout their distribution ranges, and their ability to adapt to shifting life conditions will be a key determinant of their performance during the coming decades (Mallory and Boyce, 2018). The effects of climate change may impact ecosystems at a very local scale and are thus not always easy to measure or predict. Many *Rangifer* populations have already declined during the past few decades, and climate change adds to the challenges these animals are already facing. *Rangifer* represent key species in their habitats, being important herbivores and prey species, and also provide an important socio-economic value to many northern people and communities. Climate change thus represents challenges to the management of wild *Rangifer* populations as well as to the reindeer herding industry.

REFERENCES

Aerts, R., and F. S. Chapin. 2000. The mineral nutrition of wild plants revisited: A re-evaluation of processes and patterns. *Adv Ecol Res* Vol 30:1–67.

Aikens, E. O., M. J. Kauffman, J. A. Merkle et al. 2017. The greenscape shapes surfing of resource waves in a large migratory herbivore. *Ecol Lett* 20:741–50. doi:10.1111/ele.12772.

Backx, A., R. Heutink, E. van Rooij, and P. van Rijn. 2009. Transplacental and oral transmission of wild-type bluetongue virus serotype 8 in cattle after experimental infection. *Vet Microbiol* 138:235–43.

Bastille-Rousseau, G., J. A. Schaefer, S. P. Mahoney, and D. L. Murray. 2013. Population decline in semi-migratory caribou (*Rangifer tarandus*): Intrinsic or extrinsic drivers? *Can J Zool* 91:820–8.

Becher, P. R., and M. O. Ramirez, S. C. Avalos et al. 2003. Genetic and antigenic characterization of novel pestivirus genotypes: Implications for classification. *Virology* 311(1):96–104.

Beck, P. S. A., and S. J. Goetz. 2011. Satellite observations of high northern latitude vegetation productivity changes between 1982 and 2008: Ecological variability and regional differences. *ERL* 6.

Bernes, C., K. A. Bråthen, B. C. Forbes, J. D. M. Speed, and J. Moen. 2015. What are the impacts of reindeer/caribou (*Rangifer tarandus* L.) on arctic and alpine vegetation? A systematic review. *Environ Evid* 4.

Bjorkvoll, E., B. Pedersen, H. Hytteborn, I. S. Jonsdottir, and R. Langvatn. 2009. Seasonal and interannual dietary variation during winter in female Svalbard reindeer (*Rangifer tarandus platyrhynchus*). *Arct Antarct Alp Res* 41:88–96.

Bjune, A. E. 2000. Pollen analysis of faeces as a method of demonstrating seasonal variations in the diet of Svalbard reindeer (*Rangifer tarandus platyrhynchus*). *Polar Res* 19:183–92.

Boggs, C. L. 2016. The fingerprints of global climate change on insect populations. *Curr Opin Insect Sci* 17:69–73.

Bokhorst, S., J. W. Bjerke, H. Tømmervik, C. Preece, and G. K. Phoenix. 2012. Ecosystem response to climate change: The importance of the cold season. *Ambio* 41(3):246–55.

Bråthen, K. A., and V. T. Ravolainen. 2015. Niche construction by growth forms is as strong a predictor of species diversity as environmental gradients. *J Ecol* 103:701–13.

Bråthen, K. A., R. A. Ims, N. G. Yoccoz, P. Fauchald, T. Tveraa, and V. H. Hausner. 2007. Induced shift in ecosystem productivity? Extensive scale effects of abundant large herbivores. *Ecosystems* 10:773–89.

Brooker, R., and R. van der Wal. 2004. Mosses mediate grazer impacts on grass abundance in arctic ecosystems. *Funct Ecol* 18:77–86.

Capinera, J. L. 2010. *Insects and Wildlife – Arthropods and Their Relationships with Vertebrate Animals.* Oxford, UK: Wiley-Blackwell, 487 pp.

Carpenter, S., A. Wilson, and P. S. Mellor. 2009. Culicoides and the mergence of bluetongue virus in northern Europe. *Trends Microbiol* 17(4):172–8.

Cederlund, G., and R. Bergström. 1996. Trends in the moose – Forest system in Fennoscandia, with special reference to Sweden. In: *Conservation of Faunal Diversity in Forested Landscapes.* Volume 6 of the series Conservation Biology. the Netherlands: Springer, pp. 265–81.

Chapin, F. S., M. Sturm, M. C. Serreze et al. 2005. Role of land-surface changes in Arctic summer warming. *Science* 310:657–60.

Christie, K. S., J. P. Bryant, L. Gough et al. 2015. The role of vertebrate herbivores in regulating shrub expansion in the Arctic: A synthesis. *Bioscience* 65:1123–33.

Corbière, F., S. Nussbaum, J. P. Alzieu et al. 2012. Bluetongue virus serotype 1 in wild ruminants, France, 2008–10. *J Wildl Dis* 48:1047–51.

Coutinho A. E., and K. E. Chapman. 2011. The anti-inflammatory and immunosuppressive effects of glucocorticoids, recent developments and mechanistic insights. *Mol Cell Endocrinol* 335:2–13.

das Neves, C. G., S. Roth, E. Rimstad, E. Thiry, and M. Tryland. 2010. Cervid herpesvirus 2 infection in reindeer: A review. *Vet Microbiol* 143(1):70–80.

das Neves, C. G., M. Tryland, and H. Li. 2013. Identification of a putative new gammaherpesvirus in reindeer (*Rangifer tarandus tarandus*) in Norway. In: *4th ESVV Herpesvirus Symposium.* Zurich, Switzerland: European Society of Veterinary Virology, p. 31.

Derksen, C., and R. Brown. 2012. Spring snow cover extent reductions in the 2008–2012 period exceeding climate model projections. *Geophys Res Lett* 39.

Díaz, J. M., A. Prieto, C. López et al. 2015. High spread of Schmallenberg virus among roe deer (*Capreolus capreolus*) in Spain. *Res Vet Sci* 102:231–3.

Doceul, V., E. Lara, C. Sailleau et al. 2013. Epidemiology, molecular virology and diagnostics of Schmallenberg virus, and emerging orthobunyavirus in Europe. *Vet Res* 44:31.

Doiron, M., G. Gauthier, and E. Levesque. 2014. Effects of experimental warming on nitrogen concentration and biomass of forage plants for an arctic herbivore. *J Ecol* 102:508–17.

Douhard, M., L. E. Loe, A. Stien et al. 2016. The influence of weather conditions during gestation on life histories in a wild Arctic ungulate. *Proc R Soc Lond B Biol Sci* 283.

Elbers, A. R., C. J. Koenraadt, and R. Meiswinkel. 2015. Mosquitos and Culicoides biting midges: Vector range and the influence of climate change. *Rev Sci Tech* 34(1):123–37.

Elmendorf, S. C., G. H. R. Henry, R. D. Hollister et al. 2012. Global assessment of experimental climate warming on tundra vegetation: Heterogeneity over space and time. *Ecol Lett* 15:164–75.

Falconi, C., J. R. López-Olvera, and C. Gortázar. 2011. BTV infection in wild ruminants, with emphasis on red deer: A review. *Vet Microbiol* 151(3-4):209–19.

Falldorf, T., O. Strand, M. Panzacchi, and H. Tømmervik. 2015. Estimating lichen volume and reindeer winter pasture quality from Landsat imagery (vol 140, pg 573, 2014). *Remote Sens Environ* 166:286.

Fernández-Pacheco, P., J. Fernández-Pinero, M. Agüero, and M. A. Jiménez-Clavero. 2008. Bluetongue virus serotype 1 in wild mouflons in Spain. *Vet Rec* 162:659–60.

Forbes, B. C., T. Kumpula, N. Meschtyb et al. 2016. Sea ice, rain-on-snow and tundra reindeer nomadism in Arctic Russia. *Biol Lett* 12.

García-Bocanegra, I., A. Arenas-Montes, C. Lorca-Oró et al. 2011. Role of wild ruminants in the epidemiology of bluetongue virus serotypes 1, 4 and 8 in Spain. *Vet Res* 42:88.

Gibbens, N. 2012. Schmallenberg virus: A novel viral disease in northern Europe. *Vet Rec* 170:58.

Guay, K. C., P. S. A. Beck, L. T. Berner et al. 2014. Vegetation productivity patterns at high northern latitudes: A multi-sensor satellite data assessment. *Global Change Biol* 20:3147–58.

Gustine, D. D., P. S. Barboza, J. P. Lawler et al. 2012. Diversity of nitrogen isotopes and protein status in caribou: Implications for monitoring northern ungulates. *J Mammal* 93:778–90.

Handeland, K., L. Qviller, T. Vikøren et al. 2013. *Ixodes ricinus* infestation in free-ranging cervids in Norway – a study based upon ear examinations of hunted animals. *Vet Parasitol* 195:142–9.

Hansen, B. B., and R. Aanes. 2012. Kelp and seaweed feeding by High-Arctic wild reindeer under extreme winter conditions. *Polar Research* 31.

Hansen, B. B., R. Aanes, I. Herfindal et al. 2011. Climate, icing, and wild arctic reindeer: Past relationships and future prospects. *Ecology* 92:1917–23.

Hansen, B. B., S. Henriksen, R. Aanes, and B. E. Saether. 2007. Ungulate impact on vegetation in a two-level trophic system. *Polar Biol* 30:549–58.

Hansen, B., K. Isaksen, R. Benestad et al. 2014. Warmer and wetter winters: Characteristics and implications of an extreme weather event in the High Arctic. *ERL 9*.

Helle, T. P. and L. M. Jaakkola. 2008. Transitions in herd management of semi-domesticated reindeer in northern Finland. *Ann Zool Fenn* 45:81–101.

Holand, Ø., P. Aikio, H. Gjøstein et al. 2002. Modern reindeer dairy farming—The influence of different milking regimes on udder health, milk yield and composition. *Small Ruminant Res* 44:65–73.

Høye, T. T., and M. C. Forchhammer. 2008. The influence of weather conditions on the activity of high-arctic arthropods inferred from long-term observations. *BMC Ecol* 8:8. doi: 10.1186/1472-6785-8-8.

Høye, T. T., E. Post, H. Meltofte, N. M. Schmidt, and M. C. Forchhammer. 2007. Rapid advancement of spring in the High Arctic. *Curr Biol* 17(12) R449–51. doi: dx.doi.org/10.1016/j.cub.2007.04.047.

Hudson, J. M. G., and G. H. R. Henry. 2010. High Arctic plant community resists 15 years of experimental warming. *J Ecol* 98:1035–41.

Hvidsten, D., F. Stordal, M. Lager et al. 2015. *Borrelia burgdorferi sensu lato*-infected *Ixodes ricinus* collected from vegetation near the Arctic Circle. *Ticks Tick Borne Dis* 6(6):768–73. doi: 10.1016/j.ttbdis.2015.07.002.

Ihl, C., and D. R. Klein. 2001. Habitat and diet selection by muskoxen and reindeer in western Alaska. *J Wildl Manag* 65:964–72.

IPCC. 2014. *Climate Change 2014: Synthesis Report.* Contribution of working groups I, II and III to the Fifth assessment report of the Intergovernmental Panel on Climate Change (Core writing team: Pachauri RK and Meyer LA (Eds.). IPCC, Geneva, Switzerland, 151 pp.

Iversen, M., P. Fauchald, K. Langeland et al. 2014. Phenology and growth forms predict herbivore habitat selection in a high latitute ecosystem. *PLoS One*, Jun 27;9(6):e100780.

Jaenson, T. G. T., D. G. E. Jaenson, L. Eisen, E. Peterson, and E. Lindgren. 2012. Changes in the geographical distribution and abundance of the tick *Ixodes ricinus* during the past 30 years in Sweden. *Parasit Vectors* 5:8 doi: 10.1186/1756-3305-5-8.

Jaenson, T. G. T., K. Värv, I. Fröjdman et al. 2016. First evidence of established populations of the taiga tick *Ixodes persulcatus* (Acari: Ixodidae) in Sweden. *Parasit Vector* 9:377. doi: 10.1186/s13071-016-1658-3.

Jean, P. O., R. L. Bradley, J. P. Tremblay, and S. D. Côté. 2015. Combining near infrared spectra of feces and geostatistics to generate forage nutritional quality maps across landscapes. *Ecol Appl* 25:1630–9.

Jefferies, R. L., A. P. Jano, and K. F. Abraham. 2006. A biotic agent promotes large-scale catastrophic change in the coastal marshes of Hudson Bay. *J Ecol* 94:234–42.

Joly, K., M. J. Cole, and R. R. Jandt. 2007. Diets of overwintering caribou, *Rangifer tarandus*, track decadal changes in Arctic tundra vegetation. *Can Field Nat* 121:379–83.

Joo, S., D. Han, E. J. Lee, and S. Park. 2014. Use of length heterogeneity polymerase chain reaction (LH-PCR) as non-invasive approach For dietary analysis of Svalbard reindeer, *Rangifer tarandus platyrhynchus. Plos One* 9.

Jore, S., S. O. Vanwambeke, H. Viljugrein et al. 2014. Climate and environmental change drives *Ixodes ricinus* geographical expansion at the northern range margin. *Parasit Vector* 7:11. doi: 10.1186/1756-3305-7-11.

Kaplan, J. O., and M. New. 2006. Arctic climate change with a 2 degrees C global warming: Timing, climate patterns and vegetation change. *Climatic Change* 79:213–41.

Kattsov, V. M., E. Källén, H. Cattle et al. 2005. Future climate change: Modeling and scenarios for the Arctic. In: *Arctic Climate Impact Assessment.* Cambridge: Cambridge University Press, pp. 99–150.

Kaunisto, S., L. Härkönen, P. Niemelä, and H. Roininen. 2011. Northward invasion of the parasitic deer ked (*Lipoptena cervi*), is there geographical variation in pupal size and development duration? *Parasitology* 138:354–63.

Kautto, A. H., S. Alenius, T. Mossing et al. 2012. Pestivirus and alphaherpesvirus infections in Swedish reindeer (*Rangifer tarandus tarandus* L.). *Vet Microbiol* 156(1–2):64–71.

Klein, D. R. 1968. The introduction, increase, and crash of reindeer on St. Matthew Island. *J Wildl Manag* 32 (2):350–67. doi: 10.2307/3798981.

Kynkäänniemi, S. M., M. Kettu, R. Kortet et al. 2014. Acute impacts of the deer ked (*Lipoptena cervi*) infestation on reindeer (*Rangifer tarandus tarandus*) behavior. *Parasitol Res* 113(4):1489–97.

Laaksonen, S., J. Pusenius, J. Kumpula et al. 2010. Climate change promotes the emergence of serious disease outbreaks of filarioid nematodes. *Ecohealth* 7(1):7–13.

Laloy, E., E. Bréard, C. Sailleau et al. 2014. Schmallenberg virus infection among red deer, France, 2010–2012. *Emerg Infect Dis* 20:131–4.

Larter, N. C., and J. A. Nagy. 2004. Seasonal changes in the composition of the diets of Peary caribou and muskoxen on Banks Island. *Polar Res* 23:131–40.

Linden, A., D. Desmecht, R. Volpe et al. 2012. Epizootic spread of Schmallenberg virus among wild cervids, Belgium, fall 2011. *Emerg Infect Dis* 18(12):2006–8.

Loe, L. E., B. B. Hansen, A. Stien et al. 2016. Behavioral buffering of extreme weather events in a high-Arctic herbivore. *Ecosphere* 7.

Macias-Fauria, M., B. C. Forbes, P. Zetterberg, and T. Kumpula. 2012. Eurasian Arctic greening reveals teleconnections and the potential for structurally novel ecosystems. *NCC* 2:613.

Mallory, C. D., and M. S. Boyce. 2018. Observed and predicted effects of climate change on Arctic caribou and reindeer. *Environ Rev* 26:13–25.

Medlock, J. M., K. M. Hansford, A. Bormane et al. 2013. Driving forces for changes in geographical distribution of *Ixodes ricinus* ticks in Europe. *Parasit Vector* 6:1. doi: 10.1186/1756-3305-6-1.

Moen, K. 2008. Climate change: Effects on the ecological basis for reindeer husbandry in Sweden. *Ambio* 37(4):304–11.

Mouchantat, S., K. Wernike, W. Lutz et al. 2015. A broad spectrum screening of Schmallenberg virus antibodies in wildlife animals in Germany. *Vet Res* 46:99.

Mullens, B. A., W. J. Tabachnick, F. R. Holbrook, and L. H. Thompson. 1995. Effects of temperature on virogenesis of bluetongue virus, serotype 11 in *Culicoides variipennis sonorensis. Med Vet Entomol* 9:71–6.

Myers-Smith, I. H., S. C. Elmendorf, P. S. A. Beck et al. 2015. Climate sensitivity of shrub growth across the tundra biome. *NCC* 5:887–91.

Myers-Smith, I. H., B. C. Forbes, M. Wilmking et al. 2011. Shrub expansion in tundra ecosystems: Dynamics, impacts and research priorities. *ERL* 6.

Myneni, R. B., C. D. Keeling, C. J. Tucker, G. Asrar, and R. R. Nemani. 1997. Increased plant growth in the northern high latitudes from 1981 to 1991. *Nature* 386:698–702.

Niedbalski, W. 2015. Bluetongue in Europe and the role of wildlife in the epidemiology of disease. *Pol J Vet Sci* 18(2):455–61.

Ogden, N. H., and L. R. Lindsay. 2016. Effects of climate and climate change on vectors and vector-borne diseases: Ticks are different. *Trends Parasitol* 32(8):646–56.

Ophof, A. A., K. W. Oldeboer, and J. Kumpula. 2013. Intake and chemical composition of winter and spring forage plants consumed by semi-domesticated reindeer (*Rangifer tarandus tarandus*) in Northern Finland. *Anim Feed Sci Technol* 185:190–5.

Pedersen, A. O., I. Tombre, J. U. Jepsen et al. 2013. Spatial patterns of goose grubbing suggest elevated grubbing in dry habitats linked to early snowmelt. *Polar Res* 32.

Post, E., and M. C. Forchhammer. 2008. Climate change reduces reproductive success of an Arctic herbivore through trophic mismatch. *Philos Trans R Soc Lond B Biol Sci* 363:2369–75.

Prach, K., and G. Rachlewicz. 2012. Succession of vascular plants in front of retreating glaciers in central Spitsbergen. *Pol Polar Res* 33:319–28.

Putkonen, J., and G. Roe. 2003. Rain-on-snow events impact soil temperatures and affect ungulate survival. *Geophys Res Lett* 30.

Ravolainen, V. T., K. A. Bråthen, R. A. Ims et al. 2011. Rapid, landscape scale responses in riparian tundra vegetation to exclusion of small and large mammalian herbivores. *BAEE* 12:643–53.

Ravolainen, V. T., K. A. Bråthen, N. G. Yoccoz, J. K. Nguyen, and R. A. Ims. 2014. Complementary impacts of small rodents and semi-domesticated ungulates limit tall shrub expansion in the tundra. *J Appl Ecol* 51:234–41.

Rominger, E. M., and J. L. Oldemeyer. 1990. Early-winter diet of woodland caribou in relation to snow accumulation, Selkirk Mountains, British-Columbia, Canada. *Can J Zool* 68:2691–4.

Rominger, E. M., C. T. Robbins, M. A. Evans, and D. J. Pierce. 2000. Autumn foraging dynamics of woodland caribou in experimentally manipulated habitats, northeastern Washington, USA. *J Ecol* 64:160–7.

Rozenfeld, S. B., A. R. Gruzdev, T. P. Sipko, and A. N. Tikhonov. 2012. Trophic relationships of musk ox (*Ovibos moschatus*) and reindeer (*Rangifer tarandus*) on Wrangel Island. *Zool Zh* 91:503–12.

Ruiz-Fons, F., A. Sánchez-Matamoros, C. Gortázar, and J. M. Sánchez-Vizcaíno. 2014. The role of wildlife in bluetongue virus maintenance in Europe: Lessons learned after the natural infection in Spain. *Virus Res* 182:50–8.

Sharma, S., S. Couturier, and S. D. Côté. 2009. Impacts of climate change on the seasonal distribution of migratory caribou. *Global Change Biol* 15:2549–62. doi: 10.1111/j.1365-2486.2009.01945.x.

Skogland, T. 1980. Comparative summer feeding strategies of Arctic and alpine *Rangifer*. *J Anim Ecol* 49:81–98.

Staaland, H., and H. Sletten. 1991. Feeding reindeer in Fennoscandia: The use of artificial food. In: *Wildlife Production: Conservation and Sustainable Development*. Eds. Renecker, L. A., Hudson, R. J., Fairbanks, Alaska: Agricultural and Forestry Experiment Station, University of Alaska Fairbanks, pp. 227–42.

Stien, A., L. E. Loe, A. Mysterud et al. 2010. Icing events trigger range displacement in a high-arctic ungulate. *Ecology* 91:915–20.

Strathdee, A. T., and J. S. Bale. 1998. Life on the edge: Insect ecology in arctic environments. *Annu Rev Entomol* 43:85–106.

Stuen, S. 2007. *Anaplasma phagocytophilum* – the most widespread tick-borne infection in animals in Europe. *Vet Res Commun* 31 Suppl 1:79–84.

Stuen, S., E. G. Granquist, and C. Silaghi. 2013. *Anaplasma phagocytophilum* – A widespread multi-host pathogen with highly adaptive strategies. *Front Cell Infect Microbiol* doi: 10.3389/fcimb.2013.00031.

Tambs-Lyche, H. 1943. Ixodes ricinus of piroplasmosen i Norge. *Nor. Vet. Tidsskr.* 55:337–66.

Tape, K. D., M. Hallinger, J. M. Welker, and R. W. Ruess. 2012. Landscape heterogeneity of shrub expansion in Arctic Alaska. *Ecosystems* 15:711–24.

Tews, J., M. A. D. Ferguson, and L. Fahrig. 2007. Potential net effects of climate change on High Arctic Peary caribou: Lessons from a spatially explicit simulation model. *Ecol Model* 207:85–98.

Thiry, J., F. Widen, F. Gregoire et al. 2007. Isolation and characterisation of a ruminant alphaherpesvirus closely related to bovine herpesvirus 1 in a free-ranging red deer. *BMC Vet Res* 3:26. doi: 10.1186/1746-6148-3-26.

Thompson, I. D., P. A. Wiebe, E. Mallon et al. 2015. Factors influencing the seasonal diet selection by woodland caribou (*Rangifer tarandus tarandus*) in boreal forests in Ontario. *Can J Zool* 93:87–98.

Trudell, J., and R. G. White. 1981. The effect of forage structure and availability on food-intake, biting rate, bite size and daily eating time of reindeer. *J Appl Ecol* 18:63–81.

Tryland, M. 2012. Are we facing new health challenges and diseases in reindeer in Fennoscandia? *Rangifer* 32(1):35–47.

Tryland, M., C. G. das Neves, M. Sunde, and T. Mørk. 2009. Cervid herpesvirus 2, the primary agent in an outbreak of infectious keratoconjunctivitis in semi-domesticated reindeer. *J Clin Microbiol* 47(11):3707–13. doi: 10.1128/JCM.01198-09.

Tryland, M., T. D. Josefsen, A. Oksanen, and A. Aschfalk. 2001. Parapoxvirus infection in Norwegian semi-domesticated reindeer (*Rangifer tarandus tarandus*). *Vet Rec* 149(13):394–5.

Tryland, M., J. S. Romano, N. Marcin, I. H. Nymo, T. D. Josefsen, K. K. Sørensen, and T. Mørk. 2017. Cervid herpesvirus 2 and not Moraxella bovoculi caused keratoconjunctivitis in experimentally inoculated semi-domesticated Eurasian tundra reindeer. *Acta Vet Scand* 59(1):23. doi: 10.1186/s13028-017-0291-2.

Turunen, M., and T. Vuojala-Magga. 2014. Past and present winter feeding of reindeer in Finland: Herders' adaptive learning of feeding practices. *Arctic* 67: 173–88.

Tveraa, T., A. Stien, B. J. Bardsen, and P. Fauchald. 2013. Population densities, vegetation green-up, and plant productivity: Impacts on reproductive success and juvenile body mass in reindeer. *Plos One* 8.

Välimäki, P., K. Madslien, J. Malmsten et al. 2010. Fennoscandian distribution of an important parasite of cervids, the deer ked (*Lipoptena cervi*), revisited. *Parasitol Res* 107(1):117–25. doi: 10.1007/s00436-010-1845-7.

van der Wal, R., and M. J. J. E. Loonen. 1998. Goose droppings as food for reindeer. *Can J Zool* 76:1117–22.

van der Wal, R., N. Madan, S. van Lieshout et al. 2000. Trading forage quality for quantity? Plant phenology and patch choice by Svalbard reindeer. *Oecologia* 123:108–15.

van der Wal, R., and A. Stien. 2014. High-arctic plants like it hot: A long-term investigation of between-year variability in plant biomass. *Ecology* 95:3414–27.

Vickers, H., K. A. Høgda, S. Solbø et al. 2016. Changes in greening in the high Arctic: Insights from a 30 year AVHRR max NDVI dataset for Svalbard. *ERL* 11.

Vors, L. S., and M. S. Boyce. 2009. Global declines of caribou and reindeer. *Global Change Biol* 15:2626–33.

Walsh, J. E., J. E. Overland, P. Y. Groisman, and B. Rudolf. 2011. Ongoing climate change in the Arctic. *Ambio* 40:6–16.

Wernike, K., F. Conraths, G. Zanella et al. 2014. Schmallenberg virus – Two years of experiences. *Preventive Vet Med* 116:423–34.

Wilson, A., and P. Mellor. 2008. Bluetongue in Europe: Vectors, epidemiology and climate change. *Parasitol Res* 103:69–77.

Wisløff, H., B. S. Nordvik, S. Sviland, and R. Tønnessen. 2014. The first documented clinical case of Schmallenberg virus in Norway: Fetal malformations in a calf. *Vet Rec* 174(5):120.

Witter, L. A., C. J. Johnson, B. Croft, A. Gunn, M. P. Gillingham. 2012. Behavioural trade-offs in response to external stimuli: Time allocation of an Arctic ungulate during varying intensities of harassment by parasitic flies. *J Anim Ecol* 81(1):284–95. doi: 10.1111/j.1365-2656.2011.01905.x.

Wittmann, E. J., and M. Baylis. 2000. Climate change: Effects on culicoides-transmitted viruses and implications for the UK. *Vet J* 160:107–17.

Wittmann, E. J., P. S. Mello, and M. Baylis. 2002. Effect of temperature on the transmission of orbiviruses by the biting midge, *Culicoides sonorensis*. *Med Vet Entomol* 16(2):147–56.

Xu, L., R. B. Myneni, F. S. Chapin et al. 2013. Temperature and vegetation seasonality diminishment over northern lands. *NCC* 3:581–6.

Index

Page numbers followed by f and t indicate figures and tables, respectively.

T - #0878 - 101024 - C550 - 254/178/24 - PB - 9781032094335 - Gloss Lamination